Modern Magnetic Resonance

Part 2

Part 1: Applications in Chemistry, Biological and Marine Sciences
Part 2: Applications in Medical and Pharmaceutical Sciences
Part 3: Applications in Materials Science and Food Science

Modern Magnetic Resonance

Part 2: Applications in Medical and Pharmaceutical Sciences

Graham A. Webb (Ed.)
Royal Society of Chemistry, London, UK

Springer

A C.I.P. Catalogue record for this book is available from the Library of Congress.

ISBN-10 1-4020-3894-1 (HB)
ISBN-13 978-1-4020-3894-5 (HB)
ISBN-10 1-4020-3910-7 (e-book)
ISBN-13 978-1-4020-3910-2 (e-book)

Published by Springer,
P.O. Box 17, 3300 AA Dordrecht, The Netherlands.

www.springer.com

Printed on acid-free paper

List of Section Editors

Subject	Name	Type of Editor
	Graham A. Webb	Editor-In-Chief
Chemistry	Hazime Saitô Himeji Institute of Technology and QuLiS, Hiroshima University Japan e-mail: hsaito@siren.ocn.ne.jp	Section Editors
	Isao Ando Department of Chemistry and Materials Science Tokyo Institute of Technology, Ookayama, Meguro-ku Tokyo 152-0033 Japan e-mail: iando@polymer.titech.ac.jp	
	Tetsuo Asakura Department of Biotechnology Tokyo University of Agriculture and Technology Koganei, Tokyo 184-8588 Japan e-mail: asakura@cc.tuat.ac.jp	
Biological Sciences	Jimmy D. Bell Molecular Imaging Group, MRC Clinical Sciences Centre Hammersmith Hospital Campus, Imperial College London London, W12 OHS UK e-mail: jimmy.bell@csc.mrc.ac.uk	Section Editor
Marine Science	M. Aursnad SINTEF Fisheries and Aquaculture Ltd N-7465 Trondheim Norway e-mail: marit.aursand@sintef.no	Section Editor
Medical Science	Carolyn Mountford Institute for Magnetic Resonance Research, and Department of Magnetic Resonance in Medicine University of Sydney, PO Box 148, St Leopards, 1590, NSW Australia email: caro@imrr.usyd.edu.au	Section Editor
	Uwe Himmelreich, PhD Max-Planck-Institute for Neurological Research In vivo NMR Group, Gleueler Str 50, Cologne, D-50931 Germany email: himmelreich@mpin-koeln.mpg.de	
	Deborah Edwards	Sub-editor

Preface

It is a great pleasure for me to Introduce the handbook of Modern Magnetic Resonance, MMR. The various techniques which comprise MMR derive essentially from three sources, all of which were produced by physicists. To-day they are widely used by scientists working in many diverse areas such as chemistry, biology, materials, food, medicine and healthcare, pharmacy and marine studies.

The first source of MMR studies is nuclear magnetic resonance, NMR. This provides details on the relative positions of nuclei, i.e. atoms, in a molecule. Consequently NMR provides structural information on samples which may be in the solid, liquid or gaseous state. Nuclear relaxation data yield dynamic information on the sample and the topology of the dynamic processes if the sample is undergoing a molecular change. Thus high and low resolution NMR studies provide information on all interesting aspects of molecular science. The protean nature of NMR is reflected in its many applications in chemistry, biology and physics which explore and characterize chemical reactions, molecular conformations, biochemical pathways and solid state materials, to name a few examples.

Magnetic resonance imaging, MRI, is the second source of MMR data. MRI provides a three-dimensional image of a substatnce, and is consequently widely employed to assess materials both *in vitro* and *in vivo*. The importance of MRI studies in many areas of science and medicine is shown by the recent award of the Nobel Prize to Lauterbur and Mansfield.

The third source of MMR results is due to electron spin resonance, ESR. This is a technique for detecting unpaired electrons and their interactions with nuclear spins in a given sample. Thus ESR data are often used to complement the results of NMR experiments.

Taken together NMR, MRI and ESR comprise the field of MMR, recent years have witnessed the fecundity of these techniques in many scientific areas. The present three volumes cover applications in most of these areas. Part 1 deals with Chemical Applications, Biological and Marine Sciences. Medical and Pharmaceutical Sciences are covered in Part 2. Part 3 provides examples of recent work in the Materials Science and Food Science.

I wish to express my gratitude to all of the Section Editors and their many contributors for their hard work and dedication in the creation of MMR. My thanks also go to Emma Roberts and the production staff at Kluwer, London, for their assistance in the realization of these volumes.

Royal Society of Chemistry
Burlington House
Piccadilldy
London, W1J OBA

G.A.WEBB
February 2005

Graham A. Webb (ed.), Modern Magnetic Resonance, VII–LV.

Foreword to Application in Chemistry

Magnetic resonance has continued to be an emerging technique, to be applied to almost all fields of pure and applied sciences, including chemistry, physics, biology, materials science, medicine, etc. during past 60 years since its discovery. The applications in chemistry of this volume covers advanced studies on chemical aspect of magnetic resonance spectroscopy and imaging dealing with the state-of-the-art developments of new techniques together with those of basic concepts and techniques, consisting of 93 articles which are grouped to 25 chapters. They are alphabetically arranged for convenience of readers: amyloids, chemical shifts and spin coupling constants, fibrous proteins, field gradient NMR, host-guest chemistry, imaging, inorganic materials and catalysis, lipid bilayers and bicelles, membrane-associated peptides, membrane proteins, new developments, NOE and chemical exchange, NQR and ESR, organometallic chemistry, paramagnetic effects, protein structures, polymer structure, polymer dynamics, polymer blends, quantum information processing, residual dipolar couplings and nucleic acids, solid state NMR techniques, structural constraints in solids, and telomeric DNA complexes. The section editors are grateful to contributors to this section for their fine contributions.

Tetsuo Asakura, Hazime Saitô and Isao Ando

Contents

PART III

List of Tables

Part 3: Applications in Materials Science and Food Science

Color Plate Section

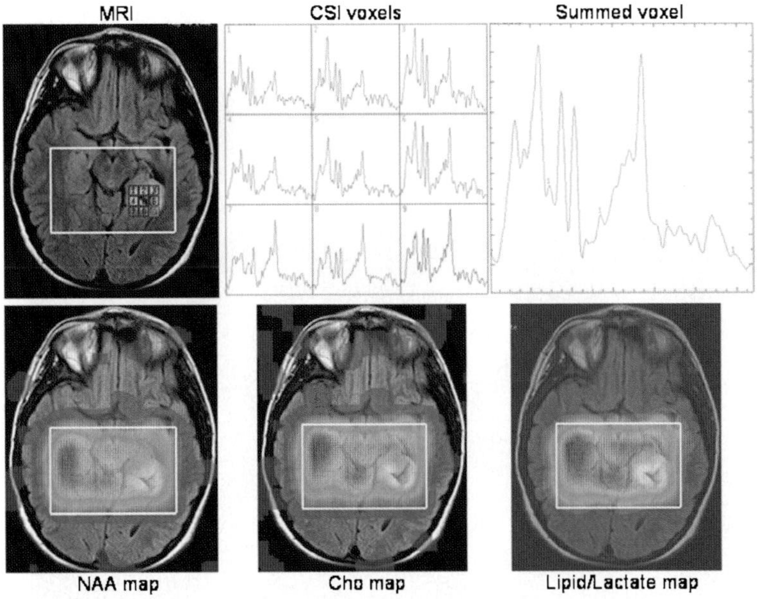

Plate 82. See also Figure 15 on page 988.

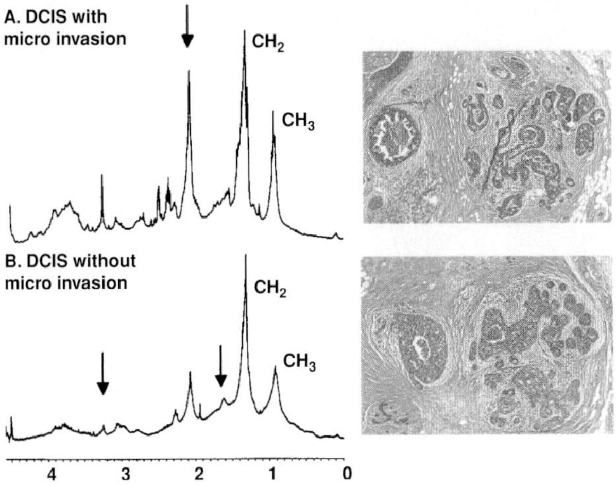

Plate 83. See also Figure 2 on page 1017.

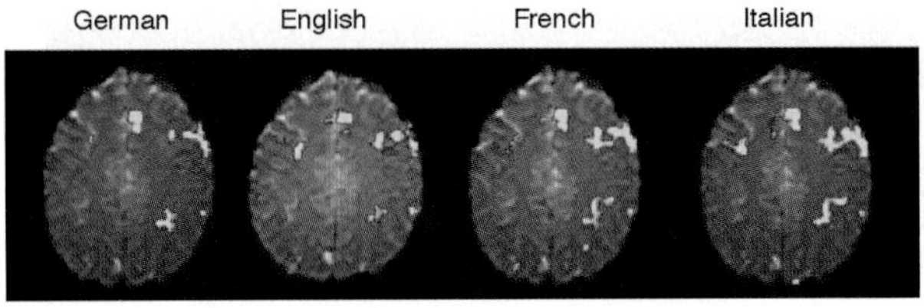

Plate 84. See also Figure 3 on page 1029.

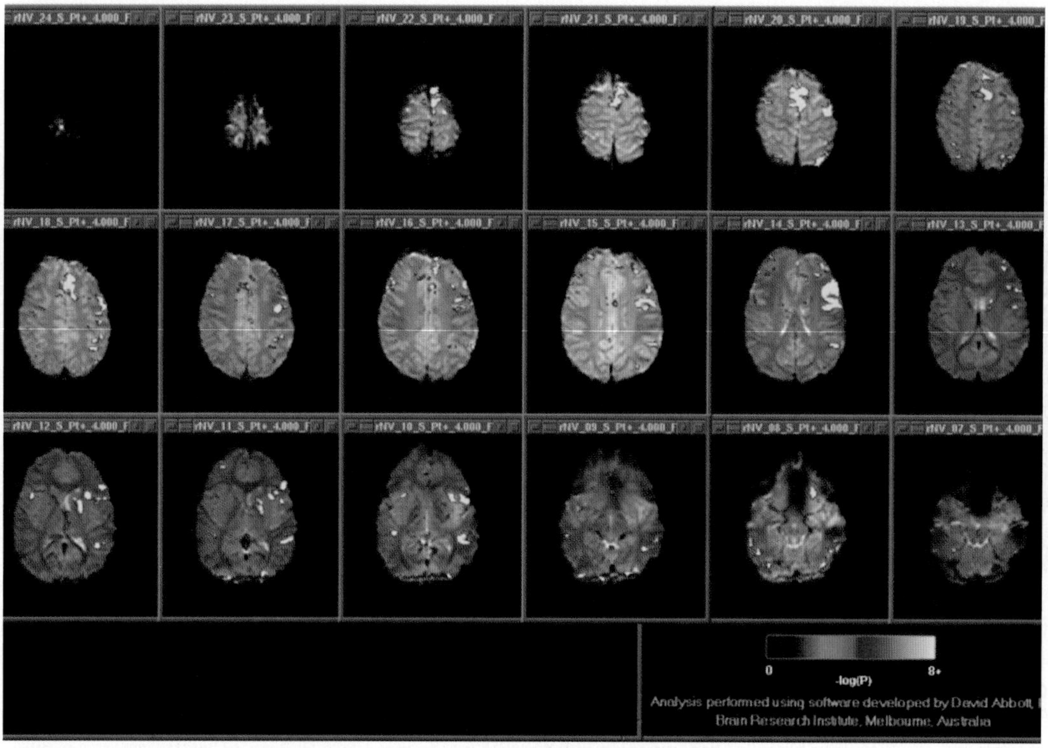

Plate 85. See also Figure 6 on page 1031.

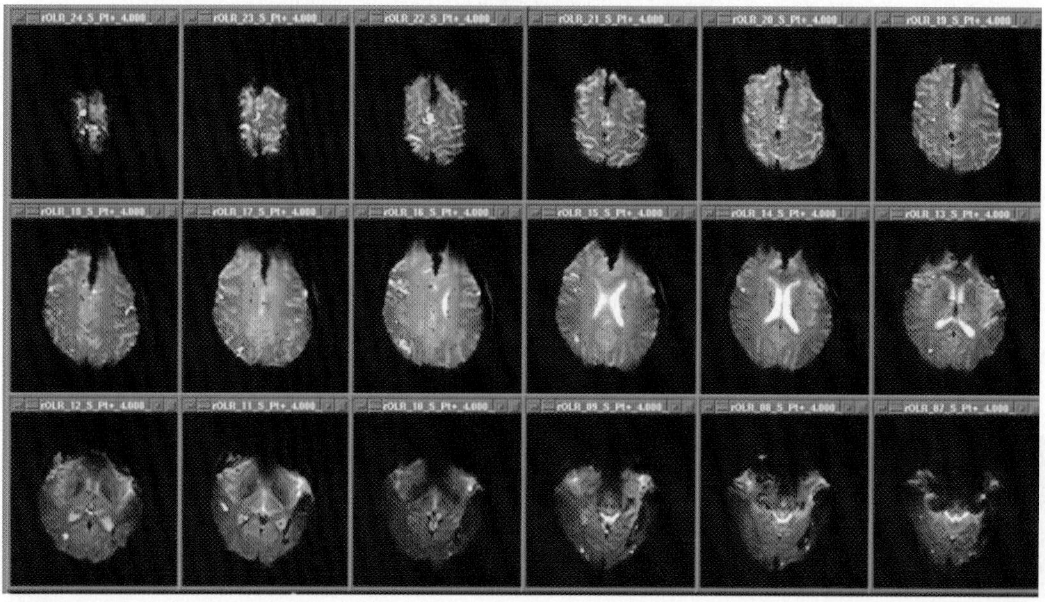

Plate 86. See also Figure 8 on page 1032.

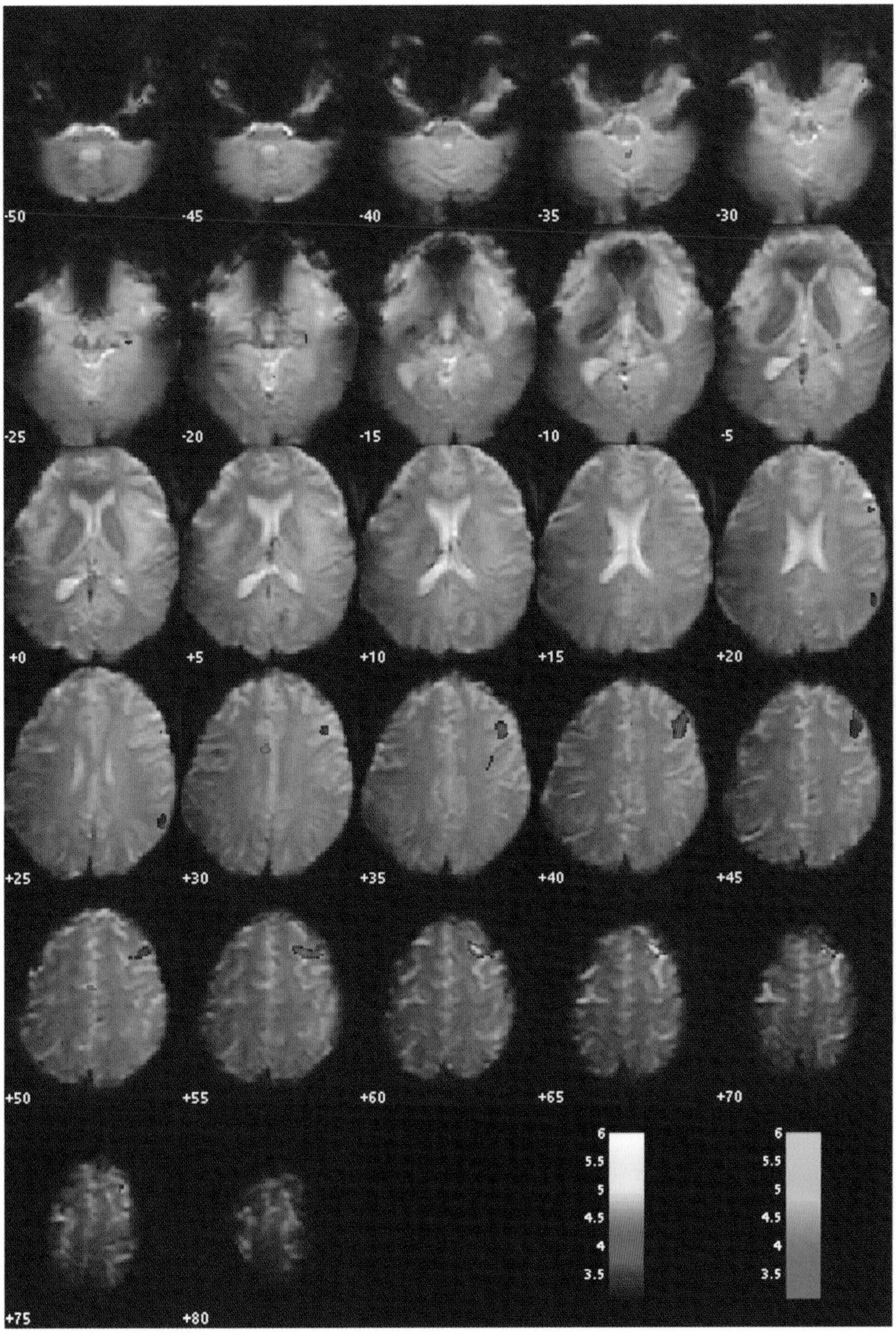

Plate 87. See also Figure 10 on page 1034.

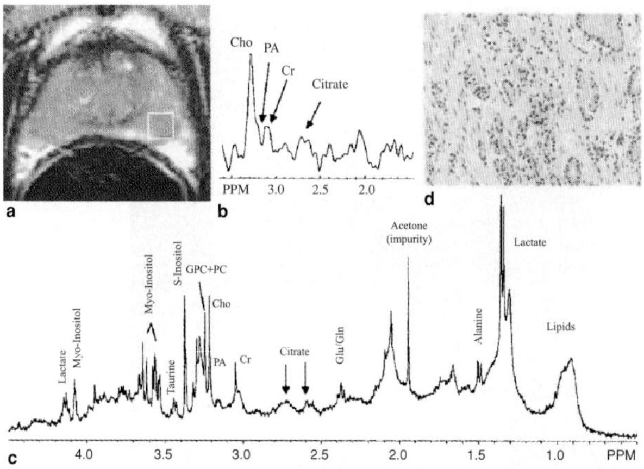

Plate 88. See also Figure 6 on page 1045.

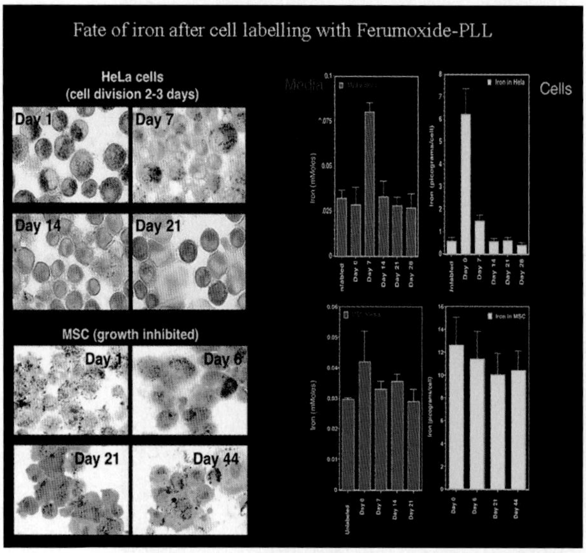

Plate 89. See also Figure 3 on page 1080.

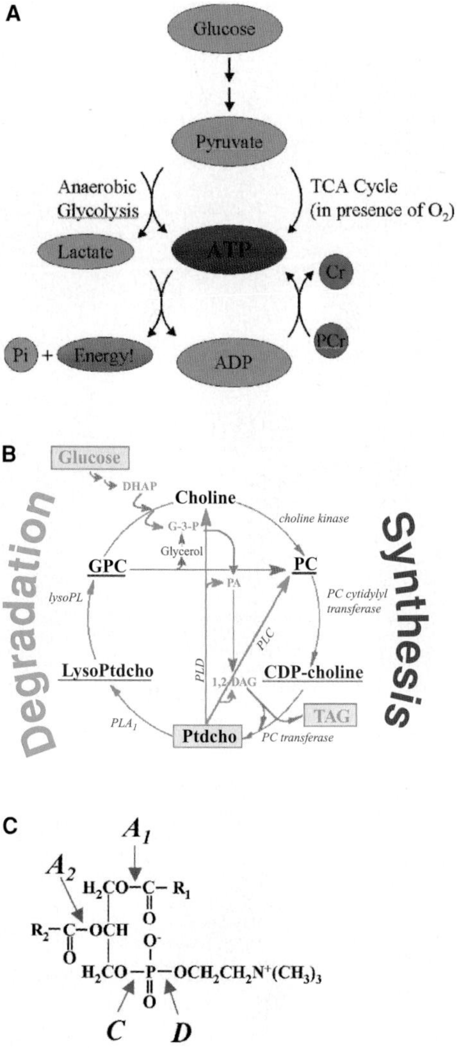

Plate 90. See also Figure 3 on page 1132.

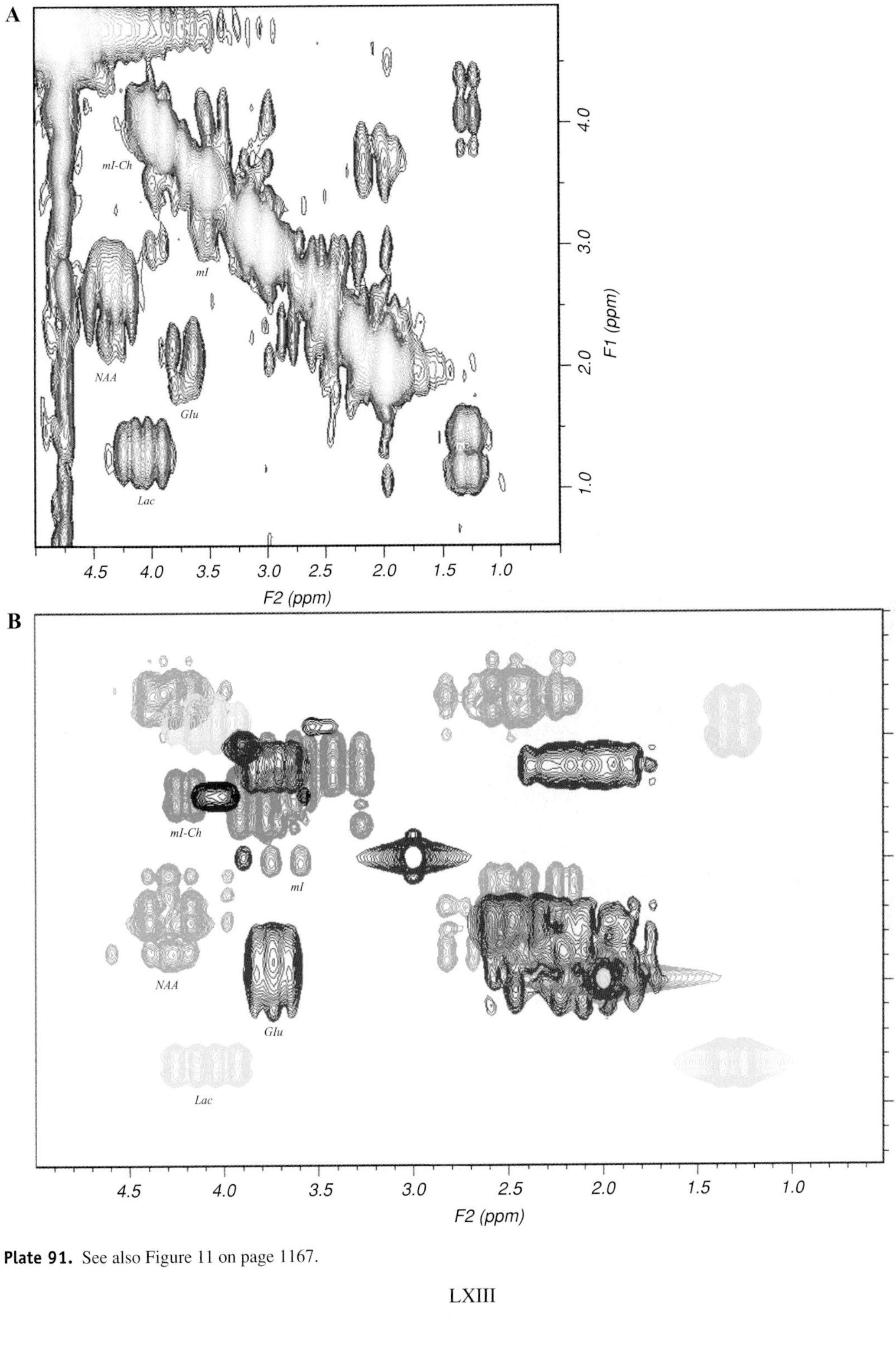

Plate 91. See also Figure 11 on page 1167.

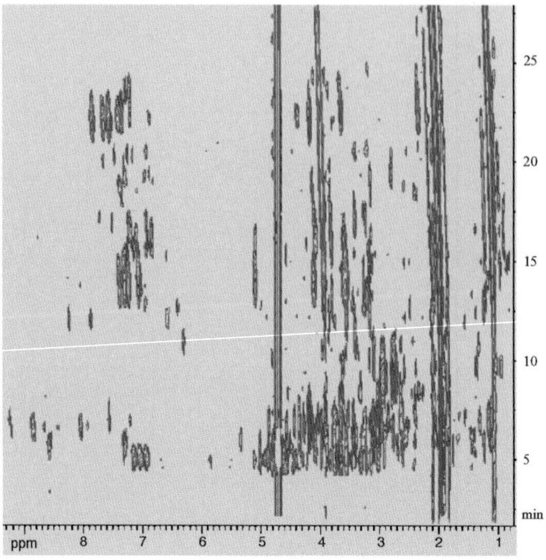

Plate 92. See also Figure 2 on page 1205.

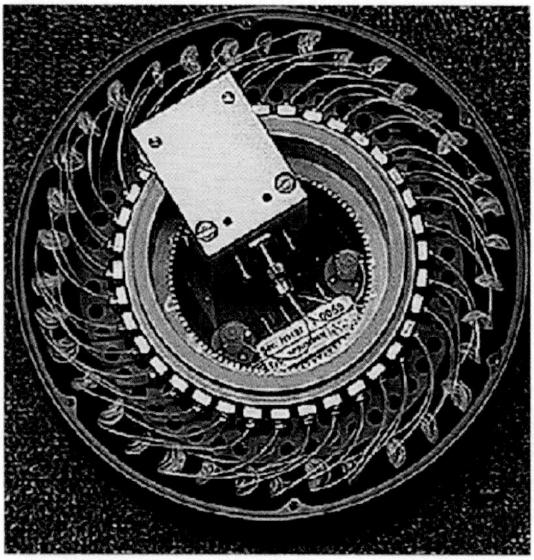

Plate 93. See also Figure 4 on page 1207.

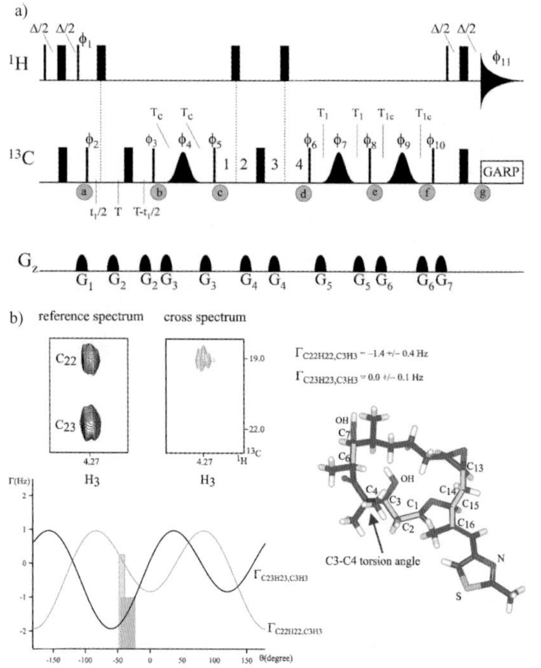

Plate 94. See also Figure 3 on page 1233.

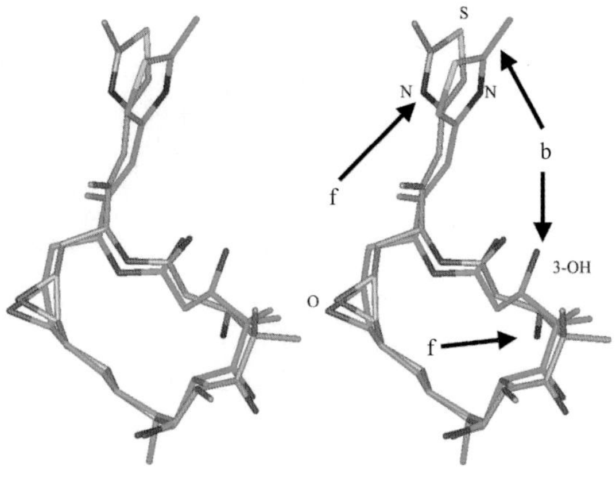

Plate 95. See also Figure 4 on page 1235.

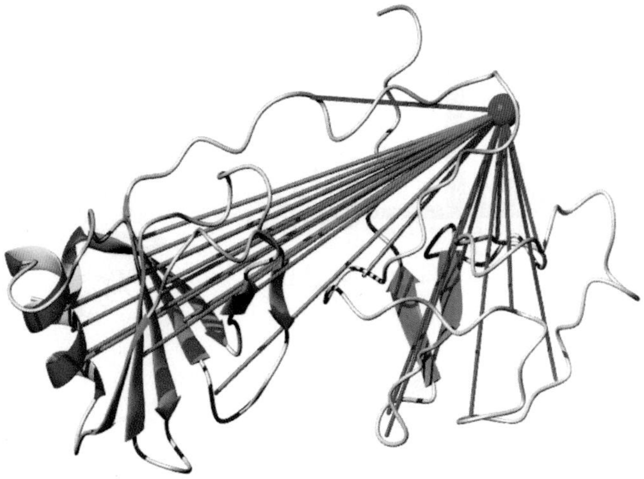

Plate 96. See also Figure 3 on page 1242.

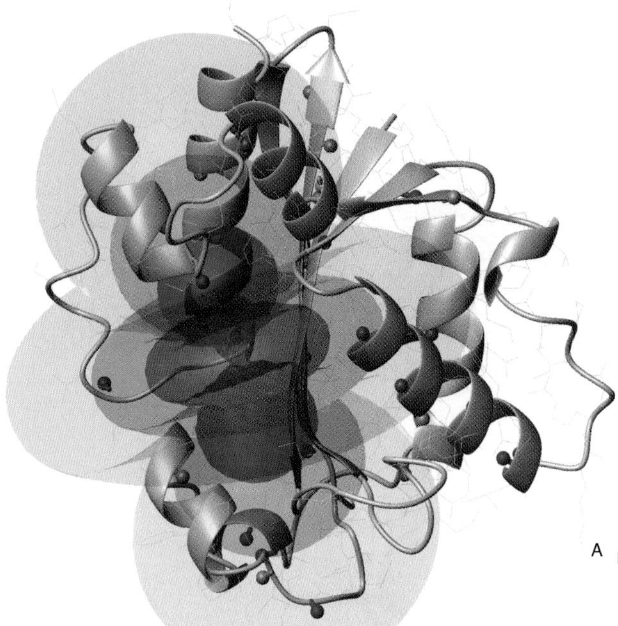

Plate 97. See also Figure 4 on page 1248.

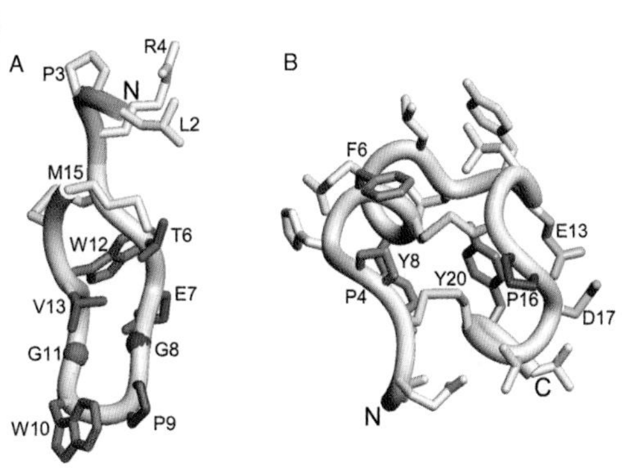

Plate 98. See also Figure 1 on page 1276.

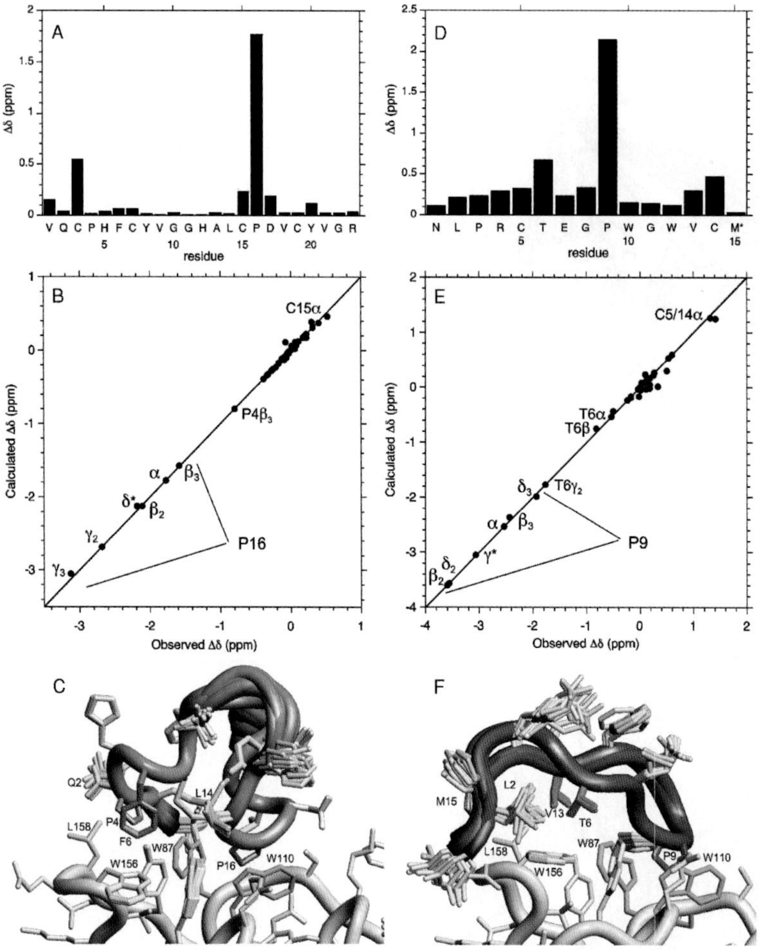

Plate 99. See also Figure 3 on page 1278.

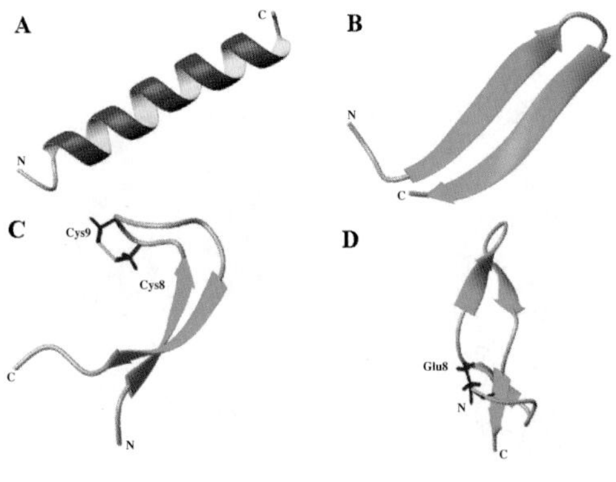

Plate 100. See also Figure 1 on page 1299.

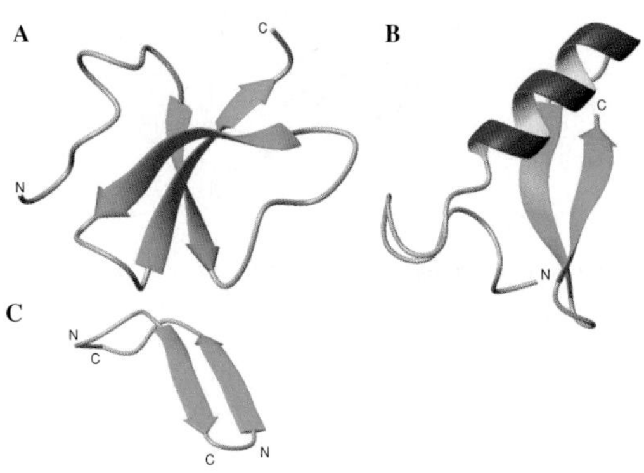

Plate 101. See also Figure 2 on page 1299.

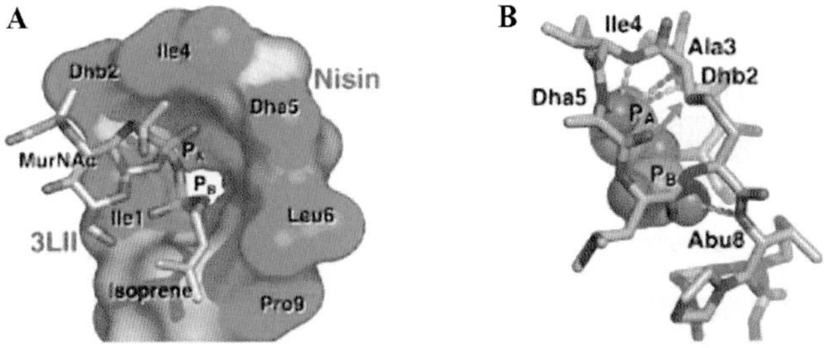

Plate 102. See also Figure 3 on page 1300.

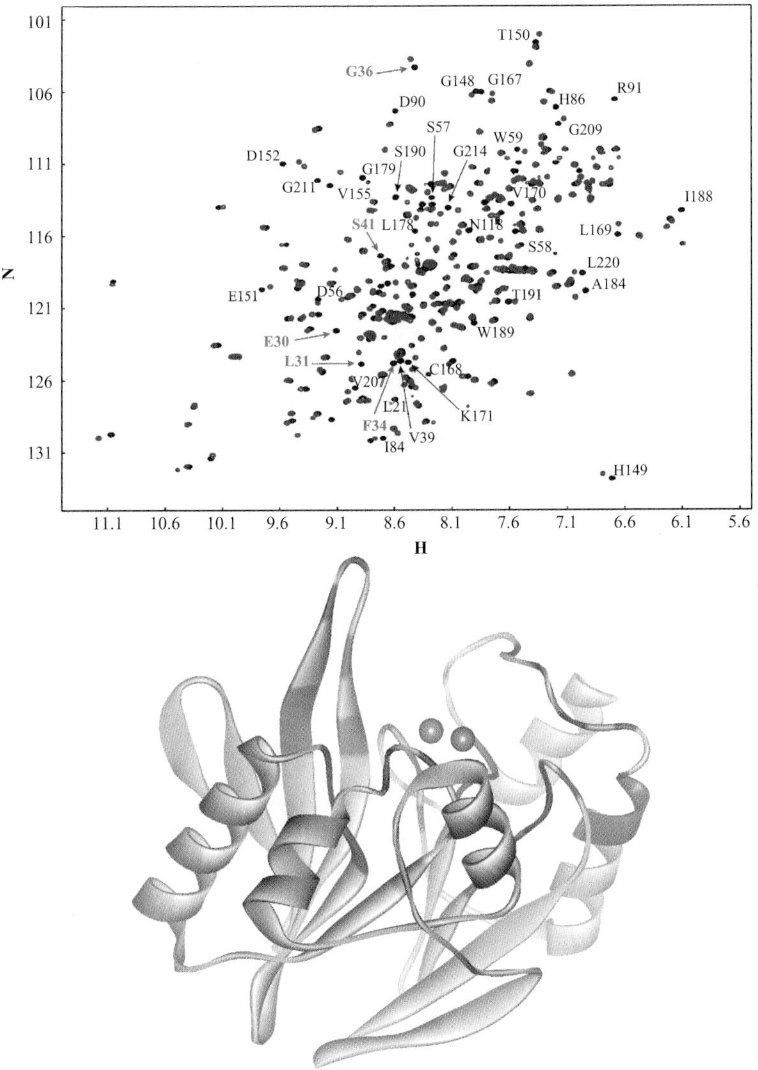

Plate 103. See also Figure 1 on page 1315.

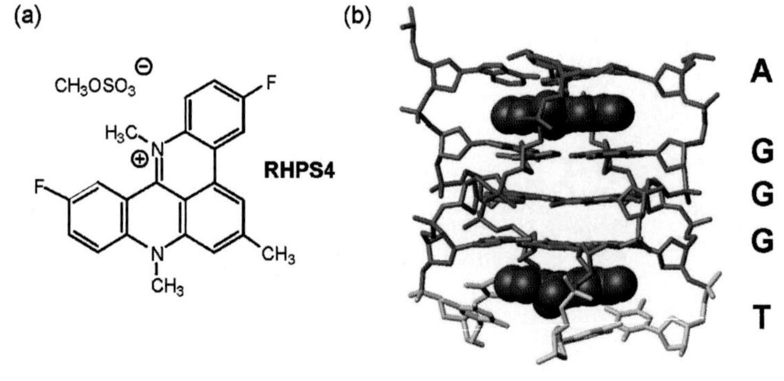

Plate 104. See also Figure 1 on page 1346.

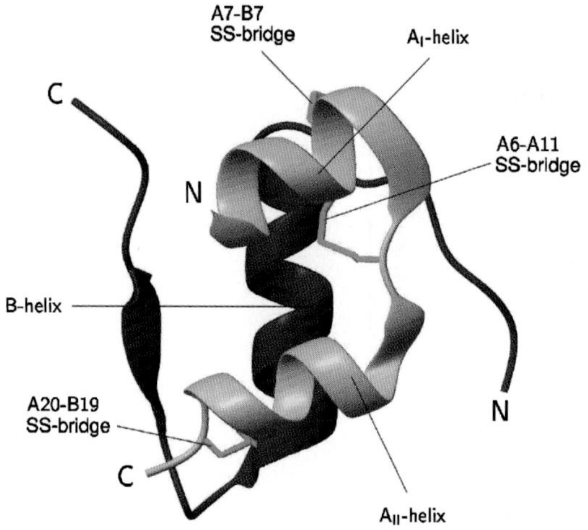

Plate 105. See also Figure 1 on page 1352.

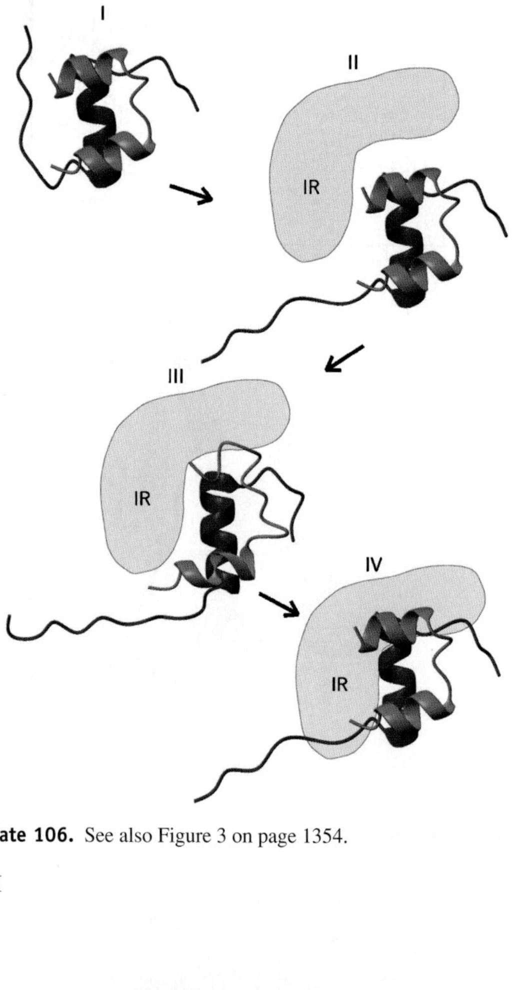

Plate 106. See also Figure 3 on page 1354.

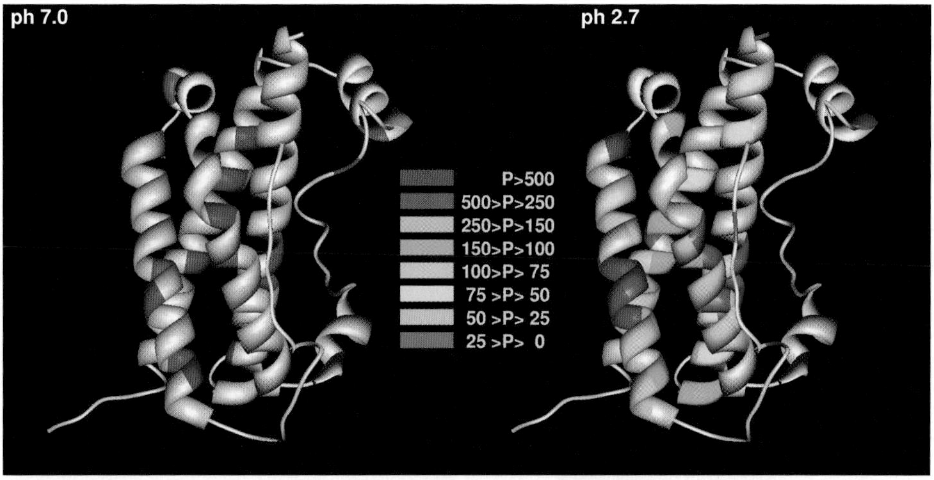

Plate 107. See also Figure 4 on page 1355.

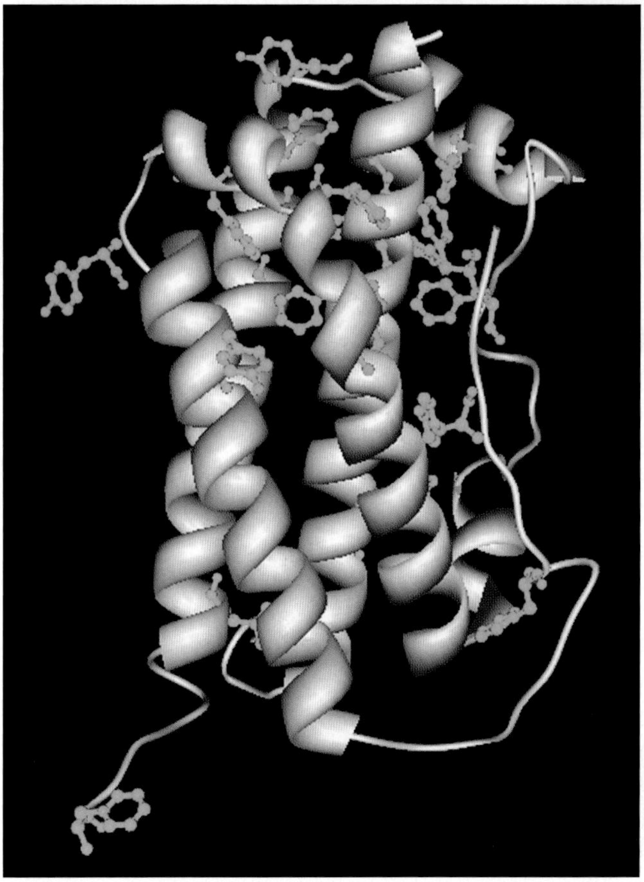

Plate 108. See also Figure 6 on page 1356.

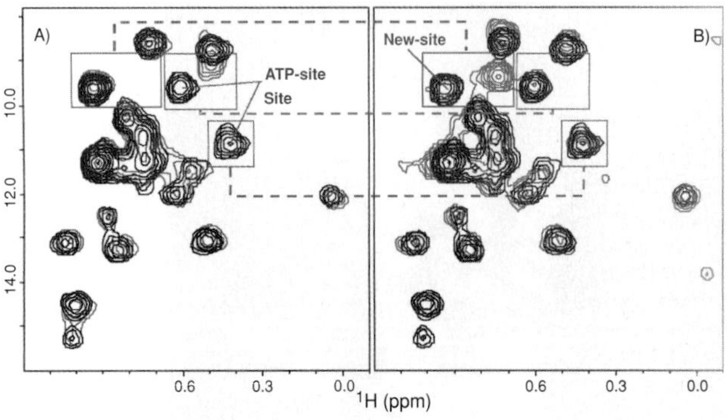

Plate 109. See also Figure 2 on page 1385.

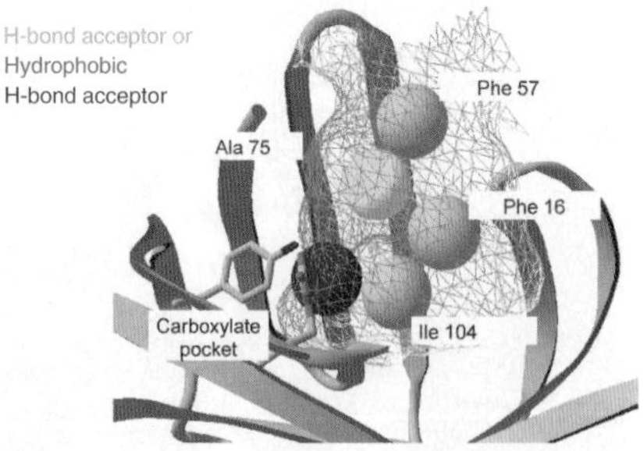

Plate 110. See also Figure 4 on page 1395.

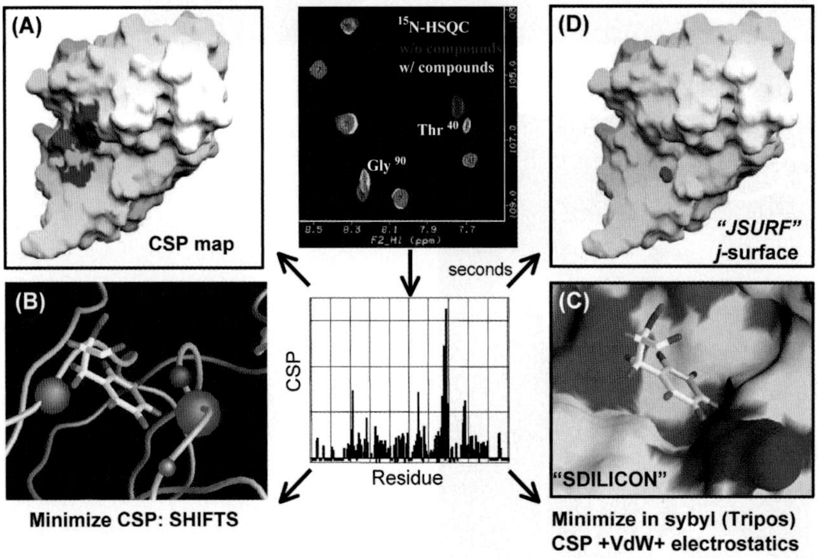

Plate 111. See also Figure 1 on page 1402.

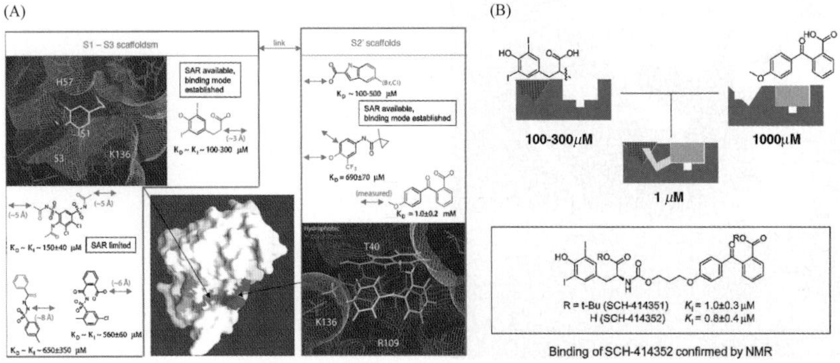

Plate 112. See also Figure 3 on page 1407.

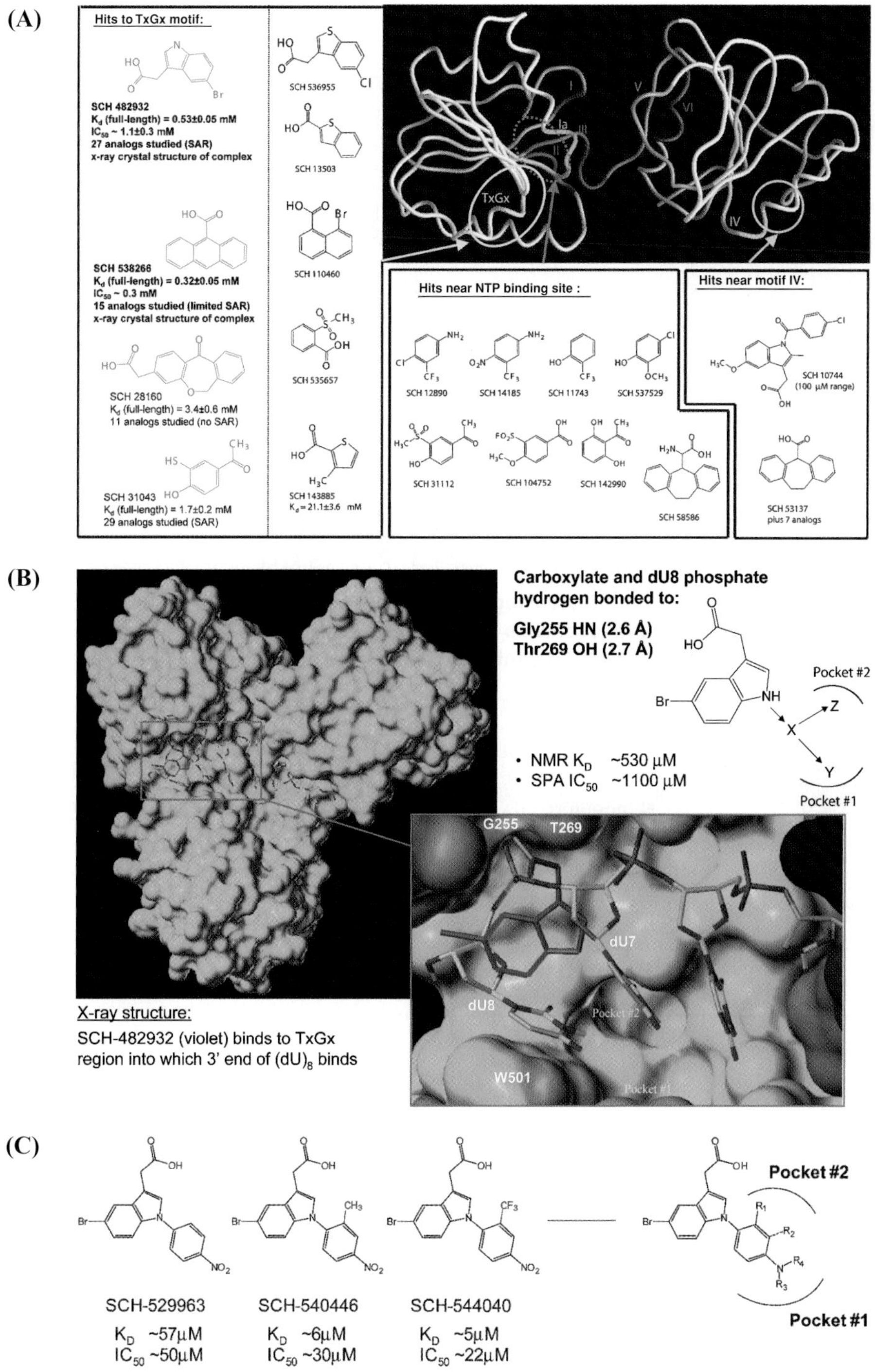

Plate 113. See also Figure 4 on page 1408.

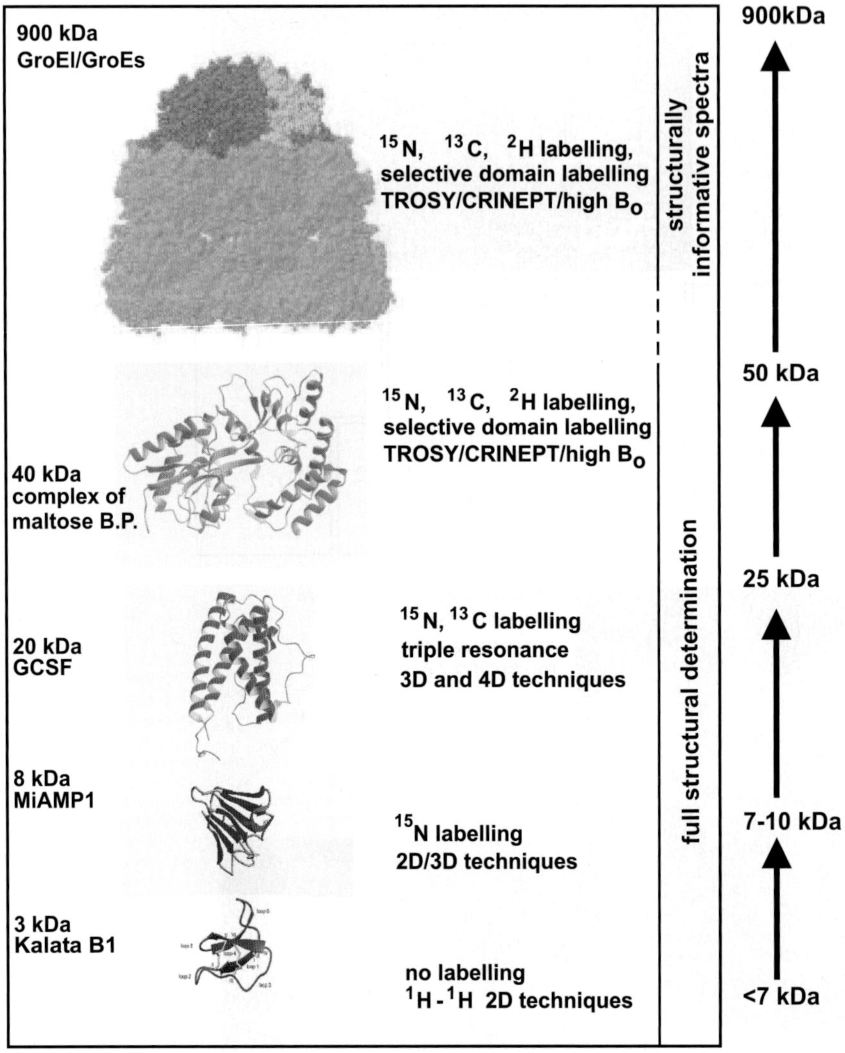

Plate 114. See also Figure 1 on page 1414.

Foreword

Magnetic resonance (MR) remains an emerging trend in technological innovation. In support is the award of four Nobel Prizes since 1991. Magnetic resonance has provided a paradigm shift in medical diagnosis and treatment planning and now plays an ever increasing role in health and medical research. Magnetic resonance imaging (MRI) is used worldwide with magnetic resonance spectroscopy (MRS), functional MRI and intraoperative MRI increasingly used in the clinic and for medical research purposes.

Each is technically difficult and requires a determined effort by those who develop, and use, the methods to obtain reproducible output at multiple sites. In the case of medical applications multidisciplinary teams are needed to develop, test and implement the method for routine clinical use in a controlled environment.

In this volume the authors, researchers from around the globe, have documented their technology in a way that will allow others to implement the methods for either research or clinical use. They share their many years of experience with the readers. We would like to thank them for taking the time to write their contributions with the end-user in mind.

We hope that the field of magnetic resonance in medicine will benefit from this volume by sharing the various experimental protocols and details. This will ultimately translate into improved medical care.

We would like to thank Professors Robert Lenkinski, Brian Tress, Drs Roger Bourne, Peter Malycha, Alan McLaughlin, Saadallah Ramadan, Robert Savoy and Peter Stanwell for their assistance reviewing the chapters.

Our special thanks to Dr Deborah Edward for the large amount of time and effort she put into editing and collating the Chapters. Thanks also to Sinead Doran for assistance with the diagrams.

Carolyn Mountford D. Phil (Oxon)
Uwe Himmelreich Ph.D.

Abbreviations

(HR)MAS: (High Resolution) Magic Angle Spinning

(S)SB: (shifted) sine bell

1D: One-Dimensional

^{1}H: MRS Proton Magnetic Resonance Spectroscopy

2D: Two Dimensional

31P MRS: phosphorus magnetic resonance spectroscopy

3D: Three-dimensional

Ac: Acetate

ADC: apparent diffusion coefficient

AQ: acquisition time

Asp: Aspartate

B1: Magnetic Field of RF Coil

BASING: Band Selective Inversion with Gradient Dephasing

BIR: B1-Insensitive Rotation

BISTRO: B1-Insensitive Train to Obliterate Signal

BPH: Benign Prostatic Hyperplasis

CABINET: Coherence transfer based spin-echo spectroscopy

CE-MRI: contrast enhanced magnetic resonance imaging

CHESS: chemical shift selective

CHESS: Chemical Shift Selective Saturation

CHESS: Chemical Shift Selective Suppression

COSY: correlated spectroscopy

CSF: Cerebral Spinal Fluid

CSI: Chemical Shift Imaging

CT-COSY: Constant time-based COSY

CT-PRESS: Constant time-based PRESS

DCIS: Ductal Carcinoma In Situ

DQF: double-quantum filtered

DWI: diffusion weighted imaging

EPI: echo planar imaging

EPSE: Echo-Planar Spin-Echo

FDA : Federal Drug Administration

FID: Free Induction Decay

fMRI: functional magnetic resonance imaging

FNAB: Fine Needle Aspiration Biopsy

FOV: Field of View

FWHM: full width at half maximum

GABA: γ-amino butyric acid

GAMMA: General Approach to Magnetic Resonance Mathematical Analysis

GB: Lorentzian-Gaussianlineshape transformation fraction

GBM : Glioblastoma Multiforme

Glc: Glucose

Gln : Glutamine

Glu: Glutamate

GM: grey matter

HCA: Hierarchical Cluster Analysis

HCO3-: Bicarbonate

HDR: haemodynamic response function

HMQC: heteronuclear multiple-quantum correlation

HPLC: High Pressure Liquid Chromatography

HR-MAS: High Resolution Magic Angle Spinning

HSQC: heteronuclear single-quantum correlation

ISI: inter-stimulus interval

ISIS: Image Selected In Vivo Spectroscopy

J: scalar coupling constant

KD: Ketogenic Diet

LASER: Localization by Adiabatic Selective Refocusing

LB: Lorentzian line broadening

LCModel: Linear Combination of Model Spectra

L-COSY : Localized Correlated Spectroscopy

LDA: Linear Discriminant Analysis

LG: Lorentzian Gaussian

MAS: magic angle spinning

MM: magnitude mode

MR: magnetic resonance

MR(S): magnetic resonance (spectroscopy)

MRI: magnetic resonance imaging

MRS: Magnetic Resonance Spectroscopy

MRSI: magnetic resonance spectroscopic imaging

NAFLD: Non-Alcoholic Fatty Liver Disease

NASH: Non-Alcoholic Steatohepatitis

NE: number of experiments

NOE: Nuclear Overhauser Effect

NOE: Nuclear Overhauser Enhancement

NS: number of transients

NVG: Noun Verb Generation

OLR: Orthographic Lexical Retrieval

OSIRIS: outer volume suppressed ISIS

OVS: Outer Volume Suppression

PA: Polyamines

PCA: Principal Component Analysis

PCG: posterior cingulated gyrus

PDE: phosphodiester

PME: phosphomonoester

PPM: Parts-Per-Million

PRESS: Point Resolved Spectroscopic Sequence

PRESS: Point Resolved Spectroscopy

PRESSCSI: Multi-voxel based PRESS

PRESSSV: single-voxel based PRESS

PS: phase-sensitive

PTC: Phosphatidylcholine

PWI: Perfusion Weighted Imaging

RF: Radio Frequency

SAR: Specific Absorption Rate

SCS: Statistical Classification Strategy

SECSY: Spin-echo Correlated Spectroscopy

SI: Spectroscopic Imaging

SNR: Signal-to-Noise Ratio

SPM: statistical parametric mapping

SS: Slice-Selective

SSB: Spinning Side Bands

STEAM: STimulated Echo Acquisition Mode

STEAMCSI: Multi-voxel based STEAM

STEAMSV: single-voxel based STEAM

STIR: Short Time Inversion Recovery

SVS: single voxel spectroscopy

T_1: spin-lattice or longitudinal relaxation time

T1: Longitudinal (Spin-Lattice) Relaxation Time

T_2: spin-spin or transverse relaxation time

T2 : Transverse (Spin-Spin) Relaxation Time

TCA: Tricarboxylic Acid

T_E: echo-time

TOCSY: total correlated spectroscopy

TPPI: time-proportional phase incrementation

T_R: repetition time

TRUS: Transrectal Ultrasound

U-FLARE: Ultra-fast Low Angle Rapid Acquisition with Relaxation Enhancement

VOI: Volume Of Interest

VOSY: Volume Localized Spectroscopy

VSS: Very Selective Suppression

WALTZ: Broadband Decoupling Scheme

W-F: water-to-fat ratio

WM: white matter

Metabolite Abbreviations

Ala: Alanine

ATP: Adenosine Triphosphate

Cho: Choline

Cr: Creatine

Eth: Ethanolamine

Glx: Glutamine/Glutamate

GPC: Glycerophosphocholine

GPE: Glycerophosphoethanolamine

Lac: Lactate

MI: Myo-Inositol

NAA: N-acetyl aspartate

NTP: Nucleotide Triphosphate (e.g., ATP)

PC: Phosphocholine

PCr: Phosphocreatine

PE: Phosphoethanolamine

Pi: Inorganic Phosphate

SI: Scyllo-Inositol

Tau: Taurine

UDP: Uridine Diphosphate

Glossary of Terms

Abscess: Focal infection that consists of an abscess capsule, containing fibrous material and chronic inflammation cells, and a central cavity filled with pus.

Absolute quantitation: Determination of concentrations of metabolites in mMoles per kg of tissue or volume.

Astrocytoma: The most common type of brain tumor in adults; astrocytomas, known for their marked potential for malignant progression and infiltrative nature, can be classified histopathologically into three grade of malignancy: WHO grade II astrocytoma, WHO grade III anaplastic astrocytoma, and WHO grade IV GBM.

B_0: Strong magnetic field, constant in time and space, generated by the superconducting magnet.

B_1: Radiofrequency (RF) magnetic field generated by *radiofrequency coils.*

Benign Prostatic Hyperplasia: Enlargement of the prostate and more specifically, overgrowth of the epithelium and fibromuscular tissue of the transition zone and periurethral area.

Bifunctional contrast agent: Contrast agent that increases the intrinsic contrast for two different imaging modalities.

calar coupling: See *J-coupling*

Contrast agent : Chemical compound that increases the intrinsic contrast in an image.

Δf: bandwidth (in Herz) of the receiver

Dendrimers: Large and complex polymers of a consistent size and a regular and highly branched architecture. They are often used for the encapsulation of smaller molecules.

Ductal Carcinoma In Situ: Noninfiltrative lesions composed of malignant epithelial cells that are confined to the mammary ducts and lobules.

Endocytosis : In endocytosis, the cell engulfs some of its extracellular medium including material dissolved or suspended in it. A portion of the plasma membrane is invaginated and pinched off forming a membrane-bound vesicle called an endosome.

Endosome : Vesicle formed during endocytosis.

η: filling factor

F: noise figure of the preamplifier

Focal infection: Infection that is results in the formation of lesions. Can be caused by bacteria, fungi, parasites or viruses.

Glioblastoma Multiforme: Highly malignant tumour with a low overall survival time for most patients. They are histologically characterised by a dense cellularity, high endothelial proliferation, and often the presence of focal tissue necrosis. These tumours are can occur in a cystic form.

Glioblastoma Multiforme: The most malignant grade of astrocytoma (WHO grade IV), which has an overall survival time of less than two years for most patients. These tumors are histopathologically characterized according to their dense cellularity, high proliferative indices, endothelial proliferation, and most importantly the presence of focal tissue necrosis.

Hetero-nuclear J-coupling: *J-coupling* between different species of spins, e.g. proton and carbon.

Homo-nuclear J-coupling: *J-coupling* between the same species of spins, e.g. proton and proton.

ISIS: Image-selected *in vivo* spectroscopy is based on a cycle of eight acquisitions which need to be added and subtracted in the right order to get a single volume. ISIS is considerably more susceptible to motion then *STEAM* or *PRESS* and is mostly used in heteronuclear studies, where its advantage of avoiding *T2-relaxation* is valuable.

J-coupling (or scalar coupling): Many resonances split into multiplet components. This is the result of an internal indirect interaction of two spins via the intervening electron structure of the molecule. The coupling strength is measured in Hertz (Hz) rather than *ppm* because it is independent of the external B_0 field strength.

k: Boltzman constant

K: numerical factor depending on the coil geometry

Lipoma: Commonly diagnosed benign tumors of adipose tissue that can develop anywhere in the body where fat is normally found.

Liposarcoma: A commonly diagnosed soft tissue sarcoma in adults which be found anywhere in the body and for which there exists several types with varying clinical outcomes.

Meningioma: Intracranial tumors that arise in the meninges and compress the underlying brain. In general, many of these tumors are benign, however, others are malignant with the capability to metastasize, both locally and distally.

Metabolite Profiling: The study of small molecules, such as lipids, peptides, amino acids and carbohydrates, which represent steady state concentrations of intermediate products or end products of cellular processes and as a result can be thought of as the ultimate response to genetic and environmental stimuli.

Mo: Magnetization

μo: Permeability of free space

Molecular imaging: Non-invasive, *in vivo* visualization of cellular and molecular events in normal and pathological processes.

Necrosis: Tissue and/or cell death.

Neoplasm: Literally meaning "new growth". An abnormal growth of tissue which may be benign or malignant.

NOE: Nuclear Overhauser effect, the magnetization of protons dipolar-coupled to ^{13}C nuclei can be used to enhance the ^{13}C signal. While the term NOE is mainly associated with ^{13}C MRS, NOE enhancement can also be observed in ^{31}P MRS and with other nuclei.

ω0: Larmor frequency

Oncogenomics: The study of genes, gene sequences, and the underlying genetic alterations that appear to be involved in oncological pathways.

Polarization transfer: Many interesting nuclei like ^{13}C suffer from low inherent sensitivity compared with proton MR. Techniques like DEPT (**d**istortionless **e**nhancement by **p**olarization **t**ransfer) and INEPT (**in**sensitive **n**uclei **e**nhanced by **p**olarization **t**ransfer) improve the ^{13}C sensitivity by transferring the higher polarization of coupled protons to the carbon nuclei. Special hardware with two RF channels is needed for polarization transfer experiments. A modification of DEPT and INEPT is reverse DEPT and inverse INEPT where the polarization is transfered back to utilize the higher sensitivity of the protons for observation (inverse detection).

PRESS: Point-resolved spectroscopy, utilizes three 180° slice selective pulses along each of the spatial directions and generates signals from the overlap in form of a spinecho.

Pus: Content of an infective lesion. It consists of dead and dying polymorphous cells (leucocytes), living and dead microorganisms, tissue debris and other components of inflammation (oedema fluid and fibrin).

Q: Quality factor of the coil

Reactive Nodes: Lymph nodes that have been immunologically challenged.

Responsive contrast agent: Contrast agent that conditionally increases the intrinsic contrast (depending on the change of environmental conditions like enzyme activation, pH change, change in ion concentration and others).

SAR: Specific absorption rate. Due to inductive and dielectric losses energy from radiofrequency pulses is absorbed by tissue and mainly transferred into rotational and translational movements of water molecules which causes an increase of tissue temperature. Limits for the human brain are established by FDA guidelines.

Schwannoma: Benign tumors that arise on peripheral nerves in general and on cranial nerves in particular, especially on the vestibular portion of the eighth cranial nerve.

Sensitivity: The term relating to the percentage of people with disease who test positive for the disease, i.e. the percentage of patients with cancer who have positive biopsy results.

SNR: Signal to noise ratio

Specificity: The term referring to the percentage of people without disease who test negative for the disease, i.e. patients without cancer with cancer negative biopsy results.

STEAM: Stimulated **e**cho **a**cquisition **m**ode, localization method utilizing three 90° slice selective pulses, along each of the spatial directions. Signal, in form of a *stimulated echo*, from the overlap is generated in a "single shot" experiment. In contrast to *PRESS*, only half of the possible signal is recovered when the same *echo time* is used.

Stem cell: Cell with broad differentiation potential that retains the capacity for unlimited self-renewal. A totipotent stem cell has the ability to differentiate to all cell types of an organism, whereas a pluripotent stem cell produces many but not all cell types.

T1-relaxation, T1-relaxation time: After the *magnetization* vector has been flipped into the transverse plane, new magnetization builds up along the z-axis. The time after 63% (1-1/e) of the equilibrium magnetization has built up is called the T1-relaxation time. T1 and T2 relaxation is caused by time-dependent fluctuations of local magnetic fields arising mostly from the motion of molecules with electric or magnetic dipoles at the site of the spins. For accurate *absolute quantitation* the relaxation times of all metabolites must be known in order to correct peak intensities appropriately.

T1-saturation: The repetition times *TR* are usually in the range of the *T1-relaxation* times. As a consequence of this, not all the *magnetization* has recovered, for example when TR = T1 only 63% of the equilibrium magnetization can be used for each scan (with the exception of the first scan) when 90° flip angles are used for excitation. This effect is called T1-saturation. The extreme case of saturation occurs when several *RF pulses* are applied within a very short time followed by dephasing *gradients*. This technique is used in localized ^{1}H MRS to remove the dominant water signal (see *water suppression*).

T2-relaxation, T2-relaxation time: The *magnetization* vector can be flipped into the transverse plane by using an *RF pulse*. The so generated transverse magnetization undergoes an exponential decay. The time after the magnetization has relaxed to 37% (1/e) of its amplitude is called the T2 or transverse or spin-spin relaxation time. See also *T1- relaxation*

Tc: probe temperature

TR (repetition time): The time between each initial excitation of the magnetization is called the repetition time.

Transfection: A method by which experimental DNA (or in our case contrast agents) can be incorporated into mammalian cells after interaction of the encapsulating transfection agent and the cell membrane.

Vc: volume of the coil

Part II

Acquiring Neurospectroscopy in Clinical Practice

Alexander P. Lin
Huntington Medical Research Institutes, Rudi Schulte Research Institute, Pasadena, CA 91105, USA

Part I: Seven Secrets to Successful Spectroscopy

Neurospectroscopy or magnetic resonance spectroscopy (MRS) has moved from the realm of academic research into that of the clinical world. All major MR manufacturers have aided in the endeavor by automating neurospectroscopy so that it no longer requires an MR physicist and is a push-button technique that can be run by technologists just as a typical MR sequence. Thousands of papers have demonstrated the clinical efficacy of neurospectroscopy and there are a dozens of medical reviews of how this technique can be applied across a wide range of neurological disorders. However, few papers address the practical issue of acquiring neurospectroscopy in a clinical practice. Based on over a decade of clinical experience on our 1.5 T scanner and applications training for technologists and radiologists at our international clinical neurospectroscopy courses, this chapter was developed to demonstrate proven protocols for clinical diagnosis and outline the strategies involved in acquiring successful clinical spectra.

Introduction

While the goals of the MR technologist are not very different between imaging and spectroscopy, the terminology and approach are slightly different. Imagers desire clear and detailed images without motion artifacts. Spectroscopists want spectra with well-defined peaks and few artifacts as well. The composition of a peak can be broken down to two parameters: (1) peak height or how readily discernible the metabolites are from noise; this in turn is governed by the principles of signal-to-noise ratio (SNR) and (2) line width or how narrow the peaks are, which is governed by magnetic field homogeneity. It is the goal of the spectroscopist to therefore maximize these two principles using a variety of different parameters discussed in this chapter. Using the same analogy, imagers use different sequences such as T_2w or FLAIR to investigate certain pathological properties; spectroscopists use different protocols to diagnose different disorders. These protocols can be broadly categorized into focal and global protocols, each of which will require a different approach to these parameters.

When we began developing and growing a clinical spectroscopy program, we soon discovered that the "secret" to a successful spectroscopy service was CONSISTENCY. Once a methodology of acquisition or guideline is established, it must be maintained. In any clinical MR unit, all technologists must follow that guideline in order to guarantee spectral reliability from patient to patient. Worldwide, the same consistency is essential to ensure "universality" of clinical diagnosis.

Signal and Homogeneity

In the following sections, we will be discussing different parameters that can be changed to optimize your spectroscopy. There are two components to optimization, namely: SNR and homogeneity. With a high SNR, the metabolite peaks (signal) are very easy to distinguish from the smaller peaks that surround it (noise) as you can see in Figure 1. High SNR is critical to the interpretability of spectra by both the radiologist and the automated processing software. Maintaining high SNR throughout all your acquisitions is a prime responsibility for spectroscopy quality control. The factors that affect SNR are (in order of impact): voxel size, number of averages, echo time (T_E), localization sequence, and repetition time (T_R).

Homogeneity refers to the uniformity of the magnetic field present in the voxel of interest (VOI: a voxel is the box which defines the area you wish to examine). A very homogeneous field has very little change in the magnetic field throughout the VOI (Figure 2B). A heterogeneous field has differing magnetic properties throughout the VOI (Figure 2A). A "sharp" spectrum at 1.5 T reflects both consistent and relatively thin line widths (<3 Hz) as shown in Figure 2D. The reason why spectroscopy is performed primarily in the brain is that the magnetic field is remarkably homogeneous throughout it. A severe example of inhomogeniety would be the magnetic field as dental braces affect it. By having a homogeneous field in your VOI, you reduce the chances of artifacts that would make the spectrum unreadable. These artifacts can affect the entire spectrum or affect just a certain area of chemical shift, however, you would like to avoid them if at all possible. The elements that cause homogeneity problems will be discussed in voxel positioning.

Graham A. Webb (ed.), Modern Magnetic Resonance, 971–989.

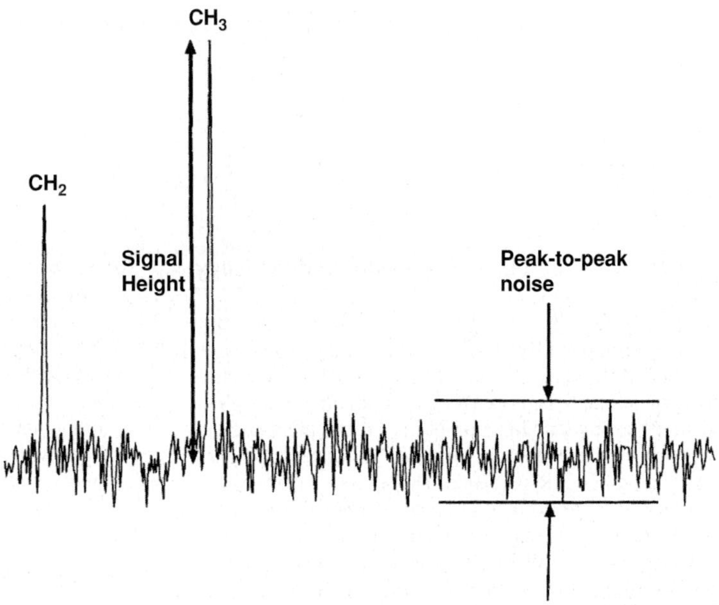

Fig. 1. Definition of signal-to-noise ratio. The large peak labeled as CH_3 is the signal and the peak-to-peak noise is indicated. Signal-to-noise ratio is simply the dividend of the height of the two measurements.

Acquisition Paradigms

As you will see throughout this chapter, there are essentially two methods of acquiring spectra: FOCAL vs. GLOBAL. The definition in terms of spectroscopic applications is as follows:

- *Focal:* Acquiring spectra from specific regions of interest that are variable from patient to patient. For example: in brain tumor spectroscopy, the location, size, and homogeneity of the tumor will differ from patient to patient. It is therefore critical to understand the underlying concepts behind each parameter such that you can use it to your advantage in different situations.
- *Global:* Acquiring spectra from a specific region of interest that is the same in all patients. A well-defined protocol is used for this type of spectroscopy. For example, in diagnosing Alzheimer's disease, the technologist will always place the voxel in the posterior cingulate gyrus (PCG) in the posterior parietal lobe of the brain. In this paradigm, very little changes from patient to patient. Therefore, great care must be taken to ensure that the protocol is followed to the letter.

Patient Positioning

Patient positioning plays a key role in maintaining consistency from patient to patient as well as from exam to exam. This is critical in use of spectroscopy for monitoring treatment or determining medication efficacy. More and more clinicians are turning to spectroscopy as a non-invasive tool to help determine how the patient is progressing in a quantitative way through neurospectroscopy. If the patient's head is not in the same position from his/her first MRS exam before treatment, it makes it all the more difficult to compare the results of the MRS exam after treatment.

- Landmark the patient at a consistent anatomical location. The eyebrow is a good location that is easily determined in each patient (Figure 3). By routinely landmarking in the same location, you can ensure that you are achieving maximum homogeneity within the magnet bore.
- Make sure the patient's head is centered. The left and right halves of the patient head should be symmetrical. This is critical for exams that compare the contralateral areas of the brain.
- Try to place the center of the coil at the landmark position. This ensures that the most homogeneous part of the coil is at the area of the brain where you will be acquiring your spectra. Depending on the anatomy of the patient, this may be difficult (i.e. short neck); however, try to do your best!
- Reduce patient movement by securing the head. Foam inserts or rolled up towels to the left and right side of the head help to reduce the chance of the patient moving their head after imaging (Figure 3). Using a strap around the forehead can also be useful but difficult to employ in claustrophobic patients.

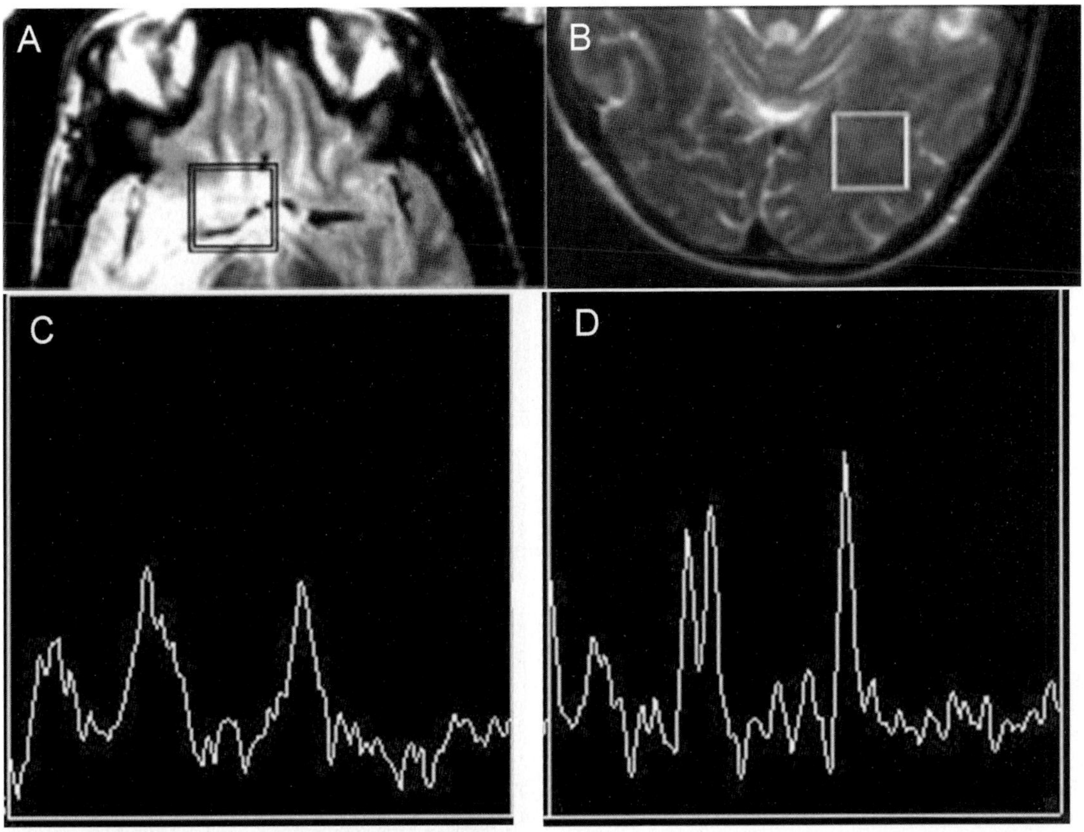

Fig. 2. Example of poor and good homogeneity. (A) This MRI image on the left demonstrates a voxel (indicated by the square) that contains different types of tissue as well as arteries. This is a very heterogeneous region. (B) The image on the right contains a voxel (square) from a very homogeneous region. (C) The long echo spectrum that was the result of the heterogeneous voxel in image A on the left has low signal as well as poor homogeneity as evidenced by the lack of splitting between Cho and Cr. (D) The long echo spectrum on the right, the result of the homogeneous voxel indicated in image B, although not of excellent quality, has enough signal to be interpreted.

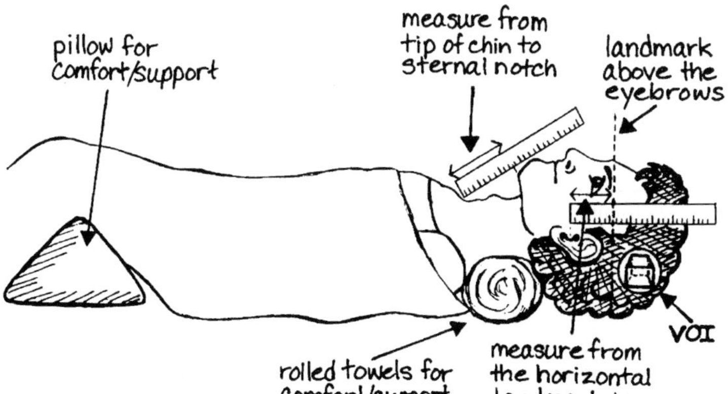

Fig. 3. Good positioning is achieved by a proper landmark, increased patient comfort, and additional measurements that can be used for repeat exams. Pillows and rolled towels provide comfort and support that prevents the patient from squirming about. Measurements such as those shown can dramatically improve spectral reproducibility.

Sequences

Manufacturers have greatly improved upon the localization sequences used in diagnostic spectroscopy. PRESS can reliably acquire spectra at short echo times and gives an almost doubled SNR advantage over STEAM. It is clearly the localization sequence of choice.

However, in many of the protocols described in available clinical texts, STEAM localization was utilized to acquire the spectra at 1.5 T. It is important to note that there are significant differences between metabolite ratio values for STEAM and PRESS that go beyond a simple scalar increase in SNR. For example, in exactly the same voxel in the same patient STEAM will give an NAA/Cr of 1.35, whereas PRESS will give a value of 1.71! Therefore, in order to use the spectra presented in other textbooks and papers as a comparison, you MUST use the exact methods described in the protocol. To avoid errors in diagnosis and in order to compare against those results, it is imperative not to mix STEAM and PRESS sequences. It is best to rigorously adhere to ONE sequence.

Echo Time

The greatest advantage of short echo neurospectroscopy is that the shorter the echo time, the greater the signal. Of course there is a practical limit to how short an echo time can be acquired. In general, echo times of 21–35 ms are used for neurospectroscopy. Another advantage of short echo times is that metabolites with short T_2 relaxation times such as *myo*-inositol, glutamate, glutamine, *scyllo*-inositol, etc. can be observed as shown in Figure 4. This adds to the diagnostic power of neurospectroscopy. For example, the diagnosis of Alzheimer's disease is less sensitive with NAA alone; the addition of mI allows neurospectroscopy to achieve excellent sensitivity and specificity. Initially, short echo neurospectroscopy was more prone to artifacts, however, at this time PRESS is robust. Because of another property of brain metabolites, termed *J* coupling, even small differences in T_E are critical: while using 27 ms may not give significantly different results, using 45 ms will!

So why use long echo time ($T_E = 135$, 144, 270, 288 ms)? Initially, it reduced the baseline "artifacts" observed sometimes with short echo spectroscopy. Also at long echo, lipid signals are significantly reduced, allowing other metabolites to be observed originally obscured by lipid. At $T_E = 135$ or 144 ms, lactate is easily observed as an inverted doublet (an admirable use of *J* coupling as described above), whereas in short echo it can often be hidden within the lipid resonance. This is all at the sacrifice of SNR of at least 40% as well as the reduction in the number of diagnostic markers. Typically, echo times of 135 or 144 ms are used. If you do decide to employ long echo times, be sure to recognize that the normal spectra heights are dramatically different in long echo than they are in short echo simply due to different intrinsic metabolite T_2 relaxation times. It is therefore imperative to stick with just ONE choice of echo time to save yourself from confusion and misdiagnosis.

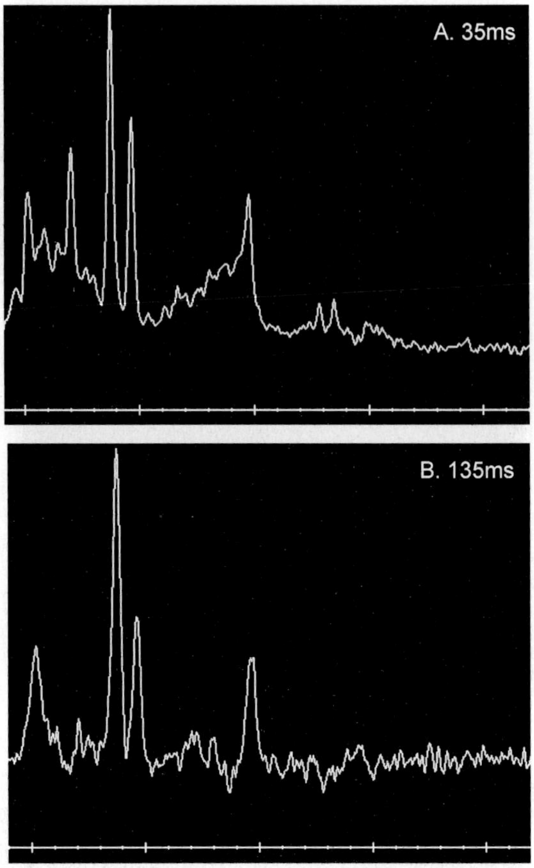

Fig. 4. Both of these spectra were acquired from the same exam using the same voxel position, voxel size, and number of averages using PRESS in a patient with recurrent brain tumor. The spectrum on top was acquired with an echo time of 35 ms, the bottom spectrum at 135 ms. Note that the Cho/Cr is vastly different and mI is absent in the bottom spectrum. Although the abnormal choline that is diagnostic of brain tumor is evident in both spectra, additional information, such as the significantly elevated mI that may indicate low-grade glioma, is missing from the long echo spectrum. It is difficult to tell the amount of signal loss since both spectra are scaled to the highest peak. However, the noise is much greater at long echo when compared to short echo to the extent that the small amount of lactate, which is diagnostic in the short echo spectrum, is not observed in the long echo spectrum.

Voxel Size

SNR is highly dependent on your voxel size. The typical voxel size of 2.0 × 2.0 × 2.0 cm (8 cc) is standard on most 1.5 T scanners. This is also an optimal voxel size for most 1.5 T applications (except notably in neonatal neurospectroscopy where smaller voxels may be required). In global diseases, the voxel sizes are well established and should not change from patient to patient. In some protocols, the default of 8 cc may not be used due to location-specific optimization; nonetheless, all patients examined with that particular protocol must have the same voxel size.

However, with focal neurospectroscopy, 8 cc may not be suitable to obtain enough partial volume of the lesion or may be too small for a large lesion. For lesions that are smaller than 8 cc, we recommend that you first obtain a spectrum from the larger volume to guarantee that you receive signal at all. A common mistake is to shrink your voxel to too small a volume as shown in Figure 5. In that case, there will not be enough signal to generate a readable spectrum. If the first voxel is successful, then at the very least the radiologist has some data to review. For a more accurate reflection of the metabolites within the lesion, after obtaining a successful first voxel, one can

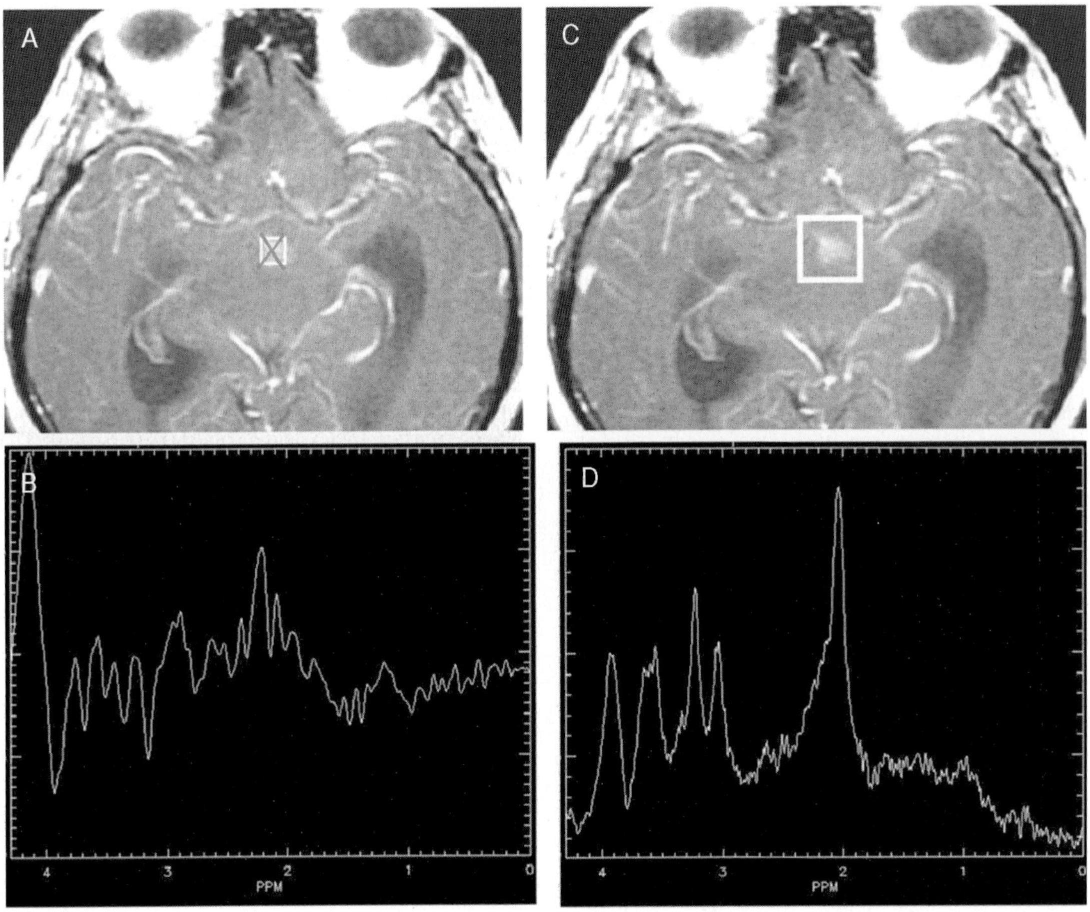

Fig. 5. Example of how voxel size affects spectral quality. In this patient case, a small hyperintense signal was discovered in the MRI that was too dangerous for surgical biopsy. Furthermore, MRI was indeterminate for neoplasm or stroke. It is therefore imperative for neurospectroscopy to provide an answer. When trying to acquire signal from a small lesion, it is common to try to shrink the voxel size to as small as possible. The result of such a small voxel (A) results in an uninterpretable spectrum (B). Although the larger voxel (C) has a small partial volume of the lesion, it produces (D) a good quality spectrum that gives a diagnostic result (high Cho) that proved to be a brain tumor.

then shrink the voxel to increase the partial volume. A good rule of thumb is that if the lesion occupies greater than 40% partial volume, there will be enough signal from tumor to generate an abnormal spectrum.

If the lesion is greater than 8 cc, then it would most likely make more sense to increase your voxel to match that volume. However, the larger the voxel, the greater the chances of observing inhomogeneity effects (such as artifacts and eddy current problems) in the volume. We recommend not obtaining a voxel larger than 30 cc. It is best to use a volume between 8 and 27 cc and change the voxel position instead of trying a larger voxel. This will be further discussed in the section regarding voxel position.

Another option when dealing with partial volume is to not assume that the voxel volume must be cubic. The advantages of using a different shape would be to maximize partial volume or minimizing the artifacts that one may get. For example, in trying to obtain a spectrum from the basal ganglia, we have observed that by using a rectangular voxel shape one can obtain a greater partial volume of the basal ganglia as well as minimizing the chances of the voxel being affected by other structures, such as the ventricles. However, it is best to keep the shape of the voxel as close to a square as possible since manufacturers most likely optimized their hardware to a cubic volume.

Number of Averages

The number of averages is directly proportional to SNR. This implies that the longer the acquisition, the more signal one can get. Obviously there are limits to the amount of time one can spend for each voxel. This is dependent not only on the patient's limits of tolerance but also due to the number of acquisitions required in a certain protocol. The total time for the scan is the prescan plus the number of averages in the scan multiplied by the repetition time.

In the global paradigm where you already have a defined size and location, it is best to set the number of averages the same for every acquisition as it does not need to be adjusted for signal and homogeneity effects. The data from the defined size and location should also always be acquired with the same number of averages.

In the focal paradigm, adjustments are dependent on the size of the voxel. Generally an 8 cc voxel requires 128 averages for a high quality spectrum. If you reduce the volume to half, you must increase the number of averages to 256 in order to maintain the same spectral quality. Therefore, if you use a very small volume such as 2 cc, you would require 512 averages, which at $T_R = 1.5$ ms would take nearly 13 min! This implies that small volumes are NOT the place to start acquiring data, hence why we recommend first acquiring data from an 8 cc volume and then reduce size accordingly. Homogeneity within the voxel should also factor into your determination of number of averages. By averaging for longer time, you reduce the artifacts that can occur in a heterogeneous acquisition.

Voxel Position

Voxel position is an inherent part of optimizing your ability to collect interpretable spectra. More importantly, the position plays a key role in the diagnostic power of the spectra. The diagnosis of diseases with neurospectroscopy is dependent on differences in metabolite peak heights as compared to normal values. If the difference you observe in a certain pathology is 15%, the reproducibility of that voxel must be 10% or better. In order to optimize reproducibility, the voxel parameters must be the same in every patient. All other parameters can be set as part of a protocol, leaving voxel position as the only factor that can affect reproducibility.

For example, in the diagnosis of Alzheimer's disease, an increase of mI/Cr of 10% is considered "significant." How does one determine if this change is significant? Utilize the standard deviation in a cohort of spectra from normal, healthy adults. In this case, the standard deviation is 5%; therefore the physician can rightly assume that the 10% change in mI/Cr is due to pathology and not due to normal variation in the human brain. In order to ensure that the standard deviation in all measurements was 5%, one must maintain reproducibility of the spectra. The voxel is placed and landmarked in the same position in every patient. The greater the reproducibility of the voxel, the less likely that the change observed has come from a different voxel location.

As with voxel size, changing voxel position can greatly alter your results when dealing with global spectroscopy. Gray and white matters contain different Cho/Cr ratios, therefore a shift in voxel position, causing it to contain more white than gray matter could lead the reader of the spectra to believe that there is a higher Cho/Cr in the voxel that he/she will attribute to pathology. This is the wrong conclusion to draw if the voxel position is incorrect and the observed change is due to greater partial volume of white matter. Although not all brains are exactly the same, using anatomical landmarks, such as the ventricles, will allow for a high measure of reproducibility. Positions for each diagnostic test will be described in detail with accompanying figures in the following section.

When dealing with the focal spectroscopy paradigm, the concerns of acquiring exact voxel position from patient to patient is disregarded because by definition, the key location will NOT be the same from patient to patient. In this case, it is more important to optimize the position such that you increase homogeneity as well as

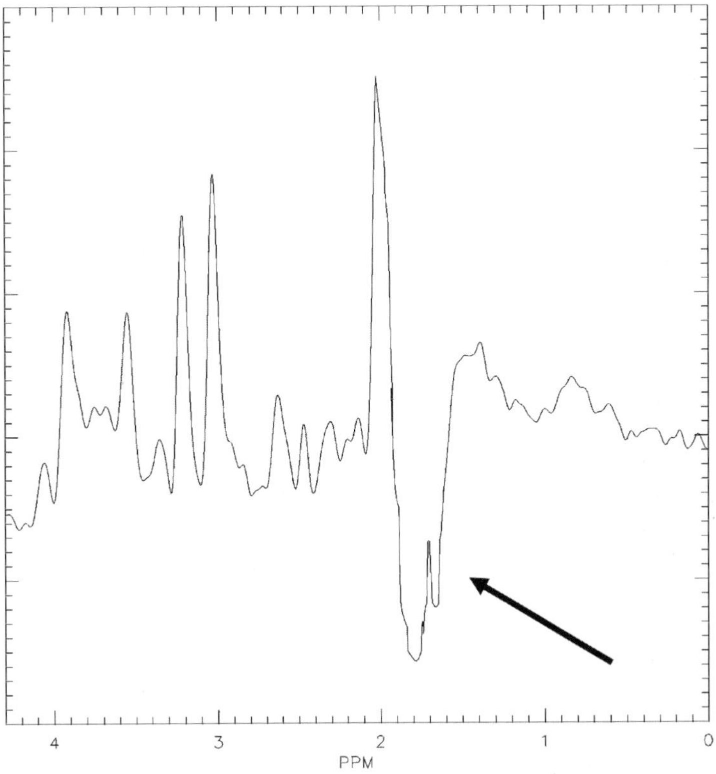

Fig. 6. Example of out-of-volume lipid artifact. The artifact is indicated by the arrow. This artifact comes about as a result of the voxel being to close to the back of the brain. The danger of this artifact is that it may impact upon the NAA signal that results in an artificially low NAA/Cr reading that would impact on spectral interpretation. This artifact is the reason why with voxel placements described in Protocols 1, 2, and 3 in Part II of this text it is critical to stay away from the skull.

signal. While in each pathology, different techniques in determining voxel position can be used; there are some areas of the brain or pathology you want to avoid:

- *Skull:* Stay away from the surfaces of the brain. The lipids that lie in the meninges of the brain, within the bones of the skull, or in the scalp can contaminate your spectra in two ways: (1) By creating an overriding lipid signal that can be misinterpreted or can dominate the entire spectra thus hiding the spectra; (2) an artifact that can result is "out-of-volume lipid artifact" (Figure 6) which is a large broad negative resonance or "dip" in the spectra around 1.8 ppm which can cause the resonance for NAA that resides at 2.0 ppm to be "dragged" down. Be sure to check all slices of the image that your voxel is located to ensure you have not come too close to the skull.
- *Necrosis:* As described above, lipids can cause severe problems with spectra. One major problem is domination of the spectra as shown in Figure 7. The spectra displayed on the scanner are scaled based on the highest peak. If there is a huge amount of lipid in the spectra, the scanner will scale the remainder of the peaks to the height of the lipid peaks. If the lipid peaks are very tall, the remaining peaks will be reduced and therefore become unreadable. In certain pathologies this is an important factor and therefore the strategy is to move away from the necrosis to the edge of the lesion.
- *Blood:* Blood contains heme-iron which is paramagnetic and will destroy signal coming from brain tissue. Post-surgical locations or highly necrotic centers can pool blood which makes spectra unobtainable or at the very least very difficult to read due to decrease in SNR.
- *Open space:* Within the brain there can be areas of open space such as in air sinuses adjacent to the temporal lobes. In cases of resection, there can be areas of the brain "missing." These should be avoided not only because of the possibility of blood; the large air spaces cause field inhomogeneity problems. You must be sure to check through all slices of the images to ensure that your voxel does not contain any of the air space.
- *Cerebral spinal fluid (CSF):* While having some CSF in the voxel can be unavoidable, it is important not to have CSF occupy more than 40% of the voxel. CSF has no brain metabolite signal, therefore, if CSF comprises

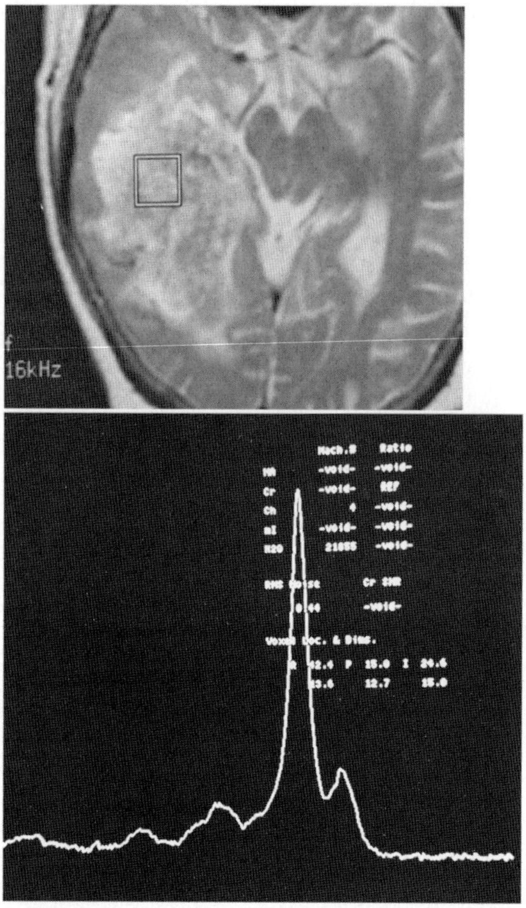

Fig. 7. Example of lipid overwhelming the spectrum. With the lipid peaks overwhelming the spectrum, it is very difficult to ascertain if there is an increase in Cho for diagnosis. Although this is a short echo acquisition, enough lipid within the voxel can negatively impact a long echo neurospectroscopy exam as well. This is a common problem when dealing with focal neurospectroscopy where the voxel is often placed in the center of a necrotic lesion as shown in the MRI at the top. The necrosis contains a great deal of lipid which would overwhelm the spectra, reducing its diagnostic value. The key to avoiding this artifact is to move the voxel to the edge of the lesion where there is not as much lipid signal from necrosis.

a large part of your voxel, the total signal is reduced causing a very "noisy" spectrum. CSF does, however, contain lactate (1–2 mM) and glucose, which could be misinterpreted as an indicator of pathology. Nevertheless, it is better to include CSF in your voxel if it helps avoid the four areas listed above.

As described in the section dealing with voxel size, partial volume is one of the attributes of focal spectroscopy that can be optimized with voxel position. This is especially the case in large areas of interest that should not or cannot be acquired with a single large voxel. In this case, one must use a smaller voxel that brings into question where to place the voxel: In the middle of the area of interest or at the edge? Studies have shown that in cases such as brain tumors, it is better to obtain spectra from the edge of the lesion away from the brain. This is due to the necrotic centers often found in tumors which as described above, can cause serious problems with your acquisition. By obtaining spectra from the edge you may not have complete partial volume but you guarantee yourself a more interpretable spectra. You should choose a voxel that has the highest partial volume while avoiding the areas stated above.

A frequently asked question is whether it is necessary to obtain a second spectrum that is contralateral to the focal voxel. This is obviously quite useful to the radiologist so that there is a comparable "normal" spectrum with which to compare the pathological one. The danger of blindly trusting such a spectra is that in some pathologies, the changes in metabolism are not limited to the region of interest. For example, when trying to distinguish radiation necrosis from tumor, it is assumed that the contralateral region to the tumor is "normal." However, studies have shown that radiation therapy itself causes global changes to the brain and therefore would affect the metabolite ratios of "normal" spectra. We recommend that one use this technique for the first generation of cases. As the radiologist becomes more familiar with the spectral patterns in different areas of the brain, it will not be necessary to provide a so-called normal contralateral spectrum.

Consistency

Once again we would like to reemphasize the importance of being consistent and reliably following protocol. When training our own technicians, we follow a set pattern that has allowed us over the course of a decade to acquire spectra with a high degree of reproducibility and very low variability. For example, one patient who has been examined every year for the past 6 years has demonstrated spectra with a reproducibility of less than 5% despite several upgrades and different technologists running each exam (Figure 8). The goal in mind is to try to achieve this level of consistency. This way, any biochemical changes can be attributed to pathology and not to changing acquisition parameters.

With the global paradigm, it is essential to be consistent. By following the protocol, you can ensure that your results will be comparable to published data utilizing the same technique. With the focal paradigm, it is not quite

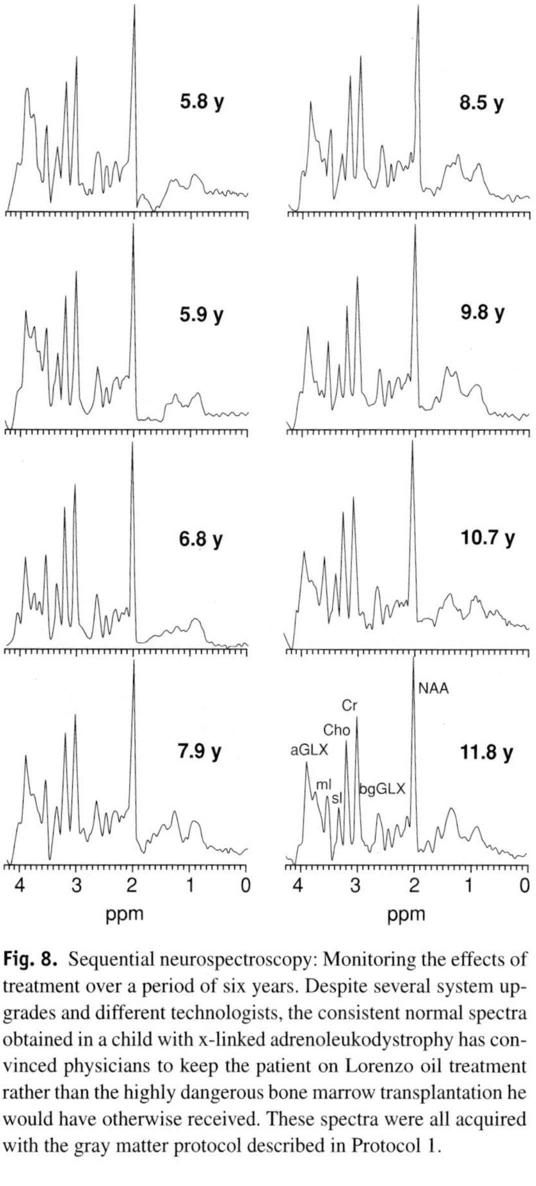

Fig. 8. Sequential neurospectroscopy: Monitoring the effects of treatment over a period of six years. Despite several system upgrades and different technologists, the consistent normal spectra obtained in a child with x-linked adrenoleukodystrophy has convinced physicians to keep the patient on Lorenzo oil treatment rather than the highly dangerous bone marrow transplantation he would have otherwise received. These spectra were all acquired with the gray matter protocol described in Protocol 1.

as simple: all of these parameters do interact with one another. In particular, voxel size and number of averages are very dependent on one another and will require adjustments if either is altered. If you decrease your voxel size by a factor of two (one-half) you increase the number of averages (to obtain the same level of SNR) by a factor of four! This in turn causes the exam to last longer which, in turn, can jeopardize the exam as a whole because the patient may become restless and move the head (moving your voxel to an entirely different location!) or worse, demand to be released from the scanner. How do you balance all of these elements?

We recommend starting with a "basic" protocol for all exams (this of course does not apply to global protocols which should not be changed) and then adjust the parameters according to the situation. Factors such as sequence and echo time should NOT be changed unless absolutely required. This is critical in maintaining consistency of spectra from patient to patient. Voxel size should be set at a default size: 8 cc is generally system's default size on most 1.5 T scanners. If you can acquire good quality spectra with 128 averages at 8 cc, then simply adjust averages according to size. Voxel position does not necessarily negatively affect the other parameters unless positioned incorrectly. Therefore, take special care in choosing the best area possible.

Multivoxel Spectroscopy

All of the rules described above also apply to multivoxel spectroscopy (a.k.a. Chemical Shift Imaging (CSI), Metabolite Maps, Spectroscopic Imaging, etc), the main difference being that instead of just an excitation volume (voxel) you must also consider the number of phase and frequency encoding steps as well as the field of view (FOV). The approach to multivoxel neurospectroscopy is more similar to imaging. It is easiest to imagine the FOV defining the image area as shown in Figure 9. The frequency and phase encode steps then divide the FOV into smaller boxes which can be reproduced as individual spectra. The increased number of encoding steps and the smaller the FOV, the higher the resolution you will have for each individual spectra. For example, a 16 × 16 cm FOV using 16 frequency and 16 phase encode steps will provide a volume of 1 cc per voxel when using a thickness of 1 cm. The same FOV, with the number of encoding steps increased to 24 will then give a resolution of 0.44 cc. This of course comes with the time penalty of increasing scan time approximately from 5 to 10 min. However, there is a signal gain by increasing the encoding steps. Depending on the resolution you require, you want to set your protocol accordingly. We give our standard protocol at the end of the next section.

The excitation volume is therefore equivalent to the voxel where the same rules in voxel positioning apply as shown in Figure 9. However, the number of averages and voxel size are no longer correlated since scan time is completely dependent on the number of encoding steps. It is a common mistake to try to use the largest volume possible (since there is no impact on time and all of the data from that region is acquired anyways), however, we have found that this affects the line width of the scan. We

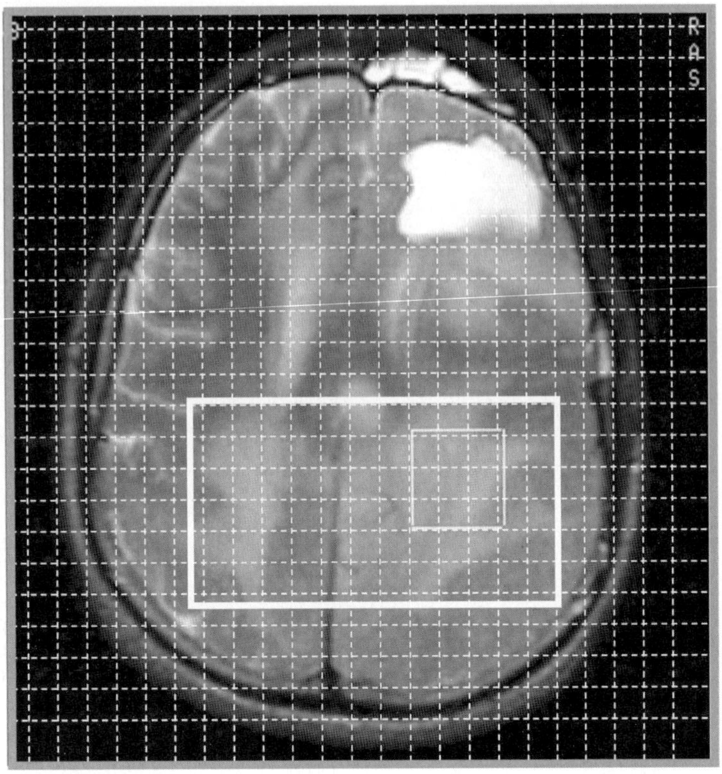

Fig. 9. Multivoxel neurospectroscopy. This is an example of short echo multivoxel neurospectroscopy in a patient with acute disseminated encephalomyelitis. The FOV is the entire image as indicated by the thick gray box. The FOV is then split into smaller voxels by phase and frequency encode steps as represented by the dotted line boxes. The excitation volume, or the voxel, is indicated by the thick white box. Note that the voxel does not acquire excitation from the entire brain but rather just from the region that is clinically relevant which are the hyperintense areas in the posterior aspect of the brain.

therefore recommend that the voxel only encompasses the region that you are interested in and no larger.

The third aspect of multivoxel neurospectroscopy that is different is the display of the data. Initially, multivoxel had the great promise of eliminating the need for interpreting spectra by using "metabolite maps"—color maps that correspond to the peak height of the metabolite over the area of excitation. However, we found that not only are the metabolite maps not reliable, they could be actually misleading. This has to do with the fact that it does not correct for overall signal strength. In other words, the high choline observed may not be due to increased Cho in the region but simply due to an increase in all MR signals from a homogeneous region. We therefore rely on the individual spectra for analysis. This presents a problem in itself since the radiologist will not want to comb over hundreds of spectra generated by each multivoxel exam. Instead we describe a "speed reading" technique where only select regions (those that are most clinically relevant) of the spectra are displayed. The region of interest is displayed with a summed spectrum and the individual spectra, usually in an easy-to-read 3 × 3 grid. This also has the benefit of providing quality assurance by generating a summed spectrum from the same voxel region as the single voxel exam. In this way, we find the most efficacious exams to include both single and multivoxel neurospectroscopy both using the same echo time.

Part II: Neurospectroscopy Protocols

Clinical Protocol Decision Matrix

All of the following protocols have been time-tested and proven to provide all of the necessary information required for the accurate diagnosis of disease. They are based on peer-reviewed published studies demonstrating efficacy with these techniques. In addition, the protocols are designed to be practical for the demands of a clinical practice.

The first step in applying these protocols is to know which diagnostic test is most efficacious for a disease process. With that in mind, the following section details a decision matrix for deciding which protocols should be used in each patient case. Although the list is not all inclusive, it provides a framework from which exam planning

Table 1: Clinical protocol decision matrix

Diagnosis	Protocol
Dementia, memory loss	Standard GM/WM
Frontal lobe Lewy body dementia	+Frontal GM
Focal Lesion (tumor, recurrence)	Focal SV-MRS, CSI
+Contrast	+T_1w, T_1w + c, MRS OK after Gad
vs. Stroke	+DWI, voxel in center of lesion and rim
vs. Abscess	Voxel in the center of lesion, maximize PV
Radiation necrosis vs. recurrence	Voxel at the edge of the lesion
Demyelinating disease, white matter disease	Standard WM
Multiple sclerosis w/o lesions	Standard WM
Multiple sclerosis w/ lesions	Lesion if >3 cc, CSI if <3 cc L or R WM in normal appearing WM
Epilepsy	Standard GM/WM
Mesial temporal lobe epilepsy	+L/R temporal lobe
Known focus	Focal SV (bilateral)
Parkinson's/Huntington's disease	Standard GM/WM, L/R BG
Hepatic encephalopathy	Standard GM/WM (earliest changes in WM)
Global hypoxia	Standard GM/WM
Trauma	SV MRS away from blood/lesions
Possible hypoxia	GM
No hypoxia	WM
Systemic diseases, unknown not focal	Standard GM/WM

can be initiated. The decision matrix also provides a basis for dynamic decision making at the time of the exam. For example, discoveries made at the time of the exam may necessitate additional tests to be run. It is important to note, however, that input from the managing physician at the time of the exam provides the greatest diagnostic precision.

Protocol 1: Standard Gray Matter (GM) or Posterior Cingulate Gyrus (PCG)

1. Position the patient in the scanner as indicated in Section 1 (Figure 3).
2. Acquire localization images for spectroscopy
   a. For single-plane acquisition scanners:
      i. Slice thickness + Interspace = 5.0 mm. That way you can place your voxel on the center slice and check two slices up and two slices down to ensure that your voxel does not fall into any bad areas.
      ii. For ease of landmarks, we suggest using the axial plane.
   b. For three-plane acquisition scanners:
      i. Choose a sequence that shows the best contrast between gray and white matters. We generally use a T_2w image.
3. Set up your neurospectroscopy scan parameters:
   a. Pulse sequence: PRESS
   b. $T_E = 35$ ms
   c. $T_R = 1500$ ms
   d. Voxel size = 20 × 20 × 20 mm
   e. Number of averages = 128
4. Place your voxel using the following landmarks:
   a. Center across falx at level 1 cm above posterior commissures of corpus callosum.
   b. Start inferior one slice (5 mm) above the slice with the (a) internal capsule, (b) angular artery in sylvian fissure, (c) occipitoparietal fissure, (d) vein of Galen, (e) internal cerebral vein, and (f) frontal horn of lateral ventricle. *Note:* If the head is tilted not all of these landmarks may be visible.
   c. By choosing a voxel shape of 20 mm in A/P and 20 mm in R/L, most of the tissue within the voxel is GM.
   d. Review above and below center position at least two slices (for 5 mm MRI slice thickness and 20 mm voxel slice thickness) for single-plane acquisitions. Check all three planes to ensure that you stay away from the skull, white matter (particularly the corpus callosum), and ventricles (Figure 10).
5. *Prescan routine:* this is different on different platforms. What is performed, however, remains the same:

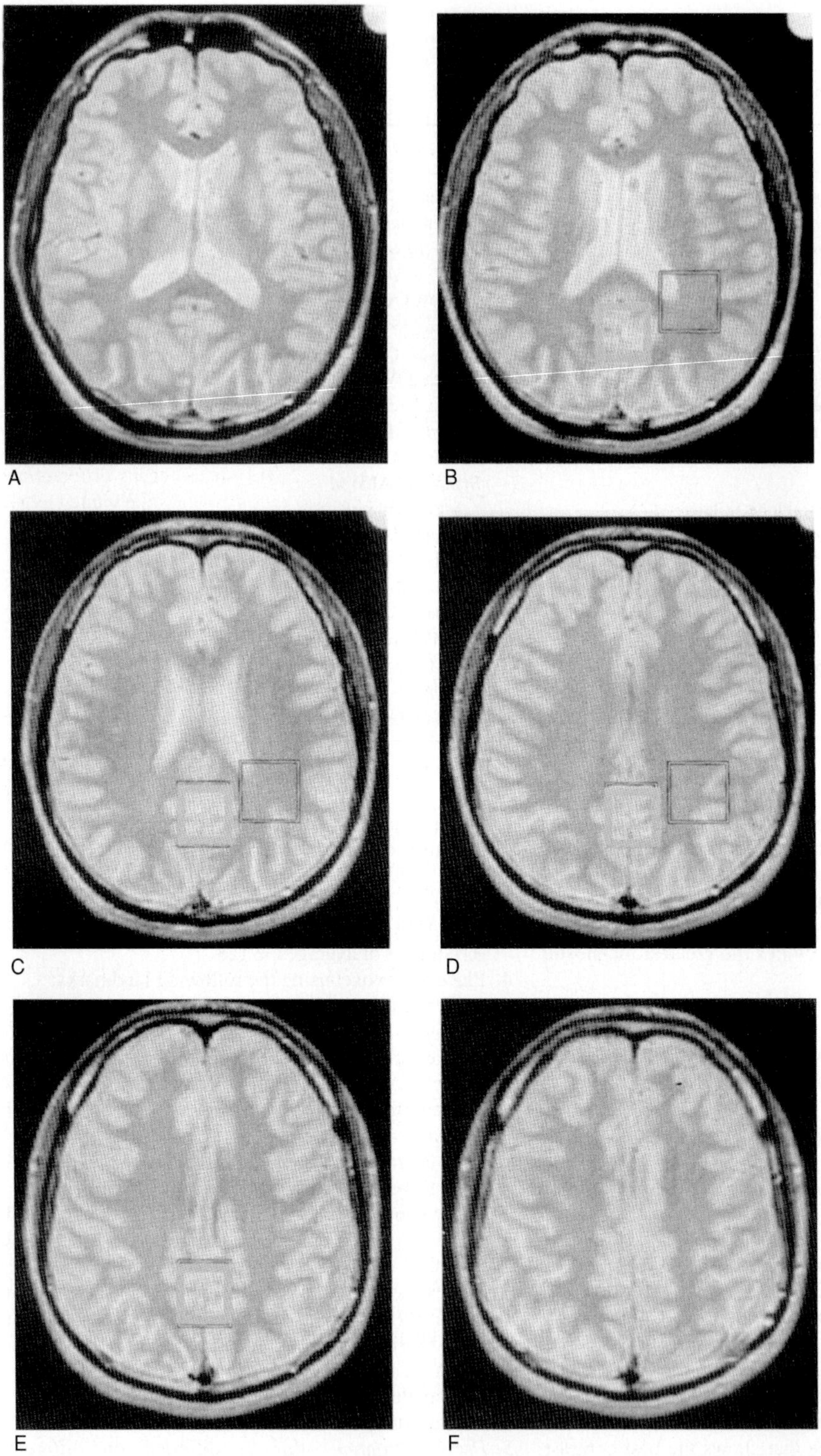

Fig. 10. Gray matter voxel position. Note that you should not only have a well-placed center slice but keep in mind all other slices as well.

   a. *TG optimization:* Optimize for the greatest amount of signal.
   b. *Water suppression:* You are better off to slightly under-suppress than to over-suppress.
   c. *Shim:* This is the most critical factor, the line width (sometimes represented as FWHM, full width half max of the water peak) should be less than 7 Hz at 1.5 T.
6. Acquire the scan.
7. Reconstruct the data. Generally this is automated, however, the routines include spectral apodization, fast Fourier transform, baseline corrections, and peak fitting.
8. Filming the data:
   a. Show the voxel location in three dimensions—"3D"
      i. Three-plane localizers, that is fairly straightforward. You simply place a voxel in each plane as shown.
      ii. For single-plane localizers, you will need to prescribe the center slice and then two slices above and below to demonstrate the voxel position in 3D.
   b. Show the reconstructed spectra. Label as GM or PCG.

Protocol 2: Standard White Matter

Repeat Protocol 1 with just a modification to step 4:

4. Place your voxel using the following landmarks:
   a. Stay in largely white matter, but allow up to 25% of GM.
   b. Landmarks are the
      i. Center posterior rim of left (right) lateral ventricle
      ii. 1–1.5 cm above posterior commissures of corpus callosum.
   c. Review above two and below at least two slices (for 5 mm MRI slice thickness and 20 mm voxel slice thickness).
   d. The S/I center position should be not more than 0–5 mm (superior) off from the GM center (Figure 11).

Protocol 3: Frontal GM

Repeat Protocol 1 with the following modifications:

4. For placement of the frontal GM voxel, use the axial T_2w MRI as shown in Fig. 12 (FGM is the box with the dotted lines, ignore the posterior boxes).
   a. Place the single voxel neurospectroscopy in the frontal GM just as you would for the PCG.
   b. You must tuck it in between the anterior skull and the anterior corpus callosum.
   c. Avoid being close to the nasal sinuses as well as CSF space.
   d. Try not to include the frontal white matter.

On GE systems: change FreqDIR to L/R instead of A/P. This is an important step as it reverses the gradient order which allows for better shims (Figure 12).

Protocol 4: Hippocampus/Temporal Lobe

1. For scanners that utilize the true coronal plane, make sure that the patient head is as straight as possible. If the head is tilted to the left or right, it will make voxel placement and/or comparison more difficult.
2. Acquire coronal localizers that can easily differentiate the hippocampal GM and temporal lobe white matter.
3. Set up the following parameters:
   a. Pulse sequence: PRESS
   b. $T_E = 35$ ms
   c. $T_R = 1500$ ms
   d. Voxel size $= 15 \times 15 \times 30$ mm (30 mm thick in the S/I direction)
   e. Number of averages $= 256$
4. Find the slice where the brainstem is apparent (as shown in Fig. 13) and place the voxel in the hippocampus. Try to capture as much of the GM of the hippocampus as possible. This will be your center slice.
   a. Check three slices up and three slices down—you may have to shift the center slice to capture as much of the hippocampus as possible.
   b. Make sure to not fall outside of the temporal lobe! Check your bottom left corner of the voxel for the left-sided voxel and bottom right corner for the right.
   c. Try not to include any of the inferior ventricles (Figure 13).
5. Prescan routine: It is absolutely essential that the line widths are less than 7 Hz. In many cases, you may need to check the line width and move the voxel to try to obtain a better line width. This is the most challenging aspect of hippocampal spectroscopy and why measurements from this region have a high variation.
6. Acquire data.
7. Prescribe a voxel that is exactly contralateral to the previously acquired voxel. This is also difficult but you must be absolutely precise in order to ensure comparability between the two spectra.
   a. Remember to take into account any tilt of the patient to ensure that the same regions are measured.
   b. For single-plane scanners, you may have to prescribe your center slice on a different slice if the patient's head is tilted to the left or right (reference the eye balls to check for head tilt).

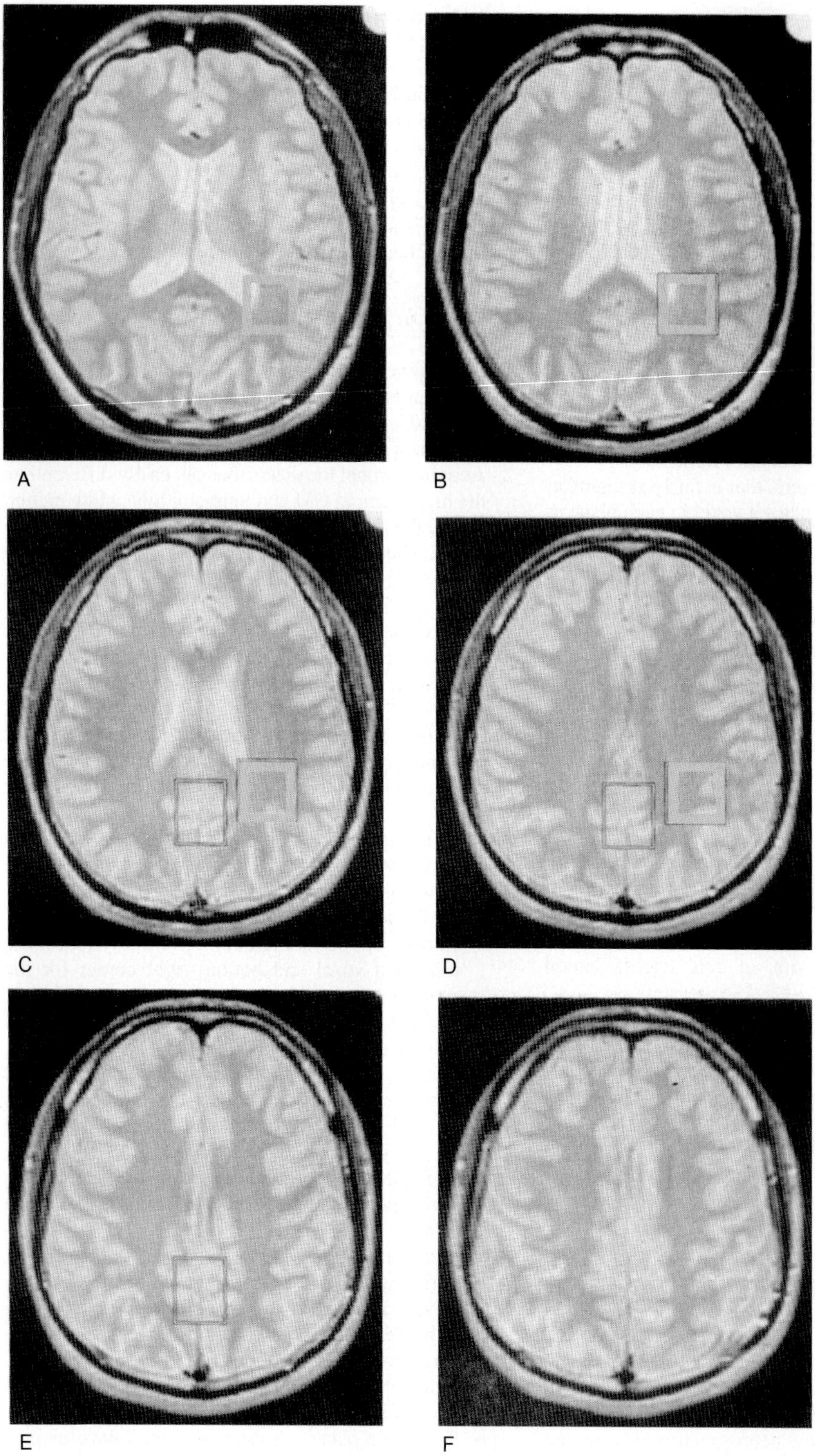

Fig. 11. White matter voxel position.

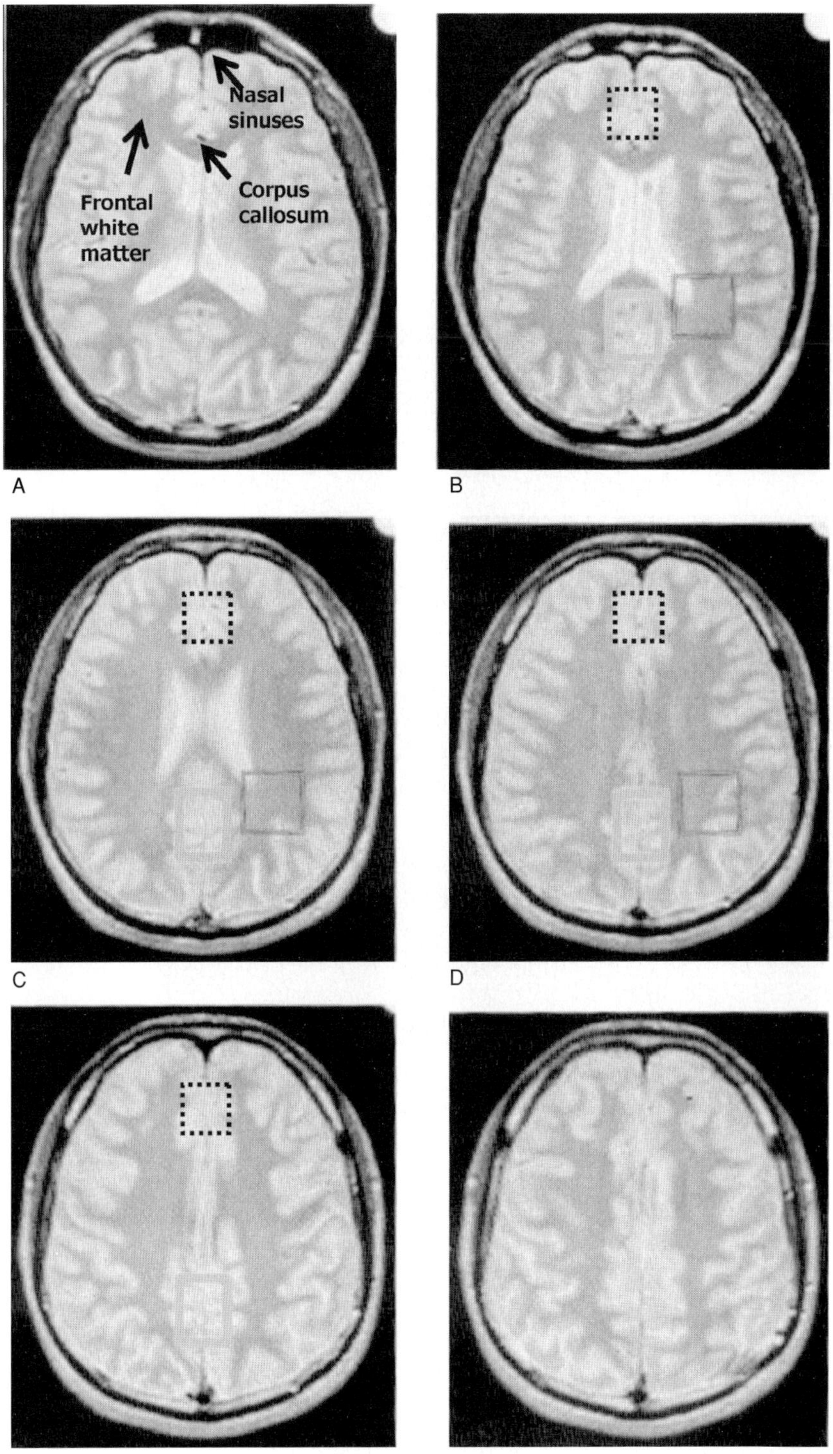

Fig. 12. Frontal gray matter voxel position. The frontal gray matter voxel is indicated in the dotted black lines. The posterior gray matter voxel is shown for reference. Be careful to avoid too much frontal white matter and corpus callosum in your voxel but at the same time stay away from the nasal sinuses.

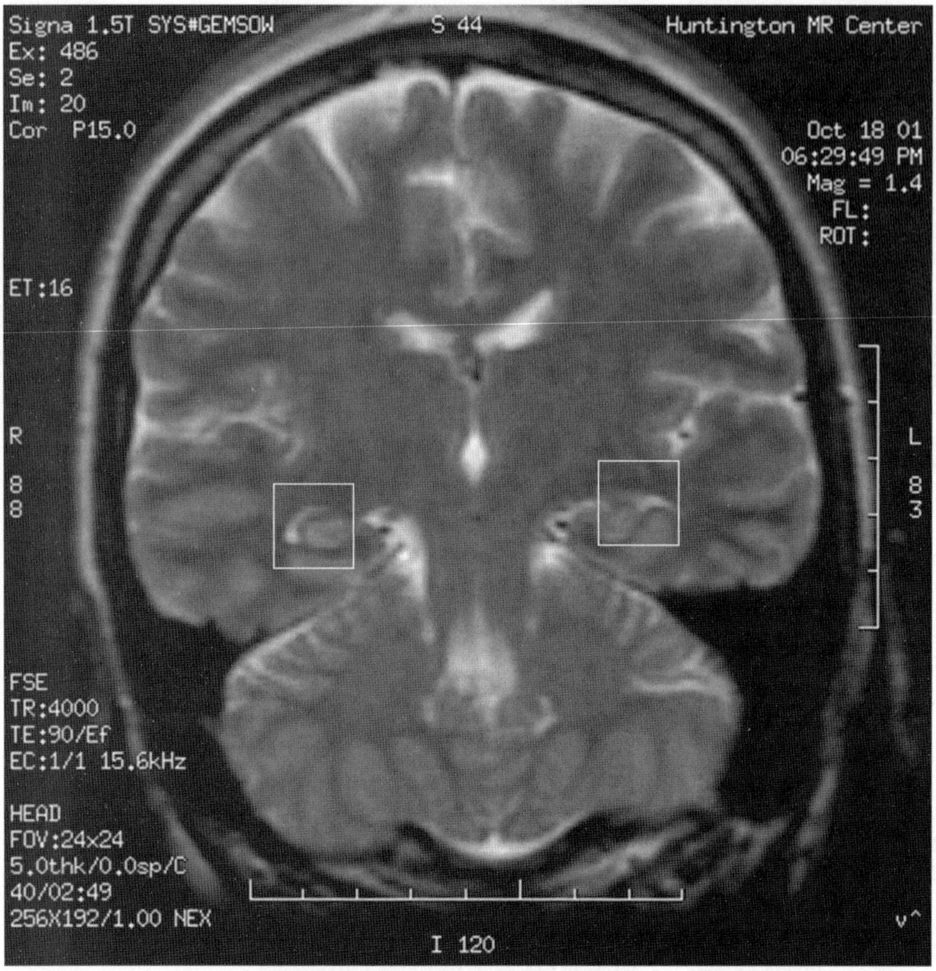

Fig. 13. Left and right hippocampal voxel position. Note how the voxel may not be on the same S/I level due to some minor head tilt. Not shown are the other slices that are 15 mm above and below this center slice.

Protocol 5: Multivoxel Neurospectroscopy (For Focal Use Only)

1. Position the patient in the scanner as indicated in Section 1 (Figure 3).
2. It is recommended that a single voxel exam precedes the multivoxel exam for the most efficacious exam. Proceed with focal techniques described in the first section.
3. Acquire localization images for spectroscopy
   c. For single-plane acquisition scanners:
      i. Slice thickness + Interspace = 5.0 mm. That way you can place your voxel on the center slice and check two slices up and two slices down to ensure that your voxel does not fall into any bad areas.
      ii. For ease of landmarks, we suggest using the axial plane.
   d. For three-plane acquisition scanners:
      i. Choose a sequence that shows the best contrast between gray and white matters. We generally use a T_2w image.
4. Set up your neurospectroscopy scan parameters:
   e. Pulse sequence: PRESS-CSI
   f. $T_E = 35$ ms
   g. $T_R = 1000$ ms

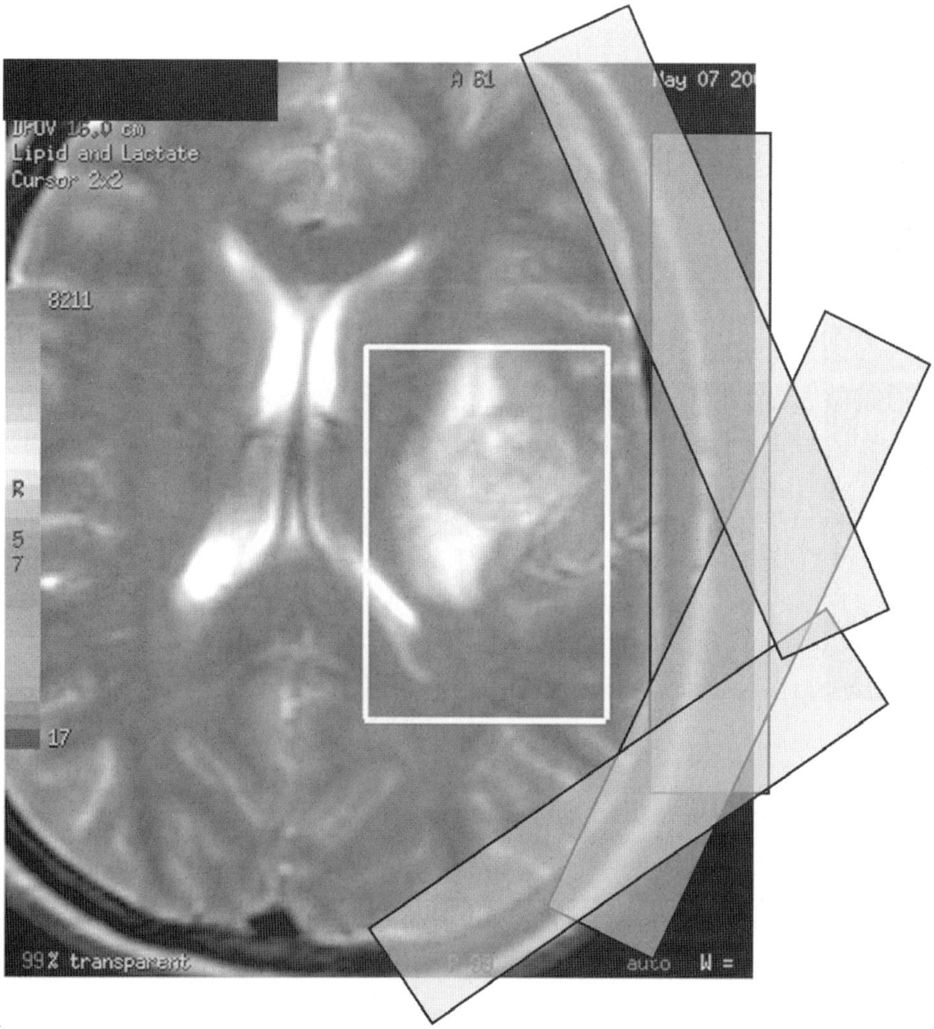

Fig. 14. Saturation band and excitation volume placement. If available on your scanner, additional saturation bands can be placed to regions close to the excitation volume to avoid lipid contamination from the skull.

h. Phase encode = 24
i. Frequency encode = 24
j. FOV = 24 × 24 cm
k. Slice thickness = 10 mm
l. NEX = 1

5. Prescribe the excitation volume voxel around the area of interest (i.e. lesion) and a bit more. Make the size relevant to what you are looking for. Use the same center slice as that of the single voxel exam. Use the same care described above in not placing the excitation volume too close to the skull and other areas of inhomogeneity.
6. If they are available on your scanner system, place saturation bands over skull closest to the voxel in order to avoid any lipid contamination. Make sure that the saturation bands do not intersect with the excitation volume. The saturation bands are used to cut out the residual water and any lipids surrounding the skull (Figure 14).
7. Prescan routine: In the case of multivoxel neurospectroscopy, since the volumes are larger, the line widths will not be as low as those for single voxel. Generally, a line width of <11 Hz is acceptable.
8. Acquire the data.

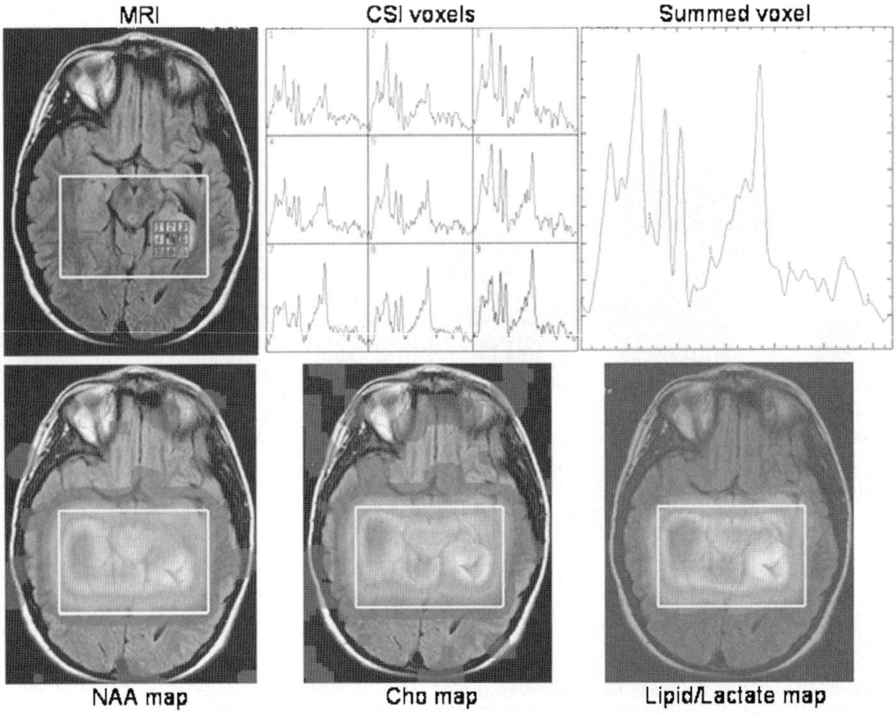

Fig. 15. Filming multivoxel neurospectroscopy for "Speed Reading." This case of a temporal lobe lesion demonstrates the value of multivoxel short echo neurospectroscopy. The region of interest is indicated on the images (top left) from which the summed spectrum is shown (top right). To benefit from the fine resolution of multivoxel neurospectroscopy without the obfuscation of hundreds of spectra, the individual spectra from the region of interest are also shown (top middle). The bottom row demonstrates the metabolite maps generated from this same exam. They are not particularly diagnostic despite excellent spectral quality as demonstrated in the top row. These spectra demonstrate a high mI signal which indicates a low-grade tumor which would have been missed if it had been acquired with long echo TE or relying on high Cho alone. The results were confirmed by autopsy. (See also Plate 82 on page LIX in the Color Plate Section.)

9. Data analysis: it is best to visualize the multivoxel as soon as possible. Users will want to check for quality first and then use the multivoxel results for planning a confirmatory single voxel exam. The reasons why a second single voxel result is recommended are a) it provides confirmation of multivoxel results, and b) the single voxel technique is much more robust and not prone to artifacts that could mislead the diagnosis.
10. When filming the multivoxel results, the "speed reading" technique is used where multiple regions from the excitation volume are selected and summed as well as individual spectra from that region are shown (Figure 15, top row). This allows for quick yet detailed inspections of multiple data points. It is important to also film the regions from which single voxel were acquired to provide quality assurance of your multivoxel data.

Summary

Once the above routines are well established, they can be enshrined in a protocol on your scanner. This limits the number of variables—and hence sources of error—as well as greatly speeding the diagnostic scan. In our clinical practice, technologists must go through a training routine before they are allowed to run patients. Initially they run "phantom" studies: acquiring spectra on a container with different concentrations of chemicals similar to those found in the brain. This can be purchased from the manufacturer or the Spectroscopy Institute (http://www.spectroscopy.org). This will enable the technologist to become familiar with the controls and the protocols. Secondly, they acquire several spectra from volunteers. This allows the new technologist to build up a working knowledge of different brain regions and their biochemistry as well as providing the group with a larger base of their own "normal values." If this is not possible, then close supervision of the acquisition on patients is necessary. Quality control is essential at this stage. Ensure that they are following protocol religiously. At the end, you will have a well-trained technologist in spectroscopy.

This chapter was written with the assumption that the reader will have a basic familiarity with clinical applications of neurospectroscopy. Recommended reading would include "Magnetic Resonance Spectroscopy Diagnosis of Neurological Diseases" by Else Danielsen and Brian D. Ross (Marcel Dekker, 1999). Additional resources include the Clinical MR Spectroscopy Course offered annually by Dr. Brian Ross and his staff at Huntington Medical Research Institutes (http://www.hmri.org) as well as on-site technologist training course offered by the Spectroscopy Institute available throughout the year (http://www.spectroscopy.org). The manufacturers such as General Electric Healthcare, Siemens Medical Systems, and Philips also offer applications training and courses that cover spectroscopy throughout the United States, Asia, and Europe.

Acknowledgments

All of the experiences detailed above would not have been possible without the guidance and support of Dr. Brian D. Ross at Huntington Medical Research Institutes. He and the many scientists that have been under his tutelage have truly brought the science to the bedside by developing neurospectroscopy as a clinical diagnostic tool. Drs. Stefan Bluml, Else Danielsen, Thomas Michaelis, Thomas Ernst, and Roland Kreis are among the many that have contributed to the clinical practice of neurospectroscopy.

Suggested Reading List for "Clinical Neurospectroscopy Protocols"

1. Danielsen E, Ross BD. Magnetic Resonance Spectroscopy Diagnosis of Neurological Diseases. Marcel-Dekker: New York, 1999.
2. Salibi NM, Brown MA. Clinical MR Spectroscopy: First Principles. Wiley-Liss: New York, 1997.
3. Mountford CE, Doran S, Lean CL, Russell P. Proton MRS can determine the pathology of human cancers with a high level of accuracy. Chem. Rev. 2004;104(8):3677–704.
4. Lin A, Bluml S, Mamelak AN. Efficacy of proton magnetic resonance spectroscopy in clinical decision making for patients with suspected malignant brain tumors. J. Neurooncol. 1999;45(1):69–81.
5. Preul MC, Caramanos Z, Collins DL, Villemure JG, Leblanc R, Olivier A, Pokrupa R, Arnold DL. Accurate, noninvasive diagnosis of human brain tumors by using proton magnetic resonance spectroscopy. Nat. Med. 1996;2(3): 323–5.
6. Ricci PE, Pitt A, Keller PJ, Coons SW, Heiserman JE. Effect of voxel position on single-voxel MR spectroscopy findings. AJNR Am. J. Neuroradiol. 2000;21(2):367–74.
7. Chang L, Miller BL, McBride D, Cornford M, Oropilla G, Buchthal S, Chiang F, Aronow H, Beck CK, Ernst T. Brain lesions in patients with AIDS: H-1 MR spectroscopy. Radiology. 1995;197(2):525–31. Erratum in: Radiology 1996; 198(2):586.
8. Shonk TK, Moats RA, Gifford P, Michaelis T, Mandigo JC, Izumi J, Ross BD. Probable Alzheimer disease: diagnosis with proton MR spectroscopy. Radiology. 1995;195(1):65–72.
9. Kantarci K, Jack CR Jr, Xu YC, Campeau NG, O'Brien PC, Smith GE, Ivnik RJ, Boeve BF, Kokmen E, Tangalos EG, Petersen RC. Regional metabolic patterns in mild cognitive impairment and Alzheimer's disease: a 1H MRS study. Neurology. 2000;55(2):210–7.
10. Ashwal S, Holshouser BA, Shu SK, Simmons PL, Perkin RM, Tomasi LG, Knierim DS, Sheridan C, Craig K, Andrews GH, Hinshaw DB. Predictive value of proton magnetic resonance spectroscopy in pediatric closed head injury. Pediatr. Neurol. 2000;23(2):114–25.
11. Lee PL, Yiannoutsos CT, Ernst T, Chang L, Marra CM, Jarvik JG, Richards TL, Kwok EW, Kolson DL, Simpson D, Tang CY, Schifitto G, Ketonen LM, Meyerhoff DJ, Lenkinski RE, Gonzalez RG, Navia BA, HIV MRS Consortium. A multicenter 1H MRS study of the AIDS dementia complex: validation and preliminary analysis. J. Magn. Reson. Imaging. 2003; 17(6):625–33.
12. Ross B, Bluml S. Magnetic resonance spectroscopy of the human brain. Anat. Rec. 2001;265(2):54–84.

Application of Magnetic Resonance for the Diagnosis of Infective Brain Lesions

Uwe Himmelreich[1] and Rakesh K. Gupta[2]

[1] *Institute for Magnetic Resonance Research and Centre for Infectious Diseases and Microbiology, Westmead Hospital at the University of Sydney, Sydney, Australia; and*
[2] *Department of Radiodiagnosis, Sanjay Gandhi Postgraduate Institute of Medical Sciences, Lucknow, India*

Abbreviations: ADC, apparent diffusion coefficient; DWI, diffusion-weighted imaging; GBM, glioblastoma multiforme; MRI, magnetic resonance imaging; MRS, magnetic resonance spectroscopy; PCA, principal component analysis; PWI, perfusion-weighted imaging; PRESS, point-resolved spectroscopy; SCS, statistical classification strategy; SNR, signal-to-noise ratio; STEAM, stimulated echo acquisition mode; TE, echo time; TR, repetition time; VOI, volume of interest

Introduction

Brain infections are often acute and potentially life-threatening diseases of immunocompromised and healthy hosts. Rapid, accurate, and safe diagnostic methods are essential to initiate optimal treatment and reduce mortality from brain infections. We will focus here on focal brain infections as an example of the diagnostic potential and limitations of magnetic resonance imaging (MRI) and spectroscopy (MRS) methods. Particular attention will be paid to distinguishing between tumors and infective lesions. Different types of infections will be discussed.

Brain lesions are usually diagnosed after biopsies are taken. A definitive diagnosis is still typically established by culture of bacteria from pus obtained after neurosurgical biopsy, or inferred by culture of infected material from other body sites. Computerized tomographic (CT) and MRI methods have reduced mortality by improving the detection and localization of brain lesions over the last decades [1,2]. Typically, necrotic or cystic tumors, tumorfactive demyelinating lesions, and cerebral abscesses show similar appearances using conventional radiological methods. The use of diffusion-weighted images, perfusion MRI and MRS have been suggested for differential diagnosis of these pathologies [3–5]. However, differential diagnosis of post-contrast ring-enhancing cerebral lesions continues to be a challenge, as such lesions include glioblastoma, metastasis, pyogenic abscess, subacute ischemic infarction, resolving hematoma, and demyelinating disease [3,6–8]. This becomes a particular problem when other indications of infections (like fever) are missing and microbial culture from body fluids is negative, as can be the case in up to 50% of infective lesions [2]. In addition, case reports are relatively sporadic and the microbial spectrum is diverse.

Pathological, clinical, and diagnostic aspects of brain infections, and their diversity, have been addressed in a number of reviews and are beyond the scope of this review [1,2,6,9]. Although detailed descriptions of the fundamentals of MR techniques have been addressed elsewhere [6,10], essential methods will be discussed and their application will be demonstrated, using the example of focal brain infections.

Scenarios that need to be addressed by non-invasive diagnostic tools include: (1) the differentiation between infective lesions and other pathologies like glioma and (2) the identification of the infection-causing microorganisms.

Magnetic Resonance Techniques

The physical basis of MRI and MRS has been reviewed by numerous authors (for example Ref. [10]). In order to access the most appropriate MR techniques one has to consider the physical nature of an infective brain lesion. A "true" abscess consists of an abscess capsule, containing fibrous material and chronic inflammation cells, and a central cavity filled with pus. Pus is of variable viscosity and consists of dead and dying polymorphous cells (leucocytes), living and dead microorganisms, tissue debris, and other components of inflammation (edema fluid and fibrin) [1,2]. In contrast to "true" abscesses, some focal infections (parasitic, fungal) lack an abscess wall. In order to distinguish between abscesses and other focal brain diseases and to distinguish between various types of focal infections the following properties are of interest for diagnosis: proton density, spin–lattice relaxation time (T_1), spin–spin relaxation time (T_2), flow, diffusion, chemical composition, and variation in uptake of contrast agents. For clinical diagnosis, it is important to select protocols that are rapid, as some of these patients are critically ill. Short protocols increase patient comfort, allow

Graham A. Webb (ed.), Modern Magnetic Resonance, 991–999.

additional studies, and decrease motion artifacts. It is also important to suggest protocols that can be implemented on various hardware platforms, including older MR systems that are common in developing countries where the cohort of abscess patients is large.

At present, two-dimensional (2D) multislice images are commonly used in routine diagnosis. The most widely used pulse sequences are based on spin-echo sequences. The main drawback of the spin-echo sequence is a relatively long acquisition time. Alternatively to spin-echo T_1-weighted images, inversion recovery results in stronger T_1 weighting due to an initial 180° pulse followed by an evolution time, during which the inverted spins relax back to $+Z$ according to T_1. By choosing a specific evolution time that suppresses lipid resonances, T_1-weighted images can be acquired with inversion recovery spin-echo sequences that suppress lipid signals. Combining inversion recovery with T_2-weighted spin-echo sequences can be used to attenuate otherwise hyperintense signals from fluids [fluid-attenuated inversion recovery (FLAIR)]. FLAIR permits the detection of CNS lesions that are adjacent to regions containing fluids (e.g. CSF). Significant improvement of lesion-to-CSF contrast has been achieved using FLAIR [11].

Reduction of repetition times (TR) is possible with gradient echo imaging. Gradient echo images are particularly sensitive to field inhomogeneities. Instead of 180° pulses, gradients are used to refocus the signal. Adjustment of pulses is therefore less critical than in spin-echo sequences. Spin-echo sequences are sensitive to T_2 effects, gradient echo sequences are sensitive to T_2^* effects. Due to the sensitivity to field inhomogeneity, the quality of gradient echo imaging is usually lower compared with spin-echo sequences.

A more rapid version of spin-echo sequences is fast-spin-echo or turbo-spin-echo imaging [12]. The total scan time is hereby reduced by traversing through several phase-encoding levels during each TR. Thus, several echoes are collected during each TR. Fast-spin-echo methods require rapid switching of gradients, imposing considerable strain on the hardware. The large refocusing angles make constraining RF power a challenge. Nutation angles of much less than 180° can overcome this problem [13]. Other rapid acquisition methods include gradient echo fast-low-angle-shot (FLASH) [14] and echo planar imaging (EPI) [15,16]. In EPI the entire matrix is acquired in one excitation with multiple phase-encoding gradient echoes. The penalty of the rapid data acquisition is a relatively low image quality. EPI is of great importance for applications that do not necessarily require a high resolution such as functional MRI or diffusion-weighted imaging (DWI) [17]. The drawback of EPI is image distortion, dropout, blurring, and signal loss [18]. EPI quality and resolution can be improved by reducing the readout time (parallel imaging or half-Fourier techniques) [19].

Acquisition time gained by the application of rapid pulse sequences can be reinvested in improving the quality of the image. This applies to increasing the signal-to-noise ratio (SNR) by averaging more transients, increasing the resolution to reduce partial volume effects by acquiring a larger data matrix or by applying stronger T_2 weighting using longer TR and TE values.

Differences in tissue molecular self-diffusion due to translational motion can be monitored by application of strong magnetic field gradients. Molecular motion between the application of these gradients results in dephasing of the MR signal. Diffusion-weighted images can be acquired by varying either the gradient strength (G), the gradient length (δ), or the separation (Δ) between two read out gradients (dephasing and rephasing of water protons) [20]. The so-called b value ($b = (\gamma G \delta)^2 [\Delta - \delta/3]$) describes these variables in comparable form. In diffusion-weighted images, regions of hyperintensity represent restricted diffusion. Maps of apparent diffusion coefficients (ADC) can be calculated by acquisition of a number of images in which either G, δ or Δ have been incremented systematically. Due to time constraints for the acquisition of a series of images, diffusion-weighted EPI is usually used to obtain ADC maps. It has become common practice to use only two b values for the calculation of ADCs. However, in order to avoid an overestimation of the ADC, multiple points (b) should be used for more accurate calculations [4,21].

Perfusion-weighted MR imaging (PWI) exploits signal changes following the passage of a contrast agent or labeled spins through the cerebrovascular system. A series of images is acquired in short intervals from which relative cerebral blood volume (rCBV) maps are calculated. The CBV is proportional to the area under the contrast agent concentration–time curve. Fast acquisition methods are required for PWI (rapid gradient echos, EPI).

Magnetization transfer (MT) has been used to increase tissue specific contrast in various diseases [22,23]. Selective saturation of signal contributions from protons with very low T_2 values (broad lines due to decreased mobility) can be achieved by off-resonance irradiation. MT from the saturated protons to mobile protons results in decreased signal intensity and produces substantial changes in contrast [24].

Localized *in vivo* MRS is probably the most powerful tool for determining the chemical composition of an object of interest non-invasively. In contrast to MRI methods, chemical shift information for the acquisition of metabolite profiles is acquired from a volume of interest (VOI), suppressing unwanted signals from outside the VOI. Although, ^{31}P and other nuclei like ^{13}C are increasingly studied (for example [25–27]), the most commonly used method for clinical diagnosis is ^1H MRS. From the relatively large number of MRS protocols [28], single VOI spectroscopy is most useful when studying lesions that are magnetically distinct from surrounding tissue. Although, chemical shift imaging (CSI) provides a matrix of spectra over relatively large regions within a short time, it is less suitable for the acquisition of high-quality spectra of focal lesions due to difficulties of shimming over a heterogeneous region. The two main pulse sequences used for single VOI MRS are point-resolved spectroscopy (PRESS) [29] and stimulated echo acquisition mode (STEAM) [30]. PRESS uses an excitation pulse ($\pi/2$) and a double echo and STEAM uses stimulated echoes. In general, PRESS results in higher SNRs but requires more precise pulse calibrations to suppress signal from outside the defined VOI. It can be distinguished between localized MRS that has been acquired with relatively short echo times (TE < 35 ms) and with long echo times (TE > 130 ms). The latter has the advantage of suppressing contributions from compounds with short T_2 relaxation times like lipids and macromolecules that would otherwise mask resonances from small metabolites in low concentrations. On the other hand, those compounds may be of diagnostic value. MR spectra can be applied for the quantification of metabolites [31,32].

More recently, methods from high-resolution (*ex vivo*) MR spectroscopy like ^{1}H, ^{1}H COSY and J-resolved MRS have been applied to *in vivo* assignment of resonances [33,34].

Contrast Enhancement

MR contrast agents modulate T_1 and T_2 relaxation times of water molecules inside the organism. Intravenous administration of gadolinium (Gd)-based contrast agents results in signal enhancement in T_1-weighted images in areas of accumulation of the substance. In the case of brain lesions, this most commonly reflects an increase in blood volume or a breakdown of the blood–brain barrier. Post-contrast hyperintensity of a perilesional ring is common in cerebral abscesses in the capsule stage as well as cystic and necrotic brain tumors [most commonly glioblastoma multiformes (GBMs)]. This makes differentiation between abscess and tumor difficult based on conventional contrast agents.

Most recently, responsive contrast agents have become more and more important. These contrast agents are usually responsive to chemical modification or specific binding to receptors, resulting in an activation or accumulation of the contrast agents at the targeted site [35,36]. One example in cancer diagnosis is the combination of iron oxide particles with ligands that bind to transferin receptors, resulting in accumulation of contrast agent in glioma cells [36]. Similarly, antibody–polylysine Gd chelates or Gd chelates targeting cell surface molecules have been suggested for specific visualization of tumor cells [37–39].

Ultra-small paramagnetic iron oxide particles (USPIOs) generate large susceptibility effects in surrounding tissue. This results mainly in shortening of T_2 and T_2^* relaxation times and subsequent hypointensity in respectively weighted MR images [40,41]. Labeling of phagocytes with USPIOs has been used for cell tracking *in vivo* [42]. Krieg *et al.* suggested the use of USPIOs for the detection of inflammatory processes as they result in the accumulation of large numbers of phagocytes [43]. This hypothesis has been proven by the utilization of *in vitro* labeled neutrophils in an animal model of induced abscesses. Intravenous administration of USPIOs was subsequently shown to differentiate between normal tissue and abscess [44]. Phagocytosis of USPIOs by macrophages and neutrophils partly contributed to a persistent signal decrease in interstitial spaces of the abscess wall and perifocal granulation tissue. It remains to be seen if the use of USPIOs is useful in clinical diagnosis of abscesses, as other diseases such as stroke also result in inflammatory response and subsequent accumulation of iron labeled macrophages [42,45].

The utilization of iron or Gd-based contrast agents linked to antigens specific for particular pathogenic microorganisms is still an unexplored, potentially useful tool for the diagnosis and etiological characterization of brain infections.

Conventional MRI of Infective Brain Lesions

Pyogenic bacterial abscesses are the most common focal brain infections. Radiologically, a brain abscess appears in CT and MRI as a mass surrounded by edema, similar to brain tumors [9]. They are hypointense on T_1- and hyperintense on T_2-weighted MR images (Figure 1). The mature abscess often has a rim that is slightly hyperintense on T_1- and hypointense on T_2-weighted images. This has partly been attributed to accumulation of macrophages [9]. Post-contrast ring enhancement in T_1-weighted MRI is common in brain abscesses (Figure 1).

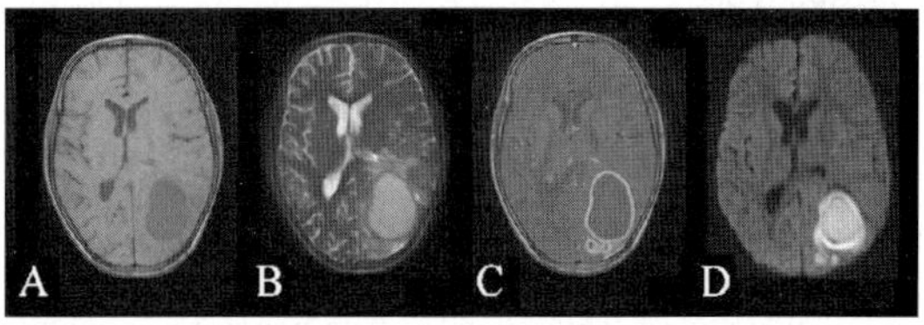

Fig. 1. Transverse MR images of a mixed bacterial abscess (*Streptococcus milleri* and *Bacteroides fragilis*). Images were acquired on a 1.5 T human MR scanner. (A) Hypointensity on T_1-weighted MR image (TR/TE = 660/15), (B) Hyperintensity and perifocal edema on T_2-weighted MR image (2200/80), (C) post-Gd contrast enhancement of a perilesional ring in T_1-weighted MR image (660/15) and (D) hyperintensity in a DWI ($b = 1000$ s/mm^{-2}) indicating restricted diffusion.

Features similar to pyogenic abscess were seen in intracranial tuberculoma [46,47] and focal infections caused by higher bacteria like *Nocardia* spp. [48,49].

Focal brain infections caused by fungi are less common. These infections can be caused by filamentous fungi (*Aspergillus* spp., *Mucor* spp.) or yeast and yeast like organisms (*Cryptococcus neoformans*, *Blastomyces dermatitides*, *Candida* spp., *Histoplasma capsulatum*). They differ from pyogenic abscesses as they do not necessarily form an abscess wall. The MR appearance of infections by filamentous fungi is often consistent with infarcts. Their hyphal elements lodge in intracerebral blood vessels, causing occlusions which result in microinfarcts and thrombosis. The actual infective lesions are not necessarily well defined. They show inhomogeneous hyperintensity on T_2-weighted images, surrounded by edema [50–53]. Hypointense peripheral rims have been reported due to blood products [51]. T_1-weighted images are often consistent with hemorrhages (hyperintense). Ring enhancement after administration of Gd contrast agent has been reported in some cases. The center of the lesion is often necrotic. Pituitary lesions due to *Aspergillus* spp. result in similar features to pituitary bacterial abscesses: a hypo- or isointense mass on T_1-weighted images, a hyperintense mass on T_2-weighted images and a peripheral contrast enhancing rim, which may also be features of pituitary tumor [54].

Similarly, yeast infections are characterized by isointensity [53,55] or hypointensity [9,53,56] on T_1-weighted images, hyperintensity on T_2-weighted images [9,52,53,55,56], and absence [53,56] or presence [52,53] of ring enhancement after Gd contrast. Some lesions show hypointense rings on T_2-weighted images due to the presence of macrophages [9].

Parasitic infections are caused by organisms that live part of their life cycle in a host. Parasites that cause focal brain infections in humans are most commonly *Taenia solium* larvae (causing cysticercosis), *Toxoplasma gondii* (toxoplasmosis), *Paragonimus* spp. (paragonimiasis), *Echinococcus* larvae (echinococcosis), amoebic infections, and others [57]. The appearance of lesions may change with development of the parasite and treatment. Many parasites can cause multiple lesions. They appear hypointense on T_1-weighted images, hyperintense on T_2-weighted images and show often no ring enhancement post-contrast [57–64]. Toxoplasmosis shows usually variable signal intensity on T_2-weighted images due to different amount of necrotic tissue, and post-contrast ring enhancement [64]. Later stages of paragonimiasis result in calcification and subsequent hypointense T_1- and T_2-weighted images [60]. Treatment of cysticercosis results in increased fluid content and immune response resulting in increased signal intensity on T_1, DW, and PW images and contrast enhancement [62].

Other MRI Methods

Diffusion

DWI provides unique information about translational molecular diffusion of water protons, which is also affected by the presence of cellular substructures [65]. Restricted diffusion within the cavity of a brain abscess, as opposed to the increased diffusion observed in cystic or necrotic tumors, was reported by Ebisu *et al.* [5]. This phenomenon was correlated with the high viscosity and cellularity of the respective pus samples. Further studies confirmed restricted diffusion in abscesses (Figure 1) [3,4,66]. However, the ADC values between and within these studies varied widely. Increased or unchanged ADCs have been reported for a number of abscesses [7,67,68]. Similar to these pyogenic abscesses, toxoplasmosis also resulted in increased ADCs [69]. These findings are not surprising as the viscosity of pus varies widely with the amount of inflammatory cells and microorganisms, the etiology and the age of the infective lesion (data not shown). In addition, restricted diffusion has also been found in cystic metastasis [7]. While decreased ADCs are highly suggestive of abscesses, they are not pathognomonic. A definite diagnosis of infective brain lesions cannot be based solely on DWI.

Perfusion

The main pitfall of PWI for the detection of brain lesions is that it can be inaccurate, where there is breakdown of the blood–brain barrier, which is the case in GBM and many infections [70]. In a low number of studies, it has been reported that intracranial infections resulted in a variable appearance in PWI, depending on the type of infection [71–73]. Pyogenic abscesses showed areas of increased rCBV in regions adjacent to the capsular rim in one study [71], suggesting that this may distinguish abscesses from solid tumors with areas of high vascularity, like high-grade GBMs and metastases [74]. Other studies have reported significantly lower rCBVs in abscesses compared to tumors, with lower rCBVs of the walls in abscesses compared to cystic/necrotic tumors [73]. However, these tumors can still be mistaken for abscesses based on PWI. Heterogeneous, hypervascular appearance of a significant number of tuberculomas has been reported, making differentiation between these lesion and tumors difficult [72].

This clearly indicates that more data are required to draw robust conclusions about the diagnostic value of PWI and DWI in distinguishing between subgroups of infective brain lesions and other pathological processes.

Magnetization Transfer

MT MRI for the diagnosis of infections is based on exchange of magnetization of protons from proteins with protons from water [22,75]. Pre-contrast MT spin-echo imaging was used for assessment of disease load in CNS tuberculosis and other infections due to better visualization of the lesions. Tuberculous lesions showed significantly lower MT ratios compared to pyogenic abscesses. Quantitative MT imaging was suggested in order to distinguish between various types of infective lesions [22]. A larger number of cases need to be evaluated in order to assess the influence of variable protein concentrations in abscesses.

Magnetic Resonance Spectroscopy

Although sophisticated MRI techniques are able to distinguish between certain infective lesions and tumors, cystic and necrotic gliomas can still be difficult to differentiate from abscesses. MRI methods are not able to distinguish between different infection-causing microorganisms. Microorganisms have diverse metabolite profiles that are distinct from mammalian metabolites [76]. This makes MR spectroscopy a suitable method for determining the etiology of an infection.

MRS chemotypes living microbial and mammalian cells by providing information on their chemical constituents rapidly and simultaneously. Studies on cell suspensions have shown that even closely related bacteria and fungi could be distinguished when MR spectra were analyzed by pattern recognition techniques [77,78]. Metabolites identified by MRS in cell cultures were also found in body fluids and biopsies from infective tissue [68,79–81]. If typical marker metabolites are present infection-causing microorganisms can be identified or at least classified without isolation and cultivation. In cases where spectroscopic patterns are similar to those from other pathologies, such as the presence of large concentrations of lipids and lactate in *Staphylococcus aureus* or *Streptococcus milleri* abscesses as well as necrotic GBMs, diagnostic "marker" metabolites can be masked. Analysis of MR spectra by statistical classification methods is still powerful enough to distinguish between such similar spectroscopic patterns [68,82,83].

In vivo MRS has been reported to distinguish the broad group of pyogenic abscesses from malignant gliomas and tuberculoma non-invasively [8,48,75,84–90]. A selection of *in vivo* MR spectra from brain abscesses is shown in Figure 2. While highly necrotic and cystic tumors and pyogenic abscesses contain both, large amounts of lipids and lactate, other organic acids like acetate and succinate have been identified in cerebral abscesses in humans and

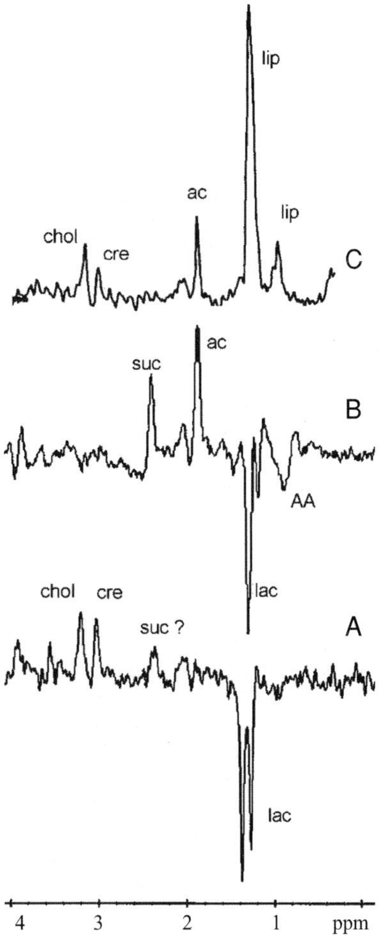

Fig. 2. MR spectra of bacterial abscesses with different etiological composition. Spectra were acquired from an 8 cm^3 VOI placed in the center of the lesion on 1.5 T human MR scanners (TR/TE = 3000/135). Bacteria causing the infections were: (A) *Streptococcus milleri*, (B) *Streptococcus milleri* and *Bacteroides fragilis* (same patient as in Figure 1), and (C) *Staphylococcus aureus, Proteus mirabilis*, and *Escherichia coli*.

utilized as bacterial "marker metabolites" for non-invasive diagnosis by MRS. Diverse profiles of organic acids are mainly characteristic for hetero-fermentative bacteria [76]. In addition, elevated concentrations of amino acids have been found in pyogenic abscesses [86,88]. This could not always be confirmed, as MRS of abscesses caused by microaerophilic bacteria like *S. milleri* and *S. aureus* did not consistently show elevated resonances from amino acids [8,68]. A remaining diagnostic problem is the differentiation between abscesses caused by exclusively

Table 1: Choline to creatine ratio determined by integration of the resonances at 3.2 and 3.0 ppm, respectively. Ratios were determined for cystic GBMs, abscesses with growth of *Streptococcus aureus* and sterile abscesses

	Cases	Mean ± SD	Range of choline:creatine ratio
Cystic GBM	11	1.8 ± 0.8	0.4–4.2
S. aureus abscess	9	0.9 ± 0.4	0–1.4
Sterile pus	5	0.2 ± 0.1	0–0.4

microaerophilic bacteria, sterile cysts, and cystic glioma. MR spectra of these lesions contain only resonances from lipids, lactate, and choline containing compounds. Attempts were made to distinguish between these groups based on increased choline levels in gliomas [8]. As Table 1 indicates, substantial overlap exists between the choline to creatine ratio in these lesions. However, statistical analysis of MRS from animal models has indicated that classification is feasible [68].

MR spectroscopy was also able to identify a characteristic pattern for parasitic infections (cysticercus and hydatid cysts) [91,92] as well as yeast infection (cryptococcoma) [56,80]. Relatively non-specific MR spectra were acquired from cerebral aspergillosis [53], nocardidosis [48], and tuberculomas [75,93].

Treatment of brain abscesses was monitored in animal models as well as humans. It was noticed that characteristic metabolites of hetero-fermentative bacteria like acetate and succinate disappeared completely [8,85,94]. Only lactate remained in these cases. These results indicate that the spectral specificity is only valid for untreated abscesses or soon after the start of treatment and that success of antibiotic treatment can be monitored by MRS.

In general, MR spectra acquired with long echo times are most informative as potentially masking intense resonances of lipids are suppressed and phase modulation due to J-coupling supports resonance assignment (lactate, amino acid residues). Unequivocal resonance assignment is currently not possible *in vivo*. First attempts to utilize localized *in vivo* ^{1}H, ^{1}H-COSY spectra for clinical diagnosis are promising and might further improve diagnostic accuracy when applied to abscesses [33,34]. It remains to be seen if the currently unsuitably long acquisition times can be reduced.

Considering the diversity of microorganisms causing brain lesions and the sparseness of cases for some of them, more MRS data are required to evaluate the whole potential of MR spectroscopy and to suggest robust strategies for its application to *in vivo* diagnosis. Based on small numbers, distinction between abscess causing microorganisms seems to be feasible [8,86,87]. This was confirmed for animal models in which a relatively large number of reproducible lesions can be evaluated by MRS [68,80].

Data Analysis

Diagnosis in radiological practice is based on interpretation of images by individuals. Using various MRI methods, this is feasible by following flow charts in the decision making process. Although the diagnostic value of MRS is established, it is little used for medical diagnosis, since interpretation of spectra is complex and few clinicians are trained to use this method. Diagnosis of abscesses is relatively easy in the presence of anaerobe microorganisms that produce unique "marker" metabolites like acetate or succinate [6,84,86,90]. However, focal infections caused by other microorganisms may result in MR spectra that appear similar to those acquired from other pathologies [8,68,86].

Metabolite assignment and quantification can be assisted by utilization of fitting routines like LCModel and other programs [32,95,96]. These semi-automated spectral evaluation tools have the disadvantage that some prior knowledge is required for data fitting, as only metabolites that are in the base set of LCModel can be quantified. Although, the latest versions of LCModel contain a relatively large set of metabolites or can be extended by additional model metabolites [32,97], this would always require prior knowledge and the accumulation of MR spectra from metabolites that are expected to be of value for the diagnosis of infections. The list of such metabolites could include organic acids like acetate, succinate, butyrate, propionate, and carbohydrates/polyols like trehalose, and mannitol. It remains to be seen if fitting routines are still reliable if extended by such a large number of metabolites.

The project International Network for Pattern Recognition of Tumours Using Magnetic Resonance (INTERPRET) aims to develop a decision support system that automates the analysis of multimodal MR data (MRI and MRS) for clinical diagnosis [98,99]. The objective of this multicenter study is the classification of brain tumors. It was demonstrated that, based on a large set of MR spectra for training and various pathologies, fully automated processing and analysis of test spectra by pattern recognition techniques was relatively robust and accurate [98].

Similarly, Poptani *et al.* were able to distinguish between tumors and non-neoplastic disorders using neural networks for the classification of MR spectra [100]. However, neural networks tend to be over-optimistic in cases of relatively small data sets.

A more robust approach was successfully used for the identification of microorganisms and classification of brain abscesses in animal models [8,68,77]. A statistical classification strategy (SCS) of MR spectra that consisted of: (1) automated processing of the spectra, (2) feature reduction by a genetic algorithm-based optimal region selection [101] and (3) the development of a set of linear discriminant analysis-based classifiers was robust and highly accurate. The validation of a classification system containing five classes of closely related pathogens resulted in less than 2% misclassifications [77]. As SCS was successful for the diagnosis of abscess animal models [68] and *in vivo* MRS data to classify human diseases, this seems to be a promising approach for the development of a multiclass classification system for the non-invasive diagnosis of focal brain infections that does not rely on the operator's ability to interpret MRS data. The limiting factor is the sparseness of data to make such a system robust.

Summary

A definitive diagnosis of infective lesions is still typically established by culture of bacteria from invasively obtained samples. Further reduction of mortality from brain abscesses requires more rapid, accurate, and safe diagnostic techniques. It is currently possible with conventional MRI to distinguish infective lesions from a broad range of other pathologies that are indistinguishable by other radiological modalities. Conventional MRI methods of typical abscesses result in hypointense lesions on T_1-weighted images, hyperintense lesions on T_2-weighted images, and ring enhancement in T_1-weighted images after administration of Gd-based contrast agents. The remaining indistinguishable pathologies (necrotic and cystic tumors, tumorfactive multiple sclerosis, and other demyelinating lesions) can be distinguished from infective lesions with relatively high probability by using DWI, PWI, MT MRI combined with MRS. DWI results typically in a reduction of ADCs in abscesses. However, variability in viscosity of pus in abscesses causes a large variability of ADCs, resulting in unchanged or even increased ADCs for some cases. MRS is the most promising method to distinguish between different etiologies of brain infections due to its ability to identify bacterial "marker" metabolites. MRS has also demonstrated its potential to assess the success of treatment. With increasing computerized data analysis and statistical data interpretation, an objective, automated data interpretation is feasible for clinical diagnosis. However, MRS fails if lesions are small (<1 mm) or if only a small cohort of cases is available. Animal models can overcome the limitations of small case numbers. Correlation between models and clinical cases has been successfully demonstrated.

Glossary of Terms

Abscess—Focal infection that consists of an abscess capsule, containing fibrous material and chronic inflammation cells; and a central cavity filled with pus.
Focal infection—Infection that is results in the formation of lesions. Can be caused by bacteria, fungi, parasites, or viruses.
Glioblastoma Multiforme—Highly malignant tumor with a low overall survival time for most patients. They are histologically characterized by a dense cellularity, high endothelial proliferation, and often the presence of focal tissue necrosis. These tumors are can occur in a cystic form.
Pus—Content of an infective lesion. It consists of dead and dying polymorphous cells (leucocytes), living and dead microorganisms, tissue debris, and other components of inflammation (edema fluid and fibrin).

References

1. Mathisen GE, Johnson JB. Clin. Infect. Dis. 1997;25:763.
2. Calfee DP, Wispelway B. Semin. Neurol. 2000;20:353.
3. Leuthardt EC, Wippold FJ, Oswood MC, Rich KM. Surg. Neurol. 2002;58:395.
4. Desbaratsa LN, Herlidoua S, de Marcoa G, Gondry-Jouetb C, Le Garsc D, Deramondb H, Idy-Peretti I. Magn. Reson. Imaging. 2003;21:645.
5. Ebisu T, Tanaka C, Umeda M. Magn. Reson. Imaging. 1996;14:1113.
6. Gupta RK, Lufkin RB. MR Imaging and Spectroscopy of Central Nervous System Infection. Kluwer Academic, Plenum Publishers: New York, 2001.
7. Hartmann M, Jansen O, Heiland S. Am. J. Neuroradiol. 2001;22:1738.
8. Himmelreich U, Dorsch N, Gomes L, Jones N, Brown G, Taylor J, Mountford CE, Sorrell TC. MAGMA. 2003;17(S7):106.
9. Haimes AB, Zimmerman RD, Morgello S, Weingarten K, Becker RD, Jennis R, Deck MDF. Am. J. Neuroradiol. 1989;10:279.
10. Young IR. Methods in Biomedical Magnetic Resonance Imaging and Spectroscopy. John Wiley and Sons: New York, 2000.
11. Rydberg JN, Hammond CA, Grimm RC, Erickson BJ, Jack CR Jr, Huston J, Riederer SJ. Radiology. 1994;193:173.
12. Hennig J, Nauerth A, Friedburg H. Magn. Reson. Med. 1986;3:823.
13. Hennig J. J. Magn. Reson. 1988;78:397.
14. Haase A, Frahm J, Matthaei D, Hänicke W, Merboldt KD. J. Magn. Reson. 1986;67:258.
15. Mansfieldv P. J. Phys. C. Solid State Phys. 1977;10:L55.
16. Stehling MK, Turner R, Mansfield P. Science. 1991;254:43.
17. DeYoe EA, Bandettini P, Neitz J, Miller D, Winans P. J. Neurosci. Methods. 1994;54:171.
18. Fischer H, Ladebeck P. Echo-planar imaging. Image artifacts. In: MS Schmitt, FR Turner (Eds). Echo-Planar

Imaging. Springer-Verlag: Berlin, Heidelberg, 1998, p 179.
19. Jesmanowicz A, Bandettini PA, Hyde JS. Magn. Reson. Med. 1998;40:754.
20. Stejskal EO, Tanner JE. J. Chem. Phys. 1965;43:3579.
21. Burdette J, Elster A, Ricci P. J. Comput. Assist. Tomogr. 1998;22:792.
22. Gupta RK, Kathuria MK, Pradhan S. Am. J. Neuroradiol. 1999;20:867.
23. Mehta RC, Pike BG, Haros PS, Enzmann DR. Radiology. 1995;195:41.
24. Tanttu JI, Sepponen RE, Lipton MJ, Kuusela T. J. Comput. Assist. Tomogr. 1992;16:19.
25. Bluml S, Moreno A, Hwang JH, Ross BD. NMR Biomed. 2001;14:19.
26. Gruetter R, Adriany G, Choi IY, Henry PG, Lei HX, Oz GL. NMR Biomed. 2003;16:313.
27. Payne GS, Al-Saffar N, Leach MO. Phosphorus magnetic resonance spectroscopy on biopsy and in vivo. In: GA Webb (Ed). Handbook of Modern Magnetic Resonance. Kluwer Academic Publisher: New York, 2006, pp 1125–43.
28. Ross B, Bluml S. Anat. Rec. 2001;265:54.
29. Bottomley PA. Ann. N.Y. Acad. Sci. 1987;508:333.
30. Frahm J, Merboldt KD, Hänicke W. J. Magn. Reson. 1987;72:502.
31. Kreis R. Prog. Nucl. Magn. Reson. Spectrosc. 1997;31:155.
32. Provencher SW. Magn. Reson. Med. 1993;30:672.
33. Thomas MA, Hattori N, Umeda M, Sawada T, Naruse S. NMR Biomed. 2003;16:245.
34. Ryner LN, Sorenson JA, Thomas MA. J. Magn. Reson. 1995;B107:126.
35. Aime S, Cabella C, Colombatto S, Crich SG, Gianolio E, Maggioni F. J. Magn. Reson. Imaging. 2002;16:394.
36. Weissleder R, Moore A, Mahmood U, Bhorade R, Benveniste H, Chiocca EA, Basilion JP. Nat. Med. 2000;6:351.
37. Gohr-Rosenthal S, Scmitt-Willich H, Ebert W, Conrad J. Invest. Radiol. 1993;28:789.
38. Sipkins DA, Cheresh DA, Kazemi MR, Nevin LM, Bednarski MD, Li KCP. Nat. Med. 1998;4:623.
39. Lemieux GA, Yarema KJ, Jacobs CL, Bertozzi CR. J. Am. Chem. Soc. 1999;121:4278.
40. Thomassen T, Wiggen UN, Gundersen HG, Fahlvik AK, Aune O, Klaveness J. Magn. Reson. Imaging. 1991;9:255.
41. Weissleder R, Elizondo G, Wittenberg J, Lee AS, Josephson L, Brady TJ. Radiology. 1990;175:489.
42. Rausch M, Baumann D, Neubacher U, Rudin M. NMR Biomed. 2002;15:278.
43. Krieg FM, Andres RY, Winterhalter KH. Magn. Reson. Imaging. 1995;13:393.
44. Gellissen J, Axmann C, Prescher A, Bohndorf K, Lodemann KP. Magn. Reson. Imaging. 1999;17:557.
45. Ramsay SC, Weiller C, Myers R, Cremer JE, Luthra SK, Lammertsma AA, Frackowiak RS. Lancet. 1992;339:1054.
46. Kioumehr F, Dadsetan MR, Rooholamini AA. Neuroradiology. 1994;36:93.
47. Gupta RK, Jena A, Sharma A, Guha DK, Khushu S, Gupta AK. J. Comput. Assist. Tomogr. 1998;12:280.
48. Murray RJ, Himmelreich U, Gomes L, Ingham NJ, Sorrell TC. Clin. Infect. Dis. 2002;34:849.
49. Pyhtinen J, Paakko E, Jartti P. Neuroradiology. 1997;39:857.
50. Ashdown BC, Tein RD, Feisberg GJ. Am. J. Roentgenol. 1994;162:155.
51. Mikhael MA, Rushovish AM, Ciric I. Comput. Radiology. 1985;9:85.
52. Stave GM, Heimberger T, Kerkering TM. Am. J. Med. 1989;86:115.
53. Kathuria MK, Gupta RK. Fungal infections. In: RK Gupta, RB Lufkin (Eds). MR Imaging and Spectroscopy of Central Nervous System Infection. Kluwer Academic, Plenum Publishers: New York, p 177.
54. Iplikcioglu AC, Bek S, Bikmaz K, Ceylan D, Gökduman CA. Acta Neurochir. 2004;146:521.
55. Kelly WM, Brandt-Zawadzki M. Radiology. 1983;149:485.
56. Himmelreich U, Sorrell TC, Dzendrowskyj T, Malik R, Mountford CE. Mircobiol. Aust. 2002;23:31.
57. Gupta RK, Chang KH. Parasitic infections. In: RK Gupta, RB Lufkin (Eds). MR Imaging and Spectroscopy of Central Nervous System Infection. Kluwer Academic, Plenum Publishers: New York, p 205.
58. Turgut M, Benli K, Eryilmaz M. J. Neurosurg. 1997;86:714.
59. Nurchi G, Floris F, Montaldo C, Mastio F, Peltz T, Coraddu M. Neurosurgery. 1992;30:436.
60. Chang KH, Cha SH, Han MH. J. Korean Radiol. Soc. 1993;29:345.
61. Dietz R, Schanen S, Kramann B, Erpelding J. J. Comput. Assist. Tomogr. 1991;15:168.
62. Jena A, Sanchetee P, Tripathi R, Jain R, Gupta AK, Sapra M. Magn. Reson. Imaging. 1992;10:77.
63. Chang KH, Lee JH, Han MH. Am. J. Neuroradiol. 1991;12:501.
64. Post MJD, Sheldon JJ, Hensley GT, Solia K, Tobias JA, Chan JC, Quencer RM, Moskowitz LB. Radiology. 1986;158:141.
65. Hajnal JV, Doran M, Hall AS. J. Comput. Assist. Tomogr. 1991;15:1.
66. Tomczak R, Wunderlich A, Görich J, Brambs HJ, Rilinger N. Radiologe. 2003;43:661.
67. Krabbe K, Gideon P, Wagn P, Hansen U, Thomsen C, Madsen F. Neuroradiology. 1997;39:483.
68. Himmelreich U, Accurso R, Malik R, Dolenko B, Somorjai RL, Gupta RK, Gomes L, Mountford CE, Sorrell TC. Radiology. 2005;236:261.
69. Camacho DLA, Smith JK, Castillo M. Am. J. Neuroradiol. 2003;24:633.
70. Bruening R, Wu RH, Yousry TA. J. Comput. Assist. Tomogr. 1998;22:104.
71. Cha S, Knopp EA, Johnson G, Wetzel SG, Litt AW, Zagzag D. Radiology. 2002;223:11.
72. Batra A, Tripathi RP. J. Comput. Assist. Tomogr. 2003;27:882.
73. Chan JH, Tsui EY, Chau LF. Comput. Med. Imaging Graph. 2002;26:19.
74. Harting I, Hartmann M, Bonsanto MM, Sommer C, Sartor K. Neuroradiology. 2004;46:189.
75. Gupta RK, Vatsal DK, Husain N, Chawla S, Prasad KN, Roy R, Kumar R, Jha D, Husain M. Am. J. Neuroradiology. 2001;22:1503.
76. Schlegel HG. Allgemeine Mikrobiologie. Georg Thieme Verlag: Stuttgart, 1992.

77. Himmelreich U, Somorjai RL, Dolenko B, Lee OC, Daniel HM, Murray R, Mountford CE, Sorrell TC. Appl. Environ. Microbiol. 2003;69:4566.
78. Bourne R, Himmelreich U, Sharma A, Mountford CE, Sorrell TC. J. Clin. Microbiol. 2001;39:2916–23.
79. Himmelreich U, Allen C, Dowd S, Malik R, Shehan BP, Mountford CE, Sorrell TC. Microbiol. Infect. 2003; 5:285.
80. Himmelreich U, Dzendrowskyj T, Allen C, Dowd S, Malik R, Shehan BP, Russell P, Mountford CE, Sorrell TC. Radiology. 2001;220:122.
81. Garg M, Misra MK, Chawla S, Prasad KN, Roy R, Gupta RK. Eur. J. Clin. Invest. 2003;33:518.
82. Torri GM, Torri J, Gulian JM, Vion-Dury J, Viout P, Cozzone PJ. Clin. Chim. Acta. 1999;279:77.
83. Dzendrowskyj T, Dolenko B, Sorrell TC, Somorjai RL, Malik R, Mountford CE, Himmelreich U. Diagn. Infect. Dis. Microbiol. 2005;52:101.
84. Remy C, Grand S, Lai ES, Belle V, Hoffmann D, Berger F, Esteve F, Ziegler A, Le Bas JF, Benabid AL. Magn. Reson. Med. 1995;34:508.
85. Dev R, Gupta RK, Poptani H, Roy R, Sharma S, Husain M. Neurosurgery. 1998;2:37.
86. Garg M, Gupta RK, Husain M, Chawla S, Chawla J, Kumar R, Rao SB, Misra MK, Prasad KN. Radiology. 2004;230: 519.
87. Himmelreich U, Dzendrowskyj T, Bourne R, Mountford CE, Sorrell TC. MAGMA. 2000;11(S1):197.
88. Grand S, Passaro G, Ziegler A, Esteve F, Boujet C, Hoffman D, Rubin C, Segebarth C, Decorps M, LeBas JF, Remy C. Radiology. 1999;213:785.
89. Kim SH, Chang KH, Song IC, Han MH, Kim HC, Kang HS, Han MC. Radiology. 1997;204:239–245.
90. Chang KH, Song IC, Kim SH. Am. J. Neuroradiol. 1998;19:401.
91. Agarwal M, Chawla S, Husain M, Jaggi RS, Husain M, Gupta RK. Neuroradiology. 2004;46:211.
92. Garg M, Chawla S, Prasad KN, Roy R, Sikora SS, Kumar R, Husain M, Khetrapal CL, Gupta RK. NMR Biomed. 2002;15:320.
93. Gupta RK, Raja R, Dev R, Husain M, Poptani H, Pandey R, Kishore J, Bhaduri AP. Magn. Reson. Med. 1996; 36:829.
94. Burtscher IM. Holtas S. Am. J. Neuroradiol. 1999; 20:1049.
95. De Graaf AA, Bovee WMMJ. Magn. Reson. Med. 1990;15:305.
96. Mierisova S, van den Boogaart A, Tkac I, van Hecke P, Vanhamme L, Liptaj T. NMR Biomed. 1998;11:32.
97. Seeger U, Klose U, Mader I, Grodd W, Nägele T. Magn. Reson. Med. 2003;49:19.
98. Tate AR, Majos C, Moreno A, Howe FA, Griffiths JR, Arus C. Magn. Reson. Med. 2003;49:29.
99. Tate AR, Julia-Sape M, Ladroue G, Murphy M, Loosemore A, Bell B, Wilkins P, Capdevila A, Majos C, Moreno A, Howe FA, Arus C, Griffiths JR. Proc. Int. Soc. Magn. Reson. Med. 2002;10:572.
100. Poptani H, Kaartinen J, Gupta RK, Niemitz M, Hiltunen Y, Kauppinen RA. J. Cancer Res. Clin. Oncol. 1999; 125:343.
101. Nikulin AE, Dolenko B, Bezabeh T, Somorjai RL. NMR Biomed. 1998;11:209.

Application of 2D Magnetic Resonance Spectroscopy to the Study of Human Biopsies

Edward J. Delikatny[1] and June Q.Y. Watzl[2]

[1] *Molecular Imaging Laboratory, Department of Radiology, University of Pennsylvania School of Medicine, Philadelphia, PA 19104, USA; and*

[2] *Department of Diagnostic Radiology, School of Medicine, Yale University, New Haven, CT 06519, USA*

Abbreviations: 2D, two-dimensional; AQ, acquisition time; COSY, correlated spectroscopy; DQF, double-quantum filtered; FID, free induction decay; GB, Lorentzian–Gaussian lineshape transformation fraction; J, scalar coupling constant; LB, Lorentzian line broadening; LG, Lorentzian–Gaussian; HMQC, heteronuclear multiple-quantum correlation; HSQC, heteronuclear single-quantum correlation; MAS, magic angle spinning; MM, magnitude mode; MRI, magnetic resonance imaging; MR(S), magnetic resonance (spectroscopy); NE, number of experiments; NS, number of transients; PS, phase-sensitive; (S)SB, (shifted) sine-bell; T_1, spin–lattice or longitudinal relaxation time; T_2, spin–spin or transverse relaxation time; TOCSY, total correlated spectroscopy; TPPI, time-proportional phase incrementation

Introduction

Proton magnetic resonance spectroscopy (^{1}H MRS) provides detailed information on cellular metabolites in whole cells and tissues enabling a non-invasive assessment of the chemical composition of these samples and the changes that occur with a disease process. As such, ^{1}H MRS can be used as an adjunct method for the pathological staging of human biopsy tissues [1–3]. Single pulse or one-dimensional (1D) proton spectra are rich with information, containing overlapping resonances from a number of mobile intracellular and extracellular metabolites and present in millimolar and submillimolar quantities. As a result of this overlap, information from metabolically important or potentially diagnostic resonances can be obscured, affecting both peak assignment and quantitative estimation of metabolite levels. One method of overcoming this problem has been the development and application of two-dimensional (2D) magnetic resonance (MR) spectroscopic methods to deliver additional chemical information than that available from 1D spectra.

2D magnetic resonance spectroscopy (MRS), from its initial proposal by Jeener in 1971 (reprinted in [4]), and first applications in the late 1970s [5] has developed into a major tool in organic and inorganic chemistry for spectral assignment and structural determination [6,7]. 2D spectroscopy relies on the collection of multiple free induction decays (FIDs) with the systematic incrementation of a delay time separating one or more pulses in the time domain of a pulse sequence. This results in a variable time-dependent evolution of spin interactions during this delay, thus effectively encoding information in a second frequency domain. 2D Fourier transformation is used to convert the 2D time-domain data into two frequency domains, which is often displayed as a contour plot with two frequency axes representing the primary spin properties of interest (e.g. chemical shift, J-coupling) and cross peaks joining regions of the spectra where spin interactions have occurred. A number of excellent review articles and textbooks have been published on 2D MRS, and the interested reader is referred to them for a detailed theoretical framework and practical information on pulse sequence design and implementation [6–11].

The most common 2D MR experiment used on biological specimens to date has been homonuclear ^{1}H–^{1}H magnitude mode correlated spectroscopy (MM-COSY). The additional chemical information obtained from the COSY experiment arises from protons that are linked or coupled through covalent bonds in a molecule, appearing in the 2D spectra as a pair of intersected resonant frequencies or cross peaks [6]. In principle, however, any homonuclear or heteronuclear 2D MRS experiment can be applied to biological samples, limited only by the imagination of the investigators. In practice, however, there are a number of limitations that arise when dealing with samples of biological material, including the stability of cells or tissues, restricted metabolite motion leading to the presence of multiple and short T_2 relaxation values, and the detection of low concentration metabolites. There are also issues associated with the low natural abundance and sensitivity of some nuclei [12].

In this chapter, we will address the practical issues associated with sample preparation, data acquisition, data processing, and interpretation of 2D spectra, concentrating primarily on the application of 2D spectroscopy to the

Graham A. Webb (ed.), Modern Magnetic Resonance, 1001–1011.

study of human biopsy material. We will place the subject in perspective by briefly reviewing the literature, followed by a presentation of some theoretical aspects relevant to 2D spectroscopy of tissues, and then discuss the technical aspects of sample preparation, 2D data acquisition, and strategies for data processing.

Application of 2D NMR Spectroscopy to Cells and Tissues

Although the technique of 2D NMR spectroscopy was first introduced in the late 1970s, it was not applied to the study of biological materials until the 1980s. The first published study used MM-COSY to identify fatty acyl chain resonances in a packed suspension of rat mammary adenocarcinoma cells [13]. It was soon recognized that the fatty acyl chains and head groups, that constituted the most intense cross peaks in many 2D spectra, arose from mobile lipid resonances, predominantly neutral lipids such as triglycerides and cholesteryl esters [14–16]. However, a number of other cross peaks were also observed in 2D spectra of cells [14–16]. The assignment of these cross peaks in 2D spectra were facilitated by the applications of techniques such as T_2 filters to selectively observe resonances with long T_2 [17], and 2D J-resolved spectra to assist in resonance assignment [18]. It was recognized that in biological samples containing metabolites with a range of T_2 relaxation values, the size of the acquisition domains and the window functions applied in processing could have dramatic effects on cross peak appearance and structure in COSY spectra [12]. As a result of these advances, over 40 cross peaks were identified in COSY spectra from intracellular metabolites. A comprehensive table of resonance assignments of lipids, amino acids, and sugars present in colorectal cancer cells was published in 1992 [19], and a similar table showing resonances observed in thyroid tissue biopsies [20], is presented in Table 1.

Around the same time, COSY and 2D J-spectroscopy were applied to investigate animal tissues such as rat brain *ex vivo* [18,21,22]. One of these studies showed one of the first 2D COSY spectra on dead rat brain *in situ* [18], thus demonstrating the feasibility for the later development of 2D MRS *in vivo* [23–27]. In another study, lactate COSY cross peak intensities were shown to be linearly correlated with lactate concentration in muscle thus providing a potential method of estimating relative metabolite concentrations [28]. The first application of total correlation spectroscopy (TOCSY) on frog muscle was published in 1990 [28]. Since these initial studies, 2D spectroscopy has become a standard spectroscopic tool for the study of cells and animal tissues, including the activation of immune cells [29–32], detection of changes associated with tumor development and progression [33,34], and the study of the actions of antitumor drugs on cultured cancer cell lines [35–37].

The first application of 2D COSY to human tissue was on a human ovarian biopsy specimen [38]. An initial study demonstrated that *ex vivo* 2D spectroscopy might be useful in discriminating the pathology of human cervical punch biopsies [39]. This was followed by the first major clinical study using 1D and 2D spectroscopy on 159 cervical biopsy specimens that showed consistently increased lipid cross peaks in malignant relative to premalignant specimens [40]. Subsequent studies were published demonstrating the usefulness of 2D COSY as an adjunct to 1D MRS and histopathology on brain tumors [41,42], colon adenomas and carcinomas [43], draining lymph nodes [44], and in samples from ovarian cancer [45] and thyroid cancer [20]. TOCSY and 2D J-resolved spectra have also been used to study smooth muscle tumors [46], myosarcomas [47], and liposarcomas [48]. The use of ^{1}H MRS on human biopsies as an adjunct to histopathology is discussed in detail elsewhere in this issue and in a number of review articles [1,2,49].

There has recently been a renaissance in 2D spectroscopy of human and animal tissues because of two innovations, reviewed elsewhere in this issue: the application of magic angle spinning (MAS) and the development of techniques for obtaining *in vivo* 2D spectra. *In vivo* 2D spectroscopy has the potential to permit the detailed examination of metabolites of diagnostic or metabolic interest *in situ* [23–25], and may extend 2D methods into the realm of true non-invasive detection of disease. MAS has allowed MR spectra of biopsy tissue to be obtained with extremely high resolution. Of particular relevance to this chapter is the application of 2D MAS experiments for high-resolution resonance assignment in animal and human tissues. MAS analogs of the homonuclear COSY, TOCSY, and J-resolved experiments have been developed for the evaluation of metabolite profiles characteristic of human disease in breast [50,51], kidney [52] and prostate tumors [53], liposarcomas [48], and draining lymph nodes [54]. Further, MAS analogs of the heteronuclear ^{1}H–^{13}C heteronuclear multiple-quantum correlation (HMQC) and heteronuclear single-quantum correlation (HSQC) experiments have been employed to assist in resonance assignment in human tissues and in animal model studies [52,55–58].

2D MR Spectroscopy

The major advantage of 2D MRS is the separation of resonances that overlap in 1D spectra. 2D spectroscopy collects data in two time domains, t_1 and t_2, which can be converted to two frequency domains, F_1 and F_2, by double Fourier transformation. The t_2 is the direct or acquisition dimension that consists of the FIDs collected in the 2D experiment, and t_1 is the orthogonal, or indirect, dimension defined by the number of FIDs acquired in the 2D

Table 1: Assignment of major cross peaks in 2D ^1H–^1H COSY MR spectra of thyroid biopsy tissue*

Molecules	Abbreviation	Coupling partners (bold face)	Cross peak coordination (ppm)
Lipid			
Fatty acyl chains	A	–(CH$_2$)$_n$–**CH$_2$**–**CH$_3$**	1.27, 0.87
	B	–CH=CH–**CH$_2$**–**CH$_2$**–	2.01, 1.30
	C	–CH$_2$–**CH$_2$**–**CH**=CH–	2.00, 5.31
	D	=CH–**CH$_2$**–**CH**=CH–	2.82, 5.32
	E	–O–C(O)–CH$_2$–**CH$_2$**–**CH$_2$**	1.58, 1.27
	F	–O–C(O)–**CH$_2$**–**CH$_2$**	2.23, 1.57
Glycerol backbone	G′ (geminal, J_{AB})	RO–**CH$_A$H$_B$**–CH(OR)–**CH$_A$H$_B$**–OR	4.07, 4.27
Amino acids			
Alanine	Ala (α, β)	–**CH**–**CH$_3$**	3.77, 1.47
Glutamate/glutamine	Glu/Gln	–CH–**CH$_2$**–**CH$_2$**–COO$^-$/–CONH$_2$	2.14, 2.45
Glutamate	Glu	–**CH**–**CH**(NH$_3^+$)–COO$^-$	2.14, 3.78
Isoleucine	Ile (β, γ)	–CH–**CH**(CH$_3$)–**CH$_2$**–CH$_3$	2.04, 0.95
Leucine	Leu (γ, δ)	–**CH**–(**CH$_3$**)$_2$	1.71, 0.95
Lysine/polyamines	Lys (δ, ε)	–CH$_2$–**CH$_2$**–**CH$_2$**–NH$_3^+$	1.67, 3.00
Threonine	Thr (β, γ)	–CH–**CH**(OH)–**CH$_3$**	4.26, 1.31
Tyrosine	Tyr (γ, δ)	o, m Protons on p-substituted ring	6.89, 7.19
Valine	Val (β, γ)	–**CH**–(**CH$_3$**)$_2$	2.28, 1.01
Phospholipid metabolism			
Choline	Cho	(CH$_3$)$_3$–N$^+$–**CH$_2$**–**CH$_2$**–OH	3.51, 4.05
Glycerophosphocholine	GPC	(CH$_3$)$_3$–N$^+$–**CH$_2$**–**CH$_2$**–OPO$_2$–CH$_2$–	3.67, 4.33
Phosphocholine	PC	(CH$_3$)$_3$–N$^+$–**CH$_2$**–**CH$_2$**–OPO$_3^{2-}$	3.56, 4.17
Phosphoethanolamine	PE	H$_3$N$^+$–**CH$_2$**–**CH$_2$**–OPO$_3^{2-}$	3.22, 3.98
Other			
Inositol	Inos	–**CH**(OH)–**CH**(OH)–	3.28, 3.64
Lactate	Lac	–**CH**(OH)–**CH$_3$**	4.10, 1.30
Taurine	Tau	H$_3$N$^+$–**CH$_2$**–**CH$_2$**–OSO$_3^-$	3.25, 3.42
Unique to thyroid†	Th1		5.10, 2.05
	Th2		5.10, 1.60

*Cross peak assignments obtained from follicular thyroid tissues ^1H–^1H MM-COSY spectra at 37 °C (table adapted from Refs. 20 and 36).
†Cross peaks unique to thyroid.

experiment. The 2D pulse sequence is often divided into four time periods: the preparation, evolution, mixing, and detection periods (Figure 1). The preparation period consists of a delay time during which equilibrium is attained and/or solvent suppression is performed, followed by a radio frequency (rf) pulse or series of pulses to prepare transverse magnetization. The evolution period consists of a variable delay time $t_1 + \Delta t_1$ during which magnetization evolves. The mixing period includes at least one rf pulse or a series of pulses and delays, to induce mixing of states and magnetization coherence transfer. The detection period is the time of FID acquisition, t_2. The indirect dimension is sampled by systematically incrementing the delay time in the evolution period, thus modulating the available magnetization during the mixing period. The size of the increment, or sampling period, in the t_1 dimension is inversely proportional to the spectral width in F_1 [59].

Choice of Pulse Sequence

In principle, any 2D pulse sequence can be applied to studies of human biopsy tissue, within the constraints defined by the stability of the sample. These time constraints make heteronuclear experiments more difficult to perform, especially for nuclei of low natural abundance. Many of these problems have been overcome using gradient-enhanced techniques with inverse detection and ^1H–^{13}C HMQC and HSQC MAS experiments have been

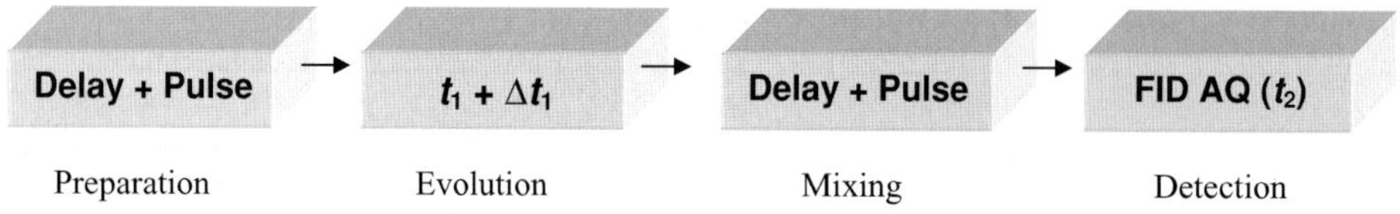

Fig. 1. The four basic periods of a 2D experiment.

reported for resonance assignment in human and rat tissues [52,55–58]. To date, however, the majority of 2D studies have been performed using ^{1}H–^{1}H homonuclear 2D spectroscopy, with COSY, 2D *J*-resolved and TOCSY experiments being by far the most common. Different and complementary information is available from these experiments, such that they may be usefully applied alone or in tandem to tissues of different origins.

The COSY sequence is the simplest and one of the most useful 2D experiments. The basic MM-COSY pulse sequence (Figure 2A) consists of two 90° pulses, the first to create magnetization in the transverse plane and the second to mix spin states, transferring coherence between scalar coupled spins. A variable delay $t_1 + \Delta t_1$ separating these pulses is incremented systematically during the evolution time and the FID is acquired during the detection time, t_2. After Fourier transformation the 2D spectrum consists of a diagonal representing spins that did not exchange magnetization during the mixing period and cross peaks or off-diagonal peaks for nuclei exchanging magnetization through scalar coupling. Since MM-COSY spectra are collected with no phase discrimination in F_1, a magnitude calculation must be performed after Fourier transformation.

Phase-sensitive double-quantum filtered COSY (PS DQF-COSY, Figure 2B) is an analog of the COSY experiment. The data are collected using time-proportional phase incrementation (TPPI) [60], in which the phase of the first pulse is incremented by 90° in successive FIDs resulting in phase discrimination in F_1. This obviates the need for the MM calculation, leading to improved apparent spectral resolution. However, the phase discrimination in the t_1 dimension means that twice as many FIDs need to be collected to achieve the same resolution. The signal to noise (S/N) is unchanged because S/N is proportional to the product of the number of transients (NS) and the number of FIDs acquired [6]. Thus, the total acquisition time (AQ) can remain constant by reducing the equivalent number of scans by a factor of two.

In the DQF pulse sequence, the first and second pulses act as in the MM-COSY. The third pulse in the DQF-COSY pulse sequence converts double-quantum magnetization into observable single-quantum magnetization, with phase cycling or gradients used to cancel unwanted single- and multiple-quantum magnetizations. This results in the diagonal signals of uncoupled nuclei, arising from single-quantum magnetization, being severely reduced as well as the conversion of the dispersive diagonal peaks into pure antiphase absorptive signals. This is useful for biological samples, since cross peaks resulting from the coupling of nuclei with similar chemical shift appearing near the diagonal can be resolved [61].

The 2D *J*-resolved sequence (Figure 2C) consists of a 90° preparation pulse, followed by the incremented delay

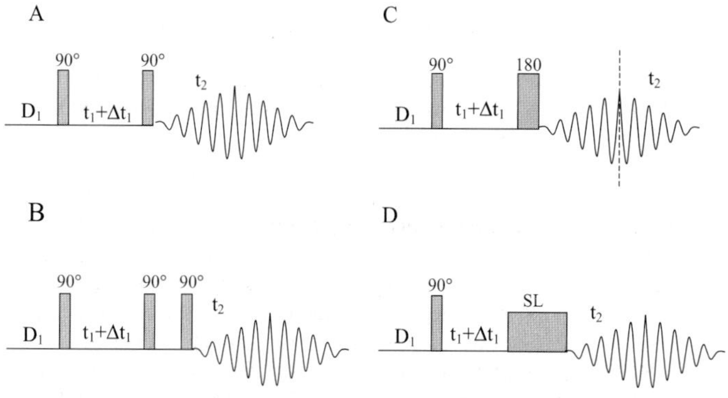

Fig. 2. Four basic 2D homonuclear pulse sequences routinely used in the study of human tissues. (A) ^{1}H–^{1}H MM-COSY, (B) ^{1}H–^{1}H DQF-PS COSY, (C) ^{1}H–^{1}H 2D *J*-resolved, and (D) ^{1}H–^{1}H TOCSY.

time and a 180° mixing pulse. The 180° pulse reverses linear spin interactions, such as chemical shift, such that at time $2(t_1 + \Delta t_1)$, chemical shift is refocused, leaving only interactions from J coupling in t_1. This results in a 2D spectrum with chemical shift on the F_2 axis and J coupling only on the F_1 axis. J-resolved spectroscopy has proven to be valuable for resonance assignment, especially in regions of the spectrum with substantial overlap.

One of the more useful methods in the 2D spectroscopic arsenal is the TOCSY (Figure 2D). In the TOCSY experiment, the mixing pulse is replaced with a spin lock field or series of pulses, often an MLEV-17 pulse train [62]. This allows coherence transfer between all coupled spins simultaneously, which ideally gives rise to cross peaks between all spins in a spin system. One advantage to the TOCSY experiment is that the coherence transfer results from in phase rather than antiphase magnetization, leading to in phase cross peaks that are not subject to cross peak cancelation.

Sample Preparation

A general harvest procedure that we have found acceptable is to collect unfixed, surgically removed tissue specimens from the operating theatre. To prevent sample decomposition, these specimens are placed immediately into vials containing phosphate-buffered saline in D_2O (PBS/D_2O), snap frozen in liquid nitrogen and stored at −70 °C. Samples can generally be stored this way for several weeks, without serious deterioration. Samples for MR experiments are cut frozen into tissue plugs of wet weight 20–50 mg and washed in PBS/D_2O, although this washing may lead to depletion or removal of certain metabolites [63]. Positioning of the sample in a standard NMR tube can be performed by placing the tissue on top of a glass susceptibility plug in a susceptibility-matched MR tube, or by suspending the tissue either in buffer or on the surface of glass wool plug. The NMR tube need not be filled with tissue to obtain reasonable spectra. Smaller cylindrically shaped samples are easier to shim and the presence of deuterated buffer is advantageous for locking purposes. Sample preparation for MAS is dealt with elsewhere in this issue.

Data Acquisition

A 2D data set consists of a series of FIDs, which are acquired and stored consecutively in a file as a 2D matrix of points. A number of operator controlled factors affect the total experiment time of the 2D data set: most notably the total repetition time, which is pulse sequence dependent, the NS collected per FID in the t_2 dimension, including dummy scans (DS) performed to put the spin system in a steady state before data accumulation, and number of experiments (NE), the number of FIDs to be collected in the t_1 dimension. The repetition time includes parameters such as the relaxation/equilibrium delay, the incrementation delay during the evolution period, and the t_2 AQ.

The relaxation delay in the preparation period is dependent on the T_1 relaxation values of the sample. Ideally, five times the longest T_1 should be used to ensure that all spins are fully relaxed before the experiment begins. In practice, a 3-h time constraint allows only about 8000 transients (for example 250 FIDs of 32 transients) to be acquired before viability drops below an acceptable level. This requires acquisition parameters that are less than ideal: a repetition time of about 1.4 s is an absolute minimum for establishing an optimal steady-state magnetization and avoiding spurious coherences and instabilities due to the presence of residual transverse magnetization before the excitation pulse [12]. In general, the AQ should be at least 3 T_2 and the repetition time have a minimum value of 1.3 T_1. For lipid resonances, T_2 and T_1 relaxation values in cells have been reported to be in the range of 120–300 and 200–600 ms, respectively [41,48,64–66].

The transmitter offsets for both t_1 and t_2 time domains should be set to the center of the spectrum. For ^1H–^1H spectra, the transmitter offsets can conveniently be set to the water frequency, which lies approximately in the center of the spectrum with spectral widths, sw_1 and sw_2 equal to 10 ppm. The size of the time increment, Δt_1, is related to the spectral width, for MM-COSY Δt_1 is equal to $1/sw_1$, and for PS-COSY to $0.5/sw_1$. The NS to be collected for each FID is generally a multiple of eight if full phase cycling is used. Gradient-assisted sequences, such as the G-COSY [67–69] can be used to reduce the number of scans in the phase cycle, but in small biopsy specimens of 20–40 mg wet weight, S/N is sufficiently limited such that often 32 or more scans must be collected to obtain 2D spectra with reasonable S/N ratios.

One major consideration in the acquisition of 2D spectra of human tissues is the time allocated to collect experimental data. This is important for two reasons, the first is that metabolite levels can change over time due to sample degradation, especially if the data are collected at temperatures above ambient. Neutral amino acids resonances from proteolytic breakdown will often appear as sharp peaks on the downfield side of the methyl resonance in 1D spectra [63]. In the case where the total experiment time is sufficiently long, the contribution of these resonances to the 2D cross peaks will almost certainly be variable over time. Secondly, it may be necessary or desirable to have histopathology performed on the same piece of tissue that was used to acquire the spectrum, in the case where only small pieces of tissue are available, or where direct comparison of spectral properties and histological diagnosis is required. This is a more stringent condition, since the tissue must remain in good enough shape for a pathologist to make a reasonable diagnosis. In cervical biopsy tissue, we found that colposcopically obtained punch biopsies kept at 37 °C for 1.5 h, remained suitable for histopathological analysis. Although these parameters may vary for different tissue types, for punch biopsy sized tissue (5–10 mg) this is not enough time for anything but the detection and assignment of major metabolite cross peaks.

The limited time available for data collection means that acquisition parameters must be established to achieve an optimum S/N ratio in the limited time available. In practice, this generally means restricting the number of FIDs collected in t_1, balancing the desired or necessary spectral resolution against S/N and sample viability [12]. For equal resolution in F_1 and F_2 in COSY or TOCSY spectra, a square 2D data set of complex data points must be collected (e.g. 512 × 2K for MM; 1024 × 2K for PS mode). In small biopsy specimens of 20–40 mg, a reasonable S/N ratio can be obtained in a total experiment time of 2–3 h. For COSY and TOCSY spectra, the collection of 200 FIDs of 32 transients will give reasonable S/N and resolution for most specimens in about 3 h. Recall that for PS data, the phase discrimination in F_1 dictates that twice the number of FIDs be collected, with half the number of scans for equivalent resolution and S/N [6]. For ^1H–^1H J-resolved spectra, where the detection of scalar coupling requires a spectral width in F_1 on the order of about 100 Hz, the number of FIDs accumulated in t_1 can be correspondingly reduced to 100–150 [55,70]. For heteronuclear MAS experiments, t_1 domains of around 200 FIDs have frequently been used to obtain reasonable resolution in human and animal tissues [55–57].

Data Processing

2D data processing converts a 2D time-domain data set to a 2D frequency domain spectrum. The general steps involved are zero filling, apodization by multiplication with a window function, and Fourier transformation. This is followed by phase correction for data collected in PS mode or magnitude calculation for MM collection. Zero filling is the procedure of supplementing the time-domain data points with a string of data points with zero amplitude prior to Fourier transformation to improve digital resolution [6]. In 2D MRS, zero filling is also used when the collected matrix of data points is not square or incomplete. In general, time constraints dictate that the number of increments collected in the t_1 domain be as few as possible. As a result zero filling of at least one level is recommended to improve resolution in t_1 [6].

Apodization and the Use of Window Functions

Since a wide range of T_2 relaxation values are encountered in biological samples, additional considerations and flexibility are required for processing 2D spectra. Different window functions may be applied to the data set to extract relevant information from different peaks [12]. A window function is a time-dependent analytical mathematical function that shapes or filters the signal envelope, providing apodization to avoid truncation artifacts created by zero filling, or when acquisition is terminated before signal has fully decayed. Window functions are also used to maximize S/N and resolution, minimize line widths or to convert the frequency domain lineshape into a more desirable form [6,7,71]. Windowing is achieved by point-by-point multiplication of the window function with each data point of the FID in the time domain, which results in a convolution of the window function with the data set in the frequency domain.

In coherence transfer 2D spectra, where the desired information is often contained in an echo not centered at the beginning of the time domain, window functions are chosen that have the form of a pseudoecho, with small initial values, peaking later in the time domain and decaying smoothly to zero. Three commonly used window functions for 2D data processing are the Lorentzian–Gaussian (LG), sine-bell (SB), and shifted sine-bell (SSB) window functions.

The unshifted SB function begins at an initial value of zero rising to a normalized value of 1 halfway through the data set (at the time point $t_i = t_m = AQ/2$), decaying smoothly to zero in both the t_1 and t_2 time domains. The unshifted SB function emphasizes sine dependent cross peaks relative to cosine dependent diagonal peaks and removes broad wings (dispersive components) from MM lineshapes. SB window functions optimize the presentation of high-resolution cross peaks from spins with long relaxation times, but spins with shorter relaxation times, those with $T_2 < AQ$ may exhibit substantial signal loss [6,12]. The SSB function is similar to the sine and can be used on samples with $T_2 > AQ$. The initial value of the SSB function is not zero but (π/n), with $n \geq 2$. However, the SSB is not as effective at canceling out dispersive wings in MM spectra [6]. The application of SB windows to a complete symmetrical 2D data set can result in a 2D spectrum in which only cross peaks from a number of free amino acids and low molecular weight metabolites appear.

The LG window function is commonly used as a window function in t_2. The LG function has two adjustable parameters, which offer increased flexibility in the position and shape of the window function within the time domain and allows optimum detection of short T_2 species such as lipids. The LG function is controlled by two parameters: a Lorentzian broadening (LB), expressed as a negative number in Hz, and a Gaussian lineshape transformation fraction (GB). GB is a decimal fraction, which positions the LG maximum at GB × AQ, scaled such that $LG_{max} = 1$. The position of the LG window is controlled by adjusting GB, while adjustment of the LB parameter controls the initial and final values and width of the window function in the time domain, but not independently of GB.

The detection of compounds with short T_2 requires that the length of the time domains be reduced before performing the 2D F T, which can be achieved either by applying the LG function with the maximum shifted to a shorter time, or by reducing the number of data points used for processing and using symmetrical window functions such as the SB. There is no particular advantage to the LG over the SB window if the maximum in the SB window is positioned by adjusting the number of time-domain points. In fact, reduction of the number of t_1 points is often the condition set by sample viability and can be achieved by reducing the number of FIDs acquired or used in processing. If short T_2 species are of most interest, the efficiency of data acquisition can be significantly improved by acquiring fewer FIDs with correspondingly more transients per t_1 value. The analysis of cross peaks from both long and short T_2 species in a single data set can only be performed by acquiring conventional data sets with long AQs and performing several transforms with processing time domains and window functions optimized for the T_2 and J values of the cross peak of interest.

Quantitation

Although 2D spectroscopy can reveal numerous differences in the metabolite profiles in normal and diseased tissues, absolute quantitation of changes in metabolite levels, as in all forms of MR, can prove to be problematic, and must be performed with care. One method of quantifying 2D coherence transfer (e.g. COSY and TOCSY) spectra is by integrating the volume enclosed by a cross peak above the base plane (noise level) of the 2D spectra. However, there are a number of factors, besides the concentration of the species that influence both the appearance and the intensities of cross peaks in 2D spectra. These factors include the molecular weight and the T_2 relaxation value of the coupling partners, the MR pulse program being used and the chosen acquisition and processing parameters. In the case of COSY spectra, cross peak intensities can be further decreased due to cross peak cancelation arising from antiphase signal components [7,59]. Cross peak cancelation is more serious for resonances with short T_2 relaxation values, or small J-couplings since the degree of cancelation depends on the ratio of J/T_2. For COSY spectra, the problem is further complicated since the sum of the integrated area for any antiphase cross peak is zero. Therefore, any non-zero cross peak volumes have to be obtained either from spectra in which a MM calculation has been performed or from the sum of positive and negative antiphase components in PS spectra. TOCSY cross peaks appear as pure absorption mode, not antiphase, and hence and are not subject to problems arising from cross peak cancelation.

As a result, of these limitations it is difficult to compare absolute concentrations of two species in the same 2D spectrum, as well as difficult to estimate the concentrations relative to internal or external standards. However, if collected and processed under identical conditions, the same cross peaks from two different 2D spectra can be compared. Assuming that the experimental treatment does not change the T_2 of the cross peak of interest, it is possible to estimate the relative concentration from cross peak integrals.

It is worth noting that the volumes of cross peaks in COSY spectra have been shown to be linearly proportional to metabolite concentrations in standard solutions [28]. We have confirmed the linear dependence of cross peak volume on concentration in samples of mixed amino acids, and find that the linear relationship holds regardless of cross peak coupling partner (Watzl and Delikatny, unpublished data). Absolute concentration estimates of lipid and lipid metabolite concentrations in activating immune cells have been obtained by comparing metabolite cross peak volumes to those of paramagnetic-doped standards of similar T_2 and known concentrations [29,30]. Quantitation of heteronuclear correlation cross peaks has been reported in bacterial supernatants [72].

Factors Affecting Cross Peak Volumes

In order to improve the diagnostic potential of 2D ^{1}H MRS, it is necessary to understand the factors controlling the resolution and intensity of cross peaks in complex 2D spectra with multiple T_2 relaxation times [12]. As discussed above, window functions commonly used to improve the spectral resolution can selectively enhance cross peaks with different T_2 relaxation times [6,12]. Moreover, in unsymmetrized COSY spectra, significant asymmetries in cross peak intensities for methyl groups (for example those from alanine, lactate, and threonine) have been observed with peaks above the diagonal being generally more intense even when time domains and windows were equal. At higher fields this effect can be quite pronounced and can lead to substantial reduction in certain cross peaks upon symmetrization [12]. For this reason, quantitation should always be performed on unsymmetrized spectra.

In Figure 3, we present two MM-COSY 2D spectra of follicular thyroid tissue, acquired from the same sample, but processed using different parameters. In the t_2 dimension, 32 transients of 2048 data points were collected, and in the t_1 dimension, 512 FIDs (NE) were collected. The spectra were processed with a LG window in t_2 and a SB window function in t_1; an NE of either

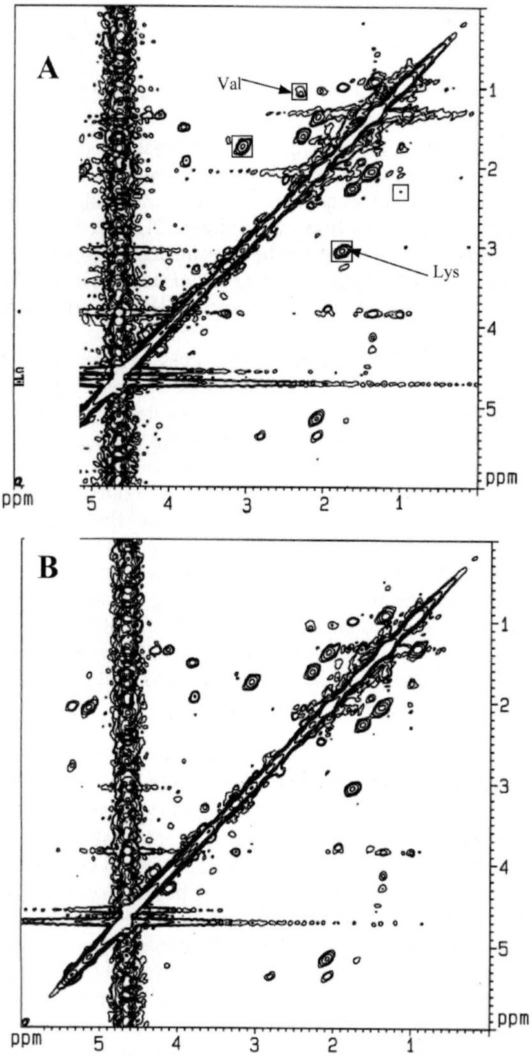

Fig. 3. Two-dimensional MM-COSY spectra of follicular thyroid tissue processed using (A) 200 and (B) 512 FIDs. In the t_2 domain, 2048 data points were collected and 512 FIDs consisting of 32 transients were collected in t_1. Data processing was performed with a sine-bell window applied in t_1 and a Lorentz-Gauss window in t_2 ($LB_2 = -30$, $GB_2 = 0.12$) before Fourier transformation. Cross peaks chosen for analysis are indicated with arrows.

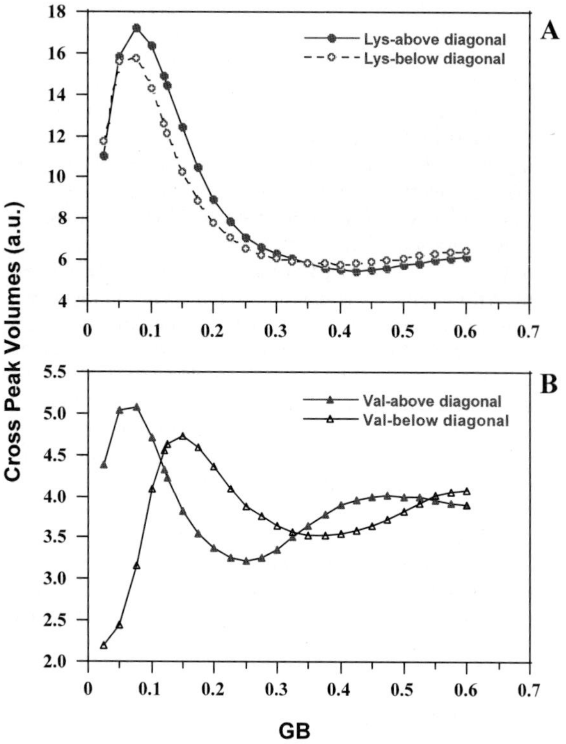

Fig. 4. Variation of (A) Lys $\mathbf{CH_2}$–$\mathbf{CH_2}$–NH_2 (1.67 and 3.00 ppm) and (B) Val $\mathbf{CH_3}$–$\mathbf{CH}$ (1.01 and 2.28 ppm) cross peak volumes as a function of GB in the t_2 time domain in magnitude-mode COSY spectra of follicular thyroid tissue processed with 200 FIDs. A sine-bell window was applied in t_1 and a Lorentz-Gauss window in t_2.

200 or 512 FIDs was used to produce Figures 3A and 3B, respectively. The volumes of two sets of amino acid cross peaks were selected from this data set (Lys, from the CH_2–$\mathbf{CH_2}$–$\mathbf{CH_2}$–NH_3^+ coupling of lysine at 1.67 and 3.00 ppm, and Val, from the $(\mathbf{CH_3})_2$–$\mathbf{CH}$–CH coupling of valine at 1.01 and 2.28 ppm; arrows in Figure 3A), and the integrated cross peak volumes on the top and bottom of the diagonal and plotted relative to systematic changes in the processing parameter GB. Figure 4 demonstrates that cross peak volumes from COSY spectra vary significantly as a function of GB. The volumes measured for the pair of symmetrically matched cross peaks on either side of the diagonal are very close in value throughout the profile for the Lys cross peak whereas for the Val pair, they are substantially different. In contrast, cross peak volume profiles generated by varying LB resulted in strict monotonic increases in cross peak volume regardless of cross peak chosen (Figure 5).

Volumes measured from cross peaks at the same chemical shifts in different spectra display similar variation with respect to processing parameters, regardless of

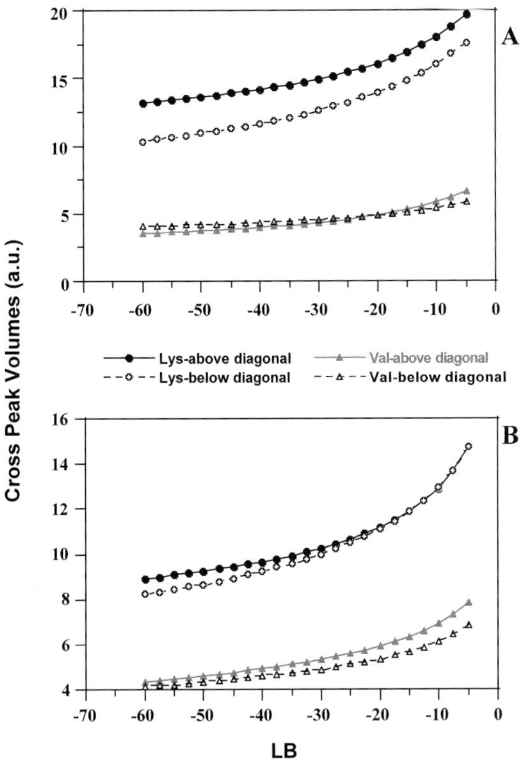

Fig. 5. Variation of cross peak volumes as a function of LB in the t_2 time domain in magnitude-mode COSY spectra of follicular thyroid tissue. A sine-bell window was applied in t_1 and a Lorentz-Gauss window in t_2. $GB_2 = 0.12$. Figure 5A and B were processed with 200 and 512 FIDs, respectively.

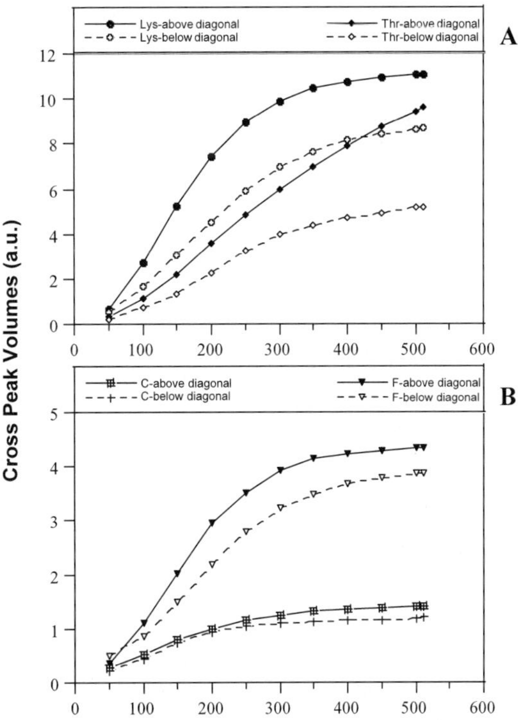

Fig. 6. Variation of cross peak volumes as a function of FIDs processed in the t_1 time domain in phase-sensitive double-quantum filtered COSY spectra of the malignant human colorectal cell line LIM 1215. A $\pi/4$ shifted sine-bell window was applied in t_1 and a Lorentzian–Gaussian window in t_2. $LB_2 = -30$, $GB_2 = 0.3$. (A) lysine and threonine, (B) F (–O–C(O)–CH_2–CH_2; 2.23, 1.57 ppm), and C (–CH_2–CH_2–CH=CH–; 2.00 and 5.31 ppm) resonances from mobile fatty acid chains.

whether the sample is a tissue specimen, cultured cells or a chemical standard [73]. Moreover, the comparison of cross peak volumes from either MM- or PS DQF-COSY spectra show similar patterns of oscillation, although they are dampened in samples with shorter T_2 relaxation values, such as tissue specimens. Similar analyses have not been performed on data obtained from other pulse sequences, e.g. TOCSY. However, this observation indicates that these processing dependent variations in cross peak volumes depend on the spin systems rather than the environment. As a general observation, cross peaks arising from symmetric AB and A_2B_2 spin systems show similar processing dependent volumes above and below the diagonal, whereas those from asymmetric A_2B, A_3B, or A_3B_2 show variation in cross peak volumes on either side of the diagonal [73].

Changing Parameters in t_1

The time constraints imposed by biological sample degradation restricts the number of t_1 increments collected during 2D experiments. The shortened t_1 time domain often leads to a restricted choice of window functions in t_1, and for COSY spectra the shifted or unshifted SB are most often used. Modulation of the phase shift of a SB window in t_1 leads to little variation in cross peak volumes for a constant number of FIDs processed (data not shown). However, for samples where large data sets have been collected, increasing the number of FIDs processed leads to increases in volumes for all cross peaks in all samples (Figure 6).

In summary, acquisition and processing parameters may be tailored to maximize the apparent S/N ratio of a cross peak and to affect the quality of 2D spectra. The disparity between volumes above and below the diagonal is controlled by the spin systems present in the sample. An understanding of the response of the spin systems and judicious choice of acquisition and processing windows allows an optimization of S/N ratio in cross peaks in 2D spectra, and allows better comparison of cross peaks between tissue samples of differing pathologies.

Concluding Remarks

2D MRS remains a powerful tool for the *in vivo* and *ex vivo* study of human cancer and other pathologies. The ability to separately identify numerous metabolites in human tissue non-invasively in a single experiment, allows a high degree of metabolic profiling to be performed that can be correlated with various disease states. The ability to assess multiple parameters through different pulse sequences gives 2D MRS extreme flexibility in experimental design. The limitations imposed by the time constraints inherent in dealing with tissue samples can be dealt with through careful experimental practice and attention to acquisition and processing details. The linking of modern 2D MR pulsed spectroscopic techniques with sophisticated technologies, such as MAS and *in vivo* spectroscopy, addressed elsewhere in this issue, hold great promise for the unraveling of metabolic parameters associated with the onset and progression of human disease.

Acknowledgments

EJD wishes to acknowledge the support of the Leo and Jenny Leukaemia Foundation, the University of Sydney Cancer Research Foundation, the University of Pennsylvania Research Foundation, the National Institutes of Health (NIH grants R21-CA-79718 and R21-EB-002537), and the Small Animal Imaging Resource Program at the University of Pennsylvania (R24-CA-83105).

JQYW wishes to thank the University of Sydney Medical Foundation for a Postgraduate Scholarship.

References

1. Mountford CE, Mackinnon WB, Delikatny EJ, Russell P. In: JD de Certaines, WMMJ Bovee, F Podo (Eds). Text of Magnetic Resonance Spectroscopy in Biology and Medicine: Functional and Pathological Tissue Characterization. Pergamon Press, Oxford 1992, p 507.
2. Mountford CE, Lean CL, Mackinnon WB, Russell P. Ann. Rep. NMR Spectrosc. 1993;27:173.
3. Mountford CE, Doran S, Lean CL, Russell P. Chem. Rev. 2004;104:3677.
4. Jeener J. In: M Goldman, M Porneuf (Eds). NMR and More in Honour of Anatole Abragam, Les editions de physique. Les Ulis cedex A: France, 1994.
5. Aue WP, Bartholdi E, Ernst RR. J. Chem. Phys. 1976;64: 2229.
6. Hull WE. In: WR Croasmun, Carlson RMK (Eds). Two-Dimensional NMR Spectroscopy: Applications for Chemists and Biochemists, 2nd ed. VCH Publishers Inc.: New York, 1994, p 67.
7. Ernst RR, Bodenhausen G, Wokaun A. Principles of Nuclear Magnetic Resonance in One and Two Dimensions, Vol. 14. Clarendon Press: Oxford, 1987, p 610.
8. Croasmun WR, Carlson RMK (Eds). Two-Dimensional NMR spectroscopy. Application for Chemists and Biochemists, 2nd edition. VCH Publishers Inc.: New York, 1994.
9. Dyson H. In: W Croasman, R Carlson (Eds). Two-Dimensional NMR Spectroscopy—Applications for Chemists and Biochemists, 2nd ed. VCH Publishers: New York, 1994, p 655.
10. Bax A. Two-Dimensional Nuclear Magnetic Resonance in Liquids. Delft University Press: Delft, Holland, 1982, p 200.
11. Ziegler A, Izquierdo M, Rémy C, Decorps M. J. Magn. Reson. B. 1995;107:10.
12. Delikatny EJ, Hull WE, Mountford CE. J. Magn. Reson. 1991;94:563.
13. Cross KJ, Holmes KT, Mountford CE, Wright PE. Biochemistry. 1984;23:5895.
14. May GL, Wright LC, Holmes KT, Williams PG, Smith ICP, Wright PE, Fox RM, Mountford CE. J. Biol. Chem. 1986;261:3048.
15. Wright LC, May GL, Dyne M, Mountford CE. FEBS Lett. 1986;203:164.
16. Holmes KT, Mountford CE. J. Magn. Reson. 1991;93:407.
17. Williams PG, Saunders JKS, Dyne M, Mountford CE, Holmes KT. Magn. Reson. Med. 1988;7:463.
18. Behar KL, Ogino T. Magn. Reson. Med. 1991;17:285.
19. Lean CL, Mackinnon WB, Delikatny EJ, Whitehead RH, Mountford CE. Biochemistry. 1992;31:11095.
20. Mackinnon WB, Delbridge L, Russell P, Lean CL, May GL, Doran S, Dowd S, Mountford CE. World J. Surg. 1996; 20:841.
21. Gillet B, Mergui S, Beloil J-C, Champagnat J, Fortin G, Jacquin T. Magn. Reson. Med. 1989;11:288.
22. Remy C, Arus C, Ziegler A, Lai ES, Moreno A, Le Fur Y, Decorps M. J. Neurochem. 1994;62:166.
23. Thomas MA, Binesh N, Yue K, DeBruhl N. J. Magn. Reson. Imaging. 2001;14:181.
24. Ryner LN, Sorenson JA, Thomas MA. Magn. Reson. Imaging. 1995;13:853.
25. Brereton IM, Galloway GJ, Rose SE, Doddrell DM. Magn. Reson. Med. 1994;32:251.
26. Ryner LN, Sorenson JA, Thomas MA. J. Magn. Reson. B. 1995;107:126.
27. Dreher W, Leibfritz D. Magn. Reson. Imaging. 1999;17:141.
28. Alonso J, Arus C, Westler WM, Markley JL. Magn. Reson. Med. 1989;11:316.
29. Dingley AJ, Veale MF, King NJC, King GF. Immunomethods. 1994;4:127.
30. Dingley AJ, King NJC, King GF. Biochemistry. 1992;31: 9098.

31. Wright LC, Groot Obbink KL, Delikatny EJ, Santangelo RT, Sorrell TC. Eur. J. Biochem. 2000;267:68.
32. King NJC, Delikatny EJ, Holmes KT. Immunomethods. 1994;4:188.
33. Mackinnon WB, Huschtscha L, Dent K, Hancock R, Paraskeva C, Mountford CE. Int. J. Cancer. 1994;59:248.
34. Ferretti A, Knijn A, Iorio E, Pulciani S, Gaimbenedetti M, Molinari A, Meschini S, Stringaro A, Calcabrini A, Freitas I, Strom R, Arancia G, Podo F. Biochim. Biophys. Acta. 1999;1438:329.
35. Delikatny EJ, Roman SK, Hancock R, Jeitner TM, Lander CM, Rideout DC, Mountford CE. Int. J. Cancer. 1996;67:72.
36. Roman SK, Jeitner TM, Hancock R, Cooper WA, Rideout DC, Delikatny EJ. Int. J. Cancer. 1997;73:570.
37. Cooper WA, Bartier WA, Rideout DC, Delikatny EJ. Magn. Reson. Med. 2001;45:1001.
38. Mountford CE, Saunders JK, May GL, Holmes KT, Williams PG, Fox RM, Tattersall MHN, Barr JR, Russell P, Smith ICP. Lancet. 1986;i:651.
39. Mountford CE, Delikatny EJ, Dyne M, Holmes KT, Mackinnon WB, Ford R, Hunter JC, Truskett ID, Russell P. Magn. Reson. Med. 1990;13:324.
40. Delikatny EJ, Russell P, Hunter JC, Hancock R, Atkinson K, van Haaften-Day C, Mountford CE. Radiology. 1993;188:791.
41. Kuesel AC, Sutherland GR, Halliday W, Smith ICP. NMR Biomed. 1994;7:149.
42. Kuesel AC, Donnelly SM, Halliday W, Sutherland GR, Smith ICP. NMR Biomed. 1994;7:172.
43. Lean CL, Newland RC, Ende DA, Bokey EL, Smith ICP, Mountford CE. Magn. Reson. Med. 1993;30:525.
44. Mountford CE, Lean CL, Hancock R, Dowd S, Mackinnon WB, Tattersall MHN, Russell P. Invasion Metastasis. 1993;13:57.
45. Mackinnon WB, Russell P, May GL, Mountford CE. Int. J. Gynaecol. Oncol. 1995;5:211.
46. Sivaraja M, Turner C, Souza K, Singer S. Cancer Res. 1994;54:6037.
47. Singer S, Sivaraja M, Souza K, Millis K, Corson JM, J. Clin. Invest. 1996;98:244.
48. Millis K, Weybright P, Campbell N, Fletcher JA, Fletcher CD, Cory DG, Singer S. Magn. Reson. Med. 1999;41:257.
49. Smith IC, Blandford DE. Biochem. Cell Biol. 1998;76:472.
50. Sitter B, Sonnewald U, Spraul M, Fjosne HE, Gribbestad IS. NMR Biomed. 2002;15:327.
51. Cheng LL, Chang IW, Smith BL, Gonzalez RG. J. Magn. Reson. 1998;135:194.
52. Moka D, Vorreuther R, Schicha H, Spraul M, Humpfer E, Lipinski M, Foxall PJ, Nicholson JK, Lindon JC. J. Pharm. Biomed. Anal. 1998;17:125.
53. Cheng LL, Wu C, Smith MR, Gonzalez RG. FEBS Lett. 2001;494:112.
54. Cheng LL, Lean CL, Bogdanova A, Wright SC Jr, Ackerman JL, Brady TJ, Garrido L. Magn. Reson. Med. 1996; 36:653.
55. Bollard ME, Garrod S, Holmes E, Lindon JC, Humpfer E, Spraul M, Nicholson JK. Magn. Reson. Med. 2000; 44:201.
56. Garrod S, Humpfer E, Spraul M, Connor SC, Polley S, Connelly J, Lindon JC, Nicholson JK, Holmes E. Magn. Reson. Med. 1999;41:1108.
57. Griffin JL, Williams HJ, Sang E, Nicholson JK. Magn. Reson. Med. 2001;46:249.
58. Chen JH, Enloe BM, Weybright P, Campbell N, Dorfman D, Fletcher CD, Cory DG, Singer S. Magn. Reson. Med. 2002;48:602.
59. Derome AE. Modern NMR Techniques for Chemistry Research. Pergamon Press: Oxford, 1989.
60. Marion D, Wuthrich K. Biochem. Biophys. Res. Commun. 1983;113:967.
61. Gray AG. In: Croasmun WR, Carlson RMK (Eds). Two-Dimensional NMR Spectroscopy: Applications for Chemists and Biochemists, 2nd ed. VCH: New York, 1994, p 1.
62. Bax A, Davis DG. J. Magn. Reson. 1985;65:355.
63. Bourne R, Dzendrowskyj T, Mountford C. NMR Biomed. 2003;16:96.
64. Mountford CE, Mackinnon WB, Bloom M, Burnell EE, Smith ICP. J. Biochem. Biophys. Methods. 1984;9:323.
65. Kotitschke K, Jung H, Nekolla S, Haase A, Bauer A, Bogdahn U. NMR Biomed. 1994;7:111.
66. Deslauriers R, Somorjai R, Geoffrion Y, Kroft T, Smith ICP, Saunders JK. NMR Biomed. 1988;1:32.
67. Hurd RE. J. Magn. Reson. 1990;87:422.
68. Brereton IM, Crozier S, Field J, Doddrell DM. J. Magn. Reson. 1991;93:54.
69. Keinlin M, Moonen CT, van der Toorn A, van Zijl PC. J. Magn. Reson. 1991;93:423.
70. Garrod S, Humpher E, Connor SC, Connelly JC, Spraul M, Nicholson JK, Holmes E. Magn. Reson. Med. 2001; 45:781.
71. Wider G, Macura S, Kumar A, Ernst RR, Wuthrich K. J. Magn. Reson. 1984;56:207.
72. Bubb WA, Wright LC, Cagney M, Santangelo RT, Sorrell TC, Kuchel PW. Magn. Reson. Med. 1999;42:442.
73. Watzl, JQY. Ph.D. thesis, Sydney University, 2001.

Correlation of Histopathology with Magnetic Resonance Spectroscopy of Human Biopsies

Carolyn Mountford[1], Ian C.P. Smith[2], and Roger Bourne[1]
[1]*Institute for Magnetic Resonance Research, Department of Magnetic Resonance in Medicine, University of Sydney, St Leonards 1590, NSW, Australia; and*
[2]*Institute for Biodiagnostics, National Research Council of Canada, Winnipeg, Manitoba, Canada R3B 1Y6*

Introduction

The pathology of human tissue can be determined by magnetic resonance spectroscopy (MRS) but it was a controversial field for over 20 years. Now MRS on human biopsies identifies disease processes, neoplastic status, and prognostic variables with high accuracy. The MRS databases were compared with careful and specialized histology [1,2]. The MR method is fast, accurate, and robust and for most organs complements routine histopathological diagnosis. For some organs such as the thyroid and esophagus it can provide information not available using a light microscope. However, rigorous procedures and quality control are required to ensure this level of accuracy and are the basis of this chapter.

The reason the MRS method is so precise is that it monitors changes to cellular chemistry during the development and progression of a disease process. This is also the reason that it was controversial. Unless the pathology was undertaken with care there were discrepancies. An early experiment, undertaken in 1978, demonstrated that MRS could identify sequential alterations in cells before they manifested frank malignancy by light microscopy [3,4]. This was an exciting step forward but it also meant that correlation with both pathology and clinical outcome was essential. Twenty-five years on the MRS method is confirmed to show pathological categories which are not documented by light microscopy, due to the absence of architectural differences. The cells do, however, have a different chemical make-up consistent with clinical outcomes [5,6].

The basis of the MRS method for identifying tumor pathology is the increase in MR visible chemical species with the onset and the development of the disease. This was clearly seen in the first clinical study on human cervix biopsies [7]. Cancer of the uterine cervix was chosen for the first clinical ^{1}H MRS study as, histologically, the distinction between the presence and the absence of malignancy is made with a high level of accuracy. As seen in Figure 1, invasive specimens are characterized by an intense MR resonance at 1.33 ppm, primarily from the methylene protons of lipid acyl chains, with additional contributions from the methyl protons of lactate and threonine. There are also methyl, choline, and olefinic resonances at 0.90, 3.2, and 5.32 ppm, respectively (Table 1). The spectrum from human papilloma virus infected tissue (Figure 1B) lacks the intense resonance at 1.33 ppm. Other visible spectral differences include increases in the broad resonances between 3.4 and 4.2 ppm in the pre-invasive specimens compared with their invasive counterpart. It can be seen by comparing these three spectra that there is a gradual increase in the MR visible chemistry associated with the development of disease. In this study, the accuracy of the MRS method was 98% [8].

It is now established that the MRS-detectable chemical information present in the primary tumor is reporting on cellular chemistry associated with tumor development and progression, as well as cellular proliferation, immunosuppression, and metastasis. For a review see Mountford *et al.* [2].

Histopathology—Strengths and Limitations

Histopathology is the gold standard of the 20th century, providing diagnostic and prognostic information for human diseases. It is a very mature discipline that draws upon the experience of medical specialists who rely on visual pattern recognition in human tissues and cells using light microscopy. Its major limitation is the restrictive range of morphological changes that tissues express. The changes are often in a continuum and from this the pathologists are expected to identify patterns for individual diseases. These patterns often overlap making it necessary to make a subjective assessment.

Of importance to the development of MRS as a mature discipline are the sampling errors inherent at several levels in routine diagnostic services. Routine hospital pathology will often examine only one or two sections per sample of tissue. This is at least partly due to the time cost of examining multiple sections of tissue. A study of prostate biopsies by Swindle *et al.* [9] demonstrated that routine hospital histopathology (i.e. one or two slices) failed to

Graham A. Webb (ed.), Modern Magnetic Resonance, 1013–1022.

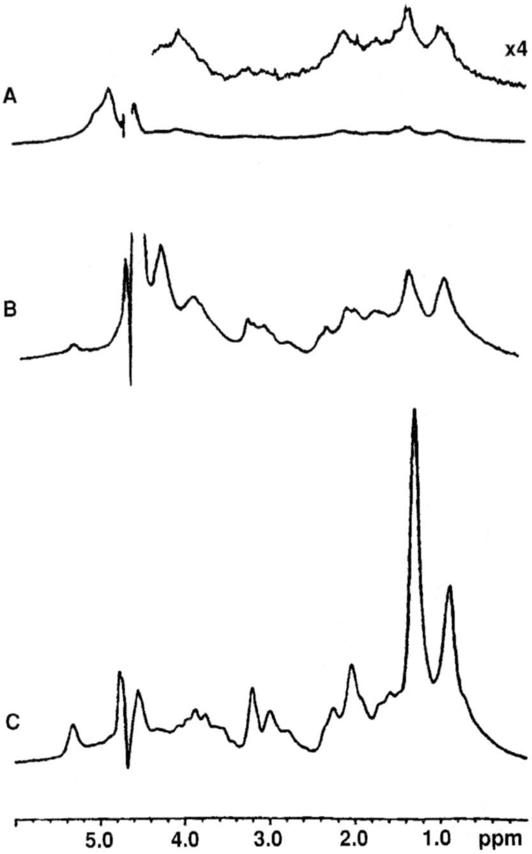

Fig. 1. ^{1}H MR spectra of cervical biopsy specimens. The histological assessment of each specimen is as follows: (A) Nonspecific chronic inflammatory disease. The inset shows the region 4.5–0 ppm with the vertical gain increased fourfold, NE = 400. (B) HPV infection but no significant dysplasia observed, 512 accumulations. (C) Cervical carcinoma, 40 accumulations. Spectra were recorded on a Bruker WM 400 (9.4 T) spectrometer at 37 °C. The water peak was suppressed by gated irradiation, sweep width 4000 Hz, acquisition time 1.024 s, and a line broadening of 3 Hz was applied in each case. *Source:* Reprinted from Magnetic Resonance in Medicine 1990;13:324–31. Mountford CE, Delikatny EJ, Dyne M, Holmes KT, Mackinnon WB, Ford R, Hunter JC, Truskett ID, Russell P. Uterine Cervical Punch Biopsy Specimens Can Be Analyzed by 1H MRS. With permission from Wiley-Liss, Inc., a subsidiary of John Wiley & Sons, Inc.

identify a number of cancers that have been detected by MRS. In order for the MRS method to achieve its potential and provide accuracies approaching 100% detailed non-routine histopathological analysis of the same piece of tissue examined by MRS is needed for correlation.

A good example of the need for serial sectioned histopathological analysis is seen in Figure 2. The spectra from breast ductal carcinoma *in situ* with and without microinvasion present are compared. Tissue sections of DCIS taken 100 μm apart are shown with microinvasion present and without microinvasion. The chemical activity seen in the MR spectrum of the cells with microinvasion is far greater. It should be noted that when both pathologies are present in a single specimen the most active chemical profile will normally be seen. The MRS method can also determine the percentage of each type of pathology but this requires a mathematical pattern recognition method. See the section on regression analysis below.

Finally, in contrast to histopathology, the MRS method has the advantage that it examines the entire piece of tissue that is placed between the transmit-receive coils thus removing any sampling errors associated with histopathological examination of only a fraction of the tissue sample.

Collection and Storage of Biopsy Specimens for Analysis

This is possibly the most important point of this chapter. If the biopsy specimens are not handled appropriately the tissue will start to degenerate and the important chemical species that have diagnostic and prognostic capabilities will alter.

For each tissue type reported to date handling protocols have been individually determined and reported [2]. Biopsy specimens can be a piece of tissue or a fine-needle aspirate biopsy (FNAB). The specimen must be immediately placed in a vial containing PBS/$D_2$0 after the biopsy is obtained and the vial placed without delay in a dewar containing liquid nitrogen. The polypropylene vials must have been tested for leakage of minute levels of chemicals which leach from the plastics and can cause cell death. This can be time consuming and often requires a number of batches of plastics to be tested. Similarly, the barrels of syringes used for FNAB need to be tested. Each tissue type, for example cervix, lung, kidney, has a finite time in which it can be stored in a freezer before the MRS examination is undertaken. It is unacceptable to store specimens for longer than 6 weeks.

The accuracy of the MRS method is dependent upon tissues and cellular pools of chemicals retaining integrity. In some of the earlier studies, tissue delivered to the pathology laboratory after surgery was then sent for MRS examination [10,11]. These spectra often displayed metabolite profiles characteristic of tissue degradation (e.g. increased lactate levels and depleted glucose). The predictive accuracy of such analyses was much lower than where the tissue was snap frozen immediately after collection.

Table 1: Resonances in one-dimensional ^{1}H MR spectra

Molecules	Abbreviation	Species	Chemical shift (ppm)
Amines			
Choline, phosphocholine, glycerophosphocholine	Chol	$-N(\mathbf{CH_3})_3$	3.2
Spermine, spermidine, polyamines	PA	$-N\mathbf{CH_2}$	3.1
Creatine, phosphocreatine	Cr	$-N\mathbf{CH_3}$	3.0
Lipids			
Triaclglycerol (fatty acyl chain)	Lip	$\mathbf{CH_3}-CH_2-CH_2-$	0.90
	Lip	$CH_3-\mathbf{CH_2}-CH_2-$	1.33
	Lip	$-OOC-CH_2-\mathbf{CH_2}$	1.6–1.7
	Lip	$CH_3-\mathbf{CH_2}-CH_2-$	2.02
	Lip	$CH{=}CH-\mathbf{CH_2}-$	2.08
	Lip	$-OOC-\mathbf{CH_2}-CH_2$	2.3
	Lip	$\mathbf{CH}{=}CH-$	5.32
Amino Acids			
Alanine	Ala	$-\mathbf{CH_3}-CH-$	1.49
Glutamate, glutamine	Glu, Gln	$-CH-\mathbf{CH_2}-CH_2-COO-/NH_{3+}$	2.2
Glutamate, glutamine	Glu, Gln	$-CH-\mathbf{CH_2}-CH_2-COO-/NH_{3+}$	2.6
Isoleucine	Ile	$-\mathbf{CH_3}$	0.97
Leucine	Leu	$-\mathbf{CH_3}$	0.97
Lysine	Lys	$H_3N^+-CH_2-\mathbf{CH_2}-CH_2$	1.7
Lysine	Lys	$H_3N^+-CH_2-\mathbf{CH_2}-CH_2$	3.03
Threonine	Thr	$\mathbf{CH_3}-CHOH-$	1.33
Fucose	Fuc	$-CH_2-\mathbf{CH_3}$	1.33
Threonine	Thr	$\mathbf{CH_3}-CHOH-$	4.25
Valine	Val	$-\mathbf{CH_3}$	1.03
Other			
Lactate	Lac	$-\mathbf{CH_3}-CH^-$	1.33
Citrate	Cit	$-OOC-\mathbf{CH_A}\mathbf{H_B}-C(OH)$	2.4–2.7
Taurine	Tau	$H_3N^+-\mathbf{CH_2}-CH_2-SO_3-$	3.25
Taurine	Tau	$H_3N^+-\mathbf{CH_2}-CH_2-SO_3-$	3.43

Care is needed to reduce or eliminate any minor and variable contamination of the sample. However, contamination with preparative agents (e.g. ultrasound gel, skin preparations) is inevitable during collection of biopsies of some tissue types. A robust pattern recognition method should be insensitive to such contamination.

Collection of a FNAB

A FNAB, which can be obtained from lymph nodes or any accessible lesion, usually comprises individual cells. It is important that for adequate signal-to-noise levels between one and five times 10^6 cells are obtained. To perform FNABs a 21/23 G needle is fitted to a 2, 5, or 10 ml syringe. After insertion of the needle tip into the tissue to be sampled the syringe plunger is withdrawn to about 1 ml to apply suction and the tip of the needle is pushed back and forth up to 20 times throughout the volume of the mass to be sampled without removal of the needle tip from the mass. The needle is withdrawn from the tissue, and the tip immersed in 300 μl PBS/D_2O in a prepared vial and the suction is then released. The buffer is aspirated into the tip of the syringe only and expelled back into the vial to wash out from the tip of the syringe and needle the cells aspirated from the tissue. No cells or liquid should be taken into the barrel of the syringe at any time. Gross contamination of the sample with blood should be noted as this may affect the quality of MR spectra. The vial is immediately placed in an upright position into liquid nitrogen and snap frozen.

Table 2: Summary of classifiers and spectral regions using SCS

Biopsy type (n)	Classification	Spectral regions (ppm)	Accuracy	References
Thyroid ($n = 107$)	Normal vs. malignant	Spectral identity not retained	99.0%	[22]
Brain ($n = 206$)	Control vs. malignant	23/55 Subregions	94.4%	[27]
Astrocytoma ($n = 91$)	High vs. low grade	0.81–0.85, 1.71–1.75, 2.16–2.20, 2.46–2.50, 2.54–2.58, 3.02–3.06, 3.51–3.55	95.7%	[28]
Breast FNA ($n = 140$)	Benign vs. malignant	1: 0.87–0.92, 1.20–1.25, 1.63–1.67, 1.79–1.82, 1.95–1.98, 2.85–2.87, 2.95–2.97, 3.19–3.33 2: 1.19–1.23, 1.32–1.38, 1.79–1.82, 1.96–2.00, 2.13–2.15, 2.70–2.76,2.92–2.94, 2.97–2.99	96.1%	[20]
	Lymph node involvement	1: 0.43–0.51, 0.64–0.77, 1.10–1.20, 1.56–1.59 2: 0.44–0.51, 0.67–0.71, 1.13–1.19, 1.56–1.60, 1.86—1.96	95.0%	[20]
Prostate ($n = 87$)	Benign vs. malignant	3.46–3.52, 3.40–3.46, 2.50–2.56, 2.14–2.20, 1.84–1.90, 1.12–1.18	96.6%	[21]
	Residual cancer after radiotherapy	0.77–0.82, 0.94–1.00, 1.32–1.37, 1.59–1.71, 2.24–2.27, 3.19–3.28	91.4%	[29]
Liver ($n = 122$)	Normal vs. HCC	1.34–1.37, 2.28–2.33, 2.83–2.85	100%	[30]
	Cirrhotic vs. HCC	3.00–3.03, 3.56–3.60, 3.66–3.68	98.4%	[30]
	Normal vs. cirrhotic	1.52–1.57, 2.03–2.08, 3.63–3.67	92.1%	[30]
Ovary ($n = 56$)	Normal vs. cancer	1.47, 1.68, 2.80, 2.97, 3.17, 3.34	98%	[31]
Esophagus ($n = 105$)	Normal vs. cancer	3.49–3.57, 3.23–3.28, 0.92–1.12	100%	[5]
	Normal vs. Barretts	3.50–3.54, 1.93–2.06, 1.37–1.43	100%	[5]
	Barretts vs. cancer	3.05–3.13, 2.51–2.57, 1.45–1.49, 1.21–1.29	98.6%	[5]
Melanoma	Metastatic vs. benign	Not yet available		[32]

Preparation of Specimens for MRS

Fine-needle Biopsy

Before ^{1}H MRS each fine-needle biopsy specimen is thawed at room temperature, then transferred to a 5 mm susceptibility-matched Shigemi MR tube. The volume is increased to 300 μl with fresh PBS/$D_2$0 when necessary.

Tissue Specimens Can Be Handled in One of Two Ways

(A) The specimen is thawed and transferred together with the storage buffer directly to a 5 mm susceptibility-matched Shigemi MR tube.

(B) The sample is suspended in a constrained environment by means of insertion into a capillary tube [12]. This tube is then placed in a Shigemi 5 mm MR sample tube containing buffer and a reference standard. The goal here was to quantify the amount of each chemical with the weighed tissue.

It is important to note that freezing, storage, and thawing of tissue biopsies collected and stored in PBS/D_2O results in very significant and unpredictable leakage of metabolites into the storage buffer immediately upon thawing [13]. Leakage of metabolites is most exaggerated in small tissue samples such as needle core biopsies. For this reason the common practice of washing thawed tissue biopsies in D_2O is not recommended for biochemical characterization of tissue. Instead, biopsies should be collected, frozen, and thawed in a small volume (e.g. 400 μl) of buffer. With efficient water suppression techniques, such as Double Watergate [14], it is not necessary to attempt to replace any of the tissue water with D_2O prior to spectroscopy.

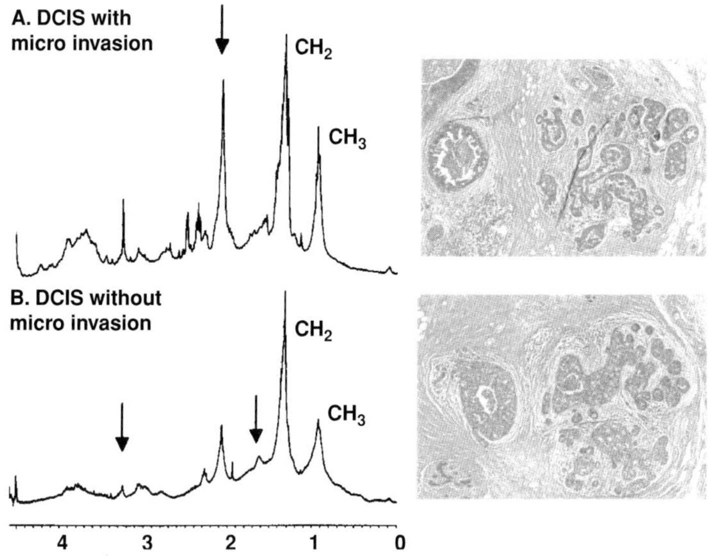

Fig. 2. ^1H MR spectra (8.5 T, 37° C) of fine-needle biopsy specimens obtained from breast ductal carcinoma *in situ*: (A) With microinvasion present and (B) without micro invasion. Tissue sections of DCIS are shown in (A) With microinvasion present and (B) without microinvasion. Spectra were acquired over a sweep width of 3597 Hz, 8192 data points, 256 accumulations, and relaxation delay of 2 s. *Source:* Reprinted from J. Women. Imaging. 2000;2:19–30. Lean CL, Russell P, Mountford CE. Magnetic Resonance Spectroscopy: A New Gold Standard for Medical Diagnosis. With permission from Lippincott Williams &Wilkins. (See also Plate 83 on page LIX in the Color Plate Section.)

All of the tissue and buffer should be within the sensitive region of the rf coils of the probe head so that diffusion of metabolites out of the tissue during the spectroscopy measurement does not result in loss of detected signal. This can be achieved by the use of Shigemi susceptibility-matched sample tubes. The maximum sample volume (tissue + buffer) that will fit within the sensitive region of the probe head will be specific to the particular probe head used. Care should be taken that all sample tubes are appropriately positioned in the probe head prior to measurement. The sample of tissue in buffer is spun at 20 Hz and shimmed optimally. In most cases a line width of 10 Hz for the residual water peak should be easily achievable. In a few cases, when examined with heterogeneous tissue samples, better shims can be obtained in the absence of sample spinning by adjustment of the x and y axis shims.

Experimental Temperature

There are several schools of thought on the appropriate temperature at which to run biopsy specimens. Much of the reported literature has MRS data collected at 37° C in an attempt to obtain data that can be correlated with *in vivo* spectroscopy. Clearly, at body temperature, the tissues degrade faster. Some groups have attempted studies at 4° C in order to reduce tissue degradation. However, it should be remembered that tissue and cell specimens contain a significant amount of lipid molecules. These lipid molecules undergo several phase transitions between 4 and 44° C [15]. It is imperative that the spectra that make up each database are collected at the same temperature.

Magnetic Field Strength

The majority of reported biopsy studies have been undertaken at the frequency of 360 MHz (8.5 T) [2]. There are some reports at 400 MHz (9.4 T) [16–18]. However, earlier work, predominantly on cultured cells, demonstrated that many of the diagnostic and prognostic spectra were frequency-dependent and thus care is needed.

MR Spectroscopy Methods

The most commonly performed measurement used for biopsy tissue characterization is a simple one-dimensional (1D) water suppressed ^1H spectrum. For a typical 50–100 mg (wet weight) tissue sample a spectrum of adequate signal-to-noise (SNR) ratio should be obtainable in 2–3 min. With manual sample changeover total preparation, spectrometer setup, and measurement time should be around 10 min per sample. Typical acquisition parameters for a 1D water suppressed ^1H MRS measurement are as follows.

Pulse angle 90° (6–7 μs), acquisition time 1 s, recycle delay 1 s, 8K data points, 128 transients, spectral width 3600 Hz, total acquisition time ca. 5 min. Water suppression can be effected by low power saturation during the recycle delay (presaturation) or, more effectively and selectively, by a selective excitation field gradient method (e.g. DPFGSE [14]).

As noted above, spectra acquired at different magnetic field strengths cannot be directly compared because, although proton resonance frequency is directly proportional to field strength, coupling constants are independent of field strength. Maximum SNR for a particular resonance of specific T_1 and T_2 is obtained under conditions of partial magnetic saturation with a specific flip angle and interpulse delay. In general, such conditions of maximum SNR will apply to, at best, a few resonances in the spectrum. It may be advantageous to choose flip angle and recycle delay to optimize SNR for specific resonances if these can be identified in advance of measurement. Flip angle, acquisition time, and recycle delay should be maintained constant for all measurements so that relative amounts of partial saturation are constant for samples with similar physical properties. For simplicity of data processing the spectral width and time domain should be maintained constant for all measurements. For samples of low cellularity the number of scans/acquisitions may be increased to obtain adequate SNR without detriment to the comparison of spectra from different samples.

After MR Spectroscopy

It is important to note that the correlation of tissue metabolite profiles by MR spectroscopy is critically dependent on rigorous histopathological assessment of the actual tissue examined by spectroscopy. To this end tissue should be transferred to fixative immediately after MR measurements are complete. The assumption that the histopathologic status of a tissue sample taken adjacent to the spectroscopy sample will be identical to that of tissue used for spectroscopy may be erroneous, especially in heterogeneous tissue types (e.g. prostate). As described earlier, routine preparation and examination of tissue for histopathological reporting may fail to detect the presence of prostate cancer in the spectroscopy sample [9]. Note that for the long-term goal of MRS-based tissue analysis to provide more rapid and sensitive diagnosis than routine histopathology it is necessary for the acquired spectra to be correlated with a more rigorous histopathological examination protocol than usual. Specifically, examination of 5×5 μm slices taken every 100 μm through the prostate tissue sample will ensure that all acini in the tissue sample are examined.

Assignments and Visual Inspection of the Data

The resonances of common chemical species found in the 1D ^{1}H MR spectra from human tissues are summarized in Table 1. These assignments have been made by different groups over 15 years. For review see [2]. Two-dimensional (2D) correlated spectroscopy has been used to make and confirm assignments [19]. See chapter on 2D methods in this Handbook. Many chemical species have been correlated with biological, diagnostic, and prognostic markers.

Typical spectra from malignant and benign breast cells obtained by aspirating the primary lesion are shown in Figure 3. The choline (3.25 ppm) to creatine (3.05 ppm) ratio clearly distinguishes the malignant and benign spectra (Figure 4). This ratio discriminates malignant and benign samples with an accuracy of 96% [20]. The four false positives, seen in Figure 4, are correctly classified when a more sophisticated pattern recognition analysis is performed [20].

The Complexity of Tumor Development and Progression

Another example of the complex and varying spectra that can occur during tumor development and progression is seen in Figure 5 where spectra from different types and pathologies of prostate tissue are seen. The simple resonance intensity ratio analysis described above for breast results in 80–90% accuracy for predicting malignancy [9]. For prostate the amount of data, spectroscopic, histological, and clinical, is considerable and difficult to analyze collectively by simple methods. Thus, a multivariate analysis method was developed [21,22] which formed the basis of the statistical classification strategy (SCS) described below.

Pattern Recognition Methods

A typical 1D MRS data set will have 4096 data points across the frequency spectrum. In contrast to simple resonance intensity ratio analysis described above where ratios of resonance intensities are compared, automated pattern recognition methods offer a more accurate and robust method with which to correlate MRS and clinicopathological characteristics. A method which combines a variety of pattern recognition procedures has been developed by Somorjai *et al.* [22] and is known as a "statistical classification strategy". The SCS method has been applied to a large number of systems and has been described in detail in a recent review [23] (see Table 1 for a summary).

The first stage of the SCS method uses an optimal region selector (ORS) to determine a small number of spectral regions that are most discriminatory [24]. To develop a robust classifier, the number of regions used to develop the classifier must be at least 5–10 times smaller than the number of spectra. The second stage uses these optimal regions and linear discriminant analysis (LDA)

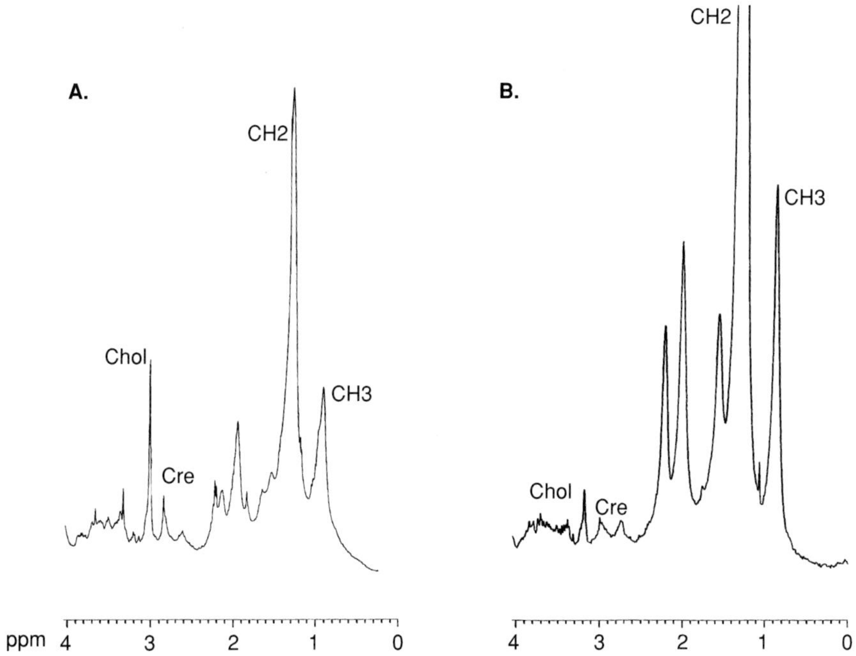

Fig. 3. [1]H MR spectra (8.5 T, 37° C) of breast FNABs with SNR >10: (A) Malignant and (B) Benign. Spectra were acquired over a sweep width of 3597 Hz, 8192 data points, 256 accumulations, and a relaxation delay of 2 s. *Source:* Reprinted from the Br. J. Surg. 2001;88:1234–40. Mountford CE, Somorjai RL, Malycha P, Gluch L, Lean C, Russell P, Barraclough B, Gillett D, Himmelreich U, Dolenko B, Nikulin AE, Smith IC. Diagnosis and Prognosis of Breast Cancer by Magnetic Resonance Spectroscopy of Fine-Needle Aspirates Analyzed Using a Statistical Classification Strategy, with permission from John Wiley & Sons Ltd on behalf of the British Journal of Surgery Society Ltd.

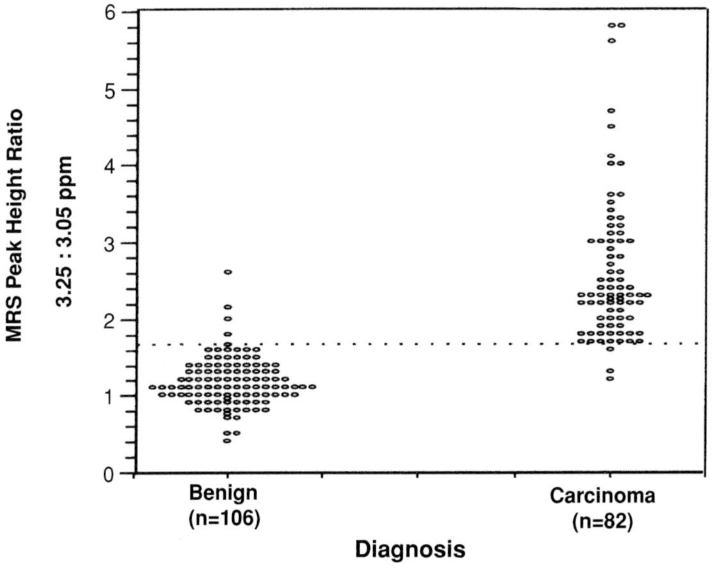

Fig. 4. Breast fine-needle biopsy MR spectroscopic (MRS) findings of unequivocally benign vs. infiltrating carcinoma. Data are grouped on the basis of the final histopathologic findings in tissue specimens from the aspiration site. *Source:* Reprinted from Radiology, 1997;204:661–6. Mackinnon WB, Barry B, Malycha P, Gillett D, Russell P, Lean CL, Doran S, Barraclough B, Bilous M, Mountford CE. Fine Needle Biopsy of Benign Breast Lesions Distinguished from Invasive Cancer by Proton Magnetic Resonance Spectroscopy, with permission from the Radiological Society of North America.

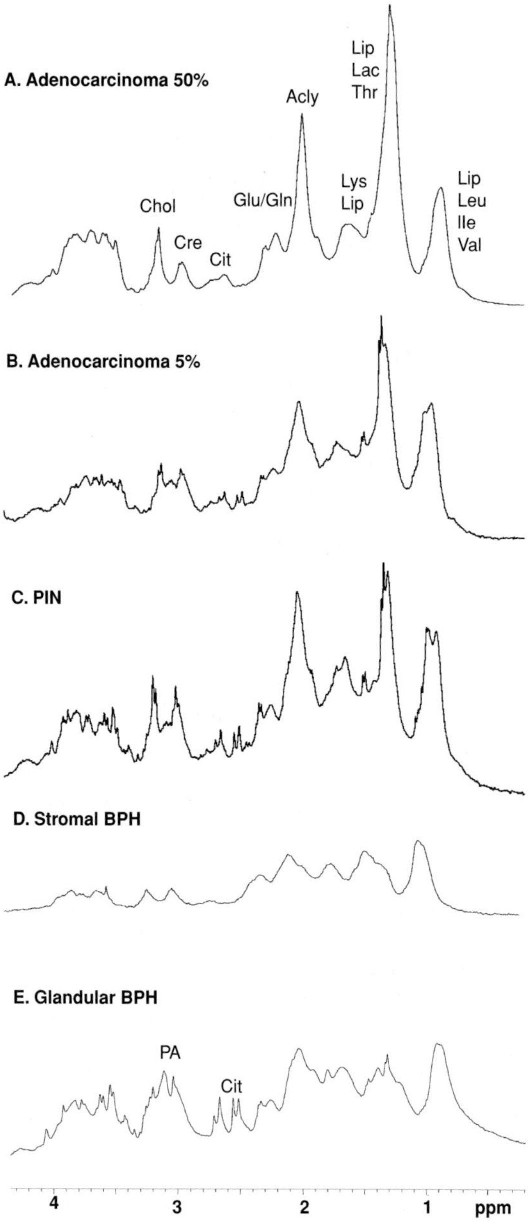

Fig. 5. ^{1}H MR (8.5 T, 37° C) spectra of prostate biopsy specimens, 256 acquisitions, sweep width 3597 Hz, and a pulse repetition time of 2.14 s. The water peak was suppressed by selective gated irradiation. (A) adenocarcinoma (50% of the tissue made up of malignant tissue), (B) adenocarcinoma (5% of the tissue made up of malignant tissue), (C) Prostatic intraepithelial neoplasia (PIN), (D) stromal BPH (95% stromal, 5% glandular), and (E) glandular BPH (85% glandular, 15% stromal) are compared. *Source:* Reprinted from Radiology 2003;228:144–151. Swindle P, McCredie S, Russell P, Himmelreich U, Khadra M, Lean CL, Mountford CE. Pathologic Characterization of Human Prostate Tissue with Proton Magnetic Resonance Spectroscopy, with permission from Radiological Society of North America.

to develop a classifier. Data are divided into training and test sets. The classifier is developed from the training set and then applied to the test set to see how well it does on data with which it has not trained. The original data are then mixed again and a different training and test set used to perform the same analysis. This is done up to 10,000 times to develop up to 10,000 different classifiers. This we call bootstrapping. The 10,000 classifiers are then used in combination to produce a final classifier. If at the end of stage 2 the accuracy of the final classifier is not sufficient, stage 3 of the SCS method is employed. Here other representations of the spectra, such as first or second derivatives, are used to make separate classifiers using stages 1 and 2. All stage 2 classifiers are then combined to yield a meta-classifier of increased accuracy [22].

In a study of FNAB from the breast, the data were subdivided into women with cancer with ($n = 29$) and without ($n = 32$) lymph node involvement [20]. Applying the SCS yielded classification of nodal involvement or not with sensitivity and specificity of 97 and 96%, respectively. Similar classification of patients with and without tumor vascular invasion yielded sensitivity and specificity of 84 and 100%. These classifications should be of great value in determining the therapeutic path for patients. More recent data from the same authors show great promise in calculating the tumor grade and estrogen and progesterone receptor status of breast tumors, also very important in planning therapy [25].

Regression Analysis

Routine histopathological reporting commonly includes a semi-quantitative description (e.g. "occasional foci," "extensive infiltration") of the amount of structural abnormality observed microscopically. Although imprecise, such semi-quantitative descriptions of observed disease burden are a factor in treatment planning. However, correlation of the MRS characteristics of tissue with histopathological characterization of the same tissue has largely been based on a simple two-class analysis in which tissue samples are classified according to the presence or absence of a single specific characteristic such as cancer. Diagnostic interpretation of the spectra from such mixed tissue voxels would ideally report not only the presence of disease, but also the extent or partial volume of disease within the volume of interest (VOI).

To address these two problems a linear regression analysis was undertaken on spectra acquired from 82 cancer-containing prostate tissue specimens. MR spectra were correlated with serial section histopathological examination of each tissue specimen for which the volume fraction of cancer tissue was estimated in 5% increments. A classifier based on five spectral features developed from a training set of 45 randomly chosen spectra was used to predict from spectra the volume fraction of cancer in a validation set of 37 tissue samples. The overall accuracy of predicted cancer volume, measured as the average accuracy of all cancer volume estimates, was 96% [26].

Future Challenges

Putting magnets into a pathology environment is a new concept. The challenge is to move this technology into a routine clinical environment. This was achieved by Liposciences for serum lipoprotein analysis of serum (www.liposcience.com). Serum specimens are easier to transport and the test is for screening purposes.

Clinical acceptance testing is required and is currently being undertaken in Sydney by a group of endocrine and urology surgeons. They need to verify that the tests provide additional information that is necessary for the management of the patient.

Automation of the software to drive the system and to allow comparison of spectra with classifiers will see the spectroscopy method for analyzing human pathology move into a new age.

Acknowledgments

We thank Deborah Edward, Brooke O'Donnell, Sinead Doran, and Raquel Baert for assistance in preparing this manuscript. Our thanks go to all present and past collaborators in particular Professor Peter Russell and Dr. Ray Somorjai, without whom none of this would have been possible.

References

1. Mountford CE, *et al.* Lancet. 1986;1(8482):651–3.
2. Mountford C, *et al.* Chem. Rev. 2004;104:3677–704.
3. Metcalf D. Recent Result Cancer Res. 1966;5:5.
4. Mountford CE, *et al.* Br. J. Cancer. 1980;41(6):1000–3.
5. Doran S, *et al.* Am. J. Surg. 2003;185(3):232–8.
6. Russell P, *et al.* Am. J. Med. 1994;96(4):383–8.
7. Mountford CE, *et al.* Magn. Reson. Med. 1990;13(2):324–31.
8. Delikatny EJ, *et al.* Radiology. 1993;188(3):791–6.
9. Swindle P, *et al.* Radiology. 2003;228:144–51.
10. Ende D, *et al.* Proc. Soc. Magn. Reson. Med. 1993;2:273.
11. Mackinnon WB, *et al.* Int. J. Gynaecol. Cancer. 1995;5:211–21.
12. Kuesel AC, *et al.* Magn. Reson. Med. 1992;27(2):349–55.
13. Bourne R, Dzendrowskyj T, Mountford C. NMR Biomed. 2003;16(2):96–101.
14. Braun S, Kalinowski HO, Berger S. 2nd ed. 150 and More Basic NMR Experiments: A Practical Course. Weinheim: Wiley-VCH, 1998, pp 473–5.
15. Baenziger JE, Jarrell HC, Smith ICP. Biochemistry. 1992;31:3377–85.
16. Cheng LL, *et al.* FEBS Lett. 2001;494(1–2):112–6.
17. Leemans CR, *et al.* Cancer. 1994;73:187–90.
18. Tzika AA, *et al.* J. Neurosurg. 2002;96(6):1023–31.
19. Cross KJ, *et al.* Biochemistry. 1984;23(25):5895–7.
20. Mountford CE, *et al.* Br. J. Surg. 2001;88(9):1234–40.
21. Hahn P, *et al.* Cancer Res. 1997;57(16):3398–401.
22. Somorjai RL, *et al.* Magn. Reson. Med. 1995;33(2):257–63.
23. Lean CL,*et al.* In: G Webb (Ed). Annual Reports NMR Spectroscopy. Guildford, UK: Academic Press, 2002, pp 71–111.
24. Nikulin A, *et al.* NMR Biomed. 1998;11:209–17.
25. Doran S, *et al.* Proc. ISMRM. Kyoto, Japan, 2004, p 2495.
26. Bourne R, *et al.* Proc. ISMRM. Toronto, Oral, 2003.

27. Somorjai RL, *et al.* J. Magn. Reson. Imaging. 1996;6(3): 437–44.
28. Somorjai RL, *et al.* J. Med. Biochem. 1999;3:17–24.
29. Menard C, *et al.* Int. J. Radiat. Oncol. Biol. Phys. 2001;50(2):317–23.
30. Soper R, *et al.* Pathology. 2002;34:417–22.
31. Wallace JC, *et al.* Magn. Reson. Med. 1997;38(4):569–76.
32. Bourne R, *et al.* Proc. ISMRM. Toronto, Canada, 2003, p 1302.

Functional MRI

Graeme D. Jackson, Regula S. Briellmann, Anthony B. Waites, Gaby S. Pell, and David F. Abbott
Brain Research Institute, Melbourne, Victoria, 3081, Australia

Principles of fMRI

In recent years, it has been demonstrated that MRI is capable of detecting changes in cerebral blood volume, flow, and oxygenation that accompany a change in neuronal activity. MRI techniques that observe these activation-induced signal changes are termed "functional MRI" (fMRI).

The most common contrast mechanism in fMRI is known as blood oxygenation level dependent (BOLD) contrast, first described by Ogawa *et al.* This contrast mechanism relies on the relationship between local changes in oxygenation and blood flow that occur with neuronal activation. The sequence of events is as follows:

1. The neuronal cells fire.
2. The small amount of energy used is recovered via oxygen metabolism, resulting in a small increase in deoxyhemoglobin in the first second or two after cell firing.
3. A vascular response also occurs, being a local increase in blood flow (perfusion).
4. In the next few seconds, the delivery of oxygen via the vascular response is far in excess of that depleted to meet the local energy demands. Therefore, paradoxically, there is a local *decrease* in the concentration of deoxyhemoglobin of a much greater magnitude than the initial increase.

BOLD contrast relies on the sensitivity of MRI to detect the changing concentration of deoxyhemoglobin. The initial small increase in deoxyhemoglobin has been observed by some groups as a very small "initial dip" in the time course of the BOLD MR signal; however, it is the subsequent larger decrease in deoxyhemoglobin concentration that provides the substantial and robust signal increase that makes the method most useful. The signal change arises because deoxygenated hemoglobin is paramagnetic, whereas oxygenated hemoglobin is diamagnetic. Paramagnetic molecules are especially visible in MRI since they exert a strong enhancing effect on the relaxation rate of the water in the vascular environment. The transverse relaxation times, T_2 and T_2^* are reduced by this mechanism. These are the contrast dependencies of spin echo and gradient echo images, respectively. Both these basic imaging techniques can therefore be used to detect BOLD contrast. Differences between these two techniques exist with regard to the extent of sensitivity to the environmental change. The T_2^* relaxation time is uniquely sensitive to the intravascular changes of water environment and therefore BOLD contrast is magnified in gradient echo images. Currently, the most commonly applied imaging technique for BOLD fMRI is echo planar imaging (EPI). It allows very rapid imaging, allowing for whole brain coverage with standard paradigms.

Another more recently developed contrast mechanism for fMRI is that of perfusion. This utilizes imaging methods that are sensitive to detecting the local change in blood flow which is the initial response to the neuronal firing. This technique acts as a more direct means of detecting the activation site. However, the perfusion imaging methods suffer from limitations in signal-to-noise ratio (SNR) in comparison with the simpler BOLD imaging techniques.

Design of fMRI Trials

One of the most important parts of an fMRI experiment is the designing of the appropriate paradigm that will allow identification of brain voxels involved in a particular cognitive or sensory–motor function. In this chapter, we will discuss two commonly used approaches, the "block design" and the "event-related design." For both approaches, it is important to keep in mind that the vascular response (which is what we can measure with BOLD) follows the neuronal activity (which is what we would like to know) with a delay of several seconds.

Block Design

The simplest form of a design for fMRI experiments is called a block design. This comprises alternating cycles of periods of task and rest conditions (Figure 1). Each task and rest period should not be too short, as the BOLD signal takes around 5–10 s to reach a maximum after a neuronal event. Shorter periods will lead to a reduced contrast between task and rest states. Very long periods also have a drawback, as they may resemble the low frequency fluctuations or drift in the scanner signal. Such a design would lead to confusion between the cognitive variation, which we wish to measure, and artifactual changes, which we

Graham A. Webb (ed.), Modern Magnetic Resonance, 1023–1036.

Fig. 1. Scheme of a block design paradigm with alternating block and task periods of identical length.

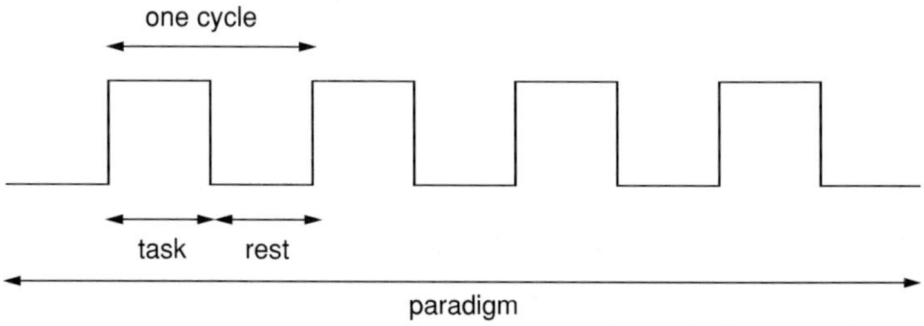

wish to minimize. The ideal design allows the BOLD signal to reach a steady state within each condition in order to maximize contrast. In practice, a 40 s to 1 min cycle time (i.e. 20–30 s task then 20–30 s rest) with about 5–8 cycles, giving a total experiment time of 4–10 min is suitable for many experiments. With more robust tasks (such as visual stimulation, which leads to large neuronal response) and higher magnetic field strengths, such as 3 T, the total experiment time may be reduced. Longer experiment times increase the chance of motion artifacts due to subject discomfort, as well as loss of concentration on the task.

Block design experiments allow maximization of SNR, but also have some disadvantages. Repeating the same task may lead to the subject anticipating the task and sometimes even the response. This may considerably confound the results. A randomized design avoids this issue since the occurrence of a particular task type is unpredictable. Further, not all cognitive studies can be adapted to a block design. For example, experiments assessing the cognitive effect of an unexpected stimulus within another task, known as an oddball paradigm, cannot be performed in a block. Similarly, studies assessing unpredictably occurring physiological events, such as myoclonic jerks, cannot be performed as block designs. Block designs also do not provide information on the shape of the blood flow change in response to the task. This vascular response to a neuronal event is referred to as the hemodynamic response function (HDR). The HDR is used in event-related designs.

Event-Related Designs

Event-related designs are a more flexible approach, and involve the placement of single events within a train of "rest state" conditions. The order of the events is often randomized, preventing anticipation of the task. Events with varying periods of rest allow measurement of different parts of the hemodynamic response, allowing the HDR to be identified. This can be useful in understanding the different response from different brain regions, which may be a result of either the different vasculature of the region, or the differences in cognitive dynamics in the different regions.

The disadvantage of such designs is the low SNR. This is due to the fact that the task state is not sustained for long periods, leading to a less intense vascular response. This can be partially overcome by using "clustered" random presentation of events, where the random pattern of task events is chosen so that there are clumps of task and clumps of rest, albeit with some variability. This is a compromise, giving some increase in SNR whilst maintaining the benefits of a random design.

Event-related analysis techniques can also be applied to data from a block-designed study. This can either take the form of modeling the block of task as a series of events, or performing a *post hoc* analysis of the data. The *post hoc* analysis involves identifying events within a block, such as correct responses to a stimulus, and comparing them to for example incorrect responses.

Inter-Stimulus Interval Selection for Event-Related Designs

It is important to consider the inter-stimulus interval (ISI) carefully. One must first decide whether a sparse or rapid event-related design is to be used. Sparse designs have a long ISI such that the hemodynamic response returns to baseline before the next event is encountered. More common are rapid event-related designs where the minimum ISI is often as little as 2 s. These designs have the advantage that they are more efficient (many more events are captured in a given period of time), and they are therefore the preferred choice if the nature of the task allows it. One must be more careful with these designs, however, as analysis methods usually rely on the assumption that the BOLD response to multiple events adds linearly. This is not always the case (especially if the ISI is much less than 2 s), and one must be careful when interpreting results in brain areas where the linearity of the BOLD response is not well known.

Other issues should also be considered when choosing the ISI. The use of an ISI equal to an integer multiple of the repetition time (TR) can lead to a missed peak in the HDR. This occurs because the stimulus and measurement are synchronized, and measurements may repeatedly occur before or after the peak of the HDR. Making matters worse, there would be a systematic difference in the time each individual slice in a single volume is acquired relative to the event, so that some slices may always be more sensitive to the event than others. If synchronous timing is to be used, it is therefore best to choose an ISI that differs from the TR such that measurements of the HDR occur at different times for many successive events. For example, if events were timed to occur each 1.3 × TR, the HDR would be effectively sampled at nine different time points before the sampling pattern began to repeat. This would be better than 1.5 × TR, for example, where only two different time points would ever be sampled. Another approach is to use jittering. Jittering involves randomly varying the ISI slightly for each event, which again leads to measurements at different points throughout the HDR.

Modeling the Hemodynamic Response in an Even-Related Study

Analysis of an event-related study requires a specific model for the HDR. If the primary goal of the experiment is to determine the precise form of the HDR, a flexible basis set is usually chosen such as a set of sine and cosine waves (a Fourier set), or a set of Gamma functions. On the other hand, if the goal is to determine where and when an event occurred, a canonical HDR is typically used. The canonical HDR is an empirically determined shape found to fit experimental data. One can improve the canonical model by including the temporal derivative (which allows flexibility in the delay of the HDR) and a dispersion term to allow a broader than typical response. This can accommodate some regional and inter-subject variation in response. The canonical model is the most powerful at detection, since it has the fewest parameters, and thus preserves the highest number of degrees of freedom. Conversely, the basis function approaches will be more sensitive if the response differs significantly from the canonical form. A combination of approaches is possible. For example, a portion of a study can be dedicated to eliciting the regional shape of the HDR in an individual (perhaps with a stimulus known to yield a large signal change), and this can then be used as a more accurate model for detection of subtle stimuli in that individual.

Principles of Experimental Design

Where to Start?

There are many factors that affect detection power including the type of experimental design, the scanner, the population, and the nature of the task being imaged. In order to identify the optimal task, a clear hypothesis will aid greatly in designing the experiment. Such a hypothesis may be based on theoretical background of a particular function, or more commonly, is an extension of an observation made outside the scanner. The development of an appropriate task for fMRI scanning is often based on published, somewhat similar paradigms. A protocol can then be further refined after a few pilot studies have been undertaken. Typically, the durations of the final protocol should be a little longer than the estimated optimum, as the subjects used for pilot studies are often highly motivated individuals who will give better results than the average study population.

Block or Event-Related Design?

The block design is optimal for determining the location of activity, but cannot readily be used to determine the nature of the HDR time course. Event-related designs could be optimized to efficiently determine the nature of the HDR time course; however they generally offer much less power to prove that a response actually exists at a particular location. There are many subtleties to possible designs, including hybrid block and event-related designs. In general a block design is preferred if there are no contraindications, as it gives higher statistical power. Alternative approaches should be taken only for questions that cannot be answered using a block design.

Baseline Considerations

Choosing the appropriate baseline is as important as choosing the optimal task. Commonly, the baseline consists of "rest," during which the subject is asked to do "nothing." This is a very poorly defined state, and some paradigms improve with a more defined, engaging baseline. Contrast in fMRI arises from a measured *change* in signal, and it is therefore important to know from what baseline state the signal changed. Our brains are always "on," so one must be careful when interpreting signal change related to a task. All we can really say about significant activation is that there is greater signal whilst a subject is performing a task of interest than when they are performing the baseline task. Rest can and should be considered as just a different task than the task of interest. Similarly, if signal decreases during the task of interest, one can regard this as an increase in activity during rest. Importantly, signal decrease or "deactivation" is not the same as inhibition, since inhibition is itself an active process that can contribute to an increase in signal. If inhibitory processes are involved, the reduction of activity that is observed may have been signaled from an active inhibitory process adjacent to or some distance from the site of "deactivation."

Performance Measures

Proper interpretation of fMRI results generally relies on knowledge of the tasks subjects were performing during the imaging session. One common assumption is that the subjects are doing what they are told. Ideally performance measures should be undertaken concurrently with the imaging, but this is not always practical. For example, in a commonly used language paradigm subjects may be asked to generate as many words as they can that begin with a provided letter. To avoid problems associated with subject motion, this task is normally performed covertly (i.e. without vocalization). This task performance cannot be directly verified. To minimize the possibility of incorrect interpretation, one should obtain some measure of task performance. This is often the measurement of a subject's performance on a similar task out of the scanner. Subjects should also be asked after their scan whether or not they encountered any difficulty in performing the task in the scanner.

Improved interpretation of functional imaging results can often be made when performance of aspects other than the task of interest are also measured. For example, it is known that changes in physiological measures such as cardiac and respiratory rates can affect the cerebral circulation, and breath-holding especially can result in very large changes in the BOLD response that can potentially confound interpretation. Measures of skin conductance and eye tracking are also possible, and these and other measures can be used to remove variance in the data that may be unrelated to the task of interest, or to explicitly model effects that may be related to the task of interest.

Principles of Analysis

There are a number of steps involved in taking raw fMRI data and turning it into colored maps of activity that can be readily interpreted. The processing pipeline contains procedures to minimize artifact, maximize signal, and aid interpretation. We will describe here a commonly used imaging pipeline for BOLD fMRI.

Data Preparation

Often the first step in an analysis pipeline is extraction of the imaging data from the scanner and conversion into a format suitable for further processing. Whilst scanner manufacturers are slowly introducing functional image analysis tools into their standard software, at the present time such software is often rudimentary and inflexible.

Pre-Processing

Pre-processing steps are those performed prior to statistical image analysis. Depending upon the type of acquisition they can include:

- *Correction for signal drop out near magnetic susceptibility inhomogeneities.* The variation in magnetic susceptibility especially near tissue–air boundaries can cause extensive areas of signal loss in typical EPI acquisitions due to the inability of scanners to shim to a uniform field in these regions. Various methods have been devised to overcome this problem, including the z-shim method that involves acquiring two or more sequential acquisitions with different imposed corrective gradients that effectively image separately the areas of drop out. A post-processing step is required to combine these multiple acquisitions.
- *Correction for geometric distortion*, e.g. Measured B_0 phase maps can be used to calculate and correct for geometric distortions that plague EPI fMRI images.
- *Slice timing correction.* Acquisition volumes most commonly consist of a series of sequentially acquired slices. Slice timing correction methods generally resample the temporal response of a voxel, applying interpolation to estimate what the response would be at the time of commencement of each acquisition volume. Of course this correction should not be used in conjunction with imaging methods that involve acquisition of an entire volume in one imaging shot.
- *Gross motion rejection and/or realignment.* The deleterious effects of subject motion can be reduced with these steps. Gross motion rejection can also remove data contaminated by acquisition failures such as a gradient misfire.
- *Global intensity normalization.* This step is necessary with some scanners that are not able to provide a stable signal from one acquisition to the next. If one assumes that the BOLD signal change due to the task is small relative to the mean signal of all within-brain voxels, one can normalize the global signal to a common value (proportional scaling). Unfortunately, the BOLD signal change is not always negligible in relation to the mean global signal, so there is a risk that this processing step will do more harm than good (for example by reducing the apparent size of real activation). If the MRI system is reasonably stable, then global intensity normalization should usually be avoided.
- *Spatial normalization.* This step involves registration of the imaging data of each subject in a study to a common template, to enable voxel-wise statistical analysis of group data and/or to allow reporting of activation locations relative to a standard atlas. In the case of a single subject analysis, the derived transformation may

be applied to the statistical result instead of the raw imaging data, so as to avoid resampling artifact. If this is done, nearest-neighbor resampling is used to preserve the fidelity of the statistical result.
- *Smoothing*. Spatial and temporal smoothing with a Gaussian kernel is often performed to improve the SNR of imaging data. This pre-processing step also ensures that the variance distribution is Gaussian, allowing the valid application of Gaussian random field theory in statistical analysis. To achieve this, the smoothing kernel should have a full width at half maximum (FWHM) of at least twice the voxel size.
- *Linear de-trending*. This is a controversial processing step as the underlying assumption that a linear trend in a voxel time course is "noise" is unlikely to be valid. There are other ways to deal with this later when statistically analyzing the images.

It is important to note that the order in which these processing steps are performed can have a substantial effect on results. The order listed above is a typical scenario, however, it may not always be the best arrangement.

Statistical Analysis

Statistical analysis of fMRI data is necessary because the signals of interest are usually of a similar magnitude to the noise in the data. By collecting many images and applying statistical analysis, one can enumerate the voxels that have an acceptable probability of being associated with the task of interest. The most common method of analysis is a general linear model based approach. The signal is assumed to comprise a linear combination of effects of interest, confounding effects, and a random noise term. The effect of interest in the model usually consists of a function representing the time course of tasks performed convolved with an estimate of the brain's HDR. Confounding effects are also often included in the model, such as the realignment parameters determined in the motion correction pre-processing stage (these are often included to model any residual motion effects such as changes in the magnetization experienced by protons that are not corrected by realignment). The statistical analysis determines, for each voxel independently, the proportion of included model terms that best fit the measured data, and provides an estimate of the probability that the coefficient of the effect of interest is non-zero (implying the effect is present). Gaussian random field theory is often used to estimate a probability corrected for the multiple comparisons made in this massively univariate data.

Other statistical analyses are also possible, including data driven approaches where no assumptions are made about the tasks performed. Rather the data are interrogated in such a way as to extract structured components that are able to describe the data in some way. Such methods include principle components analysis and independent components analysis, the latter generally being more useful in the neuroimaging context. Detailed descriptions of these approaches are beyond the scope of the current discussion. However, it is worth noting that these approaches are beneficial when the nature of the expected response is not known. Results of data driven analyses can be difficult to interpret however, and model based approaches are generally more powerful when a good estimate of the expected response is available.

Software Packages

There are now a number of software packages available to assist with functional imaging analysis. Probably the package in most widespread use is SPM, a powerful and flexible package for performing "statistical parametric mapping." A comprehensive competitor with some unique advantages is FSL. Both of these packages are available free of charge. There are many other packages that are also worth considering, including several specialized utilities that have been designed to perform particular aspects of processing, analysis, and/or visualization. Often the best analysis approach will take advantage of several packages. Selection can often be a matter of determining the packages that are already in use in your geographical area, since an experienced user is often the best source of help for inexperienced users.

Artifacts and Pitfalls

Since fMRI studies usually require a relatively long series of scans in order for the activation-induced signal changes to build up to a detectable level, gross subject motion may occur. Image processing techniques exist to at least partially correct for motion, as we discussed earlier, however prevention is better than cure. For example, head immobilization via straps, vacuum-bean bags, and/or use of bite-bars can help minimize head motion. Immobilizing the upper part of the extremities during sensory–motor tasks can help minimize task-related motion. A further confounding factor can be task-associated physiological effects resulting in changes in brain activation, such as subconscious breath-holding during challenging tasks. Such artifacts may lead to misinterpretation of the study results, and every effort should be undertaken to minimize their occurrence and to monitor for their presence.

Another common problem with fMRI is geometric distortion. The most common rapid imaging technique for fMRI is EPI. With this technique, images of the brain can be acquired in times of 100 ms or less; however the technique is particularly prone to exhibit geometric distortion.

This is especially severe in areas of high susceptibility differences such as the air–tissue boundary in the temporal lobes. Correct interpretation and detection of fMRI changes in these areas is often difficult. Even in relatively susceptibility-free regions of the brain, EPI images will display subtly different distortions in comparison with slower structural scans and this complicates the overlay of function and structure that is a necessary for interpretation of results.

Even when all these problems are minimized, and an appropriate activation map has been generated, one has to be careful in the interpretation of the fMRI results. A common pitfall in group studies is an inadequate number of participants, leading to group results that do not survive correction for multiple comparisons. Whereas such results may be important pilot data, directing further experiments, one has to consider that they are at best a "trend" and provide no statistical proof of an effect greater than noise. A further common problem is the interpretation that a significant "activation" is equivalent to preserved function. There are many examples in the literature, demonstrating that the degree of activation does not linearly follow performance of a function. A very poor performer and an excellent performer may both activate relatively few areas, whereas a moderate performer may show a relatively higher degree of activation. Task-related activation in a particular area may not represent "the site" of a particular function, but rather represent part of a network of areas used to perform this task, or even an associated function, such as attention to the task or eye movements in relation to the paradigm. Furthermore, just because an area is active does not necessarily mean that it is essential for the performance of the task.

Practical Applications

All the fMRI examples shown here were acquired on a GE Signa 3 Tesla MR scanner (GE Medical Systems). Higher magnetic field strength is associated with a higher SNR, which can be translated into shorter imaging times for a given resolution, or higher resolution for a given imaging time, or combination of both. At the time of writing 1.5 T scanners are most common. Compared to 1.5 T, at 3 T the same fMRI paradigm could be run for a shorter period of time, or paradigms evoking typically a very small response may give sufficient signal, whereas they may not be useful at lower field strength. However, the higher field strength has also some disadvantages. There is the potential for greater energy deposition into tissue and increased susceptibility-related artifacts.

Functional images are always acquired together with other sequences. The T_2^*-weighted sequence used for fMRI has a low resolution, so T_1 or T_2-weighted images are usually acquired in the same planes as the T_2^*-weighted images. Further, angiogram/venogram images are often acquired to localize vessels in the brain, in order to avoid confusion of a signal arising from a vessel with the BOLD response. fMRI in the examples below was performed using a whole brain (22 slices) gradient EPI technique (TE = 40 ms, TR = 3600 ms, acquisition matrix = 128 × 128, flip angle = 60°, and 4 mm thick slices, 1 mm gap).

Language fMRI

Language is a cognitive brain function that involves a network of brain areas. In the majority of healthy subjects, these language-related brain areas are predominantly in the left hemisphere. Since the first reported use of fMRI for the study of language in 1993 by McCarthy *et al.*, fMRI has become established for the purpose of assessing language lateralization. The determination of "language dominance" is an important investigation step in patients undergoing brain surgery. Many different language fMRI tasks have been developed. Some are designed to activate particular parts of the language network. In our Institute, we have developed and validated two language paradigms, both performed as a block designs.

For the orthographic lexical retrieval (OLR) paradigm a letter is presented visually, and the subject asked to retrieve silently as many words as possible beginning with that letter. After 18 s a second letter is presented, giving a 36-s task period. For the noun–verb generation (NVG) paradigm, the subject is visually presented with a common noun (e.g. fish) and is asked to generate an appropriate verb (e.g. swim). A new noun is presented every 4 s, with a task period of 36 s. For both tasks the baseline consists of visual fixation of a cross hair, this rest period also has a duration of 36 s (Figure 1). Both paradigms have four cycles with an additional rest period at the start of the paradigm; the total duration of the paradigm is 5.4 min.

Determination of Language Lateralization

Image processing, as described above, will result in maps of activation; these can be used to quantify the lateralization of a function. The laterality index (LI) is based on the number of activated pixels in language-associated areas. The LI is calculated using the following equation: LI = (left-hemispheric pixel count—right-hemispheric pixel count)/(left-hemispheric pixel count + right-hemispheric pixel count).

Examples of fMRI Language in Controls

We recently performed a study in multilingual subjects using a NVG task in four different languages. For reference to this publication, see figure legend. We hypothesized that the degree of proficiency in each language would be related to the extent of functional activity measured in a region of interest analysis. Therefore, the degree of

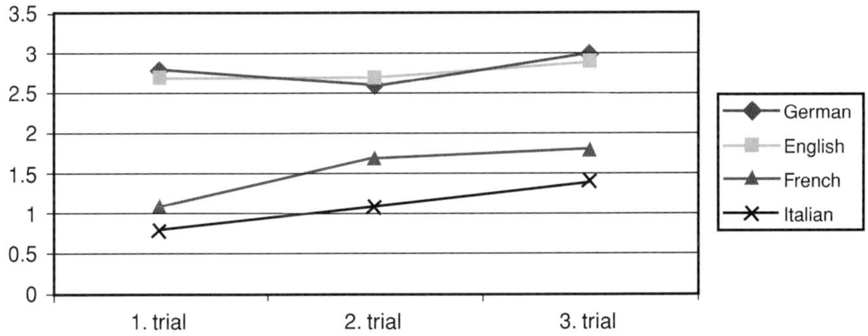

Fig. 2. This diagram shows an example of an out-of-scanner performance test in a quadrilingual subject. The subject was given a cartoon with a word-free story, and was asked to produce a discourse based on this cartoon. This was repeated three times, and then a new cartoon was given to assess the discourse in another language. Results are given for the words/second produced at all three trials. Note the overall increase in word production, particularly noticeable for languages with poorer proficiency. This subject had two "good" and two "poor" languages. *Source:* Adapted from Briellmann RS, Saling MM, Connell AB, Waites AB, Abbott DF, Jackson GD. A high-field functional MRI study of quadri-lingual subjects. Brain Lang. 2004;89:531–42.

proficiency in each language had to be quantified in detail. We used several tests, including a discourse test, in which the subjects were asked to produce a story based on a cartoon, which displayed a word-free story. Figure 2 shows the results in one subject, demonstrating a clear difference between languages with good performance (in this subject German and English), and languages with poor performance (Italian and French). The fMRI showed that all four languages activated overlapping brain areas, corresponding to the major language regions (Figure 3). Interestingly, the number of activated voxels correlated with proficiency, so that the activated volume increased for languages in which a subject had poorer proficiency. This observation highlights an important issue in the design and interpretation of fMRI experiments. In some cases better performance is associated with more brain activity, whilst in others the reverse is true. Correct interpretation of a novel task therefore requires some independent measure of subject performance or response.

Examples of fMRI Language in Patients

Atypical language lateralization is more frequently found in patients with epilepsy and some other focal and chronic brain disorders. In a series of 30 healthy controls and 30 epilepsy patients, we confirmed the increased frequency of atypical language in epilepsy patients (Figure 4). Such atypical lateralization may represent reorganization due to early disturbances of the left hemisphere, such as a perinatal vascular event. fMRI is not only used to determine language lateralization, it has been used to assess the

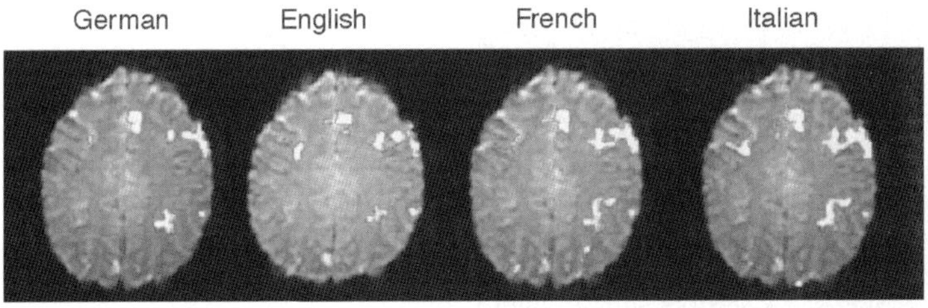

Fig. 3. Functional MRI of language of a quadrilingual subject. The subject performed the NVG task in four different languages. The figure shows the activation during task in comparison to rest in all four languages. Note that all languages activate similar brain areas, however, there is a subtle increase in the amount of activation in poor compared to good languages. *Source:* Image taken from Briellmann RS, Saling MM, Connell AB, Waites AB, Abbott DF, Jackson GD. A high-field functional MRI study of quadri-lingual subjects. Brain Lang. 2004;89:531–42. (See also Plate 84 on page LIX in the Color Plate Section.)

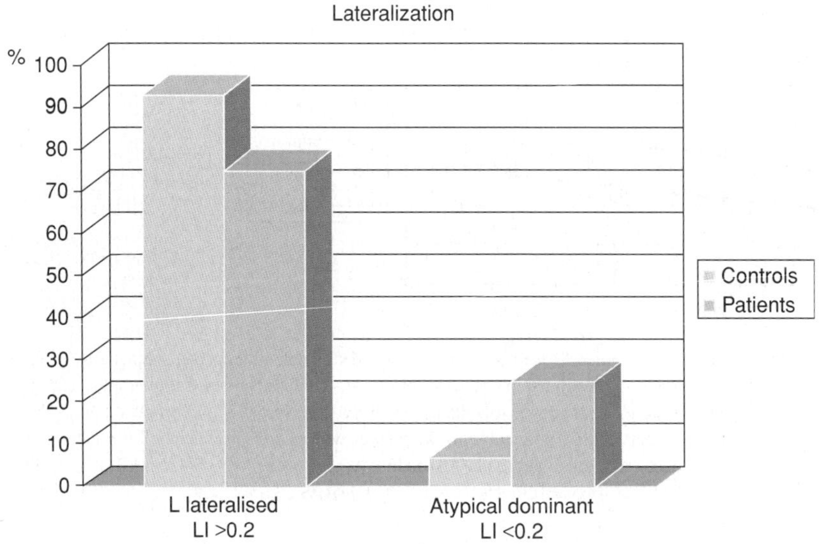

Fig. 4. This diagram shows the distribution of typical and atypical language lateralization in a group of 30 healthy controls and 30 patients with refractory partial epilepsy. Note the predominance of left-lateralized language in both groups, but the relatively higher proportion of atypical language in the patient population.

spatial relationship between the seizure focus and brain function (for review see Sunaert and Yousry 2001). This is of particular interest in the planning of neurosurgical resection.

Figures 5 and 6 show findings of a 9-year-old girl who was investigated for possible epilepsy surgery. Her epilepsy was first noted at the age of 5 years, when she developed complex partial seizures, occurring in clusters, often out of sleep. Her structural MR investigation showed a left mesial temporal lesion, suggestive of a low-grade tumor (Figure 5). Her language fMRI shows clear left-lateralized language dominance (Figure 6). Therefore, in this case language and seizure focus were co-lateralized to the same hemisphere.

A different situation is present in the patient described in Figures 7 and 8. These figures show the findings of a 35-year-old woman, who is left-handed. Her epilepsy started at the age of 12 years. Her EEG showed two independent, left-hemispheric epileptogenic foci, one in the occipital region, the other in the mesial temporal lobe. Her structural MR shows a left occipital lesion, suggestive of an old hemorrhagic lesion (Figure 7).

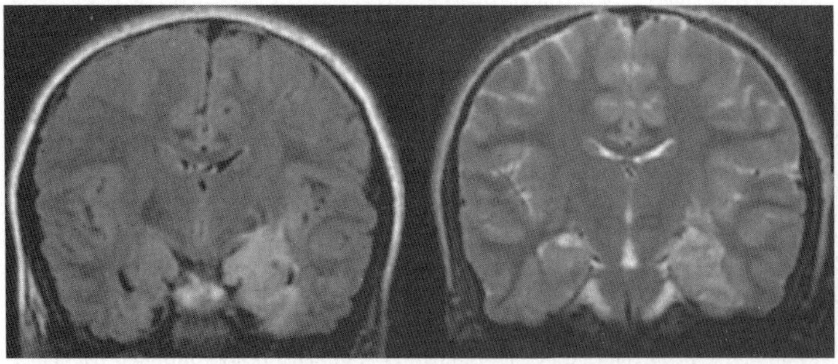

Fig. 5. This figure shows the structural MR examination in a 9-year-old girl with temporal lobe epilepsy. Representative coronal slices in the area of the mesial temporal lesion are shown. On both the FLAIR (left) and T_2-weighted (right) scans, the lesion is characterized by heterogeneously increased signal intensity, affecting mainly the mesial temporal area, but also partially involving the stem of the temporal lobe. This pattern is indicative for a low-grade tumor, such as dysembryoplastic neuroepithelial tumor.

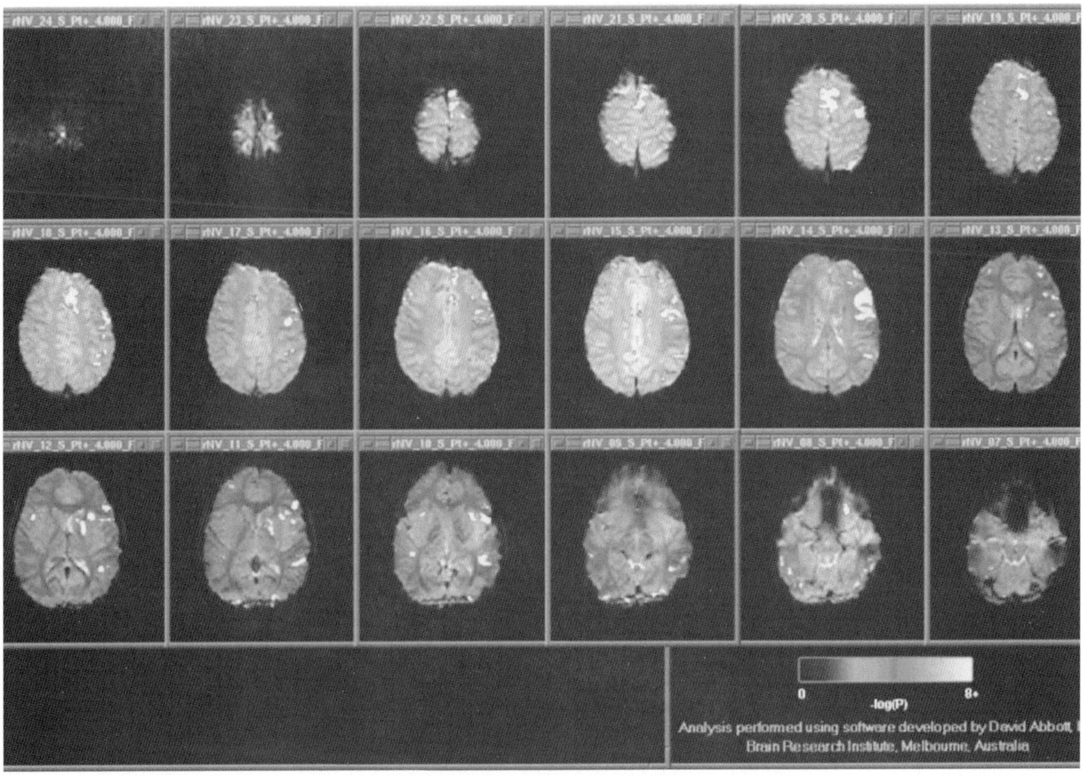

Fig. 6. Language fMRI in the same patient as shown in Figure 5. There is strong left-lateralized language dominance, with particularly strong activation in the frontal lobe language areas (middle frontal gyrus and inferior frontal gyrus). There is also activation in the superior temporal gyrus, which is not involved in the lesion. Therefore, language and seizure focuses were co-localized to the same hemisphere. Note: the subject's left is displayed on the image right (i.e. radiological convention). (See also Plate 85 on page LX in the Color Plate Section.)

Based on the clinical symptomatology, this lesion was considered the primary seizure focus. Additionally, there were also signs of ipsilateral hippocampal sclerosis, suggestive for secondary damage. Her fMRI of language shows right-lateralized language dominance (Figure 8). Therefore, in this case language and seizure focus were in different hemispheres, and given the suspected early left-hemispheric damage, shift of the language function to the undamaged right hemisphere may have taken place.

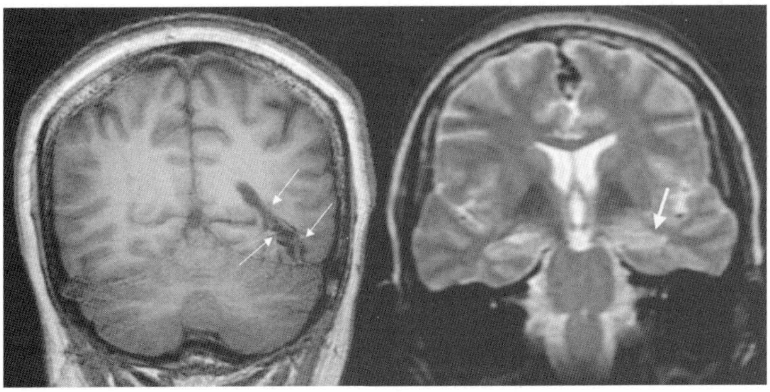

Fig. 7. Structural MRI of a 35-year-old women with refractory partial epilepsy. The image on the left shows a T_1-weighted, coronal slice at the level of the left occipital abnormality (highlighted by thin arrows). The image on the right shows a T_2-weighted, coronal slice at the level of the hippocampus (highlighted by thick arrow). There is increased T_2-weighted signal in the body of the hippocampus (114% of mean normal value, whereas the empirically determined cut-off indicating an abnormally increased value is 105% of mean normal value).

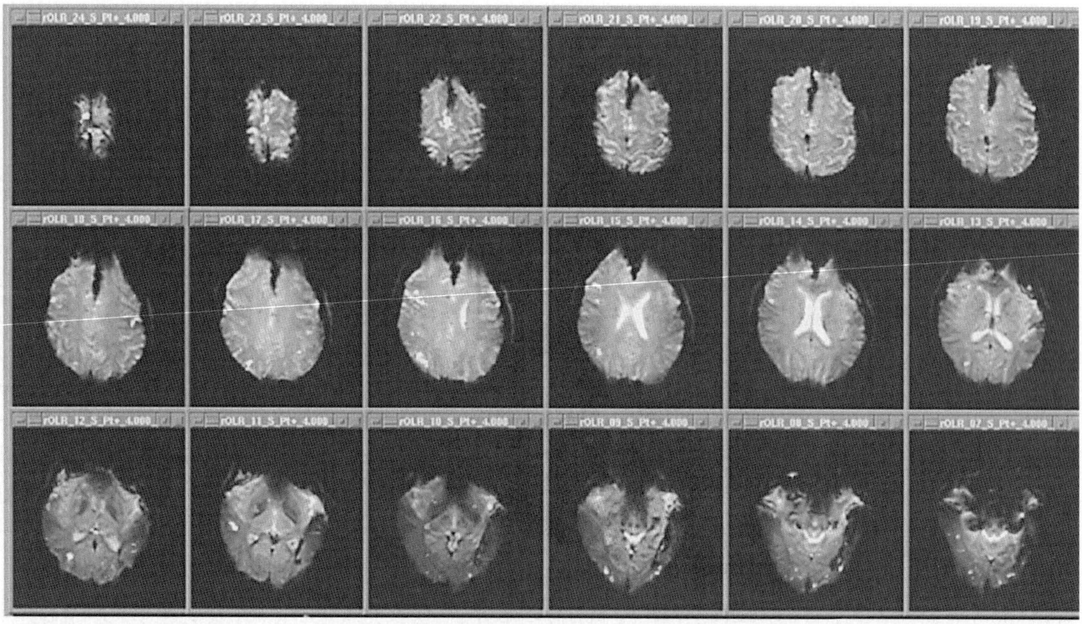

Fig. 8. Language fMRI in the same patient as shown in Figure 7. There is strong right-lateralized language dominance, involving both frontal and temporal lobe areas. In this left-handed patient, language and seizure focus were not co-localized (see text for more details). Note: the subject's left is displayed on the image right (i.e. radiological convention). (See also Plate 86 on page LX in the Color Plate Section.)

FMRI/EEG

Some neurological diseases are associated with changes on the electroencephalogram. Probably the best-known example is the epileptiform discharge in patients with epilepsy. These discharges can occur without associated clinical symptomatology, and are termed interictal discharges. These discharges can be used as "events-of-interest" in an fMRI paradigm. This technique has a great practical potential for the investigation of patients with focal epilepsy. Interictal focal discharges typically originate from around the seizure focus, and can therefore be used as a localization indicator. However, whereas the EEG gives excellent temporal information on when these discharges occur, it has rather poor spatial resolution. The combination of EEG and fMRI overcomes this limitation. Until relatively recently, recording of EEG within the bore of the magnet has been regarded as too difficult since the EEG signal is impaired due to factors including induced currents from slight movement of leads in a large static magnetic field, changing magnetic field gradients during the scanning procedure and the radio frequency (RF) energy emitted by the transmitter coil. The MR image itself can be further impaired by RF interference from conductors and electrical devices within the magnet. However, these problems have been gradually overcome, and an increasing number of centers are reporting successful EEG recording within the scanner. This approach allows studying fMRI activation in association with subclinical events.

The quality of the EEG during scanning is of crucial importance for the fMRI/EEG analysis. Our system is now of a quality that allows discrimination of the interictal discharges during continuous fMRI scanning. Figure 9A demonstrates the high quality of the EEG obtained during scanning and Figure 9B shows further improvement of the EEG using post-processing methods (Figure 9). In suitable patients, fMRI/EEG can show focal activation in the area of the presumed seizure focus, as demonstrated in Figure 10.

Conclusion

fMRI is a relatively young investigation method. Nevertheless, it has proven to be a very useful diagnostic and research tool. fMRI is safe, non-invasive and can be repeated as many times as needed. Due to the need for active participation of the subject it can, however, only be applied to cooperative subjects. Typically, this is difficult to achieve

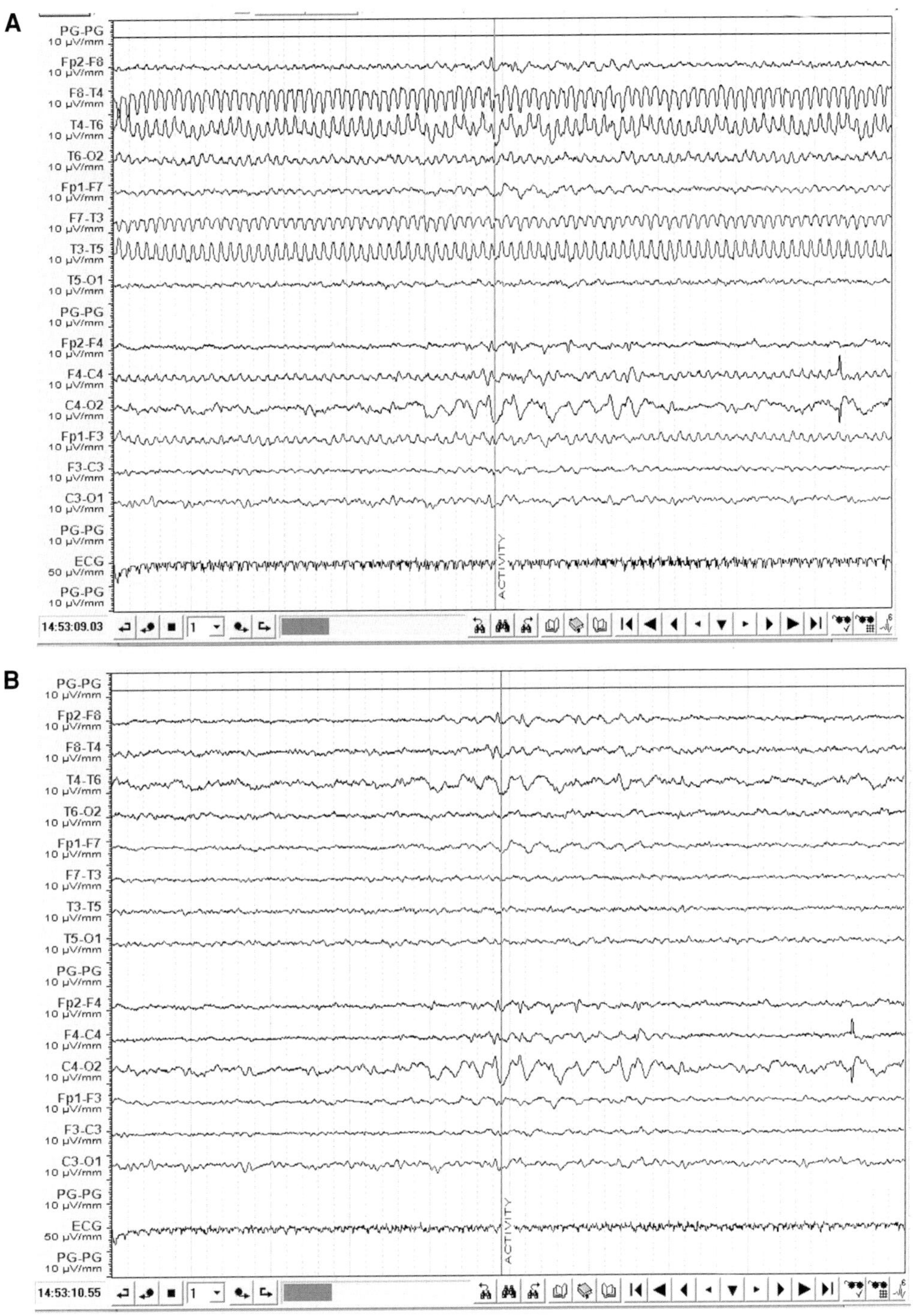

Fig. 9. Example of EEG recording during scanning, at the time of an epileptiform discharge. (A) The image shows the EEG before post-processing, demonstrating the quality of our EEG during scanning. A run of rhythmic theta is seen with maximum over C4. (B) The image shows the same trace after filtering. Note the disappearance of residual gradient artifact after filtering, allowing one to appreciate the sharp transients associated with the theta waves. For the analysis of this study, see Figure 10.

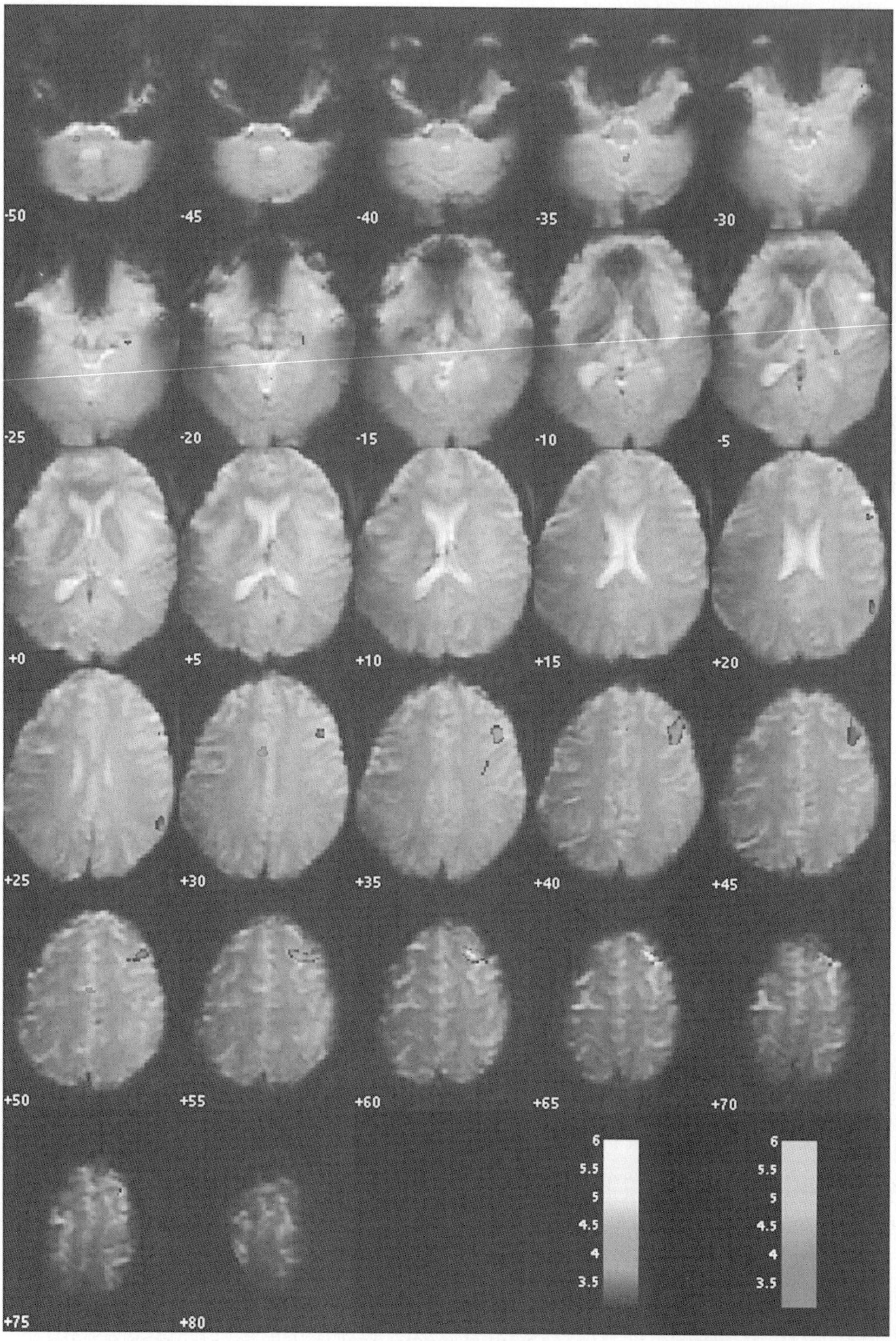

Fig. 10. This figure shows the analysis of an fMRI/EEG study in a 39-year-old woman with refractory partial seizures, arising from a small dysplastic lesion in the right frontal lobe. The figure shows positive BOLD response associated with six interictal discharges, captured during 35 min of fMRI acquisition. The fMRI analysis was performed using SPM99. The areas of activation are shown in warm colors, and the areas of deactivation in cold colors. The threshold was set at $p = 0.001$, uncorrected for multiple comparisons. There is a focal signal increase in the right frontal lobe in close proximity to the suspected seizure focus. Note: Here the subject's left is displayed on the image left (i.e. neurological convention). (See also Plate 87 on page LXI in the Color Plate Section.)

in children younger than 6 years (although we have occasionally successfully performed fMRI in 5-year-olds), or in the severely ill or disabled patient. If applied and interpreted correctly, fMRI has the great promise to further improve our knowledge of the function of the human brain, by many applications in brain research. It is beginning to simplify patient investigation and monitoring, and in some centers is already replacing some of the more invasive investigation techniques.

Abbreviations

fMRI, functional magnetic resonance imaging; BOLD, blood oxygenation level dependent; EPI, echo planar imaging; SNR, signal-to-noise ratio; HDR, hemodynamic response function; ISI, inter-stimulus interval; TR, repetition time; FWHM, full width at half maximum; SPM, statistical parametric mapping; OLR, orthographic lexical retrieval; NVG, noun–verb generation.

Further Reading

1. Abbott D, Jackson GD. iBrain®—Software for analysis and visualisation of functional MR images. NeuroImage. 2001;13:59.
2. Abbott DF, Opdam HI, Briellmann RS, Jackson GD. Brief breath-holding may confound functional magnetic resonance imaging studies. Human Brain Mapping. 2005;24(4):284–90.
3. Aghakhani Y, Bagshaw AP, Benar CG, *et al.* fMRI activation during spike and wave discharges in idiopathic generalized epilepsy. Brain. 2004;127(Pt 5):1127–44.
4. Al-Asmi A, Benar CG, Gross DW, *et al.* fMRI activation in continuous and spike-triggered EEG–fMRI studies of epileptic spikes. Epilepsia. 2003;44(10):1328–39.
5. Baxendale S. The role of functional MRI in presurgical investigation of temporal lobe epilepsy patients: a clinical perspective and review. J. Clin. Exp. Neuropsychol. 2002; 24:664–76.
6. Binder JR, Swanson SJ, Hammeke TA, *et al.* Determination of language dominance using functional MRI: a comparison with the Wada test. Neurology. 1996;46:978–84.
7. Boxerman JL, Hamberg LM, Rosen BR, Weiskoff RM. MR contrast due to intravascular magnetic susceptibility perturbations. Magn. Reson. Med. 1995;34:555–66.
8. Constantine G, Shan K, Flamm SD, Sivananthan MU. Role of MRI in clinical cardiology. Lancet. 2004;363(9427): 2162–71.
9. Dale AM. Optimal experimental design for event-related fMRI. Hum. Brain Mapp. 1999;8:109–14.
10. Desmond JE, Glover GH. Estimating sample size in functional MRI (fMRI) neuroimaging studies: statistical power analyses. J. Neurosci. Methods. 2002;118:115–28.
11. Detre JA, Wang J. Technical aspects and utility of fMRI using BOLD and ASL. Clin. Neurophysiol. 2002;113(5): 621–34.
12. Frackowiak R, Friston K, Frith C, Dolan R, Price C, Ashburner J, Penny W, Zeki S (Ed). Human Brain Function, 2nd ed. Academic Press: London, 2004.
13. Friston KJ, Zarahn E, Josephys O, Henson RNA, Dale AM. Stochastic designs in event-related fMRI. Neuroimage. 1999;10:607–19.
14. Fujiwara N, Sakatani K, Katayama Y, Murata Y, Hoshino T, Fukaya C, Yamamoto T. Evoked-cerebral blood oxygenation changes in false-negative activations in BOLD contrast functional MRI of patients with brain tumors. Neuroimage. 2004;21(4):1464–71.
15. Gaillard WD, Hertz-Pannier L, Mott SH, Barnett AS, LeBihan D, Theodore WH. Functional anatomy of cognitive development: fMRI of verbal fluency in children and adults. Neurology. 2000;54:180–5.
16. Glover GH. Deconvolution of impulse response in event-related BOLD fMRI. Neuroimage. 1999;9(4):416–29.
17. Golay X, Hendrikse J, Lim TC. Perfusion imaging using arterial spin labeling. Top Magn. Reson. Imaging. 2004; 15(1):10–27.
18. Gusnard DA, Raichle ME. Searching for a baseline: functional imaging and the resting human brain. Nat. Rev. Neurosci. 2001;2:685–94.
19. Hampson M, Peterson BS, Skudlarski P, Gatenby JC, Gore JC. Detection of functional connectivity using temporal correlations in MR images. Hum. Brain Mapp. 2002;15: 247–62.
20. Himberg J, Hyvarinen A, Esposito F. Validating the independent components of neuroimaging time series via clustering and visualization. Neuroimage. 2004;22(3): 1214–22.
21. Jackson GD, Connelly A, Cross JH, Gordon I, Gadian DG. Functional magnetic resonance imaging of focal seizures. Neurology. 1994;44:850–856.
22. Lazeyras F, Zimine I, Blanke O, Perrig SH, Seeck M. Functional MRI with simultaneous EEG recording: feasibility and application to motor and visual activation. J. Magn. Reson. Imaging. 2001;13:943–8.
23. Lemieux L, Salek-Haddadi A, Josephs O, *et al.* Event-related fMRI with simultaneous and continuous EEG: description of the method and initial case report. Neuroimage. 2001;14(3):780–7.
24. Logothetis NK, Pauls J, Augath M, Trinath T, Oeltermann A. Neurophysiological investigation of the basis of the fMRI signal. Nature. 2001;412(6843):150–7.
25. Logothetis NK. The neural basis of the blood-oxygen-level-dependent functional magnetic resonance imaging signal. Philos. Trans. R. Soc. Lond. B. Biol. Sci. 2002;357 (1424): 1003–37.
26. McKeown MJ, Hansen LK, Sejnowsk TJ. Independent component analysis of functional MRI: what is signal and what is noise? Curr. Opin. Neurobiol. 2003;13(5):620–9.
27. Menon RS. Imaging function in the working brain with fMRI. Curr. Opin. Neurobiol. 2001;11:630–6.
28. Ogawa S, Lee TM, Kay AR, Tank DW. Brain magnetic resonance imaging with contrast dependent on blood oxygenation. Proc. Natl. Acad. Sci. U.S.A. 1990;87:9868–72.
29. Prichard JW, Rosen BR. Functional study of the brain by NMR. J. Cereb. Blood Flow Metab. 1994;14:365–72.

30. Rosen BR, Buckner RL, Dale AM. Event-related functional MRI: past, present, and future. Proc. Natl. Acad. Sci. 1998;95:773–80.
31. Schenck JF. MR safety at high magnetic fields. Magn. Reson. Imaging Clin. N. Am. 1998;6:715–30.
32. Schlosser R, Hutchinson M, Joseffer S. Functional magnetic resonance imaging of human brain activity in a verbal fluency task. J. Neurol. Neurosurg. Psychiatry 1998;62: 492–8.
33. Springer JA, Binder JR, Hammeke TA, *et al.* Language dominance in neurologically normal and epilepsy subjects: a functional MRI study. Brain. 1999;122:2033–45; Sunaert S, Yousry TA. Clinical applications of functional magnetic resonance imaging. Neuroimaging Clin. N. Am. 2001;11:221–36.
34. Wilke M, Holland SK, Myseros JS, Schmithorst VJ, Ball WS Jr. Functional magnetic resonance imaging in pediatrics. Neuropediatrics. 2003;34(5):225–33.

High Resolution Magic Angle Spinning (HRMAS) Proton MRS of Surgical Specimens

Leo L. Cheng[1,2], Melissa A. Burns[1] and Cynthia L. Lean[3,4]
[1]*Departments of Pathology;*
[2]*Departments of Radiology Massachusetts General Hospital, Harvard Medical School Boston, Massachusetts 02114, USA;*
[3]*Institute for Magnetic Resonance Research, NSW 1590, Australia; and*
[4]*Department of Magnetic Resonance in Medicine, University of Sydney, NSW 2006, Australia*

List of Abbreviations

BPH	Benign prostatic hyperplasis
DCIS	Ductal carcinoma *in situ*
FNAB	Fine needle aspiration biopsy
GBM	Glioblastoma multiforme
HCA	Hierarchical cluster analysis
^{1}H MRS	Proton magnetic resonance spectroscopy
HPLC	High pressure liquid chromatography
(HR)MAS	(High resolution) Magic angle spinning
NAA	N-acetyl aspartate
PA	Polyamines
PC	Phosphocholine
PCA	Principal component analysis
PTC	Phosphatidylcholine
SCS	Statistical classification strategy
SSB	Spinning side bands

Introduction

Over the last two decades, a large body of *ex vivo* work and some *in vivo* work has demonstrated the utility of proton magnetic resonance spectroscopy (^{1}H MRS) in detecting and monitoring cellular chemical alterations associated with the development and progression of human malignant diseases [1–3]. Reports of conventional *ex vivo* ^{1}H MRS studies of human tissues, i.e. analysis of intact tissues with liquid-state MRS probes, have demonstrated that diagnosis, and for some organs, prognosis of malignant disease using metabolite ratios measured from spectra may reach 95% for both sensitivity and specificity. The accuracy can further be improved using an objective pattern recognition technique, statistical classification strategy (SCS) [2]. While tissue conventional MRS is fast and has shown the ability to diagnose accurately various human malignancies, it is limited by poor spectral resolution caused by the magnetic susceptibility effects of heterogeneous structures of the sample, and therefore detailed identification of individual metabolites is difficult. Metabolite profiling involving the measurement and quantification of tissue metabolites [4] is of increasing interest in the era of genomics and proteomics due to the direct involvement of tissue metabolites in tumor development and progression. High-resolution magic angle spinning (HRMAS) was developed to improve the spectral resolution of MR spectra of intact tissue by the reduction of susceptibility induced broadening such that individual metabolites may be identified and correlated with disease states.

Characterization of Human Malignancies

A malignant lesion exhibits unregulated growth characteristics both at primary and secondary (metastatic) sites. Initial attempts to characterize and understand tumor development and progression hinged therefore on lesion morphology and, as such, on histopathology using the light microscope, which has been the medical diagnostic "gold standard" for much of the 20th century. With the discovery of DNA and the advent of molecular genetics it became evident that tumor morphology represented only a single variable in the characterization of malignancies and could not report on risk factors or on the biological potential of a lesion. Predicting tumor behavior is essential for determining optimal disease management protocols and it is proteomics and spectroscopic metabolite profiling that have the potential to provide the information necessary for this endeavor. Metabolite profiling data, thus far, has been generated predominantly by studies involving MR spectroscopy and mass-spectrometry.

Specific Limitations of Histopathology

The principal limitation of histopathology is the restricted range of morphological changes that tissues can express. Pathologists attempt to extract a sophisticated pattern specific for an individual disease process from a continuum of morphological changes. These patterns overlap,

Graham A. Webb (ed.), Modern Magnetic Resonance, 1037–1050.

are susceptible to subjective assessment, and may be altered by sampling error. Patient assessment involves histological grading and clinicopathological staging. Tumor grading attempts to establish the aggressiveness of the tumor based on the degree of differentiation, or anaplasia, of tumor cells. However, such a distinction is subjective and often inconclusive. Staging of cancer determines the extent and spread of the disease, but it is unable to distinguish between recent but aggressive primary tumors and older but more slowly growing ones since clinical presentation of patients does not usually reveal how long a neoplasm has been present. Adequate staging of disease is often problematic. For instance, histopathological assessment of lymph nodes is subject to observer and sampling errors due to the large volumes of tissue to be assessed and the time and resource constraints that prevent thorough and complete examination. A retrospective study revealed that: "Serial sectioning of lymph nodes judged to be disease-free after routine examination revealed micrometastases in an additional 83 (9%) of 921 breast cancer subjects" [5].

Current histopathology has shown both conceptual and procedural inadequacies in providing optimal medical diagnosis. The development of new modalities capable of improving the accuracy of disease diagnosis is needed.

Cancer Pathology Determined by Conventional ^{1}H MRS

Conventional MRS measures cellular chemicals that are mobile on the MR time scale and their variations with changes in physiological or pathological function [6]. Although restricted by achievable spectral resolutions due to susceptibility induced broadening of resonances, many promising results have been obtained from conventional proton MRS studies of human malignant tissue and fine needle aspiration biopsies. The potential for diagnoses of malignant diseases in many organs has been demonstrated and more importantly, metabolic changes have been reported that were not morphologically manifested [7,8]. A study of aspiration biopsies from primary breast tumors not only determined diagnosis, but also reported on nodal involvement and tumor vascularization [9]. The sensitivity of MRS was demonstrated in a study using a rat model for lymph node metastasis where malignant cells in lymph nodes were detected with a greater sensitivity than histology. Micrometastases were detected that were not apparent even when the entire node was serially sectioned for examination by histology. The MR diagnoses were confirmed to be correct by xenografting nodal tissue into nude mice [10].

Historically, the limitation in spectral resolution of conventional proton MRS of tissue samples was realized as an obstacle to the identification of detailed cellular metabolism. A number of procedures, including sample packing and chemical extraction of tissue were investigated as ways to improve spectral resolution [11]. While solutions of tissue extracts did produce high-resolution spectra, the process was destructive and was found to alter the spectroscopic results to an unknown degree, depending on the procedure used and the thoroughness of extraction. Therefore, neither broad-line tissue analyses nor high-resolution measurements of extracts have been able to successfully evaluate malignant disease or exemplify the advantages of proton MRS in metabolite identification.

Methodology

Limited spectral resolution is a common problem in solid-state MRS for which many techniques have been proposed and tested. Among these magic angle spinning (MAS), a line-narrowing technique, was invented after the realization of the angle-dependent characteristics of the so-called "solid effects" that broaden the resonance lines [12,13]. These "effects" classically include dipolar coupling and chemical shift anisotropy, which always exist in solids. It was later shown that MAS could also reduce resonance line-broadenings caused by bulk magnetic susceptibility [14]. Specifically, in solids, spectral broadening due to these effects follows $(3\cos^2\theta - 1)$, where θ is the angle between the static magnetic field and the internuclear vector. Therefore, if a sample is spun mechanically at the "magic angle ($54°44'$, i.e. $3\cos^2(54°44') - 1 = 0$)," and at a rate faster than the broadening originating from these effects, the contribution of these effects to the spectral broadening can be reduced [15].

The HRMAS Method

The term HRMAS is currently used to refer to proton MRS of non-solution samples, such as biological tissues and cells, obtained with the application of MAS. It is important to note, however, that there are conceptual differences between classical MAS and tissue HRMAS used with proton spectroscopy [16]. The targets of MAS in classical proton solid-state studies are chemical shift anisotropies and dipolar couplings that are in ordinary solids >50 kHz. The application of MAS in solid-state studies has involved the use of high spinning rates and strong radio-frequency pulses in an attempt to decouple these homonuclear interactions. This is not the case, however, for the MAS study of cellular metabolites in biological tissues. Here, water-soluble molecules reside in the cytoplasm wherein their motion is restricted by magnetic susceptibilities caused by various interfaces with other cell structures and by inherent viscosity. Although classical "solid-effects" exist, for instance within cell membranes, they would contribute only to an almost invisible spectral background that is too broad (~50 kHz) to be measurable in a typical HRMAS spectrum of metabolites (~5 kHz, or 10 ppm on a 500 MHz spectrometer). Overall, the magnitude of line-broadenings due to magnetic susceptibility is approximately 10^3 less than that caused by "solid-effects," and can be greatly reduced by MAS.

Experimental

Sample Preparation

Fresh and previously frozen samples may be measured directly with HRMAS MRS [17–21]. For the purpose of preserving the metabolite concentrations within tissues, samples should not be collected or stored in any liquid medium. In cases where the size of frozen tissue exceeds the sample size required for HRMAS MRS analysis, samples must be cut on a frozen surface to avoid multiple freezing and thawing. Such a surface can be made with a thin metal plate, covered with gauze and placed on top of dry ice.

Tissue samples may be washed briefly with D_2O prior to spectroscopy if they contain visible amounts of blood. However, exposure to D_2O should be brief to minimize the possible loss of cellular metabolites [22]. We have observed complete depletion of metabolites for human prostate samples (~10 mg blocks) submerged in ~2 ml D_2O for approximately 10 min.

As always, when working with human materials of potential biohazard, universal precautions need to be practiced at all times. In particular, tissue samples will undergo spinning and even at "slow spinning speeds" of less than 1 kHz, tissue fluid can leak from the rotor cap if the cap is not in tight-fit with the rotor (Figure 1). Thus the compatibility and seal between rotor and cap should be tested

Fig. 1. Photograph of the rotor and a typical biopsy tissue sample that will be placed in the rotor using the tweezers for HRMAS MRS analysis.

using a D_2O solution prior to tissue analysis. The test can be done by comparing the weights of the rotor with solution before and after HRMAS using the same experimental conditions (temperature and spinning speed) used for tissue analysis.

To maximize spectral resolution, tissue samples should be limited to the physical boundaries of the receiving coils; for example by using Kel-F inserts to create a spherical sample when measuring samples on a Bruker spectrometer (Bruker BioSpin Corp., Billerica, MA). The use of a spherical sample is recommended for it minimizes shimming efforts as well as reduces the effect of the magnetic field inhomogeneity on the broadening of spectral lines.

Spectrometer Settings

Before tissue measurements, the HRMAS probe should be adjusted for its magic angle with potassium bromine (KBr) following the manufacturer's protocol for solid-state MRS. Ideally, measurements should be made at a low temperature (e.g. 4 °C) to reduce tissue degradation during acquisition.

The optimal spinning rate should be decided after consideration of several factors, including tissue type, metabolites of interest, and the plan for the tissue after HRMAS MRS analysis. Generally, the higher the spinning rate, up to 10 KHz as reported in the literature, the better the spectral resolution [23]. However, since high spinning rates can function as a centrifuge that can potentially disrupt tissue structures, if subsequent histopathological evaluation of the tissue is critical to the study, spinning rates must be reduced to limit any structural damage of the tissue that may interfere with histopathology. It is important to note that different tissue types endure different levels of stress. For instance, for the same spinning rate, skin tissue may be perfectly preserved in structure, while brain tissue can be completely destroyed. On the other hand, if less mobile metabolites such as lipids are the focus of HRMAS MRS evaluation, faster spinning rates may be necessary. The preservation of tissue architectures during HRMAS MRS is often critical and, as such, a number of studies have explored HRMAS MRS tissue analysis using moderate to slow spinning conditions [24–27]. These studies aimed to suppress spinning side bands (SSB) that overlapped with metabolite spectral regions of interest, using spinning rates that were not fast enough to "push" the 1st SSB beyond these regions. Interested readers should test these reported techniques for applicability to their specific tissue systems.

Optimal probe shimming is another critical factor that directly affects achievable spectral resolution. We have found that shimming on the lock or tissue water signal was not as sensitive as shimming on the splitting of the lactate doublet at 1.33 ppm, which fortunately presents in most excised biological tissues. However, in order to shim interactively on the degree of splitting, it is necessary to work with the frequency domain. This should not present a challenge to most current spectrometers. Furthermore, it should be possible to establish autoshimming protocols based on this criterion.

It is now accepted that malignancy related cellular marker metabolites may not present simply as present or absent, but rather as continuous changes in intensity throughout disease development and progression. Quantification of these metabolites can be extremely important if they are to accurately diagnose and characterize stages of disease. Metabolite concentrations may be estimated from HRMAS MR spectra by using either the intensity of tissue water signals or an external standard (e.g. a small piece of silicone rubber) permanently attached to the inside of the rotor or attached to the rotor inserts. Such an external standard can be calibrated with known compounds of known concentrations. Interested readers can make such compounds by dissolving known amounts of relevant metabolites in agarose gel.

HRMAS MR spectra may be acquired with or without water suppression, depending on whether water intensities are required for the estimation of metabolite concentrations. A rotor-synchronized CPMG sequence may be applied to achieve a flat spectral baseline if there are undesired broad resonances from the probe background.

Histopathology

The clinical utility of high-resolution tissue metabolite profiles obtained with HRMAS MRS needs to be investigated and validated by means of accurate and detailed correlation with serial-section tissue pathologies. Such correlations are particularly important for studies of human malignancies due to the heterogeneity that may be inherent in the disease. Tissue pathology can vary greatly from region to region within a single tumor, and intertumor differences may be even more pronounced. An obvious advantage of HRMAS MRS is its preservation of tissue for subsequent histopathological analyses, allowing the establishment of correlations between metabolite profiles and quantitative pathologies [28–30].

Routine clinical histopathology data is most often inadequate. Routine histopathology, in particular 5 micron slices of the tissue biopsy, provides information about the presence or absence of certain features. These features may vary from slice to slice and therefore information from a small percentage of the tissue may not correlate with the spectral profile, which consists of the weighted sum of the profiles from all tissue slices or indeed the entire piece of tissue. Hence, for certain types of tissues, particularly neoplasm samples with known heterogeneities, histopathology needs to be evaluated quantitatively. This can be achieved laboriously by the histopathologist's examining serially sectioned tissue slices or possibly with the assistance of computer image analysis of these same sections. The sectioning frequency differs with tissue type and should be determined in consultation with the pathologist. For instance, the optimal sectioning frequency for human prostate was found to be between $200 \sim 400$ μm.

Data Analysis

Histopathological analysis in cancer diagnosis relies on the observation of variations in colors and shapes using a light microscope. Importantly, to suspect disease the pathologist must observe widespread rather than isolated changes in color and/or shape, a process requiring keen pattern recognition. Similarly, pattern recognition methods are most often required to allow the diagnosis of malignancy from MRS tissue metabolite profiles. Development and progression of malignant disease involves the simultaneous evolution of many metabolic processes. Therefore, although some individual metabolites have been reported to correlate with disease types and stages [1,2] it is more likely that the overall metabolite profile rather than changes in single metabolites will be sensitive and specific for disease diagnosis [31,32]. Sophisticated statistical classification strategies (SCS) have been developed and applied to the analysis of conventional MRS of malignant tissues [33]. To date, however, principal component analysis (PCA) using readily available statistical programs has been sufficient for revealing accurate diagnostic information from HRMAS MRS data [34,35]. However, the application of SCS may further improve these accuracies.

HRMAS MRS of Human Surgical Specimens

Over recent years, HRMAS MRS has been applied biomedically to the analysis of human surgical samples, research animal tissues, and cultured cells. The scope of the methodology presented in this section will be limited to HRMAS MRS studies of human tissues and will be presented where possible in context with preceding studies of the same tissues using conventional MRS. Although presented in this context, comparison of the sensitivities and specificities obtained using conventional and HRMAS MRS are not at this stage warranted as unlike the mature discipline of conventional MRS, HRMAS MRS is still in its infancy and has reported only studies with restricted patient numbers. Although the capability of HRMAS has been clearly demonstrated in generating high-resolution spectra from which individual metabolites can be measured and such measurements were impossible with conventional methods, studies of large patient populations, as with conventional MRS studies, that allow evaluation of the sensitivity and specificity of the method have not yet appeared in literature. Similarly, it is not as yet possible to provide the reader with one concise and optimal method for undertaking HRMAS MRS. Detailed methods are provided for each of the studies presented and these methods discussed in terms of the study aims. In an emerging discipline such as HRMAS MRS care must be taken to rigorously address methodological variables with respect to the type of tissue being analyzed and the specific information required. Increased spectral quality will most often be obtained with higher spinning speeds but this will be at the expense of tissue preservation allowing subsequent histopathological assessment of the tissue. Some tissues are less easily destroyed than others by high spinning speeds and thus preliminary experiments need to be undertaken to determine optimal parameters.

MRS analyses on tissue extracts have been studied for many years on many diseases, and have formed a large body of literature. Results from these studies are worthy of close examination and review but will not, however, be included in this review due to the following: the pathology of the tissue samples most often remains incomplete and/or the degree of extraction cannot be certain.

Brain

Human brain was the first study reported using proton HRMAS MRS to determine tissue pathology. Due to the relative motion stability and homogeneity compared to other organs, brain has dominated the development of *in vivo* MRS, and accordingly has inspired many *ex vivo* studies, primarily including neurodegenerative diseases and tumors, aimed at understanding metabolism and defining brain tissue chemistry.

The first HRMAS MRS human studies on brain tissues from autopsies described semi-quantitative evaluation of the pathology of a neurodegenerative disease, specifically Pick disease [36]. Through MRS measurements and traditional neurohistopathology, direct and semi-quantitative correlations were found to exist between the levels of N-acetyl aspartate (NAA) and the amount of surviving neurons in varying regions of examined brain as seen in Figure 2. The study also, for the first time, demonstrated that the spectral resolution of HRMAS proton MRS was comparable with that measurable with conventional proton MRS of tissue

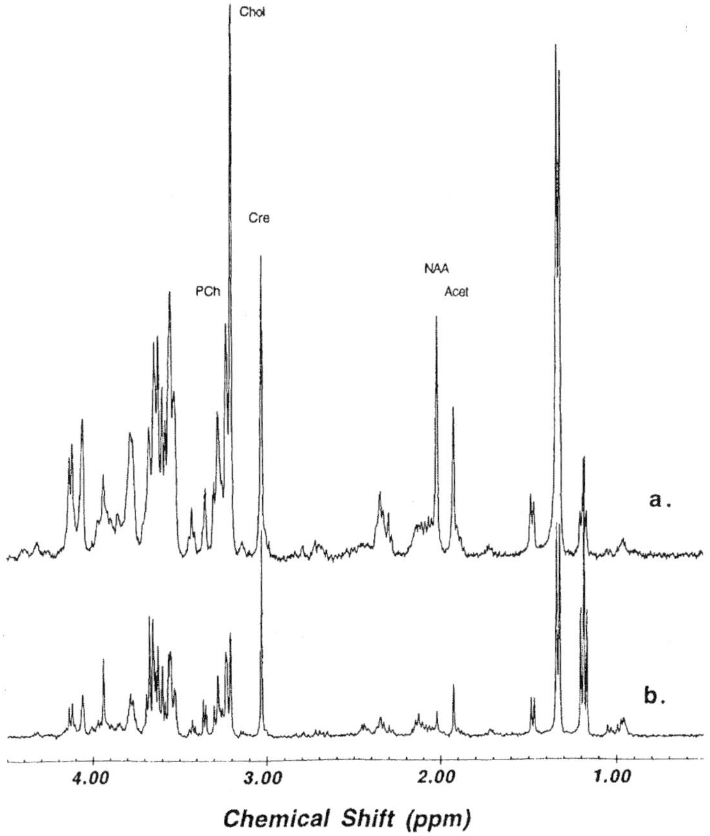

Fig. 2. Comparison of HRMAS MR spectra of brain tissue from the relatively unaffected primary visual cortex region (a) and the severely Pick disease affected rostral inferior temporal gyrus region (b), showing a marked decrease in NAA concentration, 8.48 μmol/g for spectrum a and 4.96 μmol/g for spectrum b. This decrease in NAA was found to correlate with an average neuronal count decrease of 33 neurons per 0.454 mm^2. The spectra were acquired at 2 °C, and were scaled according to concentration of creatine at 3.03 ppm for enhanced visualization. *Figure 4 from ref. [36]*

Fig. 3. Comparison of human brain proton MRS acquired with (a) HRMAS on intact tissue and (b) conventional method on extract solution. The spectral resolution was comparable while relative intensities for certain metabolites varied.

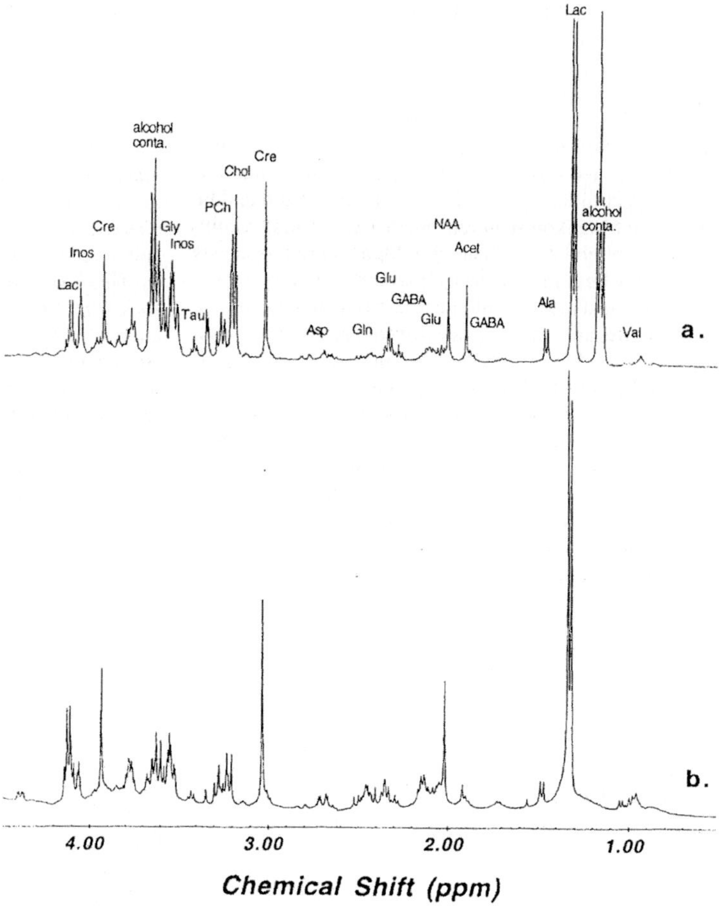

extracts (Figure 3), and superior to that obtained using conventional MRS of intact tissue (Figure 4). Another early study, also on autopsy brain tissues, correlated MRS data and stereological pathology for Alzheimer's disease and confirmed NAA concentration to be proportional to neuronal density (Figure 5). In this study, 7 human brains were examined, 3 of which were Alzheimer diseased and 4 which presented as normal human control brains. Figure 5 demonstrates the quantitative nature of the NAA and neuronal count relationship, as the correlation intercepted at zero (–0.29 ± 1.15 μmol/g) [37].

The MRS study of brain tumor intact tissues using conventional MRS suggested the diagnostic importance of lipid and lipid metabolites. Kuesel and colleagues reported correlations between MR lipid signal intensities and the amount of necrosis in astrocytomas [38]. With 42 cases, they showed that the intensity of the mobile fatty acyl –CH=CH- resonance at 5.3 ppm, differentiated 0, 1–5 and 10–40% of necrosis with statistical significance [39]. These results represented a great potential use for *ex vivo* MRS for astrocytoma diagnosis, in particular for differentiating Grade III and Grade IV tumors, which have radically different prognoses. The technique also ensured that necrotic foci often missed by clinical pathology were identified. Working with conventional MR spectra of brain tumor tissues, Rutter *et al.* attempted to categorize tumors according to 1D peak ratios (3.1–3.4 vs. 1.1–1.5 ppm), T_2 values of peaks at 1.3 ppm and cross-peaks on 2D COSY spectra (0.9 ppm with 1.35 ppm, representing methyl-methylene couplings; and 1.3 ppm with 2.05 ppm for couplings between methylene groups in fatty acids) [40]. With 38 samples studied from 33 subjects (including normal tissue, astrocytoma, GBM, meningiomas, and metastases), they were able to use peak ratios to

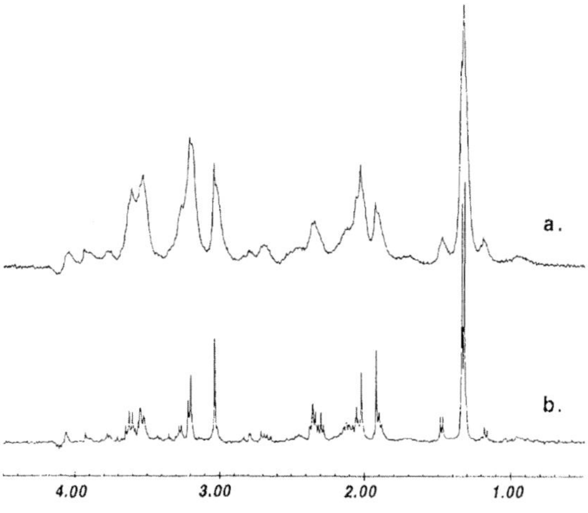

Fig. 4. Comparison of human brain proton MRS acquired with (a) conventional MRS and (b) HRMAS MRS. *Figure 2 from ref. [36].*

differentiate GBM from astrocytomas and normal tissues. They also found that the T_2 of the 1.3 ppm peak could be fitted by a double exponential function and that the long fraction of T_2 values could be used to group both GBM and metastasis from normal tissue. However, tissue spectra measured with conventional method of low resolution

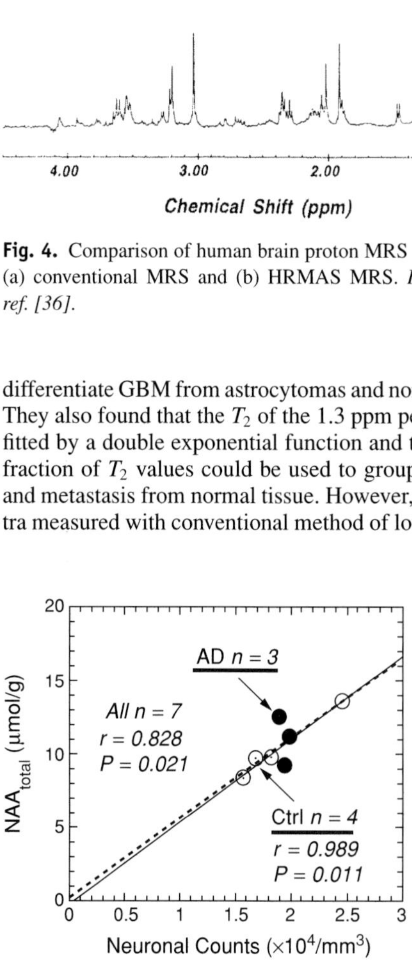

Fig. 5. A statistically significant linear correlation was found to exist between the number of neurons and the concentration of NAA_{total}, measured from 3 Alzheimer diseased and 4 normal control brains (the dotted line, $r = 0.828$, $P = 0.021$). The solid line represents the linear correlation obtained when only normal control brains were included ($r = 0.989$, $P = 0.011$). *Figure 3 from ref. [37].*

prevented these studies from observation of individual brain metabolites other than identification of broad lipid peaks.

HRMAS MRS of brain tumors allowed for the first-time detailed profiles of water-soluble metabolites and lipids to be obtained from the same spectra [31]. In a study of 19 brain tumors, including astrocytomas, GBM, meningiomas, Schwannomas, and normal brain, metabolite concentrations both in absolute units and relative ratios normalized to the creatine resonance at 3.03 ppm were reported and T_2 values for these metabolites were measured *ex vivo* for the first time (Table 1). Metabolite concentrations from intact tissues and tissue extracts from the same tumors were compared to each other and to the literature values. While the concentrations for some metabolites, measured by HRMAS MRS, were similar to those in extracts, others showed much higher concentrations in tissue than in extracts. Metabolite concentrations and T_2 values accurately differentiated tumor types based on clinical data, but detailed histopathology of the tissue samples was not performed due to the false belief at the time that HRMAS MRS damaged tissue such that subsequent histopathological analysis was not possible.

The important fact that accurate histopathological evaluation of tissue samples after HRMAS MRS is possible and the necessity of performing quantitative pathology on the same tissues after MRS was reported by Dr. Anthony and colleagues [28]. Tissue pathologies from both HRMAS MRS analyzed tissue and adjacent tissue that had not undergone HRMAS MRS before histopathological evaluation were analyzed semi-quantitatively for each region of the brain tumor. The quantitative histopathological data obtained from the study showed that adjacent specimens from the same tumor region shared similar histopathological features. Although quantitative differences were noted, these differences were most likely due to extensive tumor microheterogeity and not the result of the HRMAS MRS procedure. Futhermore, correlations between the amount of tumor necrosis and the concentrations of mobile lipids ($R^2 = 0.961$, $p < 0.020$) and lactate ($R^2 = 0.939$, $p < 0.032$), as well as between the numbers of glioma cells and the ratio of phosphocholine (PC) to choline resonances ($R^2 = 0.936$, $p < 0.033$) were observed. The strong linear correlation between tissue necrosis (%area) and lipids (mM) indicated that the amount of tissue necrosis can be estimated using the measured concentration of lipids from HRMAS MRS, and that according to the long T_2s these lipids are relatively mobile consistent with them being products of cell membrane degradation. Additionally, the results of the correlation obtained between the number of glioma cells and the phospocholine to choline resonances suggested the importance of measuring and quantifying these two resonances separately, which is difficult with both *in vivo* and *ex vivo*

Table 1: Matrix of selected brain metabolite concentrations measured with HRMAS MRS for differentiation between different pathological specimens NAA, in the table, includes both measured resonances of NAA at 2.01ppm and acetate at 1.92 ppm (see text for details); Numbers in parentheses represent resonance chemical shift in ppm. The resonance at 3.93 is tentatively assigned to the Cr metabolite. As an example of the use of this matrix, the Chol resonance can be used to differentiate low-grade/anaplastic astrocytomas from GBMs with a significance of $p < 0.05$. Similarly, the glycine resonance (Gly) can be used to distinguish GBMs from Schwannomas with a $p < 0.005$

	Normal	LG and AA[a]	GBMs	Schwannomas	Meningiomas
Normal		NAA[b] Lac (1.33)[c]	Cr (?, 3.93)[c] Gly (3.55)[d] Chol (3.20)[b] Cr (3.03)[b] NAAf Lac (1.33)[b]	Chol (3.20)[b] NAA[b] Lac (1.33)[d]	Cr (?, 3.93)[b] Chol (3.20)[e] Glu (2.35)[d] Ala (1.48)[b]
LG&AA			Chol (3.20)[e]		Glu (2.35)[b]
GBMs				Gly (3.55)[d]	Gly (3.55)[b]
Schwannomas					
Meningiomas					

[a] LG and AA, low-grade/anaplastic astrocytomas
[b] $p < 0.05$, two-tailed student's t test.
[c] $p < 0.0005$, two-tailed student's t test.
[d] $p < 0.005$, two-tailed student's t test.
[e] $p < 0.05$, calculated according to one-tailed student's t test, based on the hypothesis that chol increases in tumors.
[f] $p < 0.00005$, two-tailed student's t test.

conventional MRS methods, but readily achievable with MAS.

It is expected that HRMAS MR spectra of tissue rather than MRS of tissue extracts will provide metabolite information more closely related to that in the *in vivo* brain. However, cautions should be exercised when comparing *in vivo* MRS and HRMAS MRS data, as *in vivo* MRS will always be broad-line in nature. Nevertheless, a number of studies, both in adult [41] and pediatric [42] brain tumors, have concluded that there is good agreement between *in vivo* and *ex vivo* tissue MR spectra with high-resolution *ex vivo* results providing both insight into which metabolites reside within the broad resonances observed *in vivo* and a link between *in vivo* MRS evaluations and neuropathologies.

Prostate

The search for marker metabolites of prostate cancer has been inspired by the current state of prostate cancer pathology, wherein more than 70% of newly diagnosed cases are categorized with similar Gleason scores (6 or 7), but for which individual patient outcomes within these tumors are drastically different and unpredictable.

Hahn and colleagues reported the first intact prostate tissue conventional MRS study of 66 benign prostatic hyperplasia (BPH) and 21 prostate cancer samples from 50 patients [43]. They divided the proton spectral region between 0.5 and 3.55 ppm into 50 equal subregions, and applied multivariate linear-discriminant analysis to the point-reduced spectra. The study found six spectral regions including those containing citrate, glutamate, and taurine to be sensitive in differentiating BPH from cancer with an overall accuracy of 96.6%. This algorithm was tested by the same research group on another group of 140 samples from 35 patients after radiotherapy to test for the sensitivity of spectroscopy analysis in differentiating cancer positive vs. cancer negative samples [44]. After eliminating 24 samples that did not have sufficient signal-to-noise ratios, they reported, with the remaining 116 spectra, the sensitivity and specificity of tissue spectra in identifying cancer samples to be 88.9 and 92%, respectively.

Van der Graaf and colleagues presented another interesting study of intact prostate tissue combining conventional MRS and high-pressure liquid chromatography (HPLC) analysis to measure the relationship between polyamines (PA) and prostate cancer [45]. Although they observed PA in the proton spectra and measured statistically significant drops of PA in cancer samples with HPLC, no correlation between PA levels measured by MRS and those determined by HPLC were presented for the same cases due to very limited number of samples analyzed. Nevertheless, the study suggested the existence

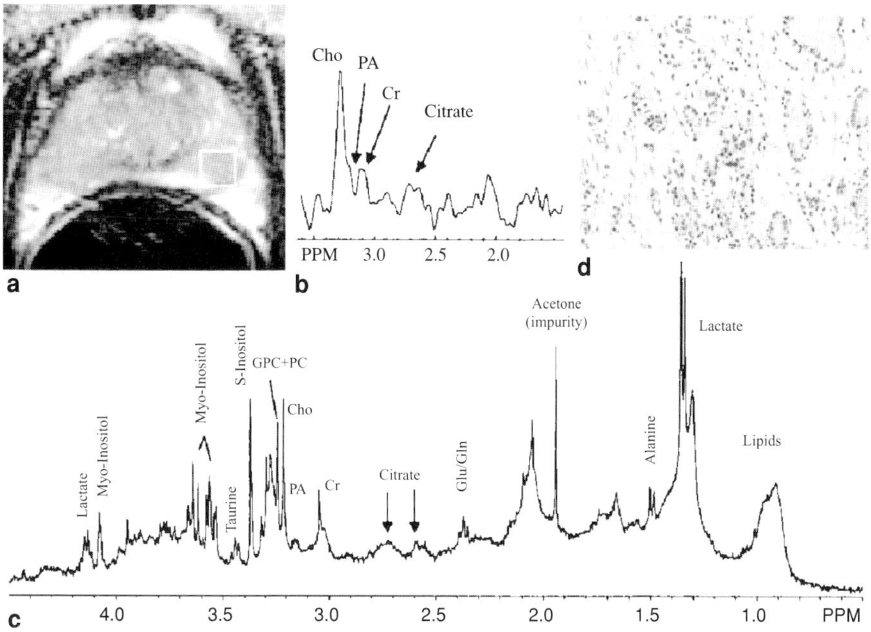

Fig. 6. Results from a presurgical 3D-MRSI of a 56-year-old prostate cancer patient were concordant with histopathologically classified malignancy of H and E stained tissue samples of excised tissue with Gleason 3 + 4 prostate cancer, **d**. Spectrum **b**, shows elevated levels of choline and low levels of citrate and polyamines relative to creatine in the 0.24 cm3 voxel of **a**. The postsurgical HRMAS MR spectrum (c) confirms the 3D-MRSI results with enhanced resolution that also identifies elevated levels of GPC + PC relative to choline. *Figure 2 from ref. [30].* (See also Plate 88 on page LXII in the Color Plate Section.)

of a potential prostate biomarker that MRS may be able to quantify, as well as a direction for future studies. More recently, Mountford *et al.* reported a conventional MRS study of 71 prostate samples from 41 patients who underwent both cancer and non-cancer prostate surgeries [46]. By using peak ratios 3.2/3.0 ppm (choline-to-creatine) and 1.3/1.7 ppm (lipid-to-lysine), they were able to differentiate malignant from benign tissues with 97% sensitivity and 88% specificity.

Tomlins and colleagues experimented with proton HRMAS MRS analysis on human prostate tissues with both 1D and 2D MR spectroscopy. In their qualitative report, they confirmed the usefulness of HRMAS MRS in producing high-resolution spectra from intact human prostate tissues [47]. A second HRMAS MRS study of human prostate tissues from 16 patients was the first to include quantitative pathology on the MRS specimens and to include multiple subjects [29]. The results of the study proved the validity of HRMAS MRS for the accurate determination of tissue histopathology. Both citrate and spermine were quantified from the tissue HRMAS MR spectra and shown to be linearly correlated with the amounts of prostate normal epithelium. In 2003, Swanson and colleagues reported an interesting study of HRMAS MRS of tissues from 26 patients harvested post surgically under the guidance of 3D-MRSI from lesions that had been analyzed using *in vivo* MRS prior to prostatectomy. Figure 6 shows the resulting spectra from a 56-year-old prostate cancer patient [30]. By combining the MRS results with quantitative pathology of the same tissue after spectroscopy, metabolite discriminators (i.e. ratios of citrate, polyamine, and choline compounds to creatine) were found to differentiate normal prostate epithelial tissue from cancer and stromal tissue. Furthermore, a correlation between the intensity of MIB-1 immunohistochemical staining and the ratio of choline to creatine resonances was reported, supporting their findings *in vivo*. These results were dependent on the assumption that the creatine concentration did not alter during the disease process, which awaits verification.

Breast

Human breast tissue was difficult to analyze using conventional proton MRS due to high lipid content. As pioneers of this work, Mountford and colleagues noted that to acquire a spectrum that was diagnostically meaningful,

31. Cheng LL, Chang IW, Louis DN, Gonzalez RG. Cancer Res. 1998;58:1825.
32. Cheng LL, Chang IW, Smith BL, Gonzalez RG. J. Magn. Reson. 1998;135:194.
33. Somorjai R, Dolenko B, Nikulin A, Pizzi N, Scarth G, Zhilkin P, Halliday W, Fewer D, Hill N, Ross I, West M, Smith I, Donnelly S, Kuesel A, Briere K. J. Magn. Reson. Imaging. 1996;6:437.
34. Tate AR, Foxall PJ, Holmes E, Moka D, Spraul M, Nicholson JK, Lindon JC. NMR Biomed. 2000;13:64.
35. Sitter B, Bathen T, Hagen B, Arentz C, Skjeldestad FE, Gribbestad IS. MAGMA. 2004;16:174.
36. Cheng LL, Ma MJ, Becerra L, Ptak T, Tracey I, Lackner A, Gonzalez RG. Proc. Natl. Acad. Sci. U.S.A. 1997;94:6408.
37. Cheng LL, Newell K, Mallory AE, Hyman BT, Gonzalez RG. Magn. Reson. Imag. 2002;20:527.
38. Kuesel A, Donnelly S, Halliday W, Sutherland G, Smith I. NMR Biomed. 1994;7:172.
39. Kuesel AC, Briere KM, Halliday WC, Sutherland GR, Donnelly SM, Smith ICP. Anticancer Res. 1996;16:1485.
40. Rutter A, Hugenholtz H, Saunders J, Smith I. J. Neurochem. 1995;64:1655.
41. Barton SJ, Howe FA, Tomlins AM, Cudlip SA, Nicholson JK, Bell BA, Griffiths JR. Magma 1999;8:121.
42. Tzika AA, Cheng LL, Goumnerova L, Madsen JR, Zurakowski D, Astrakas LG, Zarifi MK, Scott RM, Anthony DC, Gonzalez RG, Black PM. J. Neurosurg. 2002;96:1023.
43. Hahn P, Smith I, Leboldus L, Littman C, Somorjai R, Bezabeh T. Cancer Res. 1997;57:3398.
44. Menard C, Smith IC, Somorjai RL, Leboldus L, Patel R, Littman C, Robertson SJ, Bezabeh T. Int. J. Radia. Oncol. Biol. Phys. 2001;50:317.
45. van der Graaf M, Schipper RG, Oosterhof GO, Schalken JA, Verhofstad AA, Heerschap A. Magma 2000;10:153.
46. Swindle P, McCredie S, Russell P, Himmelreich U, Khadra M, Lean C, Mountford C. Radiology 2003;228:144.
47. Tomlins A, Foxall P, Lindon J, Lynch M, Spraul M, Everett J, Nicholson J. Analy. Common. 1998;35:113.
48. Mackinnon WB, Barry PA, Malycha PL, Gillett DJ, Russell P, Lean CL, Doran ST, Barraclough BH, Bilous M, Mountford CE. Radiology, 1997;204:661.
49. Mountford CE, Delikatny EJ, Dyne M, Holmes KT, Mackinnon WB, Ford R, Hunter JC, Truskett ID, Russell P. Magn. Reson. Med. 1990;13:324.
50. Delikatny E, Russell P, Hunter J, Hancock R, Atkinson K, van Haaften-Day C, Mountford C. Radiology, 1993;188:791.
51. Moka D, Vorreuther R, Schicha H, Spraul M, Humpfer E, Lipinski M, Foxall P, Nicholson J, Lindon J. Analy. Common. 1997;34:107.
52. Moka D, Vorreuther R, Schicha H, Spraul M, Humpfer E, Lipinski M, Foxall PJ, Nicholson JK, Lindon JC. J. Pharm. Biomed. Anal. 1998;17:125.
53. Millis KK, Maas WE, Cory DG, Singer S. Magn. Reson. Med. 1997;38:399.
54. Millis K, Weybright P, Campbell N, Fletcher JA, Fletcher CD, Cory DG, Singer S. Magn. Reson. Med. 1999;41:257.
55. Lean CL, Bourne R, Thompson JF, Scolyer RA, Stretch J, Li LX, Russell P, Mountford C. Melanoma Res. 2003;13:259.

Intraoperative MRI

Isabelle Latour and Garnette R. Sutherland
Division of Neurosurgery, Seaman Family MR Research Centre, Foothills Hospital, Calgary, AB, Canada T2N 1N4

Nothing is certain and everything changes
——Leonardo da Vinci

Contemporary neurosurgeons depend on precise lesion localization to maximize surgery and clinical outcome. Advances in neurosurgical technique and instrumentation evolved as surgical corridors became more precise. Magnetic resonance imaging (MRI) has considerably improved the diagnosis, treatment, and follow-up of patients with neurological disease. With the widespread use of MRI, it is difficult to conceive that, before the introduction of imaging technology, neurosurgery relied on precarious tools.

Historical Milestones in Neurology

Cranial trephination was used thousands of years ago by Neolithic man to liberate individuals thought to be possessed by evil spirits [1,2]. It resurfaced as a treatment for infantile convulsions and was still performed as late as the mid-19th century for both trauma and epilepsy [3]. Franz Joseph Gall proposed that intellectual faculties are highly compartmentalized in the brain and that examination of the surface of the skull reflects the extent to which an aptitude is developed [4]. The use of scientific evidence to determine cortical compartmentalization of function was first initiated by Paul Broca [5]. The development of imaging techniques began with the discovery of X-rays by Wilhelm Röntgen in 1895. In 1918, Walter Dandy made the serendipitous discovery of intraventricular air in a patient who had suffered cranial trauma [6]. This observation led to the development of pneumoencephalography (Figure 1) as a method for localizing intracranial pathology, based on ventricular displacement by cerebral mass lesions. A decade later, Moniz introduced cerebral angiography, a method still used to visualize vascular abnormalities [7]. Ultrasound was subsequently developed as a non-invasive method of localizing central nervous system (CNS) pathology [8]. Originating from ideas of Cormack [9] and Hounsfield [10], brain imaging was revolutionized in the mid-1970s by the invention of computerized tomography (CT). This discovery significantly enhanced diagnostic accuracy and improved surgical planning and outcome. The subsequent development of MRI by Lauterbur [11] and Mansfield [12] provided enhanced tissue contrast and remains the best imaging modality for diagnosis and determining the effect of treatment on both the lesion and the brain.

Principles of Intraoperative Imaging

Intraoperative imaging allows acquisition of near real-time images in the operating room (OR). As surgery often disrupts the brain environment, acquiring instant images at any time during and after completion of the surgery represents a considerable advantage. CT imaging was introduced into the OR in 1982 [13]. The first report of intraoperative MRI (iMRI), by Black *et al.* was in 1997 [14]. The use of intraoperative CT declined following the introduction of iMRI, in part due to lower tissue contrast and image artifact due to surgery and instrumentation. MR technology confers non-invasive, radiation-free, multiplanar, real-time, high-resolution images of the nervous system. MRI systems that include magnets of ≥ 1.5 T and high performance gradients provide a unique method for the evaluation of diffusion and perfusion tensor imaging, as well as a way to assess brain function. Diffusion images measure the mobility of water molecules in tissue. As pathological states can disrupt barriers that normally restrict water motion, diffusion imaging reveals information about tissue integrity [15]. Perfusion images represent the rate at which blood is delivered to the tissue, and can therefore be used to identify areas displaying an increase or decrease in blood flow [16]. Functional MRI allows in a non-invasive way the precise identification of brain areas undergoing a sharp increase in oxygen supply induced by a certain cognitive process or task [17]. Essential components of an iMRI system are the magnet, gradients, radio frequency (RF) coil designs, the operating table, and the RF shielding. Real-time, three-dimensional (3D) navigation software allows the neurosurgeon to localize surgical instruments and intracranial targets in 3D space. This improves localization of intracranial pathology from which more minimalist approaches may be directed.

Hardware and Configuration

Three different types of magnets can be used for MRI: permanent, resistive, and superconducting. iMRI has utilized

Graham A. Webb (ed.), Modern Magnetic Resonance, 1051–1062.

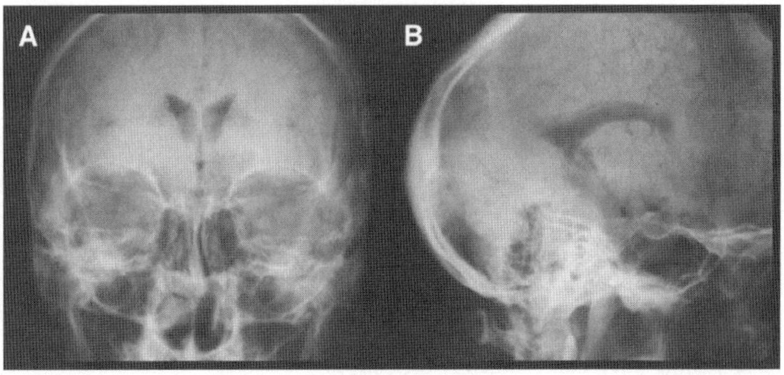

Fig. 1. Coronal (A) and lateral (B) pneumoencephalography. The lateral ventricles are visible and show no evidence of displacement.

magnets of 0.12–1.5 T, with gradient strength ranging from 12 to 33 mT/m (Table 1). Gradient strength and performance is important for rapid imaging sequence, in particular echo-planar imaging (EPI), diffusion, perfusion, and fMRI. For iMRI, inductively coupled [18] or volume coils [19] can be used. The full spectrum of imaging sequences including T_1, T_2, MR angiography, diffusion, perfusion, and fluid attenuated inversion recovery (FLAIR) have been applied to the evaluation of the utility of iMRI as an adjunct to surgery. The first iMRI system, manufactured by General Electric, consisted of a 0.5 T vertical, biplanar double doughnut magnet configuration [14]. A significant advantage of this setup is that real-time images can be acquired frequently, without interrupting surgery nor having to move the patient or the magnet. Although this initial system represented a major advance in intraoperative imaging technology, it has limitations. The system requires positioning the patient and surgeon between two magnetic poles. This results in a constrained working environment, restricted patient positioning, and the requirement for MR-compatible instrumentation. Furthermore, the magnetic field strength and gradient performance are below that needed for contemporary EPI. To partly overcome these issues, 0.2 and 0.3 T, horizontal, biplanar imaging systems were produced by Siemens and Hitachi, respectively [20,21]. For these, the surgical area is located outside of the magnetic field and the patient is moved into the magnet for image acquisition. Physically separating surgery and imaging resolved the problem of operating within a magnetic field. However, this requires interruption of surgery and moving the anesthetized, monitored patient from the surgical site to the magnet. In need of better image resolution and to maintain a patient focused environment, a movable, ceiling-mounted wide bore 1.5 T magnet was developed and manufactured by IMRIS [22–24]. With this system, conventional neurosurgical procedures are maintained and the magnet is moved to the patient for imaging (Figure 2). The higher magnetic field and improved gradient performance provides high-resolution and rapid imaging including echo planar. The system includes a versatile MR-compatible OR table accommodating various patient positioning including prone, supine, and lateral. Philips and Siemens have recently developed and installed 1.5 T iMRI systems based on stationary magnets. In the Siemens configuration, surgery takes place in the fringe field and, for imaging, the OR

Table 1: iMRI systems with main advantage and disadvantage

Manufacturer	Magnet strength	Gradient performance	Main advantage	Main disadvantage
Odin	0.12	25 mT/m	Portable	Image resolution
Siemens	0.2	20 mT/m	Unrestricted instrumentation	Image resolution
Hitachi	0.3	21 mT/m	Unrestricted instrumentation	Need to move patient for imaging
General Electric	0.5	12 mT/m	Real-time imaging	Requires MR-compatible tools
Philips	1.5	33 mT/m	Image resolution	Need to move patient for imaging
Siemens	1.5	30 mT/m	Image resolution	Need to move patient for imaging
IMRIS	1.5	20 mT/m	Image resolution	Interruption of surgery for imaging

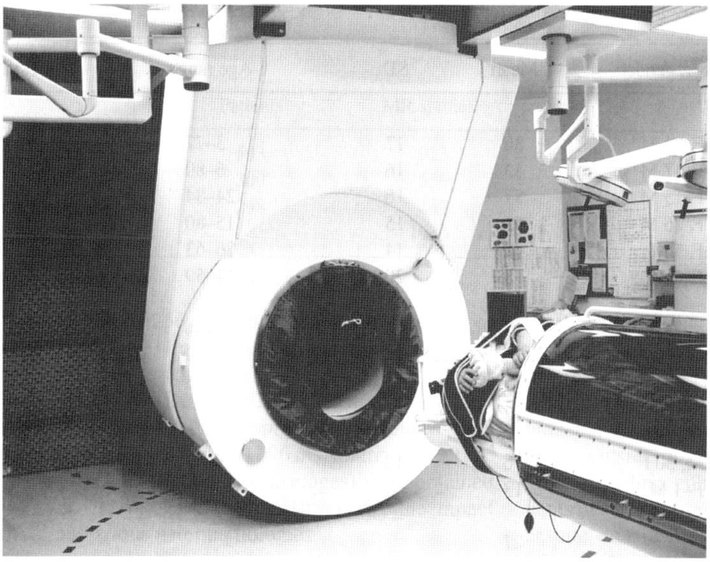

Fig. 2. Ceiling-mounted 1.5 T magnet being moved into the surgical field, together with the OR table and local RF shielding.

table is rotated to the magnet [25]. Philips designed a 1.5 T interventional/neurosurgery suite where the patient is moved by a floating tabletop from the operative table to the imaging platform for image acquisition [26]. Using the Philips system, surgeons have performed stereotaxy by reaching into the magnet, using iMRI for accurate instrument placement. To decrease the impact of iMRI in the OR environment, a small highly portable 0.12 T iMRI system, positioned under the OR table, was developed by Odin [27]. While this represented a unique innovation, imaging quality, and patient positioning are compromised by the low magnetic field and close opposition of the two magnetic poles.

Clinical Applications of iMRI

The development of iMRI has been driven by neurosurgical desire to have intraoperative imaging of CNS pathology following the induction of anesthesia and patient positioning, at various stages of the surgical dissection and as a measure of quality assurance prior to reversal of anesthesia. Despite these advantages, instrument size, weight, magnetic field, RF interference, and cost continue to pose significant challenges for widespread acceptance. Also, there has been no randomized controlled trial demonstrating superiority of iMRI-assisted neurosurgery over standard neurosurgical procedures. This in part reflects traditional education bias, the difficulty of recruiting patients for trials, crossover, and the relatively recent introduction of iMRI technology. Nevertheless, case studies and existing databases of patients who underwent iMRI procedures represent important sources of information to assess potential efficacy. Non-randomized retrospective studies matching patients based on prognostic factors are considered valuable. Examples where the use of iMRI led to modifications of the surgical procedure are also accepted as supportive evidence of utility. To fully evaluate the benefit of iMRI, randomized clinical trials, with blinded outcome assessment and long-term follow-up, are needed. At the University of Calgary, iMRI has been used to study surgery in 530 patients, 61 of which were pediatric (4 months to 18 years), with various CNS pathologies (Table 2). For each condition, surgical planning iMRI was performed after induction of anesthesia and patient positioning; interdissection iMRI was performed at various stages of the resection and quality assurance iMRI was acquired after wound closure, but before emergence from anesthesia. On average, an imaging session required 30 min. This included re-registration of surgical navigation. A total of 1930 MR studies were performed (Table 3). In 225 patients, trajectories were established by coupling surgical planning images with frameless stereotaxy (BrainLAB, Heimstetten, Germany). Time required for imaging depended on the specific MRI sequences used. Scout images (field of view 30 cm; $T_R = 35$ ms; $T_E = 15$ ms; matrix size 128×256; slice thickness 10 mm; one average) were obtained in 15 s. T_1-weighted images were obtained with spin-echo sequences (field of view 26 cm; $T_R = 400$ ms; $T_E = 12$ ms;

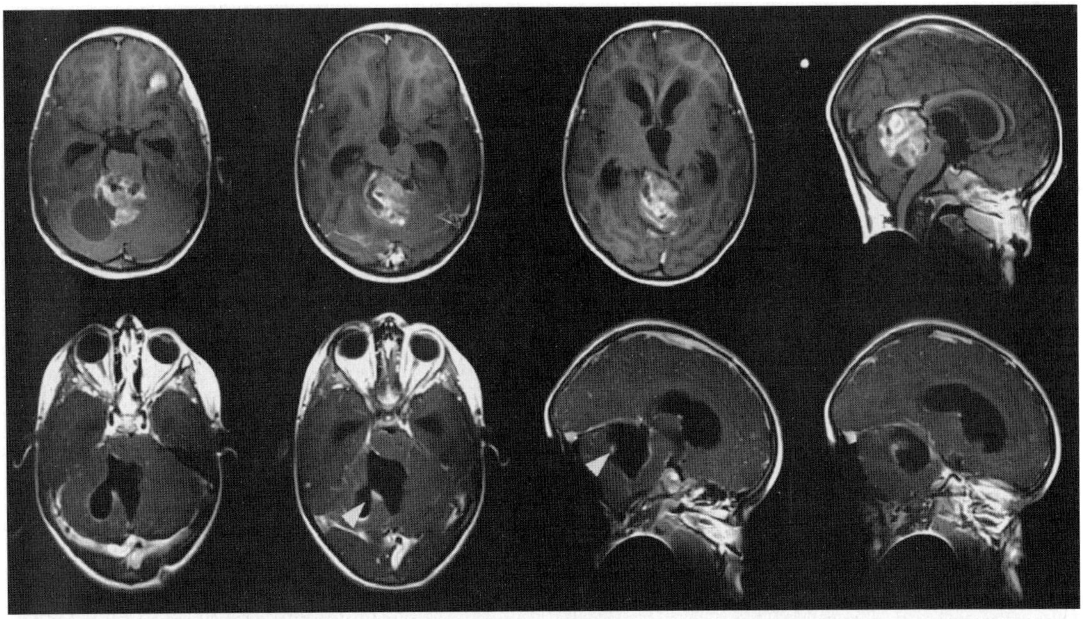

Fig. 5. Surgical planning T_1-Gd iMRI obtained from a 5-year-old boy with pilocytic astrocytoma show the large tumor with associated brain shift (upper). Interdissection images (lower) revealed unsuspected residual tumor (arrows).

[30]. Despite the fact that this procedure is routinely performed, anatomical targets can become unclear during microdissection. When approaching through the mid temporal gyrus, the amygdala may be hard to define, while with trans-sylvian dissections, it is often difficult to determine the extent of hippocampal removal. In 19% of patients with temporal lobe epilepsy, interdissection iMRI, showed unsuspected residual tissue, which was then removed (Figure 7). Notwithstanding these encouraging results, a randomized trial is necessary to both confirm the observations and determine the impact of iMRI on clinical outcome.

The assessment of cortical dysplasia requires the signal-to-noise ratio of 1.5 T or higher-field MR systems. In 4 patients with cortical dysplasia, extent of the abnormality was well assessed with the 1.5 T iMRI system, enhancing craniotomy placement and resection control.

Vascular disorders

Surgical planning iMRI significantly improved craniotomy placement for patients with arteriovenous malformation (AVM) and cavernous angioma, making lesion exposure relatively straightforward. Among the 60 patients with cerebral vascular pathology, 18 had AVM and 27 had cavernous angioma. We have, as of yet, not used diffusion/perfusion MR imaging to evaluate the effect of resecting AVM on the adjacent brain. The literature suggests that AVM resection may result in hyperperfusion of the adjacent brain. It is this consequence that is thought to contribute to the relatively high frequency (10%) of postoperative hemorrhage [31]. There were 12 patients with aneurysm and 3 had extracranial carotid occlusion. For all vascular procedures, interdissection iMRI and quality assurance iMRI confirmed complete resection of the target or treatment of the abnormality.

For patients with aneurysm, intraoperative MR angiography accurately revealed vascular anatomy and clip placement. However, clip-induced susceptibility and/or eddy current artifact interfered with evaluation of the aneurysm neck. Intraoperative assessment of the aneurysm neck will require the development of MR-invisible clips. EPI was used to identify diffusion/perfusion abnormality in the distribution of distal vascular branches, including perforators. The presence of such abnormalities would allow clip repositioning prior to irreversible brain injury (Figure 8). We have not encountered such changes and none of the aneurysm patients have suffered deficit.

Patients with carotid occlusion were symptomatic with recurrent transient ischemic attacks, and diagnostic MRI showed diffusion/perfusion mismatch. Extracranial to intracranial bypass resulted in immediate correction of the

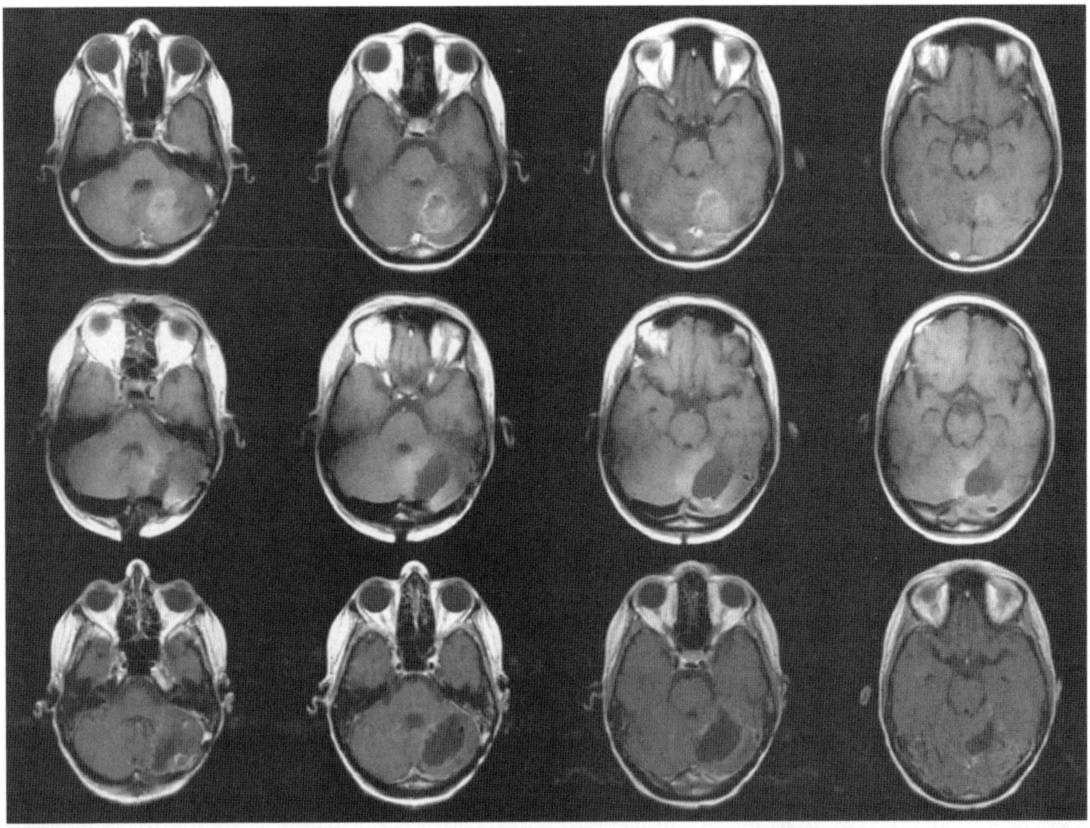

Fig. 6. Surgical planning (upper), interdissection (middle), and day 2 postoperative (lower) T_1-Gd iMRI of a patient with recurrent medulloblastoma. Diffusion of contrast well beyond the suspected tumor margin can be observed on interdissection but not on follow-up images.

perfusion deficit as shown with iMRI. Diffusion-weighted EPI quality requires a homogenous magnetic field, excellent eddy current compensation, strong gradients with fast switching times, and excellent shimming [32].

Spine

iMRI was used as an adjunct to spine surgery in 21 patients. The procedures included anterior cervical discectomy, cervical corpectomy with instrumentation, and the transoral resection of various C1/2 pathologies. iMRI with surgical navigation provided precise targeting of spinal pathology particularly lesions involving the C1/2 region, that are difficult to visualize through surgical corridors designed to limit trauma associated with extensive exposure. Interdissection images accurately determined the extent of decompression and showed the relationship of the surgical dissection to neural vascular structures (Figure 9). Extending iMRI capability to the thoracic and lumbar spine will be increasingly important as spinal surgery becomes reliant on microsurgical technique including restricted surgical corridors. The modified phased array spine RF coils are used to maximize images of the spine. The volume coil is used to transmit the RF power while the phased array coils work in the receive only mode.

Future Directions

Technology

While imaging techniques have greatly enhanced diagnostic accuracy and improved surgical outcomes, technical limitations necessitate further iMRI improvement. Increasing the magnetic field would further enhance image

Fig. 7. Surgical planning T_1-weighted iMRI from a patient with intractable temporal lobe epilepsy, showing the targeted amygdala and hippocampus (upper). Interdissection images (middle) indicate unsuspected residual amygdala (arrow). Quality assurance iMRI reveals complete dissection of the target (bottom).

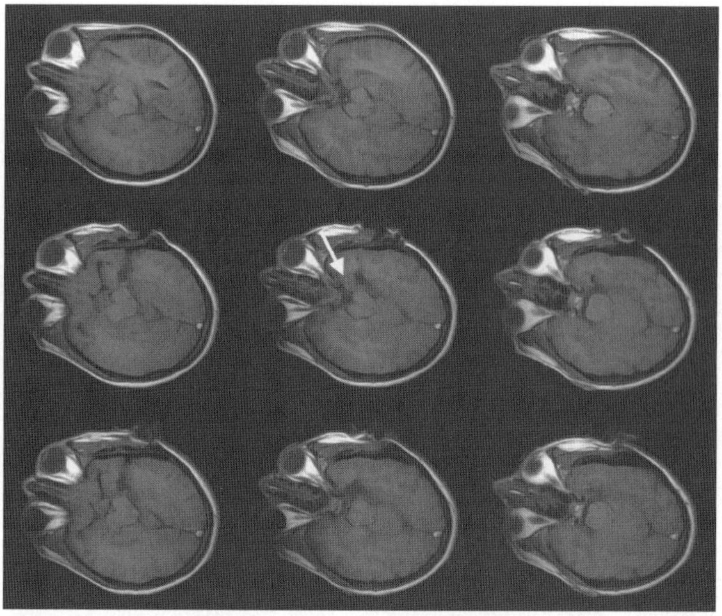

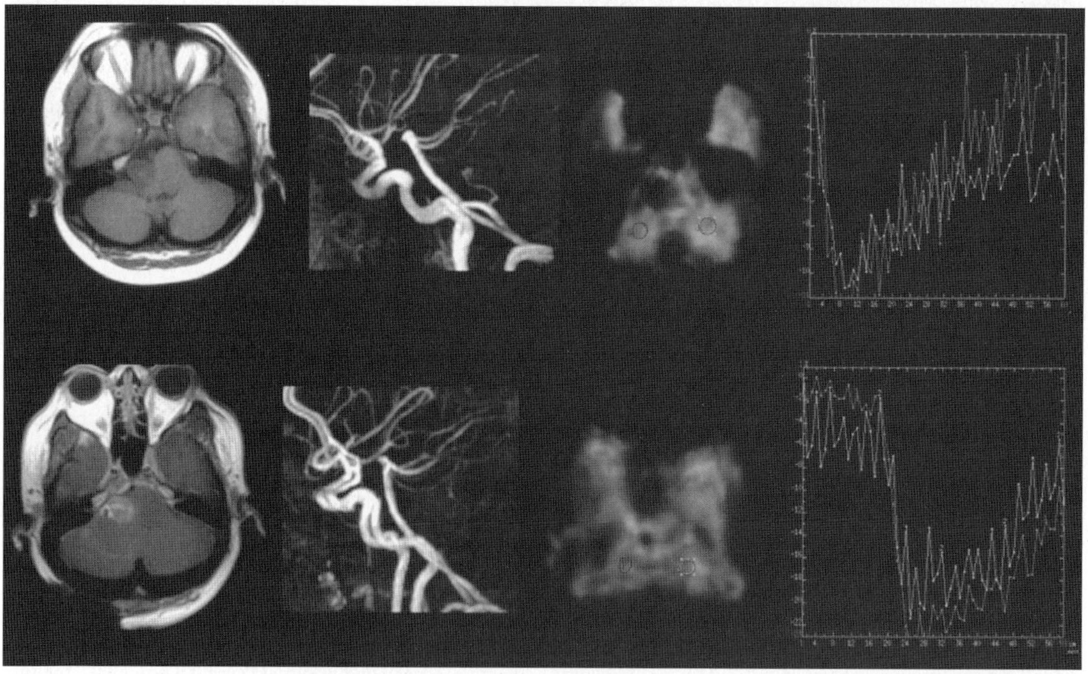

Fig. 8. Surgical planning (upper) and interdissection (lower) T_1-weighted, MR angiography, diffusion, and perfusion iMRI studies obtained from a patient with a giant left PICA aneurysm. Interdissection studies obtained 5–10 min following clip ligation of PICA show flow change within the aneurysm, parent vessel occlusion, aneurysm obliteration, and no diffusion or perfusion abnormality.

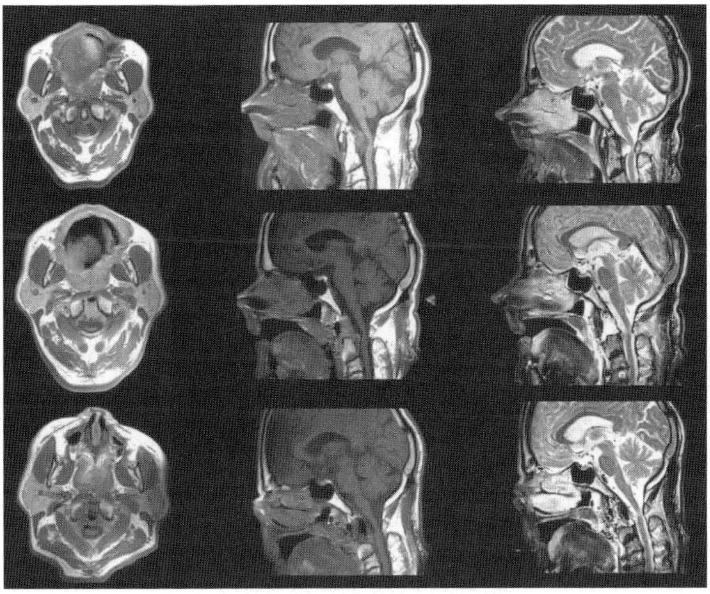

Fig. 9. T_1- and T_2-weighted iMRI from patient with retro-odontoid synovial cyst. Surgical planning iMRI (upper) show significant spinal cord compression. Inter-dissection iMRI show unsuspected residual spinal cord compression at the C2 level (middle). Quality assurance iMRI show optimal decompression (lower).

resolution by improved signal-to-noise ratio, although this might also lead to an amplification of susceptibility artifacts in certain areas. Improvements in RF coil design need to take advantage of improved technology while not limiting surgery. Larger RF coils, not restricting access to the operative field, would be an asset. Additional reduction in total acquisition time could be achieved by the application of phased array technology that allows the acquisition of parallel images. Enlarging the magnet bore facilitates patient positioning and the introduction of surgical technologies. To take full advantage of an iMRI system and maximize its use, the magnet should ideally be movable and sited for use in two or more rooms.

Bioinformatics

Software development has made possible 3D reconstruction of iMRI including lesion segmentation. 3D reconstruction augments lesion localization, diagnosis, and surgery planning. MRI and spectroscopic characteristics may correlate with outcome [33,34]. Multimodal fusion of MR images provides information as to the relationship of eloquent brain and vasculature to the planned surgical corridor and target.

Robotics

The introduction of iMRI has provided an imaging adjunct to surgery. Despite this and other technologies, such as microscopy, surgeons operate much like they did 100 years ago, depending to a large extent on hand–eye coordination. Robotic advances, such as tremor filtration and motion scaling permit maximal use of magnification and enable precise, tremor-free tool manipulation during surgery. Integrating robotic technology with iMRI would allow surgeons to manipulate images in real time and reduce limitations imposed by iMRI, particularly those related to a restricted working environment (GE double doughnut configuration) and disruptions in surgery related to image acquisition (1.5 T IMRIS, Siemens or Philips systems).

At the University of Calgary, we are developing an MRI-compatible, image-guided, ambidextrous robot called neuroArm [35]. Our industrial collaborator, MD Robotics (Brampton, Ontario, Canada), has successfully developed safety-critical robotic systems for the space shuttle and International Space Station. NeuroArm consists of a neurosurgical robot, main controller, and a workstation. The robot has two MR-compatible articulated arms with dextrous mechanical manipulators that grasp and move surgical tools. Each arm has 8 degrees of freedom (DOF) and are small enough to work within the 68 cm working diameter of the 1.5 T IMRIS magnet and capable of 30 μm precision (Figure 10). A 3-DOF strain gauge sensor system was integrated to provide haptic feedback to the surgeon. The workstation incorporates a computer processor, two hand controllers to manipulate the robot arms, a controller for positioning the microscope and lights, and three types of display and data recorders.

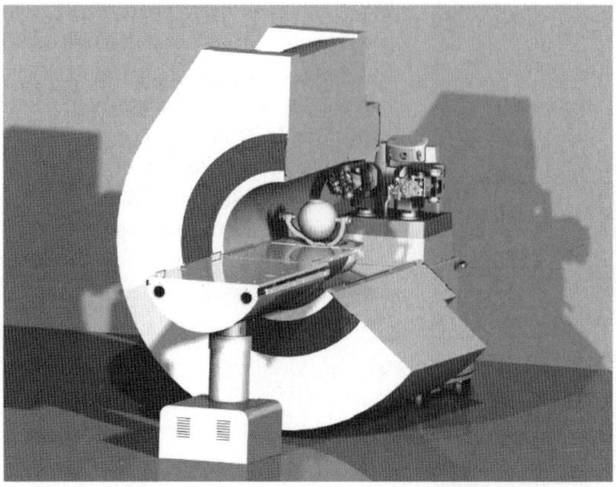

Fig. 10. NeuroArm in position for stereotaxy. The base is equipped with an extension platform (not shown), allowing the manipulator arms to reach inside the magnet bore.

The interface maximizes ergonomic comfort and minimizes surgeon fatigue.

The robotic platform and imaging environment include surgical simulation capability. This will be of increasing importance as advances in surgical technologies and associated techniques increase case complexity and produce a tyranny of choice. Risk-free rehearsal and development of patient specific interventional recipes will allow surgeons to safely take full advantage of iMRI and robotic technologies.

Acknowledgments

The authors would like to thank Boguslaw Tomanek and John Saunders for their kind suggestions. The iMRI program has been supported by a grant from the Canadian Foundation for Innovation. The iMRI system is being marketed by IMRIS (info@IMRIS.com).

References

1. Clower WT, Finger S. Neurosurgery. 2001;49:1417.
2. Piek J, Lidke G, Terberger T, von Smekal U, Gaab MR. Neurosurgery. 1999;45:147.
3. Barker FG II. J. Neurosurg. 1993;79:948.
4. Gall FJ. Sur les fonctions du cerveau et sur celles de chacune de ses parties avec des observations sur la possibilité de reconnaitre les instincts, les penchans, les talens, ou les dispositions morales et intellectuelles des hommes et des animaux, par la configuration de leur cerveau et de leur tête, Vol 1. J.B. Baillière: Paris, 1822.
5. Broca P. Bull. Soc. Anat. Paris. 1861;6:398.
6. Dandy WE. Ann. Surg. 1919;70:397.
7. Moniz E. Rev. Neural. 1927;2:72.
8. Fry WJ, Fry FJ. IRE Trans. Med. Electron. 1960;ME-7:166.
9. Cormack AM. Phys. Med. Biol. 1973;18:195.
10. Hounsfield GN. Br. J. Radiol. 1973;46:1016.
11. Lauterbur PC. Philos. Trans. R. Soc. Lond. B. Biol. Sci. 1980;289:483.
12. Mansfield P, Maudsley AA. Br. J. Radiol. 1977;50:188.
13. Lunsford LD. Appl. Neurophysiol. 1982;45:374.
14. Black PM, Moriarty T, Alexander E III, Stieg P, Woodard EJ, Gleason PL, Martin CH, Kikinis R, Schwartz RB, Jolesz FA. Neurosurgery. 1997;41:831.
15. Le Bihan D, Mangin JF, Poupon C, Clark CA, Pappata S, Molko N, Chabriat H. J. Magn. Reson. Imaging. 2001; 13:534.
16. Keston P, Murray AD, Jackson A. Clin. Radiol. 2003;58:505.
17. Le Bihan D. J. Neuroradiol. 1996;23:1.
18. Hoult DI, Saunders JK, Sutherland GR, Sharp J, Gervin M, Kolansky HG, Kripiakevich DL, Procca A, Sebastian RA, Dombay A, Rayner DL, Roberts FA, Tomanek B. J. Magn. Reson. Imaging. 2001;13:78.
19. Steidle G, Graf H, Schick F. Magn. Reson. Imaging. 2004;22:171.
20. Steinmeier R, Fahlbusch R, Ganslandt O, Nimsky C, Buchfelder M, Kaus M, Heigl T, Lenz G, Kuth R, Huk W. Neurosurgery. 1998;43:739.
21. McPherson CM, Bohinski RJ, Dagnew E, Warnick RE, Tew JM. Acta Neurochir. Suppl. 2003;85:39.
22. Sutherland GR, Kaibara T, Louw DF, Hoult DI, Tomanek B, Saunders J. J. Neurosurg. 1999;91:804.
23. Kaibara T, Saunders JK, Sutherland GR. Neurosurgery. 2000;47(1):131.
24. Sutherland GR, Kaibara T, Wallace C, Tomanek B, Richter M. Intraoperative assessment of aneurysm clipping using magnetic resonance angiography and diffusion-weighted imaging: technical case report. Neurosurgery. 2002;50(4): 893.
25. Nimsky C, Ganslandt O, von Keller B, Fahlbusch R. Acta Neurochir. Suppl. 2003;88:21.
26. Hall WA, Liu H, Martin AJ, Truwit CL. Top. Magn. Reson. Imaging. 2000;11:203.
27. Schulder M, Jacobs A, Carmel PW. Stereotact. Funct. Neurosurg. 1998;76:151.
28. Susil RC, Camphausen K, Choyke P, McVeigh ER, Gustafson GS, Ning H, Miller RW, Atalar E, Coleman CN, Menard C. Magn. Reson. Med. 2004;52:683.
29. Dick EA, Joarder R, de Jode M, Taylor-Robinson SD, Thomas HC, Foster GR, Gedroyc WM. Clin. Radiol. 2003; 58:112.
30. Olivier A. Can. J. Neurol. Sci. 2000;27(Suppl 1):S68.
31. Schaller C, Urbach H, Schramm J, Meyer B. Neurosurgery. 2002;51(4):921.
32. Fischer H, Ladebeck R. Echo-Planar Imaging: Theory, Technique and Application. Springer: New York, 1998, pp 179–200.
33. Ekinci G, Akpinar IN, Baltacioglu F, Erzen C, Kilic T, Elmaci I, Pamir N. Eur. J. Radiol. 2003;45(2):99.
34. Wu WC, Chen CY, Chung HW, Juan CJ, Hsueh CJ, Gao HW. Am. J. Neuroradiol. 2002;23(10):1775.
35. Louw DF, Fielding T, McBeth PB, Gregoris D, Newhook P, Sutherland GR. Neurosurgery. 2004;54(3):525.

In Vivo Magnetic Resonance Spectroscopy in Breast Cancer

Uma Sharma and N.R. Jagannathan
Department of NMR, All India Institute of Medical Sciences, New Delhi 110029, India

Abbreviations: MR, magnetic resonance; MRI, magnetic resonance imaging; MRS, magnetic resonance spectroscopy; ^{1}H MRS, proton magnetic resonance spectroscopy; ^{31}P MRS, phosphorus magnetic resonance spectroscopy; CE-MRI, contrast-enhanced magnetic resonance imaging; VOI, volume of interest; CSI, chemical shift imaging; MRSI, magnetic resonance spectroscopic imaging; STEAM, stimulated echo acquisition mode; PRESS, point resolved spectroscopy; CHESS, chemical shift selective; W–F, water-to-fat ratio; PME, phosphomonoester; PDE, phosphodiester; PCr, phosphocreatine; SV, single-voxel.

Introduction

Breast cancer is the most common cancer affecting women and is a significant health care problem, worldwide [1–3]. Breast cancer is classified as either ductal or lobular, depending on its morphological origin and similarities to the normal cellular components. Considerable heterogeneity among breast tumors demands highly accurate diagnostic techniques. Methods like X-ray mammography, ultrasound, and physical examination are often limited in sensitivity and specificity, especially in young women [4–6]. Magnetic resonance (MR) mammography and contrast-enhanced magnetic resonance imaging (CE-MRI) methodologies have greatly improved the ability to differentiate malignant from benign breast tumors [7–11] (see chapter by Kaiser). However, the reported specificity of the above studies is widely variable [8] and does not provide any metabolic information.

Recently, there has been an increasing interest in augmenting breast MRI with magnetic resonance spectroscopy (MRS) investigations by using the biochemical information to increase the specificity in differentiating benign from malignant tumors. Results of *in vivo* proton (^{1}H) [12–25] and phosphorus (^{31}P) [26–28] MRS, *ex vivo* and *in vitro* [29–32], and cell line [33,34] studies reported to date suggest that malignant breast tumors exhibit elevated levels of choline-containing compounds (total choline, TCho). In the case of MRS applied to fine needle aspiration biopsies (FNAB), the creatine to choline ratio can provide the pathological diagnosis with accuracies in the late 90th percentile [32,35]. This chapter reviews the potential of *in vivo* MRS (both ^{1}H and ^{31}P) in breast cancer especially in relation to the various methodological issues involved, technical limitations, diagnostic specificity, clinical applications, and future directions.

In vivo Localization in MRS

The basic requirement of *in vivo* localized MRS is to acquire the signal from a particular volume of interest (VOI) with optimal sensitivity. In the last 20 years, the availability of gradients has led to the development of localization techniques to obtain spectra from specific VOI. Magnetic field gradients in either the B_0 field as in imaging or the B_1 (radio frequency) field are used for localization of a particular volume. *In vivo* MRS began with the analysis of isolated tissues and surface regions. Most ^{31}P studies on breast cancer use surface coils, which are one of the earliest methods of localization. Surface coils provide rough localization. However, spatial selectivity can be achieved by varying the radio frequency pulse length. The disadvantages include extremely inhomogeneous transverse magnetic field, difficulty in assessing VOI that is below the surface, and contamination of signals from extraneous tissues. For these reasons, surface coils are used with other techniques like depth resolved spectroscopy, DRESS [36], which uses B_0 magnetic field gradients and a frequency selective pulse for localization. Depth resolved spectroscopy is also possible by using B_1 changes [37].

Currently employed image guided localization methods use proton images acquired in three orthogonal planes (transverse, sagittal, and coronal) to guide the placement of VOI. The magnetic field gradients spatially encode the resonance frequencies and the frequency selective pulse excites the spin distribution within the sample. Position, size of VOI, and the actual shape depend on the slice profile of the selective pulses. Localization methods acquire spectra either as single-voxel (SV) or as multi-voxel [chemical shift imaging (CSI) or magnetic resonance spectroscopic imaging (MRSI)]. MRSI offers many advantages over SV techniques, such as localized spectra from many locations and small VOIs can be acquired simultaneously which can be used to obtain the metabolite

Graham A. Webb (ed.), Modern Magnetic Resonance, 1063–1072.

images. The pixel intensity in these images is proportional to the relative concentrations of the metabolites which is useful in visually assessing the spatial variation in the metabolite concentrations. However, the technique suffers from some disadvantages, such as signal from a voxel is prone to contamination from outside the VOI. This contamination occurs due to discrete and finite sampling, since the number of phase-encoding steps is typically more limited than normal imaging acquisitions. In addition, shimming the magnetic field over large volume especially in breast due to heterogeneity of tissues and long acquisition times are other disadvantages. In recent times, echo-planar imaging-based MRSI sequences which offer less image acquisition times have also become available. The most commonly used pulse sequences in breast cancer MRS are image selected *in vivo* spectroscopy (ISIS) [38], stimulated echo acquisition mode (STEAM) [39,40], point resolved spectroscopy (PRESS) [41], and CSI or MRSI [42,43]. In breast ^{1}H MRS, detection of resonances from metabolites with lower concentration in the presence of a large water and lipid signals is a major problem. Chemical shift selective (CHESS) radio frequency pulses that excite a limited narrow band (~60 Hz) of frequencies corresponding to the water signal are used to suppress the water signal [44]. Sequences that simultaneously suppress both water and lipid resonances are also being developed (*vide infra*).

^{31}P MR Spectroscopy

^{31}P MR spectra provide information on cellular metabolism by measuring various phosphorus metabolites and have been reviewed recently [26,27]. Most studies were carried out at 1.5 T with the patient positioned either supine or prone with a dual ^{1}H/^{31}P surface coil covering the area of the tumor. Shimming is carried out at ^{1}H frequency. The measurement methodology adopted by different groups varies in relation to the use of repetition time, size of coil used, and other experimental parameters. Figure 1A shows the ^{31}P MR spectrum obtained from a normal volunteer, and reflects the level of various

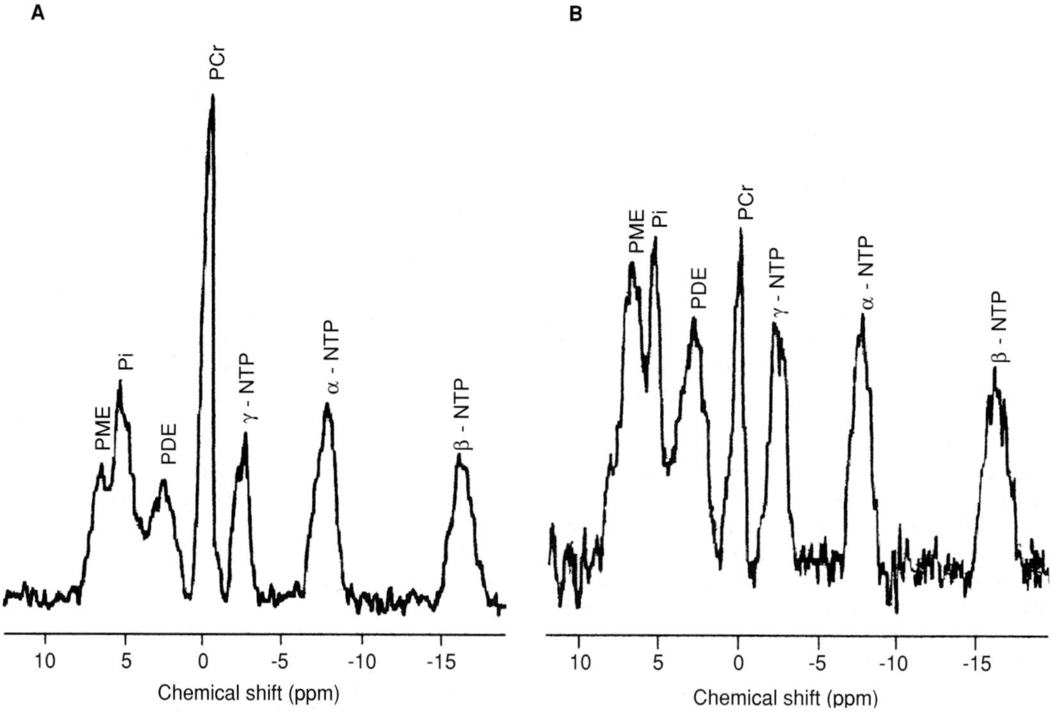

Fig. 1. (A) ^{31}P MR spectrum from normal breast tissue of a volunteer. Peak assignment corresponds to adenosine triphosphate (ATP), phosphodiesters (PDE), phosphomonoesters (PME), phosphocreatine (PCr), and inorganic phosphate (Pi). (B) ^{31}P MR spectrum from a patient suffering from infiltrating duct carcinoma of the breast. Spectra were obtained using a 10 cm surface coil with a single RF pulse.

phosphorus containing metabolites. In general, considerable variations in the ^{31}P metabolite profile of the normal breast were observed during the menstrual cycle [45,46]. Pre-menopausal women showed a rise in phosphomonoester (PME) signal at the late follicular stage of the menstrual cycle and less signal at the early luteal phase [46]. Phosphodiester (PDE) was maximal at the early follicular phase, and minimal at the late follicular stage. Twelves *et al.* [45] have reported lower PME and a high PDE/PME ratios during the second week of menstrual cycle. It is also reported that PME is increased in lactating breast compared with non-lactating post-menopausal women [45,47]. In general, ^{31}P MR signals are lower in post-menopausal women compared with pre-menopausal women [45,46].

Figure 1B shows the ^{31}P MR spectrum of a patient suffering from infiltrating duct carcinoma with elevated levels of PME and PDE, similar to that reported by other researchers [26,28,48–51]. In general, breast tumors have higher total phosphate content compared to the normal breast tissues of non-menstruating women [26,48]. The potential of ^{31}P MRS in distinguishing benign lesions from malignant has also been reported on the basis of elevated PME [49,50]. Biopsy results demonstrate that carcinomas contain three times more adenosione triphosphate and PME than benign breast lesions [29]. Recently, the utility of *in vivo* ^{31}P MRS for patients with various malignant and benign diseases has been evaluated [28]. Significant differences in PME and the ratio of PDE to phosphocreatine in malignant tumors were observed compared to normal breast tissue. However, no difference was observed between malignant and benign tumors suggesting that ^{31}P MRS is not helpful in differential diagnosis between benign and malignant lesions [28]. The use of ^{31}P MRS to monitor the response of breast tumors to chemotherapy has also been reviewed [26]. Serial measurements of localized ^{31}P MRS of breast cancer patients using the image guided *in vivo* spectroscopy sequence [26] was carried out both before and after chemotherapy showing decreased PME that was found to be associated with the response of the disease, while an increase in PME was associated with the disease progression [12,45,51–53].

Even though ^{31}P MRS studies were promising, they gave diverse results [28]. Moreover, its use in characterizing the tumors *in vivo* is hampered by lower MR sensitivity for detecting ^{31}P signals (sensitivity 1/10 of ^{1}H MRS). To achieve a similar signal-to-noise ratio of metabolites detected by ^{1}H MRS, a ^{31}P MRS study requires a voxel that is about 10 times larger than that used in ^{1}H MRS and hence can be performed only on large sized tumors. In addition, ^{31}P MRS requires special hardware that may not be available with all clinical scanners. By contrast, ^{1}H MRS examination can easily be integrated into a routine MRI examination. In view of this, considerable interest has been shown in recent years in evaluating the role of *in vivo* ^{1}H MRS in breast cancer.

^{1}H MR Spectroscopy of Breast

Presently seven groups are involved in breast *in vivo* ^{1}H MRS [12–25]. Katz-Brull *et al.* [54] have reviewed these reports and carried out pooled analysis of the existing ^{1}H MRS data. However, several methodological issues need to be addressed and these are outlined below.

Methodological Issues Related to Proton *in vivo* Breast MRS

Most *in vivo* ^{1}H MRS of human breast cancer studies reported to date were performed at 1.5 T except for the work of two groups [24,55,56]. The patient is usually positioned prone and this has some technical advantages, in that a tumor may be relatively distant from other structures that provide a substantial signal, such as chest wall muscle. However, positioning a surface coil is difficult in prone position and thus, a relatively large volume coil is often employed which has the disadvantage of potentially reducing the signal-to-noise ratio. In the supine position, a surface coil can be used, but contamination from chest wall and motion artifacts may increase. An additional disadvantage is the loss of signal from lesions that are not close to the surface coil of the breast.

A dedicated single or bilateral surface receive coil (custom/commercial make) is used for optimal signal reception with the body coil as transmitter. Automatic tuning and electronic decoupling is used for optimal signal reception. Some groups have used a transmit–receive breast coil (usually custom made) for their study [15,23,24]. Patients are positioned prone with the breast fitting into the cup of the coil. Followed by localizer images, T_1-weighted images in sagittal plane are obtained using a spin-echo sequence. To identify precisely the full extent of the irregular, speculated border of malignant tumors, lipid saturated proton MR images in the transverse, and coronal planes are acquired. Some centers use CE-MRI to localize the tumor region. Kvistad *et al.* used non-contrast MRI, when the tumor was easy to localize, otherwise CE-MRI was used for positioning the voxel [16]. Caution should be exercised when MRS is performed after using contrast agents to identify the tumor, since gadolinium has been reported to influence the *in vivo* detection of TCho [57]. Depending on the tumor size, voxels of appropriate dimension are chosen and positioned well within the tumor area.

Magnetic field shimming must be carried out both globally and over the voxel region to optimize the field homogeneity and for good water-suppression. In our experience, the line width after voxel shimming typically should correspond to 10–25 Hz for the lipid peak in case of normal breast tissue and 5–20 Hz for the water peak in patients with breast tumors. If this is not achieved, the quality of water-suppression will be poor leading to broad or no detectable choline signal. Solvent suppression pulses can be optimized to have narrow bandwidth excitation to minimize spectral perturbations outside the desired area. Water-suppressed and un-suppressed localized spectra with an adequate number of scans are acquired using an appropriate echo time (TE = 30/135/270/350 ms). Chemical shifts are reported using water as the internal standard at 4.7 ppm. The total study time per patient ranges between 60 and 75 min including MRI and MRS. Table 1 summarizes details of the pulse sequence, acquisition parameters, etc. used in various studies reported in the literature.

Table 1: Summary of experimental details used in various ^{1}H MRS studies

Study and reference	Patients, number of cases studied, and other details	Sequence used and SV/MRSI	Water or lipid suppression sequence used	Field, coil used, methods, acquisition parameters, sensitivity, and specificity
Roebuck *et al.* [13]	Malignant, $n = 10$ (pre-therapy); benign, $n = 7$	STEAM; SV	Water; CHESS	1.5 T; single breast multicoil receive; CE-MRI; TR = 2 s; TE = 31, 270 ms; voxel size = 0.7–9.8 ml; sensitivity = 70%; specificity = 86%
Jagannathan *et al.* [14,17]	Malignant, $n = 17$ (pre-therapy); $n = 34$ (post-therapy); benign, $n = 2$; controls, $n = 14$	STEAM; SV	Water; CHESS	1.5 T; bilateral surface receive coil; non-CE-MRI; TR = 3 s; TE = 135 ms; voxel size = 1–8 ml
Kvistad *et al.* [16]	Malignant , $n = 11$ (pre-therapy); $n = 1$ (post-therapy); benign, $n = 11$; control, $n = 11$; lactating, $n = 7$	PRESS; SV	Water frequency selective inversion	1.5 T; single breast receive coil; non-CE and CE-MRI; TR = 2 s; TE = 135, 350, 450 ms; voxel size = 1–25.2 ml; sensitivity = 82%; specificity = 82%
Jagannathan *et al.* [22]	Malignant, $n = 46$ (pre-therapy); $n = 35$ (post-therapy); benign, $n = 14$; control, $n = 16$; lactating, $n = 1$	STEAM; SV	Water; CHESS	1.5 T; bilateral surface receive coil; non-CE-MRI; TR = 3 s; TE = 135 ms; voxel size = 1–27 ml; sensitivity = 78%; specificity = 86%
Cecil *et al.* [18]	Malignant, $n = 23$ (pre-therapy); benign, $n = 15$	STEAM; SV	Water; CHESS	1.5 T; phased array multicoil receive; non-CE and CE-MRI; TR = 2 s; TE = 31, 270 ms; voxel size = 1–3.4 ml; sensitivity = 83%; specificity = 87%
Yeung *et al.* [19]	Malignant, $n = 24$ (pre-therapy); benign, $n = 6$	PRESS; SV	Water frequency selective inversion recovery	1.5 T; double breast receive coil; CE-MRI; TR = 2 s; TE = 38, 135, 270 ms; voxel size = 1–95 ml; sensitivity = 92%; specificity = 83%
Gribbestad *et al.* [15]	Malignant, $n = 12$ (pre-therapy); control, $n = 10$	PRESS; SV	Water frequency selective inversion	1.5 T; double breast transmit-receive coil; non-CE-MRI; TR = 2 s; TE = 136 ms; voxel size = 8–27 ml
Gribbestad *et al.* [61]	Malignant, $n = 6$ (pre-therapy)	PRESS; SV	Water and lipid using MEGA	1.5 T; standard coil; CE-MRI; TR = 1.6 s; TE = 270 ms; voxel size = 3.4 ml
Stanwell *et al.* [23]	Malignant, $n = 21$ (pre-therapy); control, $n = 43$; lactating , $n = 3$	PRESS; SV	VSS and CHESS	1.5 T; single breast transmit–receive coil; non-CE-MRI; TR = 2 s; TE = 135, 350 ms; voxel size = 3.4 ml

Table 1: *(Continued)*

Study and reference	Patients, number of cases studied, and other details	Sequence used and SV/MRSI	Water or lipid suppression sequence used	Field, coil used, methods, acquisition parameters, sensitivity, and specificity
Bolan *et al.* [24]	Malignant, $n = 86$ (pre-therapy); $n = 14$ (post-therapy); control, $n = 5$	LASER; SV	VAPOR	4 T; single breast transmit–receive coil; CE-MRI; TR= 3 s; TE = 45–196 ms; voxel size = 0.4–6 ml
Thomas *et al.* [66]	Malignant, $n = 8$; control, $n = 9$	2D, SV CABINET	No suppression	1.5 T; phased array coil; TR = 2 s; TE = 30 ms; voxel size = 1 ml
Jacobs *et al.* [25]	Malignant, $n = 8$ (pre-therapy); benign, $n = 7$	PRESS; MRSI	Water and lipid; CHESS and STIR	1.5 T; phased array coil; TR = 2 s; TE = 272 ms; voxel size = 1 ml
Meisamy *et al.* [56]	Malignant, $n = 16$; MRS both pre- and post-therapy	SV, localization with adiabatic selective refocusing	VAPOR	4 T; single breast quadrature transmit–receive coil; TR = 3 s; TE = 45–196 ms; voxel size = variable dimensions according to tumor size

Note: TR = time for repetition; TE = time for echo; LASER = localization by adiabatic selective refocusing; STIR = short inversion time recovery; VAPOR = variable pulse power and optimized relaxation; CABINET = coherence transfer based spin-echo spectroscopy; MEGA = MEscher GArwood (named after the authors); VSS = very selective suppression.

In our work, the observation of TCho was based on strict experimental criteria [22], namely: (i) the line width of the un-suppressed water peak to be around 5–20 Hz and (ii) the ratio of the water-suppression factor $\geq$20. If these criteria were not met, the data were discarded [22]. It is our experience that these criteria are essential since the observation of choline may be affected by poor shimming, the relative position of the voxel in relation to the surface coil, coil loading, and partial volume effects [22,24,54]. Moreover, SV-MRS of breast is sensitive to the size and placement of the voxel because of the heterogeneous distribution of TCho in the breast [24].

In addition, quantification of metabolite levels in breast is more difficult because of the heterogeneous distribution of the glandular and adipose tissues. Absolute concentration of TCho in breast lesions has also been reported using a phantom containing 1 mmol/l solution of choline placed in the multicoil array [13]. Recently, Bolan *et al.* used water as an internal reference peak for quantification of TCho [24]. This approach compensates for the partial volume of adipose tissue in the voxel and leads to a molal (mol/kg) concentration of water-soluble metabolites. Use of an appropriate echo time in breast *in vivo* spectroscopy involves a trade-off between signal intensity (high with short echo time) and signal contrast (the ability to resolve the composite choline signal from the lipid signal, which is higher with long echo time). The use of long echo times ($\geq$135 ms) is preferred, despite the loss of signal intensity, for improved visibility of TCho signal because of a decreased overlap with the lipids [13,19,22]. Recently, Stanwell *et al.* have demonstrated the importance of referencing and optimized post-acquisitional data processing to improve the spectral resolution of MR spectra [23]. This allows the resolution of the composite choline resonance into its constituent components and helps in improving the specificity of *in vivo* MRS method [23]. In addition, water-to-fat (W–F) (lipid) ratio can also be calculated using the respective peak areas (using area under the water and the major lipid peak) from the un-suppressed MR spectra [12,14,17].

Interpretation of ^{1}H and ^{31}P spectra should include the age and the menopausal status of the patient. Postmenopausal women have more adipose tissue compared to glandular tissue. In pre-menopausal women, glandular tissue can give an adequate signal, and this needs to be considered both in the measurement strategies as well as in interpreting the spectra obtained. In a study by Stanwell *et al.* composite choline signal was observed in three non-lactating normal volunteers [23]. The chemical shift of this resonance was found to differ from that recorded from patients with malignant disease. When ^{1}H MRS on two of these volunteers was repeated once a week for a period of 8 weeks [23], in one volunteer the choline resonance was absent at different stages throughout the menstrual cycle but re-appeared in the next cycle. In the other volunteer, the resonance was always detectable but appeared to fluctuate during the 8-week period. However, long-term follow-up of these individuals has revealed no disease.

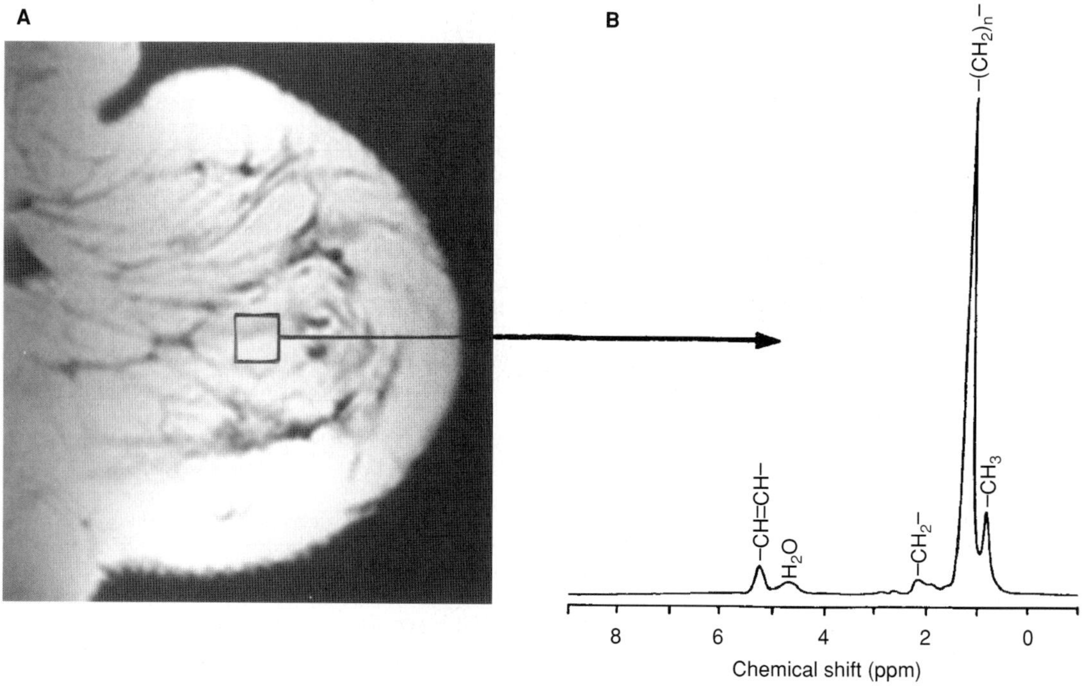

Fig. 2. (A) T_1-weighted axial image from a volunteer using spin-echo sequence with the following parameters: TR = 600 ms, TE = 15 ms, and slice thickness = 4 mm. (B) ^{1}H MR spectrum acquired from an 8 ml voxel localized in the normal breast tissue at TE = 135 ms using PRESS sequence.

Comparison of ^{1}H MRS of Normal and Malignant Breast Tissues

Water-to-Fat Ratio

A typical ^{1}H MR spectrum obtained from our laboratory using STEAM sequence (without water-suppression) from the normal breast tissue of a volunteer is shown in Figure 2B while the corresponding voxel location in Figure 2A. Lipid resonance dominates and the assignment of the various lipid resonance peaks are as shown. A ^{1}H spectrum without water-suppression from the tumor of a patient suffering from infiltrating duct carcinoma is shown in Figure 3B and the voxel location in Figure 3A. In tumor spectrum, the water peak dominates with low contributions from protons of the fatty acid chains in comparison to the normal breast tissue of controls and from the unaffected contralateral breast of the patient. Elevated W–F ratio characterizes the breast tumor. Similar findings have been documented by several others [12,14,17]. Observation of elevated W–F ratio in breast cancer patients is in agreement with the generally established trend that tumors have considerably higher water content [12,58,59]. Mackinnon *et al.* [60] reported changes in lipid content with tumor development and progression. In patients receiving chemotherapy, resulting in the reduction of primary tumor size, the W–F ratio showed a statistically significant ($p < 0.01$) decrease compared with the pre-therapy value, thus providing a non-invasive indicator of favorable clinical outcome of chemotherapy [14,17]. The method thus provides the potential for non-invasively monitoring and assessing the response of breast cancer to chemotherapy. However, due to overlap of W–F values between benign and malignant lesions it has limited utility [59] and hence a search for other biochemical/s, which can be used as marker for malignancy, was undertaken.

Choline Signal

Water-suppressed ^{1}H MR spectra of breast tumor showed, in addition to the residual water and lipid peak, a peak at 3.2 ppm due to TCho as shown in Figure 3C. In general, the observation of TCho from breast tumors is hampered by the presence of huge water and lipid signals. However, if the strict experimental criteria outlined earlier are followed, then the observation of TCho is feasible. Evaluation of a pulse sequence to suppress simultaneously both the water and lipid signals to improve the detection

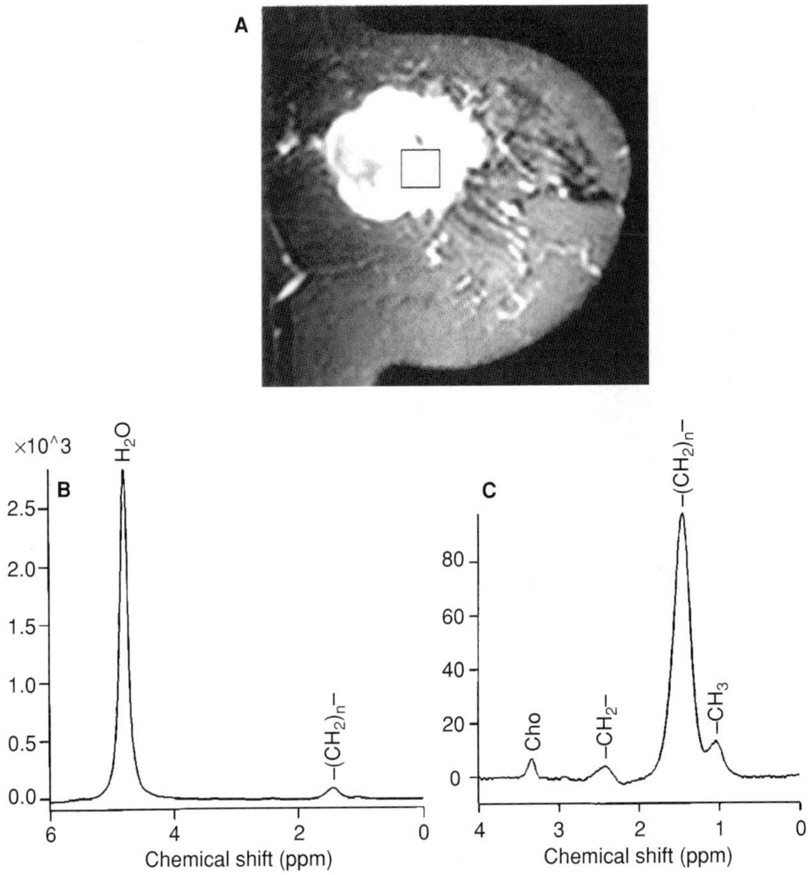

Fig. 3. (A) T_2-weighted lipid-suppressed sagittal MR image of a patient suffering from infiltrating duct carcinoma of the right breast, the tumor is seen as hyper-intense area. ^{1}H MR spectrum from an 8 ml voxel positioned in the tumor of the same patient obtained at TE = 135 ms using PRESS sequence: (B) without water-suppression and (C) with water-suppression.

of choline resonance is currently underway in our laboratory (see Figure 4). Lipid/water signal reduction is obtained using the spectral saturation method, by which transverse magnetization is dephased selectively before and after the second slice selective 180° spin-echo pulse. This dephasing is defined to affect both the water and lipid signals from 0.7 to 2 ppm, following the procedure of Mescher *et al.* [61]. Recently, Gribbestad *et al.* have also used a similar sequence to suppress the signals from water and lipid [62].

Of the various TCho (Cho, glycerophosphocholine, and phosphocholine) that contribute to the *in vivo* peak at 3.2 ppm, an increase in phosphocholine has been documented [23,33,63]. Elevation of TCho levels in malignant breast tumors compared to benign cases from *in vitro* MRS of FNAB and tissue has also been reported [30,31,32]. Results of ^{1}H *in vivo* breast MRS investigations to date have shown that the TCho resonance is specific to malignant tissue and can be used to differentiate cancerous from benign tissue. Combined analysis of the published reports reveal that the overall sensitivity and specificity of MRS is 83 and 85%, respectively [54]. The details are as presented in Table 1. In younger women, differentiation of benign from malignant lesion is important since the incidence of benign breast disease in young women is high [64,65] and mammography examination has lower sensitivity. Results from Cecil *et al.* [18], Yeung *et al.* [19], and Roebuck *et al.* [13] show that in younger patients ($\leq$40 years of age), ^{1}H MRS has a sensitivity of 100% and a specificity of 89–100% in detecting malignancy.

Several factors that limit the sensitivity of ^{1}H MRS were discussed earlier and the false negative results reported in many studies are ascribed mainly to technical limitations. For example, if the tumor size is small,

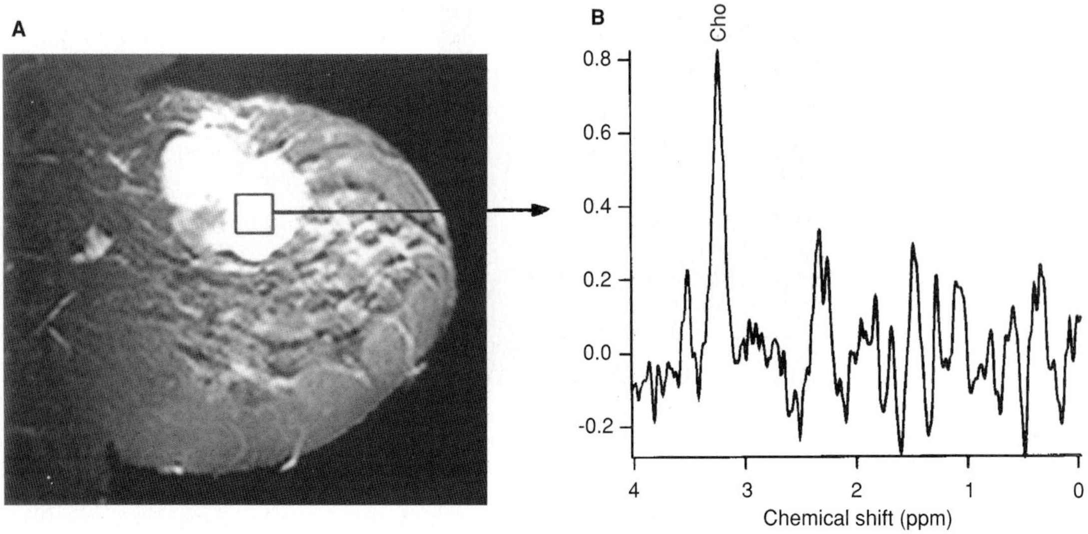

Fig. 4. (A) T_2-weighted lipid-suppressed sagittal MR image of a patient suffering from infiltrating duct carcinoma. (B) ^{1}H MR spectrum from an 8 ml voxel positioned in the tumor of the same patient obtained at TE = 270 ms using PRESS sequence with both the water and the lipid peak simultaneously suppressed.

relatively long acquisition time is required resulting in the patient becoming restless leading to movement artifacts. This also could lead to incorrect sampling of the tumor, thereby leading to contributions from surrounding fatty tissue. Moreover, movement of a patient would also lead to incorrect sampling and hence to false diagnosis. Incorrect sampling may also play a role in the detection of the choline signal. Few benign cases also showed TCho signal [16,18,19,22]. Detection of choline from the normal breast tissue of lactating women has also been reported [16,22,23]. This observation implies that the interpretation of spectra from such patients needs to be evaluated with caution and also points out the limitation of using TCho as a marker of malignancy. Stanwell *et al.* have documented that the main contribution to the TCho resonance observed in lactating breast is from the glycerophosphocholine [23]. They have documented that optimized post-processing of the spectra resolved a resonance at 3.22 ppm, consistent with phosphocholine, in patients with cancer. In contrast, the spectra recorded for three false positive volunteers and the three lactating mothers had a resonance centered at 3.28 ppm (possibly taurine, *myo*-inositol, or glycerophosphocholine).

Recently Jacobs *et al.* [25] have reported on a proton MRSI of breast cancer patients; their results show that TCho from malignant tissue was significantly elevated compared to benign breast tissue. The potential advantages of MRSI over SV-MRS include the ability to assess multiple lesions and tissues with normal appearance, as well as to distinguish lesion borders and infiltration into the surrounding tissue. Another major concern of *in vivo* breast MRS is the spectral overlap, in particular the dominant peaks due to lipids. To address this issue, Thomas *et al.* have carried out localized two-dimensional correlation spectroscopy experiments [66] and have shown that the W–F ratio can be calculated using two-dimensional spectral peak volumes. In addition, peaks due to unsaturated fatty acids can also be differentiated from saturated fatty acids. The TCho signal can also be used to monitor tumor response to different therapeutic regimens [22,56]. In a study carried out in our laboratory on 44 infiltrating duct carcinoma patients, the TCho peak was either reduced or absent after the third and/or sixth cycle of neoadjuvant chemotherapy [22].

Future Directions and Conclusions

Results obtained to date indicate that the sensitivity of MRS may be limited by various technical factors rather than by intrinsic properties of the tumors. Therefore, any improvement in signal-to-noise ratio that will effectively enhance the detection of TCho resonance may increase the sensitivity and improve the diagnostic potential especially in relation to ^{1}H MRS. Moreover, the detection of choline signal can be increased by performing MRS at higher fields [24,55]. Use of special radio frequency pulse sequences, like simultaneous suppression of water and lipid signals that are specifically optimized for detection of the signal at 3.2 ppm, would also improve the sensitivity of detection of TCho. Use of respiratory-gated radio frequency sequences would improve motion related artifacts. In addition, advances in the design of MR coils should improve detection of TCho. Use of metabolic imaging will also allow exploration of tumor heterogeneity and characterization.

Currently, breast MRS is not routinely performed as part of a breast MRI examination in many centers, in part because it is technically challenging and time consuming. More research focusing on early diagnosis with improved sensitivity and specificity and evaluation of response to chemotherapeutic agents in breast cancer patients would be clinically useful. Presently, MRS and MRI are complementary tools to histology, mammography, and other accepted techniques. Increasing use of these methods is expected for basic research, clinical investigations, and ultimately for patient diagnosis. The sensitivity and specificity of *in vivo* MR particularly for small lesions needs to be improved before MRS can be incorporated into clinical practice.

Acknowledgments

The financial grant (SP/S0/B27/95 and SP/S0/B21/2001) to NRJ from the Department of Science and Technology, Government of India, is gratefully acknowledged. We wish to thank Prof. G. Govil for critical evaluation of the manuscript and for many useful suggestions. The authors thank Drs. Mahesh Kumar, V. Seenu, O. Coshic, S.N. Dwivedi, Profs. A. Srivastava, P. K. Julka, G.K. Rath, Dr. Stefan Roell, and Mr. Virendra Kumar for their support and help.

References

1. Greenlee RT, Hill-Harmon MB, Murray T, Thun M. CA Cancer J. Clin. 2001;51:15.
2. National Cancer for Health Statistics. SEER Cancer Statistics Review (1973–1995). Bethesda, MD: US National Cancer Institute, 1998.
3. National Cancer Registry Programme (NCRP): Consolidated Report (1990–1996). An Incidence and Distribution of Cancer. Indian Council of Medical Research, New Delhi, 2001.
4. Stavros AT, Thickman D, Rapp CL, Dennis MA, Parker SH, Sisney GA.Radiology. 1995;196:123.
5. Fenlon HM, Phelan N, Tierney S, Gorey T, Ennis JT. Clin. Radiol. 1998;53:17.
6. Birdwell R, Ikeda D, O'Shaughnessy KF, Sickles EA. Radiology. 2001;219:192.
7. Schnall MD, Orel SD. Magnetic Resonance Imaging Clinics of North America: Breast MR Imaging, Vol. 9. WB Saunders Co.: Philadelphia, 2001.
8. Orel SG, Schnall MD. Radiology. 2001;220:13.
9. Sinha S, Sinha U. Ann. N. Y. Acad. Sci. 2002;980:95.
10. Kneeshaw PJ, Turnbull LW, Drew PJ. Br. J. Cancer. 2003;13:4.
11. Schnall MD. Radiol. Clin. N. Am. 2003;41:43.
12. Sijens PE, Wijrdeman HK, Moerland MA, Bakker CGJ, Vermeulen JW, Luyten PR. Radiology. 1988;169:615.
13. Roebuck JR, Cecil KM, Schnall MD, Lenkinski RE. Radiology. 1998;209:269.
14. Jagannathan NR, Singh M, Govindaraju V, Raghunathan P, Coshic O, Julka PK, Rath GK. NMR Biomed. 1998;11:414.
15. Gribbestad IS, Singstad TE, Nilsen G, Fjosne HE, Engan T, Haugen OA, Rinck PA. J. Magn. Reson. Imaging. 1998;8:1191.
16. Kvistad KA, Bakken IJ, Gribbestad IS, Ehrnholm RJB, Steiner L, Fjosne HE, Haraldseth O. J. Magn. Reson. Imaging. 1999;10:159.
17. Jagannathan NR, Kumar M, Raghunathan P, Coshic O, Julka PK, Rath GK. Curr. Sci. 1999;76:777.
18. Cecil KM, Schnall MD, Siegelman ES, Lenkinski RE. Breast Cancer Res. Treat. 2001;68:45.
19. Yeung DKW, Cheung HS, Tse GMK. Radiology. 2001;220:40.
20. Lee JK, Tsai SC, Ho YJ, Chanclai SP, Kao CH. Anticancer Res. 2001;21:1481.
21. Bakken IJ, Gribbestad IS, Singstad TE, Kvistad KA. Magn. Reson. Med. 2001;46:189.
22. Jagannathan NR, Seenu V, Coshic O, Dwivedi SN, Julka PK, Srivastava A, Rath GK. Br. J. Cancer. 2001;84:1016.
23. Stanwell P, Gluch L, Clark D, Tomanek B, Baker L, Giuffre B, Lean C, Malycha P, Mountford CE. Eur. Radiol. 2005;15:1037.
24. Bolan PJ, Meisamy S, Baker EH, Lin J, Emory T, Nelson M, Everson LI, Yee D, Garwood M. Magn. Reson. Med. 2003;50:1134.
25. Jacobs MA, Barker PB, Bottomley PA, Bhujwalla Z, Bluemke DA. J. Magn. Reson. Imaging. 2004;19:68.
26. Leach MO, Varrill M, Glaholm J, Smith TAD, Collins DJ, Payne GS, Sharp JC, Ronen SM, McCready VR, Powles TJ, Smith IE. NMR Biomed. 1998;11:314.
27. Ronen SM, Leach MO. Breast Cancer Res. 2001;3:36.
28. Park JM, Park JH. Korean J. Radiol. 2001;2:80.
29. Degani H, Horowitz R, Itzchak Y. Radiology. 1986;161: 53.
30. Gribbestad IS, Fjosne HE, Haugen OA, Nilsen G, Krane J, Petersen SB, Kvinnsland S. Anticancer Res. 1993;13: 1973.
31. Gribbestad IS, Petersen SB, Fjosne HE, Kvinnsland S, Krane J. NMR Biomed. 1994;7:181.
32. Mackinnon WB, Barry PA, Malycha PL, Gillett DJ, Russell P, Lean CL, Doran ST, Barraclough BH, Bilous M, Mountford CE. Radiology. 1997;204:661.
33. Glunde K, Jie C, Bhujwalla ZM. Cancer Res. 2004;15:4270.
34. Ting YT, Sherr D, Degani H. Anticancer Res. 1996;16:1381.
35. Mountford CE, Somorjai RL, Malycha P, Gluch L, Lean C, Russell P, Barraclough B, Gillett D, Himmelreich U, Dolenko B, Nikulin AE, Smith ICP. Br. J. Surg. 2001;88:1234.
36. Bottomley PA. U.S. Patent 4. 1984;480:228.
37. Bendall MR, Gordon RE. J. Magn. Reson. 1983;53:365.
38. Ordidge RJ, Connelly A, Lohman JAB. J. Magn. Reson. 1986;66:283.
39. Frahm J, Merboldt KD, Hanicke W. J. Magn. Reson. 1987;72:502.

40. Frahm J, Bruhn H, Gyngell ML, Merboldt KD, Hanicke W, Sauter R. Magn. Reson. Med. 1989;9:79.
41. Orididge RJ, Bendall MR, Gordon RE, Connelly A. Volume selection for *in vivo* biological spectroscopy. In: G Govil, CL Khetrapal, A Saran, AS Tata (Eds). Magnetic Resonance in Biology and Medicine. McGraw Hill: New Delhi, 1985, p 387.
42. Brown TR, Kincaid BM, Ugurbil K. Proc. Natl. Sci. U.S.A. 1982;79:3523.
43. Maudsley AA, Hilal SK, Perman WH, Simon HE. J. Magn. Reson. 1983;51:147.
44. Haase A, Frahm J, Hanicke W, Matthei D. Phys. Med. Biol. 1985;30:341.
45. Twelves CJ, Porter DA, Lowry M, Dobbs NA, Graves PE, Smith MA, Rubens RD, Richards MA. Br. J. Cancer. 1994;69:1151.
46. Payne GS, Dowsett M, Leach MO. Breast. 1994;3:20.
47. Twelves CJ, Lowry M, Porter DA, Dobbs NA, Graves PE, Smith MA, Richards MA. Br. J. Radiol. 1994;67:36.
48. Ng TC, Grundfest S, Vijaykumar S, Baldwin NJ, Majors AW, Karalis I, Meaney TF, Shin KH, Thomas FJ, Tubbs R. Magn. Reson. Med. 1989;10:125.
49. Merchant TE, Gierke LW, Meneses P, Glonek T. Cancer Res. 1988;48:5112.
50. Merchant TE, Meneses P, Gierke LW, Otter WD, Glonek T. Br. J. Cancer. 1991;63:693.
51. Kalra R, Wade KE, Hands L, Styles P, Camplejohn R, Greenall M, Adams GE, Harris AL, Radda GK. Br. J. Cancer. 1993;67:1145.
52. Glaholm J, Leach MO, Collins DJ, Mansi J, Sharp JC, Madden A, Smith IE, Mc Cready VR. Lancet. 1989;1:1326.
53. Redmond OM, Stack JP, O'Connor NG, Carney DN, Dervan PA, Hurson BJ, Ennis JT. Magn. Reson. Med. 1992;25: 30.
54. Katz-Brull R, Lavin PT, Lenkinski RE. J. Natl. Cancer Inst. 2002;94:1197.
55. Roebuck JR, Lenkinski RE, Bolinger L, Schnall MD. Proc. Int. Soc. Magn. Reson. Med. 1996;4:1246.
56. Meisamy S, Bolan PJ, Baker EH, Bliss RL, Gulbahce E, Everson LI, Nelson MT, Emory TH, Tuttle TM, Yee D, Garwood M. Radiology. 2004;233:424.
57. Sijens PE, vander Bent MJ, Nowak PJ, van Dijk P, Oudkerk M. Magn. Reson. Med. 1997;37:222.
58. Bakker CJG, Vriend J. Phys. Med. Biol. 1983;28:331.
59. Jagannathan NR, Seenu V, Kumar M. Radiology. 2002; 223:281.
60. Mackinnon WB, Huschtscha L, Dent K, Hancock R, Paraskeva C, Mountford CE. Int. J. Cancer. 1994;59: 248.
61. Mescher M, Tannus A, Johnson MO'N, Garwood M. J. Magn. Reson. 1996;A123:226.
62. Gribbestad IS, Singstad TE, Kristofferson A, Kvistad KA, Johnson IT, Lundgren S, Roell S. Proc. Int. Soc. Magn. Reson. Med. 2004;11:2044.
63. Katz-Brull R, Margalit R, Bendel P, Degani H. MAGMA 1998;6:44.
64. Johnstone PA, Moore EM, Carrillo R, Goepfert CJ. Cancer. 2001;91:1075.
65. Wang J, Shih TT, Hsu JC, Li YW. Clin. Imaging. 2000;24: 96.
66. Thomas MA, Binesh N, Yue K, DeBruhl N. J. Magn. Reson. Imaging. 2001;14:181.

In Vivo Molecular MR Imaging: Potential and Limits

Mathias Hoehn and Uwe Himmelreich
In-vivo-NMR-Laboratory, Max-Planck-Institute for Neurological Research, Cologne, Germany

Introduction

The recent years have seen dramatic progress in molecular imaging across the board of various techniques. Molecular imaging, in the context of the authors' goals in this chapter, is termed as the non-invasive, *in vivo* visualization of cellular and molecular events in normal and pathological processes. In particular, optical and *in vivo* bioluminescence imaging (BLI) techniques [1,2] but also positron emission tomography (PET) [3,4] have set the pace. All these high-resolution imaging modalities are based on the same general idea that the presence of specific contrast agents, incorporated or intracellularly induced by cellular activities, will provide the necessary contrast of the cells or compartments of interest against the background of the host tissue [1,5].

For the purpose of specific contrast generation, either tracers or contrast agents are incorporated in cells or attached via selective antigen–antibody bindings. Alternatively, cells are, in close cooperation with molecular biologists, transfected with the goal of expressing selected markers. These markers are considered "reporter genes," as the genes responsible for the marker expression are coupled to other predefined genetic expression patterns. Thus, the reporter genes can be chosen to be expressed constitutively (i.e. permanently) or conditionally (i.e. dependent on specific gene activities, linked to particular cellular dynamics or functional activity).

Magnetic resonance imaging (MRI) has joined this group of imaging modalities with cellular/molecular focus only very recently. The reason for this "late appearance" as a player in the field of molecular imaging, despite its high spatial three-dimensional (3D) resolution, is based on the fact that labeling strategies of cells for this imaging technique are not as intuitively available and not directly accessible. For example, optical techniques exploit very successfully the existence of naturally occurring chromophores, genetic code of which is transfected in a selected group of cells. The most successful single chromophore to date is probably the green fluorescent protein "GFP," originally found in certain jellyfish.

For MRI applications of microscopic resolution and for the study of cellular dynamical processes, cells can be effectively loaded with iron oxide nanoparticles. This cell loading generates a pronounced contrast in T_2^*-weighted images. This approach is by now well established and has already led to a variety of applications in diverse medical but also in developmental biological areas. In the latter field, individual cells of an embryo have been labeled [6,7] and, consequently, the fate of the labeled cells during further cell division and selective cell migration was followed *in vivo*. In the medical arena, there are two major application fields: one dealing with the monitoring of macrophage activity during inflammatory processes in various organs including liver [8,9] as well as brain [10–12] and another field dealing with stem cell implantation in the hope of successful tissue regeneration [13–16] (Figure 1).

This chapter is not aiming to review the broad application field. For this purpose, the reader is referred to a couple of excellent recent reviews [17,18]. The basis for the success of these applications in animal models rests in all cases on fundamental considerations concerning the reliable detectability of the cells of interest and the monitoring of their fate in space and time. The present chapter will review the methodological and technical aspects of the ultrahigh-resolution MRI technique, and, most of all, the key tool of this strategy: cell labeling. The goal of the present chapter is to provide the reader with an insight into the methodological requirements for successful *Molecular MRI*, and to prepare an overview of the various strategies from which the reader is then free to determine an optimal strategy for particular applications.

Detectability

Intrinsic contrast is in most cases insufficient to monitor cells *in vivo*. Cell-specific contrast must be generated in order to distinguish between targeted cells and other tissue or cells. Generation of sufficient contrast depends on (a) the availability of high-affinity contrast agents, (b) the ability of these contrast agents to overcome biological delivery barriers (e.g. cell membranes; blood–brain barrier), (c) minimal disturbance of the cellular microenvironment (low toxicity), (d) amplification strategies that result in appropriate concentrations, and (e) the availability of sensitive, high-resolution imaging techniques. Cells can be labeled *in vivo* or by pre-labeling, *ex vivo*.

Paramagnetic MRI contrast agents modulate the spin–lattice relaxation time (T_1) of rapidly exchanging water

Graham A. Webb (ed.), Modern Magnetic Resonance, 1073–1084.

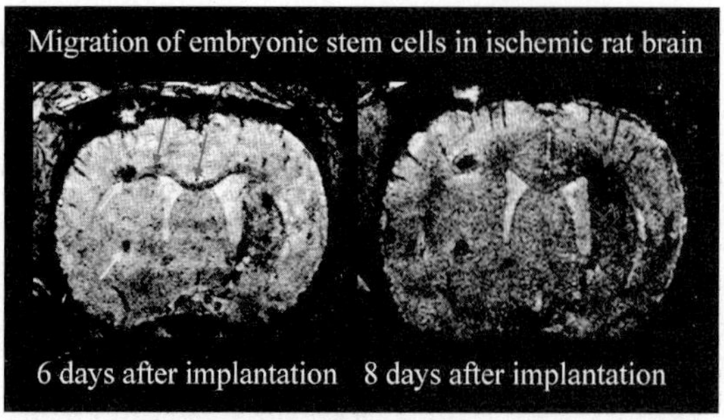

Fig. 1. Coronal sections through a rat brain at 6 and 8 days following implantation of embryonic stem cells into the left hemisphere, contralateral to the induced transient 60 min focal cerebral ischemia. 3D data sets were recorded with isotropic spatial resolution of 78 μm with a scan time of approximately 70 min. The primary implantation sites in the cortex and the striatum are indicated by white arrows. Note at 6 days, the discrete dark line (dark arrow) along the corpus callosum between the cortical implantation site and the ventricular wall showing cells migrating toward the lesioned hemisphere. At 8 days postimplantation, a dark region becomes visible in the dorsal part of the lesioned territory reflecting the first arrival of USPIO-labeled cells.

molecules and alter the local homogeneity of the magnetic field (T_2 and T_2^* relaxation times). Cells labeled with contrast agents generate *hypo*intense contrast in T_2-weighted (T_2w) images (labeled e.g. with iron oxide particles) or *hyper*intense contrast in T_1-weighted (T_1w) images (labeled e.g. with Mn or lanthanide chelates). These contrast agents most commonly consist of nanoparticles (iron oxide, T_2w), inorganic chelates (Gd, Eu, and Mn chelates; T_1w), or particles contained in a polymer sphere (iron oxide microspheres).

A precondition for a contrast agent is that the relaxivity of exchanging/surrounding protons is sufficiently modified. The large magnetic susceptibility of small paramagnetic particles can affect much larger regions of the image than is suggested by the actual size of the particle [19]. The limits of cell detectability are defined by the spatial resolution, the signal-to-noise ratio (SNR), and the contrast-to-noise ratio (CNR).

Contrast Agents for Molecular Imaging

The suitability of magnetic markers for molecular MR imaging is influenced by the change of the MR signal caused by alterations in the magnetic susceptibility and expressed by the relaxivity (R). Contrast agents for molecular imaging must provide a high relaxivity *in vivo* without compromising the cellular microenvironment. This sets certain limitations for the concentrations of contrast agents. The choice of contrast agents also depends on the intrinsic background signal of the respective application.

Dextran coated superparamagnetic iron oxide particles (SPIOs) [20] and ultrasmall SPIOs (USPIOs) have magnetic moments more than three orders of magnitude larger than those of clinically applied contrast agents [21]. Iron oxide particles can be used for T_1- [22], T_2- [12], and T_2^*- [23] weighted imaging. Due to the large magnetic moment of (U)SPIOs, the relaxivities R_2 and R_2^* are several thousand times larger than R_1 [24]. (U)SPIOs are well suited for cell tracking because of the long-range susceptibility effects. Relaxivity (R_2/R_2^*) of iron particles can easily be tested *in vitro* by comparison of signal quench in the presence of respective contrast agents. R_2^* relaxivity increases with increasing concentrations of the iron oxide particles but will be saturated (complete quench of signal) [24–26]. Saturation of magnetization occurs once the mass of iron per cell exceeds a certain threshold [24]. When normalized to iron content, larger particles greatly decrease T_2^* and enhance the susceptibility effect when compared to smaller particles [27].

The T_2^* effect of iron oxide based contrast agents is also larger than T_1 effects of Mn- and lanthanide-chelates because water molecules must be in close proximity of the paramagnetic center to generate T_1 contrast [28]. Crich *et al.* estimated the minimum detectable number of Gd-HPDO3A labeled cells in a phantom at approximately 10^3 cells [29]. Visualization is attainable when cellular concentrations of Gd-contrast agents are in the order of 10^{-4}–10^{-5} molar [30,31].

Magnetic properties that are most commonly modified for the design of contrast agents that predominantly increase R_1 relaxivity (hyperintensity in T_1w MR images) include:

(a) hydration number (q)—water bound to agent
(b) exchange rate of water molecules (τ_m)
(c) rotational correlation time of complex (τ_r).

In order to achieve sufficient contrast with T_1w contrast agents *in vivo*, accumulation of contrast agent to millimolar concentrations is usually required [32]. A system that entraps several units of a Gd chelate is apoferritin, resulting in an increase of relaxivity by almost 20 times compared to the free contrast agent [33,34]. Other approaches to concentrate Gd chelates at the site of interest include self-assembling aggregates [30,35,36]. It was suggested in recent studies that even picomolar binding is sufficient to achieve detectable CNR if high loads of particles were delivered to the cell [37].

The rotational correlation time of a contrast agent can be altered by changing the viscosity of the environment or by interaction with large molecules [38]. The latter has resulted in the design of contrast agents that specifically bind to certain proteins [39,40].

Resolution

Low image resolution can attenuate the relaxation effects through partial volume effects and thereby reduce detectability of labeled cells. The spatial resolution is limited by the strength of the gradients and the matrix size. The acquisition of MR images with a resolution in the order of tens of microns is currently possible within a reasonable time frame. The trade-off of high resolution is a reduced SNR.

A precondition for detection of single labeled cells by MRI is that the number of voxels is higher than the number of individual cells in the respective volume. Dodd *et al.* have demonstrated a good correlation of individual cell distribution between MRI and microscopic methods, proving the detectability of single cells by MRI [41]. Larger susceptibility effects, like for micron-size iron oxide particles, result in higher contrast and permit detection at lower resolution [27,42,43]. These microspheres can be considered a point source that creates a localized field larger than the field created by the evenly distributed, smaller (U)SPIOs, resulting in detectability of single cells with a spatial resolution as low as 200 μm [42].

Signal- and Contrast-to-Noise Ratio

The dilemma of cell tracking in animal models is that acquisition time is limited and resolution has to be as high as possible in order to meet the requirements set by working with anaesthetized animals and being able to detect small numbers of cells *in vivo*. This limits the options to maximize SNR and CNR. Apart from longer acquisition times, the SNR can be increased by stronger magnetic fields, smaller RF coils, and cryoprobes [44,45]. Pulse sequences, repetition time (T_R), echo time (T_E), and pulse length have to be chosen carefully and adjusted for each animal model. Optimization of MR parameters has been reviewed extensively [46,47]. The CNR is influenced by the efficiency of the labeling strategy. Visually apparent contrast is defined as CNR ≥ 5, with CNR $= |(I_A - I_B)N^{-1}|$ (I_A, I_B—signal intensities of the two adjacent regions, N—noise level). Using the relaxivity of the respective contrast agents and T_1/T_2—values of the tissue, acquisition parameters (T_R, T_E) can be optimized to maximize CNR [31,37].

It was demonstrated that detection of single cells *in vitro* is feasible [41,42,48].

Quantification

Cells can be quantified *in vitro* by the construction of simple gelatin phantoms [25,49] (Figure 2). If a sufficient spatial resolution is provided in 3D MR images and homogeneous phantoms are utilized individual cells labeled with iron oxide particles can be counted in high dilutions [49].

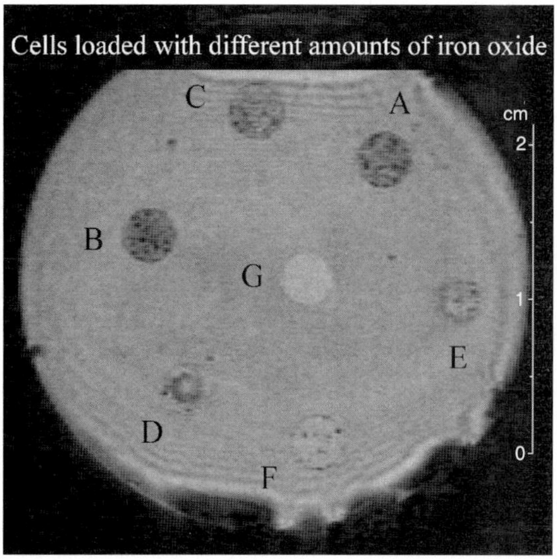

Fig. 2. Agarose gel phantom containing embryonic stem cells transfected with iron oxide particles. The holes A–E and G contain approximately 10^4 cells and the hole F contains 5×10^2 cells, which were suspended in 50 μl agarose. Cells were incubated with different concentrations of iron oxide nanoparticles prior to suspension (amount of iron oxide per 5×10^5 cells: A, 40 mg; B, 18 mg; C and F, 5 mg; D, 1 mg; E, 0.2 mg; and G, no iron oxide particles). The 3D gradient-echo MR image (FLASH) was acquired with the following parameters: field strength = 7 T, matrix = 256K × 128K × 64K data points; T_R = 250 ms, T_E = 20 ms, FOV = 4 cm × 4 cm).

Bowen *et al.* have addressed the quantification of iron oxide concentrations in cell suspensions based on relaxivities [24]. Relaxation rates of compartmentalized (U)SPIOs correlated with predicted values using the static dephasing theory [50]. Best correlations were achieved using (R_2^*–R_2) under the dilute pertuber assumption, indicating the suitability for quantification for cases where volume fractions of cells containing iron oxide are relatively small, as is the case, e.g. for implanted cells *in vivo* [24]. This model has only been tested *in vitro*. Experimentally, detection of cell clusters of tens to hundreds of (U)SPIO-labeled cells was demonstrated in animal models [14,51]. It was suggested that the utilization of micron-size iron oxide particles are suitable to detect single cells *in vivo* [52]. An additional advantage of these latter particles is that single particles or cells can be detected *in vitro* even with relatively low resolution (due to the large diameter and magnetic moment), and that the label is not diluted beyond the minimum observable threshold by cell division but remains within the cell for up to 1 year [27,43,52].

If the relaxivity of lanthanide contrast agents is known model-based predictions of local concentrations of T_1 contrast agents can be reliably performed by measurement of T_1 values [31,37]. If particle loading of cells and relaxivity of the contrast agent remain constant after implantation this could result in quantitative MRI in *in vivo* models. Contrary to these results, Crich *et al.* have found that cell pellets with variable Gd-chelate loading show a decrease of R_1 relaxivity if concentrations exceed 0.1 μmol/mg protein [29]. Possible explanations include competing T_2^* effects, resulting in signal broadening.

Cell Labeling

Route of Uptake

Conventional MR cell labeling relies on surface attachment of contrast agents to the cell [53–56]. Contrast agents are often chemically modified so that they bind unspecifically or to specific cell types. A common strategy involves linking a contrast agent with a monoclonal antibody that is recognized by receptors on the cell surface [55]. For review see Ref. [30]. These strategies are generally not suitable for *in vivo* applications, where interactions with other cells have to be avoided, as for stem cell tracking. Extracellular labeling may result in detachment of the label or transfer to other cells. We will therefore focus on strategies to internalize contrast agents.

Probably the most effective route to internalize compounds into cells is by endocytosis. Hereby, intracellular vesicles (endosomes) are formed after introflection of portions of the cell membrane and fusion of its extremities. The endosomes contain compounds that are either bound to the membrane or present in the extracellular medium. Endocytosis is most effective in phagocytic cells (also called phagocytosis). Paramagnetic iron oxide particles but also Gd chelates have been successfully internalized into macrophages, monocytes, mononuclear T-cells, and oligodendrocyte progenitors [13,26,29,57–61]. In order to internalize high quantities of a contrast agent, receptor-mediated endocytosis is sometimes the method of choice. Hereby, the substrate in the extracellular medium interacts with receptor molecules on the membrane (like contrast agents linked with a monoclonal antibody) and are internalized in high concentrations [26,62].

Hinds and co-workers [27,43] have demonstrated that even micron-scale particles (0.9–2.8 μm) of an aggregate of iron oxide magnetite that is encapsulated in a polymer, are incorporated into perinuclear endosomes by stem and progenitor cells [48]. This labeling strategy has been highly efficient (almost 100%). It was also found that concentrating (U)SPIOs into cells increases the relative R_2^* sensitivity, compared to uniformly distributed particles [26,63].

In order to increase the uptake of contrast agents that do not spontaneously accumulate in endosomes, magnetoliposomes [64], and lectins [65] have been used in several experiments. A method developed in molecular biology for transfection of oligonucleotides, lipofection has been successfully adopted for labeling of cells with contrast agents [14]. Commercially available polycationic transfection agents encapsulate (U)SPIOs through electrostatic interaction [66]. The transfection agents shuttle the (U)SPIOs into the cell and form endosomes that finally merge into liposomes. This procedure has successfully been applied for labeling T-cells, stem cells, and other mammalian cells [14,66–69]. A similar approach is the labeling of cells with magnetodendrimers, which consist of (U)SPIOs coated with carboxylated dendrimers [51]. These compounds are also used as transfection agents and result in the formation of endosomes [70]. A broad range of cells has been labeled with dendrimers [15,51,71,72].

Others have produced transgenic cells expressing, for example, the transferrin receptor and are thus susceptible to the uptake of conjugates of transferrin and monocrystalline iron oxide nanoparticles [73,74].

Many larger probes are membrane impermeable and fail to attain intracellular concentrations required for MRI. Peptides or proteins mediating membrane translocation signals (MTS) have been coupled with contrast agents for internalization. MTS from the HIV tat peptide have been linked to SPIOs and lanthanide chelates for transport into several cell types [75–78]. Hereby, the contrast agents most likely accumulate in the cytoplasm [77]. Allen and Meade have suggested a relatively simple system for internalization that consists of a polyarginine-modified Gd DOTA system [79].

Microinjection of contrast agents was used in embryology for cell tracking [6,7]. Increased contrast was detectable in the developed embryo compared to controls, indicating sufficient contrast with cell division. However, this method has only applications limited to injection of large single cells in developing embryos. It is not practical for tracking populations of cells due to its labor-intensive preparation and the dilution of the label with cell division [7].

Zhang *et al.* adopted another method from molecular biology, where so-called biolistic "gene-guns" are used to introduce oligonucleotides into cells [80,81]. Biolistic labeling was performed with ferromagnetic particles and resulted in a labeling efficiency of 95% of surviving cells. Similarly to microinjection, cell labeling remained stable.

Electroporation is a process that permeabilizes cell membranes by exposure to an external electric field. Rapid resealing of the membrane pores traps internalized compounds inside the cell. Widely used to introduce foreign DNA into cells, it has also been used for drug delivery [82]. Paturneau-Jouas *et al.* have used electroporation for the uptake of low-molecular weight Gd chelates *in vivo* [83]. It remains to be seen how efficient this labeling method is and how stressful it proves for the respective cells.

In vivo and *ex vivo* Labeling

In vivo labeling of cells is possible if phagocytes take up contrast agents that were administered either intravenously or after injection in the targeted organ. This labeling strategy is based on naturally occurring or targeted, mediated endocytosis (mainly macrophages). Hereby, macrophages accumulate the iron oxide particles from the blood, whereas the remaining iron oxide in the blood (or other organs) is washed out. It was first introduced for hepatic imaging [84]. Subsequent development of clinically approved USPIOs have resulted in labeling strategies to distinguish between normal tissue and tumor metastases in lymph nodes [85] and bone marrow [86] based on the uptake of these particles by macrophages in those organs.

The uptake of iron oxide by macrophages has also been utilized in some disease models to highlight increased macrophage activity in areas of inflammation. Such applications include visualization of atherosclerotic plaques [87], visualization of increased macrophage activity in the kidney after injury and organ rejection [88,89], and macrophage infiltration in brain lesions following ischemia [10,90].

Direct injection of micron-size iron oxide particles into various regions of the sub-ventricular zone of the rat brain resulted in *in vivo* labeling of neuronal stem cells by endocytosis, allowing their migration to be tracked along the rostral migratory stream to the olfactory bulb [52]. Coupling of MTS peptides with nanoparticles and antibody or receptor binding ligands has been suggested as a possible targeted *in vivo* labeling strategy [91]. Paturneau-Jouas *et al.* have used electroporation after intraperitoneal injection of Gd chelates for internalization into muscle cells *in vivo* [83].

So far, targeted *in vivo* labeling was mainly demonstrated for phagocytes. In all other cases, pre-labeling of cells according to the procedures listed above is necessary. Hereby, *in vitro* pre-labeled cells are transferred to the targeted organ or tissue by injection.

Responsive Contrast Agents

Contrast agents have been developed with the aim of reporting on the physiological status and metabolic activity of cells [30,92]. Most of these contrast agents are Gd based chelates with one or more potential coordination sites blocked in the inactive state, preventing free access of exchanging water molecules to the paramagnetic ion.

Lowe *et al.* synthesized an MR contrast agent that is responsive to changes in pH, generating no contrast at high pH but high contrast at low pH [93]. The Gd DO3A derived agent contains a sulfonamide nitrogen, which is protonated at low pH. Therefore, it cannot coordinate at low pH and provides free access to exchanging water molecules. Meade and co-workers developed a probe that is responsive to intracellular changes in calcium concentrations [94,95]. Similarly, two carboxyl groups of the chelate coordinate either to the Gd^{3+} ion (inactive) or to the Ca^{2+} ion (active). Probes responsive to enzyme activity (e.g. β-galactosidase) have been developed [96] and applied for cell visualization *in vivo* [6]. Contrast agents responsive to pO_2 were developed on the basis of the BOLD effect [97] and coupling an increase in τ_r with and redox switch [98].

Other contrast agents utilize the so-called chemical exchange saturation transfer (CEST) agents to be responsive to variable concentrations of targeted molecules [99,100]. Aime *et al.* have suggested an insoluble contrast agent that is activated by intracellular enzymatic solubilization [30]. However, the effective incorporation of the contrast agent in its insoluble state is still an unresolved issue.

In vivo MRI Experiments

Cell tracking *in vivo* requires the acquisition of isotropic 3D MR images of living animals at near-cellular resolution in order to minimize partial volume effects. Temporal constraints are set by working with live animals so that spatial resolution is typically in the range of tens up to a hundred of microns. The principle challenge is to obtain sufficient SNR. Hardware improvements often target the reduction of thermal noise voltages from the sample by using small surface RF coils [101]. Design of the receiver coil aims for low coil resistance. This can be achieved by cryo-cooled probes [102], which have been applied to micro-imaging in animal models [44,45]. RF coils made of high temperature superconducting material have also been demonstrated to reduce coil loss substantially *in vitro* and in animal models [103,104].

Animal models require usually anaesthetized animals. Apart from the obvious associated disadvantages, it has the advantage of decreasing movement artifacts in MR images as the animal can be firmly fixed in an animal holder.

MRI of animal models is usually performed at a high field strength (>4 T). While this is an advantage for improving the SNR it has the disadvantage of an increase in the T_1/T_2 ratio, resulting in an increase in the acquisition time and in a loss of intrinsic contrast in the brain. The increase in the magnetic field strength, compared to clinical systems, also results in more pronounced magnetic susceptibility effects. While this can lead to unwanted artifacts in the MR images, it has also the advantage that magnetically labeled cells can be more easily visualized. Gradient-echo sequences tend to be more susceptible to those artifacts than spin-echo experiments.

Data acquisition methods utilized for cell tracking are similar to those employed for clinical MRI (spin-echo, gradient-echo, diffusion-weighted imaging, inversion recovery, and magnetization transfer sequences). Rapid acquisition methods are commonly used for animal models, namely gradient-echo fast-low-angle-shot (FLASH) [105] and echo planar imaging (EPI) [106]. The utilization of MR microscopy for animal models was reviewed by Ahrens *et al.* [107].

Pulse sequences that emphasize signal loss due to field inhomogeneities are beneficial for (U)SPIOs. Cell tracking in animals is usually performed using 3D gradient-echo sequences (FLASH) for T_2/T_2^* effects [14,51,62] and two-dimensional (2D) multi-slice or 3D spin-echo sequences for T_1 effects [6,29,57]. Unless very strong T_1 weighting is used with T_Rs of approximately 100 ms, spin-echo sequences are not fast enough to permit the acquisition of a high-resolution 3D image within the time frame required for anaesthetized animals. 2D multi-slice images have the disadvantage of strong partial volume effects. A resolution of 100 to 200 μm is usually sufficient for the visualization of small cell clusters. The detection of single cells is not realistic as naturally occurring movements like breathing and muscle twitching are in the order of hundreds of microns, resulting in blurring effects. Additional blurring effects are caused by diffusion [41]. The acquisition of 3D data sets is, at any rate, a requirement for the co-registration of data sets in longitudinal studies and for the co-registration with independently recorded, non-MR data sets.

The effects of gradient strength, T_R and T_E to optimize SNR and CNR in MRI have been studied extensively [41,46,47]. For conditions of large magnetic moments due to (U)SPIOs loaded cells in animal models, the greatest sensitivity was observed with R_2^*, rather than R_2 or R_1[24]. Low gradient strengths through the use of low acquisition bandwidths will enhance susceptibility effects. This is also beneficial in terms of SNR, due to reduced diffusion loss.

An intrinsic problem for tracking iron oxide labeled cells in animal models are other areas of susceptibility induced field inhomogeneities (signal loss in T_2/T_2^*w images), namely from airspaces, but more importantly, from blood clots/hemorrhages and tissue interfaces. Seppenwoolde *et al.* have developed a technique that exploits the natural dipolar field surrounding paramagnetic markers by introducing a dephasing gradient on the slice-select axis to spoil signal across all the sample except in regions where the local gradients surrounding the paramagnetic marker are of the right amplitude and orientation to combine constructively with the added dephasing gradient to result in refocusing of the lost signal [108]. Coristine *et al.* have applied this "white-marker" technique to distinguish iron oxide labeled cells from other sources of field inhomogeneities [109].

The optimal acquisition parameters need to be determined for each combination of cell type, animal model, and contrast agent.

Biological Aspects of Cell Labeling

Reproducible and efficient visualization of cells with MRI does not only require optimized MR properties of contrast agents but also biological properties that result in stable labels without affecting other cellular processes.

Toxicity

Many contrast agents are administered in coated (polymer coating) or chelated (in particular the lanthanides) form, in order to reduce toxicity. Commercially available and clinically tested contrast agents are often used for molecular imaging, for which short and medium term toxicity is minimal. However, one has to consider that contrast agents for molecular imaging are often applied for longitudinal labeling, and in concentrations much higher than clinically applied. Immediate toxic effects can easily be tested by growth and viability studies in cell suspensions. It is more difficult to assess long-term stability (and toxicity) of coated contrast agents and the effect of contrast agents on cellular processes like differentiation.

It has been reported that very high concentrations of (U)SPIOs for cell labeling (up to 2 mg iron/ml) can cause decreased proliferation and even cell death [110]. This is in particular true after the formation of reactive oxygen species [111]. Some dextran coatings for iron oxide particles are biodegradable [112]. Released iron enters into the cellular iron metabolic pathways. Iron in high concentrations is known to interact with a number of other metabolic pathways [113,114]. Overloading can result in increased oxidative stress. Although iron oxide can be released by biodegradation, one has to consider that this will occur gradually resulting in non-toxic free iron concentrations. Stroh *et al.* have found that decomposition of the coating of iron oxide particles does not affect cell growth but causes oxidative stress [113]. These authors have suggested the application of antioxidants or iron chelators to prevent toxic effects of leaking iron.

T-cells labeled with SPIOs by endocytosis retain their function [58,115]. Sapiro *et al.* have shown in longitudinal studies that even large iron oxide loaded microspheres do not influence cell proliferation and differentiation at low concentrations [48,52]. Hill *et al.* found a dose-dependent alteration in cell proliferation but not in differentiation for mesenchymal stem cells [43]. Also, magnetodendrimers and (U)SPIOs did not cause differences between labeled and unlabeled cells in terms of proliferation, viability, and differentiation of neural stem cells in short term *in vitro* experiments [14,51,71]. Frank and co-workers have studied the effect of superparamagnetic iron oxide nanoparticles and transfection agents extensively [67–69,116]. No long-term effect on cell viability, rate of apoptosis, phenotype, proliferation, activation, and differentiation was found.

The toxicity of lanthanide contrast agents has also been studied extensively [38]. As lanthanides are known to be highly toxic, stable chelates are required for protection. Gd-HPDO3A is one of the best tolerated contrast agents with an $LD_{50} > 10$ mmol/kg in rats [38]. Although well tolerated in clinical applications, high dose toxicity for molecular imaging application requires further investigations. Several studies have confirmed that Gd-dextrans remain intact in cells [29] and are tolerated by cells in culture and *in vivo* after implantation [13,29,57].

In addition to metal ions, the interactions of chelates, transfection agents, or antibodies have to be considered for the assessment of toxicity. Although concern has been raised about potential release of reactive iron species from tat-labeled cells [67], functional tests with and without HIV tat peptide based contrast agents have shown no difference between labeled and unlabeled cells [77]. When used alone, some transfection agents are reported to be toxic [116]. However, iron oxide complexed with transfection agents does not affect viability, differentiation, or proliferation [68,69,116]. It was found in one study that the differentiation of stem cells can be influenced by transfecting agents [117]. This highlights the need to study possible biological side effects for each labeling system and cell type.

In addition to toxicological aspects of the contrast agent, one has to consider that labeling procedures like electroporation, microinjection, or biolistic "gene-guns" can cause mechanical damage to cells.

Assessment and Stability of Label

It is crucial for *in vivo* cell tracking to estimate the efficiency of cell labeling, the internalization of the label, and for how long these cells are detectable (i.e. dilution of the label; Figure 3). Various approaches have been taken to assess *in vitro* labeled cells by using agar or gelatin based phantoms (Figure 2).

It was found that uptake of (U)SPIOs as well as Gd chelates by endocytosis is linear with dose [24,29]. Linearity was also shown for (U)SPIO uptake with incubation time [24,118]. Interestingly, uptake is increased with particle size [118], which also explains the excellent uptake of micron-size microspheres [27,48].

The labeling efficiency can be assessed by NMR relaxometry [29,30,7,78] or by other techniques like atomic absorption spectroscopy to determine the iron content of USPIO-labeled cells [14]. Internalization of contrast agents (Figure 3) can also be verified by staining techniques in combination with (electron) microscopy.

The visibility of cells with time depends primarily on the dilution of the contrast agent due to cell proliferation and on degradation processes of the coating of contrast agents. Stem cells labeled with iron oxide particles have been monitored *in vivo* for up to 7 weeks in rat brains [14] and for up to 3 months in a spinal cord implantation model [72]. Embryonic stem cells were detectable, with contrast condition comparable to the originally labeled culture, after 20 cycles of cell division, which is an increase of cell numbers by a factor of 10^6 [14,25,49]. In another study, the complete disappearance of iron particles has been observed due to dilution after proliferation within 2–3 weeks [69].

An advantage of large micron-size particle is that a single particle per cell may result in sufficient contrast. Hill *et al.* were able to detect mesenchymal stem cells for up to 3 months [43]. Shapiro and co-workers monitored microsphere labeled cells for 1 year [52].

Cell labeling with lanthanide chelates is more affected by dilution after cell division due to its lower sensitivity. Modo and co-workers were able to visualize stem cells for up to 7 days after labeling with a Gd chelate [13].

Histological evaluation at the endpoint of longitudinal MR studies is necessary to evaluate MR tracking of cells *in vivo*. Other causes for loss of label are exocytosis (Figure 3) and immune reaction with a subsequent transfer of contrast agents to macrophages. Co-location can be estimated by co-registration of MR images with stained histological sections. Due to the invasiveness of histological methods, co-registration with MR images will always be problematic due to swelling, distortion, etc. Possible approaches include corroboration of MRI detection of labeled cells by using Prussian blue staining against iron [62], validation with X-gal staining [51], or electron-dense inclusions for visualization by electron microscopy [61]. In addition, dual staining detects co-localization of contrast agent with certain cell types. An example is the Prussian blue and $ED1^+$ staining to discriminate USPIO-labeled macrophages [10,119].

The use of bifunctional contrast agents allows co-registration of histology and MRI without recourse to other methods. Dual labeling strategies can be classified in methods that use dual contrast agents and methods that utilize cells that can be visualized by specific modalities (e.g. cells that express certain proteins). Modo labeled stem cells with a Gd-Rhodamine dextran (GRID) that is

They are often used for the encapsulation of smaller molecules.

Endocytosis—In endocytosis, the cell engulfs some of its extracellular medium including material dissolved or suspended in it. A portion of the plasma membrane is invaginated and pinched off forming a membrane-bound vesicle called an endosome.

Endosome—Vesicle formed during endocytosis.

Molecular imaging—Non-invasive, *in vivo* visualization of cellular and molecular events in normal and pathological processes.

Stem cell—Cell with broad differentiation potential that retains the capacity for unlimited self-renewal. A totipotent stem cell has the ability to differentiate to all cell types of an organism, whereas a pluripotent stem cell produces many but not all cell types.

Transfection—A method by which experimental DNA (or in our case contrast agents) can be incorporated into mammalian cells after interaction of the encapsulating transfection agent and the cell membrane.

References

1. Weissleder R, Mahmood U. Radiology. 2001;219:316.
2. Edinger M, Cao YA, Hornig YS, Jenkins DE, Verneris MR, Bachmann MH, Negrin RS, Contag CH. Eur. J. Cancer. 2002;16:2128.
3. Blasberg R, Tjuvajev JG. Mol. Imaging. 2002;1:160.
4. Gambhir SS. Nat. Rev. Cancer. 2002;9:683.
5. Weissleder R, Ntziachristos V. Nat. Med. 2003;1:123.
6. Louie AY, Hüber MM, Ahrens ET, Rothbächer U, Moats R, Jacobs RE, Fraser SE, Meade RJ. Nat. Biotechnol. 2000;18:321.
7. Jacobs RE, Fraser SE. Science. 1994;263:681.
8. Weissleder R. Radiology. 1994;193:593.
9. Stark DD, Weissleder R, Elizondo G, Hahn PF, Saini S, Todd LE, Wittenberg J, Ferrucci JT. Radiology. 1988;168:297.
10. Saleh A, Wiedermann D, Schroeter M, Jonkmanns C, Jander S, Hoehn M. NMR Biomed. 2004;17:163.
11. Saleh A, Schroeter M, Jonkmanns C, Hartung HP, Mödder U, Jander S. Brain. 2004;127:1670.
12. Dousset V, Delalande C, Ballarino L, Quesson B, Seilhan D, Coussemacq M, Thiaudiere E, Brochet B, Canioni P, Caille JM. Magn. Reson. Med. 1999;41:329.
13. Modo M, Cash D, Mellodew K, Williams SCR, Fraser SE, Meade TJ, Price J, Hodges H. Neuroimage. 2002;17: 803.
14. Hoehn M, Küstermann E, Blunk J, Wiedermann D, Trapp T, Wecker S, Föcking M, Arnold H, Heschler J, Fleischmann BK, Schwindt C, Bührle C. Proc. Natl. Acad. Sci. U.S.A. 2002;99:16267.
15. Bulte JWM, Ben-Hur T, Miller BR, Mizrachi-Kol R, Einstein O, Reinhartz E, Zywicke HA, Douglas TD, Frank JA. Magn. Reson. Med. 2003;50:201.
16. Veizovic T, Beech JS, Stroemer RP, Watson WP, Hodges H. Stroke. 2001;32:1012.
17. Bulte JWM, Modo M, Hoehn M. Mol. Imaging. 2005;4:143.
18. Bulte JWM, Duncan ID, Frank JA, Cereb. J. Blood Flow Metab. 2002;22:899.
19. Lauterbur PCM, Bernardo ML Jr, Mendonca Dias MH, Hedges LK. Proc. Int. Soc. Magn. Reson. Med. 1986;5:229.
20. Fahlvik AK, Klaveness J, Stark DD. J. Magn. Reson. Imaging. 1993;3:187.
21. Bulte JWM, Brooks RA, Moskowitz BM, Bryant LH Jr, Frank JA. Magn. Reson. Med. 1999;42:379.
22. Moore A, Marecos E, Bogdanov A Jr, Weissleder R. Radiology. 2000;214:568.
23. Furman-Haran E, Margalit R, Grobgeld D, Degani H. J. Magn. Reson. Imaging. 1998;8:634.
24. Bowen CV, Zhang X, Saab G, Gareau PJ, Rutt BK. Magn. Reson. Med. 2002;48:52.
25. Küstermann E, Wiedermann D, Bührle C, Arnold H, Föcking M, Trapp T, Hoehn M. Proc. Int. Soc. Magn. Reson. Med. 2002;10:1255.
26. Weissleder R, Cheng HC, Bogdanowa A, Bogdanow A Jr. J. Magn. Reson. Imaging. 1997;7:258.
27. Hinds KA, Hill JM, Shapiro EM, Laukkanen MO, Silva AC, Combs CA, Varney TR, Balaban RS, Koretsky AP, Dunbar CE. Blood. 2003;102:867.
28. Lauffer RB. Chem. Rev. 1987;87:901.
29. Crich SG, Biancone L, Cantaluppi V, Duo D, Esposito G, Russo S, Camussi G, Aime S. Magn. Reson. Med. 2004;51:938.
30. Aime S, Cabella C, Colombatto S, Crich SG, Gianolio E, Maggioni F. J. Magn. Reson. Imaging. 2002;16:394.
31. Ahrens ET, Rothbächer U, Jacobs RE, Fraser SE. Proc. Natl. Acad. Sci. U.S.A. 1998;95:8443.
32. Wickline SA, Lanza GM. J. Cell Biochem. 2002;39:90.
33. Harrison PM, Arosio P. Biochim. Biophys. Acta. 1996;1275:161.
34. Aime S, Crich SG, Frullano L. Angew. Chem. Int. Ed. Engl. 2002;114:1059.
35. Aime S, Botta M, Fedeli F, Gianolio E, Terreno E, Anelli PL. Chem. Eur. J. 2001;7:5261.
36. Bogdanov A, Matuszewski L, Bremer C, Petrovsky A, Weissleder R. Mol. Imaging. 2002;1:16.
37. Morawski AM, Winter PM, Crowder KC, Caruthers SD, Fuhrhop RW, Scott MJ, Robertson JD, Abendschein DR, Lanza GM, Wickline SA. Magn. Reson. Med. 2004;51: 480.
38. Caravan P, Ellison JJ, McMurry TJ, Lauffer RB. Chem. Rev. 1999;99:2293.
39. Nivorozhkin AL, Kolodziej AF, Caravan P, Greenfield MT, Lauffer RB, McMurry TJ. Angew. Chem. Int. Ed. Engl. 2001;40:2903.
40. De Leon-Rodriguez LM, Ortiz A, Weiner AL, Zhang S, Kovacs Z, Kodadek T, Sherry T. J. Am. Chem. Soc. 2002;124:3514.
41. Dodd SJ, Williams M, Suhan JP, Williams DS, Koretsky AP, Ho C. Biophys. J. 1999;76:103.
42. Sapiro EM, Skrtic S, Hill JM, Dumbar CE, Koretsky AP. Proc. Int. Soc. Magn. Reson. Med. 2003;11:229.
43. Hill JM, Dick AJ, Raman VK, Thompson RB, Yu ZX, Hinds KA, Pessanha BSS, Guttman MA, Varney TR, Martin BJ, Dunbar CE, McVeigh ER, Lederman RJ. Circulation. 2003;108:1009.
44. Wecker S, Küstermann E, Radermacher B, Hoehn M. Proc. Int. Soc. Magn. Reson. Med. 2002;10:878.

45. Wright AC, Song HK, Wehrli FW. Magn. Reson. Med. 2000;43:163–169.
46. Vinitski S, Griffey R, Fuka M, Matwiyoff N, Prost R. Magn. Reson. Med. 1987;5:278.
47. Haacke EM. Magn. Reson. Med. 1987;4:407.
48. Shapiro EM, Koretsky AP. Proc. Int. Soc. Magn. Reson. Med. 2004;12:1734.
49. Wiedermann D, Küstermann E, Blunk J, Wecker S, Bührlez C, Schwindt W, Trapp T, Föcking M, Hescheler J, Hoehn M. Proc. Int. Soc. Magn. Reson. Med. 2002;10:1257.
50. Yablonskiy DA, Haacke EM. Magn. Reson. Med. 1994;32:749.
51. Bulte JWM, Douglas T, Witwer B, Zhang SC, Strable E, Lewis BK, Zywicke H, Miller B, van Gelderen P, Moskowitz BM, Duncan ID, Frank JA. Nat. Biotechnol. 2001;19: 1141.
52. Shapiro EM, Skrtic S, Koretsky AP. Proc. Int. Soc. Magn. Reson. Med. 2004;12:166.
53. Safarik I, Safarikova M. J. Chromatogr. 1999;B722:33.
54. Högemann D, Basilion JP. Eur. J. Mol. Med. 2002;29:400.
55. Sipkins DA, Cheresh DR, Kazemi MR, Nevin LM, Bednarski MD, Li KCP. Nat. Med. 1998;4:623.
56. Lemieux GA, Yarema KJ, Jacobs CL, Bertozzi CR. J. Am. Chem. Soc. 1999;121:4278.
57. Modo M, Mellodew K, Cash D, Fraser SE, Meade TJ, Price J, Williams SCR. Neuroimage. 2004;21:311.
58. Yeh TC, Zhang W, Ildstad ST, Ho C. Magn. Reson. Med. 1993;30:617.
59. Sipe JC, Filippi M, Martino G, Furlan R, Rocca MA, Rovaris M, Bergami A, Zyrozz J, Scotti G, Comi G. Magn. Reson. Imaging. 1999;17:1521.
60. Moore A, Weissleder R, Bogdanov A Jr. J. Magn. Reson. Imaging. 1997;7:1140.
61. Franklin RJM, Blaschuk KL, Mearchell MC, Prestoz LLC, Setzu A, Brindle KM, ffrench-Constant C. Neuroreport. 1999;10:3961.
62. Bulte JWM, Zhang SC, van Gelderen P, Heryneki V, Jordan EK, Duncan ID, Frank JA. Proc. Natl. Acad. Sic. U.S.A. 1999;96:15256.
63. Majumdar S, Zoghbi SS, Gore JC. Magn. Reson. Med. 1989;10:289.
64. Bulte JWM, Ma LD, Magin RL, Kamman RL, Hulstaert CE, Go KG, The TH, de Leij L. Magn. Reson. Med. 1993;29:32.
65. Bulte JWM, Laughlin PG, Jordan EK, Tran VA, Vymazal J, Frank JA. Acad. Radiol. 1996;3:S301.
66. Kalish H, Arbab AS, Miller BR, Lewis BK, Zywicke HA, Bulte JWM, Bryant LH, Frank JA. Magn. Reson. Med. 2003;50:275.
67. Frank JA, Miller BR, Arbab AS, Zywicke HA, Jordan EK, Lewis BK, Bryant LH Jr, Bulte JWM. Radiology. 2003;228:480.
68. Arbab AS, Yocum GT, Kalish H, Jordan EK, Anderson SA, Khakoo AY, Read EJ, Frank JA. Blood. 2004.
69. Arbab AS, Bashaw LA, Miller BR, Jordan EK, Lewis BK, Kalish H, Frank JA. Radiology. 2003;229:838.
70. Zhang ZY, Smith BD. Bioconjug. Chem. 2000;11:805.
71. Walter GA, Cahill KS, Huard J, Feng H, Douglas T, Sweeney HL, Bulte JWM. Magn. Reson. Med. 2004;51:273.
72. Lee IH, Bulte JWM, Schweinhardt P, Douglas T, Trifunovski A, Hofstetter C, Olson L, Spengera C. Exp. Neurol. 2004;187:509.
73. Moore A, Josephson L, Bhorade RM, Basilion JP, Weissleder R. Radiology. 2001;221:244.
74. Weissleder R, Moore A, Mahmood U, Bhorade R, Benveniste H, Chiocca EA, Basilion JP. Nat. Med. 2000;6:351.
75. Bhorade R, Weissleder R, Nakakoshi T, Moore A, Tung CH. Bioconjug. Chem. 2000;11:301.
76. Lewin M, Carlesso N, Tung CH, Tang XW, Cory D, Scadden DT, Weissleder R. Nat. Biotechnol. 2000;18:410.
77. Dodd CH, Hsu HC, Chu WJ, Yang P, Zhang HG, Mountz JD Jr, Zinn K, Forder J, Josephson L, Weissleder R, Mountz JM, Mountz JD. J. Immunol. Methods. 2001;256:89.
78. Josephson L, Tung CH, Moore A, Weissleder R. Bioconjug. Chem. 1999;10:186.
79. Allen MJ, Meade TJ. J. Biol. Inorg. Chem. 2003;8:746.
80. Zhang RL, Zhang L, Zhang ZG, Morris D, Jiang Q, Wang L, Zhang LJ, Chopp M. Neuroscience. 2003;116:373.
81. Zhang ZG, Jiang Q, Zhang R, Zhang L, Wang L, Zhang L, Arniego P, Ho KL, Chopp M. Ann. Neurol. 2003; 53:259.
82. Rols MP, Delteil C, Golzio M, Dumond P, Cros S, Teissie J. Nat. Biotechnol. 1998;16:168.
83. Paturneau-Jouas M, Parzy E, Vidal G, Carlier PG, Wary C, Vilquin JT, de Kerviler E, Schwartz K, Leroy-Willig A. Radiology. 2003;228:768.
84. Stark DD, Weissleder R, Elizondo G, Hahn PF, Saini S, Todd LE, Wittenberg J, Ferrucci JT. Radiology. 1988; 168:297.
85. Weissleder R, Heautot JF, Schaffer BK, Nossiff N, Papisov MI, Bogdanov A Jr, Brady TJ. Radiology. 1994;191:225.
86. Seneterre E, Weissleder P, Jaramillo D, Reimer P, Lee AS, Brady TJ, Wittenberg J. Radiology. 1991;179:529.
87. Ruehm SG, Corot C, Vogt P, Kolb S, Debatin JF. Circulation. 2001;103:415.
88. Beckmann N, Cannet C, Fringeli-Tanner M, Baumann D, Pally C, Bruns C, Zerwes HG, Andriambeloson E, Bigaud M. Magn. Reson. Med. 2003;49:459.
89. Zhang Y, Dodd SJ, Hendrich KS, Williams M, Ho C. Kidney Int. 2000;58:1300.
90. Rausch M, Baumann D, Neubacher U, Rudin M. NMR Biomed. 2002;15:278.
91. Wunderbaldinger P, Josephson L, Weissleder R. Bioconjug. Chem. 2002;13:264.
92. Meade TJ, Taylor AK, Bull S. Curr. Opin. Neurobiol. 2003;13:597.
93. Lowe MP, Parker D, Reany O, Aime S, Botta M, Castellano G, Gianolio E, Pagliarin R. J. Am. Chem. Soc. 2001;123:7601.
94. Li WH, Parigi G, Fragai M, Luchinat C, Meade TJ. Inorg. Chem. 2002;41:4018.
95. Li WH, Fraser SE, Meade TJ. J. Am. Chem. Soc. 1999;121:1413.
96. Moats RA, Fraser SE, Meade TJ. Angew. Chem. Int. Ed. Engl. 1997;36:726.
97. Burai L, Scopelliti R, Toth E. Chem. Commun. 2002;20:2366.
98. Aime S, Botta M, Gianolio E, Terreno E. Angew. Chem. Int. Ed. Engl. 2002;39:747.
99. Aime S, Castelli DD, Terreno E. Angew. Chem. Int. Ed. Engl. 2002;41:4334.
100. Aime S, Botta M, Mainero V, Terreno E. Magn. Reson. Med. 2001;47:10.

101. Hoult DI, Lauterbur PC. J. Magn. Reson. 1979;34:425.
102. Styles P, Soffe NF, Scott CA, Cragg DA, Row F, White DJ, White PCJ. J. Magn. Reson. 1984;60:397.
103. Black RD, Early TA, Roemer PB, Mueller OM, Mogro-Campero A, Turner LG, Johnson GA. Science. 1993;259:793–795.
104. Miller JR, Hurlston SE, Ma QY, Face DW, Kountz DJ, MacFall JR, Hedlund LW, Johnson GA. Magn. Reson. Med. 1999;41:72.
105. Haase A, Frahm J, Matthaei D, Hnicke W, Merboldt KD. J. Magn. Reson. 1986;67:258.
106. Stehling MK, Turner R, Mansfield P. Science. 1991;254:43.
107. Ahrens ET, Narasimhan PT, Nakada T, Jacobs RE. Prog. Nucl. Magn. Reson. Spectrosc. 2002;40:275.
108. Seppenwoolde JH, Viergever MA, Bakker CJG. Magn. Reson. Med. 2003;50:784.
109. Coristine AJ, Foster P, Deoni SC, Heyn C, Rutt BK. Proc. Int. Soc. Magn. Reson. Med. 2004;12:163.
110. van den Bos EJ, Wagner A, Mahrholdt H, Thompson RB, Morimoto Y, Sutton BS, Judd RM, Taylor DA. Cell Transpl. 2003;12:743.
111. Emerit J, Beaumont C, Trivin F. Biomed. Pharmacother. 2001;55:333.
112. Weissleder R, Stark DD, Engelstad BL. Am. J. Roentenol. 1989;152:167.
113. Stroh A, Zimmer C, Gutzeit C, Jakstadt M, Marschinke F, Jung T, Pilgrimm H, Grune T. Free Radiat. Biol. Med. 2004;36:976.
114. Pouliquen D, le Jeune JJ, Perdrisot R, Ermias A, Jallet P. Magn. Reson. Imaging. 1991;9:275.
115. Yeh TC, Zhang W, Ildstad ST, Ho C. Magn. Reson. Med. 1995;33:200.
116. Arbab AS, Yocum GT, Wilson LB, Parwana A, Jordan EK, Kalish H, Frank JA. Mol. Imaging. 2004;324.
117. Kostura L, Mackay D, Pittenger MF, Kraitchman DL, Bulte JWM. Proc. Int. Soc. Magn. Reson. Med. 2004;12:167.
118. Pratten MK, Lloyd JB. Biochim. Biophys. Acta. 1986;881:307.
119. Schroeder M, Saleh A, Wiedermann D, Hoehn M, Jander S. Magn. Reson. Med. 2005;52:403.
120. Pautler RG, Silva AC, Koretsky AP. Magn. Reson. Med. 1998;40:740.
121. Saleem KS, Pauls JM, Augath M, Trinath T, Prause BA, Hashikawa T, Logothetis NK. Neuron. 2002:34:685.
122. van der Linden A, Verhoye M, van Meir V, Tindemans I, Eens M, Absil P, Balthazart J. Neuroscience. 2002;112:467.
123. Cohen B, Dafni H, Meir G, Neeman M. Proc. Int. Soc. Magn. Reson. Med. 2004;11:1707.

In vivo ^{13}C MRS

Stefan Bluml

Childrens Hospital Los Angeles, Department of Radiology, USC Keck School of Medicine Los Angeles, CA

Abbreviations: MRS = Magnetic Resonance Spectroscopy; MRI = Magnetic Resonance Imaging; SNR = Signal-to-noise ratio; RF = Radiofrequency; FDA = federal drug administration; SAR = specific absorption rate; B_1 = magnetic field of RF coil; NOE = nuclear Overhauser effect; VOI = volume of interest; LCModel = linear combination of model spectra; TCA = tricarboxylic acid; Glc = glucose; Glu = glutamate; Gln = glutamine; Asp = aspartate; NAA = N-acetylaspartate; GABA = γ-amino butyric acid; Lac = lactate; Ala = alanine; HCO_3^- = bicarbonate, Ac = acetate; KD = ketogenic diet.

Introduction

The application of ^{13}C Magnetic Resonance Spectroscopy (MRS) for basic research/medical applications is challenging and, despite being available for more than 25 years, only a few groups have attempted ^{13}C MRS *in vivo*. However, the few studies undertaken illustrate the great promise of *in vivo* ^{13}C MRS. Interest in ^{13}C MRS has grown considerably in recent years. The potential of ^{13}C arises from its biggest handicap: Low natural abundance [and the compromised sensitivity ($\approx$1/50 of ^{1}H)] renders ^{13}C spectroscopy *in vivo* very difficulty due to the inherently very low signal-to-noise ratio (SNR). This low natural abundance, on the other hand, is the key to new, exciting applications of *in vivo* MRS: Investigation of metabolic pathways and the measurement of flux rates *in vivo* in animal and humans after ^{13}C enriched substrate infusion. The interested reader is referred to a recent issue of *NMR in Biomedicine* [1] exclusively dedicated to the application of ^{13}C MRS to study biological systems for a more detailed update on progress in ^{13}C MRS than given below. Instead, in this chapter, basic experimental procedures of *in vivo* ^{13}C MRS, with emphasis on experimental work in humans, are illustrated and discussed for the researcher interested in conducting similar experimental work in the future.

Methods

Equipment

A list of additional equipment needed beyond that required for MR imaging or ^{1}H MRS is provided in Table 1.

Radiofrequency (RF) Coils

Even for "niche" manufacturers, ^{13}C RF coils are too "exotic" and most groups conducting *in vivo* ^{13}C research have built their own coils. Volume coils are more flexible and have a special advantage when quantitation of cerebral metabolites is the goal. Nevertheless, surface coils have exclusively been used for ^{13}C MRS since they furnish the much-needed extra signal required for *in vivo* ^{13}C MRS. Because all ^{13}C methods either apply direct detection with proton-decoupling, polarization transfer, or heteronuclear editing, dual-tuned coils (^{1}H and ^{13}C frequencies) are required. Two concentric circular surface coils, a small ^{13}C coil and a larger ^{1}H coil, have been used by several investigators. However, the need to block the ^{1}H flux of the inner ^{13}C coil resulted in poor performance of the ^{1}H coil in the sensitive volume of the ^{13}C coil. A much improved assembly of two semi-orthogonal surface ^{1}H coils, operated in quadrature mode, and a single linear ^{13}C coil (Figure 1) was first introduced by Adriany and Gruetter [2] and has since been adopted by other groups [37]. The quadrature ^{1}H coil assembly offers twofold improved efficiency over linear-polarized coils in the power requirement. This is important to avoid excessive power RF absorption by tissue during *proton-decoupling* (see below).

Broadband Excitation/Detection

Even if dual-tuned RF coils are available, not all magnetic resonance imaging (MRI) scanners can progress beyond ^{1}H excitation and detection. Broadband amplifiers covering the frequency of ^{13}C are essential. Similarly, RF receivers tuned to the ^{13}C resonance frequency must be provided along with several other hardware features.

Graham A. Webb (ed.), Modern Magnetic Resonance, 1085–1098.

Table 1: Equipment needed for ^{13}C MRS beyond that required for standard MR imaging

Hardware	Purpose	Specifications/comments
RF Coil	^{13}C and ^{1}H excitation, receive, editing, decoupling	Dual-tuned ^{13}C-1H coil, ^{13}C surface coil/surface coil assembly, quadrature ^{1}H coil for efficient decoupling
Broadband capabilities		
Broadband receiver	To digitalize, filter, and store RF signal at ^{13}C frequency	Low noise figure
Transmit/receive switch for ^{13}C with pre-amp	Protects ^{13}C receiver from high power RF during excitation, boosts the gain of the ^{13}C signal during receive	Low noise figure ($<$0.60 dB), high gain ($\sim$35 dB) pre-amp. May be supplemented with a simple gain block to provide further enhancement of the signal. A proton trap or filter (for ^{13}C a low-pass design is acceptable) placed directly before the pre-amp to limit saturation of the device and subsequent reduction in ^{13}C SNR.
Broadband exciter[a]/amplifier	Create/amplify RF pulses at ^{13}C frequency	Minimum output power to drive the ^{13}C coil described above is 2 kW and 8 kW at 1.5 T and 3.0 T, respectively. 4 kW amplifiers on some 3.0 T systems may be barely adequate.
2nd RF channel or "Stand-alone" decoupler		
^{1}H exciter, ^{1}H amplifier[b]	Proton-decoupling	The decoupling amplifier should be capable of delivering $>$50 W of CW RF power. The output of the amplifier should be high-pass filtered to suppress ^{13}C frequencies.
RF source and modulation capabilities	Generation of a decoupling sequence at the ^{1}H frequency	Typical operating requirements: e.g. WALTZ-4 sequence with a 90 degree pulse range from 0–9.9 ms (in steps of 0.1 ms), bilevel (NOE/decoupling) operation, and manual control of the WALTZ-4 pulse width, attenuation control of output RF power over a 50 dB range in 1 dB steps.
Power meter	Power monitoring	Able to measure forward/reflected RF power. Values (in Watts or dB) should be displayed in real-time, e.g. on a computer screen. In normal operations, hard power limits should force the automatic shutdown of the amplifier if the output exceeds the limits. In addition, it should be possible to establish soft limits that generate warning signals or messages.
Miscellaneous		
Phantom	System/sequence performance evaluation	A spherical phantom containing one or more ^{13}C compounds suitable to test system performance/demonstrate decoupling.

[a]The manufacturers all have specifications for exciter output levels, linearity, phase stability, etc.— the specs for broadband are always a little lower than those for the narrow band imaging exciter.
[b]With amplifiers capable of high CW output, it is advisable that some safety interlock is available to insure that a power limit, e.g. 100 W, cannot be exceeded in normal operation. Typical amplifier specifications: 1000 W (peak)/100 W (CW) into a 50 Ohm load. The RF power amplifier should be electrically connected to the MR scanner so that all automatic, emergency electrical power shutdown procedures apply.

Proton-Decoupler and Power Monitoring

Along with the capability to perform heteronuclear (other than ^{1}H) MRS comes the need to enhance sensitivity and specificity of chemical analyses by *proton-decoupling* and *heteronuclear polarization transfer* and *editing*. This involves excitation at different frequencies, and accordingly requires a second RF channel and amplifier. Since it appears that there is a renewed interest in advanced multinuclear spectroscopy, the major manufacturers of clinical MR scanners are all offering broadband capabilities and second RF channels. There are two basic options for decoupling; a fully integrated, second RF exciter channel

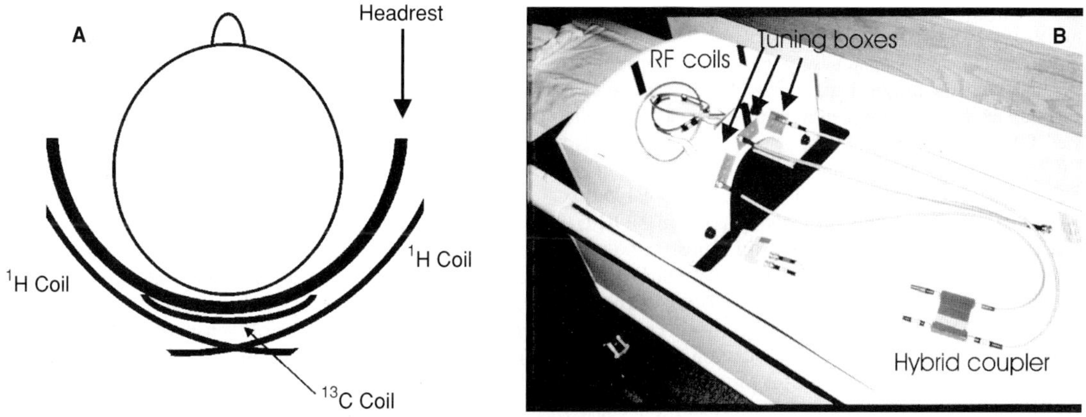

Fig. 1. Dual-tuned RF coil for in vivo ^{13}C MRS of human brain. (**A**) Schematic of the "half-volume" coil assembly as originally proposed by Adriany and Gruetter [2] and since then adopted by other groups. The two orthogonal ^{1}H coils can be used for MRI, ^{1}H MRS, as well as for proton-decoupling during ^{13}C signal reception. A single ^{13}C surface coil is mounted immediately adjacent to the head rest. The overlap of the ^{1}H coils and the position of the individual coils needed to be adjusted to minimize interaction between the different coils. It is also important to minimize the distance between ^{13}C coil and the region of interest to optimize SNR. (B) Picture of the coil assembly used at Huntington Medical Research Institutes, Pasadena, California. A hybrid coupler is necessary to operate the two ^{1}H coils 90° out of phase for quadrature excitation and detection. "Tune and match" is accomplished for each coil individually with three separate tuning boxes. Coils, cables, and connectors are "color-coded" to avoid mistakes when connecting the coil assembly with the MR scanner.

that is part of the MR scanner, or an independent or "Stand-alone" sequence generator with simple timing signals provided by the MR scanner. The fully integrated channel requires pulse sequence programming but gives the user direct access to the decoupling sequence and parameters. Standalone decouplers exist in several flavours, ranging from simple pre-programmed decoupling sequences with hardware control of all parameters to more complex signal generating devices under computer control.

Excessive RF power deposition during decoupling can result in potentially damaging heating of tissue. Federal drug administration (FDA) guidelines limit the deposited energy below levels which are considered to be of any risk [8,9]. Manufacturers have developed decouplers with built-in power monitoring and automatic shutdown should a certain threshold for the average power be exceeded. It should be noted that since for decoupled ^{13}C MRS ^{1}H surface coils are being used, the *local* specific absorption rate (SAR) may vary with the magnetic field profile of the proton coils (B_1). It is possible that the *average* SAR over the whole sensitive volume is below the FDA approved limit, whereas the *local* SAR of tissue adjacent to the ^{1}H decoupling coils may be above that threshold [2,10].

Proton-Decoupling and NOE

Proton-decoupling and the nuclear Overhauser effect (NOE) are essential tools for ^{13}C MRS. It is not advised to attempt ^{13}C MRS without decoupling/NOE. Even the simplest application, such as measuring the lipid profile of a human leg [11], would be quite demanding without the improved SNR and resolution facilitated by decoupling and NOE.

Proton-Decoupling

Heteronuclear scalar couplings between ^{1}H and ^{13}C nuclei result in a splitting of ^{13}C resonance lines in to multiplets. The goal of *proton-decoupling* is to suppress these couplings and to collapse multiplets into singlets, simplifying the spectrum and improving the signal intensity. This is achieved by irradiation of decoupling proton RF pulses during the readout of ^{13}C signal over the whole range of the spectrum (= *broadband* decoupling). The most commonly used proton-decoupling RF sequence for *in vivo* ^{13}C MRS is WALTZ [12–15]. Several WALTZ decoupling schemes have been developed named WALTZ-4, WALTZ-8, and WALTZ-16 and they all appear to work well for *in vivo* applications.

NOE—The NOE is based on changes in the polarization, which occur in coupled systems when the population of one of the systems is manipulated. In contrast to *scalar* coupling, which is facilitated indirectly by the electron bond, it is the *dipolar* coupling between ^{13}C and ^{1}H nuclei, which generates the NOE. In fluids, dipolar coupling does not result in a line splitting due to "Motional Narrowing," however, it is still an effective relaxation mechanism resulting in changes of signal intensities. While it was necessary to irradiate at the proton frequency *during* the readout of ^{13}C signal for proton decoupling, NOE enhancement is generated by irradiation at the proton frequency *before* the ^{13}C signal is detected. Excitation of protons coupled to ^{13}C has an impact on the polarization of ^{13}C. In particular, if the protons are saturated an increase of the polarization of ^{13}C can be observed. Since the maximum NOE enhancement for two nuclei A and B in fluids is given by, $NOE = 1 + \frac{1}{2}\,\gamma_B/\gamma_A$, a threefold signal enhancement can be achieved for ^{13}C MRS ($\gamma_{1H} \approx 4\gamma_{13C}$). To have maximum NOE enhancement for all resonances it is necessary to irradiate RF over the full chemical shift range of dipolar coupled protons. As for proton decoupling, a WALTZ-*n* cycle can be employed, however, it is usually sufficient to irradiate at a much lower power levels ($\approx$1/10 of decoupling power). The optimum choice for the center frequency and the bandwidth for NOE may be different from what works best for decoupling.

Pulse Sequences for *in vivo* ^{13}C MRS

More detailed discussions of pulse sequences and data acquisition strategies than provided below can be found elsewhere [16,17] and in references therein.

^{13}C Excitation and ^{13}C Detection

Direct detection ^{13}C MRS is the *least sensitive* method and spectra with sufficient quality can only be acquired from very large volumes [18] or without any localization [3]. *Gradient localized* direct ^{13}C MRS is further compromised by considerable chemical shift displacement errors due to the large chemical shift dispersion of ^{13}C spectra. On the other hand, spectra acquired with this method offer high spectral resolution and an impressive number of ^{13}C resonances, all measured simultaneously under identical conditions (Figure 2). In particular when a simple "pulse and acquire" sequence is used, the impact of acquisition parameters on the spectral pattern is small and sequence parameter adjustment is quick. These features may be important in a clinical setting where experimental simplicity is important [19,20]. Direct ^{13}C excitation and detection requires proton-decoupling and NOE to improve the spectral quality.

Polarization Transfer

A better SNR can be achieved with heteronuclear polarization transfer. The basic concept of polarization transfer is to excite protons coupled to ^{13}C and, as the polarization is transferred to the adjacent ^{13}C nucleus during the evolution period, to detect the signal at the ^{13}C frequency. Because the heteronuclear J-coupling constants between ^{1}H and ^{13}C nuclei are very similar ($\approx$120 Hz for methyl groups, $\approx$160 Hz for aromatic groups), sequence parameters, in particular the delay for coupling evolution, can be adjusted to the optimum detection of all signals. There are two distinct features of this method. (i) Proton excitation can be combined with *localization* such as image-selected *in vivo* spectroscopy (ISIS) [21] greatly reducing the chemical shift displacement error [4,22]. (ii) ^{13}C detection retains the large dispersion and thus spectral resolution of ^{13}C (Figure 3). This method also allows recovering the maximum fourfold sensitivity gain due to the higher magnetization of protons. During detection at the ^{13}C frequency, proton decoupling is applied to generate a simplified spectrum for better interpretation. Polarization transfer becomes less advantageous in situations where the resonance of interest has a very short T_2 (e.g. glycogen). If T_2 is short relative to the time delay necessary to facilitate polarization transfer, a substantial loss of signal can occur [17].

Indirect Detection

Indirect detection can improve greatly the sensitivity by detecting protons attached to ^{13}C utilizing heteronuclear scalar couplings. In the simplest case, protons coupled to a ^{13}C nucleus will form two resonances symmetrically placed around the center of the singlet generated by a proton attached to a ^{12}C nucleus. In practice however, the ^{1}H spectra of chemicals such as glutamate and glutamine are complex because of home- and heteronuclear couplings and resonances overlap due to the small chemical shift dispersion of ^{1}H MRS. This renders the detection of ^{13}C coupled protons difficult. It is therefore necessary to apply *editing* pulses at the ^{13}C frequency, which will result in a modulation of a spectrum, e.g. an inversion of the signal from ^{13}C bound protons relative to ^{12}C bound protons. When unedited and edited spectra are analyzed, the difference spectra will represent protons attached to ^{13}C while the added spectrum represents protons bound to ^{12}C. This approach is referred to as *proton-observed, carbon-edited* (POCE) [23]. Indirect detection offers the highest SNR among the methods discussed here and has the advantage of localization at the proton frequency, reducing chemical shift displacement artifacts. Another advantage of indirect detection is that only hardware for the *transmission* of RF at the ^{13}C frequency is required and that the proton channel of MR systems, used for signal detection, is usually well engineered, guaranteeing good performance (Figure 4). A disadvantage is that the high spectral resolution of ^{13}C is sacrificed. While this is acceptable for simple spectra with selective isotopic enrichment, it may

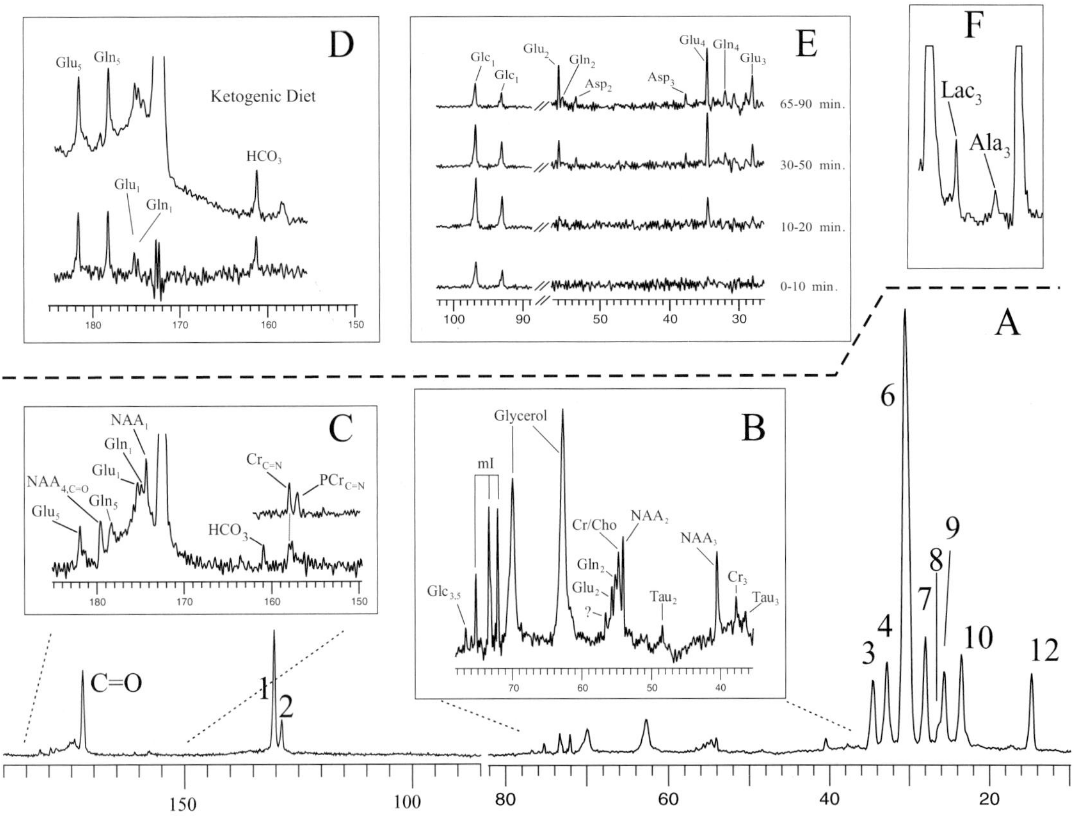

Fig. 2. *In vivo* natural abundance and ^{13}C MR spectra after substrate infusion of the human brain acquired with "pulse-and-acquire" and direct ^{13}C detection. (A–C) Shown is a natural abundance ^{13}C MR spectrum of a patient with Canavan disease. B + C are expansions of A to allow a more detailed inspection. Lipid signal originating from the skull and subcutaneous fat dominates the spectrum due to non-localized acquisition. (D) ^{13}C spectrum (upper trace) obtained from ketogenic diet (KD) patients 30–80 min after infusion [1-^{13}C] acetate. The lower trace is the difference spectrum computed by subtraction of the baseline spectrum, acquired before infusion start, (not shown) to directly illustrates ^{13}C label accumulation. ^{13}C label incorporation into bicarbonate (HCO_3), Glu C_5 and Gln C_5 and, Glu C_1 and Gln C_1 can be observed. (E) The time course of glucose uptake and metabolism to glutamate and other products in the brain in a normal adult subject after i.v. [1-^{13}C] glucose infusion. Difference spectra were calculated to subtract prominent lipid peaks. ^{13}C enriched peaks of Glu $C_{1,2,3,4}$, Gln $C_{1,2,3,4}$, Asp $C_{2,3}$, NAA $C_{2,3}$ can be observed [36]. (F) Although lactate (Lac) and alanine (Ala) concentrations are too low to be detected with natural abundance ^{13}C MRS in normal brain, ^{13}C enriched resonances are readily observed after [1-^{13}C] glucose infusion. All spectra were acquired with the dual-tuned half-volume head coil shown in Figure 1 on a GE, 1.5 T clinical scanner.

be more problematic when high resolution of complex spectra with many components is the goal.

Checking System Performance

Once the required hardware is installed and an MR sequence developed, the next essential step is the thorough evaluation of the system performance.

Low SNR

Low overall SNR of ^{13}C MRS may be caused by several reasons. A poorly designed or tuned ^{13}C coil will not provide adequate SNR. Also, poor SNR may be caused by site specific factors. RF noise at ^{13}C frequency may be generated by other MR equipment used for MRI and not filtered appropriately. Also, equipment not related with MR may emit RF at the frequency of ^{13}C and adversely

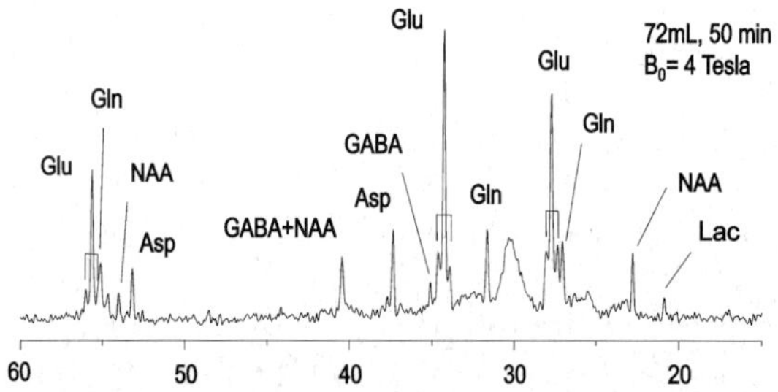

Fig. 3. Localized ^{13}C MR spectrum of the human occipital lobe The spectrum was acquired from 72-ml volume 60–110 min after start of [1-^{13}C] glucose infusion on a 4 T system using polarization transfer, proton excited carbon detected, for improved SNR and localization (Figure provided by Rolf Gruetter, Ph.D. and reproduced with permission from Dev Neurosci 1998 S. Karger AG).

interfere with the quality of a spectrum. There are many tests that can be performed to isolate the cause for insufficient SNR, however, the questions the new investigator faces first are: "How does a good spectrum look like?" "Is the performance of my system the best possible?" For

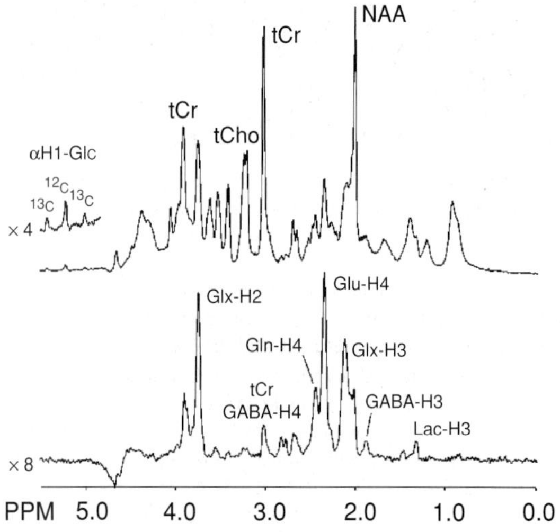

Fig. 4. Indirect detection of ^{13}C label accumulation The spectrum was acquired from rat brain at 9.4 T (180 ul volume, T_R = 4000 ms, T_E = 8.5 ms, 512 averages). It was acquired between 110 and 140 min following the intravenous infusion of [U-^{13}C6]-glucose (i.e. uniformily labelled glucose). The upper trace is the regular ^{1}H spectrum, while the bottom spectrum is the ^{1}H-[^{13}C] edited fraction (i.e. it only shows resonances from protons attached to ^{13}C). The sequence is described in de Graaf *et al.* [37] (figure provided by Robin de Graaf, Ph.D. and reproduced with permission from Wiley-Liss. Inc., 2004).

that reason, it is recommended that the investigator should team-up with a reference site with experience with ^{13}C MRS. The coil, a phantom, and specific instruction on how to perform a test should be sent to a collaborating site with the request to acquire data for comparison. Even better, a new investigator may want to visit an established site for ^{13}C MRS to witness a quality scan and to ensure that scans at the reference site and at the local site are performed under identical conditions. Once a problem has been identified or narrowed down, the manufacturer of the MR system or the RF coil can be approached with a specific request to fix a problem.

Crosstalk

Crosstalk between ^{1}H and ^{13}C coils during decoupling is caused by the transmission of RF at the proton frequency with relatively high power while the small ^{13}C signal is received. It can cause two unwanted effects. (i) Although, the ^{1}H signal is suppressed by subsequent filtering, the ^{13}C pre-amplifier usually placed immediately after the ^{13}C coil, sees both ^{13}C and ^{1}H signal. Both signals may be amplified. Since the ^{1}H signal is much larger than the ^{13}C signal it may *saturate* the pre-amplifier. This results in an overall scaling (down) of the decoupled ^{13}C spectrum. It may not affect the SNR but is nevertheless unwanted since an absolute comparison of scans becomes very difficult. (ii) Unacceptable artifacts are caused when decoupling causes spikes at the ^{13}C frequency. Detecting of these two problems is relatively straightforward. If the maximum amplitude of decoupled spectra becomes smaller after further increasing decoupling power, saturation may have occurred. Spikes are readily detectable in the spectrum because of the dramatic adverse impact on the SNR. The problems caused by crosstalk of coils can be minimized by additional filtering of the ^{13}C signal and by using coil assemblies with efficient built-in proton traps. Reducing the decoupling power is useful only when it does not result in incompletely decoupled spectra.

Incomplete Decoupling

Insufficient decoupling power results in a "hybrid" spectrum, partially non-decoupled and partially decoupled. These spectra are not quantifiable and should not be interpreted.

Instabilities

Scanner and coil instabilities are often another concern. Electronic components may heat-up during a pro-longed acquisitions and change the tuning of a coil, an amplifier may not deliver the same power over an extend period of time, etc. Usually these variations are small, however, in particular when difference spectroscopy is attempted or the time course of a signal is observed over a long time, the quality of the results and the accuracy of the interpretation may be compromised. Scanner and coil instabilities can readily be checked in vitro by scanning phantoms.

Reproducibility

The nature of a study defines when insufficient reproducibility (= different results from the same subject on different occasions) is a problem. Hardware imperfections can cause insufficient reproducibility. However, often the positioning of a subject, selection of region of interest, adjustment of acquisition parameters, shimming, etc. is not consistent thus limiting the reproducibility. These problems can be minimized/avoided by generating macros (if possible), dedicated protocols, and clear written instruction on how to conduct a scan. Macros, protocols, and instructions should be tested by scanning different phantoms. Sequence parameters should not be "guessed" when starting an acquisition. Ideally, good and reproducible spectra should be obtained without the need to "manually" adjust scan parameters in each cause. If the lack of reproducibility is small compared with the postulated differences between two groups of subjects, it only complicates the study by the need to examine more individuals in order to reach significance. It is always a good strategy to conduct several studies in the same subject to investigate this matter. There are of course limitations to that when a study is very complex such as a lengthy examination with substrate infusion.

Data Processing

High quality spectra are of no use if the information they contain is not or incorrectly extracted. Therefore appropriate data processing is as important as the acquisition of the spectra. In general, data processing should be as *user independent* as possible to prevent unavoidable operator bias.

Since the signal generated by a specific nucleus is proportional to the number of equivalent nuclei in the volume of interest (VOI), *in vivo* concentrations of a metabolite M (in mmol/kg tissue or mmol/l) can be determined by $[M] = C_{\mathrm{MRS}} A_{\mathrm{M}}$, where C_{MRS} comprises all correction factors (T_1 saturation, T_2 relaxation, VOI, number of equivalent nuclei, receiver gains, etc.) and A_{M} is the total area under a curve. Resonances in decoupled ^{13}C spectra are usually well resolved and line fitting of individual resonances is relatively straightforward. However, when quantifying spectra of low SNR (^{13}C spectra are compromised by low SNR), the investigator may want to keep in mind that not all parameters of a curve are measured with the same accuracy. In particular the determination of the position of a resonance on the chemical shift axis and the amplitude can be measured with a much higher accuracy than the width (or area). Therefore, most of the time, it is advantageous to measure peak amplitudes instead of areas. On the other hand, the amplitude is affected by the homogeneity whereas the area is not. Nevertheless, in a well-designed study, spectra are acquired from comparable regions of interest and the variation of the homogeneity across subjects should be very small and there should be no correction necessary. In cases where small systematic variations could be expected (e.g. when a comparison of spectra from different brain region is attempted), it is still better to perform an adjustable lineshape correction (see below) and to measure amplitudes rather than areas. If due to the nature of a study, large systematic differences of the field homogeneity is expected then areas rather than amplitudes should be evaluated.

Lineshape Transformation

For spectra of low SNR and partially overlapping peaks it is generally advisable to perform a lineshape transformation. A common lineshape transformation is the Lorentzian-to-Gaussian transformation. The first step is a negative Lorentzian linebroadening to create a spectrum with a standardized linewidth. E.g. a spectrum with resonances with 4 Hz linewidth will be corrected by −3 Hz whereas a spectrum with resonances with 4.5 Hz linewidth will be corrected by −3.5 Hz to result in standardized spectra with a 1 Hz linewidth. The second step is a positive Gaussian linebroadening to improve SNR. Although this lineshape transformation is not linear (= the relative areas of peaks with different linewidths are not the same after transformation), in practice, this imperfection is more than compensated for by the fact that Gaussian lines can be fitted much easier than Lorentzian lines because of the reduced overlap of adjacent resonances.

LCModel for ^{1}H MRS

^{13}C spectra are usually well resolved and fitting of the individual resonances can be achieved relatively straightforward. Should the user decide to use inverse detection, proton spectra need to be processed. Proton spectra suffer from the narrow chemical shift range, complex pattern due to homonuclear couplings, and a broad baseline due to macromolecules. On the other hand, good processing software for ^{1}H MRS is commercially available. LCModel (Linear combination of model spectra, Stephen Provencher Inc., Oakville, Ontario, Canada) is a well-developed, robust, and at the same time flexible software package [24]. LCModel fits the *in vivo* data by finding the best linear combination of spectra obtained from phantoms with known concentrations. LCModel is most suited for standard ^{1}H MRS as carried out in a clinical environment (e.g. 1.5 T, 2.0 T, 3.0 T, PRESS, STEAM) with pre-measured basis spectra available from the vendor. For applications with complex acquisition sequences and unusual sequence parameters the user may need to prepare his/her own phantoms to acquire the appropriate model spectra the specific experimental conditions. Although this can be tedious when many metabolites are being analyzed, it is time well spent.

Modeling and Determination of Flux Rates

Infusion of a ^{13}C labeled substrate allows the metabolism of the substrate to be followed by sequential ^{13}C MRS. A *mathematical model*, basically a set of differential equations, of the metabolism is then needed to determine the *in vivo* flux rates (Figure 5). It needs to be pointed out that the formulation of the differential equations for a biological model is the easy part—finding the right model is the tricky part. Assumptions and simplifications are necessary, even if the correct model is known [e.g. for the tricarboxylic acid (TCA) cycle] because only a limited number of metabolites can be measured with sufficient SNR. E.g. the measurement of the TCA-cycle rate after [1-^{13}C] labeled glucose infusion is based to a large extent on the time course of the glutamate C_4 enrichment. However, glutamate is NOT a metabolite of the TCA-cycle. Only because the pool sizes of TCA-cycle intermediates are small and the exchange rate between α-ketoglutarate and glutamate is fast (there are some controversies in the literature on how fast "fast" is), it is reasonable to assume that glutamate C_4 acts as a trap for ^{13}C label entering

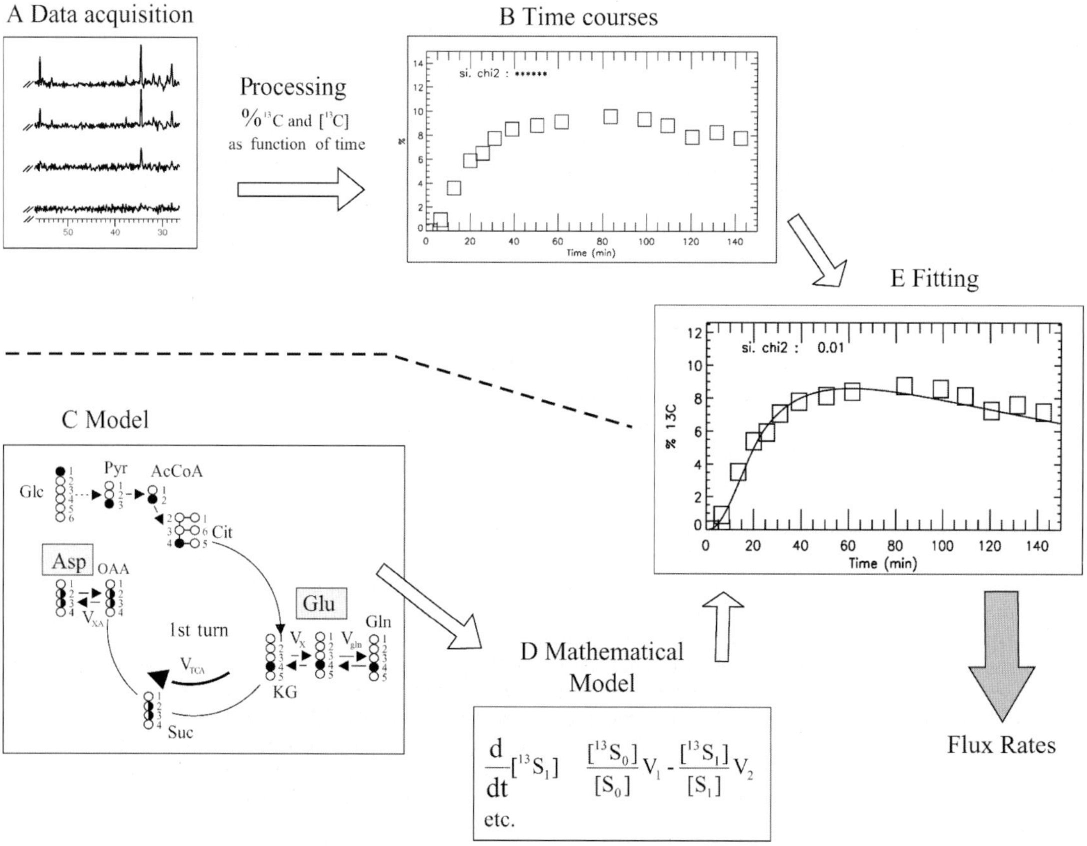

Fig. 5. (A) Dynamic ^{13}C MRS involves the sequential acquisition of ^{13}C spectra after substrate infusion to ^{13}C label accumulation. (B) The next step is the processing and quantitation of spectra to generate time courses of ^{13}C concentration (or ^{13}C enrichment). (C + D) A set of differential equations is derived from the metabolic model applied for an experiment. (E) By iteratively varying flux parameters (and pool concentrations) of the mathematical model to optimize the fit with experimental data, flux rates (and pool concentrations) can be determined.

the TCA-cycle at acetyl-Coa [25]. It is beyond the scope of this chapter to get involved in the discussion of current complex models of glucose and acetate metabolism in animal and human brain and it is recommended that the interested reader studies recently published articles [1]. Sometimes it is possible to simplify the mathematical model considerably to answer very specific questions. Examples are the determination of the N-acetyl-asparate synthesis rate in humans *in vivo*, the glutamine synthesis rate in human hepatic encephalopathy, or the measurement of the astro-glial TCA-cycle rate after [1-^{13}C] acetate infusion [20,26,27].

A simple two Compartment Model—Mathematical Description

$$S_0 \xrightarrow{V_1} S_1 \xrightarrow{V_2} S_2$$

In this model S_0 is the initial substrate and S_1, S_2 are intermediates of S_0 metabolism. [$^{13}S_0$], [$^{13}S_1$], [$^{13}S_2$] are the ^{13}C enriched pools; e.g. before infusion $[^{13}S_1] = 1.1\%$ [S_1]. V_1 and V_2 are the fluxes from S_0 to S_1 and from S_1 to S_2. Assuming that the total concentration of S_1 remains constant ($V_1 = V_2$), the ^{13}C flux into and out of pool S_1 is described by:

$$\frac{d}{dt}[^{13}S_1] = \frac{[^{13}S_0]}{[S_0]}V_1 - \frac{[^{13}S_1]}{[S_1]}V_2. \qquad (1)$$

In the special case that the fractional enrichment of substrate S_0 is constant ($[^{13}S_0]/[S_0] = E_0 =$ constant), the ^{13}C enrichment of S_1 will exponentially approach that of the precursor:

$$[^{13}S_1](t) = E_0[S_1] \times \left(1 - \exp V_1 t/[S_1]\right) \qquad (2)$$

Since the time constant for the pool to turn over is the ratio of pool size and flux rate ($[S_1]/V_1$), large metabolic pools take longer to turn over whereas small pools will quickly have the same enrichment as the precursor. For this hypothetical model, the concentration of ^{13}C enriched S_1, would be measured by sequential MRS. Once the total pool size is known (e.g. from literature, quantitative natural abundance ^{13}C, or ^{1}H MRS), the flux rate V_1 can be obtained by fitting an exponential function to the measured time course for [$^{13}S_1$].

Real biological systems, although analytical differential equations for each metabolic pool can be written readily, are too complex for analytical solutions and numerical solutions need to be found by a step-by-step integration. Software (CWave, Graeme Mason, Ph.D., Yale University) for the design of models and analysis of ^{13}C-labeling studies is available after signing a license agreement [28]. To obtain a copy contact Graeme Mason, Ph.D., at the Magnetic Resonance Center, Yale University, School of Medicine.

Miscellaneous

Even when all technical obstacles have been resolved, the investigator still needs to plan any study very carefully. *In vivo* ^{13}C studies, in particular of human requiring large amounts of ^{13}C labeled substrates, are expensive, lengthy, and the subsequent data analysis requires a considerable effort. To avoid frustration and jeopardizing the success of a ^{13}C experiment, all "peripheral" steps of a study need to be planned with equal due diligence than the acquisition of the ^{13}C spectra. Depending on the biological question asked, it needs to be decided what substrate should be infused for how long and in what fashion. For some applications oral administration may be appropriate [29,30,26] which would simplify the procedure considerably because one inter-venous (i.v.) infusion line could be eliminated. In particular for studies of humans, i.v. administered substrates should be prepared with care, kept refrigerated, and used within a couple of days. It is usually necessary to determine fractional ^{13}C enrichment of the substrate in plasma. Therefore the drawing, storage, and analysis of blood samples need to be planned. A new investigator is advised to carefully read the "Methods" sections of previous publications [22,25,28] and references therein.

Applications

Although there are undoubtedly applications for *in vivo* natural abundance ^{13}C MRS, to harvest the full potential of ^{13}C MRS, enrichment of metabolites with ^{13}C via intravenous or oral administration of ^{13}C labeled substrates is necessary. Glucose is the principal substrate for energy metabolism for both neurons and glia cells in the brain and also facilitates the *de novo* synthesis of many neurochemicals. In normal human adults, i.v. infused [1-^{13}C] labeled glucose (Glc) passes the blood-brain barrier and is readily metabolized. ^{13}C enrichment of individual carbon atoms of glutamate (Glu), glutamine (Gln), aspartate (Asp), N-acetylaspartate (NAA), γ-amino butyric acid (GABA), lactate (Lac), alanine (Ala), and bicarbonate (HCO_3^-) follows [5,19,20,22,25,31]. From repeated ^{13}C MR spectra acquired during studies over 2–3 h, the *in vivo* rates of several of the principal bioenergetic pathways of normal adult brain have been determined [22,25,4].

The role of ^{13}C glucose MRS in *diseased brain* remains a matter of speculation as there are only a few studies

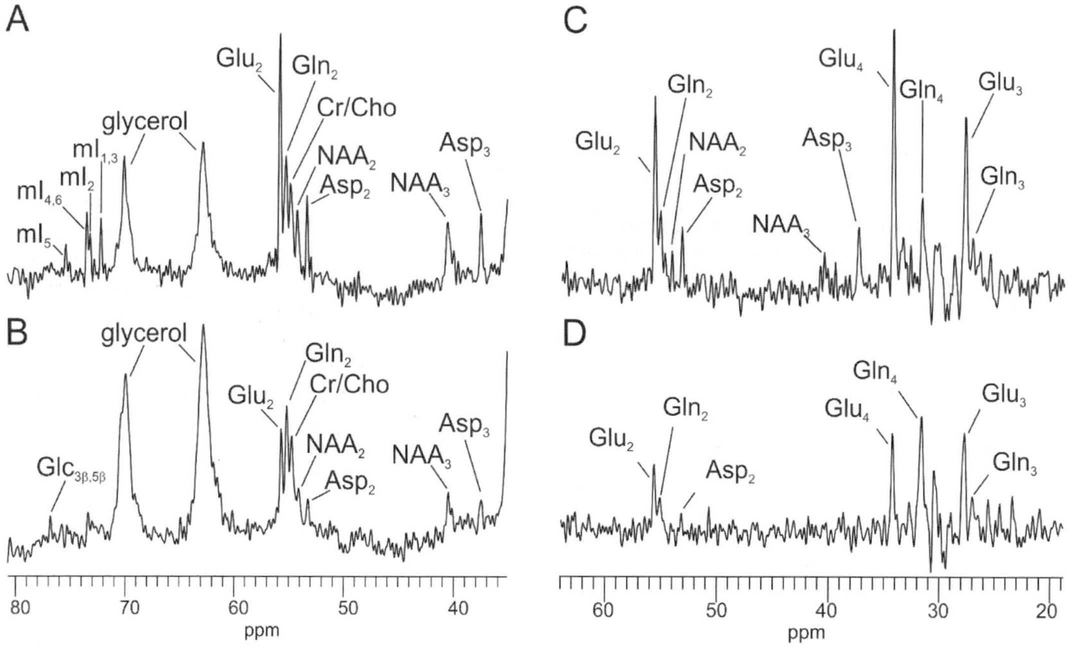

Fig. 6. ^{13}C MRS of human hepatic encephalopathy (HE) Hepatic encephalopathy is a metabolic brain disorder caused by liver dysfunction and incomplete removal of ammonia and other neurotoxins from the blood. (**A+B**) Spectra acquired over a period of 60–120 min after a 20 min infusion of [1-^{13}C] glucose of an adult control (A) and a patient with severe chronic HE (grade III–IV) are shown. ^{13}C enrichment of Glu C_2, Asp C_2, Asp C_3 is reduced while ^{13}C accumulation in Gln C_2 is comparable to the control. Note the absence of natural abundance *myo*-inositol (mI) signal in HE. (**C+D**) Difference spectra, to enable inspection of ^{13}C accumulation in Glu C_4, Gln C_4, Glu C_4, and Gln C_4, of a control and a chronic HE patient (grade I-II) acquired 80–100 min after infusion start, are shown. Spectra were scaled to represent absolute concentrations. Note the apparent reduction of ^{13}C incorporation into Glu C_4, C_2, Asp $C_{2,3}$ in HE. Even though affected by the subtraction analysis to a larger extent in HE than in control, the Gln C_4 resonance appears strikingly prominent in spectra of the patient. Label accumulation in NAA $C_{2,3}$ observed in the control was not detected in HE. The pattern of label accumulation in HE is consistent with an overall reduction of glucose oxidation and altered glutamate/glutamine cycling.

conducted. Striking abnormalities in glucose metabolism and glutamate and glutamine label accumulation were observed in patients with chronic hepatic encephalopathy (Figure 6). Glucose oxidation was also reduced in a juvenile with hypoxic injury and in a premature infant. Abnormalities were also detected in pediatric patients with leukodystrophies and in children with mitochondrial disorders [19,20]. The information obtained from following the fate of ^{13}C labeled glucose goes beyond that of providing a rate for energy production. A tight coupling between cerebral glucose metabolism and glutamate neurotransmitter flux in humans has been proposed by Magistretti *et al.* [32]. Aspartate, (a neurotransmitter?), can be studied *in vivo* in humans by its ^{13}C label accumulation. The role of NAA in mammalian brain, a neuronal/axonal marker which is central for its diagnostic power in ^{1}H MRS, is incompletely understood. NAA synthesis can be measured with ^{13}C MRS after glucose infusion in a clinical setting [26] (Figure 7).

The application of ^{13}C MRS after substrate infusion is by no means limited to glucose. Glucose is convenient because of its rapid oxidation and the fast appearance of ^{13}C label in its metabolites and its non-toxicity even at extremely high concentrations. However, the use of other substrates may further enhance the potential of ^{13}C MRS as a research and diagnostic tool in human brain disease. The candidate next best to glucose appears to be acetate [7,27]. Acetate (Ac) is metabolized to acetyl-CoA only in the glial compartment [33]. Using the same MR technique as for [$1-{}^{13}$C] glucose, glial acetate metabolism can be investigated in normal and diseased brain. ^{13}C label accumulation in HCO_3, in Glu C_5 and Gln C_5 (first turn of the TCA-cycle), and in Glu C_1 and Gln C_1 (second turn) can be observed after [$1-{}^{13}$C] Ac in fusion. The rate of Ac

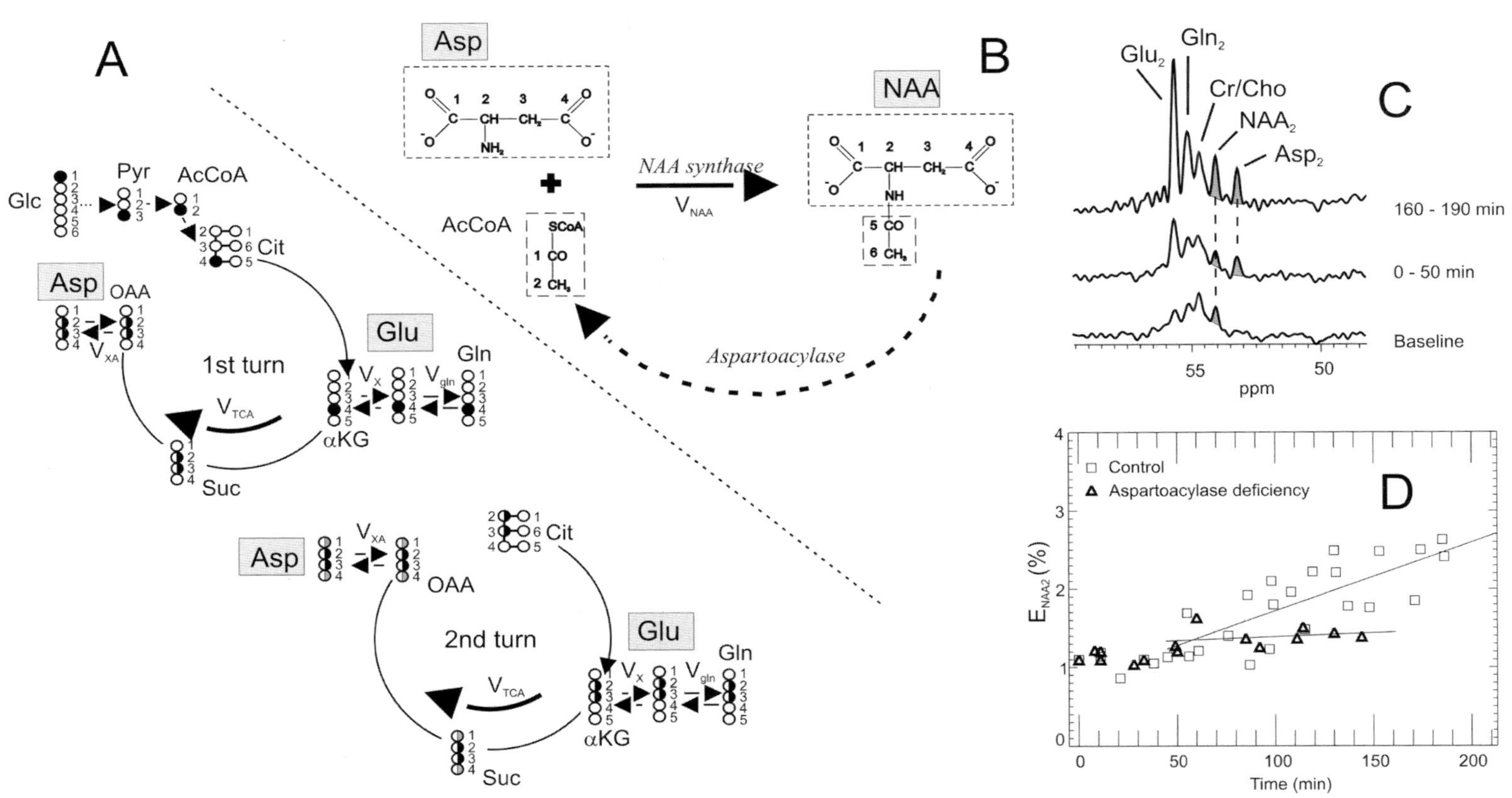

Fig. 7. Determination of N-acetyl-aspartate synthesis rate, V_{NAA}, in humans *in vivo* (A) ^{13}C labeling scheme used to predict the transfer of ^{13}C label from Glu_4 to $Asp_{2,3}$. Only relevant TCA-cycle intermediates are shown (B). N-acetyl-aspartate (NAA) synthesis by condensation of aspartate and Acetyl CoA. (C) Human brain natural abundance (baseline) and ^{13}C enriched spectra after [1-^{13}C] glucose infusion. The time intervals indicate the time of acquisition relative to the commencement of [1-^{13}C] glucose infusion. Peak intensities of Asp_2 and NAA_2 (as well as Glu C_2 and Gln C_2) increased progressively as ^{13}C label accumulates. (D) Fractional ^{13}C enrichment of NAA (E_{NAA2}) versus time was lower in patients with aspartoacylase deficiency (Canavan Disease) than in control subjects. Linear regression lines are superimposed on data points $t > 50$ min in each group. The calculated mean rate of *in vivo* NAA synthesis in normal subjects was 9.2 ± 3.9 nmol/min/g. In patients with Canavan Disease statistically significant reductions of [Asp] and [Glu] were observed. NAA pool size was increased and V_{NAA} was significantly lower (3.6 ± 0.1 nmol/min/g, $p < 0.001$) than in controls. V_{TCA} = TCA-cycle rate, V_x = α-ketoglutarate/glutamate exchange rate. V_{gln} = glutamine synthesis rate, V_{xa} = oxaloacetate/aspartate exchange rate, Cit = citrate, Suc = succinate, Pyr = pyruvate, AcCoA = Acetyl CoA.

oxidation in human brain was estimated to be ≈20% of the total neuronal/glial TCA-cycle rate in fasted human brain [7,27]. Recently it was demonstrated that Glu C_5 and Gln C_5, which are primarily derived from acetate in the glial compartment, accumulated more in patients on ketogenic diet (KD) than in controls, whilst accumulation of bicarbonate was similar [34] (Figure 8). These results are consistent with altered glutamate-glutamine neutrotransmitter cycling and adaptation to ketogenic diet with up-regulation of acetate oxidation relative to glucose oxidation. KD is a treatment option for epileptic patients, which is particularly effective in children. The biochemical mechanisms why KD improves seizure activity are incompletely understood. *In vivo* ^{13}C MRS may elucidate a possibly biochemical basis for reduction in seizures during KD therapy. These studies indicate that ^{13}C MRS is an appropriate tool to investigate diseases which are believed to originate in glial cells. Together with data from glucose infusion experiments this could result in a more complete understanding of cerebral metabolism in normal and diseased human brain.

Hyperpolarized ^{13}C Compounds

Recently, dynamic nuclear polarization was used to obtain highly polarized ^{13}C nuclei in endogenous substances such as isotopically enriched ^{13}C urea [35]. Methods have been developed to perform the polarization process outside the body and a nuclear polarization of 50% for ^{13}C appears to be possible [35]. This polarization corresponds with an enhancement of 500,000 (!) compared with the thermal equilibrium polarization at 1.5 T for ^{13}C. Hyperpolarization vanishes due to longitudinal relaxation (T_1) and spectroscopy studies with the goal to investigate metabolic pathways face two major obstacles. (i) The hyperpolarized substrate needs to reach the organ of interest quickly and (ii) turnover rates need to be sufficiently fast to allow the observation of metabolites of a hyperpolarized substrate. Therefore, the application of hyperpolarized substrates to study normal and abnormal metabolism depends mainly on the capability to prepare metabolically active, non-toxic substrates with long T_1-times. Once that is achieved, the promise of ^{13}C MRS with endogenous substances is enormous. Already, Golman *et al.* [35] report a T_1-time of 20 s for ^{13}C urea *in vivo*. The polarization of such a substrate would still be ≈10 times higher than the thermal equilibrium polarization 4 min after injection. A hyperpolarized compound with a T_1 of 40 s would have a ≈1000 (!) higher polarization and a hypothetical 12 day acquisition time could be shortened to 1 s without compromising SNR! As these compounds become available, detailed studies of chemicals of low concentrations in tissue, first in animals and then in humans, may be feasible. In particular, the study of neurotransmitters, until now out of reach for spectroscopist may become a reality.

Acknowledgments

The author likes to thank Thomas Raidy, Ph.D. and Mark Albers, BS for discussions and for using their expertise with MR hardware. The author also likes to thank Rolf Gruetter, Ph.D. and Robin A. de Graaf, Ph.D. for providing figures.

Glossary

Absolute quantitation—Determination of concentrations of metabolites in mMoles per kg of tissue or volume.
B_0—Strong magnetic field, constant in time and space, generated by the superconducting magnet.
B_1—Radiofrequency (RF) magnetic field generated by *radiofrequency coils*.
Hetero-nuclear J-coupling—*J-coupling* between different species of spins, e.g. proton and carbon.
Homo-nuclear J-coupling—*J-coupling* between the same species of spins, e.g. proton and proton.
ISIS—**I**mage-**s**elected *in vivo* **s**pectroscopy is based on a cycle of eight acquisitions which need to be added and subtracted in the right order to get a single volume. ISIS is considerably more susceptible to motion then *STEAM* or *PRESS* and is mostly used in heteronuclear studies, where its advantage of avoiding *T_2-relaxation* is valuable.
J-coupling (or scalar coupling)—Many resonances split into multiplet components. This is the result of an internal indirect interaction of two spins via the intervening electron structure of the molecule. The coupling strength is measured in Hertz (Hz) rather than *ppm* because it is independent of the external B_0 field strength.
NOE—Nuclear Overhauser effect, the magnetization of protons dipolar-coupled to ^{13}C nuclei can be used to enhance the ^{13}C signal. While the term NOE is mainly associated with ^{13}C MRS, NOE enhancement can also be observed in ^{31}P MRS and with other nuclei.
Polarization transfer—Many interesting nuclei like ^{13}C suffer from low inherent sensitivity compared with proton MR. Techniques like DEPT (**d**istortionless **e**nhancement by **p**olarization **t**ransfer) and INEPT (**i**nsensitive **n**uclei **e**nhanced by **p**olarization **t**ransfer) improve the ^{13}C sensitivity by transferring the higher polarization of coupled protons to the carbon nuclei. Special hardware with two RF channels is needed for polarization transfer experiments. A modification of DEPT and INEPT is reverse DEPT and inverse INEPT where the polarization is transfered back to utilize the higher sensitivity of the protons for observation (inverse detection).
PRESS—Point-resolved spectroscopy, utilizes three 180° slice selective pulses along each of the spatial directions

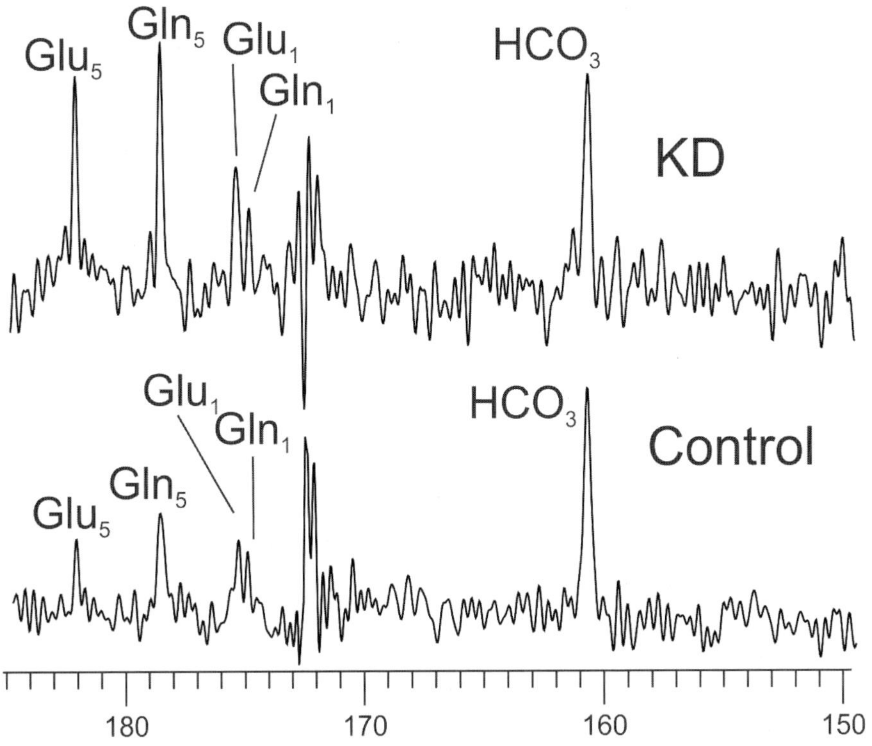

Fig. 8. Impact of ketogenic diet (KD) on astroglial acetate oxidation (A) ^{13}C difference spectra obtained from ketogenic diet patients (average of three) and (B) controls after infusion of [1-^{13}C] acetate. ^{13}C label incorporation into Glu C_5 and Gln C_5 was more pronounced in patients. In addition, Glu C_1 and Gln C_1 enrichment was more prominent, whereas equivalent production of HCO_3 was observed.

and generates signals from the overlap in form of a spinecho.

SAR—Specific absorption rate. Due to inductive and dielectric losses energy from radiofrequency pulses is absorbed by tissue and mainly transferred into rotational and translational movements of water molecules which causes an increase of tissue temperature. Limits for the human brain are established by FDA guidelines.

Scalar coupling—See *J-coupling*

STEAM—Stimulated echo acquisition mode, localization method utilizing three 90° slice selective pulses, along each of the spatial directions. Signal, in form of a *stimulated echo*, from the overlap is generated in a "single shot" experiment. In contrast to *PRESS*, only half of the possible signal is recovered when the same *echo time* is used.

***T_1*-relaxation, *T_1*-relaxation time**—After the *magnetization* vector has been flipped into the transverse plane, new magnetization builds up along the *z*-axis. The time after 63% (1-1/e) of the equilibrium magnetization has built up is called the T_1-relaxation time. T_1 and T_2 relaxation is caused by time-dependent fluctuations of local magnetic fields arising mostly from the motion of molecules with electric or magnetic dipoles at the site of the spins. For accurate *absolute quantitation* the relaxation times of all metabolites must be known in order to correct peak intensities appropriately.

***T_1*-saturation**—The repetition times T_R are usually in the range of the T_1*-relaxation* times. As a consequence of this, not all the *magnetization* has recovered, for example when $T_R = T_1$ only 63% of the equilibrium magnetization can be used for each scan (with the exception of the first scan) when 90° flip angles are used for excitation. This effect is called T_1-saturation. The extreme case of saturation occurs when several *RF pulses* are applied within a very short time followed by dephasing *gradients*. This technique is used in localized ^{1}H MRS to remove the dominant water signal (see *water suppression*).

***T_2*-relaxation, *T_2*-relaxation time**—The *magnetization* vector can be flipped into the transverse plane by using an *RF pulse*. The so generated transverse magnetization undergoes an exponential decay. The time after the magnetization has relaxed to 37% (1/e) of its amplitude is called the T_2 or transverse or spin-spin relaxation time. See also T_1- *relaxation*.

T_R (repetition time)—The time between each initial excitation of the magnetization is called the repetition time.

References

1. NMR in Biomed, 2003;16(6–7):301–457. Special issue: ^{13}C NMR studies of cerebral metabolism.
2. Adriany G, Gruetter R. J. Magn. Reson. 1997;125(1):178.
3. Bluml S. J. Magn. Reson. 1999;136(2):219.
4. Shen J, Petersen KF, Behar KL, Brown P, Nixon TW, Mason GF, Petroff OA, Shulman GI, Shulman RG, Rothman DL. Proceedings of the National Academy of Sciences of the United States of America. 1999;96(14), 8235.
5. Pan JW, Stein DT, Telang F, Lee JH, Shen J, Brown P, Cline G, Mason GF, Shulman GI, Rothman DL, Hetherington HP. Magn. Reson. Med. 2000;44(5):673.
6. Chhina N, Kuestermann E, Halliday J, Simpson LJ, Macdonald IA, Bachelard HS, Morris PG. J. Neurosci. Res. 2001;66(5):737.
7. Lebon V, Petersen KF, Cline GW, Shen J, Mason GF, Dufour S, Behar KL, Shulman GI, Rothman DL. J. Neurosci. 2002;22(5):1523.
8. FDA. Federal Register, 1988;53:7575.
9. FDA. US Department of Health and Human Services, Food and Drug Administration, http://www.fda.gov/cdrh/ode/mri340.pdf(http://www.fda.gov/cdrh/ode/mri340.pdf) (1998).
10. van den Bergh AJ, van den Boogert HJ, Heerschap A. Magn. Reson. Med. 1998;39(4):642.
11. Hwang JH, Bluml S, Leaf A, Ross BD. NMR Biomed. 2003; 16(3):160.
12. Levitt MH, Freeman R. J. Magn. Reson. 1981;43:502.
13. Levitt MH. J. Magn. Reson. 1982;48:234.
14. Shaka AJ KJ, Frenkiel AJ, Freeman R. J. Magn. Reson. 1983;52:335.
15. Shaka AJ KJ, Freeman R. J Magn Reson, 1983;52:313.
16. de Graaf RA, Mason G, Pantel AB, Behar KL, Rothman DL. NMR Biomed. 2003;16:339.
17. Gruetter R, Adriany G, Choi IY, Henry P-G, Lei H, Oz G. NMR Biomed. 2003;16:313.
18. Gruetter R, Rothman DL, Novotny EJ, Shulman RG. Magn. Reson. Med. 1992;25(1):204.
19. Bluml S, Moreno A, Hwang JH, Ross BD. NMR in Biomed. 2001;14(1):19.
20. Bluml S, Moreno-Torres A, Ross BD. Magn. Reson. Med. 2001;45(6):981.
21. Ordidge RJ, Connelly A, Lohman JAB. J. Magn. Reson. 1986;66:283.
22. Gruetter R, Seaquist ER, Kim S, Ugurbil K. Dev. Neurosci. 1998;20(4–5):380.
23. Rothman DL, Behar KL, Hetherington HP, den Hollander JA, Bendall MR, Petroff OA, Shulman RG. Proc. Natl. Acad. Sci. U.S.A. 1985;82:1633.
24. Provencher SW. Magn. Reson. Med. 1993;30(6):672.
25. Mason GF, Gruetter R, Rothman DL, Behar KL, Shulman RG, Novotny EJ. J. Cereb. Blood. Flow. Metab. 1995;15(1):12.
26. Moreno A, Ross BD, Bluml S. J. Neurochem. 2001;77(1): 347.
27. Bluml S, Moreno-Torres A. Shic F, Nguy CH, Ross BD. NMR Biomed. 2002;15(1):1.
28. Mason GF, Falk Petersen K, de Graaf RA, Kanamatsu T, Otsuki T, Shulman GI, Rothman DL. Brain Res. Brain Res. Protoc. 2003;10(3):181.
29. Watanabe H, Umeda M, Ishihara Y, Okamoto K, Oshio K, Kanamatsu T, Tsukada Y. Magn. Reson. Med. 2000;43(4): 525.
30. Moreno A, Bluml S, Hwang JH, Ross BD. Magn. Reson. Med. 2001;46(1):39.
31. Beckmann N, Turkalj I, Seelig J, Keller U. Biochemistry. 1991;30(26):6362.
32. Magistretti PJ, Pellerin L, Rothman DL. Shulman RG. Science. 1999;283(5401):496.
33. Muir D, Berl S, Clarke DD. Brain Res. 1986;380(2):336.
34. Bluml S, Shic F, Lai L, Yahya K, Lin A, Ross BD. In Proceedings, 10th International Society of Magnetic Resonance in Medicine, Honolulu, 417 (2002).
35. Golman K, Ardenkjaer-Larsen JH, Petersson JS, Mansson S, Leunbach I. Proc. Natl. Acad. Sci. U.S.A. 2003;100(18): 10435.
36. Bluml S, Hwang JH, Moreno A, Ross BD. J. Magn. Reson. 2000;143(2):292.
37. de Graaf RA, Brown PB, Mason GF, Rothman DL, Behar KL. Magn. Reson. Med. 2003;49(1):37.

Magnetic Resonance Spectroscopy and Spectroscopic Imaging of the Prostate, Breast, and Liver

Mark G. Swanson, Susan M. Noworolski, and John Kurhanewicz
Center for Molecular and Functional Imaging, Department of Radiology, University of California, San Francisco, CA 94107, USA

Metabolite Abbreviations: Ala, Alanine; ATP, Adenosine triphosphate; Cho, Choline; Cr, Creatine; Eth, Ethanolamine; Glx, Glutamine/Glutamate; GPC, Glycerophosphocholine; GPE, Glycerophosphoethanolamine; Lac, Lactate; MI, *myo*-Inositol; NAA, N-acetyl aspartate; NTP, Nucleotide triphosphate (e.g. ATP); PC, Phosphocholine; PCr, Phosphocreatine; PE, Phosphoethanolamine; Pi, Inorganic phosphate; SI, *scyllo*-Inositol; Tau, Taurine; UDP, Uridine diphosphate.

Introduction

In this chapter, we describe the current and potential role of magnetic resonance spectroscopy (MRS) and magnetic resonance spectroscopic imaging (MRSI) in organs of the body with an emphasis on the technical aspects of applications in prostate and breast cancer, and diseases of the liver. In contrast to anatomical magnetic resonance imaging (MRI), which detects changes in the relaxivity or density of bulk tissue water, spectroscopy detects small molecular weight metabolites within the cytosol of cells or within extracellular spaces such as glands or ducts. The addition of spectroscopy has been shown to improve the ability (i.e. sensitivity, specificity, and accuracy [1]) of conventional MRI to detect and stage prostate and breast cancer [2–4] and has shown promise in the evaluation of primary and metastatic liver tumors [5] and other liver diseases [6,7]. However, the potential of spectroscopy is even greater because multiple metabolic markers may be combined to provide an independent assessment of disease state, cancer aggressiveness, and therapeutic response [8–10]. The clinical use of spectroscopy as an adjunct to MRI has expanded dramatically over the past several years. This has been due to both the need to answer clinically relevant questions and recent technical advances in hardware and software that have provided improvements in the spatial and time resolution of the spectral data and have resulted in the incorporation of this technology on commercial MR scanners. These breakthroughs have allowed the routine addition of spectroscopy sequences to clinical MRI exams, and have led to spectroscopy being factored into the clinical decision process.

Historically, ^{31}P and ^{1}H have been the nuclei of choice for *in vivo* MRS in the body and each has distinct advantages and disadvantages. The major advantage of ^{1}H spectroscopy is its high sensitivity, which is necessary to achieve high spatial resolution (< 1 cm^3) in a clinically reasonable amount of time. Because the sensitivity of ^{31}P is only 6.6% that of ^{1}H, much larger voxel sizes (typically >8 cm^3) must be used to achieve the same sensitivity in the same amount of time. ^{31}P MRS also suffers from long T_1 and short T_2 relaxation times relative to ^{1}H. The inherently low sensitivity of ^{31}P MRS can be improved by broadband ^{1}H decoupling (e.g. WALTZ) during the acquisition, which sharpens signals by collapsing multiplets and produces a large nuclear overhauser enhancement (NOE) [11], and through the use of higher magnetic field clinical MR scanners. The major advantages of ^{31}P MRS are that no water or lipid suppression is needed and there is less spectral overlap because a relatively small number of metabolites are dispersed over a large spectral window (~25 ppm[1]). However, both ^{1}H and ^{31}P spectroscopy require additional hardware, software, and post-processing and display tools, much of which can now be purchased in the form of spectroscopy packages from the major MR manufacturers General Electric, Siemens, and Philips.

Several metabolites are present in high enough concentrations (>1 mM) to be detected by ^{1}H or ^{31}P MRS, although many have not yet been detected or fully exploited *in vivo* in the body. The ^{1}H MR spectrum spans a frequency range of about 10 ppm and is centered around water at 4.8 ppm. To date, the resonances

[1] Parts-per-million (ppm) is a dimensionless unit used to describe differences (usually from zero) in chemical shift (δ) independent of the applied frequency (B_0). Frequencies in Hertz (Hz) are converted to ppm by dividing the value by the carrier frequency in MHz and multiplying by 10^6. For example, a magnetic field strength of 1.5 T corresponds to a ^{1}H carrier frequency of ~63 MHz, such that 1 ppm ≈ 63 Hz, whereas at 3.0 T, 1 ppm ≈ 126 Hz.

Graham A. Webb (ed.), Modern Magnetic Resonance, 1099–1111.

upfield of water have received the most attention, including lactate (Lac, $\delta = 1.33$ and 4.12 ppm), alanine (Ala, $\delta = 1.48$ and 3.78 ppm), glutamine/glutamate (Glx, $\delta =$ 2.04, 2.11, 2.35, and 3.76 ppm), taurine (Tau, $\delta = 3.26$ and 3.43 ppm), *myo*-inositol (mI, $\delta = 3.28$, 3.54, 3.60, and 4.05 ppm), *scyllo*-inositol (sI, $\delta = 3.35$ ppm), creatine/phosphocreatine (Cr/PCr, $\delta = 3.04$ and 3.93 ppm), the choline containing compounds choline (Cho, $\delta =$ 3.21, 3.55, and 4.07 ppm), phosphocholine (PC, $\delta = 3.23$, 3.62, and 4.18 ppm) and glycerophosphocholine (GPC, $\delta = 3.24$, 3.68, and 4.34 ppm), and the ethanolamine containing compounds ethanolamine (Eth, $\delta = 3.15$ and 3.80 ppm), phosphoethanolamine (PE, $\delta = 3.22$ and 3.99 ppm), and glycerophosphoethanolamine (GPE, $\delta =$ 3.30 and 4.12 ppm). Healthy prostate tissue is unique in that citrate ($\delta = 2.55$ and 2.71 ppm) [12] and polyamines (predominantly spermine, $\delta = 3.11$, 2.09, and 1.78 ppm) [13] are also present in very high concentrations and can be readily observed by ^{1}H MRS. There is also much interest in observing glucose (3.43, 3.80, and 5.23 ppm) and uridine diphosphate (UDP) sugars (5.5–6.1 ppm), which resonate very close to water, because of the role of increased glycolysis in cancer [14]. It has also been demonstrated that amide proton transfer from the downfield exchangeable amide protons of proteins and peptides ($\sim$7.8–8.8 ppm) to water can improve sensitivity by several orders of magnitude and provide a novel imaging mechanism [15].

The major contributors to the ^{31}P MR spectrum include inorganic phosphate (Pi, $\delta = 2.26$ ppm), phosphocreatine (PCr, $\delta = -2.89$ ppm), the phosphomonoesters (PC and PE, $\delta = 3.76$), the phosphodiesters (GPC and GPE, $\delta = 0.11$ and 0.74 ppm), diphosphodiesters (e.g. UDP sugars, $\delta = -11.07$ and -12.76 ppm), and nucleotide phosphates [e.g. adenosine triphosphate (ATP), $\delta = -5.24$, -10.37, and -19.02 ppm] [16]. ^{31}P chemical shifts are typically referenced to either *in vivo* PCr or external 85% H_3PO_4 (as listed here). It should be noted that when ^{31}P chemical shifts are reported with PCr set to 0.00 ppm, the values are 2.89 ppm greater than the corresponding values referenced with H_3PO_4 set to 0.00 ppm. It is also widely believed that ^{13}C spectroscopy, which to date has been primarily limited to the brain [17,18], will play a major role in the future of body MRS as higher field human scanners, ^{13}C labeled substrates (e.g. glucose, acetate, and pyruvate), and commercial ^{13}C hyperpolarizers [19] become more widely available and approved for human studies. However, because ^{13}C MRS is described elsewhere in this book, it will not be further discussed in this chapter.

Previous MR studies on prostate, breast, and liver tumors have identified elevated levels of phosphomonoesters and phosphodiesters (detected by ^{31}P MRS) [5,16,20–24] and elevated levels of the composite Cho resonance (detected by ^{1}H MRS) relative to normal healthy tissues [4,12,20,25–29]. Although the *in vivo* ^{1}H signal that is attributed to Cho contains contributions from Cho, PC, GPC, Eth, PE, GPE, Tau, *m*I, and *s*I, the Cho head group contains nine equivalent protons; consequently, a small increase in concentration results in a large increase in signal intensity. In cancer, the observed increases in Cho and Eth containing compounds have been primarily associated with increased cell membrane synthesis and degradation [24]. However, changes in cell density and altered phospholipid metabolism with cancer evolution and progression also contribute to the observed changes in phospholipid metabolites [27,30].

In spectroscopy it is also very important to identify markers for healthy or normal tissue that can be used for quantitation purposes. In ^{31}P MRS, metabolite ratios are often calculated relative to Pi, PCr, or NTP. In studies of cancer, before and after therapy, the phosphomonoester to phosphodiester ratio is particularly useful because it compares markers of proliferation (PC and PE) to markers of cellular breakdown or apoptosis (GPC and GPE). ^{1}H MRS has been highly successful in the prostate and brain because, in addition to increased Cho, there are unique markers for healthy tissue that decrease in cancer [31]. Consequently, the ratios of Cho to citrate in the prostate and Cho to N-acetyl aspartate (NAA) in the brain are significantly greater in regions of cancer compared to surrounding healthy tissues [2,32]. However, because citrate and NAA are not observed in any other tissues of the body, novel markers for healthy tissue, which may or may not change with malignancy or disease, are needed.

Techniques for Spectroscopy and Spectroscopic Imaging of the Body

Spatial Localization

The most common localization schemes for single voxel spectroscopy are image selected *in vivo* spectroscopy (ISIS) [33], stimulated echo acquisition mode (STEAM) [34], and point resolved spectroscopy (PRESS) [35]. ISIS consists of a series of selective inversion (i.e. 180°) pulses, which are turned on and off according to an eight-step encoding scheme, in the presence of magnetic field gradients. Because the magnetization remains along the "*z*" axis prior to the read pulse, ISIS is relatively insensitive to T_2 relaxation, and therefore has historically been popular for ^{31}P MRS. However, ISIS is particularly sensitive to motion because the eight transients must be added and subtracted to achieve spatial localization. Consequently, STEAM and PRESS, which are capable of three-dimensional (3D) localization in a single acquisition, are preferred for ^{1}H MRS studies in the body.

The STEAM radio frequency (rf) pulse sequence can be represented as $90^\circ - t_1 - 90^\circ - t_2 - 90^\circ - t_1-$acquire, where t_1 and t_2 are inter-pulse delay times. STEAM generates three FIDs, four spin echoes, and one "stimulated echo" at distinct temporal positions, depending on t_1 and t_2, and the unwanted coherences are removed by applying crusher gradients. The desired stimulated echo appears at time $2 \times t_1 + t_2$ and corresponds to the signal from the volume of interest. The PRESS rf pulse sequence can be represented as $90^\circ - t_1 - 180^\circ - t_1 - t_2 - 180^\circ - t_2-$acquire, where t_1 and t_2 are inter-pulse delay times. When the first 180° pulse is applied after time t_1, a spin echo forms at time $2 \times t_1$. When a second 180° pulse is applied after time t_2, the spin echo formed at $2 \times t_1$ is refocused into a second spin echo at time $2 \times t_1 + 2 \times t_2$, producing the signal that corresponds to the volume of interest. As in STEAM, crusher gradients are used in PRESS to remove the unwanted coherences.

Although STEAM can be used with shorter echo times for the observation of short T_2 metabolites, PRESS offers a factor of two times greater signal-to-noise, is less sensitive to motion and diffusion, and is not susceptible to the effects of multiple quantum coherence [36]. The use of longer echo times with PRESS also improves water and lipid suppression, but with improved gradient technology, echo times of the PRESS sequence can be reduced to that of STEAM to exploit the increased chemical information that can be obtained at shorter echo times. Recently, modified PRESS sequences for single voxel MRS and MRSI with very short echo times have been described using asymmetric rf pulses as well as optimized design and timing of the PRESS sequence [37].

Improved Volume Selection and Outer Volume Suppression

Spectroscopy studies in the body are critically dependent on accurate volume selection, since the region of interest is often adjacent to regions of lipid or air-tissue interfaces, which can significantly impair spectral quality. A recent technical advance for ^{1}H MRS has been the substitution of optimally shaped rf pulses, e.g. Shinnar-Le Roux pulses [38], in place of conventional sinc-shaped pulses for improved volume selection in PRESS acquisitions. Although low tip angle pulses can produce reasonably good slice profiles, optimized pulses are essential for 90° and especially 180° excitations [38]. Water saturation performance can also be improved using shaped pulses; however, due to the imperfect excitation profiles of the PRESS spin-echo pulses, even with Shinnar-Le Roux pulses, significant contamination from lipids outside the PRESS selected region can still occur. Several groups have used outer volume suppression (OVS) sequences to better conform the volume of interest [39–43]. These sequences utilized optimized pulses or special excitation schemes to shape the excitation volume to the region of interest. However, due to the non-rectangular suppression profiles of these pulses residual unsuppressed water and lipid signals at the band edges often rendered large portions of the spectral array unusable.

Quadratic phase pulse designs, e.g. very selective suppression (VSS) pulses [44], can provide excellent spatial selectivity, high effective bandwidths, and improved B_1 and T_1 insensitivity compared to conventional OVS pulses. VSS pulses can be inserted just before the PRESS excitation pulses and are used to better define the edges of the PRESS box. Additional VSS pulses can also be graphically prescribed in order to shape the selected volume to the region of interest to exclude regions of lipid or air tissue interfaces. Because of the imperfect PRESS excitation profile, the effects of chemical shift misregistration and the rounded edges of the PRESS selected volume can be dramatically reduced by overprescribing the PRESS selection by ~20–30% and applying the VSS pulses to define the desired dimensions of the box as shown in Figure 1. The use of graphical VSS pulses to exclude periprostatic lipids in the prostate is illustrated in Figure 2A and B.

Water and Lipid Suppression

In order to detect the resonances of biological interest, the 110 molar water resonance must be suppressed by approximately 1000–10,000 fold and spectral contamination from lipids outside the volume of interest must be minimized as much as possible. Good shimming is absolutely essential for water and lipid suppression. Although automated shimming routines are often adequate and should be used as a starting point, it is well worth the extra time to manually shim and visually assess the shape of the water resonance and the FID. The ability to obtain a narrow water line width (<10 Hz) is also dependent upon the proper placement of the PRESS or STEAM box. When there are large differences in magnetic susceptibility, which can be caused by air-tissue or bone-tissue interfaces or the presence of radioactive seeds, it may be impossible to obtain adequate shim. If an adequate shim cannot be obtained in less than 5 min, the volume of interest should be re-prescribed and shimmed again, or the spectroscopy exam should be aborted. Prior to starting the MRS sequence, the water and lipid suppression pulses should be turned on in the pre-scan window to ensure that adequate suppression is being achieved.

closer placement to regions of interest, and specialized breast coil arrays which circle the breasts and extend into the axilla for the supine patient. Many surface coils are now commercially available from a number of vendors (e.g. USA Instruments Inc., MRI Devices Corporation, and Medrad Inc.).

While surface coils and phased arrays of surface coils can provide much better sensitivity than the body coil, their sensitivity varies with position and decreases with increasing distance from the coil. If interpretation of spectra is to be based on ratios of resonance intensities or areas then the inhomogeneous reception profile of the surface coil can generally be neglected. However, the absolute amplitudes of individual metabolite peaks cannot be directly compared without taking into account the surface coil reception profiles. Like MR images, spectra can be at least partially corrected for inhomogeneous reception profiles by numerical evaluation of the Biot–Savart law [51]. Simplistically, the coil is modeled as a series of finite elements of various lengths from which a theoretical reception profile is calculated and the spectra are then divided by the theoretical profile. In practice, such correction algorithms need to take into account the anatomical location of the coil with respect to the location of the voxels to be corrected. The quality of the coil correction can then be evaluated by applying the same algorithm to the corresponding MR images, and by comparing the corrected images to the coil map. If the location of the coil map is incorrect, the images will contain very distinct hypointense (dark) or hyperintense (bright) spots, and the position of the coil should be re-determined and the algorithm reapplied. An alternative approach is to acquire proton density weighted images, which demonstrate the sensitivity profile of the coil, and use these to correct the data [52].

In heteronuclear and high field (e.g. 4 and 7 T) MRS, combined transmit/receive surface coils are required and inhomogeneous excitation profiles become a major issue. Specifically, an inhomogeneous excitation profile results in different locations experiencing different nutation (flip) angles, which causes poor or inadequate spatial selection and chemical shift misregistration. One way of overcoming problems associated with inhomogeneous excitation profiles is to excite with a much larger surface coil than that used for signal reception. Another way to overcome the problems with inhomogeneous excitation profiles of surface coils is to use adiabatic rf pulses, such as B_1 insensitive rotation (e.g. BIR-4) pulses [53]. Adiabatic pulses produce a uniform flip angle, typically 90° or 180°, across a region of interest despite variations in B_1, provided that the rf power is above a minimum threshold value. Adiabatic pulses are highly versatile and can be exploited for spatial localization [e.g. localization by adiabatic selective refocusing (LASER) [54]], as well as water suppression and OVS [e.g. B_1-insensitive train to obliterate signal (BISTRO) [43,55]].

Motion

Respiratory and peristaltic motion can also be major problems in body spectroscopy. In the prostate, peristaltic motion is reduced by the use of an inflatable endorectal coil. Because the prostate is directly beneath the bladder, as shown in Figure 2B, additional motion may be caused by the bladder filling up during the course of the examination. To minimize this, patients are asked to refrain from drinking (especially caffeinated beverages) for 2–3 h prior to the exam. In the breast, motion is reduced by having the patient lay supine on a breast coil, supported by the chest, resulting in the breasts remaining relatively free of respiratory motion. Nonetheless, one study showed that in 20 breast and abdominal tumors, 30% moved 6–23 mm, while the diaphragm and fatty tissues of the gut typically moved ~15–20 mm [56]. Recently, it has been reported that breathheld MRS can significantly reduce phase and frequency shifts and outer voxel contamination due to respiratory motion [57], while the use of navigator echoes can aid with retrospective motion correction [58].

Data Processing and Display

Single voxel MRS data can often be processed and displayed using commercial software packages designed for conventional NMR spectroscopy, provided that the header can be interpreted or removed. Basic data processing involves Lorentzian and/or Gaussian apodization of the FID to enhance resolution and/or signal-to-noise, baseline correction, zero-filling, and Fourier transformation of the data. Phasing and frequency referencing are then performed manually on the resulting spectra. Because MRSI data may contain hundreds of useable spectra, completely automated and robust data processing algorithms are essential. In addition to the basic data processing steps just described, the data must be reconstructed to correctly reproduce the spatial dependence of the data [59] and can also be spatially zero filled as shown in Figure 2C and D. After the spectra have been Fourier transformed, automated baseline, phase, and frequency corrections can then be applied using water as a reference or by using prior knowledge of the approximate relative positions of the major peaks in the spectrum. Peak areas may be estimated by integration across fixed frequency ranges, by fitting baseline subtracted data as a sum of components with particular lineshapes [60–63], or using linear combinations of model *in vitro* spectra [64].

Several different approaches have been used to display the information from multi-dimensional localized spectra and to correlate spatial variations in metabolites with the anatomy [50,65–68]. These include superimposing a grid on the MR image and plotting the corresponding arrays of spectra, and calculating images of the spatial distribution of metabolites to overlay on the corresponding MR images. These formats provide an excellent summary of the spatial distribution of different metabolites enabling rapid identification of regions of suspected abnormal signal and facilitating correlation with the anatomy. Additionally, since 3D volume MRI and MRSI data are collected, the data can be viewed in any plane (axial, coronal, or sagittal), and the spatial position of spectroscopic voxels can be selected retrospectively via "voxel-shifting," using the appropriate mathematical weighting of the raw data based upon the translation property of the Fourier transform [69,70]. This method of interactive analysis will be the way that MRI/MRSI data is used in the future and should reduce interpretative errors associated with the overlap of normal and abnormal tissues.

Data Interpretation

Interpretation of spectroscopy and spectroscopic imaging data requires both knowledge of what constitutes a clinically interpretable spectrum and an understanding of the underlying biochemistry and morphology that result in the observed changes. ^{1}H spectra are considered clinically interpretable if they are not contaminated by insufficiently suppressed water or lipid and have resolvable metabolite peaks with peak area to noise ratios of greater than 5 to 1. Metabolic criteria must then be established to distinguish abnormal from normal metabolism and then validated using a pathologic "gold standard." In the prostate, the (Cho + Cr)/Cit ratio discriminates prostate cancer from benign glandular tissues with high specificity [12]. In the other organs of the body, similar metabolic criteria must still be established. *Ex vivo* high-resolution magic angle spinning (HR-MAS) spectroscopy and quantitative pathologic analysis of intact surgical or biopsy tissues can aid in understanding the relationship between metabolism and tissue composition [71] and thereby help with the identification of the appropriate metabolic criteria. Finally, the metabolic criteria used to identify residual or recurrent disease often change following therapy [8]. Treatment effects typically result in an overall reduction in the signal-to-noise of all metabolites, which underscores the need to be able to distinguish an absence of metabolites (termed "metabolic atrophy") from a technically failed study.

Applications in the Prostate, Breast, and Liver

To date, *in vivo* MRS has been applied to the prostate [12,22], breast [4,25,72,73], liver [74,75], kidney [76,77], colon [78], heart [79–81], skeletal muscle [82–84], sarcomas [85,86], and non-Hodgkin's lymphomas [10,87,88]. In the following sections, specific applications in the prostate, breast, and liver are described and ^{1}H MRS examples are shown.

Prostate Cancer

The earliest MRS studies of the human prostate involved ^{31}P spectroscopy using a dual tuned (^{31}P/^{1}H) transmit/receive endorectal probe [22,89,90]. The proton frequency was used to image the location of the coil and obtain a homogeneous field of view, prior to performing phosphorus spectroscopy. These studies demonstrated the ability of ^{31}P MRS to detect metabolic differences between normal, hyperplastic, and malignant prostate tissues. Specifically, the ^{31}P MR spectra taken from regions of prostate cancer were characterized by increases in the phosphomonoester to β-NTP ratio and decreases in the PCr to β-NTP ratio relative to healthy prostate tissues [22,89,90]. Additional studies using murine models of prostate cancer also identified ^{31}P spectral characteristics that may be related to the hormone sensitivity, radiation sensitivity, and metastatic potential of the cancer [16]. However, these early ^{31}P MRS studies of the prostate were limited by coarse spectral localization and spatial resolution due to the inherent insensitivity of ^{31}P MRS. Therefore, most of the clinical prostate studies performed to date have utilized a combination of high spatial resolution MRI and ^{1}H MRSI.

High-resolution T_2-weighted imaging, using combined endorectal and pelvic phased array coils, has demonstrated good sensitivity but relatively poor specificity for identifying prostate cancer, because numerous other conditions, including prostatitis, benign prostatic hyperplasia, and treatment effects can all mimic cancer. Prior to therapy, the addition of 3D-MRSI to anatomical MRI has been shown to improve the localization [2] and staging [3] of prostate cancer, provide a measure of prostate tumor volume [91], and provide an assessment of prostate cancer aggressiveness [92]. The ability of combined MRI/MRSI to identify residual or recurrent prostate cancer after hormone deprivation [8,9] and radiation therapy [93,94] has also been described.

Healthy glandular prostate tissue demonstrates two unique metabolic markers, citrate and spermine, that are produced by highly specialized epithelial cells and secreted into the prostatic ducts that empty into the ejaculate. Both citrate and spermine are reduced or absent in regions of prostate cancer, due to both biochemical changes and a loss of the prostate's normal ductal morphology [95–98]. Although the citrate resonance is completely resolved, its detection is complicated due to strong coupling. At 1.5 T, under good shimming conditions, polyamines are seen as a hump resonating between Cho and creatine, while in regions of cancer, the reduction or loss of polyamines is observed as an increase in the discrimination between Cho and creatine.

At our institution, axial high-resolution T_2-weighted images (3 mm slice thickness, no intersection gap) are used to select a PRESS volume (typically 50–100 cm^3) that encompasses the entire prostate, but excludes periprostatic lipids, the seminal vesicles which contain very high levels of GPC, and the air tissue interface of the rectum. The PRESS box is overprescribed by 20–30% in all three dimensions and then six VSS pulses are used to define the desired edges of the PRESS box. Six additional VSS pulses are graphically prescribed to eliminate contamination from surrounding lipids (Figure 2A and B). With a typical voxel size of ~7 mm on a side (0.34 cm^3), 16 × 8 × 8 step phase encoding (112 × 56 × 56 mm^3 field of view) is performed with a 1 s repetition time for a total acquisition time of ~17 min.

Spectral data from a MRI/MRSI staging exam acquired under these conditions are shown in Figure 2. Figures 2A and B show the location of the PRESS box and the locations of the graphical VSS bands. Figure 2C shows a portion of the 3D spectral grid, which has been zero filled in the anterior/posterior direction, and Figure 2D shows the corresponding MRSI spectral data. As can be seen, the right side of the gland (left side of the image/array) demonstrates fairly normal metabolism with high levels of citrate and low levels of Cho. However, the left side of the gland (right side of the image/array) demonstrates very elevated levels of Cho and in some areas a complete absence of citrate, consistent with the presence of an aggressive prostate tumor. The patient subsequently went on for a transrectal ultrasound (TRUS) guided biopsy and was diagnosed with Gleason 4 + 3 prostate cancer.

The interpretation and utility of prostate MRSI can be complicated by the presence of chronic inflammation (prostatitis) [99], which appears metabolically similar to prostate cancer. Additionally, post-biopsy hemorrhage can persist for 6–8 weeks or more after biopsy and demonstrates a reduction or absence of citrate and polyamines due to the disruption of prostatic ducts, and in worse cases results in a complete absence of all prostatic metabolites [100]. Regions of hemorrhage are usually identified as bright areas, and less frequently dark areas, on axial T_1-weighted images and these can be used to exclude suspected regions from the MRSI data analysis. Following therapy, the time course of treatment-induced metabolic changes must be taken into account when interpreting the data. For example, hormone deprivation has a very fast impact on prostate metabolism and often results in a total loss of citrate and polyamines within 16 weeks [8]. Conversely, radiation therapy has a much slower impact on prostate metabolism and may take 1–3 years to achieve metabolic atrophy. In either case, the presence of elevated Cho relative to creatine appears to be the best current indicator of recurrent prostate cancer.

Breast Cancer

For breast cancer, one of the critical clinical questions is whether the lesion detected on screening is benign or malignant. About 75% of the breast lesions detected by mammography and about 50% of the enhancing lesions detected by contrast-enhanced MRI are pathologically benign [101]. Sonographic classification of benign and malignant tumors has low specificity (about 30%) as well [102]. Recent advances in contrast-enhanced MRI methodology and interpretation have greatly improved the ability to differentiate malignant from benign breast tumors. However, there still remains a clinical need for improved specificity [103,104], which spectroscopy may be able to provide [4].

Breast spectroscopy has lagged behind prostate and brain spectroscopy because of the presence of mobile lipids and a lack of multiple metabolic markers. Early ^{31}P MRS studies typically observed increased phosphomonoester, phosphodiester, and sometimes PCr levels in breast cancer vs. normal tissues [105]. However, because of the poor sensitivity of ^{31}P MRS combined with the decreasing size of breast tumors due to early detection, ^{1}H MRS is now being primarily used. Typical breast MRS studies use a single voxel ($\sim 1 - 27$ cm^3) [4] localization technique such as STEAM or PRESS with the intent of selecting signals from the lesion of interest and excluding signals from the surrounding adipose tissue and normal parenchyma [4]. However, due to the magnitude of lipid resonance in the breast, gradient induced sidebands of the lipid resonance can cause both positive and negative artifacts in the Cho region of the spectrum. Recently, a technique called "T_E-averaging," based upon oversampled 2D J-resolved spectroscopy [106], has demonstrated the ability to separate lipid induced sidebands and provide increased sensitivity for the study of small or irregularly shaped lesions [72]. An example of the spectral improvement provided by T_E-averaging is shown in Figure 3.

Currently, the interpretation of breast ^{1}H MRS data is relatively simple and mainly involves determining whether Cho is present (malignant) or absent (benign) [4]. This observation is consistent with the high PC content of human breast cancer cells, which is 10-fold higher than that of normal human mammary epithelial cells [107]. However, Stanwell *et al.* found that 20% of normal volunteers also demonstrated detectable Cho, leading to an overall sensitivity and specificity of 80 and 86%, respectively, when Cho presence alone is used to define cancer [108]. Additionally, smaller tumors tend to be diagnosed as benign (false–negative) because of the lack of a detectable composite Cho signal [4]. Other complications in the interpretation of breast spectra have primarily been technical and associated with poor spectral quality [4,109]. The utility and robustness of breast spectroscopy can be improved by increasing the Cho signal-to-noise

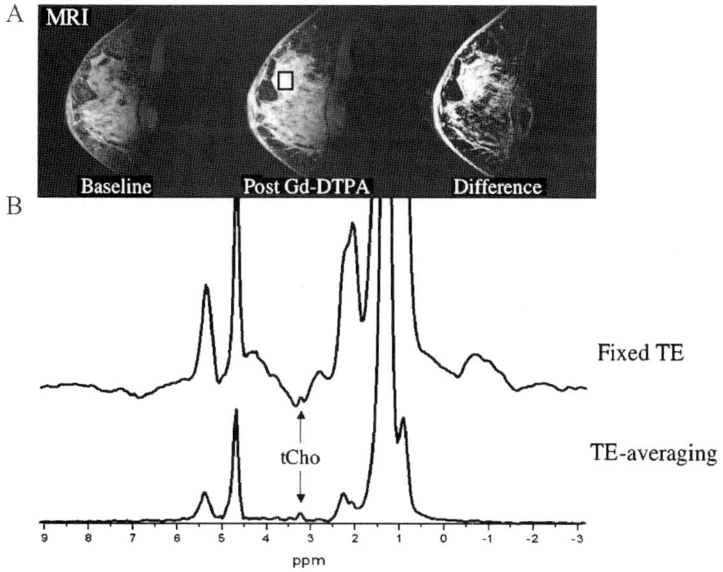

Fig. 3. (A) T_1-weighted image, dynamic contrast-enhanced image, and difference image of a poorly differentiated adenocarcinoma of the breast acquired at 4.0 T. (B) Single voxel ^{1}H MR spectra localized using LASER, demonstrating the improvement in lipid sideband suppression obtained by T_E-averaging (data acquired at 4.0 T). Figure courtesy of Drs Patrick J. Bolan and Michael Garwood, University of Minnesota School of Medicine.

through the use of higher sensitivity coils, the addition of new metabolic markers, and the use of improved pulse sequences and higher field human scanners.

MRS of the Liver

The liver is the most biochemically complex organ in the human body, and in addition to being the second most common site of cancer metastasis (after lymph nodes) [110], the liver is subject to a variety of non-malignant diseases including inflammation, hepatitis, cirrhosis, and fatty liver disease. The majority of liver MRS studies have looked at fat content and have shown that proton MRS of liver fat correlates well with *ex vivo* liver fat measurements [111]. Lipid levels have been shown to change with cancer [28,29], metabolism [112,113], and non-alcoholic fatty liver disease (NAFLD) and non-alcoholic steatohepatitis (NASH) [7,114,115]. Example single voxel ^{1}H MR spectra from a healthy individual and from a patient with fatty liver disease are shown in Figure 4. While the body mass index of these two individuals were similar (26 vs. 27, respectively), the lipid to water ratio is much greater in the subject with fatty liver disease.

Because of the problem of motion in the abdomen, liver MRS is typically performed as a single voxel technique, with a surface coil phased array used for reception. In the case of diffuse disease (e.g. fatty liver disease), where the location within the organ is not critical, the voxel should be placed at least 1 cm from all edges of the liver to reduce the chance of contamination from signals outside the liver. In the superior/inferior direction, the voxel should be placed such that it will remain in the liver throughout respiration. This can be ascertained by comparing breathheld images acquired at end-expiration to those at end-inspiration. During pre-scanning, the signal amplitude should also remain relatively constant from pulse to pulse. To account for small shifts in position over time, MR spectra can be saved individually before averaging. These individual MR spectra can then be phase and frequency aligned before summation to account for some respiratory motion.

Summary

MR spectroscopy and spectroscopic imaging are promising techniques for the metabolic assessment of cancer and other diseases in the body. However, spectroscopy in the body is challenging due to motion and the deep location of many organs. The critical considerations when performing spectroscopy in the body include the choice of: nucleus (^{1}H is preferred at 1.5 T due to high sensitivity, but ^{31}P and ^{13}C may be valuable at higher field); single voxel vs. multi-voxel (CSI or MRSI) approach; localization scheme (e.g. PRESS, STEAM, LASER); water and lipid suppression (e.g. BASING, CHESS, STIR) and OVS techniques (e.g. VSS pulses); surface coils; and data display and analysis tools. Proper spectral interpretation based upon a pathologic "gold standard" and knowledge of the impact of therapy on metabolism are both critical for evaluating the clinical utility of new MRS applications in the

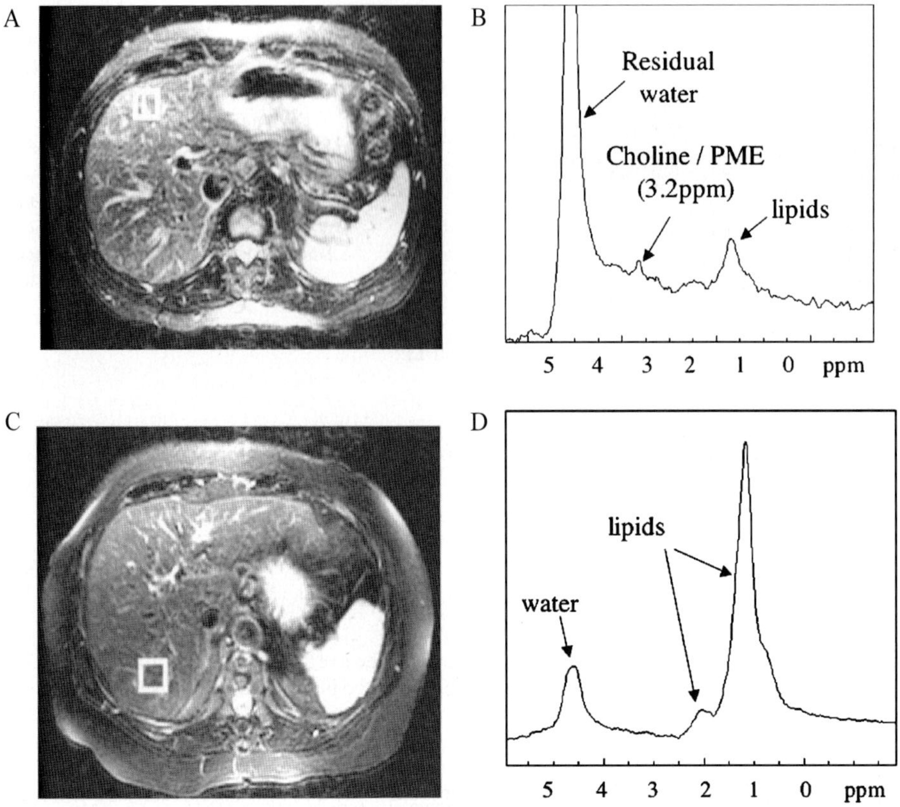

Fig. 4. Axial T_2-weighted image and $2 \times 2 \times 2$ (8 cm^3) single voxel ^1H spectra acquired from the liver of a normal volunteer (A and B) and a patient with NAFLD (C and D). Data were acquired at 1.5 T using PRESS with VSS pulses and CHESS water suppression (no lipid suppression), with 128 transients, $T_R = 2500$ ms, and $T_E = 35$ ms. Voxels were prescribed >1 cm from the edges of the liver, avoiding vessels, and spectra were individually phased and frequency shifted before averaging.

body. The increased sensitivity provided by higher field MR scanners, improved pulse sequences and technology, and the ability to include additional metabolic markers will also have a major impact on the clinical potential of body spectroscopy in the future.

Acknowledgments

The authors wish to acknowledge and thank Drs. Michael Garwood and Patrick J. Bolan, Center for Magnetic Resonance Research, University of Minnesota School of Medicine, Minneapolis, for providing the breast figure presented in this chapter, and Dr. Aliya Qayyum, Department of Radiology, University of California, San Francisco, for providing the liver figures presented in this chapter.

Glossary of Terms

BASING	Band selective inversion with gradient dephasing
BIR	B_1-insensitive rotation
BISTRO	B_1-insensitive train to obliterate signal
CHESS	Chemical shift selective saturation
CSI	Chemical shift imaging
EPSE	Echo-planar spin-echo
FID	Free induction decay
HR-MAS	High-resolution magic angle spinning
ISIS	Image selected *in vivo* spectroscopy
LASER	Localization by adiabatic selective refocusing
MRI	Magnetic resonance imaging
MRS	Magnetic resonance spectroscopy

MRSI	Magnetic resonance spectroscopic imaging
NAFLD	Non-alcoholic fatty liver disease
NASH	Non-alcoholic steatohepatitis
NOE	Nuclear overhauser enhancement
OVS	Outer volume suppression
PPM	Parts-per-million
PRESS	Point resolved spectroscopy
STEAM	Stimulated echo acquisition mode
STIR	Short time inversion recovery
T_1	Longitudinal (spin–lattice) relaxation time
T_2	Transverse (spin–spin) relaxation time
TRUS	Transrectal ultrasound
VSS	Very selective suppression
WALTZ	Broadband decoupling scheme

References

1. Obuchowski NA. Radiology. 2003;229:3.
2. Scheidler J, Hricak H, Vigneron DB, Yu KK, Sokolov DL, Huang LR, Zaloudek CJ, Nelson SJ, Carroll PR, Kurhanewicz J. Radiology. 1999;213:473.
3. Yu KK, Scheidler J, Hricak H, Vigneron DB, Zaloudek CJ, Males RG, Nelson SJ, Carroll PR, Kurhanewicz J. Radiology. 1999;213:481.
4. Katz-Brull R, Lavin PT, Lenkinski RE. J. Natl. Cancer. Inst. 2002;94:1197.
5. Bell JD, Bhakoo KK. NMR Biomed. 1998;11:354.
6. Cho SG, Kim MY, Kim HJ, Kim YS, Choi W, Shin SH, Hong KC, Kim YB, Lee JH, Suh CH. Radiology. 2001;221: 740.
7. Heiken JP, Lee JK, Dixon WT. Radiology. 1985;157:707.
8. Mueller-Lisse UG, Swanson MG, Vigneron DB, Hricak H, Bessette A, Males RG, Wood PJ, Noworolski S, Nelson SJ, Barken I, Carroll PR, Kurhanewicz J. Magn. Reson. Med. 2001;46:49.
9. Mueller-Lisse UG, Vigneron DB, Hricak H, Swanson MG, Carroll PR, Bessette A, Scheidler J, Srivastava A, Males RG, Cha I, Kurhanewicz J. Radiology. 2001;221:380.
10. Arias-Mendoza F, Smith MR, Brown TR. Acad. Radiol. 2004;11:368.
11. van Sluis R, Payne GS, Leach MO. Magn. Reson. Med. 1995;34:893.
12. Kurhanewicz J, Vigneron DB, Hricak H, Narayan P, Carroll P, Nelson SJ. Radiology. 1996;198:795.
13. Cheng LL, Wu CL, Smith MR, Gonzalez RG. FEBS Lett. 2001;494:112.
14. Stubbs M, Bashford CL, Griffiths JR. Curr. Mol. Med. 2003; 3:49.
15. Zhou J, Lal B, Wilson DA, Laterra J, van Zijl PC. Magn. Reson. Med. 2003;50:1120.
16. Kurhanewicz J, Dahiya R, Macdonald JM, Jajodia P, Chang LH, James TL, Narayan P. NMR Biomed. 1992;5:185.
17. Ross B, Lin A, Harris K, Bhattacharya P, Schweinsburg B. NMR Biomed. 2003;16:358.
18. Morris P, Bachelard H. NMR Biomed. 2003;16:303.
19. Ardenkjaer-Larsen JH, Fridlund B, Gram A, Hansson G, Hansson L, Lerche MH, Servin R, Thaning M, Golman K. Proc. Natl. Acad. Sci. U.S.A. 2003;100:10158.
20. Cornel EB, Smits GA, Oosterhof GO, Karthaus HF, Deburyne FM, Schalken JA, Heerschap A. J. Urol. 1993;150:2019.
21. Kalra R, Wade KE, Hands L, Styles P, Camplejohn R, Greenall M, Adams GE, Harris AL, Radda GK. Br. J. Cancer. 1993;67:1145.
22. Kurhanewicz J, Thomas A, Jajodia P, Weiner MW, James TL, Vigneron DB, Narayan P. Magn. Reson. Med. 1991;22:404.
23. Ronen SM, Leach MO. Breast Cancer Res. 2001;3:36.
24. Daly PF, Lyon RC, Faustino PJ, Cohen JS. J. Biol. Chem. 1987;262:14875.
25. Cecil KM, Schnall MD, Siegelman ES, Lenkinski RE. Breast Cancer Res. Treat. 2001;68:45.
26. Kurhanewicz J, Dahiya R, Macdonald JM, Chang LH, James TL, Narayan P. Magn. Reson. Med. 1993;29:149.
27. Aboagye EO, Bhujwalla ZM. Cancer Res. 1999;59:80.
28. Kuo YT, Li CW, Chen CY, Jao J, Wu DK, Liu GC. J. Magn. Reson. Imaging. 2004;19:598.
29. Soper R, Himmelreich U, Painter D, Somorjai RL, Lean CL, Dolenko B, Mountford CE, Russell P. Pathology. 2002;34: 417.
30. Schiebler ML, Tomaszewski JE, Bezzi M, Pollack HM, Kressel HY, Cohen EK, Altman HG, Gefter WB, Wein AJ, Axel L. Radiology. 1989;172:131.
31. Kurhanewicz J, Vigneron DB, Nelson SJ. Neoplasia. 2000; 2:166.
32. Vigneron D, Bollen A, McDermott M, Wald L, Day M, Moyher-Noworolski S, Henry R, Chang S, Berger M, Dillon W, Nelson S. Magn. Reson. Imaging. 2001;19:89.
33. Ordidge RJ, Connelly A, Lohman JAB. J. Magn. Reson. 1986;66:283.
34. Frahm J, Merboldt KD, Hanicke W. J. Magn. Reson. 1987;72:502.
35. Bottomley PA. Ann. N.Y. Acad. Sci. 1987;508:333.
36. Moonen CT, von KM, van ZP, Cohen J, Gillen J, Daly P, Wolf G. NMR Biomed. 1989;2:201.
37. Geppert C, Dreher W, Leibfritz D. MAGMA. 2003;16:144.
38. Pauly J, Le Roux P, Nishimura D, Macovski A. IEEE Trans. Med. Imaging. 1991;10:53.
39. Shungu DC, Glickson JD. Magn. Reson. Med. 1993;30: 661.
40. Posse S, Tedeschi G, Risinger R, Ogg R, Le Bihan D. Magn. Reson. Med. 1995;33:34.
41. St Lawrence K, Lee TY, Henkelman M. Magn. Reson. Med. 1998;40:944.
42. Le Roux P, Gilles RJ, McKinnon GC, Carlier PG. J. Magn. Reson. Imaging. 1998;8:1022.
43. Luo Y, de Graaf RA, DelaBarre L, Tannus A, Garwood M. Magn. Reson. Med. 2001;45:1095.
44. Tran TK, Vigneron DB, Sailasuta N, Tropp J, Le Roux P, Kurhanewicz J, Nelson S, Hurd R. Magn. Reson. Med. 2000;43:23.
45. Haase A, Frahm J, Hanicke W, Matthaei D. Phys. Med. Biol. 1985;30:341.
46. Bydder GM, Young IR. J. Comput. Assist. Tomogr. 1985; 9:659.
47. Star-Lack J, Nelson SJ, Kurhanewicz J, Huang LR, Vigneron DB. Magn. Reson. Med. 1997;38:311.

48. Schricker AA, Pauly JM, Kurhanewicz J, Swanson MG, Vigneron DB. Magn. Reson. Med. 2001;46:1079.
49. Brown TR, Kincaid BM, Ugurbil K. Proc. Natl. Acad. Sci. U.S.A. 1982;79:3523.
50. Maudsley AA, Hilal SK, Simon HE, Wittekoek S. Radiology. 1984;153:745.
51. Moyher SE, Vigneron DB, Nelson SJ. J. Magn. Reson. Imaging. 1995;5:139.
52. Liney GP, Turnbull LW, Knowles AJ. J. Magn. Reson. Imaging. 1998;8:994.
53. DeGraaf RA, Nicolay K. Concepts Magn. Reson. 1997;9:247.
54. Garwood M, DelaBarre L. J. Magn. Reson. 2001;153:155.
55. de Graaf RA, Luo Y, Garwood M, Nicolay K. J. Magn. Reson. B. 1996;113:35.
56. Schwarz AJ, Leach MO. Phys. Med. Biol. 2000;45:2105.
57. Katz-Brull R, Rofsky NM, Lenkinski RE. Magn. Reson. Med. 2003;50:461.
58. Tyszka JM, Silverman JM. Magn. Reson. Med. 1998;39:1.
59. Nelson SJ. Magn. Reson. Med. 2001;46:228.
60. Spielman D, Webb P, Macovski A. Magn. Reson. Med. 1989;12:38.
61. Nelson SJ, Brown TR. J. Magn. Reson. 1987;75:229.
62. Nelson S, Brown T. J. Magn. Reson. 1989;84:95.
63. Derby K, Hawryszko C, Tropp J. Magn. Reson. Med. 1989;12:235.
64. Provencher SW. NMR Biomed. 2001;14:260.
65. Tropp JS, Sugiura S, Derby KA, Suzuki Y, Hawryszko C, Yamagata H, Klein JE, Ortendahl DA, Kaufman L, Acosta GF. Radiology. 1988;169:207.
66. Twieg DB, Meyerhoff DJ, Hubesch B, Roth K, Sappey MD, Boska MD, Gober JR, Schaefer S, Weiner MW. Magn. Reson. Med. 1989;12:291.
67. Lenkinski RE, Holland GA, Allman T, Vogele K, Kressel HY, Grossman RI, Charles HC, Engeseth HR, Flamig D, MacFall JR. Radiology. 1988;169:201.
68. Vigneron DB, Nelson SJ, Murphy BJ, Kelley DA, Kessler HB, Brown TR, Taylor JS. Radiology. 1990;177:643.
69. Bracewell RN. The Fourier Transform and Its Applications. McGraw-Hill: New York, 1978, p 104.
70. Marcei TH, Booker HR. J. Magn. Reson. 1984;57:157.
71. Swanson MG, Vigneron DB, Tabatabai ZL, Males RG, Schmitt L, Carroll PR, James JK, Hurd RE, Kurhanewicz J. Magn. Reson. Med. 2003;50:944.
72. Bolan PJ, DelaBarre L, Baker EH, Merkle H, Everson LI, Yee D, Garwood M. Magn. Reson. Med. 2002;48:215.
73. Bolan PJ, Meisamy S, Baker EH, Lin J, Emory T, Nelson M, Everson LI, Yee D, Garwood M. Magn. Reson. Med. 2003;50:1134.
74. Li CW, Negendank WG, Murphy-Boesch J, Padavic-Shaller K, Brown TR. NMR Biomed. 1996;9:141.
75. Gonen O, Murphy-Boesch J, Li CW, Padavic-Shaller K, Negendank WG, Brown TR. Magn. Reson. Med. 1997;37:164.
76. Dixon RM, Frahm J. Magn. Reson. Med. 1994;31:482.
77. Kim DY, Kim KB, Kim OD, Kim JK. J. Korean Med. Sci. 1998;13:49.
78. Kasimos JN, Merchant TE, Gierke LW, Glonek T. Cancer Res. 1990;50:527.
79. Bottomley PA, Hardy CJ, Roemer PB. Magn. Reson. Med. 1990;14:425.
80. Deslauriers R, Kupriyanov VV. Biochem. Cell. Biol. 1998;76:510.
81. Neubauer S, Horn M, Hahn D, Kochsiek K. Mol. Cell. Biochem. 1998;184:439.
82. Kreis R, Boesch C. J. Magn. Reson. B. 1996;113:103.
83. Boska MD, Nelson JA, Sripathi N, Pipinos II, Shepard AD, Welch KM. Magn. Reson. Med. 1999;41:1145.
84. Meyerspeer M, Krssak M, Moser E. Magn. Reson. Med. 2003;49:620.
85. Li CW, Kuesel AC, Padavic-Shaller KA, Murphy-Boesch J, Eisenberg BL, Schmidt RG, von Roemeling RW, Patchefsky AS, Brown TR, Negendank WG. Cancer Res. 1996;56:2964.
86. Sijens PE. NMR Biomed. 1998;11:341.
87. Negendank WG, Padavic-Shaller KA, Li CW, Murphy-Boesch J, Stoyanova R, Krigel RL, Schilder RJ, Smith MR, Brown TR. Cancer Res. 1995;55:3286.
88. Griffiths JR, Tate AR, Howe FA, Stubbs M. Eur. J. Cancer. 2002;38:2085.
89. Narayan P, Vigneron DB, Jajodia P, Anderson CM, Hedgcock MW, Tanagho EA, James TL. Magn. Reson. Med. 1989;11:209.
90. Narayan P, Jajodia P, Kurhanewicz J, Thomas A, MacDonald J, Hubesch B, Hedgcock M, Anderson CM, James TL, Tanagho EA *et al.* J. Urol. 1991;146:66.
91. Coakley FV, Kurhanewicz J, Lu Y, Jones KD, Swanson MG, Chang SD, Carroll PR, Hricak H. Radiology. 2002;223:91.
92. Kurhanewicz J, Swanson MG, Nelson SJ, Vigneron DB. J. Magn. Reson. Imaging. 2002;16:451.
93. Pickett B, Ten Haken RK, Kurhanewicz J, Qayyum A, Shinohara K, Fein B, Roach M. Int. J. Radiat. Oncol. Biol. Phys. 2004;59:665.
94. Coakley FV, Qayyum A, Swanson MG, Lu Y, Roach M, Pickett B, Shinohara K, Vigneron DB, Kurhanewicz J. Radiology. 2004;233:441.
95. Costello LC, Franklin RB. Prostate. 1991;18:25.
96. Costello LC, Franklin RB. Prostate. 1991;19:181.
97. Heby O. Differentiation. 1981;19:1.
98. Heston WD. Cancer Surv. 1991;11:217.
99. Shukla-Dave A, Hricak H, Eberhardt SC, Olgac S, Muruganandham M, Scardino PT, Reuter VE, Koutcher JA, Zakian KL. Radiology. 2004;231(3):717.
100. Kaji Y, Kurhanewicz J, Hricak H, Sokolov DL, Huang LR, Nelson SJ, Vigneron DB. Radiology. 1998;206(3):785.
101. Orel SG, Schnall MD. Radiology. 2001;220:13.
102. Buchberger W, Niehoff A, Obrist P, DeKoekkoek-Doll P, Dunser M. Semin. Ultrasound CT MR. 2000;21:325.
103. Degani H, Gusis V, Weinstein D, Fields S, Strano S. Nat. Med. 1997;3:780.
104. Esserman L, Hylton N, George T, Weidner N. Breast J. 1999;5:13.
105. Leach MO, Verrill M, Glaholm J, Smith TA, Collins DJ, Payne GS, Sharp JC, Ronen SM, McCready VR, Powles TJ, Smith IE. NMR Biomed. 1998;11:314.
106. Hurd RE, Gurr D, Sailasuta N. Magn. Reson. Med. 1998;40:343.
107. Gribbestad IS, Petersen SB, Fjosne HE, Kvinnsland S, Krane J. NMR Biomed. 1994;7:181.

108. Stanwell P, Gluch L, Clark D, Tomanek B, Baker L, Giuffre B, Lean C, Malycha P, Mountford C. Eur. Radiol. 2005;15:1037.
109. Star-Lack J, Vigneron DB, Pauly J, Kurhanewicz J, Nelson SJ. J. Magn. Reson. Imaging. 1997;7:745.
110. Pickren J, Tsukada Y, Lane WW. In: L Weiss, HA Gilbert (Eds). Analysis of Autopsy Data. GK Hall and Company: Boston, 1982, p 2.
111. Thomsen C, Becker U, Winkler K, Christoffersen P, Jensen M, Henriksen O. Magn. Reson. Imaging. 1994;12:487.
112. Ryysy L, Hakkinen AM, Goto T, Vehkavaara S, Westerbacka J, Halavaara J, Yki-Jarvinen H. Diabetes. 2000;49:749.
113. Sutinen J, Hakkinen AM, Westerbacka J, Seppala-Lindroos A, Vehkavaara S, Halavaara J, Jarvinen A, Ristola M, Yki-Jarvinen H. AIDS. 2002;16:2183.
114. Rosen BR, Carter EA, Pykett IL, Buchbinder BR, Brady TJ. Radiology. 1985;154:469.
115. Szczepaniak LS, Babcock EE, Schick F, Dobbins RL, Garg A, Burns DK, McGarry JD, Stein DT. Am. J. Physiol. 1999;276:E977.

first, Babylonian confusion and a broad range of opinions about the usefulness of MRM [22–72], especially before 1995.

3. Using MRM small lesions of a few millimeters could be detected, which were often not seen by X-ray or ultrasound. Therefore, a need for the development of interventional MR procedures was rising, at first for positioning of markers, later for biopsy and therapeutic removal [73–80]. These techniques currently require an additional MR image to be taken later, i.e. a second measurement and a repeated use of the MR-device and contrast injection. At present single breast biopsy coils in the so-called "closed" as well as in "open" MR-devices are tested. A bilateral coil for combined imaging and simultaneous intervention has been used in clinical tests since 1997 [81].

Pathophysiological Background of MRM

The secret of the very high sensitivity of MRM for the detection of breast lesions is the detection of tumor-based angiogenesis [82–84]. It is well-established that any tumor of more than 2 mm needs an increased perfusion to selectively direct nutrients towards the malignant tumor. Fifty years ago it was found that the implantation of breast cancer cells induces new micro-vessels after 3 days post implantation. Meanwhile numerous publications especially from Judah Folkman, have clarified that human breast cancers, like other malignant tumors, possess the angiogenentic power as well [85–91]. Over 20 years ago [92] it was established that, in a "far" distance of 3 cm from the cancer, an increased concentration of tumor angiogenetic factors can be measured. Probably induced by

A

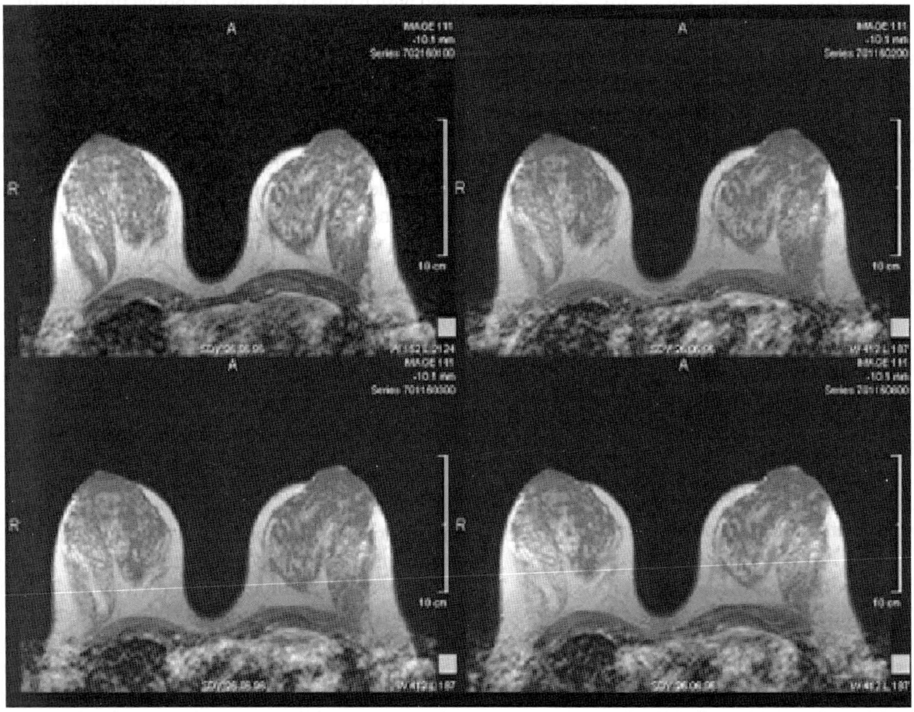

Fig. 1. Normal enhancement, Inflow-phenomenon (40-year-old patient, dense breast) (A) Original dynamic scans showing the uptake of contrast over 7 min (upper left = precontrast, upper right = 1-min-postcontrast, lower left = 2-min-postcontrast, lower right = 7 min postcontrast), slow progressive enhancement especially in the outer portions of the parenchyma, bilateral, symmetric enhancement. The contrast usually arrives via lateral thoracic artery in the outer upper quadrant at first (inflow phenomenon) (arrows). (The left breast is displayed on the right side of the image, the right breast is displayed on the left side of the images.) (B) The inflow phenomenon is best delineated in the corresponding subtraction images (arrows) (upper left = subtraction 1-min-postcontrast minus precontrast, upper right = subtraction 2-min-postcontrast minus precontrast, lower left = subtraction 3-min-postcontrast minus precontrast, lower left = subtraction 7-min-postcontrast minus precontrast). (C) So-called "mosaic-image" (upper left = precontrast, upper right = 1-min-postcontrast, lower left = subtraction 1-min-postcontrast minus precontrast, lower left = T_2w-TSE image; all images in identical sclice positions).

B

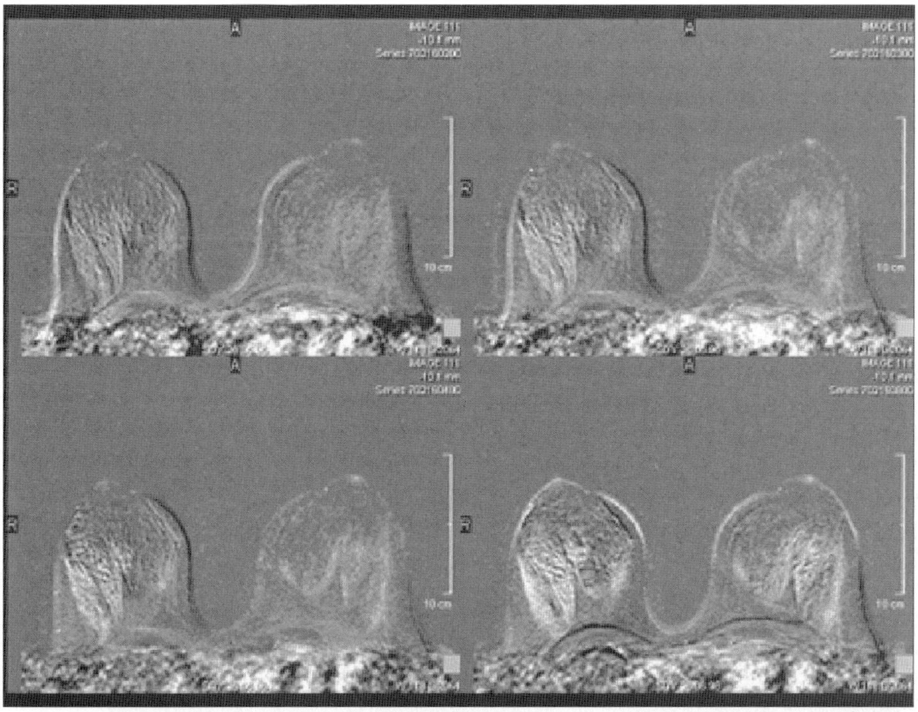

C

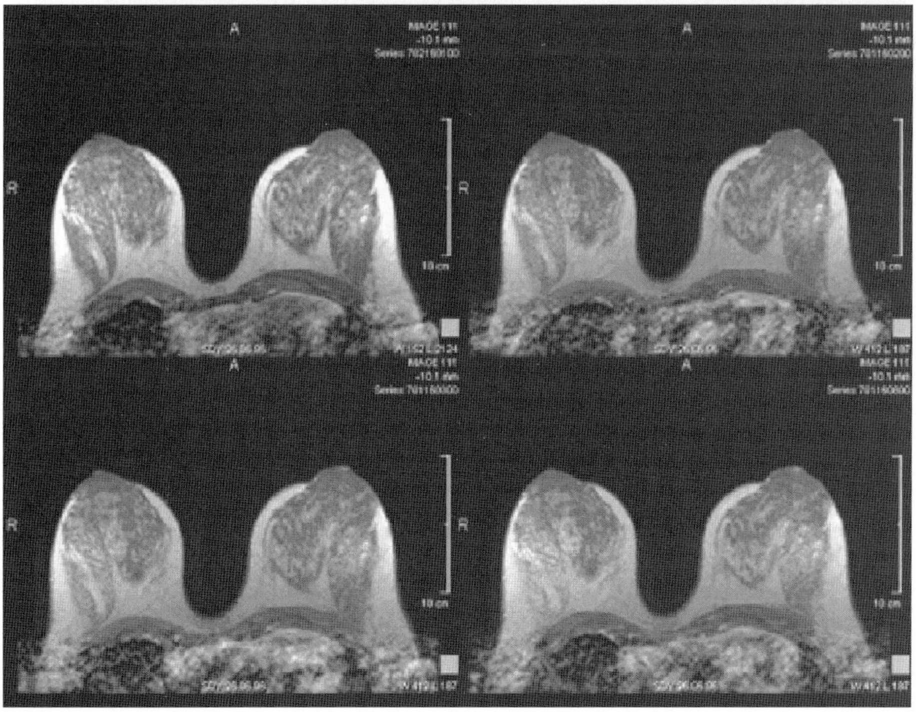

Fig. 1. *(Continued)*

hypoxia, acidosis, and many tumor angiogenetic factors, and eventually combined by an inactivation of the p53 suppressor gene, the so-called "tumor angiogenetic switch" is induced and new vessels arise through the sprouting of capillaries from pre-existing micro-vessels. Host capillaries dilate and become hyperpermeable, fibrin exudes through the leaky capillaries, proteinases, and collagenases break basal membranes, endothelial cells migrate across ruptures, form loops and canalize and finally a "functional neovasculature" called "tumor angiogenesis" is built.

Unique properties of tumor vessels are increased tumor blood volume, increased arterio-venous shunt formation, altered capillary bed transit time, increased interstitial pressure, and increased capillary permeability.

Among more than fifty so-called "tumor-angiogenetic factors" the vascular permeability factor (VPF), vascular endothelial growth factor (VEGF), and basic fibroblastic growth factor (bFGF) are most important.

Various pharmacokinetic models revealed that, in particular, differences in tumor vessel permeability are the

A

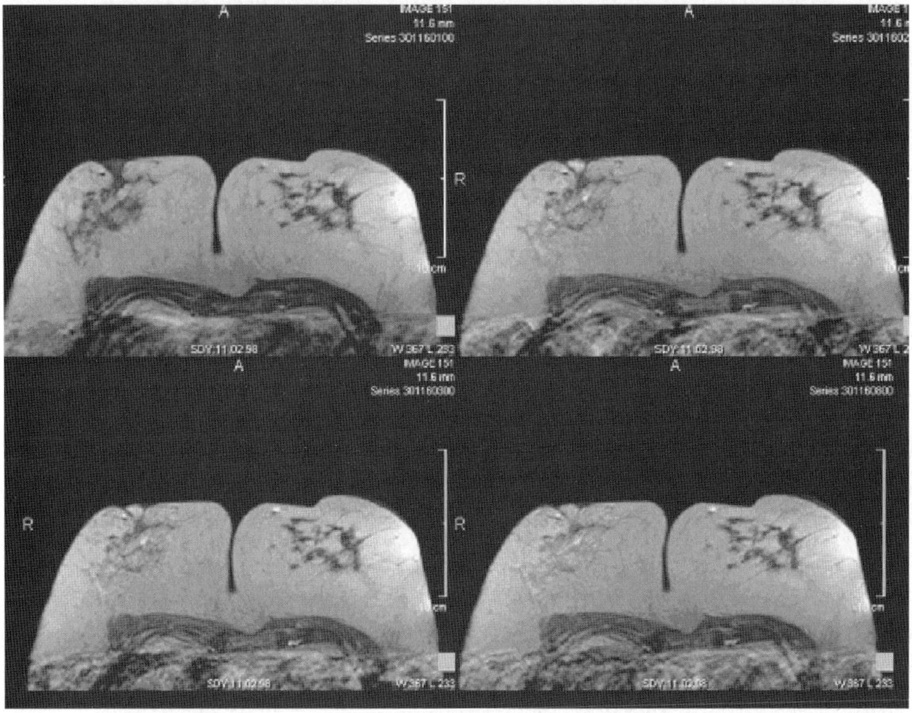

Fig. 2. Invasive lobular cancer and Lobular cancer in situ (LCIS, 78-year-old patient). (A) Original dynamic scans showing the uptake of contrast over 7 min (upper left = precontrast, upper right = 1-min-postcontrast, lower left = 2-min-postcontrast, lower right = 7 min postcontrast), strikingly enhancing area in a diameter of 1.3 cm behind the nipple in the right breast (arrow) with a maximum signal rise within the first minute after the injection of contrast followed by a decrease of signal intensity ("washout-phenomenon") which is very typical for a cancer. In addition to that, nearly the complete right parenchyma shows a reticular, network-like enhancement compared to the other non-affected left breast (ellipse). (B) This reticular enhancement (ellipse) can clearly be detected in the corresponding subtraction images besides the focal mass enhancement of the invasive cancer (arrow). Most of this network-loke enhancement representing LCIS is steadily continuously enhancing with confluent irregular lesions in late scans (upper left = subtraction 1-min-postcontrast minus precontrast, upper right = subtraction 2-min-postcontrast minus precontrast, lower left = subtraction 3-min-postcontrast minus precontrast, lower left = subtraction 7-min-postcontrast minus precontrast). (C) The "mosaic-image" (upper left = precontrast, upper right = 1-min-postcontrast, lower left = subtraction 1-min-postcontrast minus precontrast, lower left = T_2w-TSE image in identical sclice positions) shows the invasive cancer dark in T_2w-scans (arrow), whereas the LCIS is unspecific in T_2w-signal intensity (ellipse). The main hallmark for invasive cancer is a striking "washin" in a typical range for cancers followed by a "washout-phenomenon;" the main hallmark for non-invasive cancer is the asymmetric enhancement in relation to the opposite breast (both the invasive and the non-invasive cancer had not been described by X-ray mammography and ultrasound in this case).

B

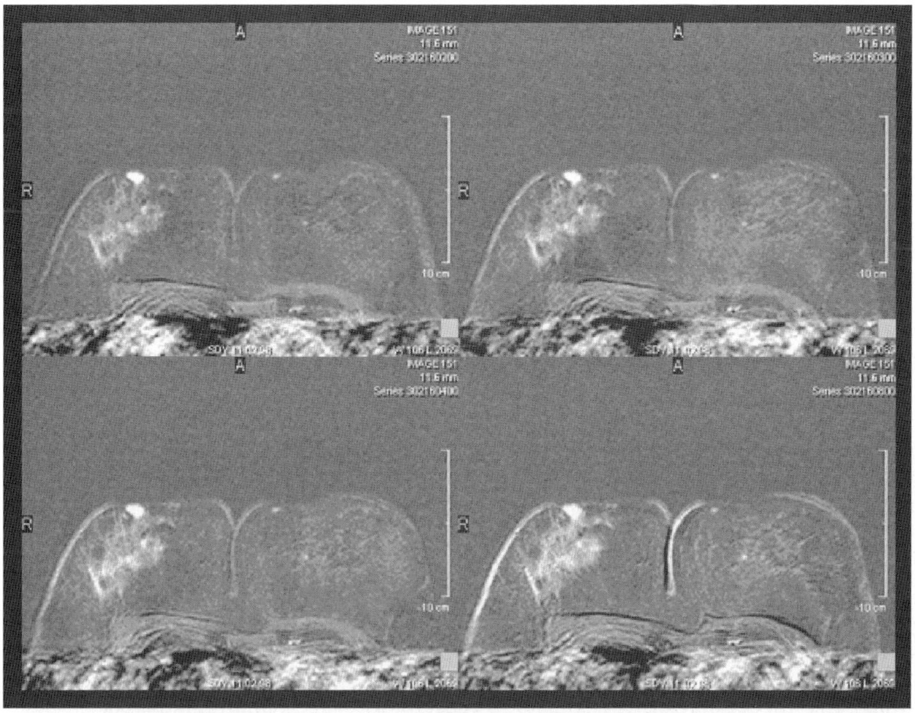

C

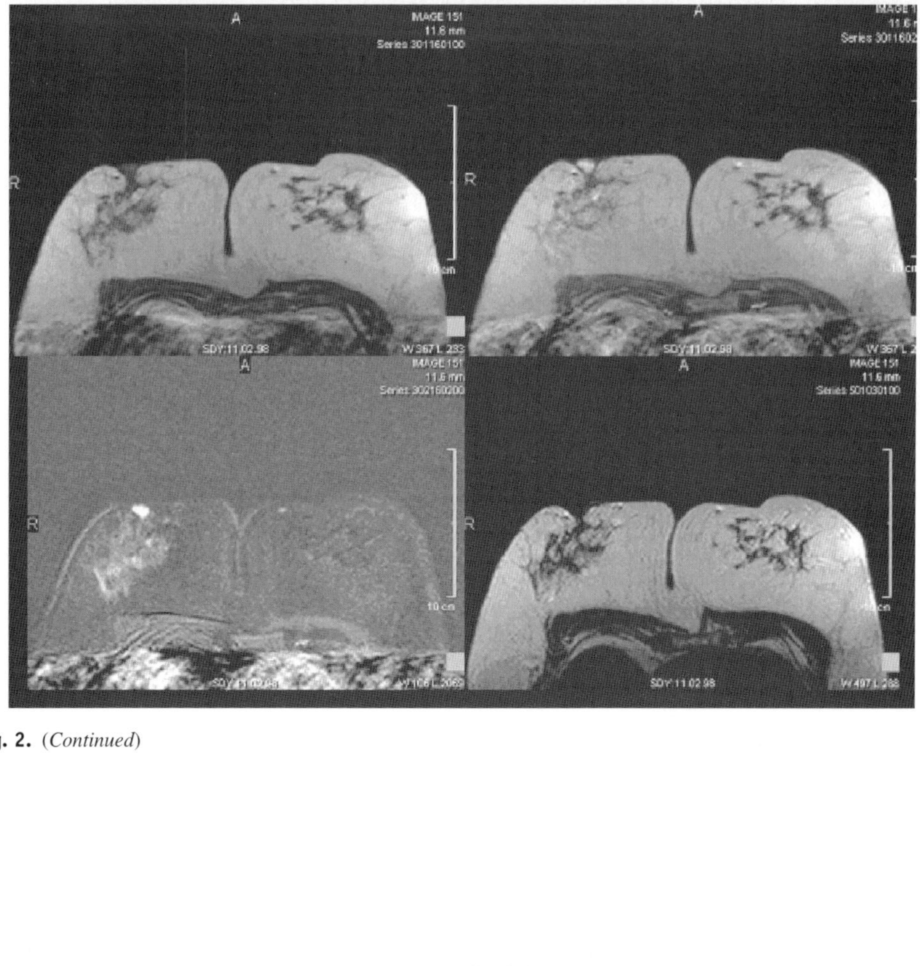

Fig. 2. (*Continued*)

defining feature of tumor angiogenetically based uptake of contrast medium in breast cancer [93–96].

As a consequence of this pathophysiological basis MRM is looking for the type, the amount and the morphology of enhancement within the neighborhood of, and in, a lesion [97–102]. As mentioned above, malignant lesions are characterized by a tumor angiogenetic network of tumor-specific micro-vessels; thus, they show a rapid initial increase of signal intensity in T_1-weighted images (T_1w-images) after the injection of contrast medium within the first 1–2 min after injection, followed by a plateau, or the so-called "washout" effect, i.e. a decrease of signal intensity. This "washout" effect is probably caused by the arterio-venous-shunts within the angiogenetic network. Most benign tumors show a significantly different type of enhancement, either a slow progressive uptake or a very rapid increase with a further signal rise after the initial enhancement. However, about 20% of benign lesions (especially myxoid fibro-adenomas) might also show a strong initial enhancement followed by a plateau or a washout effect.

It is vital that the kinetic analysis is drawn only from the non-necrotic part of the tumor and that the maximum signal increase is used; a tumor contains all types of cells and compartments including haemorrhage, necrosis, and fibrosis besides vital malignant tumor cells. The so-called "wash-in"-phenomenon (i.e. the initial enhancement before the bending of the curve) is the most sensitive

A

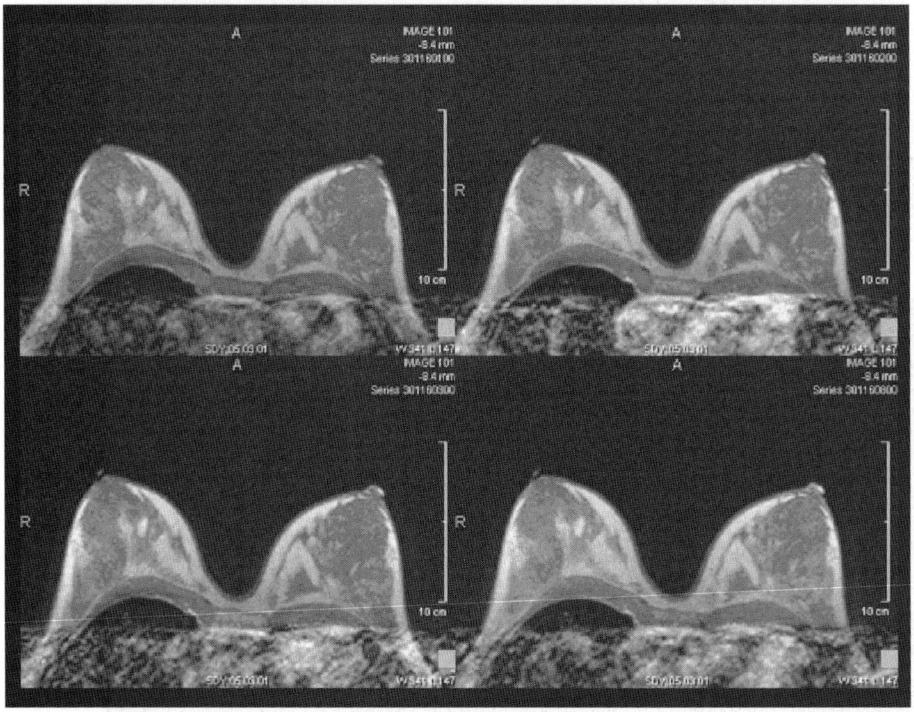

Fig. 3. Fibrous fibroadenoma (41-year-old patient with a lump in the lateral part of the right breast. The left breast is displayed on the right side of the image, the right breast is displayed on the left side of the image). (A) The original dynamic scans show a slow steady area of enhancement in the lateral portion of the right breast (arrow) without any cancer-like washin or washout-phenomenon. (upper left = precontrast, upper right = 1-min-postcontrast, lower left = 2-min-postcontrast, lower right = 7 min postcontrast). (B). The corresponding subtraction images (upper left = subtraction 1-min-postcontrast minus precontrast, upper right = subtraction 2-min-postcontrast minus precontrast, lower left = subtraction 3-min-postcontrast minus precontrast, lower left = subtraction 7-min-postcontrast minus precontrast) display this slow enhancement (arrow) better than the original images. In addition a slight motion artefact in the left breast. (C) The "mosaic-image" (upper left = precontrast, upper right = 1-min-postcontrast, lower left = subtraction 1-min-postcontrast minus precontrast, lower left = T_2w-TSE image in identical sclice positions). The T_2w-scan show the lesion in a dark signal intensity (arrow). Typical signal change of a fibrous fibroadenoma, i.e. a slow, steady enhancement without cancer signs, dark in precontrast and dark in T_2w-scans.

B

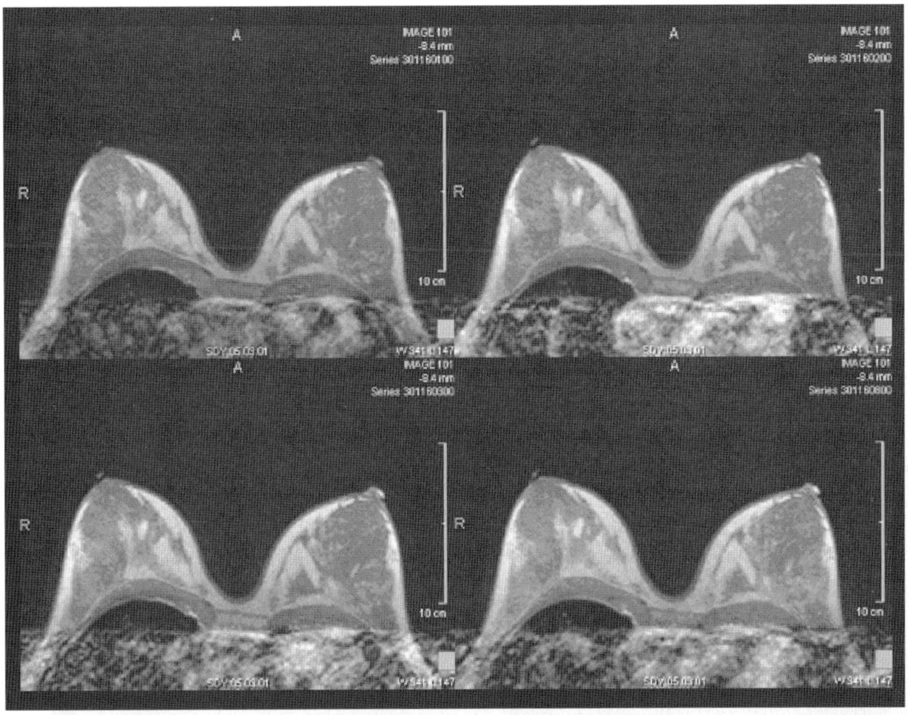

C

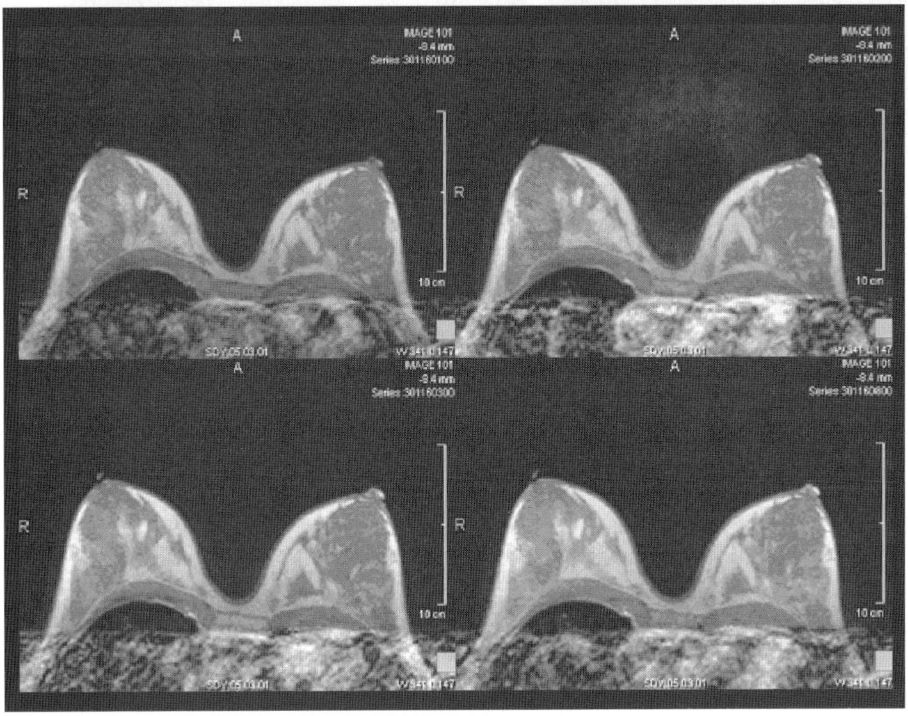

Fig. 3. *(Continued)*

lesion. After more than 20 years of clinical research a significant number of publications describe a very high sensitivity for MRM for detecting malignant lesions in the range of more than 95–100%. Even indications which had previously been described as "difficult", like the identification of Ductal carcinoma *in situ* (DCIS), are now accepted indications for MRM [99].

Technique

High temporal and spatial resolution is important for the detection and distinction between benign and malignant lesions. Today high-resolution dynamic techniques, i.e. the repeated imaging of the same slices before and in short time intervals after injection of contrast are used. The patient is placed in a prone position with the breasts hanging into a bilateral phased-array double breast coil after an iv-cannula had been placed in the cubital vein. The series of sequences in a dynamic MRM examination at 1.5 T includes (further technical details about the MRM examination are described elsewhere [101]:

1. a scout sequence (to clarify the position of the patient and breasts)
2. a coronal T_1-weighted Gradient-Echo-Sequence (T_1w-GRE-sequence) in a large field-of-view (FOV) including the axillary region and the neck region (to detect enlarged lymph nodes)
3. a transversal T_2w-TSE sequence.
4. a transversal T_1w-GRE sequence (i.e. the precontrast scan)
5. an iv injection of contrast (0.1 mmol/kg Gd-DTPA)
6. a sevenfold repetition of the precontrast scan (=nr. 4, in identical imaging parameters) every minute for the next 7 min
7. an automatic image subtraction
8. a coronal post-contrast scan (identical scan as nr. 2, so that any lesion in the breasts or the lymph nodes can be evaluated in two pre- and post-contrast orientations)
9. in the case of a breast implant, a silicone-selective sequence is added (focusing on the resonance frequency of silicone, which is 100 Hz lower than the resonance frequency of fat at 1.5 T.

The important T_1-weighted dynamic gradient-echo-sequences before and in short time intervals after the injection of contrast are performed as two dimensional-(2D) gradient-echo-sequences (TR/TE/flip-angle = 100 ms/4.7 ms/80°) in our institution. The FOV for scan nr. 4 is 350 mm, the slice thickness 3 mm for 36 slices in a matrix of 512 × 512 in a parallel imaging data aquisition. The aquisition time for sequence nr. 4 is 60 s.

Alternatively a dynamic three dimensional-(3D) gradient-echo-sequence can be performed which allows a better spatial resolution with thinner slices. However, the homogeneity of 3D-sequences in some machines is limited and artefacts caused by inhomogeneity deteriorate the kinetic data evaluation.

Another alternative is a dynamic fat-saturation-sequence (fat-sat-sequence) which usually also requires a following subtraction.

After the measurements, the kinetic data evaluation (signal-intensity-time-curves) in addition to the morphological analysis is made.

Modern MR-machines are able to acquire images in 512 × 512 matrices in time distances of each minute. Today parallel imaging techniques in fast gradient-echo-sequences (FLASH, FFE, GRASS) fulfill these prerequisites. Some teams prefer the so-called "fat-sat"-techniques by applying saturation pulses towards fat signal; these techniques suffered from signal inhomogeneities, longer measurement times and often single-breast examinations in the past and therefore could not be successfully applied in Europe [101].

Indications for MRM

The scientific results gained so far recommend MRM for the following indications:

1. The definite identification and/or exclusion of breast cancer in cases of unclear mammographic or sonographic breasts.
2. Clarification of number, size, and location of breast lesions, i.e. clarification of multifocality and/or multicentricity; thus enabling more precise planning for treatment.
3. Differentiation between scars and/or recurrent/residual tumors after previous operation/radiation.
4. Delineation of implants, implant-ruptures, and/or lesions around implants like inflammations or benign or malignant tumors.
5. Detection of primary tumors in the case of an unknown primary tumor following a positive lymph node.
6. Delineation of diseases in the contralateral breast before operation/radiation.
7. Detection of the early effect of chemotherapy after the first doses of anticancer drugs.
8. Examination of high-risk patients (family history, genetic risk like BRCA1 and BRCA2).

Discrepancies and Pitfalls

Despite striking results for the sensitivity the value of MRM is still in question because of variations in specificity, the high cost of the machine and examination, its inability to detect micro-calcifications, its role in detecting DCIS, questions about different dosages of contrast etc. As with other MR-examinations, patients with ferromagnetic particles, pacemakers, and operations that have taken place within the weeks prior to the examination, and patients undertaking hormone-replacement-therapy should not be included.

The major pitfalls consist of the following:

1. Signal-inhomogeneities in the image due to the rf-profile of the double-breast coil.
2. Signal-inhomogeneities and saturation artefacts in fat-sat images
3. Application of an inappropriate dose and type of contrast medium.
4. Measurement in an inappropriate ("opposed") echo-time where fat and water protons are oriented in different directions.
5. Inappropriate positioning of the breast using high compression of the breast and/or bent arms and lack of control of contrast uptake.
6. Incomplete injection of contrast medium and/or no detection of incomplete injection.
7. Movement of patient during the examination due to insufficient information.
8. Lack of kinetic analysis.
9. No inclusion of T_2-weighted-scans (T_2w-scans)
10. False data evaluation
11. Susceptibility artefacts due to small bleedings because of previous punctures and/or biopsies.
12. Omittance of normal enhancements and "inflow" effects.

Since the time intensity curve is (in addition to morphological evaluations) very important for the diagnosis, the signal intensity in the first post-contrast images, in particular, is absolutely crucial for diagnosis. This critical first contrast signal intensity image depends on exact measurement parameters. Injection time, injection site, sodium chloride bolus, length of plastic tube, arm position, tuning parameters are some of the major factors affecting this critical image signal intensity.

In our experience an increase of more than 90% in the first 90 s after the beginning of contrast injection ("90–90 rule") is a typical threshold for a malignant enhancement at 1.5T-2D-dynamic MRM. To reach this critical threshold measurement using a region of interest procedure (ROI) has to be performed carefully: Only the "vital" tumor areas show a critical initial enhancement followed by a plateau phenomenon or a washout effect. The inclusion of necrotic areas of the tumor or surrounding normal glandular or fatty tissue will deteriorate and falsify the effect of kinetic data evaluation. It is necessary to look for the fastest enhancing portion of the lesion.

It is also important not to include any enhancing focus and to describe it as malignant. Vessels are cut in different orientations in the slice and are therefore displayed in a round, oval or comet shaped manner. The detection of inflow phenomenon, the delineation of further vessels in other slices or other orientations or, if you have still any doubts, an additional MR angiography sequence clarifies the tiny enhancing area as a vessel or not.

Fat saturation techniques are in relatively widespread use in the United States. They allow a high spatial resolution, but have a limited value for kinetic data evaluation. Since a pre-pulse in a low bandwidth is necessary, the adjustment parameters in pre- and post-contrast scans are not identical. Current fat-sat techniques usually have a relatively high signal inhomogeneity, so that a quantitative evaluation is difficult and subtractions are difficult, though often necessary. The measurement time is in most cases longer than in non-fat saturation techniques and the "diagnostic window" for the detection of differences between benign and malignant enhancement criteria is restricted.

Fat suppression methods may be by temporal subtraction, which is ideal for fat suppression, since only enhancing structures are delineated as high signal intensity lesions. An alternative is fat suppression by pre-pulses with the above-mentioned limitations.

The injection of contrast medium is made in a bolus (3 ml/s) either manually or by an automatic injector followed by a post-contrast bolus of physiological sodium chloride of 20 ml (cubital vein) or 30 ml (lower arm/hand). The procedure should be explained to the patient before positioning in the machine. No movement of, or discussion with, the patient is recommended during injection. The dosage of contrast medium should not be higher than 0.1 mmol/kg. A higher dosage shortens the already short "diagnostic window", i.e. the time difference of only 1–2 min where the contrast uptake between malignant and benign lesions is sufficiently different in order to enable a differential diagnosis.

The evaluation begins by examining fatty tissue on both breasts to assess if the signal intensity of fat is constant. If this signal intensity of fatty tissue as an "internal standard" varies, consider changing receiver adjustment or field inhomogeneity effects or artefacts. In these cases a quantitative evaluation is difficult or impossible, since the technical problems overlap any morphological information.

Dynamic MRM should be performed in transversal scans, because in this orientation the identification of vessels, nipple, posterior borders of implants, fatty layer between parenchyma and muscle, signal homogeneity of the coil, and signal correlation with the aorta is possible. However, if the lateral border of the parenchyma towards the axillary region and/or lymph nodes is especially important, a dynamic MRM in coronal scans is advantageous. A sagittal orientation should be avoided because of doubled measurement time and the lack of correlation with the other breast, and the difficulty of detecting inflow phenomenon and hormone effects. This image orientation might be acceptable if sagittal images can be acquired simultaneously from both breasts.

If only the axillary region is important, the rotation of the phase encoding direction towards a dorso-ventral orientation is recommended to keep the lateral axillary region free of phase encoding artefacts.

It is of crucial importance to use echo times where fat and water protons are in "in-phase" conditions, because Gd has a signal-*increasing* effect on the water protons in the voxel only in these echo times. Particularly tiny reticular structures like DCIS or lobular cancers can be imaged only in these "in-phase" images. "In-phase" conditions for 1.5 T are even numbers of 2.4 ms, i. e. 4.8, 9.6 ms etc. Odd numbers of 2.4 ms, for example 7.2 ms are "forbidden" because of the "opposed effect" of water and fat protons. This effect is field-strength dependent and should be adjusted according to the field strength which is used.

During the dynamic examination it is essential not to change tuning parameters between scans and to guarantee identical pulses in pre- and post-contrast scans. No change of pulses should be allowed between pre- and postcontrast scans. Only under these conditions is a correct signal intensity/time evaluation possible.

The pre-contrast signal intensity of parenchyma should be in a typical range for this sequence. If it is too low, this may be caused by an inappropriate receiver adjustment or inappropriate coil design; try to repeat the receiver adjustment. If this pre-contrast signal intensity is too high, it could also be an inappropriate receiver adjustment or a medical/biological reason (bleeding after puncture/biopsy, hormone effect, pregnancy, previous operation, and/or radiation).

Diagnosis of breast lesions in MRM is always made by a combined morphological and kinetic analysis. It is the main feature of MRM that tumor typical "tumor-angiogenesis" can be detected by signal-intensity-time-curves. A typical malignant lesion shows the striking initial increase (wash-in phenomenon) within the first 1–2 min followed by either a constant signal (plateau phenomenon) or—more specifically—a decrease in signal intensity (washout effect). The sudden increase and the following constancy or washout effect make the "cancer-corner" of the signal-time-curve. Benign lesions normally show a slower initial increase which is continuously rising over the complete dynamic examination. The distribution of a lesion is described as regional or patchy or diffuse or symmetric; the lesion itself is morphologically described as a focal mass with sharp or non-sharp margins, linear, linear-branched or segmental aspects. The initial and late enhancement is described as being homogeneous, heterogeneous, rim enhancement, bright or dark septations after contrast and centripetal or centrifugal filling effect over time. A centrifugal filling (from inside to outside) is directed mainly towards a benign lesion, a centripetal filling (from outside to inside) towards a malignant lesion.

Another major mistake is the inclusion of patients under hormone-replacement-therapy for MRM, because hormone effects cause bilateral, patchy, continuously enhancing spots which are confluent on late scans [103] and thus might be difficult to separate from bilateral DCIS or lobular cancers. However, in most patients DCIS and lobular cancers show unilateral enhancements and, in two-thirds, also a cancer-like wash-in and washout effect.

Future Challenges

At present MRM is classified as an additional diagnostic tool in all cases of unclear mammographic and sonographic situations. The medical community needs teaching, training, increased experience, and consensus. MRM will allow a precise detection of cancers in a size as small as 3 mm. For this purpose special techniques providing high spatial and temporal resolution are needed.

New specific contrast agents might enable a selective display of specific cancers and differentiate them from other lesions in the breast and in the whole body as well.

The combination of MRI and MR-spectroscopy will probably increase the overall accuracy of MRM.

Today modern "high-field MR-devices" (high-field for imaging means above 1.0 T) do not allow direct access to the breast during data acquisition. The patient has to be in the isocenter of the magnet field during imaging and must be moved outside the machine for a following intervention. The position of a wire marker, a core biopsy or a laser fibre has to be checked by a following MR-image, thus increasing measurement time, artefacts, and pitfalls. A MR-compatible robotic system for simultaneous imaging and immediate biopsy/intervention at high field strength is in clinical testing.

Combining MRI and therapy in one session—at least for lesions smaller than 2 cm—might enable a "one-stop-shop" procedure of detection and simultaneous removal in one single examination [104–105]. Preliminary tests applying laser therapy, radio frequency therapy, or cryotherapy have been promising so far [106–107] as has receptor-specific near-infrared-imaging *in vitro* [108].

References

1. Damadian R. Science 1971;171:1151–1153.
2. Lauterbur PC. Nature 1973;242:190–191.
3. Mansfield P, Morris PG, Ordidge R. *et al.* Brit. J. Radiol. 1979;52:242–243.
4. Ross RJ, Thompson JS, Kim K, Bailey RA. Radiology 1982;143:195–205.
5. El Yousef SJ, Alfidi RJ, Duchesneau RH. J. Comput. Assist. Tomogr. 1983;7:215–218.
6. El Yousef SJ, Duchesneau RH, Alfidi RJ, Haaga JR, Bryan PJ, LiPuma JP. Radiology 1984;150:761–766.
7. Kaiser W. Arch. Int. de Physiol. et de Biochim. 1985;93:67–76.
8. Fritschy P, Müller E, Sauter R, Kaiser W. Radiology 1984;153(P):243–244.
9. Stelling CB, Wang PC, Lieber A, Mattingly SS, Griffen WO, Powell DE. Radiology 1985;154:457–462.
10. Kaiser W, Zeitler E, Fortschr. Röntgenstr. 1986;144.5:459–465.
11. Kaiser W, Zeitler E, Fortschr. Röntgenstr. 1986;144.5:572–579.
12. Heywang SH, Fenzl G, Edmaier M, RÖFO 1985;143:207–212.
13. Weinmann HJ, Laniado M, Mützel W. Physiol. Chem. Phys. Med. NRM. 1984;16:167–172.
14. Eskin BA, Parker JA, Bassett JG, George DL. Obstet. Gynecol. 1974;44:398–492.
15. Chang CHJ, Sibala JL, Lin F, Fritz SL, Gallagher JH, Dwyer-ill SJ, Templeton AW. Am. J. Roentgenol. 1978;131:459–464.
16. Heywang SH, Hahn D, Schmidt H, Krischke I, Eiermann W, Bassermann RJ, Lissner J. J. Comput. Assist. Tomogr. 1986;10:199–204.
17. Haase A, Frahm J, D Matthaei *et al.* J. Magn. Res. 1986;67:258–268.
18. Kaiser WA, Oppelt A, KST der Mamma mit Schnellbildverfahren. MR'87, 2. Internationales Kernspintomographie Symposium, 29.1.–1.2.87, Garmisch-Partenkirchen, Referateband, 303–310, Schnetztor-Verlag, Konstanz (1987).
19. Kaiser WA, Zeitler E. Radiology 1989;170:681–686.
20. Heywang SH, Wolf A, Pruss E, Hilbertz T, Eiermann W. Radiology 1989;171:95–103.
21. Kaiser WA, Kess H. Fortschr. Röntgenstr. 1989;151:103–105.
22. Kaiser WA. Eur. Front. Radiol. 1990;7:39–68.
23. Ercolani P, Giovagnoni A, Giuseppetti G, Radiol. Med. Torino. 1991;82:422–426.
24. Hachiya J, Seki T, Okada M, Radiat. Med. 1991;9:232–240.
25. Hess T, Knopp MV, Brix G, Zentralbl. Radiol. 147, 969 (1993).
26. Stack JP, Redmond OM, Codd MB. Radiology, 1990;174:491–494.
27. Boetes C, Strijk SP, Holland R, Barentsz JO, Van Der Sluis RF, Ruijs JH. Eur. Radiol. 1997;7:1231–1234.
28. Bone B, Aspelin P, Bronge L, Isberg B, Perbeck L, Veress B, Acta. Radiol. 1996;37(2):208–213.
29. Buadu LD, Murakami J, Murayama S *et al.* Radiology 1996;200:639–649.
30. Daldrup HE, Roberts TP, Muhler A, Gossmann A, Roberts HC, Wendland M, Rosenau W, Brasch RC, Radiologe 1997;37(9):733–740.
31. DeAngelis GA, de Lange EE, Miller LR, Morgan RF. Radiographics 1994;14:783–94.
32. el Kwae EA, Fishman JE, Bianchi MJ, Pattany PM, Kabuka MR. J. Digit. Imaging. 1998;11(2):83–93.
33. Farria DM, Gorczyca DP, Barsky SH, Sinha S, Bassett LW. Am. Roentgenol. J. 1996;167(1):187–189.
34. Fischer U, von-Heyden D, Vosshenrich R, Vieweg I, Grabbe E. Rofo. 1993;158(4):287–292.
35. Fobben ES, Rubin CZ, Kalisher L, Dembner AG, Seltzer MH, Santoro EJ, Radiology 1995;196(1):143–152.
36. Frankel SD, Sickles EA, Radiology 1997;202(3):633–634.
37. Gilles R, Guinebretiere JM, Lucidarme O *et al.* [published erratum appears in Radiology 1994 Oct;193(1):285], Radiology 1994;191:625–631.
38. Gilles R, Guinebretiere JM, Shapeero LG *et al.* Radiology 1993;188:473–478.
39. Gilles R, Zafrani B, Guinebretiere JM *et al.* Radiology 1995;196:415–419.
40. Gorczyca DP, DeBruhl ND, Ahn CY, Hoyt A, Sayre JW, Nudell P, McCombs M, Shaw WW, Bassett LW. Radiology 1994;190(1):227–232.
41. Gorczyca DP. Radiology 1993;186(3):906–907.
42. Graham RA, Homer MJ, Sigler CJ *et al.* Am. J. Roentgenol. 1994;162:33–36.
43. Gribbestad IS, Nilsen G, Fjosne HE, Kvinnsland S, Haugen OA, Rinck PA. J. Magn. Reson. Imaging 1994;4(3), 477–480.
44. Harms SE, Flamig DP. Magn. Reson. Imaging. Clin. Am. N. 1994;2(4):573–584.
45. Hochman MG, Orel SG, Powell CM, Schnall MD, Reynolds CA, White LN. Radiology 1997;204:123–129.
46. Hoffmann U, Brix G, Knopp MV, Hess T, Lorenz WJ, Magn. Reson. Med. 1995;33(4):506–514.
47. Hylton NM, Frankel SD, Magn. Reson. Imaging. Clin. Am. N. 1994;2(4):511–525.
48. Knopp MV, Brix G, Junkermann HJ, Sinn HP, Magn. Reson. Imaging. Clin. Am. N. 1994;2(4):633–658.
49. Kuhl CK, Kreft BP, Hauswirth A, Elevelt A, Steudel A, Reiser M, Schild HH. Rofo. 162(5):381–389 (1995).
50. Kuhl CK, Seibert C, Sommer T, Kreft B, Gieseke J, Schild HH. Rofo. 1995;163(3):219–224.
51. Kuwabara M, Nippon Igaku Hoshasen Gakkai Zasshi, 1991;51(11):1366–1374.
52. Morris EA, Schwartz LH, Dershaw DD, van Zee KJ, Abramson AF, Liberman L. Radiology 1997;205(2):437–440.
53. Muller RD, Barkhausen J, Sauerwein W, Langer R, J. Comput. Assist. Tomogr. 1998;22(3):408–412.
54. Muller-Schimpfle M, Ohmenhauser K, Stoll P, Dietz K, Claussen CD. Radiology 1997;203(1):145–149.
55. Mumtaz H, Hall-Craggs MA, Davidson T *et al.* Am. Roentgenol. J. 1997;169:417–424.

56. Nunes LW, Schnall MD, Siegelman ES *et al.* Am. J. Roentgenol. 1997;169:409–415.
57. Orel SG, Hochman MG, Schnall MD, Reynolds C, Sullivan DC, Radiographics 1996;16:1385–1401.
58. Orel SG. Semin Ultrasound MCT R. 1996;17(5):476–493.
59. Pedevilla M. Radiology 1997;205(2):580–581.
60. Peller M, Stehling MK, Sittek H, Kessler M, Reiser M, MAGMA 1996;4(2):105–113.
61. Piccoli CW. Magn. Reson. Imaging. Clin. Am. N. 1994;2(4):557–571.
62. Pierce WB, Harms SE, Flamig DP, Griffey RH, Evans WP, Hagans JE. Radiology 1991;181(3):757–763.
63. Rieber A, Zeitler H, Rosenthal H, Gorich J, Kreienberg R, Brambs HJ, Tomczak R. Br. Radiol. J. 1997;70(833):452–458.
64. Rodenko GN, Harms SE, Pruneda JM *et al.* Am. Roentgenol. J. 1996;167:1415–1419.
65. Seki T, Hachiya J, Nitatori T, Yokoyama K, Fukushima H, Uchigasaki S, Nippon Geka Gakkai Zasshi 1996;97(5): 347–56.
66. Soderstrom CE, Harms SE, Copit DS, Evans WP, Savino DA, Krakos PA, Farrell RS Jr, Flamig DP. Radiology 1996; 201(2):427–432.
67. Soo MS, Kornguth PJ, Walsh R *et al.* J. Magn. Reson. Imaging. 1997;7:724–730.
68. Stelling CB. Radiol. Clin. North. Am. 1995;33(6):1187–11204.
69. Van Goethem M, Van Breusegem L, Ceulemans R, De Schepper A, de Moor J. J. Belge. Radiol. 1995;78(1):6–10.
70. Weinreb JC, Newstead G, Controversies in Breast MRI 1994;10(2):67–83.
71. Weinreb JC, Newstead G. Radiology 1995;196(3):593–610.
72. Kaiser WA, Fortschr. Röntgenstr. 1996;165(5):425–427 (Editorial).
73. Hussman K, Renslo R, Phillips JJ, Fischer HJ, Khalkhali I, Braslau DL, Sinow RM. Work in Progress, Radiology. 1993;189:915–917.
74. Kuhl CK, Elevelt A, Leutner CC, Gieseke J, Pakos E, Schild HH. Radiology. 1997;204:667–675.
75. Thiele J, Schneider JP, Franke P, Lieberenz S, Schmidt F. Fortschr Röntgenstr. 1998;168:374–379.
76. Doler W, Fischer U, Metzger I, Harder D, Grabbe E. Radiology 1996;200:863–864.
77. Mahfouz AE, Rahmouni A, Zylbersztejn C, Mathieu D, Am. Roentgenol. J. 1996;167:167–169.
78. Wurdinger S, Noras H, Straube K, Michaelsen S, Kaiser WA, ECR'97, Eur. Congr. Radiol. 2–7. March 1997, European Radiology 1997;7. S243.
79. de Souza NN, Kormos DW, Krausz T, Coutts GA, Hall AS, Burl M, Schwieso JE, Puni R, Vernon C. J. Magn. Reson. Imaging. 1995;5:525–528.
80. Silverman SG, Collick BD, Figueira MR, Khorasani R, Adams DF, Newman RW, Topulos GP. Radiology. 1995;197:175–182.
81. Pfleiderer SO, Reichenbach JR, Azhari T, Marx C, Wurdinger S, Kaiser WA, Invest. Radiol. 2003;Jan 38(1):1–8.
82. Knopp MV, Weiss E, Sinn HP, Mattern J, Junkermann H, Radeleff J, Magener A, Brix G, Delorme S, Zuna I, G van Kaick. J. Magn. Reson. Imaging. 1999;Sep 10(3):260–266.
83. Brix G, Schreiber W, Hoffmann U, Guckel F, Hawighorst H, Knopp MV. Radiologe 1997;Jun 37(6):470–480.
84. Buckley DL, Drew PJ, Mussurakis S, Monson JR, Horsman A. J. Magn. Reson. Imaging. 1997;May to Jun 7(3):461–664.
85. Algire GH. J. Int. Chir. 1953;Jul to Aug 13(4):381–384 (includes translations).
86. Gimbrone MA Jr, Leapman SB, Cotran RS, Folkman J. J. Exp. Med. 1972;Aug 1, 136(2):261–276.
87. Folkmann J, Long DM Jr, Becker FF. 1963;Apr 16:453–467.
88. Ottinetti A, Sapino A. Breast Cancer Res. Treat. 1988;Jul 11(3):241–248.
89. Samejima N, Yamazaki KA. Jpn. J. Surg. 1988;May 18(3):235–242.
90. Folkman J. Adv. Cancer. Res. 1985;43:175–203.
91. Weidner N, Semple JP, Welch WR, Folkman J, Engl N. Med J. 1991;Jan 3, 324(1):1–8.
92. Jensen HM, Chen I, DeVault MR, Lewis AE. Science 1982;Oct 15, 218(4569):293–295.
93. Port RE, Knopp MV, Brix G. Magn. Reson. Med. 2001;Jun, 45(6):1030–1038.
94. Lucht RE, Knopp MV, Brix G. Magn. Reson. Imaging 2001;Jan 19(1):51–57.
95. Knopp MV, Himmelhan N, Radeleff J, Junkermann H, Hess T, Sinn HP, Brix G. Radiologe 2002 Apr ;42(4):280–290.
96. Buckley DL, Drew PJ, Mussurakis S, Monson JR, Horsman A. J. Magn. Reson. Imaging 1997;May to Jun 7(3), 461–464.
97. Kaiser WA. Magnetic-Resonance-Mammography (MRM), Springer-Verlag, 1993.
98. Kaiser WA. In: PL Davis (Ed). MRI-Clinics of North America, Vol 2.4, Saunders, Philadelphia, 1994, pp 539–555.
99. H Neubauer, Li M, Kuehne-Heid R, Schneider A, Kaiser WA. Br. Radiol. J. 2003 Jan;76(901):3–12.
100. Fischer DR, Baltzer P, Malich A, Wurdinger S, Fresmeyer MG, Marx C, Kaiser WA. Eur. Radiol. 2004 Mar; 14(3):394–401, Epub 2003 Sep 27 (2004).
101. Kaiser WA. Breast, Unit 21. In: EM Haake, W Lin, NYC Cheng, CP Ho, WA Kaiser, JS Lewin, ZP Liang, SK Mukhery, RC Semelka, KR Thulborn, PK Woodard (Eds). Current Protocols in Magnetic Resonance Imaging. Wiley & Sons: New York, 2004.
102. Ikeda DM, Hylton NM, Kinkel K, Hochman MG, Kuhl CK, Kaiser WA, Weinreb JC, Smazal SF, Degani H, Viehweg P, Barclay J, Schnall MD. J. Magn. Reson. Imaging 2001;Jun 13(6):889–895.
103. Pfleiderer SO, Sachse S, Sauner D, Marx C, Malich A, Wurdinger S, Kaiser WA. Breast Cancer Res. 6(3):R232–238, Epub, Mar 16 (2004).
104. Kaiser WA, Selig M, Jung R, Vagner J, S Schönherr: Patentanmeldung: Manipulator für einen geschlossenen Magnetresonanztomographen (MRT). Patent Application Nr. 19941019.4; 28.08.1999, German Patent Office.
105. Pfleiderer SO, Reichenbach JR, Azhari T, Marx C, Malich A, Schneider A, Vagner J, Fischer H, Kaiser WA. 2003 Apr;17(4):493–498.

106. Pfleiderer SO, Freesmeyer MG, Marx C, Kuhne-Heid R, Schneider A, Kaiser WA. Eur. Radiol. Dec 12(12):3009-3014, Epub 2002 Jun 21.
107. Jeffrey SS, Birdwell RL, Ikeda DM, Daniel BL, Nowels KW, Dirbas FM, SM Griffey. Arch. Surg. 1999 Oct.;134(10): 1064–1068.
108. Hilger I, Leistner Y, Berndt A, Fritsche C, Haas KM, Kosmehl H, Kaiser WA. Eur. Radiol. 2004 Jun 14(6); 1124-1129, Epub 2004, Apr 30.

Internet Resources

http://www.mediteach.de: List of detailed courses in MRM.
http://www.mrisafety.com: List of safe or unsafe materials for a quick overview.

Phosphorus Magnetic Resonance Spectroscopy on Biopsy and *In Vivo*

Geoffrey S Payne, Nada Al-Saffar, and Martin O Leach
Cancer Research UK Clinical Magnetic Resonance Research Group, Royal Marsden Hospital and the Institute of Cancer Research SUTTON, Surrey, SM2 5PT, UK

Features of ^{31}P MRS in Tissues

General Properties of ^{31}P

^{31}P is one of the most sensitive of the NMR nuclei. It has spin $1/2$, a gyromagnetic ratio of 1.083×10^8 rad/s/T, and a natural abundance of 100%, giving a sensitivity of 6.63% relative to ^1H. Since many metabolites of biochemical interest are phosphorylated, ^{31}P MRS is a useful probe of tissue biochemistry. Spectra *in vivo* are relatively uncrowded as the number of ^{31}P-containing metabolites of sufficient concentration are relatively few. Properties of the individual metabolites of major interest are given below, while the abbreviations used for the various metabolites referred to in this chapter are given in Table 1. Note that there are nucleoside triphosphates (NTP) in addition to adenosine triphosphate (ATP), so in general we have followed the convention of referring to the corresponding peaks as NTP rather than as ATP.

Spectral Dispersion. In general ^{31}P gives rise to spectral peaks covering a wide range of frequencies. However in tissues most peaks of interest lie within a 30 ppm range (see Figure 1). Chemical shift is often referenced relative to phosphocreatine (PCr) at 0 ppm *in vivo*, or to H_3PO_4 in tissue extracts *in vitro*.

Dependence of Chemical Shift on pH. Many ^{31}P metabolites have a pKa in the range that makes the chemical shift dependent on tissue pH. The classic example of this is the equilibrium between the two forms of inorganic phosphate (Pi) in tissue [1];

$$H_2PO_4^- \Leftrightarrow HPO_4^{2-} + H^+ \qquad pKa \cong 6.7$$

Since this is a dynamic equilibrium between two forms of phosphate with different chemical shifts, the net shift of the observed Pi peak depends on the relative concentrations of the two species and hence on the hydrogen ion concentration (pH). Measurement of the chemical shift of the Pi peak relative to a peak that does not shift significantly with pH (usually PCr is used) provides a non-invasive way to measure intracellular pH, as most phosphate is intracellular. Some other ^{31}P metabolites also demonstrate sensitivity to pH, in particular phosphocholine (PC) and phosphoethanolamine (PE) (see Figure 2).

Other Factors Affecting Chemical Shift. Chemical shift is also affected by factors such as ionic strength and metal ion binding. In particular the β and γ peaks of ATP are sensitive to the magnesium concentration in the tissue [2,3].

Homonuclear J-Coupling. The main example of homonuclear coupling is found in NTP. In ATP this leads to α- and γ-ATP being observed as doublets, while β-ATP is a triplet, with a coupling constant of about 16 Hz. Homonuclear decoupling is not generally required. However, allowance for coupling is necessary when measuring the T_2 of coupled metabolites, such as ATP [4].

Heteronuclear J Coupling. The phosphomonoesters (PE, PC) have weak (6–7 Hz) 3-bond coupling to ^1H nuclei, forming a triplet, while the phosphodiesters couple to two-proton pairs, producing quintets (Table 1). Nearly all preclinical systems and a few of the more advanced clinical scanners have the facility to perform ^1H-decoupling to collapse these multiplets back to singlets (see below). A few research systems can also perform more complex double resonance measurements for further sensitivity enhancement.

Metabolites Observed in Tissues

Most of the metabolites detectable with ^{31}P MRS in tissues can be seen in Figure 1, while the structures are shown in Table 1. Metabolites are involved in either energy production (high energy phosphates—ATP, PCr, sugar phosphates; NADH) or in membrane metabolism (phosphomonoesters—PC, PE; phosphodiesters—GPC, GPE; membrane phospholipids). The reactions involved are summarized in Figure 3. For more details a biochemistry textbook should be consulted [5,6].

Graham A. Webb (ed.), Modern Magnetic Resonance, 1129–1147.

Table 1: Summary of main metabolites detected by ^{31}P MRS *in vivo*

Metabolite	Abbreviation	Structure	Chemical shift relative to PCr (ppm)	*J* coupling[a] (Hz)
Phosphoethanolamine	PE	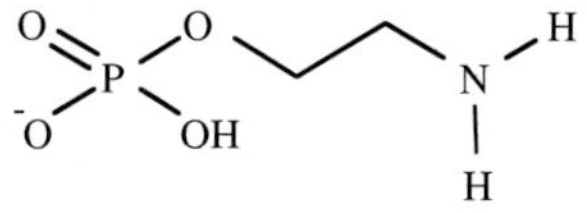	5–7	7.19
Phosphocholine	PC		4.2–6.5	6.25
Inorganic phosphate	Pi	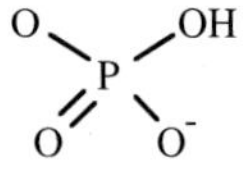	3.5–5.5	Singlet
Glycerophosphoethanolamine	GPE	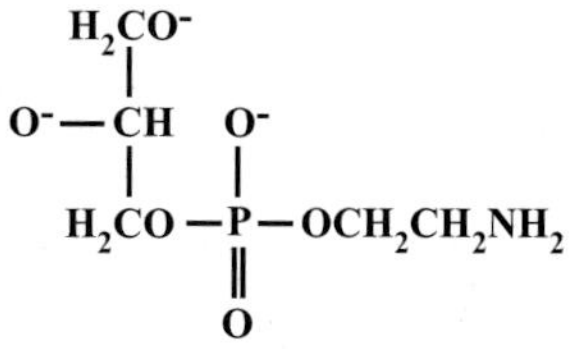	3.6	b
Glycerophospocholine	GPC	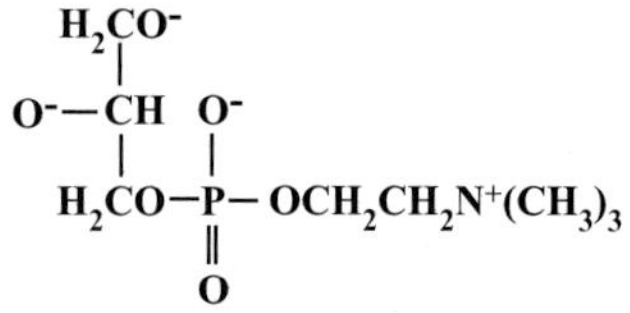	2.96	6.03
Phosphocreatine	PCr	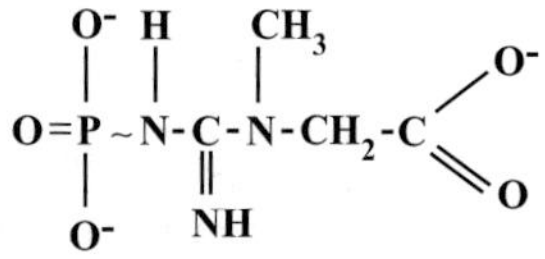	0	Singlet
Adenosine triphosphate	ATP	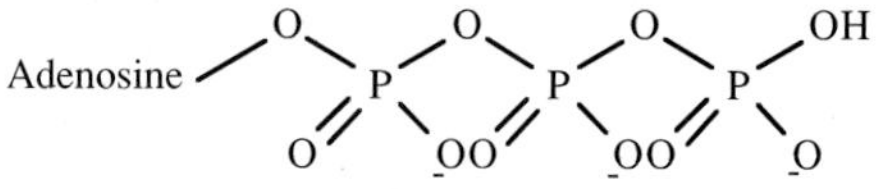	α-7.5 ppm β-16–19[c] γ-2–7.5[c]	

Notes: [a]The coupling quoted is the main $^3J_{PH}$ coupling. Additional J_{HH} couplings are often present. [138]; [b]This was not measured. It will be similar to that of GPC. [c]Depends on magnesium concentration.

Additional Notes on Specific Metabolites

Phosphomonoesters—phosphocholine and phosphoethanolamine. These metabolites are involved in membrane synthesis and degradation (Figure 3B [7]). PC is also involved in cell signalling [8,9]. The chemical shifts of both are pH sensitive (Figure 2).

Phosphodiesters—glycerophosphocholine (GPC) and glycerophosphoethanolamine (GPE). These are produced by the action of phospholipases A_1 and A_2, or phospholipase *B* on the corresponding phospholipids (Figure 3C). The chemical shift does not depend on pH. When interpreting a "PDE" peak one must be careful to distinguish phosphodiesters from the peak of the broader phospholipid component (see below).

Broad Resonances from Phospholipids and Bone. In unlocalized spectra of brain a large broad (~750 Hz

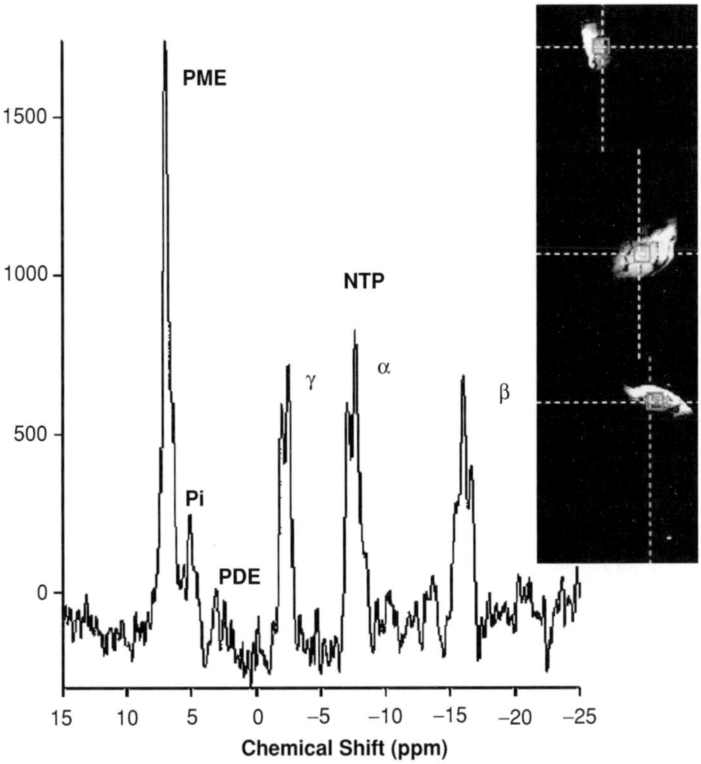

Fig. 1. Example of a ^{31}P MR spectrum from a non-Hodgkins lymphoma in the groin of a patient, measured at 1.5 T. Note the absence of PCr at 0 ppm.

width) symmetric resonance is observed due to immobile phosphates in bone [10]. In localized spectra of brain, liver, and kidney a broad asymmetric resonance is observed that decreases in amplitude with increasing magnetic field strength, and therefore suggesting a strong influence of chemical shift anisotropy. It has been demonstrated that the peak is produced by membrane lipid bilayers and that it may be eliminated by off-resonance presaturation [11].

Nucleoside Triphosphates. In most systems the dominant contribution is from ATP. As mentioned above, the chemical shifts of the β and γ peaks are sensitive to magnesium concentration.

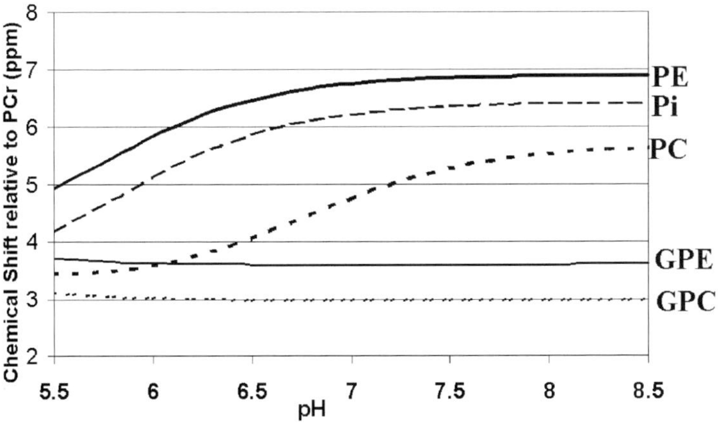

Fig. 2. Chemical shift of several metabolites relative to PCr as a function of pH (calculated from .[137]).

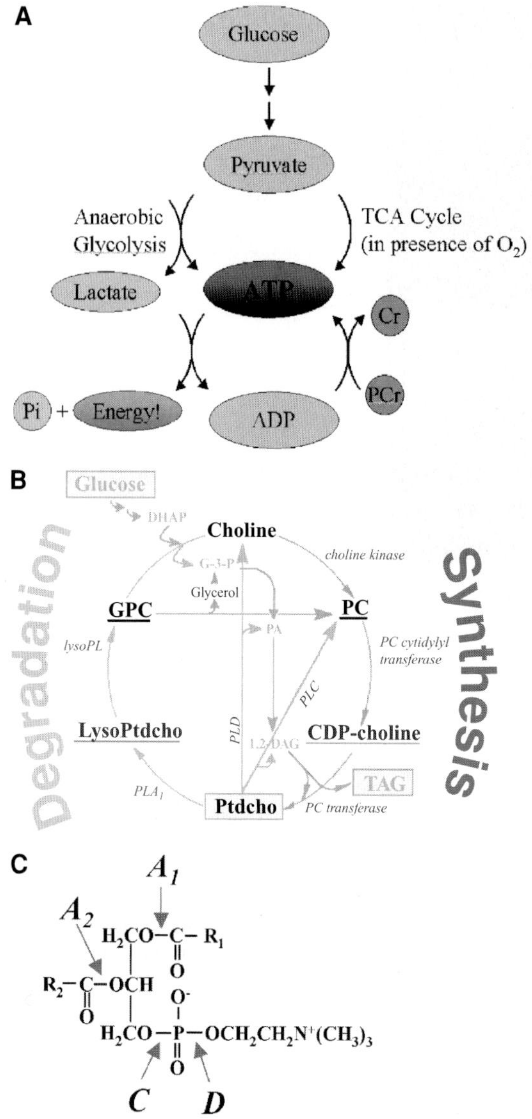

Fig. 3. Simplified biochemical pathways of main compounds visualized by ^{31}P MRS in tissues. (A) Metabolism of high energy phosphates. (B) Synthesis and degradation pathways of main membrane phospholipid. (C) Sites of action of the phospholipases on phosphatidylcholine. Phospholipase B can also bring about the successive removal of both fatty acids. (See also Plate 90 on page LXII in the Color Plate Section.)

Adenosine Diphosphate (ADP). The concentration measured by NMR is much lower than that measured biochemically after freeze-clamping of the tissue. This suggests that much of the ADP is tightly bound to larger structures leading to rapid transverse relaxation such that the peaks become invisible by NMR.

Metabolite Concentrations and Relaxation Times

Some published measurements of concentrations and relaxation times of various metabolites in different tissues are summarized in Tables 2 and 3.

^{31}P MRS of Tissue Biopsy Samples

Methods

Extraction Methods

The type of extraction method to be selected depends greatly on the nature of the investigation and the analytical methods involved. To analyze cellular lipids and/or their water-soluble metabolites, it is necessary to use extraction procedures that quantitatively recover the metabolites of interest. Currently available methods include the following.

Perchloric Acid (PCA). This is widely used for extraction of water-soluble cell components [12].

Modified Folch's Method. The extraction of cellular lipids using organic solvents, usually chloroform and methanol [12].

A Combined Extraction Technique. The above two methods performed on the same cell batch [13].

The Dual Phase Extraction Method (DPE). The recovery of both the lipid and water-soluble fractions from the same biological sample by a single extraction method [12].

Chemical Shift and Concentration Referencing for ^{31}P MRS

The separation of resonance frequencies from an arbitrary chosen reference frequency is termed the "chemical shift", and is expressed in terms of the dimensionless unit parts per million (ppm). Depending on the phase of the extracts and the type of nucleus detected, different solvent systems and references (internal or external) are used.

Aqueous Fraction. An example of stock solution used is D_2O with 10 mM ethylenediaminetetraacetate (EDTA) adjusted to pH 8.2 with tris (hydroxymethyl)-methylamine (Tris) (0.5 M, pH 10.4), and methylene-diphosphonic acid (MDPA) ($\delta = 16.8$ ppm) calibrated against GPC ($\delta = 0.49$ ppm @ pH 7.2) for referencing.

Table 2: Some measured concentrations of metabolites detected by ^{31}P MRS in different human tissues (units of mM)

Tissue	Meth	PE	PC	PME	Pi	GPE	GPC	PDE	MP	PCr	ATP (NTP)	Reference
Kidney	NMR			2.32 ± 0.50	0.66 ± 0.25			2.14 ± 0.91		None	1.6 ± 0.26	[95]
Kidney	NMR			2.6	1.6			4.9		None	2.0	[67]
Brain cerebrum	NMR			3.2 ± 0.3	1.0 ± 0.2			10.8 ± 2.0		2.9 ± 0.3	2.9 ± 0.2	[71]
Brain—cerebellum				4.0 ± 0.6	1.0 ± 0.2			9.4 ± 3.4		3.9 ± 0.4	2.6 ± 0.2	[71]
Brain cortical GM	NMR			3.1 ± 0.4	1.0 ± 0.2			10.1 ± 2.5		3.1 ± 0.3	2.9 ± 0.2	[71]
Brain cortical WM	NMR			4.3 ± 0.8	1.3 ± 0.3			14.2 ± 2.6		2.9 ± 0.4	2.9 ± 0.3	[71]
Brain	NMR			3.6 ± 0.8	1.5 ± 0.5			10.2 ± 1.9		3.8 ± 0.7	2.1 ± 0.5	[139]
Brain	NMR			2.3 ± 0.4	1.2 ± 0.2			7.0 ± 0.8		3.1 ± 0.5	2.3 ± 0.2	[140]
Brain tumor	NMR			0.9 ± 0.3	0.9 ± 0.1			2.3 ± 0.4		1.2 ± 0.1	1.3 ± 0.2	[140]
Brain (left parieto-occipital cortex)		.81 ± .21	.46 ± .14			.74 ±. 30	1.15 ±. 43		1.54 ± .95			[141]
Brain (right parieto-occipital cortex)		.94 ± .16	.46 ±. 17			.83 ± .22	1.14 ± .40		1.26 ± .78			[141]
Left thalamus		.69 ±. 18	.42 ± .16			.63 ± .20	1.05 ± .42		0.93 ± .56			[141]
Right thalamus		.68 ±. 24	.34 ±. 18			.60 ± .23	1.09 ± .36		.74 ±. 48			[141]
Muscle	Bioch									17.4	5.5	[142]
Muscle	Bioch									17.3	5.4	[143]
Muscle	Bioch				10					17.4	5.5	[144]
Muscle	NMR				2.7					25.6	5.5	[145]
Muscle	NMR				2.6					23.9	(5.5)	[146]
Muscle	NMR			2.5	3.6					24.7	(5.5)	[147]
Muscle	NMR				2.9					17.7	8.2	[148]
Muscle	NMR			2.2	4.7			4.2		32.0	6.3	[149]
Muscle	NMR				3.8					24.5	4.9	[150]
Muscle	NMR			2.0	2.9			3.8		22.0	5.7	[67]
Liver	NMR			0.9	2.1			5.4			2.2	[69]
	NMR			3.3	1.9			8.4			2.9	[67]
Liver (alcoholic hepatitis)	NMR			0.6 ± 0.3	1.4 ± 0.4			2.9 ± 1.1			1.5	[151]
Liver (alcoholic cirrhosis)	NMR			0.7 ± 0.5	1.1 ± 0.3			3.3 ± 1.8			1.65	[151]
Liver (viral hepatitis)	NMR			1.9 ± 0.6	2.1 ± 0.5			5.4 ± 2.0			2.8	[151]
Heart	NMR									8.82 ± 1.30	5.69 ± 1.02	[119]
Heart	NMR									9.0 ± 1.2	5.3 ± 1.2	[173]

Table 3A: Published values of T_1 relaxation times in different human tissues

Tissue	Field strength	PME	Pi	PDE	PCr	γ-NTP	α-NTP	β-NTP	Reference
Brain	1.9 T				4.0 ± 0.4	1.2 ± 0.2			[152]
Brain	2.0T	4.0	2.5	2.0	3.0	0.7	0.7	1.0	[56]
Brain (neonate)	2.35	4.9 ± 1.0	1.8 ± 0.6	1.7 ± 0.4	4.4 ± 1.2	1.3 ± 0.3	1.5 ± 0.4	1.6 ± 0.5	[153]
	2.35	6.4 ± 1.0	3.4 ± 1.1	2.1 ± 0.7	3.9 ± 1.0	2.2 ± 0.6	2.7 ± 1.0	1.7 ± 0.6	[153]
Brain (infants)	2.35	2.9 ± 0.6	1.8 ± 0.5	1.6 ± 0.2	3.2 ± 0.3	1.5 ± 0.3	1.7 ± 0.3	1.3 ± 0.4	[153]
Brain (infants)	2.35	4.9 ± 0.6	2.2 ± 0.6	2.0 ± 0.6	3.7 ± 1.1	2.2 ± 0.6	2.3 ± 0.8	2.5 ± 0.7	[153]
Muscle	1.5	3.20.5	3.3 ± 0.2	3.6 ± 0.4	4.6 ± 0.2	2.9 ± 0.4	2.2 ± 0.3	2.7 ± 0.3	[67]
Muscle	1.5		3.5 ± 0.4		5.0 ± 0.6	4.1 ± 0.4	2.9 ± 0.5	3.6 ± 0.3	[154]
Muscle	1.5		4.2 ± 0.5		6.1 ± 0.3	4.6 ± 0.3	3.2 ± 0.5	3.7 ± 0.6	[155]
Muscle	1.5		4.0 ± 0.5		5.6 ± 0.5	4.5 ± 0.4	3.4 ± 0.6	3.8 ± 0.8	[155]
Muscle	2.0		5.4 ± 1.7		6.0 ± 0.5	3.5 ± 1.0	3.9 ± 0.8	3.9 ± 0.8	[156]
Muscle	2.0		4.6 ± 0.5		6.5 ± 1.1	4.8 ± 0.6	3.5 ± 0.6	3.6 ± 0.8	[156]
Muscle	1.5		4.0 ± 0.4		6.5 ± 0.1	4.3 ± 0.5	3.2 ± 0.2	4.1 ± 0.3	[157]
	1.5		4.7 ± 0.3		6.5 ± 0.7	3.9 ± 1.3	4.2 ± 0.6	4.1 ± 1.3	[158]
	1.5		4.1 ± 0.5		6.6 ± 0.5	5.0 ± 0.7	3.8 ± 0.5	3.0 ± 0.5	[150]
	1.5		5.3 ± 0.5		6.9 ± 0.6	5.0 ± 0.5	3.2 ± 0.5	4.1 ± 0.4	[159]
	1.5		4.0 ± 0.8		5.5 ± 0.2	4.8 ± 0.5	3.6 ± 0.5	4.3 ± 0.6	[160]
Muscle	1.5	3.7 ± 1.6	3.9 ± 1.9	3.1 ± 1.9	5.8 ± 0.7	3.4 ± 0.1	2.8 ± 0.3	2.8 ± 0.4	[161]
Rhabdo Myosarcoma	1.5	5.2 ± 3.9	2.9 ± 1.3	2.0 ± 0.3	1.3 ± 0.3	1.3 ± 0.4	2.1 ± 0.4	1.6 ± 1.1	[161]
Muscle	3T		5.2 ± 1.0		6.4 ± 0.2	4.5 ± 0.3	2.6 ± 0.9	3.5 ± 1.1	[162]
Heart	1.5T			5.0 ± 1.0	6.1 ± 0.5	5.4 ± 0.5	5.0 ± 0.5	5.8 ± 1.0	[163,164]
Liver	1.5		0.41 ± 0.1	1.4 ± 0.13	none	0.43 ± 0.18	0.68 ± 0.09	0.40 ± 0.13	[158]
Liver	1.5	1.6 ± 0.4	0.8 ± 0.2	1.9 ± 0.4	none	0.5 ± 0.2	0.9 ± 0.3	0.5 ± 0.1	[67]
Liver	2.0	0.84 ± 0.26	0.97 ± 0.25	1.36 ± 0.37	none	0.35 ± 0.06	0.46 ± 0.09	0.35 ± 0.05	[69]
Liver	1.9	0.74 ± 0.1	0.44 ± 0.06			0.33 ± 0.04			[165]

Table 3B: Published values of T_2 relaxation times in different human tissues.

Tissue	Field strength	PME	Pi	PDE	PCr	γ–NTP	α–NTP	β–NTP	NTP	Reference
Muscle	1.5		240 ± 48		425 ± 21	93 ± 3	74 ± 1	75 ± 2		[4]
Muscle	1.5				61^a	66^a	69^b			[166]
Muscle	1.5				95^b	74^b	75^b			[4]
Muscle	1.5		205 ± 14		424 ± 21	$(16 \pm 5)^c$	$(22 \pm 6)^c$	$(8 \pm 2)^c$		[167]
Muscle	3T		148 ± 17		334 ± 30	78 ± 13	55 ± 7	55 ± 10		[162]
Brain	1.5					89 ± 9	84 ± 6	62 ± 3		[168]
Brain	2.0	70	80	20	150	30^c	30^c	20^c		[56]

Notes: ^{a}Spin echo with $T_E = 1/J$; ^{b}Selective echo 90-T_E/2-2(6)6(2)-Te/2-acquire; ^{c}These measurements have not allowed for modulation owing to homonuclear J coupling.

Lipid Fraction. A 2:1 mixture of $CDCl_3$ and 40 mM methanolic EDTA (200 mM EDTA in water adjusted to pH 6.0 with CsCl, and further diluted fivefold with absolute methanol) [14]. Examples of internal references are: GPC ($\delta = -0.13$ ppm) which is soluble in chloroform solutions, although it can be extracted with water and triphenylphosphine oxide ($\delta = 33.5$ ppm when PtdCho $\delta = 0.0$ ppm) or phosphocholine ($\delta = 3.5$ ppm when PtdCho $\delta = 0.0$ ppm) which is used as an external reference.

In all cases, metabolic contents are quantified by integration, normalized relative to the reference peak, and corrected for the number of live cells extracted and cell size in cases when treatment is expected to modify cellular size.

Acquisition Hardware and Condition

Measurements are usually performed in vertical superconducting magnets, with field strengths in the range 7–14 T. Manufacturers provide a wide range of RF probes, that are usually equipped with a deuterium lock to ensure the field does not drift over the time of the measurement, and with a separate ^{1}H coil in order to perform ^{1}H decoupling. Most measurements will use 5 mm NMR tubes, although 2.5 mm and even 1 mm tubes are becoming available.

Acquisition conditions will depend to some extent on the system and concentration being studied. The following conditions are an example of those used for cellular extracts of $0.5 - 2.0 \times 10^8$ cells measured on a 500 MHz spectrometer [15,16]. Some example spectra are given in Figure 4.

Spectra are acquired at room temperature with proton broad-band decoupling to eliminate ^{1}H-^{31}P NMR multiplets (decoupler power "on" during acquisition and "off" during the interpulse delay in order to prevent the building of NOE enhancement), spectral width of 100 ppm (aqueous) and 20 ppm (lipid), 32K data points and 4K transients accumulation. The sample spinning rate is 20 Hz. Exponential multiplication to give a line-broadening of 1–2 Hz is applied before Fourier transformation.

Spectral Analysis

Most spectrometers include software for processing and analysis of data. For off-line processing several packages are available, such as MestRe-C software (available free from Javier Sardina Research Group, University of Santiago de Compostela, Spain (http://qobrue.usc.es/jsgroup/MestRe-C/MestRe-C.html)). In addition to the measurement of peak areas directly, when large numbers of samples are available it is possible to look for patterns characteristic of different sample groups using automated pattern recognition techniques [17].

MRS of Intact Tissue Samples

While virtually all MRS measurements of biopsy samples are performed on tissue extracts, it has recently become possible to perform high resolution MRS directly on samples of intact tissue using the technique of magic angle spinning [18]. While most studies have used ^{1}H MRS, high resolution ^{31}P MR spectra may also be obtained [19]. Unlike extract studies the enzymes are still active, so it is necessary to perform measurements at low temperature (usually about 4° C). The high-energy phosphates are usually not visualized, but the metabolites involved in membrane metabolism (phosphomonoesters and phosphodiesters) are found to be reasonably stable.

Applications using ^{31}P MRS of Tissue Extracts

In practice, ^{31}P MRS of biopsy samples *ex vivo* is not as widely used as ^{31}P MRS *in vivo*, owing to the wide variety of alternative analytical techniques available for evaluating tissue samples. However the high resolution permits the identification of many more metabolites than is possible *in vivo*, for example, 14 different phospholipids were identified in a study of malignant breast tumors, several of which (sphingomyelin, PtdCho, PtdSer, phosphatidic

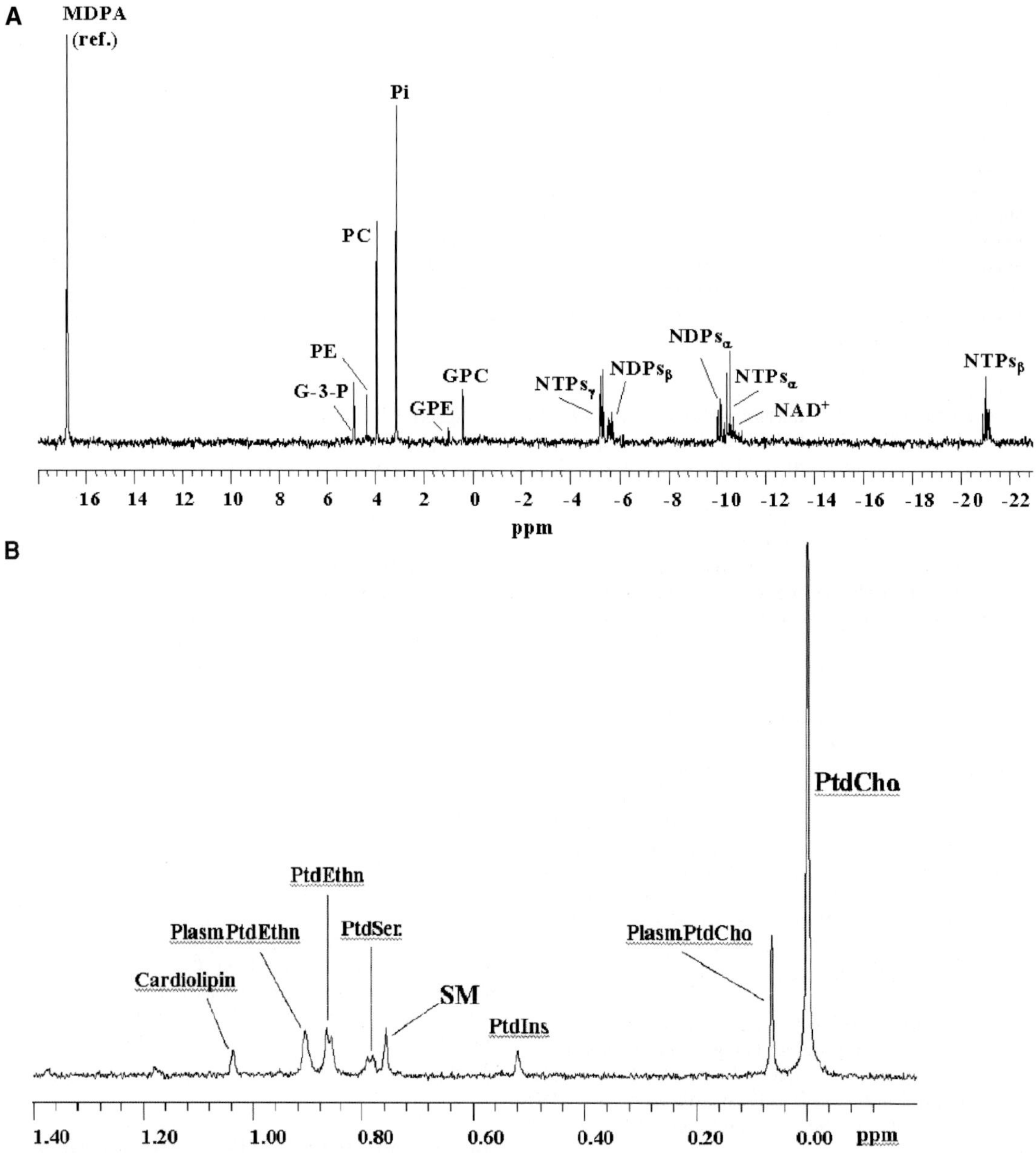

Fig. 4. Example spectra from cell extracts. All measurements were performed at 9.4 T. Spectra were plotted and analyzed using MestRe-C version 1.5.1 software. (A) ^1H decoupled ^{31}P MR spectrum of the water soluble fraction of Jurkat T-cell extracts (*G-3-P* = glycerol-3-phosphate, PE = phosphoethanolamine, PC = phosphocholine, Pi = inorganic phosphate, GPE = glycerol-3-phosphoethanolamine, GPC = glycerol-3-phosphocholine, NDPs = nucleotide diphosphates, NTPs = nucleotide triphosphates. NAD^+ = nicotinamide adenine dinucleotide). Spectrum is the result of 1280 scans plotted with line broadening of 2 Hz. (B) ^1H decoupled 31P MR spectrum of the lipid fraction of Jurkat T-cell extracts. Spectrum is the result of 1280 scans plotted with line broadening of 0.1 Hz and zero filling to 64 K data points (Plasm.PtdEthn = plasmalogen phosphatidylethanolamine, PtdIns = phosphatidylinositol, Plasm.PtdCho = plasmalogen phosphatidylcholine, SM = sphingomyelin, PtdCho = phosphatidylcholine, PtdEthn = phosphatidylethanolamine).

Table 4: Relative merits of tissue extracts and *in vivo* ^{31}P MRS measurements

	Tissue Extracts	*In vivo*
Tissue biopsy required?	Yes	No
Sample preparation required?	Yes	No
Max measurement time	Unlimited	Approx. 1 h
Measurement co-registered with anatomical image?	Not usually	Yes
System disturbed?	Yes	No
Capacity to follow kinetics	No	Yes
Spectral resolution	~0.01ppm	~0.1 ppm
Sensitivity to motion	N/A	Depends on tissue and method
Complexity of NMR sequence	Decoupled pulse-acquire	Requires localization
Field strength generally available	7–14 T	1.5–3 T
Coil design / availability	Good range of commercial coils	Few commercial coils; most need to be custom-built
Safety issues	Biopsy	MR Hazards (esp. RF heating from surface coils)

acid, phosphatidylglycerol and alkylacylphosphatidylcholine) predicted tumor characteristics such cellular infiltration, lymphatic invasion and necrosis [20].

In other studies biopsy specimens from patients with chronic ductopenic rejection [21] and liver cirrhosis [22] showed elevated PE and PC, and reduced GPE and GPC, reflecting altered phospholid metabolism [21]. In malignant hyperthermia-susceptible patients PCA extracts of vastus medialis muscle showed elevated GPC/(PCr+Pi) but no histological differences, suggesting GPC could be a marker of impairment [23]. PCA extracts of transurethral prostate samples show statistically different ratios of PE/totP, PC/totP and GPC/totP in prostate cancer samples compared with those from patients with benign prostatic hyperplasia [24]. In extracts of breast tumors no correlation was found between levels of GPC, GPE, PC, and PE and with indices of tumor proliferation, but higher PC levels were found in high grade tumors [25].

PCA and lipid extracts from kidney show that oncocytomas (benign tumors) have a biochemical composition between healthy and malignant tissue [26].

Blood samples are easier to obtain than tissue samples. One application is in thyroid cancer, where plasma levels of sphingomeylin and PtdCho have been found to be reduced in patients with remnants of thyroid tissue, and are even lower in patients with metastatic thyroid cancer, compared with patients in remission [27].

^{31}P MRS *In Vivo*

Introductory Comments

Given the difficulty and limited patient acceptance of obtaining tissue biopsy samples, the option to perform the spectroscopy measurements on the tissue *in vivo* is clearly very welcome. While the sensitivity and spectral resolution can never approach those obtained *in vitro*, the possibility of obtaining biochemical information from tissues non-invasively, repeatedly if necessary, provides the opportunity to monitor aspects of tissue biochemistry and response to treatment. Table 4 lists some of the main differences between measurements *in vivo* and *ex vivo*. A direct comparison of spectra of breast tumors from extracts and *in vivo* is given in [28].

Magnets

Most ^{31}P MRS studies on human subjects are performed in clinical NMR systems using horizontal bore superconducting magnets at a field strength of between 1.5 and 3T. Since sensitivity increases with field strength even higher fields would be advantageous but are not yet widely available. Since clinical scanners are normally designed for operation at a single frequency (for ^{1}H MR imaging) it is necessary to add or modify the RF hardware (coils, frequency source, transmit amplifier, receiver preamplifier, filters etc.) to permit operation at the ^{31}P resonant frequency.

RF Coils for *In Vivo* Studies (see also Chapter "Radiofrequency Coils for Magnetic Resonance Spectroscopy")

Appropriate RF coils of high sensitivity are essential for acquiring good ^{31}P MRS data. Because *in vivo* ^{31}P MRS is largely a research activity, with users having different specifications and applications, at present the choice of commercial coils is very limited. Unlike ^{1}H coils, for which the vendors of the MR systems provide a good range of options, commercial ^{31}P coils are often supplied by specialist coil companies, while many researchers design and build their own. Some good reviews on coil design principles can be found in [29–32]. Again in contrast to ^{1}H coils, almost all ^{31}P coils are used in transmit/receive mode—principally because there is no ^{31}P body resonator to provide a uniform transmit field. Occasionally a small surface coil receiver (to optimize sensitivity) is used together with a larger transmit coil (to optimize transmit field uniformity).

Surface Coils

Apart from studies in the brain, ^{31}P MRS measurements are usually performed with surface coils because of their superior sensitivity in superficial tissues [33]. There is a huge variety of designs, with the main differences being shape (e.g. circular or rectangular; mounting on a flat or curved surface), number of turns, material for construction (e.g. wire or foil), type of tuning network (e.g. simple LC circuit or balanced match), and whether there is a complementary ^{1}H coil to provide imaging and dual resonance capability. Surface coils are characterized by a spatial variation in the RF field. During transmit this leads to non-uniform excitation within the field of view of the coil. For small regions of interest that are not too close to the coil conventional amplitude-modulated RF pulses are often used, calibrated to give the required flip angle at the centre of the volume. However larger volumes are often of interest. In this case it is necessary to use adiabatic RF pulses to achieve uniform excitation over the region of interest (see below).

The non-uniform transmit field of surface coils also leads to non-uniform RF power deposition, which can be very high close to the coil itself. This is hard to measure experimentally, so sophisticated models (e.g. using finite element analysis) are required to calculate the expected power deposition [34]. International guidelines specify limits to RF power deposition for local and whole-body irradiation [35]. In practice the high RF power deposition near surface coils does limit the minimum repetition time or pulse amplitude in some studies, especially where double-resonant techniques are being used (see below).

Volume Coils

Volume coils are characterized by a more uniform RF field than surface coils, but usually at the expense of lower sensitivity. The most common designs for clinical studies are co-axial pairs of surface coils (often loosely called "Helmholtz pairs" but in fact having a separation larger than the radius), or birdcage coils [36]. Birdcage coils can be circularly polarized (which reduces RF power and improves receive sensitivity) and also double-resonant ($^{1}H/^{31}P$) [37].

Double-Resonance Coils

When performing ^{31}P MRS studies a ^{1}H capability is also usually required to obtain localization MR images, to shim, and sometimes to acquire complementary ^{1}H MR spectra (sequential or interleaved) or for double resonance studies (NOE, decoupling, polarization transfer). The hardware required depends on whether both ^{1}H and ^{31}P need to be used simultaneously.

Double Resonance Coil Options

(a) Two separate coils—such as ^{1}H body coil and ^{31}P surface coil or co-axial surface coils [38,39]. It is necessary to ensure they do not interact, in particular that ^{1}H transmit fields do not induce currents in the ^{31}P coil (leading to signal drop-out and local heating).
(b) Single coil, either with wide tuning range to permit sequential tuning to ^{1}H and then ^{31}P (e.g. 63 MHz and 25 MHz at 1.5 T), or that has two independent tuning networks may be switched to tune to each frequency.
(c) Single dual-resonant coil, for which the tuning circuit has resonances at both ^{1}H and ^{31}P. Depending on the design, the ^{1}H and ^{31}P connections may be on the same input [40] or different ones [31].

System Options

(a) Single channel with wide frequency range to permit transmit and receive operation at ^{1}H followed by ^{31}P in separate measurements. Some systems permit operation at specific frequencies only (e.g. at ^{1}H and ^{31}P). No double resonance experiments are possible.
(b) As (a) but with rapid switching of the transmit frequency. This permits irradiation at the ^{1}H frequency during the recovery time of ^{31}P measurements, and hence ^{31}P signal enhancement by the nuclear overhauser effect (NOE).
(c) Two-channel system that is able to transmit at one frequency (e.g. ^{1}H) while receiving on another (e.g. ^{31}P). Such systems can perform ^{1}H -decoupling (see below) as well as NOE studies.
(d) Two-channel system with full transmit and receive capability on two channels simultaneously. This permits both sequential acquisition at different frequencies [41–43] and also more complex double-resonance measurements such as polarization transfer [44,45].

Other Hardware for Double Resonance Studies

If there are no transmit pulses during signal acquisition then the only requirement is that the coils and frequencies do not interact. However for decoupling studies, where ^{1}H irradiation is applied during ^{31}P receive, it is necessary to apply a high level of isolation (~80 dB) between the channels. If this is not done the main problems are (1) saturating the receiver, leading to lower receiver gain, and (2) introducing extra noise into the receive channel, which again reduces the signal-to-noise ratio. Isolation is normally achieved in three steps, (i) use a coil that has a sharp resonance at the frequency of interest (e.g. ^{31}P) and negligible response at the other frequency (e.g. ^{1}H) (ii) a narrow band-width ^{31}P preamplifier, to minimize impact of ^{1}H irradiation, (iii) narrow bandwidth filters on both ^{1}H transmit and ^{31}P receive paths.

Adiabatic RF Pulses for *In Vivo* Studies with Surface Coils

Adiabatic RF pulses have the useful property that once a threshold transmitter amplitude is exceeded they produce uniform excitation over the sensitive region of the RF coil. In general they are characterized by a modulation of the frequency (or phase) as well as of the amplitude. For example the sech/tanh pulse ("hyperbolic secant pulse") widely used for slice-selective inversion [46] is described by

$$\text{Amp} = \Omega_0 \operatorname{sech}(\beta t)$$

$$\text{Freq} = \mu\beta \tanh(\beta t)$$

A variety of adiabatic pulses have been developed. Helpful reviews are available [47]. Other useful pulses include the tanh/tan pulse for 90° excitation [48] and the BIR4 or BIRP pulses for excitation by smaller flip angles to reduce partial saturation effects [48,49]. While there are adiabatic pulses suitable for many purposes some applications are particularly difficult, for example slice-selective excitation (for which the subtraction of two measurements is required [50]) and spin-refocusing [51].

The major disadvantages of adiabatic RF pulses compared with conventional amplitude-modulated pulses is that they are often (but not always) longer and/or requiring of a higher peak or total power.

Localization Methods for *In Vivo* Studies

For most studies *in vivo* it is necessary to select signal from a specified volume, usually identified from MR images acquired on the same occasion. A schematic representation of the three classes of technique is shown in Figure 5. An ideal acquisition method would have a sharp edge profile matching the requested target volume on the MR images, acquire signal from the selected volume with 100% efficiency, and include no contamination at all from elsewhere. The relative merits of the different acquisition strategies available are discussed briefly below.

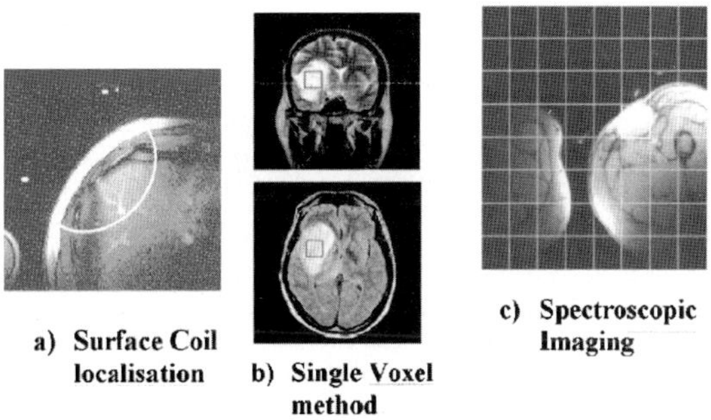

Fig. 5. Schematic illustrating the different methods of spectral localization.

Surface-Coil Localization

The relative sensitivity of an RF coil in receive mode at a particular point in space is proportional to the relative RF field produced at the same position by the coil in transmit mode [52]. Thus the sensitivity falls away with distance from a surface coil, limiting the sensitive volume. An indication of this region is shown in Figure 5, although in practice the boundary is not well defined. However, the localization method is simple and has the maximum available sensitivity for a given coil. It is normally only used alone (i.e. without further localization) when the sensitive region is almost entirely occupied with the target tissue, and when the signal losses associated with additional forms of localization cannot be tolerated.

Single Voxel Localization Methods

Single voxel methods make use of a series of frequency-selective RF pulses in combination with slice-selective B_0 field gradients. These define three intersecting slices that are usually but not necessarily orthogonal. Of the wide range of strategies that have been developed over the years, only three are in widespread use—PRESS [53], STEAM [54] and ISIS [55]. ^{31}P MRS has been demonstrated to work with STEAM [56] but ISIS has been the method of choice. While PRESS and STEAM have the great advantage of achieving localization in a single measurement, they both include echo times during which significant losses occur for nuclei which are characterized by short transverse time constants, such as are observed in most ^{31}P metabolites. As technology advances and shorter minimum echo times become possible this will become less of a problem.

PRESS and STEAM. PRESS is conceptually the simplest single voxel method and uses a 90° − 180° − 180° double spin echo sequence (Figure 6A). The first 90° RF pulse excites a slice of spins (e.g. transverse). The first 180° refocus pulse affects an intersecting slice (e.g. sagittal), so that only spins with the column defined by the intersection

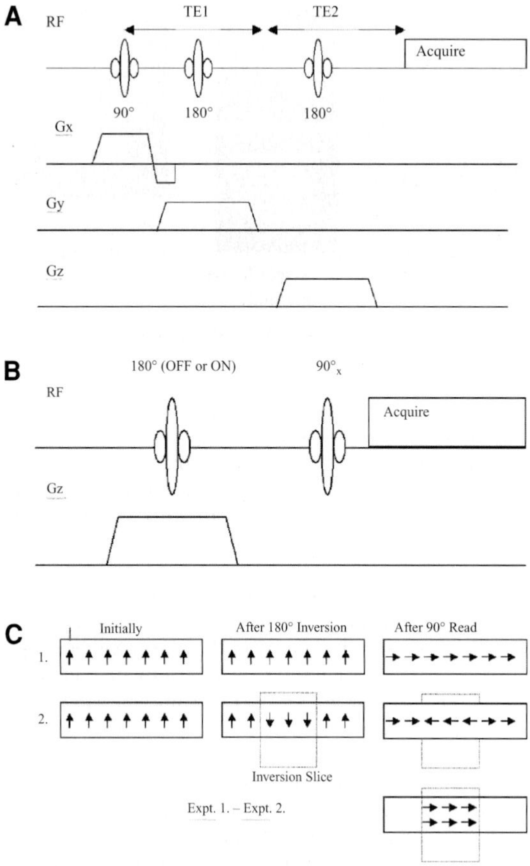

Fig. 6. (A) The PRESS localization sequence. (B) The ISIS localization sequence in 1 dimension (C) Schematic illustrating the effect on spins inside and outside the selected slice in the two measurements when performing ISIS localization in one dimension. The difference of the two measurements leads to signal addition within the slice and signal cancellation outside the slice.

of the two slices are refocused. The second 180° pulse affects a third slice that intersects the other two, so that spins within the intersecting cuboid are refocused while other spins are dephased.

The STEAM sequence is similar to PRESS except that it uses a $90° - 90° - 90°$ stimulated echo sequence instead of a double spin echo, with corresponding changes in timings and gradient patterns. It also localizes in a single measurement. However the maximum signal obtainable is 50% of the available magnetization.

ISIS. Localization with ISIS in one dimension requires two acquisitions (Figure 6B). In the first, a broadband RF pulse excites signal from the whole sample. In the second measurement the excitation pulse is preceded by a slice-selective inversion pulse. Subtraction of the two measurements causes cancellation of signal from outside the slice, but addition of signal from spins within the slice (Figure 6C). Three dimensional (3D)-Localization requires the appropriate addition/subtraction of eight separate measurements with all possible on/off permutations of slice-selective inversion pulses in the three intersecting slices.

Table 5: A summary of the relative merits of STEAM, PRESS, and ISIS for single voxel acquisition of ^{31}P MR spectroscopy data.

	STEAM	PRESS	ISIS
Surface coils	Poor	Poor	Good
Volume coils	Good	Good	Poor
Short T_2 metabolites	Moderate	Poor	Good
Single-shot	Yes	Yes	No
Edge definition	Good	Improving	Good
Sensitivity	<50%	Good	Good
Use use for shimming	Yes	Yes	Need other method

Relative Merits of the Different Single Voxel Localization Schemes for ^{31}P Acquisition. A summary of the relative merits of the three main single-voxel sequences is given in Table 5. The main advantage of the PRESS and STEAM schemes is that they are single-shot, thus reducing the problems of subtraction artefacts. However because the spins spend significant time in the transverse plane these methods are sensitive to signal loss from T_2 relaxation. With ISIS all magnetization is longitudinal till the final read pulse, thus reducing this problem. ISIS also has the advantage that it can be used easily with transmit surface coils by incorporating adiabatic RF pulses (usually hyperbolic secants); both STEAM and PRESS require slice-selective RF pulses, for which good adiabatic pulses are not available.

Spectroscopic Imaging (Chemical Shift Imaging)

Spectroscopic imaging (SI, also known as Chemical Shift Imaging—CSI) produces spectra for each member of a 2d or 3d grid, using a single excitation RF pulse followed by phase-encoding gradients [57]. For 2D-SI a slice-selective RF pulse is required, while for 3D-SI a non-selective pulse is used, which is better for use with surface coils. In principle it is possible to measure peak areas for a given metabolite in each cell and create a metabolite map. However with the voxel sizes often used for ^{31}P SI (e.g. 3 cm) this is not always useful.

Unlike single-voxel techniques the grid can be shifted after acquisition to make a cell line up with a target tissue

Table 6: Comparison of CSI and Single voxel techniques

	Single voxel	CSI
Specification of VOI		
Number of volumes	1	Typically ~512 (i.e. 8 × 8 × 8)
Voxel specification	Must be specified before measurement	1. Many voxels available 2. Grid shift can be performed retrospectively
Voxel shape	Dimensions adjustable; also tilt	(Usually) orthogonal grid e.g.(2cm)^3 voxels
Voxel Integrity		
Edge definition	That of slice	Point spread function effects
Chemical shift displacement artefact	Yes (in each direction)	No
Sensitivity		
Relaxation losses	T_2 (STEAM, PRESS)	Negligible (basic implementation)
Other losses	Imperfect RF pulse profiles and flip angles. STEAM loses 50%.	Phase-encoding loses 13% in each dimension
Conformation to target	Usually good	Sometimes poor. Addition of small voxels does not recover SNR of the larger corresponding voxel.
Other aspects		
Minimum number of acquisitions	1	Many
Use with Tx/Rx surface coils	ISIS OK; STEAM and PRESS not good	OK with adiabatic excitation pulse

of interest. However the edges of the cells are blurred by a process known as Fourier bleed, which leads to signal leakage into adjacent voxels [58]. In the presence of motion this can cause leakage into quite distant voxels [59,60]. Some of these effects can be mitigated by suitable apodisation before the spatial Fourier transform [61] or by weighting of the number of acquisitions for each phase-encoding step [62].

For an N × M array it is necessary to acquire N × M individual transients into a data matrix, from which the spatial information is recovered by a 2D (or 3D) Fourier transform. Thus the minimum number of transients for an 8 × 8 × 8 matrix is 512, giving an acquisition time of about 8.5 minutes if TR = 1s.

A summary of the relative merits of single voxel methods and spectroscopic imaging for ^{31}P acquisition is given in Table 6.

Double-Resonance Acquisition Schemes

Double resonance methods excite spins of a different species (e.g. ^{1}H) in addition to those that are being detected (^{31}P) within the same measurement sequence. In the context of *in vivo* spectroscopy the aim is usually to improve the signal-to-noise ratio, although they can also be used for spectral editing. Not all scanners are capable of performing these studies. The main classes of sequence are outlined in Table 7; the references should be consulted for a more detailed description.

Table 7: Features of double resonance techniques

Name of technique	Method	References	Mechanism	Result
Nuclear overhauser effect	Saturation of ^{1}H spins during ^{31}P recovery	[169,170]	Dipolar coupling	Signal enhancements of up to 50%
Decoupling	Saturation of ^{1}H spins during ^{31}P acquisition	[171,172]	*J* coupling	Collapse multiplets to singlets e.g. phosphocholine improves resolution and sensitivity
Polarization transfer	Simultaneous pulsing of ^{1}H and ^{31}P (large range of different schemes, e.g. INEPT, DEPT, cross-polarization etc.)	[44,45]	*J* coupling	Signal enhancements on coupled spins

PCr/ATP is reported to be reduced with age [118], dilated cardiomypathy [119], multiple sclerosis [120] and aortic stenosis [119]. The PCr/ATP ratio recovers towards control values in the latter group following surgical valve replacement [121]. In hypertensive patients the unstressed PCr/ATP ratio is variously reported as being normal [119] or reduced [122]. No reduction in PCr/ATP is observed in the non-infarcted septal myocardium of patients compared with controls [123]. Heart transplant patients studied within 24 h of a biopsy showed reduced PCr/ATP compared with controls, but with no strong correlation either with biopsy scores or with subsequent biopsies [124].

Stress tests show that cardiac patients with some forms of ischaemia showed a decrease in PCr/ATP even under light exercise, while no decrease was observed in patients whose hearts were non-ischaemic [122]. (*Note:* ^{31}P MRS measurements in the heart are often contaminated by signal from blood. In particular signal from 2,3 diphosphoglycerate overlaps with Pi signal making it hard to quantify.)

Tumors

A good review of the use of ^{31}P MRS in cancer is given by Negendank [125]. In particular ^{31}P MRS has been important in demonstrating that, contrary to expectation, the intracellular space of many cancers is alkaline [125]. In addition PE and PC have been shown to be elevated in many cancers. The ratio of PME/NTP is now being evaluated as an early indicator of response. There is also increasing evidence that both PME/NTP [126] and NTP/Pi [127] in the pretreatment spectrum may be able to predict response to treatment. Separate studies in patients with non-resectable soft tissue sarcoma have shown that changes in PME/ATP predict response to isolated limb perfusion to permit limb sparing [128].

Patients with non-small cell lung cancer have infusion of ATP to restrict weight loss; an increase in liver ATP ($8.8 \rightarrow 12.2\%$) is observed [104]. PME levels have also been observed to be higher in the weight-loss patients [129].

In breast cancer, there is an association between transformation and an increase in the phosphomonoesters (PMEs), while a decrease in PME content after treatment is associated with response to treatment as assessed by tumor volume [130]. In normal premenopausal breast a consistent pattern of changes is seen in the ^{31}P MR spectrum during the menstrual cycle, suggesting a potential application in assessing effects of exogenous hormones such as contraceptives or hormone replacement therapy which may increase breast cancer risk [131].

^{31}P MRS can also be used as marker of drug action (based on results of model systems and leading into clinical trials, e.g. 17-AAG [132]).

^{31}P MRS has been used extensively in experimental studies on effects of drugs, optimization strategies (reversing pH gradient, carbogen breathing), radiation etc. on cell systems.

Drug Detection

While ^{31}P MRS is normally used for study of endogenous compounds it also has the potential for monitoring uptake and metabolism *in vivo* of drugs that contain ^{31}P nuclei. The sensitivity is much lower than that of radionuclide labels, but the method has the advantages of being able to follow the metabolism of the drug, is not limited by a radioactive half-life, and makes use of the native compound without further chemistry required to perform the measurement. One example application has been the investigation of the alkylating agent ifosfamide. ^{31}P MRS studies have examined ifosfamide metabolites in body fluids [133], preclinical model systems [134,135], and in patients [136].

References

1. Moon R, Richards J. J. Biol. Chem. 1973;248:7276.
2. Gadian DG. Nuclear Magnetic Resonance and Its Applications to Living Systems. Oxford University Press, 1995.
3. Cohn M, Hughes TR Jr. J. Biol. Chem. 1962;237:176.
4. Jung WI, Straubinger K, Bunse M, Schick F, Kuper K, Dietze G, Lutz O. Magn. Reson. Med. 1992;28:305.
5. Cox M, Nelson DL. Lehninger Principles of Biochemistry. Palgrave Macmillan, New York, 2004.
6. Stryer L, Berg JM, Tymoczko JL. Biochemistry W H Freeman, New York, 2002.
7. Podo F. NMR Biomed. 1999;12:413.
8. Cuadrado A, Carnero A, Dolfi F, Jimenez B, Lacal JC. Oncogene. 1993;8:2959.
9. Jimenez B, del Peso L. Montaner S. Esteve P. Lacal JC. J. Cell. Biochem. 1995;57:141.
10. Gonzelez-Mendez R, Litt L, Koretsky AP, Voncolditz J, Weiner MW, James TL. J. Magn. Reson. 1984;57:526.
11. Murphy EJ, Rajagopalan B, Brindle KM, Radda GK. Magn. Reson. Med. 1989;12:282.
12. Tyagi RK, Azrad A, Degani H, Salomon Y. Magn. Reson. Med. 1996;35:194.
13. Henke J, Willker W, Engelmann J, Leibfritz D. Anticancer Res. 1996;16:1417.
14. Meneses P, Glonek T. J. Lipid. Res. 1988;29:679.
15. Al-Saffar NM, Titley JC, Robertson D, Clarke PA, Jackson LE, Leach MO, Ronen SM. Br. J. Cancer. 2002;86:963.
16. Ronen SM, DiStefano F, McCoy CL, Robertson D, Smith TA, Al-Saffar NM, Titley J, Cunningham DC, Griffiths JR, Leach MO, Clarke PA. Br. J. Cancer 1999;80:1035.
17. Tate AR, Griffiths JR, Martinez-Perez I, Moreno A, Barba I, Cabanas ME, Watson D, Alonso J, Bartumeus F, Isamat F, Ferrer I, Vila F, Ferrer E, Capdevila A, Arus C. NMR Biomed. 1998;11:177.
18. Cheng LL, Lean CL, Bogdanova A, Wright SC, Ackerman JL, Brady TJ, Garrido L. Magn. Reson. Med. 1996;36:653.

19. Payne GS, Chung Y-L, Vaidya SJ, Leach MO. In International Society for Magnetic Resonance in Medicine: Twelth Annual Meeting. Kyoto, Japan, 2004, p 68.
20. Merchant TE, Kasimos JN, Vroom T, de Bree E, Iwata JL, de Graaf PW, Glonek T. Cancer Lett. 2002;176:159.
21. Taylor-Robinson SD, Sargentoni J, Bell JD, Thomas EL, Marcus CD, Changani KK, Saeed N, Hodgson HJ, Davidson BR, Burroughs AK, Rolles K, Foster CS, Cox IJ. Gut. 1998;42:735.
22. Taylor-Robinson SD, Sargentoni J, Bell JD, Saeed N, Changani KK, Davidson BR, Rolles K, Burroughs AK, Hodgson HJ, Foster CS, Cox IJ. Liver. 1997;17:198.
23. Payen JF, Fouilhe N, Sam-Lai E, Remy C, Dupeyre R, Mezin P, Halsall J, Stieglitz P. Anesthesiology. 1996;84:1077.
24. Cornel EB, Smits GA, Oosterhof GO, Karthaus HF, Deburyne FM, Schalken JA, Heerschap A. J. Urol. 1993;150:2019.
25. Smith TA, Bush C, Jameson C, Titley JC, Leach MO, Wilman DE, McCready VR. NMR Biomed. 1993;6:318.
26. Tosi MR, Tugnoli V, Bottura G, Lucchi P. Applied Spectroscopy 2001;55:908.
27. Moka D, Dietlein M, Raffelt K, Hahn J, Schicha H. Am. J. Med. 2002;112:634.
28. Smith TA, Glaholm J, Leach MO, Machin L, Collins DJ, Payne GS, McCready VR. Br. J. Cancer. 1991;63:514.
29. Schnall M. In: P Diehl, E Fluck, H Gunther, R Kosfeld, J Seelig (Eds). NMR Basic Principles and Progress 26. Springer-Verlag: Berlin Heidelberg, 1992, p 33.
30. Link J. In: P Deihl, E Fluck, H Gunther, R Kosfeld, J Seelig (Eds). NMR Basic Principles and Progress 26. Springer-Verlag: Berlin Heidelberg, 1992, p 1.
31. Hoult DI. Prog. NMR Spec. 1978;12:41.
32. Jin J. Electromagnetic Analysis and Design in Magnetic Resonance Imaging. CRC Press LLC, London, 1999.
33. Ackerman JJ, Grove TH, Wong GG, Gadian DG, Radda GK. Nature. 1980;283:167.
34. Prock T, Collins DJ, Leach MO. Phys. Med. Biol. 2002;47:1805.
35. IEC. In: International Electrotechnical Commission (IEC), 2002.
36. Hayes CE, Edelstein WA, Schenck JF, Mueller OM, Eash M. J. Magn. Reson. 1985;63:622.
37. Murphy-Boesch J, Srinivasan R, Carvajal L, Brown TR. J. Magn. Reson. Series B 1994;103:103.
38. Klomp DW, Collins DJ, van den Boogert HJ, Schwarz A, Rijpkema M, Prock T, Payne GS, Leach MO, Heerschap A. Magn. Reson. Imaging. 2001;19:755.
39. Murphy-Boesch J, He L, Elsasser C, Brown TR. In: International Society for Magnetic Resonance in Medicine: Forth Annual Meeting. New York, USA 1996, p 1430.
40. Collins DJ, Prock T, Leach MO. In: International Society for Magnetic Resonance in Medicine: Eighth Annual Meeting, 2000, p 1417.
41. Yongbi NM, Payne GS, Leach MO. Magn. Reson. Med. 1994;32:768.
42. Arias-Mendoza F, Javaid T. Stoyanova R, Brown TR, Gonen O. NMR Biomed. 1996;9:105.
43. Gonen O, Hu J, Murphy-Boesch J, Stoyanova R, Brown TR. Magn. Reson. Med. 1994;32:104.
44. Gonen O, Mohebbi A, Stoyanova R, Brown TR. Magn. Reson. Med. 1997;37:301.
45. Payne GS, Leach MO, Magn. Reson. Med. 2000;43:510.
46. Silver MS, Joseph RI, Hoult DI. Physical Review A 1985;31:2753.
47. Garwood M, Ugurbil K. In: P. Diehl, E. Fluck, H. Gunther, Kosfeld R, Seelig J (Eds). NMR: Basic Principles and Progress, Springer-Verlag, Berlin Heidelberg, 1992, p 109.
48. Garwood M, Ke Y. J. Magn. Reson. 1991;94:511.
49. Bottomley PA, Ouwerkerk R. J. Magn. Reson. Series A 1993;103:242.
50. Johnson AJ, Garwood M, Ugurbil K. J. Magn. Reson. 1989;81:653.
51. Ugurbil K, Garwood M, Rath AR, Bendall MR. J. Magn. Reson. 1988;78:472.
52. Hoult DI, Richards RE. J. Magn. Reson. 1976;24:71.
53. Bottomley P. Ann. NY Acad. Sci. 1987;508:333.
54. Frahm J, Merboldt K-D, Hanicke W. J. Magn. Reson. 1987;72:502.
55. Ordidge R, Connelly A, Lohman J. J. Magn. Reson. 1986;66:283.
56. Merboldt KD, Chien D, Hanicke W, Gyngell ML, Bruhn H, Frahm J. J. Magn. Reson. 1990;89:343.
57. Brown T, Kincaid B, Ugurbil K. Proceedings of the National Academy of Sciences U.S.A. 1982;79:3523.
58. Koch T, Brix G, Lorenz WJ. J. Magn. Reson. Series B 1994;104:199.
59. Doyle VL, Howe FA, Griffiths JR. Phys. Med. Biol. 2000;45:2093.
60. Schwarz AJ, Leach MO. Phys. Med. Biol. 2000;45:2105.
61. de Graaf RA, In vivo NMR spectroscopy, John Wiley & Sons, 1998.
62. Pohmann R, von Kienlin M. Magn. Reson. Med. 2001;45:817.
63. Madden A, Leach MO, Collins DJ, Payne GS. Magn. Reson. Med. 1991;19:416.
64. Knijn A, de Beer R, van Ormondt D. J. Magn. Reson. 1992;97:444.
65. ven den Boogaart A, Van Hecke P, Van Huffel S, Graveron-Demilly D, de Beer R. In: European Society for Magnetic Resonance in Medicine and Biology: 13th Annual Meeting, Prague, p. 318, 1996.
66. Provencher SW. Magn. Reson. Med. 1993;30:672.
67. Buchli R, Meier D, Martin E, Boesiger P. Magn. Reson. Med. 1994;32:453.
68. Tofts PS, Wray S. NMR Biomed 1988;1:1.
69. Meyerhoff DJ, Karczmar GS, Matson GB, Boska MD, Weiner MW. NMR Biomed 1990;3:17.
70. Buchli R, Martin E, Boesiger P. NMR Biomed. 1994;7:225.
71. Buchli R, Duc CO, Martin E, Boesiger P. Magn. Reson. Med. 1994;32:447.
72. Murphy -Boesch J, Jiang H, Stoyanova R, Brown TR. Magn. Reson. Med. 1998;39:429.
73. Doyle VL, Payne GS, Collins DJ, Verrill MW, Leach MO. Phys. Med. Biol. 1997;42:691.
74. Argov Z, Lofberg M, Arnold DL. Muscle Nerve 2000;23:1316.
75. Heerschap A, Houtman C, in 't Zandt HJ, van den Bergh AJ, Wieringa B. Proc. Nutr. Soc. 1999;58:861.
76. Newcomer BR, Larson -Meyer DE, Hunter GR, Weinsier RL. Int. J. Obes. Relat. Metab. Disord. 2001;25:1309.

77. Tartaglia MC, Chen JT, Caramanos Z, Taivassalo T, Arnold DL, Argov Z. Muscle Nerve 2000;23:175.
78. Schunk K, Pitton M, Duber C, Kersjes W, Schadmand-Fischer S, Thelen M. Invest. Radiol. 1999;34:116.
79. Mattei JP, Bendahan D, Roussel M, Lefur Y, Cozzone PJ. FEBS Lett. 1999;450:173.
80. Smith SA, Montain SJ, Matott RP, Zientara GP, Jolesz FA, Fielding RA. J. Appl. Physiol. 1999;87:116.
81. Schunk K, Losch O, Kreitner KF, Kersjes W, Schadmand-Fischer S, Thelen M. Invest. Radiol. 1999;34:348.
82. Lodi R, Kemp GJ, Muntoni F, Thompson CH, Rae C, Taylor J, Styles P, Taylor DJ. Brain 122 (Pt 1) 121 (1999).
83. Park JH, Niermann KJ, Ryder NM, Nelson AE, Das A, Lawton AR, Hernanz-Schulman M. Olsen NJ. Arthritis Rheum 2000;43:2359.
84. Hsu AC, Dawson MJ. Magn. Reson. Med. 2003;49:626.
85. Roussel M, Mattei JP, Le Fur Y, Ghattas B, Cozzone PJ, Bendahan D. J. Appl. Physiol. 2003;94:1145.
86. Ross BD, Radda GK, Gadian DG, Rocker G, Esiri M, Falconer-Smith J, Engl N. J. Med. 1981;304:1338.
87. Kemp GJ, Roberts N, Bimson WE, Bakran A, Harris PL, Gilling-Smith GL, Brennan J, Rankin A, Frostick SP. J. Vasc. Surg. 2001;34:1103.
88. Pipinos, II, Shepard AD, Anagnostopoulos PV, Katsamouris A, Boska MD. J. Vasc. Surg. 2000;31:944.
89. Niekisch MB, Von Elverfeldt D, Saman AE, Hennig J, Kirste G. Transplantation 2004;77:1041.
90. Moller HE, Gaupp A, Dietl K, Buchholz B, Vestring T. Magn. Reson. Imaging. 2000;18:743.
91. Seto K, Ikehira H, Obata T, Sakamoto K, Yamada K, Kashiwabara H, Yokoyama T, Tanada S. Transplantation 2001;72:627.
92. Jonsson O, Lindgard A, Fae A, Gondalia J, Aneman A, Soussi B. Scand J. Urol. Nephrol. 2003;37:450.
93. Ogimoto G, Sakurada T, Imamura K, Kuboshima S, Maeba T, Kimura K, Owada S. Mol. Cell. Biochem. 2003;244: 135.
94. Sala E, Noyszewski EA, Campistol JM, Marrades RM, Dreha S, Torregrossa JV, Beers JS, Wagner PD, Roca J. Am. J. Physiol. Regul. Integr. Comp. Physiol. 280 R1240 (2001).
95. Kugel H, Wittsack HJ, Wenzel F, Stippel D, Heindel W, Lackner K. Acta Radiol. 2000;41:634.
96. Sullentrop F, Moka D, Neubauer S, Haupt G, Engelmann U, Hahn J, Schicha H. NMR Biomed. 2002;15:60.
97. Mairiang E, Hanpanich P, Sriboonlue P. Magn. Reson. Imaging. 2004;22:715.
98. Bak M, Thomsen JK, Jakobsen HJ, Petersen SE, Petersen TE, Nielsen NC. J. Urol. 2000;164:856.
99. Bell JD, Bhakoo KK. NMR Biomed. 1998;11:354.
100. Lim AK, Patel N, Hamilton G, Hajnal JV, Goldin RD, Taylor-Robinson SD. Hepatology 2003;37:788.
101. Changani KK, Jalan R, Cox IJ, Ala-Korpela M, Bhakoo K, Taylor-Robinson SD, Bell JD. Gut 2001;49:557.
102. Mann DV, Lam WW, Magnus Hjelm N. So NM, Yeung DK, Metreweli C, Lau WY. Gut 2002;50:118.
103. Leij-Halfwerk S, Dagnelie PC, Van Den Berg JW, Wilson JH, Sijens PE. Clin. Sci. (London) 2000;98:167.
104. Leij-Halfwerk S, Agteresch HJ, Sijens PE, Dagnelie PC. Hepatology 2002;35:421.
105. Mann DV, Lam WW, Hjelm NM, So NM, Yeung DK, Metreweli C, Lau WY. Ann. Surg. 2002;235:408.
106. Mann DV, Lam WW, Hjelm NM, So NM, Yeung DK, Metreweli C, Lau WY. Hepatology 2001;34:557.
107. Jensen JE, Miller J, Williamson PC, Neufeld RW, Menon RS, Malla A, Manchanda R, Schaefer B, Densmore M, Drost DJ. Br. J. Psychiatry. 2004;184:409.
108. Bluml S, Tan J, Harris K, Adatia N, Karme A, Sproull T, Ross B. J. Comput. Assist. Tomogr. 1999;23:272.
109. Hamakawa H, Murashita J, Yamada N, Inubushi T, Kato N, Kato T. Psychiatry. Clin. Neurosci. 2004;58:82.
110. Yildiz A, Demopulos CM, Moore CM, Renshaw PF, Sachs GS. Biol. Psychiatry. 2001;50:3.
111. Silverstone PH, Rotzinger S, Pukhovsky A, Hanstock CC. Biol. Psychiatry. 1999;46:1634.
112. Kato T, Murashita J, Kamiya A, Shioiri T, Kato N, Inubushi T. Eur. Arch. Psychiatry Clin. Neurosci. 1998;248:301.
113. Estilaei MR, Matson GB, Payne GS, Leach MO, Fein G, Meyerhoff DJ. Alcohol Clin. Exp. Res. 2001;25: 1213.
114. Rango M, Bozzali M, Prelle A, Scarlato G, Bresolin N. J. Cereb. Blood. Flow. Metab. 2001;21:85.
115. Bluml S, Seymour KJ, Ross BD. Magn. Reson. Med. 1999;42:643.
116. von Kienlin M, Beer M, Greiser A, Hahn D, K Harre, H Kostler, W Landschutz, T Pabst, Sandstede J, Neubauer S. J. Magn. Reson. Imaging. 2001;13:521.
117. Schocke MF, Zoller H, Vogel W, Wolf C, Kremser C, Steinboeck P, Poelzl G, Pachinger O, Jaschke WR, Metzler B. Magn. Reson. Imaging. 2004;22:515.
118. Schocke MF, Metzler B, Wolf C, Steinboeck P, Kremser C, Pachinger O, Jaschke W, Lukas P. Magn. Reson. Imaging. 2003;21:553.
119. Beer M, Seyfarth T, Sandstede J, Landschutz W, Lipke C, Kostler H, von Kienlin M. Harre K, Hahn D, Neubauer S. J. Am. Coll. Cardiol. 2002;40:1267.
120. Beer M, Sandstede J, Weilbach F, Spindler M, Buchner S, Krug A, Kostler H, Pabst T, Kenn W, Landschutz W, von Kienlin M, Toyka KV, Ertl G, Gold R, Hahn D. Rofo Fortschr Geb Rontgenstr Neuen Bildgeb Verfahr 2001;173: 399.
121. Beer M, Viehrig M, Seyfarth T, Sandstede J, Lipke C, Pabst T, Kenn W, Harre K, Horn M, Landschutz W, von Kienlin M, Neubauer S, Hahn D. Radiologe. 2000;40:162.
122. Butterworth EJ, Evanochko WT, Pohost GM. Ann. Biomed. Eng. 2000;28:930.
123. Beer M, Sandstede J, Landschutz W, Viehrig M, Harre K, Horn M, Meininger M, Pabst T, Kenn W, Haase A, von Kienlin M, Neubauer S, Hahn D. Eur. Radiol. 2000;10: 1323.
124. Buchthal SD, Noureuil TO, den Hollander JA, Bourge RC, Kirklin JK, Katholi CR, Caulfield JB, Pohost GM, Evanochko WT. J. Cardiovasc. Magn. Reson. 2000;2:51.
125. Negendank W. NMR Biomed. 1992;5:303.
126. Arias-Mendoza F, Smith MR, Brown TR. Acad. Radiol. 2004;11:368.
127. Zakian KL, Shukla-Dave A, Meyers P, Gorlick R, Healey J, Thaler HT, Huvos AG, Panicek DM, Koutcher JA. Cancer Res. 2003;63:9042.
128. Kettelhack C, Wickede M, Vogl T, Schneider U, Hohenberger P. Cancer 2002;94:1557.
129. Leij -Halfwerk S, van den Berg JW, Sijens PE, Wilson JH, Oudkerk M, Dagnelie PC. Cancer Res. 2000;60:618.

130. Ronen SM, Leach MO. Breast Cancer Res. 2001;3:36.
131. Payne GS, Dowsett M, Leach MO. The Breast 1994;3:20.
132. Chung YL, Troy H, Banerji U, Jackson LE, Walton MI, Stubbs M, Griffiths JR, Judson IR, Leach MO, Workman P, Ronen SM. J. Natl. Cancer Inst. 2003;95:1624.
133. Martino R, Crasnier F, Chouini-Lalanne N, Gilard V, Niemeyer U, De Forni M, Malet-Martino. J. Pharmacol. Exper. Therap. 1992;260:1133.
134. Rodrigues LM, Maxwell RJ, McSheehy PMJ, Pinkerton CR, Robinson SP, Stubbs M, Griffiths JR, Br. J. Cancer 1997;75:62.
135. Mancini L, Payne GS, Dzik-Jurasz AS, Leach MO. Magn. Reson. Med. 2003;50:249.
136. Payne GS, Pinkerton CR, Bouffet E, Leach MO. Magn. Reson. Med. 2000;44:180.
137. Robitaille P-M, Robitaille P, Brown GJ,. Brown G. J. Magn. Reson. 1991;92:73.
138. Govindaraju V, Young K, Maudsley AA. NMR Biomed. 2000;13:129.
139. Deicken RF, Calabrese G, Raz J, Sappey-Marinier D, Meyerhoff D, Dillon WP, Weiner MW, Fein G. Biol. Psychiatry 1992;32:628.
140. Hubesch B, Sappey-Marinier D, Roth K, Meyerhoff DJ, Matson GB, Weiner MW. Radiology 1990;174:401.
141. Jensen JE, Drost DJ, Menon RS, Williamson PC. NMR Biomed. 2002;15:338.
142. Harris RC, Hultman E, Nordesjo LO. Scand. J. Clin. Lab. Invest. 1974;33:109.
143. Bylund-Fellenius AC, Walker PM, Elander A, Holm S, Holm J, Schersten T. Biochem. J. 1981;200:247.
144. Dawson M, Allan RJ, Watson TR. Br. J. Clin. Pharmacol. 1982;14:453.
145. Taylor DJ, Styles P, Matthews PM, Arnold DA, Gadian DG, Bore P, Radda GK. Magn. Reson. Med. 1986;3:44.
146. Rees D, Smith MB, Harley J, Radda GK. Magn. Reson. Med. 1989;9:39.
147. Newham DJ, Cady EB. NMR Biomed. 1990;3:211.
148. Venkatasubramanian PN, Mafee MF, Barany M. Magn. Reson. Med. 1988;6:359.
149. Cady EB. J. Magn. Reson. 1990;87:433.
150. Roth K, Hubesch B, Meyerhoff DJ, Naruse S, Gober JR, Lawry TJ, Boska MD, Matson GB, Weiner MW, J. Magn. Reson. 1989;81:299.
151. Meyerhoff DJ, Boska MD, Thomas AM, Weiner MW. Radiology. 1989;173:393.
152. Blackledge MJ, Styles P. J. Magn. Reson. 1989;83:390.
153. Gruetter R, Fusch C, Martin E, Boesch C. Magn. Reson. Med. 1993;29:7.
154. Brown TR, Stoyanova R, Greenberg T, Srinivasan R. Murphy-Boesch J. Magn. Reson. Med. 1995;33:417.
155. Newcomer BR, Boska MD. Magn. Reson. Med. 1999;41:486.
156. Bottomley PA, Ouwerkerk R. J. Magn. Reson. Series B 1994;104:159.
157. Yongbi NM, Leach MO, Payne GS, Collins DJ. NMR Biomed. 1992;5:95.
158. Buchthal SD, Thoma WJ, Taylor JS, Nelson SJ, Brown TR. NMR Biomed. 1989;2:298.
159. Luyten PR, Groen JP, Vermeulen JW, den Hollander JA. Magn. Reson. Med. 1989;11:1.
160. Thomsen C, Jensen KE, Henriksen O. Magn. Reson. Imaging. 1989;7:231.
161. Moller HE, Vermathen P, Rummeny E, Wortler K, Wuisman P, Rossner A, Wormann B, Ritter J, Peters PE. NMR Biomed. 1996;9:347.
162. Meyerspeer M, Krssak M, Moser E. Magn. Reson. Med. 2003;49:620.
163. Schindler R, Krahe T, Neubauer S, Hillenbrand H, Entzeroth C, Horn M, Lackner K, Ertl G. Rofo Fortschr Geb Rontgenstr Neuen Bildgeb Verfahr 1992;157:452.
164. Neubauer S, Krahe T, Schindler R, Hillenbrand H, Entzeroth C, Horn M, Bauer WR, Stephan T, Lackner K, Haase A *et al.* Magn. Reson. Med. 1992;26:300.
165. Blackledge MJ, Oberhaensli RD, Styles P, Radda GK. J. Magn. Reson. 1987;71:331.
166. Straubinger K, Jung WI, Bunse M, Lutz O, Kuper K, Dietze G. Magn. Reson. Imaging. 1994;12:121.
167. Thomsen C, Jensen KE, Henriksen O. Magn. Reson. Imaging. 1989;7:557.
168. Jung WI, Widmaier S, Bunse M, Seeger U, Straubinger K, Schick F, Kuper K, Dietze G, Lutz O. Magn. Reson. Med. 1993;30:741.
169. Noggle JH, Schirmer RE. The Nuclear Overhauser Effect, Academic Press: New York and London, 1971.
170. Bachert P, Bellemann ME. J. Magn. Reson. 1992;100:146.
171. Murphy-Boesch J, Stoyanova R, Srinivasan R, Willard T, Vigneron D, Nelson S, Taylor JS, Brown TR. NMR Biomed. 1993;6:173.
172. Luyten PR, Bruntink G, Sloff FM, Vermeulen JW, van der Heijden JI, den Hollander JA. Heerschap A. NMR Biomed. 1989;1:177.
173. Meninger M, Landschutz W, Beer M, Seyfarth T, Horn V, Pabst T, Haase A, Hahn D, Neubauer S, von Kienlin M. Magn. Reson. Med. 1999:41:657.

Radio Frequency Coils for Magnetic Resonance Spectroscopy

Boguslaw Tomanek
Institute for Biodiagnostics, Winnipeg, Manitoba, Canada

The Requirements

Localized magnetic resonance spectroscopy (MRS) as well as magnetic resonance imaging (MRI) require both maximal signal-to-noise ratio (SNR) and radio frequency (rf) field (B_1) homogeneity. While MRI allows the application of post-processing methods, such as correction of image intensity to correct B_1 inhomogeneities, MRS is more demanding from this point of view leaving very little opportunity for post-processing and requiring mostly hardware improvement. In addition, due to the smaller volume of interest (VOI) and particularly low metabolite concentration, there is less signal available in MRS than in MRI. The time required to obtain good spectra is long pushing the SNR requirements to the limits.

The Issues

Unfortunately, when considering rf probe construction the requirements of high SNR and good B_1 homogeneity usually conflict. The SNR depends on many factors such as the strength of the magnetic field, geometry, resistance and quality factor (Q) of the coil, sample size, filling factor of the coil, type of the reception (linear or circular), etc. The B_1 field distribution depends on the geometry of the coil, its type (transmit and receive or transmit/receive), and the material used for the coil construction.

When designing an rf coil all of these factors plus patient access to the coil and the direction of the magnetic field relative to B_1 must be considered. Therefore whilst a coil may work very well, providing good SNR and B_1 homogeneity, at a low frequency (say 1.5 T) it may not work so well (or even not at all) at a higher frequency (say 3 T). A trade-off amongst all these factors is inevitable and there is no simple, consistent recipe for making a perfect rf coil that fulfills all the requirements for MRS.

With this knowledge of the problems we shall consider some of the requirements for the MRS coil: SNR, B_1 distribution, coil geometry, method of reception as well as some other factors. For this discussion we will be ignoring other potential problems associated, for example, with relaxation times, pulse parameters or even eddy currents induced in the coil, gradients coil, or in the cryostat due to the short TE (fast gradient switching) required frequently in MRS.

The first expression describing the NMR signal was derived by Abragam [1], who showed that the electromagnetic force (emf) induced by the nuclear magnetization in the receiving coil can be expressed as:

$$\Psi_{\mathrm{rms}} \propto K\eta M_0 \left(\mu_0 Q\omega_0 V_c / 4FkT_c\Delta f\right)^{1/2} \qquad (1)$$

where K is numerical factor depending on the coil geometry; η is the filling factor, i.e. a measure of the fraction of the rf field volume occupied by the sample; M_0 is the magnetization (proportional to the magnetic field strength); μ_0 is the permeability of free space; Q is the quality factor of the coil; ω_0 is the Larmor frequency; V_c is volume of the coil; F is the noise figure of the preamplifier; k is Boltzmann constant; T_c is the probe temperature; Δf is the bandwidth (in Hertz) of the receiver. As one can notice, the SNR is proportional to the 3/2 power of the magnetic field ($M_0 \sim \omega_0$). This is true assuming that other factors are independent of frequency. This is an important statement that will be analyzed in detail later.

The above revelation explains the historical tendency to introduce stronger and stronger magnets in an effort to increase SNR. Therefore, magnets generating a magnetic field of 20 T were produced for *ex vivo* MRS, while whole body MRI systems have reached 7 T, and even higher field magnets are under construction.

However, a closer analysis of the dependence of noise mechanism with frequency reveals a less favorable SNR relationship. The rule of thumb is a linear relationship between the SNR and the magnetic field strength. While the Equation (1) gives a reasonable estimate of the SNR, it does not consider other factors such as rf coil shape, preamplifier noise, and other electronic components whose performance depends on, and usually degrades with, the frequency. Thus, innocent engineers constructing rf coils for MRI systems are often blamed for insufficient increase in SNR as expected from MR theory.

Probably the most comprehensive summary of the SNR and its relationship with MR hardware was given by Hoult *et al.* [2,3], who applied reciprocity theory and

Graham A. Webb (ed.), Modern Magnetic Resonance, 1149–1156.

considering the sources of noise they obtained:

$$\psi_{\text{rms}} = \frac{K(B_1)_{xy} V_s N \gamma \bar{\lambda}^2 I(I+1)}{7.12 k T_s} \cdot \left(\frac{p}{F k T_c l \zeta \Delta f} \right)^{1/2} \times \frac{\omega_0^{7/4}}{[\mu \mu_0 \rho(T_c)]^{1/4}} \tag{2}$$

where $K(B_1)_{xy}$ is the effective field over the sample volume produced by unit current flowing in the receiving coil, p is the perimeter of the conductor, l is the length of the conductor. The above factors are dependent only on coil geometry. T_c is the temperature of the coil, $\rho(T_c)$ is the resistivity of the material from which the coil is made, ζ is the proximity factor, F is the quality of the preamplifier.

From the above equation one can see that SNR seems to be proportional to $\omega_0^{7/4}$. However, in the above equation Q and η have been replaced by a function ζ, that is also ω_0 dependent. Hence, SNR is not always proportional to the 7/4 power of the frequency. A careful reader will also notice that SNR depends on the type and geometry of the coil (factors: K, l, ρ).

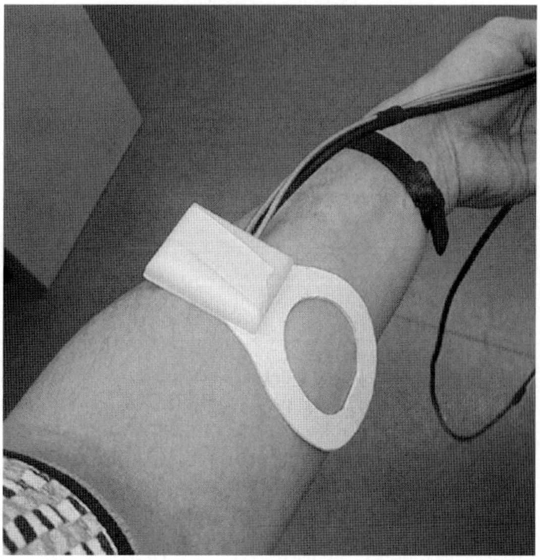

Fig. 1. Surface coil for 3 T MR system (National Research Council, Institute for Biodiagnostics, Winnipeg, Canada).

The Solenoid Coil and Saddle-Shape Coil

As seen from Equation (1) SNR can be improved by an increase of the Q factor, which depends on the coil resistance and inductance. The Q of a solenoidal coil is proportional to $\omega_0^{1/2}$ only well below self-resonance frequency of the coil, but when approaching the self-resonance frequency the Q drops away from the $\omega^{1/2}$ line and prevents further increase of the SNR as one could expect from Equation (2). The only solution is then to decrease the self-resonance frequency of the coil by the reduction of the coil inductance or/and resistance. This can be achieved by reducing the number of turns for a solenoidal coil. A detailed analysis of solenoidal and saddle-shape coils [2], the most common coils producing very homogenous rf field, yield to the conclusion that the solenoidal coil performance is about three times better that the saddle-shape coil with comparable dimensions. As shown above, a solenoidal coil would be a very good candidate for the rf probe providing good B_1 homogeneity and good sensitivity within a reasonable range of frequencies. However, further consideration should be given when the direction of the main magnetic field is considered. To produce the NMR phenomenon (to flip the magnetization), the direction of the rf field must be perpendicular to the main magnetic field. This creates additional constraints when selecting an rf coil for MRS and MRI. Unfortunately, the direction of the rf field produced by the solenoidal coil (along its long axis) prevents almost all applications for horizontal bore magnets. Such a coil could not be made for a human head, for example, since it would require the head to bend by 90°.

However, a single loop solenoidal probe, namely a surface coil (Figures 1 and 2), overcomes this obstacle. In addition, in MRS we are interested in small voxels to detect small abnormalities. Thus a volume coil is usually not required or even not welcome. A surface coil (Figure 2A) provides a very good SNR produces but also a very inhomogenous B_1 field (Figure 2B). Along the axis the B_1 field can be calculated, neglecting interaction of the tissue with the rf field, based on Bio-Savart law:

$$B_1(x) = \frac{\mu_0 I}{2} \frac{a^2}{\sqrt{(a^2 + x^2)^3}} \tag{3}$$

where a is the radius of the coil, x is the axial distance from the center of the coil and I is the current flowing through the coil.

Surface Coil

So far we have concluded that the best solution from the point of view of the SNR is the application of the surface coil. Further improvement in SNR by up to $\sqrt{2}$ can be achieved by the application of the quadrature surface coil [4,5]. However, the perfect quadrature rf coil requires the creation of equal and orthogonal B_1 fields (Figure 3), which is difficult to achieve for a surface coil. The most common configuration for the quadrature surface coil [so-called circular polarized (CP) coil] is the use of a shaped butterfly coil and a standard surface coil to

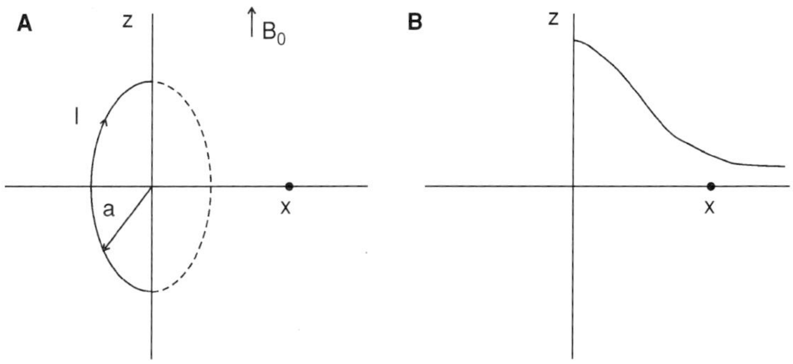

Fig. 2. (A) Geometrical model of the surface coil: A current I flows in the ring of a radius. (B) The B_1 field magnitude generated by the surface coil along the x-axis.

create two orthogonal or, more often—nearly orthogonal rf fields. While such a coil provides larger field of view (FOV) it gives only modest (less than $\sqrt{2}$) increase in SNR due to the lack of perfect perpendicularity or/and unequal amplitude of B_1 fields. The quadrature surface coil requires, however, less power when compared to the linear coil (LP coil). Let us now answer a question: what is the optimum diameter of the surface coil to maximize SNR within the specific VOI? In attempt to find the answer we will take a closer look at the sources of noise. Neglecting radiation losses, the noise picked up by the receiver (N) is proportional to the sum of the effective resistance of the sample (R_s) and the resistance of the coil itself (R_c):

$$N \propto \sqrt{R_s + R_c} \tag{4}$$

The resistance of the sample comes from two sources: dielectric losses (that can be minimized using distributed capacitance and proper capacitor shielding) and inevitable inductive losses associated with rf induced eddy currents within the sample. Hoult and Lauterbur [3] showed that

$$R_s \propto \omega_0^2 B_1 b^5 \tag{5}$$

where ω_0 is the Larmor frequency, B_1 is the rf field produce by unit current, b is the sample radius.

One can see that the sample losses increase with frequency and become predominant at a higher frequency. It can be shown [5,6] that in such a case SNR per unit volume depends on the surface coil radius a as:

$$SNR \propto 1/a \tag{6}$$

This equation, as explained in detail by Hayes and Axel [6] is exact only at the surface of the coil, however, it gives a good estimate of the dependence of the SNR of the coil radius. Thus, the smaller the coil the higher SNR hence surface coils which are so popular in MRS. The diameter of a surface coil may be chosen to give maximum SNR at a particular point within the sample. If sample loses are dominant (above 1T), maximum SNR can be obtained for $a = d/\sqrt{5}$ where a is the coil radius and d is the distance from the coil center [7,8]. When the coil losses are dominant [9] then the optimum SNR occurs for the coil radius $a = d\sqrt{2}$. The exact optimum diameter depends on the frequency and the sample. For example, at 3T and 1 cm^3 VOI placed at $d = 2.5$ cm from the coil the maximum SNR occurs, as found experimentally, when the coil diameter is 4.0 cm. This diameter lies between the coil loss dominant case (7.1 cm) and the sample loss dominant case (2.25 cm).

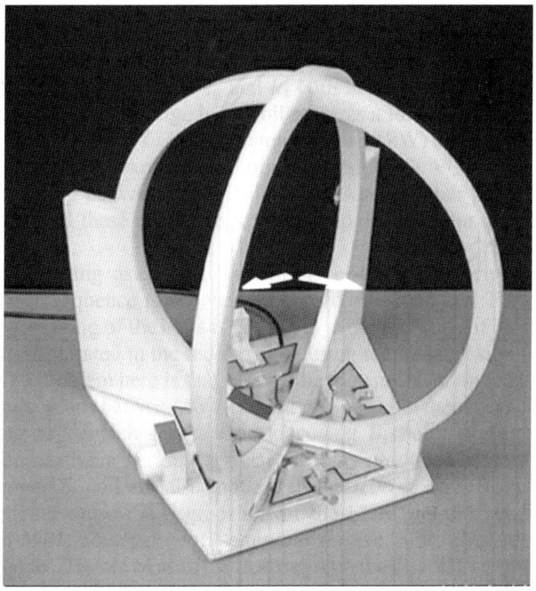

Fig. 3. Two surface coils assembled into the quadrature probe. The two perpendicular B_1 fields are marked with arrows (National Research Council, Institute for Biodiagnostics, Winnipeg).

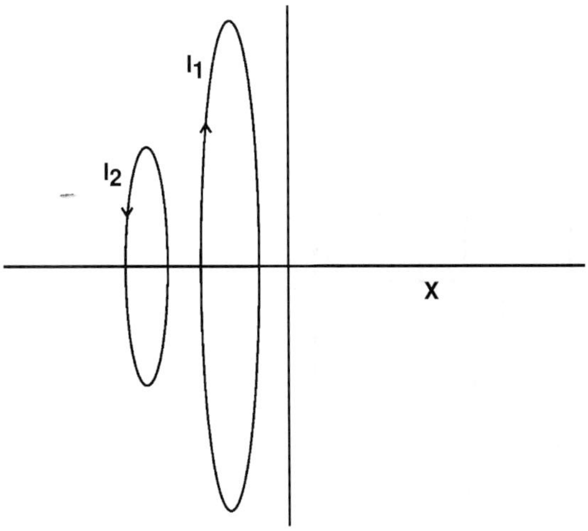

Fig. 7. Multi-ring surface coil assembly. Currents I_1, I_2 flow in opposite directions to compensate B_1 inhomogeneity of each separate surface coil.

Local rf Coils with Improved B_1 Homogeneity and SNR

An interesting solution [33,34] to generate identical fields for both excitation and receiving periods was proposed: the combination of surface coil sensitivity with volume coil homogeneity. This idea uses two or more inductively coupled rings of different diameter, carrying current in opposite directions (Figure 7). Such an assembly produces homogenous, identical excitation and receiving B_1 in the desired region of interest while providing SNR comparable with the surface coil. While this coil seems to be perfect for MRS its major disadvantage is fixed VOI, what in practice means that for each VOI another rf coil should be constructed. The possible applications of this coil include skin or thyroid MRS, where the VOI varies little between subjects.

Another solution to increase SNR is the application of rf shields [35]. This concept works particularly well for breast MR imaging and spectroscopy. If the B_1 field is perpendicular to the chest wall, a Helmholtz pair probe gives good field homogeneity over the breast and good chest penetration, but significant chest power dissipation with concomitant loss of SNR. Weak B_1 fields covering the large volume could generate substantially larger losses than bigger fields confined to a smaller volume. Thus, the combination of the Helmholtz pair with the rf shielding of the chest wall away from the breast brought a substantial improvement of up to a factor of two when compared with the standard breast rf coil.

Implanted rf Coils

For imaging of some small, deeper organs, for example, the spinal cord, the rf surface coils are used but due to their field drop with the distance from the coil, they bring even more challenges in the coil design. Because of that the SNR within the VOI is compromised due to the low and inhomogenous B_1 field. The noise source from the "disturbing" tissue (e.g. fat) between the coil and the VOI, where the B_1 is strong, is yet another source of noise. One of the possible ways of dealing with this problem is the application of implanted rf coils [36,37]. Briefly, the implanted coil consists of a loop of wire (the "actual" coil generating and receiving rf signal) tuned to the Larmor frequency, and a matching ring inductively coupled to the loop. Thus, no electrical connections between the implanted coil and the external matching network are required. Such an implanted rf coil provides better SNR than the surface coils, thanks to their close proximity to the VOI. It also brings some disadvantages: the procedure can no longer be considered non-invasive, and as such its application to human MRS becomes questionable. In addition, implanted coils are vulnerable to various failure modes including mechanical and electrical. The blood and body fluids can change the coil loading or even its position, thus SNR during the MRS experiment. In some cases matching also becomes difficult due to the restricted access of the matching coil and the loop.

Microcoils

When using ultra-high fields and ultra-low samples ($\sim$1 μl) new challenges arise such as susceptibility, low filling factors, increased coil noise, skin depth, etc. As shown [38,39] for the coils of diameter of order of millimeters or less the major source of the noise comes from the coil. Therefore, special attention must be placed to the resistance of the coil. A so-called single and multilayer rf microcoil, using metallic foil rather than wires was used to enable very high field *ex vivo* MRS [40].

The performance of the standard rf surface coil can also be surpassed by a microstrip rf surface coil for extremely high-field MRI as reported by Zhang *et al.* [41]. This design is based on the microstrip transmission line concept that does not require the use of lumped elements. The microstrip surface coil can be used up to 500 MHz. Such a coil gives about 20% better SNR when compared to the standard surface coil [41]. This concept can also be used with any other configurations such as quadrature volume or phase array rf coils. Such an assembly resonates at the frequency $f = c/2L\sqrt{\varepsilon_{\mathrm{eff}}}$ where c is the speed of light, L length of the coil, $\varepsilon_{\mathrm{eff}}$ dielectric constant of the material. This concept has not yet found a clinical use but may be promising.

Dual Frequency rf Coils

Quantification of the tissue metabolites by heteronuclear MRS can be achieved using dual frequency rf coils. The dual frequency coils allow simultaneous measurement of two nuclei frequencies, without disruptions associated with exchanging coils for imaging and spectroscopy examinations. One is usually ^1H while the other is phosphorus (^{31}P), sodium (^{23}Na), or other nuclei (^{23}Na). The proton frequency is often used for high-resolution imaging that allows lesion localization while the other frequency is used for spectroscopic lesion analysis. The proton frequency also provides the sensitivity needed for shimming. Dual frequency coils have to deliver SNR of the single frequency rf coil as well as produce homogenous rf field over the sample volume. To achieve these needs double-tuned quadrature birdcage coils were developed. One example is a design based on two coaxial birdcage coils resonating at ^{31}P and ^1H frequencies each. Such a coil provides excellent ^{31}P performance although lower SNR for ^1H when compared to a single frequency ^1H coil. This design allows altering the ^{31}P coil length thus can provide better SNR sacrificing however B_1 [42]. Another solution is to use the same sized coils and tune every other leg of the coil to the desired frequency, using trap circuits and inductive matching to optimize the coil [43,44].

An elegant design based on transmission line resonators for a volume quadrature, double-tuned coil was also proposed by Vaughan [45]. In our work to even further increase SNR for two or more frequencies, a dual frequency surface coil was designed using frequency splitting circuitry [46]. This construction allows concurrent ^1H and ^{31}P MRS and provides the same B_1 homogeneity for both frequencies almost without reducing SNR when compared to the single frequency rf coils. Such a design was further improved by combining a multi-ring coil concept [47] with a splitting circuit. This coil provides better SNR than a volume coil within the specified VOI, but about 70% of the SNR of the single frequency surface coil, however, it creates homogenous and identical B_1 fields for both frequencies over the desired VOI. The dual frequency coils require dual frequency receive and transmit channels, which are still not standard in clinical MRI systems, hence their limited application.

Summary

This chapter describes the most common rf coils for MRS as well as possible methods of improving SNR and B_1 homogeneity. As concluded, there is no perfect rf coil for MRS. Compromises must be made among SNR, B_1 homogeneity, voxel size, total experiment time, size, and position of VOI. The choice of the coil depends on the application, field strength, available MRI system, and the patient position. The volume coil—transmit and surface or local coil receive-only seem to be the most effective, yet not perfect rf coil configuration.

Part II

Glossary of Terms

SNR	signal-to-noise ratio
B_1	rf field
M_0	magnetization
μ_0	permeability of free space
Q	quality factor of the coil
K	numerical factor depending on the coil geometry
η	filling factor
ω_0	Larmor frequency
V_c	volume of the coil
F	noise figure of the preamplifier
k	Boltzmann constant
T_c	probe temperature
Δf	bandwidth (in Hertz) of the receiver

References

1. Abragam A. The Principles of Nuclear Magnetism. Clarendon Press: Oxford, 1961, p 82.
2. Hoult DI, Richards RE. J. Magn. Reson. 1976;24:71.
3. Hoult DI, Lauterbur, PC. J. Magn. Reson.1979;34:425.
4. Chen CH, Hoult DI, Sank VJ. J. Magn. Reson. 1983;54: 324.
5. Hyde JS, Jesmianowicz A, Grists TM, Froncisz W, Kneeland JB. Magn. Reson. Med. 1987;4:179.
6. Hayes CE, Axel L. Med. Phys. 1985;12(5):604.
7. Edelstien WA, Foster TH, Schenk JF. The relative sensitivity of surface coils to deep lying tissue. Proceedings of the SMRM 4th Annual Meeting, London, 1985, p 964.
8. Wang J, Reykowski A, Dickas J. IEEE Trans. Biomed. Eng. 1995;42:908.
9. Chen CN, Hoult DI. Biomedical Magnetic Resonance Technology. Adam Hilger: Bristol, 1989, p 161.
10. Hill HDW. Solid State Commun. 1997;102:169.
11. Darrasse L, Ginefri JC. Biochemie. 2003;85:915.
12. Roemer PB, Edelstein WA, Hayes CE, Souza SP, Mueller OM. Magn. Reson. Med. 1990;16:192.
13. Sodickson DK, Manning WJ. Magn. Reson. Med. 1997;38: 591.
14. Pruessmann KP, Wieger M, Scheidegger MB, Boesiger P. Magn. Reson. Med. 1999;42:952.
15. Bottomley PA. Selective volume method for performing localized NMR spectroscopy. US Patent 4,480,228, 1984.
16. Frahm J, Merboldt KD, Hanicke W. J. Magn. Reson. 1987;72:502.
17. van Zijl PCM, Moonen CTW, Alger JR, Cohen JS, Chesnick SA. Magn. Reson. Med. 1989;10:256.
18. Lawry TJ, Kaczmar GS, Weiner MW, Matson GB. Magn. Reson. Med. 1989;9:299.
19. Moonen CTW, van Zijl PCM. J. Magn. Reson. 1990;88:28.
20. Lawry TJ, Kaczmar GS, Weiner MW, Matson GB. Magn. Reson. Med. 1989;9:299.

21. de Graaf RA, Nicolay K, Garwood M. Magn. Reson. Med. 1996;35:652.
22. de Graaf RA, Nicolay K, Garwood M. J. Magn. Reson. B. 1996;113:35.
23. Bendall MR. Chem. Phys. Lett. 1983;99(4):310.
24. Edelstein WA, Hardy CJ, Mueller OM. J. Magn. Reson. 1986;67:156.
25. Haase A. J. Magn. Reson. 1985;61:130.
26. Barberi EA, Gati JS, Rutt BK, Meonon RS. Magn. Reson. Med. 2000;43:284.
27. Hoult DI. J. Magn. Reson. Imaging. 2000;12:46.
28. Crowley MG, Evelhoch JL, Ackerman JJH. J. Magn. Reson. 1985;64:30.
29. Styles P, Smith MB, Briggs RW, Radda GK. J. Magn. Reson. 1985;62:397.
30. Chen CH, Hoult DI. Biomedical Magnetic Resonance Technology. Adam Hilger: Bristol, 1989, p 157.
31. van Zijl PCM, Moonen CTW, Alger JR, Cohen JS, Chesnick SA. Magn. Reson. Med. 1989;10:256.
32. Crozier S, Field J, Brereton IM, Moxon LN, Shannon GF, Doddrell DM. J. Magn. Reson. 1991;94:123.
33. Tomanek B, Ryner L, Hoult DI, Kozlowski P, Saunders JK. Magn. Reson. Imaging. 1997;15: 1199.
34. King S, Ryner L, Tomanek B, Sharp J, Smith I. Magn. Reson. Med. 1999;42:655.
35. Tomanek B, Hoult DI, Chen X, Gordon R. Magn. Reson. Med. 2000;43:917.
36. Silver X, Xu W, Mercer EV, Beck BL, Bossart EL, Inglis B, Mareci TH. Magn. Reson. Med. 2001;46:1216.
37. Arnder LL, Shattuck MD, Black RD. Magn. Reson. Med. 1996;35:727.
38. Cho Z, *et al.* Med. Phys. 1998;15:815.
39. Peck T, Magin R, Lautenbur P. J. Magn. Reson. B. 1995;108: 114.
40. Grant SC, Murhpy LA, Magin RL, Friedman G. IEEE Trans. Magn. 2001;37(4).
41. Zhang X, Ugurbil K, Chen W. Magn. Reson. Med. 2001;46: 443.
42. Fitzsimmons JR, Beck BL, Brooker HR. Magn. Reson. Med. 1993;30:107.
43. Matson GB, Vermathen P, Hill TC. Magn. Reson. Med. 1999;42:173.
44. Shen GX, Boada FE, Thulborn KR. Magn. Reson. Med. 1997;38:717.
45. Vaughan JT, Hetherington HP, Out JO, Pan JW, Pohost GM. Magn. Reson. Med. 1994;32:206.
46. Schnall MD, Subramanian H, Leigh JS, Chance B. J. Magn. Reson. 1985;65:122:
47. Volotovskyy V, Tomanek B, Corbin I, Buist R, Tuor UI, Peeling J. Magn. Reson. Eng. Concepts in Magnetic Resonance, Part B 2003;17B:11.

Spatially Resolved Two-Dimensional MR Spectroscopy *in vivo*

M. Albert Thomas[1], Amir Huda[1,2], Hyun-Kyung Chung[1], Nader Binesh[1], Talaignair Venkatraman[1], Art Ambrosio[1], and Shida Banakar[1]

[1]*Department of Radiological Sciences, University of California, Los Angeles, CA 90095, USA; and*
[2]*Department of Physics, California State University, Fresno, CA 93740, USA*

Introduction

Acquisition of three-dimensional (3D) spatially localized, water-suppressed one-dimensional (1D) ^{1}H MR spectra has been optimized in human tissues over the last two decades [1–6]. Introduction of automated MRS a decade ago has enabled a 1–2 min pre-scan to calibrate the flip-angle using the transmitter power, to adjust the static field homogeneity (B_0) using the linear gradient shim coils, and to optimize water suppression, which used to be an at least 15 min operation, facilitating the integration of clinical MRS protocols into routine MRI examinations [7]. In ^{1}H MR spectroscopy of human brain, spectral resonances due to methyl, methylene, and methine protons of N-acetylaspartate (NAA), glutamate/glutamine (Glx), creatine (Cr), choline (Ch), *myo*-inositol (mI), GABA, aspartate (Asp), N-acetylaspartyl-glutamate (NAAG), and other metabolites have been identified in the region of 0–4.5 ppm [2–12].

One-dimensional MRS *in vivo* is hampered by the fact that it is difficult to resolve a multitude of peaks existing over a small spectral range of approximately 300 Hz at 1.5 T. Spectral overlap of macromolecules, GABA, and glutathione with the methyl resonances of NAA, Cr, NAAG, and methylene resonances of Glx has also been reported [8,14–15]. Attempts have been made to quantify Glx [16], glucose [17], and other metabolites with only minimal success due to the difficulty in extracting this information from a region with many overlapping resonances. Using spectral-editing techniques, it is possible to select a particular J-coupled metabolite [2,14,18–20]. However, one major drawback of editing techniques is that only one metabolite is optimized at a time assuming that the multiplets of the J-coupled metabolites are well separated. Using the LC-Model post-processing algorithm, quantification of several metabolites in human brain *in vivo* has been reported recently using a basis-set of 10–20 different metabolite spectra *in vitro* [21–23].

Unlike the spectral editing techniques, which target one metabolite at a time, two-dimensional (2D) MRS can unambiguously resolve many overlapping peaks non-selectively *in vitro* as shown by Ernst and co-workers two decades ago using the vertical bore NMR spectrometers [24–25]. 2D-MR spectroscopy enables converting a crowded overlapping 1D MR spectrum to a better-resolved 2D spectrum through the addition of a spectral dimension. Instead of a standard 1D spectrum plotting intensity vs. a single-axis (i.e. chemical shift), 2D MR spectroscopy techniques produce a 2D spectrum plotting intensity vs. two axes, the dimensions of which depend on the specific 2D MR technique [25]. Better dispersion of several metabolite peaks and improved spectral assignment make the proposed technique more attractive. It seems natural to explore the clinical potentials of 2D MRS techniques of human tissues *in vivo*.

There have been several attempts during the last 15 years in the implementation and evaluation of 2D NMR spectroscopy on *in vivo* MRI scanners and MR spectrometers [26–30]. The aim of this chapter is to discuss the recent progress on the implementation of selected localized 2D MRS sequences on the whole body 1.5 T and higher field MRI scanners. This includes a summary of 1D spectroscopy with and without chemical shift imaging (CSI), followed by single- and multi-voxel based 2D spectroscopy dealing with coherent interactions such as chemical shift and J-coupling. A brief overview of potential artefacts in 2D spectroscopy and simulation is also included.

Single- and Multi-voxel Based 1D ^{1}H MR Spectroscopy

Clinical 1D MR spectroscopy has some important differences and challenges compared to MR imaging. Often how the challenges are resolved define the particular technique of spectroscopy with its limitations and advantages. Some of these basic challenges are pertinent to water and lipid suppression, spatial localization of the volume of interest (VOI) including single vs. multiple volume elements (voxel) as shown in Figure 1.

Graham A. Webb (ed.), Modern Magnetic Resonance, 1157–1169.

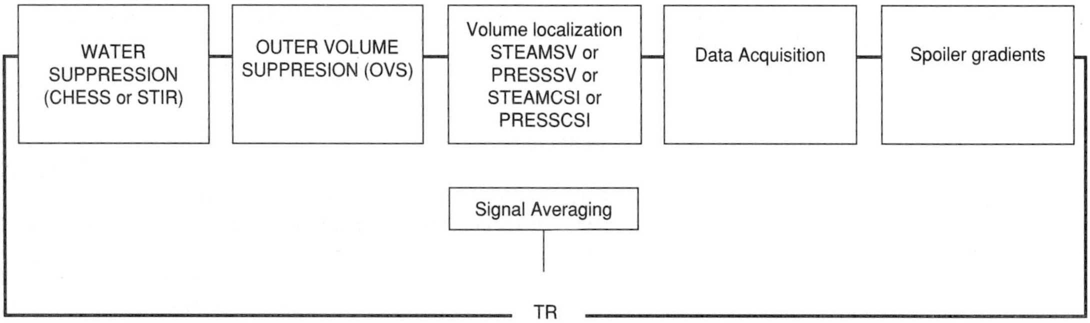

Fig. 1. Different modules of 1D MRS protocol.

Water and Lipid Suppression

What makes MR imaging quite sensitive is the large concentration of water in biological tissue. However, the dominance of water limits spectroscopy since the metabolites are tens of thousands of times lower in concentration than water. Thus, in order to obtain a visible signal from the metabolites, MRS techniques must incorporate ways of minimizing or eliminating the water signals. This water suppression can be extended from changes in hardware for data acquisition (digitizers), pulse sequences, to post-processing software. The primary contribution stems from an addition of three frequency selective (water as the target frequency) radio frequency (rf) pulses with arbitrary flip-angles followed by a dephasing gradient to the pulse sequence. These pulses are commonly called the chemical shift selective saturation (CHESS) pulses [5,31]. Instead of water saturation using CHESS, an inversion recovery scheme can also be employed with a 180° frequency selective pulse on water such that excitation of the metabolites can be prolonged until the crossing of the water signal at the null point with known T_1 values [32].

Quite often the lipid signals from either within the VOI or outside of the VOI hold a similar challenge like water. In order to suppress the lipids, the same two approaches utilized in water suppression in terms of pulse sequence modification can also be applied here [33]. The most common approach is the use of frequency selective (targeted at lipids) pulses in conjunction with spatial selection around the edges of the brain called the outer volume saturation or suppression (OVS) slabs. Of course, post-processing software can also contribute to elimination of lipid signals.

Spatial Localization of the Volume

There is a fundamental difference between the localization done in imaging and spectroscopy. Imaging relies on acquisition of the entire proton signal (primarily water and fat) and the differences in frequencies of the signal are attributed to the spatial shifts introduced *via* the application of the frequency encoding gradients during signal sampling to reproduce the selected slice within a plane or a data point in volumetric MRI. A frequency encoding gradient cannot be used in spectroscopy during sampling because the differences in frequencies of metabolites sought after are due to the small chemical shifts and a frequency encoding gradient would corrupt this basic information. However, the demand for volume definition remains, whether single voxel or multiple voxel spectroscopy is done since the metabolite concentration information is sought from a particular voxel and not the whole brain. The most common localization procedure is the use of slice selection and phase encoding gradients. These serve to demarcate the dimensions of the voxel and are operated for a very short duration. The spins then return to the influence of the external magnetic field.

Single Voxel Spectroscopy

The two most popular pulse sequences for defining a small (ranging from 1 cm^3 to 8 cm^3) volume within the anatomy are termed as STEAM and PRESS [3,9,34]. Both PRESS ($90°_{ss}-180°_{ss}-180°_{ss}$) and STEAM ($90°_{ss}-90°_{ss}-90°_{ss}$) use frequency selective RF pulses along with magnetic field gradients to isolate a single volume, called STEAMSV and PRESSSV; however, the timing and the sequence of the pulses has consequences on the sensitivity and the strength of the signal [35].

Multivoxel Spectroscopy or Chemical Shift Imaging

This technique known as spectroscopic imaging (SI) or CSI is challenged with distilling both chemical shift as well as spatial information from the acquired region of interest [36–37]. The data processing in this case requires both spatial and chemical shift dimensions. For CSI, the

PRESS and STEAM sequences described above can be used including the field gradients modified for spatial encoding (STEAMCSI and PRESSCSI). The phase encoding gradients are repeated several times between the remainder of the rf pulses each time repeating the entire sequence but with the phase encoding gradient amplitude changing slightly. This successive change is done in a stepwise fashion with the gradient amplitude changing; thus, several samples of a frequency are produced. The data can then be Fourier transformed to allocate signal intensities of different frequencies to appropriate voxels.

Single Volume Localized 2D ^{1}H MR Spectroscopy

For MR spectroscopy of a particular nucleus, namely hydrogen (^{1}H), one counts on the differences in internal spin interactions which are due to different inter- and intra-molecular environments. These differences in the molecular environment are of several types and are, hence, responsible for different features of a spectrum or an image. For example, the translational molecular motion is exploited in a diffusion weighted image. While the chemical shift expresses the local electronic environment and is predominantly an intra-molecular property, molecular motion also influences this shift significantly. A primary manifestation and measure of the chemical bonds between the nuclear spins, i.e. the indirect spin-spin coupling communicated through the covalent bonds, is the exclusively intra-molecular J-coupling, which is independent of the applied magnetic field. The motive for introducing the 2D MRS sequences to whole body MR spectroscopy is simple—the extraction of more information out of the MR spectrum to aid in the biochemical analysis of human tissue *in vivo*. 2D spectroscopy can do this by converting a crowded, overlapping 1D MR spectrum to a better-resolved 2D spectrum through the addition of a spectral dimension [24,25]. The problems due to the overlap in the 1D spectrum introduced by the multiplicity of various peaks due to J-coupling could be removed by separating the interactions due to chemical shift and indirect spin-spin coupling (J) along the two dimensions in 2D MR spectroscopy. Instead of a standard 1D spectrum plotting intensity vs. a single-axis (i.e. chemical shift), 2D NMR spectroscopy techniques produce a 2D spectrum plotting intensity vs. two axes, the dimensions of which depend on the specific 2D NMR technique.

The Nobel prizes were awarded to Prof. Richard Ernst in 1991 and Prof. Kurt Wuthrich in 2002 for their innovative contributions focused on coherent and incoherent intra- and inter-molecular interactions using multidimensional MR spectroscopy *in vitro* [38]. Exploitation of this interaction is the crucial link to the molecular chemistry of the metabolites in human tissues. Before applying these multidimensional MR techniques in human tissues, each sequence needs to be converted first into localizing a VOI using three orthogonal slice-localizing rf pulses. A second task is to minimize the number of rf pulses essential for creation of Hahn's spin-echo or coherence transfer echoes. Similar to the 1D MRS protocol, shown in Figure 2 are different compartments of 2D MRS protocol. In addition to repeated spectral acquisition for signal averaging, the sequence has to be repeated multiple times for encoding the second spectral dimension.

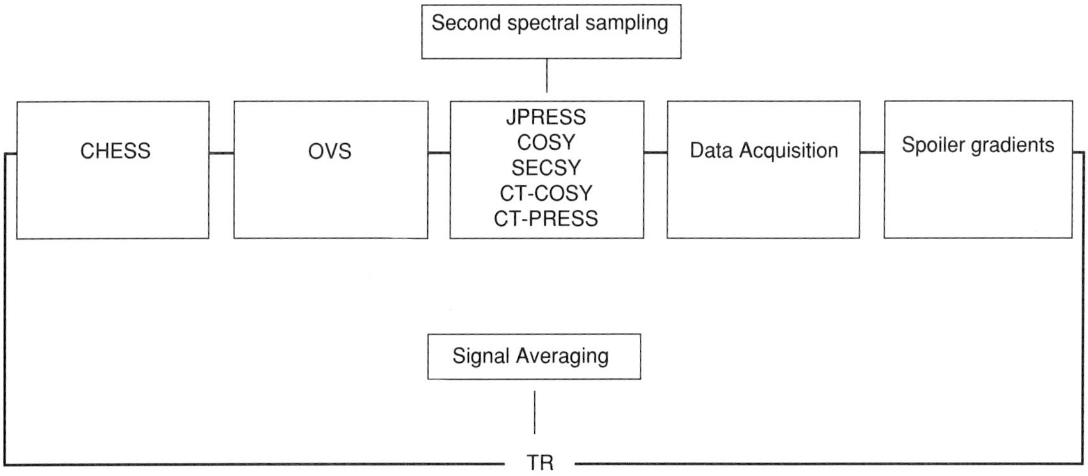

Fig. 2. Different modules of 2D MRS protocol.

2D J-Resolved MR Spectroscopy (JPRESS)

As discussed earlier, the conventional PRESS sequence uses three slice-selective rf pulses (90°, 180°, and 180°) along three orthogonal directions with the second spin-echo originating from a voxel at the intersection of three orthogonal slices. The original 2D J-resolved homonuclear MR spectroscopic sequence is a modified Hahn spin-echo sequence [90°-$t_1/2$-180°-$t_1/2$-acquire(t_2)], where signal is acquired during t_2 and the incremental period t_1 is the additional time domain to sample the second spectral dimension of a 2D MR spectrum [25]. In 1995, Ryner *et al.* coded the first 3D-localized 2D J-resolved sequence (JPRESS) by modifying the PRESS sequence to allow simultaneous incrementation of the two periods before and after the last 180° rf pulse [39,40] as shown in Figure 3. The localized 2D JPRESS sequence has been implemented on both GE and Siemens 1.5 and 3 T MRI scanners [39–46]. Shown in Figure 4 are 2D JPRESS spectra of a 100 mM Lac phantom using a Siemens 3T Trio MRI scanner before and after rotating the entire 2D dataset by 45°. A voxel size of 27 ml and a head transmit/receive coil were used. The time period Δ was kept to the minimum necessary to play out rf and gradient pulses. The evolution time (t_1) was incremented in 10-ms steps in order to achieve a spectral window of ±50 Hz that is well within the range of homonuclear (^{1}H–^{1}H) J-coupling. A digital resolution of 1024 complex points along t_2 over a sweep width of 2000 Hz and T_R of 2000 ms was used. The total duration was 4.3 min using two averages per t_1 and 64 increments. However, a longer acquisition time of 17 min will be necessary for signal averaging the metabolite signals at the physiological concentrations. Water suppression was achieved using CHESS. The refocusing of the chemical shift by the last 180° RF pulse resulted in only J-coupling information along the F_1 dimension. A 45° rotation of the 2D data matrix results in J-couplings and chemical shifts along orthogonal axes F_1 and F_2as shown in Figure 4B. As reported earlier [40,41], additional cross peaks due to strong coupling have been observed even at 3 T.

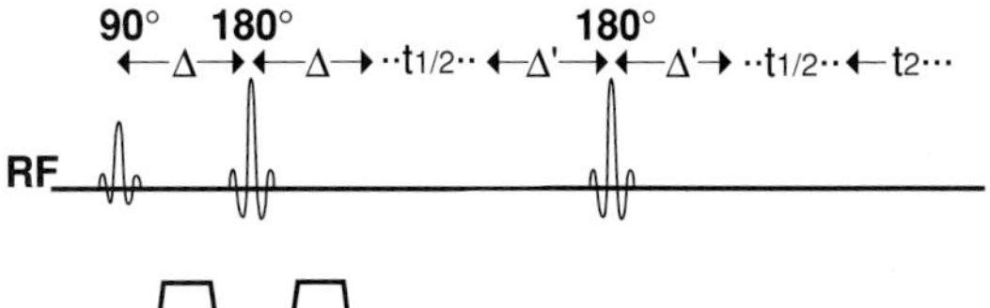

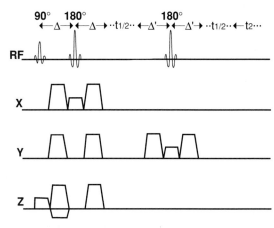

Fig. 3. Localized 2D J-resolved MRS sequence (JPRESS) consisting of three slice-selective rf (90°, 180°, 180°) pulses.

The peaks due to the non-coupled spins are also retained in the 2D J-resolved spectrum unlike the 1D edited spectra. Another advantage of this sequence is that 100% of the available magnetization is retained, neglecting losses caused by T_2 relaxation and spin diffusion. Hence, it is more advantageous than some of the filtered spectral editing techniques, in which 50% or more of the available magnetization is lost. The 2DJ-resolved pulse sequence is versatile such that it can be used for the acquisition of a standard 1D PRESS spectrum by simply removing the incrementation of the t_1-period. A homonuclear broadband decoupled ^{1}H spectrum can also be retrieved after projecting the 2D JPRESS cross peaks onto the F_2 dimension as demonstrated in Figure 4B. This technique has been further extended to human brain and prostate tumours recently to select certain metabolites [42–43]. An over-sampled 2D J-resolved sequence without any water suppression has also been demonstrated [44–46].

Chemical Shift Correlated MR Spectroscopy (L-COSY)

Figure 5 shows a localized 2D correlated spectroscopy (L-COSY) sequence [47]. Similar to JPRESS, the VOI was localized in 'one-shot' by a combination of three slice-selective rf pulses (90°-180°-90°), a new MRS volume localization sequence called CABINET (coherence transfer based spin-echo spectroscopy). The last slice-selective 90° RF pulse acts also as a coherence transfer pulse for the 2D spectrum

In addition to the slice-selective 180° rf pulse, there are B_0 gradient crusher pulses before and after the last 90° rf pulses. The incremental period for the second dimension can be inserted at two different locations: first, immediately after the formation of the Hahn spin-echo (t_1) and second, after the last gradient crusher column (k^*t_1), where k^*t_1 can be 4 µs or minimum allowed by the scanner hardware. The L-COSY sequence can be considered a single-shot technique in terms of simultaneous volume localization and coherence transfer. In contrast to PRESS, the CABINET sequence retains only 50% net signal from the localized volume due to a selection of only N-type echo, enabled by the B_0 gradient crusher pulses [47]. If a surface coil is used for both transmission/reception, one has to consider the influence of flip-angle errors due to the delivery of the inhomogeneous rf pulses delivered by a surface coil. In the CABINET sequence with the

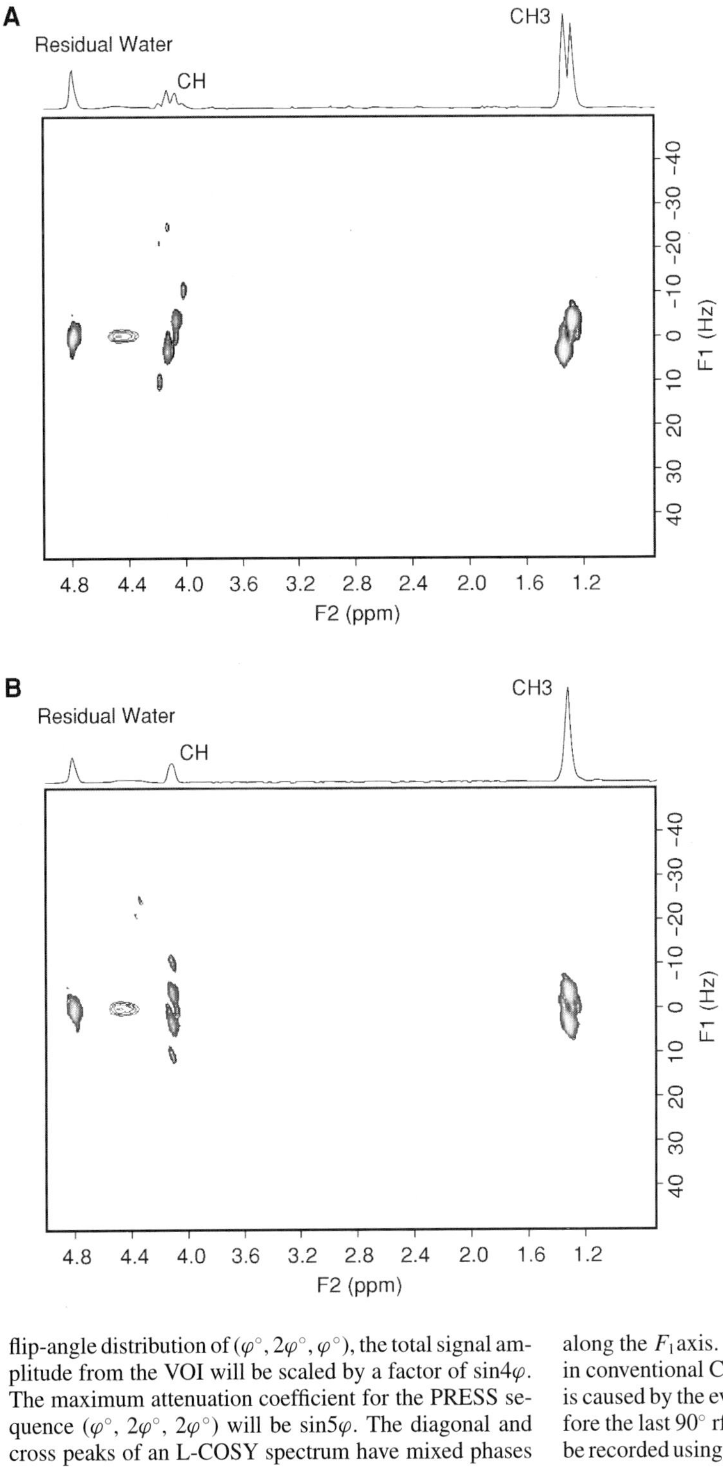

Fig. 4. Localized 2D JPRESS spectra of 100 mM Lac phantom. (A) before and (B) after 45° rotation of the 2D data matrix.

flip-angle distribution of ($\varphi°$, $2\varphi°$, $\varphi°$), the total signal amplitude from the VOI will be scaled by a factor of sin4φ. The maximum attenuation coefficient for the PRESS sequence ($\varphi°$, $2\varphi°$, $2\varphi°$) will be sin5φ. The diagonal and cross peaks of an L-COSY spectrum have mixed phases along the F_1 axis. In contrast to the amplitude modulation in conventional COSY, the phase modulation in L-COSY is caused by the evolution during the B_0 gradient pulse before the last 90° rf pulse. Pure phase L-COSY spectra can be recorded using a quadrature detection method along the

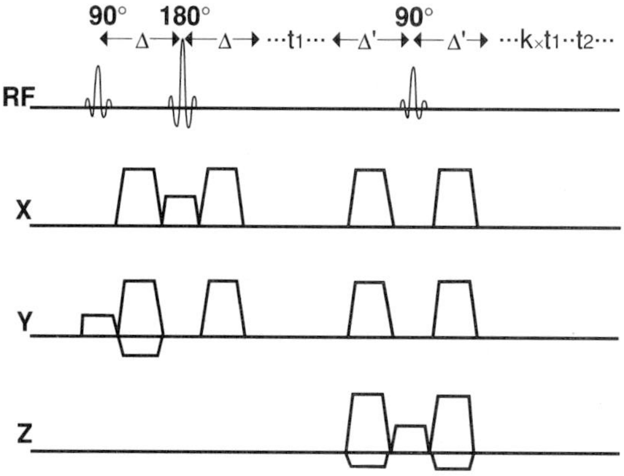

Fig. 5. The 2D L-COSY sequence consisting of three slice-selective rf (90°, 180°, 90°) pulses.

F_1 axis described by Doddrell and co-workers [48,49] that will require two separate P- and N- type spectral acquisition and recombination of the two datasets. Relaxation during the gradient pulses before the last 90° rf pulse will cause further losses in signal intensities.

Two-dimensional L-COSY has been implemented recently on GE and Siemens 1.5 and 3 T MRI scanners [41,47,50] and the test-retest reliability has been demonstrated *in vitro* as well as *in vivo* [41,51]. Shown in Figure 6 are the L-COSY spectra recorded in a) a grey matter brain phantom using a Siemens 3 T MRI scanner and b) the occipito-parietal white matter region of a 38-year-old healthy human subject using a GE 3 T MRI scanner. The brain phantom had the following metabolites at physiological concentrations: 8.9mM NAA, 0.7 mM GABA, 2.1 mM Asp, 0.9 mM Ch, 7 mM Cr, 1 mM Glc, 12.5 mM Glu, 2.5 mM Gln, 2 mM GSH, 4.4 mM mI, 0.4 mM Lac, 3.6 mM PCr, 0.6 mM PCh, 1.8 mM Tau, 0.3 mM Thr, 1 mM PE, and 1 mM DSS in a phosphate buffer solution to maintain pH of 7.2. 2D L-COSY spectra were recorded using the following parameters: 27 ml voxel, $T_E = 20$–30 ms, $T_R = 2$ s, total number of scans = 768 (96 Δt_1 increments and 8NEX/Δt_1). The total duration for each 2D scan with water suppression was approximately 24 min. Typical experimental parameters used for 2D L-COSY are listed in Table 1. Applications of this technique in brain and breast tumours, and brain disorders have also been reported recently [52–54]. Other researchers have also reported the implementation of different versions of 2D COSY [30,49,55–57]. A gradient enhanced COSY in combination with volume localized spectroscopy (VOSY) used the steam sequence for volume localization [49]. Due to additional signal losses, a human brain 2D COSY spectrum recorded in a gross occipital volume of $5 \times 6 \times 8$ cm^3 using a 2 T MRI scanner took a total sampling duration of 1 h and 42 min.

Simultaneous acquisition of COSY from multiple volumes of interest was proposed by Delmas and co-workers (55) and two-voxel localized COSY spectra of rat brain were recorded in a total duration of 42 min using a Bruker 4.7 T MRI scanner. A different pulse-sequence (ISIS-COSY) was also proposed by Welch *et al.* (57), where the volume was localized by outer-volume suppressed ISI (OSIRIS). Two major drawbacks of ISIS-COSY compared to L-COSY were: (1) eight shots are necessary to achieve the VOI and (2) five rf pulses are necessary to record the single voxel localized COSY leading to more dependence on the B_1-field inhomogeneity.

Spin-Echo Correlated MR Spectroscopy (SECSY)

A second variation of COSY, namely spin echo correlation spectroscopy (SECSY) was originally reported by Nagayama *et al.* [58]. Localized SECSY works the same way as L-COSY as shown in Figure 3 except the second incremental period (k*t_1) should be kept the same as the first period with $k = 1$. Compared to L-COSY, the diagonal peaks of SECSY lie on ($F_2 = F_1 = 0$) and the J-cross peaks are symmetrically disposed above and below the diagonal. A localized SECSY spectrum recorded in the anterior cingulate of a 24-year-old healthy subject is shown in Figure 7. 1024 complex points along the t_2 dimension, 64 points along the t_1 dimensions, a 27 ml voxel and 16 averages per t_1 were acquired. The resultant 2D spectrum was displayed in the magnitude mode. Typical experimental parameters for 2D SECSY are included in Table 1. An advantage of SECSY over COSY is that a smaller sweep width is needed along F_1, however there is additional T_2^*-weighting during the second incremental period.

Constant-Time Based Point Resolved Spectroscopic Sequence (CT-PRESS)

Constant time (CT)-PRESS based on constant time (T_c) chemical shift encoding was recently implemented by Dreher*et al.* [59]. In addition to three rf pulses required for PRESS, an additional 180° rf pulse was used. 2D CT-PRESS spectra were recorded in rat brain with improved spectral resolution using a Bruker 4.7 T MRI scanner. Two different versions of modified CT-PRESS were recently implemented on a whole body 1.5 T MRI scanner including only three-slice selective (90° – 180° – 180°) rf pulses for volume localization and the CT chemical shift encoding as an integral part of volume localization [60]. In the first version, CT was inserted prior to the first spin-echo and in the second version, prior to the second spin-echo. A 2D CT-PRESS spectrum recorded in a phantom containing six metabolites, namely 12.5 mM NAA, 10 mM Cr, 3 mM Ch, 7.5 mM mI, 12.5 mM Glu and 5 mM Lac at pH = 7.3 using a GE 1.5 T MRI scanner is shown

A

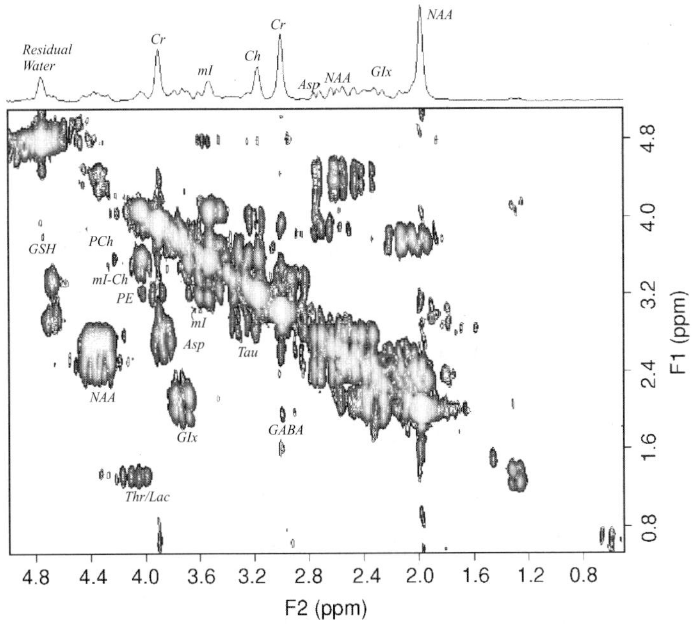

B

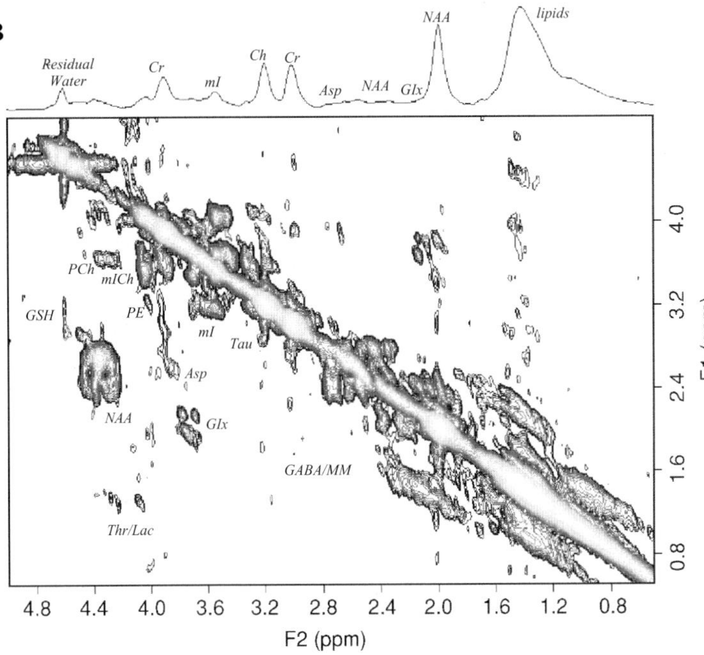

Fig. 6. 2D L-COSY spectra of (A) a grey-matter brain phantom (Siemens) and (B) occipito-parietal white matter region of a 38-year-old healthy human brain (GE) using 3 T MRI scanners.

Table 1: Experimental parameters for 2D MRS

T_R/T_E	2000/20–30 ms
Number of averages/Δt_1	8–16
Number of increments (Δt_1)	64–128
RF pulse widths (90°/180°)	3.2/5.2 ms
Digital resolution (t_2)	Complex 1024–2048
Voxel size	18–27 ml
Duration of crusher gradients	4–6 ms
Spectral width (F_2/F_1)	(2000–5000)/(625–1250) Hz

in Figure 8. Global water suppression was achieved using CHESS.

One major advantage of 2D CT-PRESS is that a homonuclear broadband-decoupled COSY-type spectrum along the F_1-dimension.can be recorded without a second rf-channel for decoupling. Also, compared to the basic localized COSY spectra, a further increase in SNR was obtained by CT-PRESS for coupled resonances since there was no coherence transfer of magnetization between the J-coupled protons leading to disappearance of cross-peaks. Two major drawbacks of the optimized 2D CT-PRESS are: 1) The signal amplitude in CT-PRESS depends on T_c. 2) The spectrum has to be acquired in the 2D mode, which requires long acquisition time.

Constant-time Based Correlated MR Spectroscopy (CT-COSY)

Constant time (CT)-COSY can be easily derived from CT-PRESS by converting the last 180° into 90° rf pulse [61]. Four versions of CT-COSY sequence with the following slice-selective (ss) rf pulses can be easily conceived as described below: (A) $90^\circ_{ss} - (T_c - \Delta t_1) - 180^\circ_{ss} - \Delta t_1 - 90^\circ_{ss}$–Acquire ($t_2$), (B) $90^\circ_{ss} - (T_c + \Delta t_1)/2 - 180^\circ_{ss} - (T_c - \Delta t_1)/2 - 90^\circ_{ss}$ – Acquire(t_2), (C) $90^\circ_{ss} - \Delta t_1 - 180^\circ_{ss} - (T_c - \Delta t_1) - 90^\circ_{ss}$ – Acquire (t_2), and (D) $90^\circ_{ss} - (T_c - \Delta t_1)/2 - 180^\circ_{ss} - (T_c + \Delta t_1)/2 - 90^\circ_{ss}$ – Acquire (t_2), where T_c is defined as the constant time between the 90°_{ss} pulses.

Shown in Figure 9 is a localized 2D CT-COSY spectrum recorded in a phantom containing six metabolites, namely 12.5 mM NAA, 10 mM Cr, 3 mM Ch, 7.5 mM mI, 12.5 mM Glu, and 5 mM Lac at pH = 7.3. The following parameters were used: $T_R = 2$ s, $T_E = 132$ ms, $T_c = 125$ ms, $\Delta t_1 = 0.8$ ms, $\Delta t_2 = 0.8$ ms, spectral widths of 2500 Hz and 625 Hz along the two spectral axes (F_2 and F_1); 1024 complex points along t_2 and 128 points along t_1 dimensions; the number of excitations (NEX) per Δt_1 between 8.

Compared to L-COSY, a fixed interval of T_c ($> t_1^{max}$) separating the preparation and mixing periods is chosen to be of the order of ($1/2J$). The precession under J-coupling is not affected by the 180° rf pulse during T_c. Hence, a

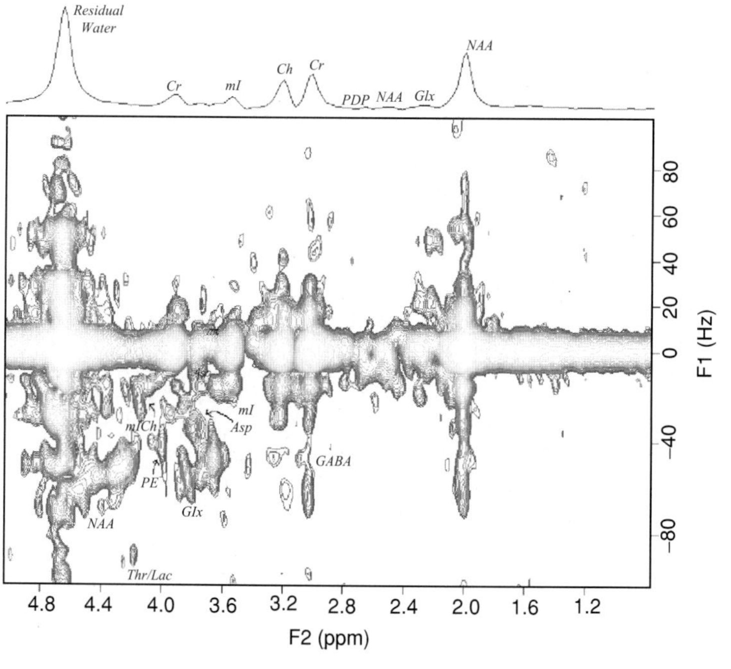

Fig. 7. 2D localized SECSY spectrum of recorded in the occipito-parietal grey matter region of a 24-year-old healthy volunteer using a 1.5 T GE MRI scanner.

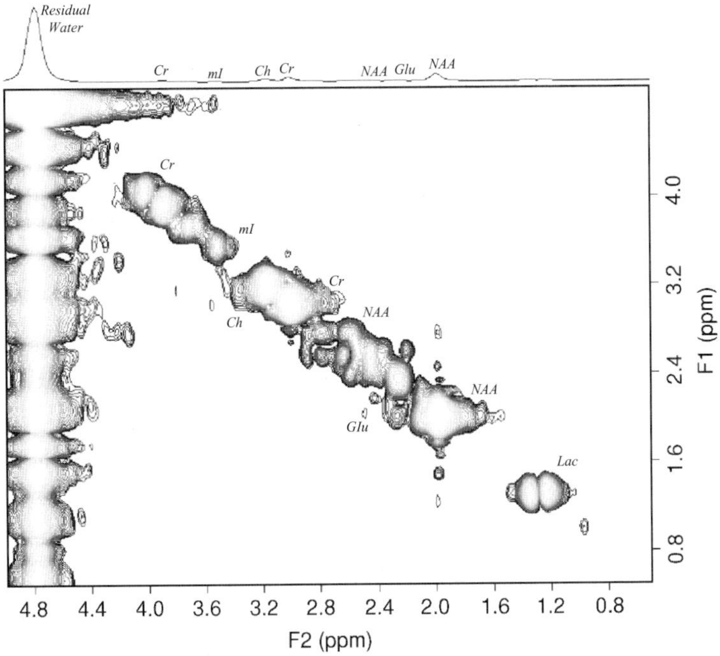

Fig. 8. A CT-PRESS spectrum of recorded in a phantom containing six metabolites using a 1.5 T GE MRI scanner.

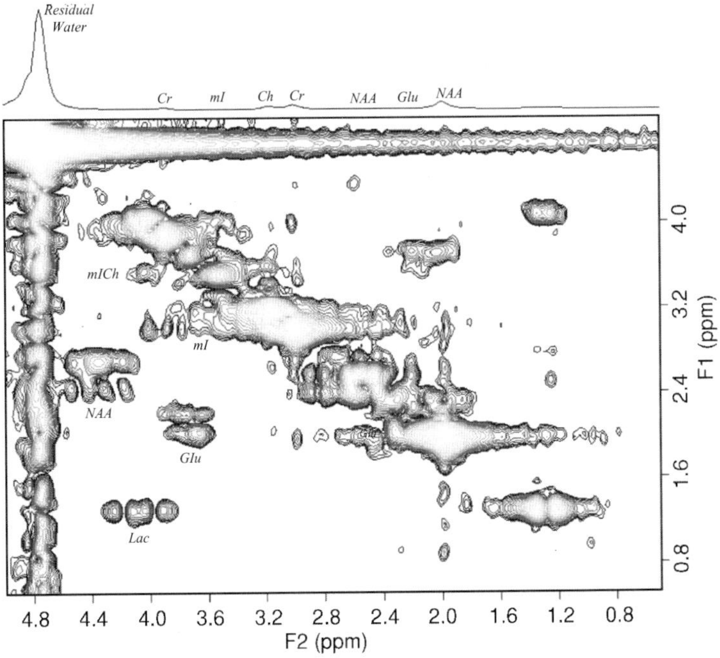

Fig. 9. A CT-COSY spectrum of recorded in a phantom containing six metabolites using a 1.5 T GE MRI scanner.

further increase in SNR was obtained by CT-PRESS for coupled resonances since there was no coherence transfer of magnetization between the J-coupled protons. Secondly, collapse of multiplets along the second dimension was evident without using a second rf channel to achieve broadband decoupling.

Artefacts in Localized 2D MRS and Simulation

In this section, a brief discussion of the commonly observed artefacts in localized 2D MR spectra in presented. First, the flip-angle errors such as an imperfect 180° rf pulse, can lead to ghost artefacts in 2D JPRESS. A four-step phase-cycling scheme, namely EXORCYCLE can eliminate such artefactual peaks [25]. Second, strong coupling effect also complicates 2D JPRESS as discussed by Ryner *et al.* [40]. These artefacts can also impact CT-PRESS and CT-COSY. Third, the asymmetric cross peaks of L-COSY/SECSY are disposed on both sides of the diagonal peaks. The asymmetry has been reported to be more severe with the upper cross peaks of NAA at ($F_2 =$ 2.5 ppm, $F_1 = 4.35$ ppm) above the diagonal [47].

A GAMMA (general approach to magnetic resonance mathematical analysis) software package can be used to simulate different metabolite spectra [62]. The GAMMA library [62] uses a density matrix description of the spin system and provides the simulation of MRS experiments. Density matrices can be calculated for each metabolite using the reported chemical shifts and J-coupling [10]. The GAMMA codes have been extended to simulate the JPRESS, L-COSY/SECSY, CT-COSY, and CT-PRESS spectra including water suppression. As shown in Figure 10, The first 90° rf pulse of the L-COSY pulse sequence is replaced by a binomial rf pulse ($1\,\bar{3}\,3\,\bar{1}$) to mimic the water suppression pulse sequence [63]; where the numerals give the relative pulse lengths, the bars indicate a 180° phase shifted pulse and equal delays τ between pulses. Thus $1\bar{3}\ 3\bar{1}$ corresponds to 11.25°(x)-τ-33.75°(-x)-τ-33.75°(x)-τ-11.25°(−x), $\tau = 1/2\Delta f$, and Δf is the offset frequency [63]. The phase shift is optimized based on the target a null at the water frequency.

Shown in Figure 11 are (A) experimental 2D L-COSY spectrum of a phantom containing NAA, Cr, Ch, mI, Lac, and Glu and (B) simulated composite 2D COSY spectra of the same metabolites using GAMMA [64]. The parameters used for simulation were identical to that of experiments (T_E of 30 ms for L-COSY, 1024 complex points along $F_2$and 128 points along F_1). Another type of water-suppression artefact is due to the t_1-ridge running vertically at $F_2 = 4.8$ ppm as evident in the experimental COSY spectrum shown in Figure 11A. This is mainly due to the scanner hardware fluctuations during the acquisition of the 2D raw matrix [25]. Also, more artefacts can arise due to wrong choice of post-processing parameters and one needs to use caution in selecting optimal apodization filters along both F_2 and F_1 dimensions during post-processing the 2D data.

Multi-Voxel Based 2D ^{1}H MR Spectroscopy

Long data acquisition has been one of the major drawbacks of single-voxel localized 2D MRS. There will be a further increase of time when conventional SI is added to 2D MRS [36–37]. In contrast to reconstructing metabolite images from the 2D spatial and 1D spectral data, the metabolite images can be reconstructed from the 2D spatial and 2D spectral data projecting the cross peak volumes into the corresponding spatial images. This methodology was termed as "cross peak imaging" (CPI) by Metzler and co-workers [65–66]. An obvious drawback of CPI is extended total acquisition time. For example, a combination of 16 × 16 spatial, 64 spectroscopic encoding steps and T_R of 1 s will result in a total duration of 4.6 h [66]. Recently, implementation of circular sampling combined

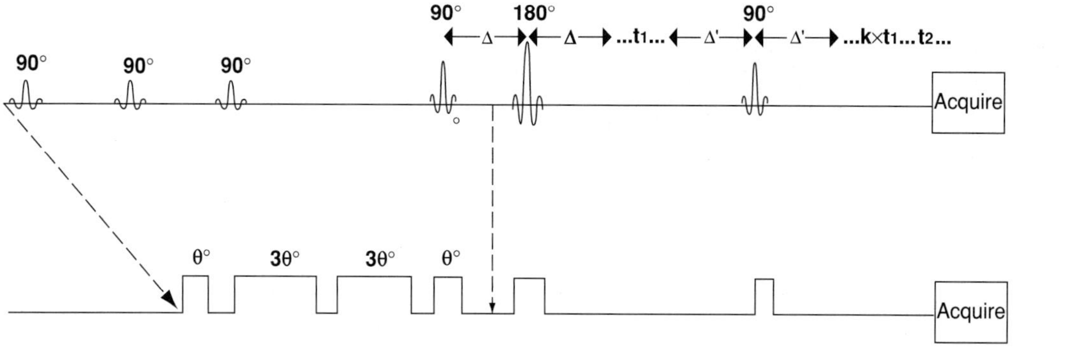

Fig. 10. A 2D COSY sequence used for simulation.

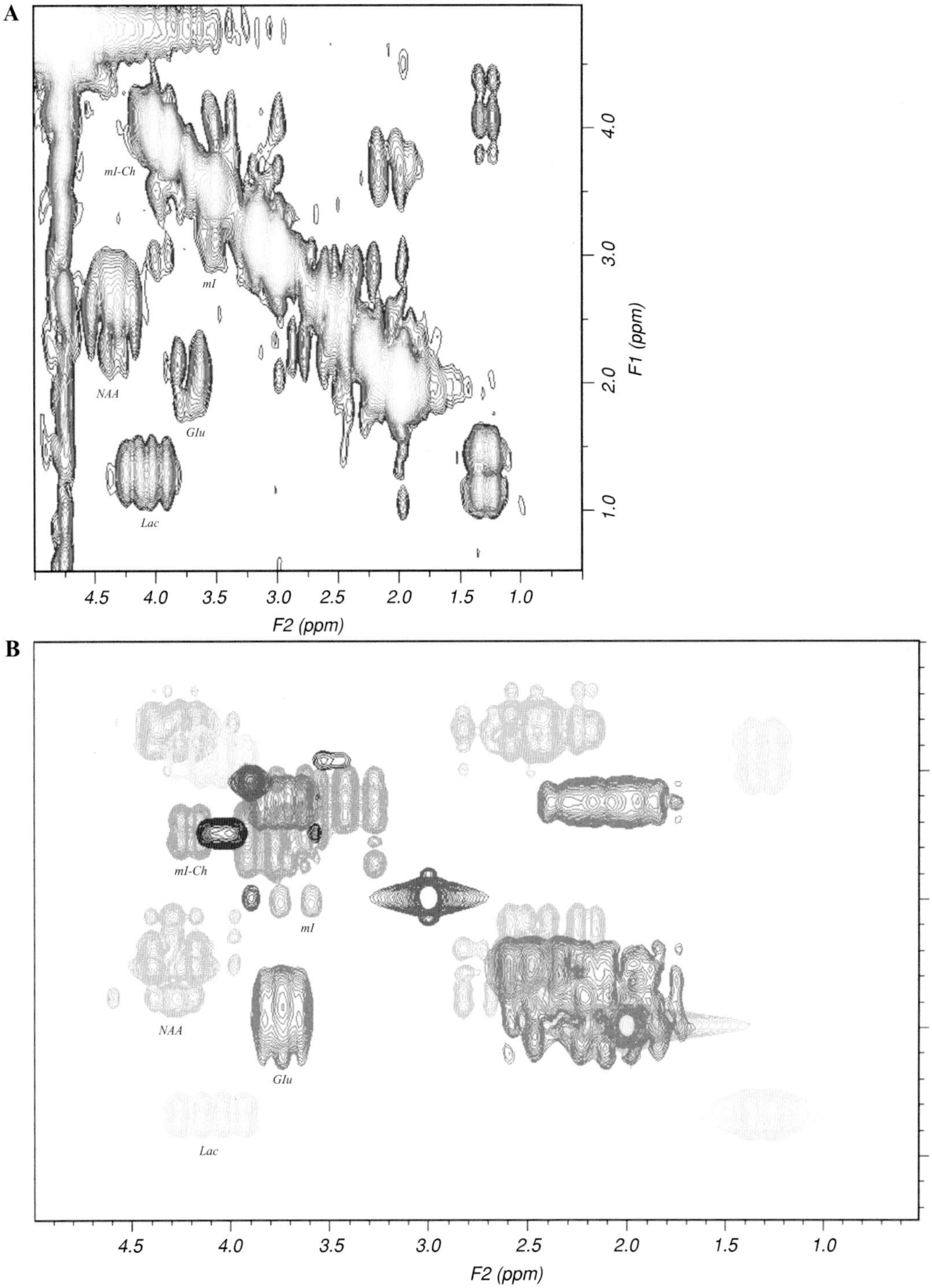

Fig. 11. 2D L-COSY spectra of (A) a brain phantom containing six metabolites and (B) simulated spectra for the same metabolites (Reproduced with permission from Banakar *et al.* [64]). (See also Plate 91 on page LXIII in the Color Plate Section.)

with 2D J-resolved spectroscopy was demonstrated by Renshaw and co-workers using a 4 T whole body MRI scanner with a total duration of 2 h [67].

A drastic reduction of total duration has been proposed by two different multidimensional MRS sequences. First, Adalsteinsson and Spielman introduced the first application of time-varying gradients during the read-out time to acquire multi-voxel based 2D J-resolved spectral data in phantom solutions containing lactate and ethanol using a 1.5 T MRI scanner [68]. Spiral-based k-space trajectories were used in conjunction with a 2D J-resolved sequence. By preceding the spiral read-out with a 2D preparation scheme, 128 spectral encoding with T_R of 2 s resulted in a total duration of 17 min.

An asymmetric variant of echo-planar spectroscopic imaging (EPSI) consisting of a train of trapezoidal read-out gradients, each followed by short refocusing pulse was implemented by Meyer *et al.* [69]. A constant-time based COSY was used and the total acquisition was 34 min. Application of this technique was demonstrated in rat brain using a 4.7 T MRI scanner and the cross peak images of mI and Tau were recorded only. In an alternative approach, COSY was combined with spectroscopic ultra-fast low angle rapid acquisition with relaxation enhancement (U-FLARE) [70]. CT along the two spectral dimensions (evolution and detection) facilitated complete effective homonuclear decoupling. A spiral imaging combined CT-PRESS was also recently proposed [71].

Summary

Localized 2D MRS is still at its infancy and different techniques discussed earlier need further improvements in terms of voxel sizes, coil design, etc. Non-selective excitation of the J-coupled metabolites is facilitated by different 2D MRS techniques. 2D cross peaks of more than ten J-coupled metabolites have been observed using L-COSY recorded in the frontal and occipital regions of human brain [41,47,54]. In contrast to ISIS-COSY [57], the 2D techniques discussed in this chapter are one-shot based. Even though it has been shown recently that 1D MR spectra processed using the LC-Model algorithm [21], 2D L-COSY shows the unambiguous presence of 2D cross peaks originating from several J-coupled metabolites. However, further work is necessary in order to ascertain the superiority of 2D L-COSY over the other methodology. Due to the differential T_2^*-weighting during the incremental duration of 2D L-COSY, absolute quantitation of several metabolites is further complicated by lack of T_2 and T_1 values in the literature for different proton groups (methyl, methylene, methine, etc.) in any given metabolite. Regarding the application of these multidimensional MRS techniques in different pathologies, targeting a specific metabolite or selected metabolites in different pathologies may be more advantageous than non-selective excitation/detection of several metabolites. Another major concern is that the 2D spectra are complicated by artefacts contributed by strong coupling of various metabolite protons unavoidable at both 1.5 and 3 T MRI scanners. As discussed earlier, the artefacts in these multi-voxel based 2D MRS sequences need to be further evaluated. Further work is necessary to demonstrate the clinical potential of both single and multi-voxel based multidimensional MR SI methods.

Acknowledgment

This work was supported by the grants from the National Institute of Health (NIH) and the US Army Breast and Prostate Cancer Research Programs. Authors acknowledge the scientific support of Drs. Kenneth Yue and Lawrence Ryner in the earlier implementation of 2D MR sequences, and Ashwin Thomas and Tai Dou during the preparation of the chapter.

References

1. Gadian DG. Nuclear Magnetic Resonance and Its Applications to Living Systems. Oxford University Press: New York, 1982.
2. Rothman DL, Behar KL, Hetherington HP, Shulman RG. Proc. Natl. Acad. Sci. U.S.A. 1984;81:6430.
3. Moonen CTW, Kienlin MV, van Zijl PCM *et al.* NMR Biomed. 1989;2:201.
4. Kreis R, Ross BD, Farrow N, Ackerman Z. Radiology 1982;182:9.
5. Michelis T, Merboldt KD, Bruhn H *et al.* Radiology 1983;187:219.
6. Gruetter R, Weisdorf SA, Rajanayagan V *et al.* J. Magn. Reson., 1998;135:260.
7. Webb PG, Sailasuta N, Kohler SJ *et al.* Magn. Reson. Med. 1994;31:365.
8. Rothman DL, Petroff OAC, Behar KL, Mattson RH. Proc. Natl. Acad. Sci. U.S.A. 1993;90:5662.
9. Bruhn H, Frahm J, Gyngell ML *et al.* Magn. Reson. Med. 1989;9:126.
10. Govindaraju V, Young K, Maudsley AA. NMR Biomed. 2000;13:129.
11. Renshaw PF, Lafer B, Babb SM *et al.* Biol. Psych. 1997;41:837.
12. Pouwels PJW, Frahm J. Magn. Reson. Med. 1998;39:53.
13. Barker PB, Hearshen DO, Boska MD. Magn. Reson. Med. 2001;45:765.
14. Trabesinger AH, Weber OM, Duc CO, Boesiger P. Magn. Reson. Med. 1999;42:283.
15. Prichard JW. Curr. Opin. Neur. 1997;10:98.
16. Pan JW, Stein DT, Telang F *et al.* Magn. Reson. Med. 2000;44:673.
17. Choi IY, Lei H, Gruetter R. J. Cereb. Blood Flow Metab. 2002;22:1343.
18. Hardy DL, Norwood TJ. J. Magn. Reson. 1998;133:70.

19. de Graaf AA, Luyten PR *et al.* Single and Double Quantum Lactate Imaging of the Human Brain at 1.5 T. 9th SMRM, New York 1990, p 138.
20. Crozier S, Brereton IM *et al.* Magn. Reson. Med. 1990;16:492.
21. Ith M, Bigler P, Scheurer E, *et al.* Magn. Reson. Med. 2002;48:915.
22. Kumar A, Thomas MA, Lavretsky H *et al.* Am. J. Psych. 2002;159:630.
23. Pfeuffer J, Tkac I, Provencher SW, Gruetter R. J. Magn. Reson. 1999;141:104.
24. Aue WP, Bartholdi E, Ernst RR. J. Chem. Phys. 1976;64:2229.
25. Ernst RR, Bodenhausen G, Wokaun A. Principles of NMR Spectroscopy in One and Two Dimensions. Oxford Publications: Oxford, 1987, pp 283–489.
26. Crozier S, Brereton IM *et al.* Magn. Reson. Med. 1990;16:492.
27. Sotak CH, Freeman DM, Hurd RE. J. Magn. Reson., 1988;78:355.
28. Desmoulin F, Seelig J. Magn. Reson. Med. 1990;14:160.
29. van Zijl PCM, Chesnick AS, Despres D. *et al.* Magn. Reson. Med. 1993;30:544.
30. Kreis R, Boesch C. J. Magn. Reson. 1996;B113:103.
31. Ogg RJ, Kingsley PB, Taylor JS. J. Magn. Reson. 1994;B103:1.
32. Patt SL, Sykes BD. J. Chem. Phys. 1972;56:3182.
33. Tran TK, Vigneron DM, Sailasuta N *et al.* Magn. Reson. Med. 2000;43:23.
34. Bottomley PA. Ann. NY. Acad. Sci. 1987;508:333.
35. Salibi N, Brown MA. Clinical MR Spectroscopy: First Principles. Wiley-Liss: New York, 1998, p 75.
36. Brown TR, Kincaid BM, Ugurbil K. Proc. Natl. Acad. Sci. U.S.A. 1982;79:3523.
37. Maudsley AA, Hilal SK, Perman WH, Simon HE. J. Magn. Reson. 1983;51:147.
38. Boesch C. J. Magn. Reson. Imag. 2004;16:177.
39. Ryner LN, Sorenson JA, Thomas MA. J. Magn. Reson. 1995;B107:126.
40. Ryner LN, Sorenson JA, Thomas MA. Magn. Reson. Ima. 1995;13:853.
41. Thomas MA, Hattori N, Umeda M, *et al.* NMR Biomed. 2003;16:245.
42. Thomas MA, Ryner LN, Mehta MP *et al.* J. Magn. Reson. Imag. 1996;6:453.
43. Yue K, Binesh N, Marumoto N *et al.* Magn. Reson. Med. 2002;47:1059.
44. Hurd RE, Gurr D, Sailasuta N. Magn. Reson. Med. 1998;40:343.
45. Swanson MG, Vigneron DB, Tran TK *et al.* Magn. Reson. Med. 2001;45:973.
46. Adalsteinsson E, Hurd RE, Mayer D *et al.* Neuroimage 2004;22:381.
47. Thomas MA, Yue K, Binesh N *et al.* Magn. Reson. Med. 2001;46:58.
48. Brereton IM, Crozier S, Field J, Doddrell DM. J. Magn. Reson. 1991;93:54.
49. Brereton IM, Galloway GJ, Rose SE, Doddrell DM. Magn. Reson. Med. 1994;32:251.
50. Binesh N, Thomas MA. Proc. Intl. Soc. Magn. Reson. Med. Kyoto, Japan, 2004, p 2303.
51. Binesh N, Yue K, Fairbanks L, Thomas MA. Magn. Reson. Med. 2002;48:942.
52. Prescot AP, Leach MO, Saran F, Collins DJ. Proc. Intl. Soc. Magn. Reson Med. 2004, p 2493.
53. Thomas MA, Yue K, Binesh N, Debruhl N. J. Magn. Reson. Imag. 2001;14:181.
54. Binesh N, Bugbee M, Fairbanks L *et al.* Proc. Intl. Soc. Magn. Reson. Med. Hawaii, 2002, p 699.
55. Delmas F, Beloeil JC, van der Sanden BPJ *et al.* J. Magn. Reson. 2001;149:119.
56. Ziegler A, Gillet B, Beloeil JC *et al.* Magn. Reson. Mat. Phys. Biol. Med. 2002;14:45.
57. Welch JWR, Bhakoo K, Dixon RM *et al.* NMR Biomed. 2003;16:47.
58. Nagayama K, Wuthrich K, Ernst RR. Biochem. Biophys. Res. Commun. 1979;90:305.
59. Dreher W, Leifbritz D. Magn. Reson. Imag. 1999;17:141.
60. Chung HK, Banakar S, Thomas MA. Proc. Intl. Soc. Magn. Reson. Med., Kyoto, Japan, 2004, p 687.
61. Chung HK, Banakar S, Thomas MA. Proc. Intl. Soc. Magn. Reson. Med. Toronto, 2003, p 1143.
62. Smith SA, Levante TO, Meier BH, Ernst RR. J. Magn. Reson. A 1994;106:75.
63. Hore PJ. J. Magn. Reson. 1983;55:283.
64. Banakar S, Venkatraman TN, Yue K *et al.* Proc. Int. Conf. Math. Engg. Tech. Med. and Biol. Sci., Las Vegas, 2002, Vol II, p 500.
65. Metzler A, Izquierdo M, Ziegler A *et al.* Proc. Natl. Acad. Sci. U.S.A. 1995;92:11912.
66. Ziegler A, Metzler A, Kockenberger W *et al.* J. Magn. Reson. 1996;B112:141.
67. Jensen JE, Frederic BD, Wang L *et al.* Proc. Intl. Soc. Magn. Reson. Med. 2004, p 2301.
68. Adalsteinsson E, Spielman DM. Magn. Reson. Med. 1999;41:8.
69. Mayer D, Dreher W, Leibfritz D. Magn. Reson. Med. 2000;44:23.
70. D. Mayer, Dreher W, Leibfritz D. Magn. Reson. Med. 2003;49:810.
71. Mayer D, Kim DH, Adalsteinsson E, Spielman DM. Proc. Intl. Soc. Magn. Reson. Med., Kyoto, Japan, 2004, p 678.

Glossary

Overview of NMR in the Pharmaceutical Sciences

Drug design

LC NMR

Ligand based screening

Metabonomics

Receptor based screening

SAR by NMR

Screening

Structure-based design

Structural genomics

Applications of Cryogenic NMR Probe Technology for the Identification of Low-Level Impurities in Pharmaceuticals

Correlation experiments

Cryogenic NMR probes

Degradation products

Metabolite identification

Metabonomic studies

Natural products

Pharmaceutical impurities

Protein characterization

Small volume NMR probes

Structure characterization

Flow NMR Techniques in the Pharmaceutical Sciences

Cold probes

DI-NMR

Direct injection NMR

FIA-NMR

Flow injection analysis NMR

Flow NMR

Flow probe

LC-NMR

LC-NMR-MS

Microcoil NMR

Developments in NMR Hyphenation for Pharmaceutical Industry

Cryogenic probes

Drug impurity profiles

LC-NMR

LC-NMR/MS

Natural product screening

Solid phase extraction

Stop flow LC-NMR

LC-NMR in Dereplication and Structure Elucidation of Herbal Drugs

Crude plant extracts

Dereplicaion

HMBC

HSQC

LC-NMR

Structure elucidation of natural products

WET pulse sequence

New Approaches to NMR Data Acquisition, Assignment and Protein Structure Determination: Potential Impact in Drug Discovery

Automated structure determination

Assignment

Computer-assisted sequential assignment

Fast multidimensional NMR

Transferred Cross-Correlated Relaxation: Application to Drug/Target Complexes

Bound conformations

Cross-correlated relaxation

Drug receptor complexes

Epothilone/tubulin complex

Transferred cross-correlated relaxation

Transferred NOE

Novel Uses of Paramagnets to Solve Complex Protein Structures

Domain orientation

J-modulation

Long-range restraints

Paramagnetic labelling

Protein structure

Pseudocontact shift

Residual dipolar coupling

Fast Assignments of ^{15}N-HSQC Spectra of Proteins by Paramagnetic Labeling

HSQC spectrum

Ligand binding

Paramagnetic labelling

PLATYPUS

Proteins

Pseudocontact shift

Resonance assignment

Phospholipid Bicelle Membrane Systems for Studying Drug Molecules

Bicelles

Conformation

Drug membrane-interactions

Model membranes

Orientation

Partial Alignment for Structure Determination of Organic Molecules

Liquid crystal phase

Partial alignment

Organic molecules

Residual dipolar couplings

Stretched polymer gels

Measurement of Residual Dipolar Couplings and Applications in Protein NMR

Aligned proteins

Conformational substrates

Echo-anti-echo manipulation

Protein folding

Protein NMR

Residual dipolar coupling

Using Chemical Shift Perturbations to Validate and Refine the Docking of Novel IgE Antagonists to the High-Affinity IgE Receptor

Chemical shift

Docking

High-affinity binding

IgE receptor

Peptides

Simulated annealing

Dual-Region Hadamard-Encoding to Improve Resolution and Save Time

Adiabatic gHMBC

Cyclosporine-A

HMBC

Hadamard spectroscopy

Region-selection

Resolution enhancement

Sensitivity enhancement

Nonuniform Sampling in Biomolecular NMR

Maximum entropy reconstruction

Nonuniform data acquisition

Sensitivity

Spectrum analysis

Resolution

Structural Characterization of Antimicrobial Peptides by NMR Spectroscopy

Antimicrobial peptides

Diffusion NMR

Disulfide cross-linking

Membrane-peptide interactions

Membrane insertion of peptides

Peptide solution structure

Solid-state NMR

Pharmaceutical Applications of Ion Channel Blockers: Use of NMR to Determine the Structure of Scorpion Toxins

Low abundance samples

Nano probes

Scorpion toxins

Small sample probes

Toxin structure

Structure and Dynamics of Inhibitor and Metal Binding to Metallo-β-Lactamases

Cadmium NMR

Cooperativity

Imidazole

Metallo-β-lactamases

Minimum chemical shift approach

Mono-cadmium

Tautomer

Thiomandelate

NMR Spectroscopy in the Analysis of Protein–Protein Interactions

Chemical exchange

Cross saturation

Deuterium labelling of proteins

Mapping protein interfaces

Large proteins

Protein complexes

Paramagnetic nuclei

Protein-protein interactions

Protein structure

Residual dipolar coupling

Titrations

TROSY NMR

Identification and Characterization of Ternary Complexes Using NMR Spectroscopy

Borate

Dihydrofolate reductase

Inorganic ions

Interligand NOE

Nuclear Overhauser Effect

Transferred NOE

Trypsin

The Transferred NOE

Fast exchange

Non-specific binding

Off rate

On rate

Saturation transfer difference (STD)

Slow exchange

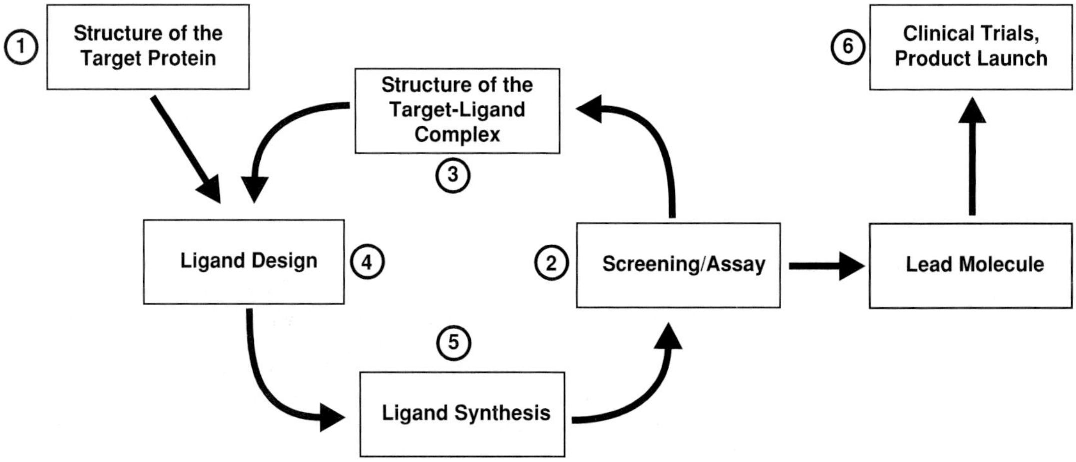

Fig. 1. Schematic outline of the process of structure-based drug design and discovery. In an iterative cycle the structures of protein targets and/or protein-ligand complexes are determined, ligands are designed using the structural information, screened for binding to the target and assayed for activity. The interactions of suitable ligands with the target proteins are structurally investigated and the information is used for improving the ligands in further iterations of this cycle until a suitable lead compound has been generated. NMR spectroscopy is involved in all aspects of this cycle: (1) Protein structure determination; (2) ligand screening; (3) structural analysis of protein ligand-complexes; (4) ligand design (as a conformational analysis tool); (5) ligand synthesis (as a structure verification tool); and (6) quality control of the pharmaceutic production process and metabonomic studies during clinical trials and after release of the drug.

bound water molecules. Finally, NMR spectroscopy can be used to determine binding constants and other thermodynamic data as well as ionization states of the protein and/or ligand.

In this overview article we will briefly discuss the applications of NMR spectroscopy techniques in drug design and discovery. More detailed discussions are presented in the individual chapters of this section of the Handbook of Modern Magnetic Resonance. A number of other review articles covering various aspects of the application of NMR spectroscopy in drug design and discovery have been published in recent years [1–8] and in addition to the specialist articles contained in this edition of Handbook of Modern Magnetic Resonance the reader is referred to these publications.

The review is divided into four sections. In the first section, we review the latest technical developments in NMR spectroscopy, ranging from hardware developments such as the use of cryoprobes and coupled techniques like LC-NMR to improvements in the acquisition and measurement of NMR spectra. The second section deals with the structural characterization of target molecules and ligands and discusses labeling strategies, NMR experiments suited for large molecules (TROSY) and the use of non-NOE-based NMR data such as residual dipolar couplings (RDC) and paramagnetic labels. The third section is dedicated to applications of NMR spectroscopy in the screening of ligands, and reviews target-based methods as well as ligand-based methods. While the first three sections focus on applications of NMR to obtain structural information, the fourth section focuses on NMR applications unique to pharmacy and pharmacology, from drug monitoring, quality control, and pharmacokinetics to the new field of metabonomics. The chapter finishes with a short section on future developments and directions.

Technical Developments

Several significant technical developments have been made in NMR in hardware, software, and experimental techniques in recent years. The most significant development in terms of NMR hardware is the introduction of cryoprobes. Cryoprobes reduce the thermal noise in the detection circuit by cooling the receiver coil and electronic parts of the probe (but not the sample) with liquid helium. Consequently, the sensitivity of a cryoprobe is two to four times higher than conventional room-temperature probes [9,10]. This sensitivity improvement benefits pharmaceutical applications in two ways: on one hand, the improved sensitivity can be utilized to investigate samples of much lower concentration than before, which is especially useful in natural product discovery or screening for low-concentration impurities in pharmaceutical samples during quality control, as reviewed in more detail in Chapter 2 by Martin [11]. On the other hand, the improved sensitivity can be utilized to acquire data of the same quality in a much shorter time—up to 16 times faster—which is especially useful in high-throughput screening applications.

Other developments on the hardware side include the integration of NMR spectroscopy with other analytical techniques such as liquid chromatography (LC) or mass spectrometry (MS) to yield techniques such as LC-NMR [12] or LC-NMR-MS [13,14]. In addition, developments such as direct-injection NMR (DI-NMR) [15] and flow-injection analysis NMR (FIA-NMR) [16] aim at utilizing NMR spectroscopy in support of pharmaceutical applications. DI-NMR has already become a routine technology in pharmaceutical laboratories and in other industrial applications [17–21]. These techniques are reviewed in Chapters 3–5 [22–24].

The most significant advance with respect to new experimental techniques is the development of transverse-relaxation optimized spectroscopy (TROSY) by Pervushin *et al.* [25], in which the line-broadening effects of transverse relaxation are eliminated by selecting specifically the line of a heteronuclear multiplet for which the effects of dipole–dipole relaxation and chemical shift anisotropy compensate each other, yielding improved spectral resolution and sensitivity. However, TROSY is applied only during evolution periods in NMR experiments, and transverse relaxation, which is still active during periods of magnetization transfer, becomes a limiting factor for molecular weights beyond 100 kDa when magnetization is transferred via spin-spin couplings with the insensitive nuclei enhanced by polarization transfer (INEPT) technique [26]. This limitation can be overcome by replacing INEPT steps with cross relaxation-enhanced polarization transfer (CRINEPT) [27], which combines INEPT with cross-correlated relaxation-induced polarization transfer [28], a highly effective transfer mechanism for molecules with a molecular weight above 200 kDa. Thus, TROSY and CRINEPT represent complementary techniques that have revolutionized NMR spectroscopy within a short period of time by making it applicable to proteins larger than the "traditional" 30 kDa limit [29]—reports of NMR data on a 900 kDa complex have been widely cited [30]. In addition, modifications of the TROSY experiment, like solvent-exposed amides with TROSY (SEA-TROSY), have been developed [31] that are specifically geared towards screening studies as part of structure-based drug design approaches. Techniques for improving the study of large proteins and protein complexes with NMR spectroscopy are reviewed in more detail in Chapter 19 [32].

Other developments with respect to experimental NMR techniques focus on modifying and speeding up data acquisition. In general, they involve replacing the uniform sampling of time-space with other sampling schedules and consequently replacement of the discrete Fourier-transformation (DFT) with other transformations. One approach is multiplexing, by encoding more than one resonance frequency in one dimension, as realized in G-matrix FT (GFT) [33,34] or Hadamard encoding [35,36]. Other approaches are projection-reconstruction NMR (PR-NMR) [37] and maximum entropy reconstruction (ME) [38]. These techniques are discussed in more detail in Chapters 6, 14, 15 [39–41].

Structure-based Design

As already outlined, the first step in the cycle of structure-based drug design is the structural characterization of the target receptor molecule. Table 1 shows a collection of

Table 1: NMR technologies used for structural characterization of receptors, ligands, and ligand–receptor complexes

Target	Technique	Information
Receptor	2-4D NMR	3D structure
	RDC	3D structure
	Relaxation	Macromolecular dynamics
Ligand	1D/2D NMR	Solution conformation
	RDC	Solution conformation
	Chemical shift	Charge/tautomeric state
	Lineshape/relaxation analysis	Solution dynamics
	Transfer NOE	Bound conformation
Receptor–ligand complex	3D/4D NMR	3D structure of complex
	RDC	Ligand orientation in complex
	Shift perturbation (HSQC)	Mapping of binding site, Stoichiometry of complex, K_D
	Line width	Binding kinetics
	Relaxation	Dynamics of complex

NMR techniques that can be employed for this purpose as well as for the structure determination of protein–ligand complexes.

Isotopic labeling is essential for the structure determination of target proteins that are not small in NMR terms (i.e. larger than 10 kDa). Labeling is also a prerequisite to harvest the benefits of the TROSY effect. The stable isotopes most commonly used for labeling are ^{15}N, ^{13}C, and ^{2}H (deuterium), and they can be employed in different labeling schemes [42,43]. Uniform labeling with ^{15}N and ^{13}C allows for sequential assignment and structure determination of the target protein with multidimensional triple resonance experiments [44]. For larger proteins, additional substitution of non-exchangeable protons with deuterium ameliorates relaxation effects. In this case, TROSY experiments are usually employed to obtain spectral information. In large multidomain proteins individual domains can be labeled by segmental labeling thereby simplifying the information in the NMR spectrum [45]. Another method of simplifying spectral information, which is especially useful for screening studies with large proteins, is the selective labeling of individual amino acids or of amino acid types [46,47]. The former approach can also be achieved with amino-acid specific pulse sequences [48,49].

After NMR spectra have been recorded, and all resonances have been assigned, the three-dimensional structure of the target protein is determined mostly from information based on NOE connectivities and dihedral restraints [50]. Recently, additional sources of structural information have been made accessible, most important among them RDCs [51] and shift and relaxation changes associated with paramagnetic nuclei [52,53]. Both methods have the distinct advantage of providing long-range constraints, which is a useful complementation for the short-range information derived from NOEs and dihedral restraints. Furthermore, they are able to provide structural information in the case of large proteins that have been extensively labeled with deuterium and thus have a reduced number of observable proton-proton NOEs. Labeling, RDC and paramagnetic methods are discussed in more detail in Chapters 8, 9, 11, 12, and 19 [32,54–57].

Similar methods can be employed to determine the three-dimensional structures of ligand molecules. Although ligand molecules are easier to examine because of their small size and can generally be studied by standard homonuclear NMR methods, the use of more advanced techniques such as RDCs has been demonstrated to be beneficial e.g. for obtaining structural information on sugars and small organic molecules as reviewed in Chapter 11 [56]. Other examples of the application of NMR spectroscopy for structure determination of small organic molecules are antimicrobial peptides and scorpion toxins as discussed in Chapters 16 and 17 [58,59].

A major milestone in structure-based drug design is a detailed determination of the three-dimensional structure of a ligand-target protein complex. If the structure of the target protein has been solved, but the ligand binding site is unknown then a quick determination of the binding site can be obtained by chemical shift mapping: a labeled sample of the target protein is titrated with the ligand, and due to the change in chemical environment associated with ligand binding amide proton resonances at, or close to, the binding site change their position in a HSQC spectrum. Mapping of the shift perturbations on the structure of the target protein reveals the ligand binding site. Similarly, in the case of a small ligand protein binding to a large receptor the binding face of the ligand can be determined by titrating labeled ligand with unlabeled receptor and chemical shift mapping.

To determine the detailed nature of protein–ligand interactions the observation of intermolecular NOEs between target protein and ligand is crucial. The aforementioned combination of labeled protein and unlabeled ligand (or vice versa) is helpful here, and for example, "filtered" experiments can be used to selectively observe NOE signals either within the protein, within the ligand, or most importantly the desired intramolecular signals between protein and ligand [60].

RDCs offer an alternative approach for determining the orientation of ligands within the target protein's binding site. Upon binding, the ligand adopts the alignment tensor of the target protein, and thus a comparison of ligand RDCs in the free and bound state makes it possible to determine the conformation of the ligand in the bound state in the complete absence of ligand-protein or ligand–ligand NOEs [61–63]. A further discussion of techniques for determining the structures of protein–ligand complexes, including some case examples, is given in Chapters 13, 18, 19, and 20 [32,64–66]. Transferred NOE and cross-correlated relaxation methods are also particularly important for determining the conformation of bound ligands, provided certain kinetic conditions are met. These approaches are described in Chapters 7 and 21 [67,68].

One of the distinct advantages of NMR spectroscopy is the possibility of obtaining information about the dynamics of a protein or a protein–ligand complex. The dynamics of the binding site is of fundamental importance in determining affinities and binding kinetics. In addition, it is a not an uncommon observation that the binding site of a protein exhibits considerable flexibility, but becomes more rigid upon binding of the ligand [69,70]. Phenomena like this can be readily investigated with NMR relaxation experiments, which can determine molecular motions on a nanosecond up to millisecond timescale [71]. In addition, lineshape analysis offers an avenue into further determining events on the millisecond timescale like chemical exchange, including the measurement of rate constants of the exchange process [72]. Finally, NMR titrations can be used to determine dissociation constants of a protein–ligand complex. These topics are further reviewed in Chapters 18, 21, 22, and 23 [66,68,73,74].

NMR Screening

The screening of libraries of compounds is becoming an increasingly useful adjunct to structure-based drug design. Table 2 gives an overview of the different screening strategies that can be employed. In general, screening techniques can either be target-based—i.e. changes in the macromolecular target protein upon ligand binding are observed, or ligand-based—i.e. changes in NMR properties of the ligand upon binding are observed.

In general target-based screening techniques are based on monitoring chemical shifts with a HSQC spectrum. For large target proteins specific labeling strategies and TROSY experiments are employed, but the principle stays the same. A recent modification of the TROSY experiment specifically geared for application in screening programs with large target proteins is the SEA-TROSY experiment [31]. The experiment is based on the concept that only groups on the solvent-exposed surface of the protein are likely to be involved in ligand binding, while residues in the protein interior are unlikely to be in contact with the ligand. SEA-TROSY observes by magnetization transfer from the bulk water only solvent-exposed amide protons in fully deuterated and ^{15}N-labeled proteins that are dissolved in water. Thus the SEA-TROSY spectrum contains significantly fewer signals than conventional spectra, alleviating the problem of spectral overlap for large target proteins.

Table 2: NMR technologies used for high-throughput screening of ligand–receptor complexes

Target	Technique	Reference
Receptor-based	SAR by NMR	[75]
	SEA-TROSY	[31]
	NMR-SOLVE	[47]
Ligand-based, Direct	Relaxation	[78,79]
	Diffusion coefficients	[78]
	STD	[83]
	ILOE	[65,85]
	Transfer NOE	[68,80]
	NOE pumping	[81,82]
	Water-LOGSY	[84]
	SHAPES	[87]
Ligand-based, Indirect	FAXS	[86]
	3-FABS	[96]
	STD with spy-molecule	[86]

The by now "classic" strategy employed in NMR target-based screening is structure–activity relationships (SAR) by NMR [75]. It aims at the identification of two ligands with moderate binding affinities that bind to two different but adjacent binding sites on the target protein. Crosslinking of the two ligands via chemical synthesis then produces a first-generation lead molecule with substantially improved binding affinity. It has been responsible for the development of numerous novel drug leads, as outlined in the article by Hajduk in this Handbook [76].

NMR-Structurally orientated library valency engineering (NMR-SOLVE) is a new target-based screening approach that has been developed specifically for the postgenomic era to provide guidance for the construction of libraries against multiple related target proteins [47]. One key element of NMR-SOLVE is to use highly selective isotopic labeling (e.g. ^{13}C-^{1}H labeled CH_3 groups in otherwise uniformly deuterated proteins) to observe only a small number of key resonances in a binding site, thus yielding information in the absence of complete assignments. The NMR-SOLVE strategy is based on the fact that several enzyme families have multiple binding sites—one that is common for all family members, and one that is specific for each protein. In a first-step NMR-SOLVE uses a known reference ligand that binds to the common binding site to map and identify the specifically labeled groups that are around this binding site. In the next steps mimics of the reference ligand that could bind to the common binding site of most family members of the protein family are investigated. Finally, in an act of core-expansion structural information is used to design a linker that could connect the core of ligand-mimetic fragments binding in the common binding site with additional ligand fragments that would bind to the secondary binding sites, unique to each protein. Thus, a library of bivalent ligands is obtained that is specifically geared towards a given protein family. The NMR-SOLVE strategy has been demonstrated with target proteins as large as 170 kDa [47]. Further discussion on target-based screening approaches can be found in Chapters 27 and 29 [76,77].

A plethora of ligand-based screening methods has been developed targeting a wide variety of observable NMR parameters, as indicated in Table 2. Parameters that have been utilized include longitudinal, transverse, and double-quantum relaxation [78,79], diffusion coefficients [78], and intramolecular and intermolecular transfer of magnetization, such as transfer NOE [68,80], NOE pumping, and reverse NOE pumping [81,82], saturation transfer experiments [83], Water-ligand observed via gradient spectroscopy (water-LOGSY) [84] and inter-ligand NOEs (ILOE) [65,85]. Individual techniques are discussed in detail in Chapters 21, 26, and 29 [68,77,86] and are reviewed here only briefly.

The main advantage of monitoring the small ligands is that there is no need for isotopic labeling of the target

Applications of Cryogenic NMR Probe Technology for the Identification of Low-Level Impurities in Pharmaceuticals

Gary E. Martin
Pfizer Global Research & Development, Michigan Structure Elucidation Group, Kalamazoo, MI 49001-0199

Abstract

Following the detection of low-level impurities in pharmaceuticals, structural characterization of these components represents a continuing challenge. Irrespective of whether the impurity in question arises from side reactions of the chemical processes being employed for the synthetic elaboration of the active pharmaceutical, through degradation chemistry processes, or through accidental contamination, the identification of low-level impurities is a mandatory component of the preparation and marketing of pharmaceuticals. The development of cryogenic NMR probes and the corresponding increases in sensitivity that they provide has significantly impacted this phase of pharmaceutical development. The role of cryogenic NMR probe technology in the characterization of low-level impurities from pharmaceuticals is discussed.

Introduction

Multiple creative approaches have been utilized in the quest for improved NMR sensitivity. NMR spectroscopy, despite the wealth of chemical structure information that it can provide, has always had an Achilles' heel in the form of the low inherent sensitivity of the technique. Approaches that have been used to improve sensitivity have included the development of new pulse sequences, such as the proton- or "inverse"-detected heteronuclear 1D-, 2D-, and 3D-NMR experiments that are now in wide use. Another approach that has been heavily exploited has been that of "shrinking" the size of NMR probes and hence the required sample [1,2]. Although small volume probes have been known for a considerable period of time, contemporary small volume probe development effectively began in 1991 with the first reports of 3 mm NMR probe applications, followed shrinking of probe diameters to 1.7 mm, then 1 mm, and finally μ-coil probes with sample volumes as small as a few μl. This area was the subject of a recent review and the interested reader is referred there for more details [3]. Cooling the rf coils of an NMR probe to cryogenic temperatures to reduce noise, thereby improving sensitivity, is also a well-established approach to enhancing NMR performance [4–7].

Unlike conventional NMR probe developments, which were commercially realized after prototype design and testing, the commercial manufacture of cryogenic NMR probes proved to be a more daunting task requiring a number of years for even the first prototypes to be delivered to customer laboratories. In part this was due to design considerations necessary to cool the internal components to temperatures for the rf coil in the range of 10–25 K and preamps housed in the probe body to temperatures near 50 K without destroying those components during thermal cycling that the probe must occasionally be subjected to. Nevertheless, numerous commercial cryogenic NMR probes have been delivered and have begun to have a significant impact on the way in which various disciplines perform NMR-based chemical structure investigations.

Cryogenic NMR Probes

Commercial examples of cryogenic NMR probes have been developed in a limited number of configurations. Typically, they have been 3 or 5 mm diameter probes and most commonly, inverse-detection triple resonance gradient designs, although a cryogenic probe for direct carbon detection has also been developed. Cryogenic NMR probes have also been developed for flow applications. Perhaps one of the most interesting of these is the interchangeable flow cell (IFC) design by Varian, which allows the user to change between NMR tubes and a flow cell without having to remove the cryoprobe from the magnet. Although some early cryogenic probes were developed and tested using an open configuration of a liquid helium dewar, to the best of the author's knowledge, all commercial examples use a closed loop system to deliver the helium refrigerant to the probe at temperatures in the vicinity of 8–10 K. The helium refrigerant is used to cool the rf coils to their operating temperature as well as the associated electronic circuitry housed in the probe body such as preamps, after which the helium refrigerant is cycled back and re-chilled. Typically,

Graham A. Webb (ed.), Modern Magnetic Resonance, 1187–1194.

the rf coil temperature for cryogenic probe installations is 25 K although in the author's laboratory 500 and 600 MHz cryogenic gradient triple resonance inverse probes are both operated at 20 K to provide improved sensitivity. From a serviceability standpoint, the closed loop helium chillers perform reliably for intervals ranging from 10 to 14 months in the author's experience before requiring a rebuild. This type of required "annual" maintenance, and the associated expense, should be taken into consideration when any NMR laboratory weighs the sensitivity advantages of a cryogenic NMR probe installation vs. a high sensitivity small volume conventional NMR probe.

Sample Preparation

In the author's experience, one factor that has a considerable impact on cryogenic NMR probe performance is the uniformity of sample preparation. While conventional NMR probes are very forgiving in terms of sample-to-sample variation in column height, etc. the same cannot be said for cryogenic NMR probes. Figure 1 shows an example of the variability of performance that can be seen simply by varying column height in a conventional NMR tube over the range from 35 to 20 mm. Given that the utilization of cryogenic NMR probes will generally be in those cases where time or sample are scarce, the impact of sample preparation on the quality of the data generated should receive adequate attention.

Another less obvious factor associated with sample preparation when using a cryogenic NMR probe is the selection of tube size. While it would seem to make sense to use a 5 mm NMR tube for samples to be run in a 5 mm cryogenic NMR probe, thereby avoiding filling factor losses, reality is quite the opposite. Almost counterintuitively, substantially better performance can be obtained with extremely scarce samples by preparing them carefully in a 3 mm NMR tube. An illustration of attainable s/n ratios for a limited quantity of the simple antibiotic clindamycin in 3, 4, and 5 mm NMR tubes is presented in Figure 2. These results can be explained when the noise sources in a cryogenic NMR probe are considered. In reality, the sample is the single largest source of noise in a cryoprobe NMR experiment. Hence, as the diameter of the NMR sample itself is decreased in going from 5 to 4 and then 3 mm sample tubes, the noise contributed to the experiment correspondingly diminishes as the isolation of the rf coil from the noise source improves with smaller diameter NMR tubes [8]. Thus, despite potential filling factor losses associated with smaller diameter NMR tubes run coaxially in a 5 mm probe, performance still improves since the impact of the sample noise is larger than filling factor losses.

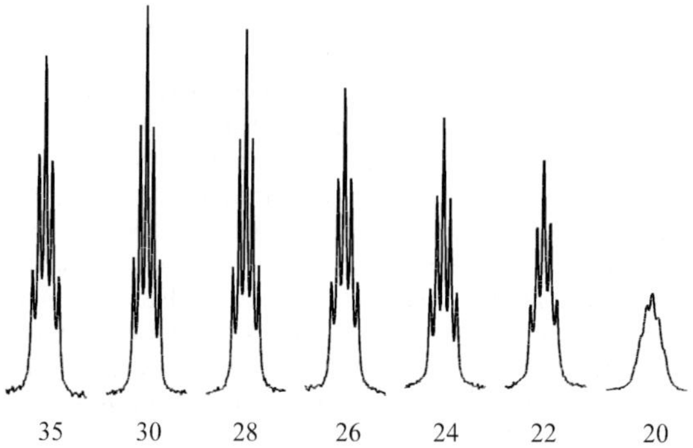

Fig. 1. The effect of solvent column height on lineshape and s/n characteristics for a 4 mm NMR tube with a sample of d_4-methanol. The spectra of the solvent multiplet were acquired in a Varian 5 mm 500 MHz triple resonance gradient-inverse detection Cold Probe^{TM}. Each column height was first gradient and then manually shimmed to afford the best possible lineshape and sensitivity.

Identification of Degradants

To illustrate what would be a typical pharmaceutical industry application of cryogenic NMR probe sensitivity, a sample of a complex alkaloid studied more than a decade ago that was kept in a sealed 5 mm NMR tube was used as a model system. The sample consisted of 2.5 mg of the complex spiro nonacyclic indoloquindoline alkaloid cryptospirolepine (**1**) that was stored in DMSO-d_6 (Scheme 1). During the life of the sample, the color changed from an initial red-orange to a dark brown. When the sample tube was cut open and the sample subjected to HPLC analysis, the chromatogram shown in Figure 3 was obtained. By LC/MS methods, there was no trace of the starting alkaloid left in the sample among the 26 degradant peaks observed in the chromatogram. An investigator dealing with a typical pharmaceutical would not be faced with total degradation. Instead, most of the drug would be intact and impurity peaks ranging from a few tenths of a percent to a few percent would be observed in the chromatogram. Nevertheless, this still affords a useful example of the capabilities engendered in cryogenic NMR probe technology.

The two largest chromatographic peaks in the HPLC of the degraded sample of **1** were identified as cryptolepinone (**2**) and cryptoquindolinone (**3**). Cryptolepinone (**2**) is a simple molecule and was readily identified from just a COSY and ^1H–^{13}C GHSQC spectra. Cryptoquindolinone (**3**) is a more complex molecule and required considerably more effort to characterize. Approximately 100 µg of **3** was isolated chromatographically and, using a 600 MHz spectrometer with a state-of-the-art 3 mm triple resonance gradient inverse-detection NMR probe afforded a good ^1H–^{13}C HSQC spectrum overnight (18 h). Given the relative differences in sensitivity of the HSQC

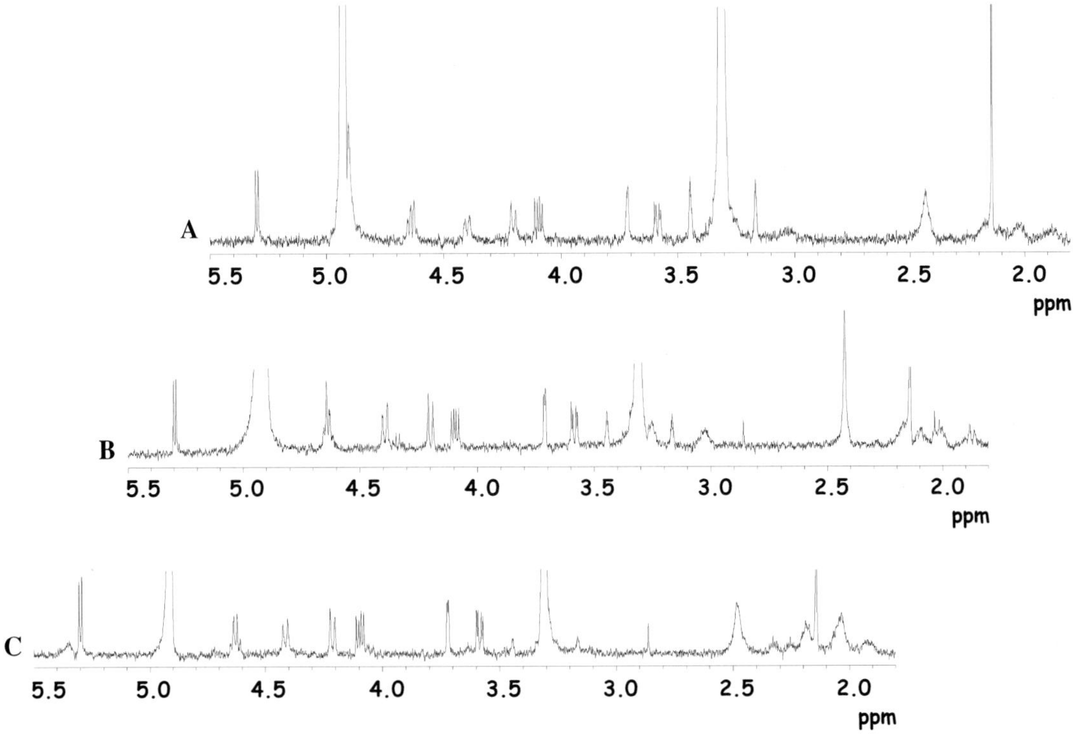

Fig. 2. Comparison spectra of 11.6 μg samples (0.0027 μmoles) of the simple antibiotic clindamycin prepared: (A) 500 μl solve in a 5 mm NMR tube; (B) 292 μl solvent in a 4 mm NMR tube; or (C) 163 μl of solvent in a 3 mm NMR tube (identical column heights of 30 mm). Each sample was gradient and then hand shimmed to afford optimum lineshape and performance, after which a spectra was recorded using 8 transients. The s/n ratio was then measured using the anomeric proton of the sugar with a 200 Hz standard "noise" window downfield of the lowfield anomeric proton doublet. The measured s/n for three samples in this limited sample scenario was: (A) 14.4:1 s/n; (B) 20.8:1; and (C) 21.5:1.

and HMBC experiments, a long-range ^{1}H–^{13}C spectrum was anticipated to require from 3 to 4 days. In contrast, using the available sample, the 6 Hz optimized HMBC [9] shown in Figure 4 was acquired overnight at 500 MHz using a 5 mm triple resonance gradient inverse-detection cryogenic NMR probe. These data were necessary to assemble the structure of the degraded molecule. Using the same sample, it was also possible over a weekend to acquire a 3–6 Hz optimized long-range ^{1}H–^{15}N CIGAR-HMBC spectrum of **3** in which correlations to three of the four nitrogens in the structure were observed, including a correlation to the N10 resonance at 230 ppm, which confirmed the presence of the C=N double bond as shown in Scheme 1. Without where indicated access to a cryogenic NMR probe, long-range ^{1}H–^{15}N data for a sample of this size would consume a week or more of spectrometer time, precluding the acquisition of these data in most laboratories.

The example just described is typical of many encountered within the pharmaceutical industry. While it is certainly possible to isolate larger samples and solve impurity and degradant structures using conventional NMR probe technology there are several disadvantages to doing so. First the chromatography required to isolate larger samples generates large volumes of chromatographic mobile phase that must be disposed of or recycled. Second, the time required to isolate larger samples for conventional spectroscopy is proportionally longer and undesirable whenever tight timelines are imposed on a pharmaceutical development effort for whatever reason.

Applications of Cryogenic NMR Probe Technology

There are several possible ways to break down applications of cryogenic NMR probe technology that have appeared in the literature thus far. One logical subdivision would be to separate applications on the basis of the type

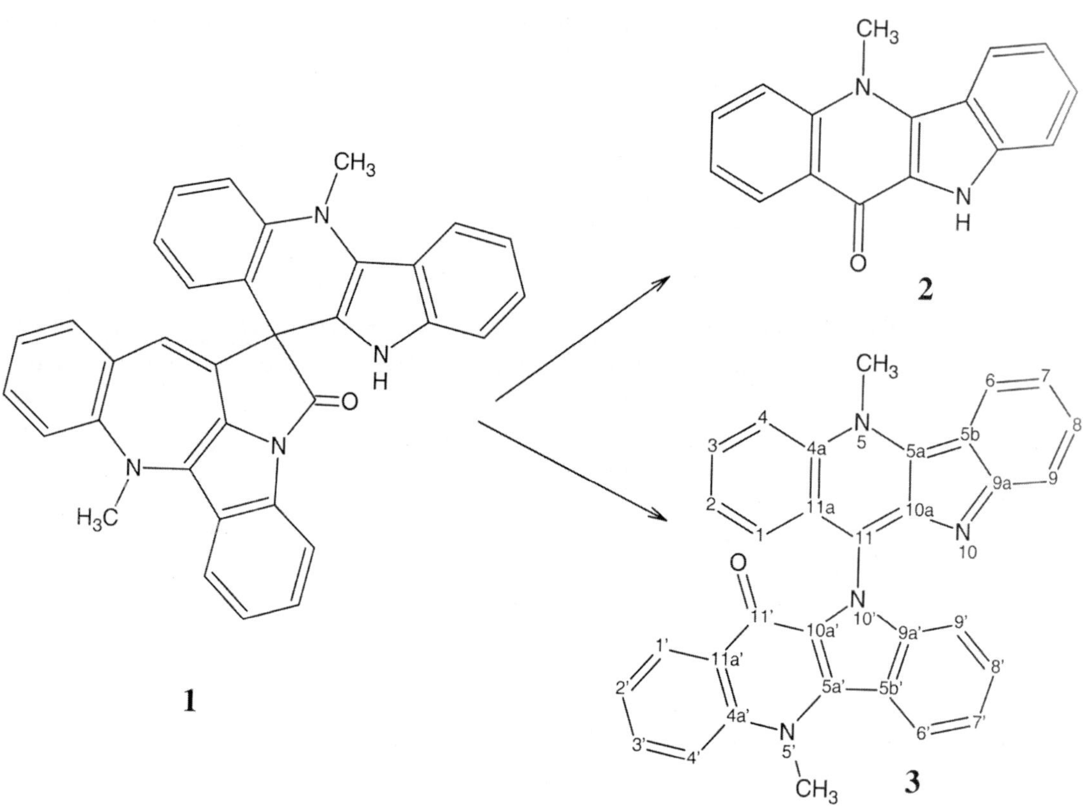

Scheme 1.

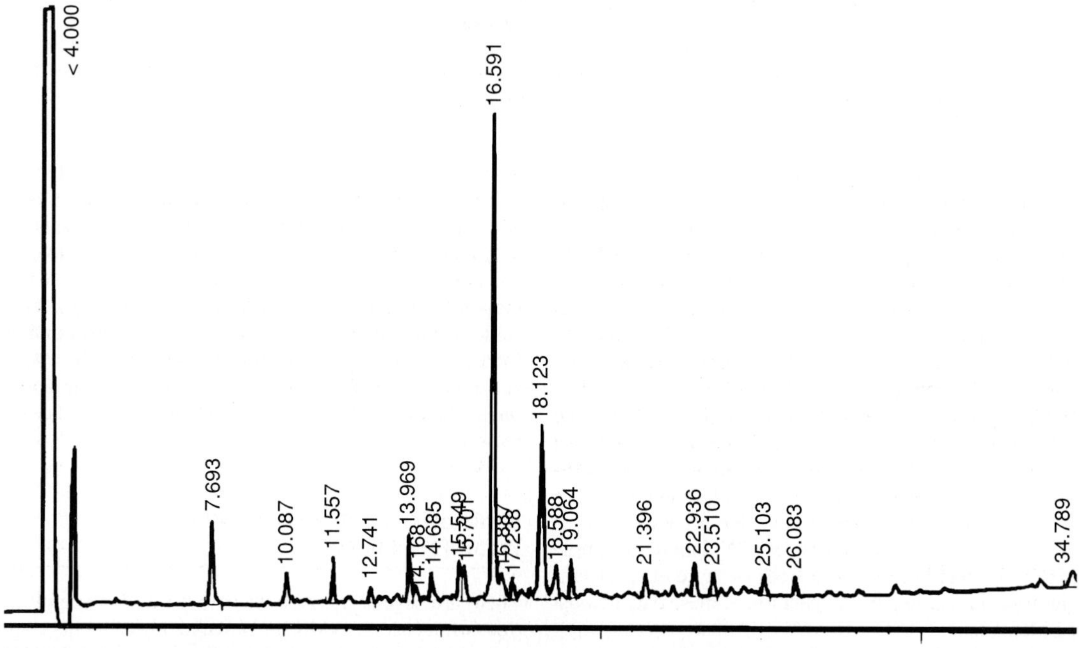

Fig. 3. HPLC chromatograph showing the result obtained after chromatographing a 2.5 mg sample of cryptospirolepine (**2**) stored in DMSO-d_6 for more than 10 years. Based on LC/MS measurements, none of the starting alkaloid remained in the sample.

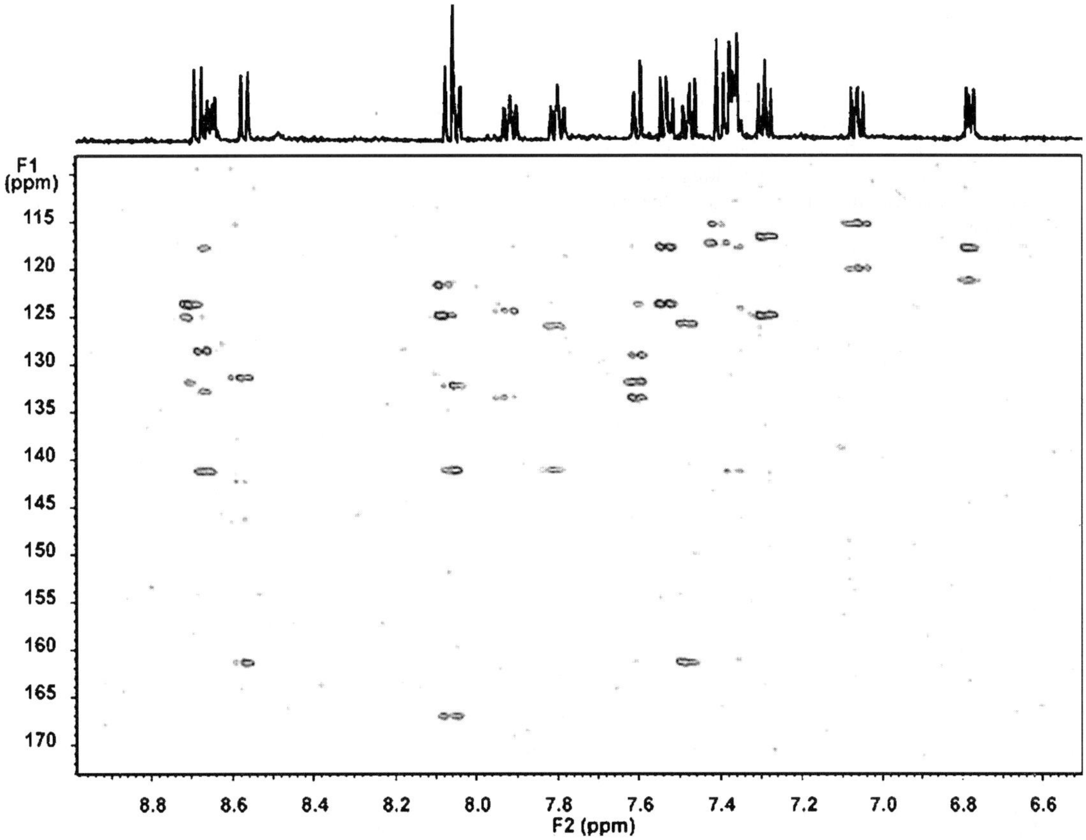

Fig. 4. Phase-sensitive 6 Hz optimized non-gradient HMBC spectrum of a sample of ~100 μg of cryptoquindolinone (**4**) dissolved in 150 μl DMSO-d_6 in a 3 mm NMR acquired overnight using a Varian 500 MHz 5 mm triple resonance gradient inverse-detection cryogenic NMR probe.

of cryoprobe employed. However, the vast majority of applications utilize inverse-detection cryogenic probes so subdivision on the basis of probe type is not particularly informative. A more logical way of sorting applications is on the basis of the type of study reported. It is this basis that will be used for this final segment of this contribution. Reports that have appeared in the literature will be subgrouped into small molecule applications, which include both natural products and pharmaceuticals; metabolic and metabonomic studies; and finally protein and biochemical applications, in which NMR screening is included.

Small Molecule Applications

Among the earliest of the small molecule applications reported in the literature was a report by Logan and co-workers in 1999 that demonstrated the results that could be obtained with a 5 mm ^{1}H cryogenic probe in the acquisition of NOE difference spectra on small samples of the natural product taxol [10]. In 2000, the author and co-workers reported a comparative study using conventional and cryogenic 3 mm gradient inverse NMR probes using a 40 μg sample of the indole alkaloid strychnine [11]. This comparison demonstrated an improvement in performance for the cryogenic probe relative to the conventional 3 mm probe by approximately a factor of 3.5X. Liu and co-workers presented the results of their application of cryogenic NMR probe capabilities in the study of several mass-limited pharmaceutical samples [12]. With samples ranging from 10 to 80 μg this group of authors was able to record data that allowed the characterization of the samples. In an example that utilized a ^{13}C detection cryogenic NMR probe, Bringmann and co-workers [13] reported the results of a study of the biosynthetic pathway leading to acetogenic isoquinoline alkaloids. The ^{13}C-^{13}C INADEQUATE experiments employed to study the incorporation of [$^{13}C_2$] acetate were greatly facilitated by the enhanced sensitivity provided by the cryoprobe.

The significant performance advantage of cryogenic NMR probes for the acquisition of long-range ^1H–^{15}N correlation data was demonstrated using the small oxazolidinone antibiotic eperezolid by the author and co-workers in 2001 [14]. Using a 2 mg sealed sample, data were acquired in both a conventional 3 mm gradient inverse triple-resonance probe and a 5 mm cryogenic probe. Data with a s/n ratio of ~100:1 were acquired in less than 30 min using the cryogenic NMR probe while it took ~4 h to achieve only half that s/n ratio in the conventional 3 mm probe.

In 2002, Gustafson and co-workers reported the elucidation of the complex polyketide-derived macrolactam pecillastrin A from a deep water Caribbean sponge [15]. It was necessary for these authors to resort to a 500 MHz gradient inverse triple resonance cryogenic probe to acquire the necessary long-range correlation data on the 800 μg sample of the C_{82} natural product to complete the structure elucidation. Later in 2002, the author and co-workers [16] reported the utilization of a 500 MHz gradient triple resonance cryoprobe in the acquisition of long-range ^1H–^{13}C and ^1H–^{15}N correlation data in the characterization of one of the isolated degradants, **3**, shown in Scheme 1. Cook and co-workers put the sensitivity enhancement of cryogenic NMR probe technology to excellent use in their study of the Boreal forest fulvinic acid [17]. This group of authors acquired both ^1H–^{13}C HSQC and HMBC on this complex mixture, which gave >300 direct correlation responses in the HSQC spectrum. In a study that represents an interesting potential technique for natural products characterization, Griesinger and co-workers reported the measurement of long-range ^1H–^{13}C coupling constants of natural products at natural abundance using orienting media [18]. Sandvoss and co-workers, exploited the sensitivity of a cryoprobe to facilitate the acquisition of HSQC-TOCSY data for limited samples of two complex new asterosaponins from the starfish *Asterias rubens* [19]. Cryogenic NMR probe capabilities were also of importance in the acquisition of HSQC-TOCSY and ^1H–^{15}N long-range data acquired during the characterization of the complex indoloquinoline dimeric alkaloid quindolinocryptotackieine, which relied on Computer-Assisted Structure Elucidation (CASE) methods because of the extensive overlap in both the proton and carbon frequency domains [20]. Another example of the use of cryogenic NMR probe capabilities for small molecules is found in the work of Exarchou and co-workers [21] who used a combination of solid phase extraction (SPE) and cryoprobe technology for the identification of constituents present in Greek oregano. Further applications are surveyed in a recently published review [8].

Metabolic and Metabonomic Applications

In the first application of cryogenic NMR probe technology of which the author is aware, Pease and co-workers reported the results obtained with a 2.5 mm cryoprobe in the characterization of several metabolites [22]. Using a spectrometer equipped with a ^{13}C detection cryogenic NMR probe, Keun and co-workers investigated the applicability of ^{13}C spectral data as a compliment to routinely utilized ^1H data for metabonomic studies [23]. Griffin [24] reported an overview of ongoing metabonomic studies of xenobiotic toxicity and disease states that relied on cryoprobe data. Corcoran and co-workers [25] characterized the sensitivity advantages afforded by cryogenic NMR flow probes in LC-NMR-MS applications in a study of acetaominophen metabolites in urine. Stöckigt and co-workers [26] described the results of an interesting study in which alkaloid metabolism in hybrid *Rauwolfia* plant cell cultures was followed using a 500 MHz cryogenic NMR probe. The authors contrast the cryoprobe results with conventional 800 MHz NMR data on the same samples. Cryogenic NMR probe capabilities were also utilized by Bertini and co-workers [27] to study the degradation of aromatic compounds by *Pseudomonas putida*. More recent applications are surveyed in a recent review [8].

Protein and Biochemical Applications

In an early application in the field of protein NMR, Dötsch and co-workers [28] reported the development of several new ^{13}C detected protein NMR experiments designed to exploit the sensitivity of a ^{13}C detection cryoprobe. Wand and co-workers [29] described their efforts in the optimization of the application of cryogenic NMR probe technology in the study of proteins. In one of the early applications of cryoprobe technology involving proteins used in NMR-based screening of pharmaceuticals, Fesik and co-workers [30] compared the results obtained in NMR screening studies using specifically ^{13}C labeled amino acids in proteins with the more traditionally employed ^1H–^{15}N HSQC NMR screening methods. The authors demonstrated that with cryogenic NMR probes, this approach gave ~3X the sensitivity of the more traditional proton-nitrogen based screening approach.

After the initial flurry of cryoprobe applications in the field of protein NMR, there was somewhat of a gap before papers again began to appear. This interval corresponds to initial vendor-investigator collaborations during the early development of cryogenic NMR probes and when they began to appear in the laboratories of the investigators. Cowburn and co-workers [31] communicated the results that they obtained for triple resonance NMR experiments of proteins using a cryogenic NMR probe in 3–4 h that would typically require 1–2 days using conventional NMR probe technology. Bertini and co-workers [32] employed ^{13}C direct detection cryoprobe capabilities to facilitate a study of broad lines in paramagnetic metalloproteins in late 2003. Another application of the use of a ^{13}C detection cryogenic probe was the early 2004 report by Bertini and colleagues [33] in the acquisition of ^{13}C–^{13}C NOESY spectra of large macromolecules. In a departure from most other cryogenic NMR probe applications, Shu and Frieden [34] described the application of a ^{19}F cryogenic NMR probe in a study of the urea-dependent unfolding of murine adenosine deaminase by measuring sequential destabilization of ^{19}F NMR. Petros and co-workers [35] employed a 500 MHz cryogenic NMR probe in a study to define the p53 DNA-binding domain/Bcl-X_L binding interface. The cryoprobe data were used to monitor perturbations in the ^{13}C shifts of labeled substrate to the unlabeled p53 DNA-binding domain. Germann and co-workers [36] used the sensitivity of a 500 MHz cryoprobe to acquire F_2-coupled ^1H–^{13}C HSQC spectra of a DNA duplex containing an α-anomeric adenosine as a part of a study to explore substrate recognition by endonuclease IV. Further examples are again found in a recent review [8]. Armengod and co-workers [37] utilized a 500 MHz cryoprobe to facilitate multinuclear NMR assignments for the 18.7 kDa GTPase domain of *E. coli* MnmE protein. The enhanced sensitivity of the cryoprobe was necessary to provide adequate signal-to-noise for the assignments of the low solubility and unstable protein sample.

Conclusions

Cryogenic NMR probes are continuing to go through what will probably be a long series of iterative design improvements that should move sensitivity incrementally higher. At present, cryogenic probes offer a very significant gain in sensitivity over conventional NMR probes, typically threefold or higher at observation frequencies ranging from 500 to 900 MHz. The pharmaceutical industry is already heavily invested in cryogenic NMR probes, which have significant reduced sample isolate requirements for the characterization of impurities and degradants of pharmaceuticals in the possessive author's—laboratories. Similar advantage has been gained in laboratories devoted to the characterization of metabolites of pharmaceuticals. As investigator access to cryogenic NMR probe-equipped spectrometers continues to improve, there will undoubtedly be many new applications reported in the literature in the fields of natural product characterization, biosynthesis, forensic sample identification, and many other disciplines.

References

1. Claridge TDW. High Resolution NMR Techniques in Organic Chemistry. Pergammon Press: Amsterdam, 1999, pp 221–58.
2. Martin GE, Hadden CE, Russell DJ. In: G Gauglitz, T Vo-Dinh (Eds). Handbook of Spectroscopy, Vol. 1. Wiley-VCH: New York, 2003, pp 234–54.
3. Martin GE. In: DM Grant, RK Harris (Eds). Encyclopedia of Nuclear Magnetic Resonance, Vol. 9. Wiley and Sons: Chichester, 2002, pp 98–112.
4. Styles P, Soffe NF, Scott CA, Cragg DA, Row F, White D, White PCJ. J. Magn. Reson. 1983;60:397.
5. Styles P, Soffe NF, Scott CA. J. Magn. Reson. 1989;84:376.
6. Anderson WA, Brey WW, Brooke AL, Cole B, Delin KA, Fuks LH, Hill HDW, Johanson ME, Kotsubo VY, Nast R, Withers RS, Wong WH. Bull. Magn. Reson. 1995;17:98.
7. Hill HDW. IEEE Trans. Appl. Superconduct. 1997;7:3750.
8. Martin GE. In: GA Webb (Ed). Annual Reports on NMR Spectroscopy, Vol. 56. Elsevier: Amsterdam, 2005, pp 1–99.
9. These data were acquired without gradients. In the author's experience, for ^1H–^{13}C long-range correlation experiments with very small samples, i.e. <50 μg, the benefits of using gradients are overshadowed by the signal losses from the experiment. Our observations have been confirmed by the work of Reynolds WF, Enriquez RG. Magn. Reson. Chem. 2001;39:531.
10. Logan TM, Murali N, Wang G, Jolivet C. Magn. Reson. Chem. 1999; 37:762. Note—this paper appears twice during 1999, an earlier version in the same journal on p 512 contained errors that were corrected in the republished form cited.
11. Russell DJ, Hadden CE, Martin GE, Gibson AA, Zens AP, Carolan JL. J. Nat. Prod. 2000;63:1047.
12. Liu Y, Pease J, Potts B, Deese A, Liu YD, O'Neil-Johnson M, Withers R, Nast R. Poster 21, SMASH meeting, Argonne National Laboratory, July 16–19, 2000.
13. Bringmann G, Wohlfarth M, Rischer H, Grüne M, Schlauer J. Angew. Chem. Intl. Ed. 2000;39:1464.
14. Crouch RC, Llanos W, Mehr K, Hadden CE, Russell DJ, Martin GE. Magn. Reson. Chem. 2001;39:555.
15. Rashid MA, Gustafson KR, Crouch RC, Groweiss A, Pannell LK, Van QN, Boyd MR. Org. Lett. 2002;4:3293.
16. Martin GE, Hadden CE, Russell DJ, Kaluzny BD, Guido JE, Duholke WK, Stiemsma BA, Thamann TJ, Crouch RC, Blinov K, Elyashberg M, Martirosian ER, Molodtsov SG, Williams AJ, Schiff PL, Jr. J. Heterocyclic Chem. 2002;39:1241.
17. Cook RL, McIntyre DD, Langford CH, Vogel HJ. Environ. Sci. Technol. 2003;37:3935.
18. Verdier L, Sakhaii P, Zweckstetter M, Griesinger C. J. Magn. Reson. 2003;163:353.
19. Sandvoss M, Preiss A, Levsen K, Weisemann R, Spraul M. Magn. Reson. Chem. 2003;41:949.

20. Blinov K, Elyashberg M, Martirosian ER, Molodtsov SG, Williams AJ, Tackie AN, Sharaf MHM, Schiff PL, Jr, Crouch RC, Martin GE, Hadden CE, Guido JE, Mills KA. Magn. Reson. Chem. 2003;41:577.
21. Exarchou V, Godejohann M, van Beek TA, Gerothanassis IP, Vervoot J. Anal. Chem. 2003;75:6288.
22. Pease J, Withers R, Nast R, Deese A, Calderon P, Mehtas S, Kelly T, Laukien F. Poster W&Th P202, 40th Experimental NMR Conference, Orlando, Florida, February 28–March 5, 1999.
23. Keun HC, Beckonert O, Griffin JL, Richter C, Moskau D, Linden JC, Nicholson JK. Anal. Chem. 2002;74: 2588.
24. Griffin JL. Curr. Opin. Chem. Biol. 2003;7:648.
25. Spraul M, Freund AS, Nast RE, Withers RS, Mass WE, Corcoran O. Anal. Chem. 2003;75:1546.
26. Hinse C, Richter C, Provenzani A, Stöckigt J. Bioorg. Med. Chem. 2003;11:3913.
27. Bertini I, Provenazni A, Viezzoli MS, Pieper DH, Timmis KN. Magn. Reson. Chem. 2003;41:615.
28. Serber Z, Richter C, Moskau D, Böhlen J-M, Gerfin T, Marek D, Häberli M, Baselgia L, Laukien F, Stern SA, Hoch JC, Dötsch V. J. Am. Chem. Soc. 2000;122:3554.
29. Flynn PF, Mattiello DL, Hill HDW, Wand AJ. J. Am. Chem. Soc. 2000;122:4823.
30. Hajduk PJ, Augeri DJ, Mack J, Mendoza R, Yang J, Betz SF, Fesik SW. J. Am. Chem. Soc. 2000;122:7898.
31. Goger MJ, McDonnell JM, Cowburn D. Spectroscopy. 2003;17:161.
32. Bermel W, Bertini I, Felli IC, Kümmerle R, Pieratelli RB. J. Am. Chem. Soc. 2003;125:16423.
33. Bertini I, Felli IC, Kümmerle R, Moskau D, Pierattelli R. J. Am. Chem. Soc. 2004;126:464.
34. Shu Q, Frieden C. Biochemistry. 2004;43:1432.
35. Petrso AM, Gunasekera A, Xu N, Olejniczak ET, Fesik SW. FEBS Lett. 2004;559:171.
36. Aramini JM, Cleaver SH, Pon RT, Cunningham RP, Germann MW. J. Mol. Biol. 2004;338:77.
37. Monleón D, Yim L, Martínez-Vicent M, Armengod ME, Celda B. J. Biomol. NMR. 2004;28:307.

Flow NMR Techniques in the Pharmaceutical Sciences

Paul A. Keifer

University of Nebraska Medical Center, Eppley Institute, 986805 Nebraska Medical Center, Omaha, NE 68198-6805, USA

Introduction

Flow NMR [1,2] is a rapidly growing subset of solution-state NMR, which itself is commonly used in the pharmaceutical sciences. Flow NMR is defined here as any solution-state NMR technique in which the sample is introduced into the NMR probe as a flowing stream. The sample may—or may not—be flowing during the time of actual data acquisition. In contrast, conventional solution-state NMR uses static samples contained in glass tubes.

All known flow NMR techniques can be separated into two categories, depending upon the primary purpose of the flow. The first category uses a separation technique, such as chromatography or electrophoresis, prior to the NMR analysis (Figure 1A). The standard examples are LC-NMR and LC-NMR-MS [3]. The second category uses the liquid flow primarily as a means of transport; as a way to conveniently move solution-state samples in and out of the NMR analysis coil (Figure 1C and 1D). The standard examples of this category are direct-injection NMR (DI-NMR) [4] and flow-injection NMR (FIA-NMR) [5].

Flow NMR has both advantages and disadvantages when compared to non-flow NMR techniques. As such, although flow NMR complements non-flow NMR, it will never replace it. One advantage of flow NMR is that it allows samples to be moved quickly and easily in and out of the flow cell in the NMR probe. One disadvantage is that some flow hardware can be prone to contamination (or even clogging), because the NMR probe's flow cell is often used for hundreds of different samples.

LC-NMR

Complex mixtures are normally analyzed by fractionation, followed by an examination of the individual components one at a time. The traditional method of doing this consists of running a chromatographic separation off-line, collecting the individual fractions, evaporating them to dryness (to remove the mobile phase), and re-dissolving them in deuterated solvents. This is followed by examination with conventional NMR methods (using specialized microprobes, submicroprobes, Nanoprobes, or microcells, as needed) [6].

LC-NMR is an alternative to this method. It combines the separation and analysis steps, and because the analysis immediately follows the separation, it is especially useful if the solutes are volatile, unstable, or air-sensitive [7].

LC-NMR was first developed in 1978 [8], but during the 1980s it was more of an academic curiosity than a robust analytical tool. It became more popular during the mid-1990s as high-sensitivity flow probes became more readily available, better solvent-suppression techniques were developed [9], and the overall robustness increased. LC-NMR is now considered to be an almost routine technique [3].

The hardware for LC-NMR, including the NMR flow probe, is shown in Figure 1A. LC-NMR experiments can be run either in an on-flow mode, in which the mobile phase moves continuously during acquisition, or in a stopped-flow mode, in which the peaks of interest are stopped in the NMR flow probe for as long as is needed for NMR data acquisition. On-flow data are more useful for preliminary or survey-mode analyses, while stopped-flow data are more useful for careful examinations of individual components, using either 1D or 2D NMR techniques.

There are at least four ways to acquire stopped-flow data. First, the chromatographic peaks of interest can be stopped in the flow probe (directly as they elute from the chromatography column) and analyzed one at a time, for as long as necessary. Second, the LC pump may be programmed to "time slice" through a chromatographic peak, stopping every few seconds to acquire a new spectrum (or pumping very slowly). This is useful both for resolving multiple components (by NMR) from within a peak that is not fully resolved chromatographically, or for verifying the purity of a chromatographic peak. A third method involves collecting the chromatographic peaks of interest into loops of tubing (off-line) and then flushing the intact fractions into the NMR flow probe (one at a time) as needed. A fourth method involves trapping the eluted peaks onto another chromatographic column and then re-eluting them with a stronger solvent into the flow probe. This method (called SPE-NMR; Figure 1B) is discussed below.

Graham A. Webb (ed.), Modern Magnetic Resonance, 1195–1201.

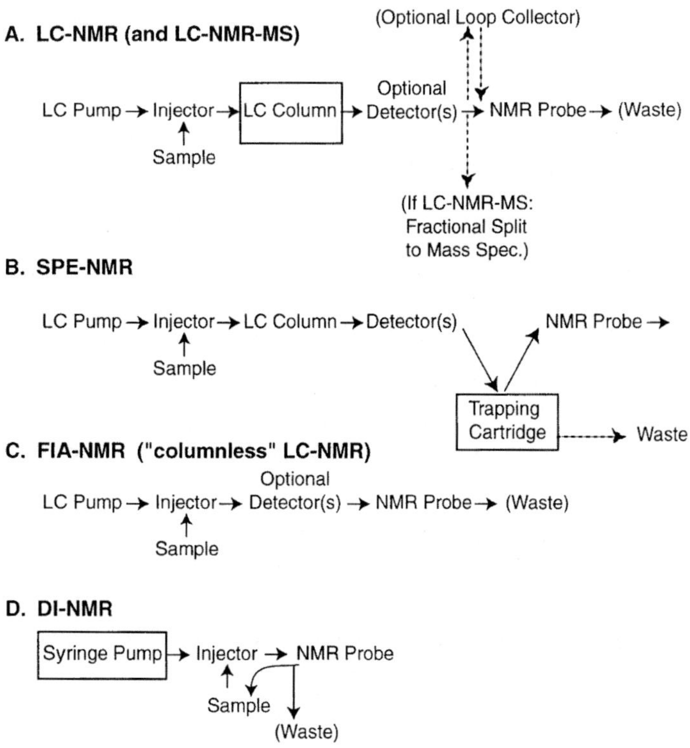

Fig. 1. Flow NMR fluid paths for: (A) LC-NMR and LC-NMR-MS, (B) SPE-NMR, (C) FIA-NMR, and (D) DI-NMR. Boxes show highly significant differences of each technique. The fluid pumps in LC-NMR and SPE-NMR are typically piston pumps capable of producing solvent gradients. The pumps in DI-NMR are typically syringe pumps. The pumps in FIA-NMR can be either (solvent gradients are not typically used nor required). The optional detectors in LC-NMR, SPE-NMR, and FIA-NMR may be any type of single (or multiple) in-line non-destructive chromatography detectors (i.e., UV, RI, radiochemical, ELSD, fluorescence). In LC-NMR, SPE-NMR, and FIA-NMR, the fluid flow is always uni-directional through the probe. In DI-NMR the flow of sample fluid is usually bi-directional (unless the flow cells are made of only uniform capillary tubing), but the additional rinsing steps may be either bi-directional or uni-directional.

The acquisition of LC-NMR data can be difficult for several reasons. First, the mobile phases in LC-NMR are solvent mixtures, and because the solvents are often non-deuterated there are usually one or more solvent resonances that need to be suppressed. Second, the resonances of the organic solvents within the mobile phase contain ^{13}C satellites, which complicate the NMR spectrum and also need to be suppressed. Third, the pulse sequences of many solvent suppression schemes (like presaturation) do not work well on samples that are flowing through the probe. Fourth, during the commonly used gradient elution, the solvent resonances that need to be suppressed are constantly changing frequency (because the solvent composition changes during the experiment). In reversed-phase LC-NMR (which uses water), the chemical shift of water can move up to 3 ppm during a gradient run. Fifth, if the mobile phase is fully protonated (non-deuterated) there will be no ^{2}H lock to keep the frequencies constant. All of these problems are solved by using WET solvent suppression combined with the SCOUT-scan technique [9]. WET uses a combination of shaped-pulse selective excitation, multi-frequency SLP pulses, PFG, and indirect-detection-style ^{13}C decoupling during the shaped pulses to properly suppress multiple solvent resonances using only a two-channel spectrometer. The FIA-NMR and DI-NMR techniques (discussed below) also benefit from using SCOUT and WET [4,5].

LC-NMR-MS

Although LC-NMR is a powerful analytical tool, it can be made even more powerful by adding mass spectrometry to create on-line LC-NMR-MS [10]. This is usually done by splitting off a fraction of the chromatographic effluent (prior to the NMR flow cell) and sending it to a mass spectrometer. Numerous papers and reviews have shown how LC-NMR-MS runs one-step analyses on problems that were either sample limited or that proved to be tough structural problems [3,11,12]. LC-NMR-MS has now been fully adopted for pharmaceutical problems that are both urgent and important. Both LC-NMR and LC-NMR-MS are now commercially available (Bruker—www.bruker.com, Varian—www.varianinc.com, and JEOL—www.jeol.com).

One advantage of incorporating a mass spectrometer into the detector chain is that the mass spectrometer output provides a selective and sensitive signal that can be used to stop the fluid flow to allow NMR signal averaging (to increase the NMR signal-to-noise). This is especially useful for drug-metabolism studies where mass information about the parent compound is already known [13].

Other Detectors in LC-NMR

In LC-NMR, although the NMR serves as one detector, additional in-line detectors (connected in series) can also be used. One or more additional detectors—such as UV, fluorescence, radiochemical, refractive index, or evaporative light scattering detectors—are commonly added (Figure 1), either to get more data or to trigger the HPLC pump to stop the flow on a specific chromatographic peak (for NMR signal averaging).

Other Chromatography in LC-NMR

Other types of chromatography (besides reversed-phase HPLC) were coupled to NMR flow probes starting around 1994. Examples include capillary electrophoresis NMR (CE-NMR), capillary electrochromatography NMR (CEC-NMR), capillary HPLC (CapLC-NMR or CHPLC-NMR), supercritical fluid chromatography (SFC-NMR), and supercritical fluid extraction NMR (SFE-NMR) [3,14–16]. Additionally, size-exclusion chromatography NMR (SEC-NMR) is used to analyze polymers [17–19], and ion-exchange chromatography NMR is used to analyze both protein mixtures [20] and polar metabolites [21].

Other Plumbing Schemes: Loop-Collection LC-NMR and Solid-Phase Extraction NMR (SPE-NMR)

As mentioned above, chromatographic peaks are sometimes collected in loops (i.e. in loops of capillary tubing controlled by a loop-collection system) prior to NMR analysis. The contents of each loop can then be individually flushed into the NMR probe as needed.

A variation upon loop collection is SPE-NMR (solid phase extraction LC-NMR), which attempts to increase solute concentration (to increase NMR sensitivity) by reversing the sample dilution that inevitably occurs during loop collection and analysis [22,23]. In SPE-NMR, the HPLC peaks that elute from the column are individually diverted to in-line solid-phase extraction cartridges (Figure 1B). Water is added to the mobile phase so that the solutes are retained on this small cartridge. Once the column trapping is complete, the solutes are eluted with a stronger solvent (which can be deuterated to ease the NMR analysis) into a concentrated plug that is pumped directly into the NMR flow cell. SPE-NMR has been used for the analysis of microsomal incubate metabolites [24] and natural products [25].

Applications of LC-NMR

LC-NMR has been readily embraced by numerous fields in pharmaceutical research, as evidenced by hundreds of research publications. LC-NMR and LC-NMR-MS are now essentially routine techniques for groups studying drug metabolism [26,27]. They are proving useful for drug-manufacturing and quality-control groups [28], for the analysis of drug impurities [29], for the analysis of drug stability [30], for identifying unstable compounds [31–33] or volatile compounds [34], for metabonomics [35,36], and for evaluating the environmental fate of pharmaceuticals [37]. LC-NMR is being used for natural product drug discovery [38–41] as well as for determining the chirality of natural products [42]. LC-NMR was also evaluated for monitoring reactions within production facilities [43] and has proven useful in evaluating compounds that do not contain UV chromophores [44]. Numerous reviews of the applications of LC-NMR and LC-NMR-MS are in the literature [3,11].

Limitations of LC-NMR

Although LC-NMR and LC-NMR-MS techniques are very powerful analytically, they tend to be slow (1–24 h) compared to LC-MS [45] or other high-throughput analysis methods (2–5 min) because of two limitations. First, NMR is less sensitive than many other techniques (i.e., mass spectrometry) and the need for signal averaging consumes time. Second, although LC-NMR can handle solvent gradients, samples for NMR need to have a "uniform magnetic susceptibility," which means that LC-NMR is not easily adaptable to ultra-fast "turbo" solvent gradients that ramp from 0 to 100% in just a few minutes. These two limitations, combined with the higher cost of an LC-NMR system (as compared to LC-MS), mean that LC-NMR and LC-NMR-MS are best suited to solving the toughest problems on the most expensive samples that cannot be solved in other ways.

Flow-Injection Analysis NMR (FIA-NMR)

The second category of flow NMR methods uses the liquid flow primarily as a means of transport to and from the NMR probe, and does not use any separation technique. This category consists of two basic techniques that differ in the type of fluid flow used: flow-injection analysis NMR (FIA-NMR) [5] and direct-injection NMR (DI-NMR) [4].

Like classical flow injection analysis [46], classical FIA-NMR [5] injects the sample as a plug into an unbroken stream of fluid (an "isocratic mobile phase"), and the resulting sample plug is pushed into the NMR detector coil (and then stopped) by the controlled motion of the fluid (Figure 1C). The fluid serves three purposes. First, it serves as a conveyer belt to carry the sample into the NMR probe. Second, it rinses out the flow cell at the end of the analysis. Third, it allows samples smaller than the volume of the flow cell to be analyzed. (The intact liquid stream reduces the NMR line broadenings caused by the magnetic-susceptibility distortions that would arise from air bubbles in a non-filled NMR flow cell, although this also causes sample dilution when small sample volumes are injected. Unlike other variants of FIA, conventional FIA does not use any air bubbles (or other devices) as separators in the liquid stream [46].)

As in all other flow NMR techniques, once the sample is placed within the NMR flow cell, the NMR acquisition itself can be readily automated. This can consist of PFG gradient shimming (if desired), automated receiver gain adjustment (if desired), and SCOUT scans (to find the solvent frequencies and create the appropriate WET-shaped pulse), followed by the acquisition of 1D or 2D NMR spectra using WET solvent suppression. The whole process (injection and acquisition), for entire queues of samples, is normally controlled by automation software running on the NMR spectrometer.

The first hyphenation of FIA and NMR was published in 1997 [47,48], where it was called "column-less LC-NMR." More recent papers have since demonstrated its viability as an analytical method [5,49,50], and it has been the subject of several reviews [6,51]. Note that there are entire books [46] and journals [52] dedicated to the technique of FIA.

Direct Injection NMR (DI-NMR)

In contrast to FIA-NMR, DI-NMR [4] injects the sample solution directly into an empty flow cell in the NMR probe (Figure 1D). The sample is not injected into a stream of liquid (because no "mobile phase" is present), which generates the maximum NMR sensitivity (because there is no sample dilution) and allows the sample to be recovered intact if desired. Sample recovery is normally accomplished by pulling the analyzed samples back out the way they went in (Figure 1D). The flow cell is then rinsed by pushing clean solvent in and out as many times as is needed to remove traces of solute. Residual rinse solvents can be a source of sample dilution, although the flow cell is emptied each time, so the dilution is usually minimal. Although samples could instead be recovered by pushing them on through the detector coil, this is usually inefficient and requires additional inefficient rinsing steps, unless the NMR flow cell has a capillary (non-standard) geometry.

Although the solvents used in LC-NMR are usually simple mixtures of water and either acetonitrile or methanol, DI-NMR has been run with a wider range of solvents and solvent mixtures, including $CHCl_3$, dimethylsulfoxide (DMSO), tetrahydrofuran (THF), biofluids, and water, as well as HPLC mobile phases. NMR software has evolved so much that it is now relatively easy to automatically acquire NMR data on samples dissolved in even complex solvent mixtures when using WET solvent suppression and the SCOUT-scan technique [9].

One disadvantage of DI-NMR is that the volume of the injected samples must be bigger than the volume of the NMR flow cell or the NMR resonances will be broadened (due to magnetic-susceptibility distortions arising from the air pockets within the flow cell). The other disadvantage of DI-NMR is that the sample cell can become contaminated from repeated contact with "raw" samples, or the connecting capillary tubing can become plugged. Both of these issues can be addressed in FIA-NMR, although FIA-NMR does not readily allow sample recovery, and does not generate the maximum possible NMR sensitivity. Note that FIA-NMR and DI-NMR are complementary, in that their advantages and disadvantages are the opposite of each other.

The injection in DI-NMR is typically performed using a robotic liquid handler controlled by the NMR spectrometer. The normal liquid handlers used for DI-NMR allow samples to be stored in a wide range of containers, including disposable vials or microtiter plates (vs. the precision glass sample tubes required in conventional solution-state NMR). DI-NMR is now commercially available [as systems called VAST (from Varian—www.varianinc.com) and BEST (from Bruker—www.bruker.com)]. These systems are normally capable of analyzing samples ranging in size from 150 to 350 ml (or more), with concentrations ranging from 1–50 mM, and stored in multiple 96-well microtiter plates. A typical analysis time may range from 2 to 7 min, depending upon the viscosity of the solvent, the concentration and volume of the sample, and the amount of rinsing desired. A complete description of DI-NMR was published in 2000 [4], which included a description of the use of DI-NMR in the analysis of combinatorial chemistry libraries, and numerous examples of different data-presentation formats.

The technique of DI-NMR was first published in 1997 [47,48], followed quickly by publications describing its use for analyzing chemical libraries [4,53–58], for analyzing biological fluids (including the commercial analysis of human plasma) [59,60], and for the screening of ligand binding with proteins and other biomolecules [36,61–63]. DI-NMR has spurred further developments in flow probes

[64], NMR pulse sequences [65], and reference standards [66,67]. The development of DI-NMR has also been the subject of several reviews [2,6,36,47,48,51,63].

Complementary Technologies

Numerous developments in other fields of NMR are proving useful to flow NMR. One important area is in probe development. "Normal" flow probes have active volumes of 30–60 ml. (Conventional flow cells require sample volumes of at least twice the active volume to obtain reasonable NMR lineshapes.) Some applications may require probes that have different volumes, to analyze either smaller-volume or lower-concentration samples. For the latter application, a larger volume (120-ml) flow probe was optimized for "SAR-by-NMR" (NMR-screening) assays that use DI-NMR to acquire ^{15}N HSQC data [64].

Microcoils

Even more interest is being generated by the work on smaller-volume "microcoil" flow probes [3,15,68–71]. The RF coils of microcoil probes are wound around capillary tubing and this produces probes with active volumes of typically 1–5 ml. These probes are usually filled using a flow technique, but not always [72,73]. There are dozens of articles now describing the use of microcoil flow probes to analyze effluents from on-line capillary HPLC, capillary electrophoresis (CE), and capillary isotachophoresis (cITP-NMR) [74–79]. As evidence of their sensitivity, a microcoil was used to analyze the 540 pmol of compound cleaved from a single 160-micron-diameter solid-phase resin bead [80]. A microcoil flow probe was also used to identify members of a natural products library using 5–50 micrograms of compound dissolved in 3 microliters of solvent [57]. Microcoil probes have now been interfaced with automated capillary liquid handlers, and are commercially available (from the MRM division of Protasis; www.protasis.com).

Interchangeable Flow Cells

Another limitation of flow NMR, particularly DI-NMR, is that repetitive injections of "raw" samples can eventually contaminate (or clog) the NMR flow cell. (LC-NMR samples have been "filtered" through the HPLC column, and FIA-NMR can incorporate in-line filters, but DI-NMR cannot easily do so.) This limitation has been addressed with the development of user-changeable flow cells to form the interchangeable flow cell (IFC) probe, which has been discussed in the literature [6,51]. User-changeable flow cells are designed to minimize instrument downtime.

Cold Probes

NMR users always need better spectral signal-to-noise. The conventional solution is to increase the signal intensity, but another way is to decrease the noise. This has been accomplished by cooling the electronics of the NMR probe (but not the sample) so as to reduce thermal noise. The resulting probe, called a cold probe, a cryogenic probe, or a cryoprobe, is usually operated at a temperature of 25 K, which provides a signal-to-noise improvement of about fourfold [81]. This technology has since been incorporated into flow probes [11,82,83].

Parallel NMR: Multiplex NMR

The goal for any high-throughput technique is to run the analyses in parallel. NMR has historically been only a serial technique, but this is changing, as ways are being developed to acquire NMR data in parallel—by placing multiple sample chambers, each with their own dedicated detection (RF) coil, all within one NMR probe (within one magnet) [84,85]. The multiple coils are usually designed so they can analyze one sample while other samples are being loaded into different coils (to reduce the percentage of analysis time lost to shuttling samples into the NMR magnet) [86,87]. Other functions of multiple-coil (multiplex) probes are to measure rapid-reaction kinetics (by placing multiple coils at different points along a flowing sample) [88], to acquire capillary electrophoresis data using solenoidal microcoils [89], or to acquire the difference spectrum between two samples in a "difference probe" [90,91].

Alternative Ways to Acquire nD-NMR Data

The filter diagonalization method (FDM) [92] is a way to rapidly acquire nD-NMR data rapidly, and store the resulting data in a much smaller space. Although the development of FDM is still in its infancy, if it can rapidly acquire 2D ^{1}H{^{13}C} or ^{1}H{^{15}N} heteronuclear correlation (HSQC) data, this could help DI-NMR either to characterize combinatorial chemistry libraries or to screen proteins for ligand binding. It may also benefit LC-NMR.

Frydman *et al.* have developed another method for rapidly acquiring nD-NMR data [93]. They have proposed using this technique in combination with LC-NMR [94]. Although its sensitivity limitations make it currently impractical for LC-NMR, it may prove useful in the DI-NMR analyses of compound libraries.

DOSY: A Possible Alternative to LC-NMR

Diffusion-ordered spectroscopy (DOSY) [95] can separate the 1D NMR spectrum of a mixture into individual 1D sub-spectra based upon the diffusion rates of the individual components. DOSY is already being used now for NMR screening (determining which compounds bind to large-molecule receptors) [96], although not in a flow mode. (It easily could be because other kinds of NMR screening are being performed using flow NMR.) DOSY has also been proposed as a possible alternative to LC-NMR [97]. The *combined* use of DOSY on LC-NMR fractions is interesting, but it is unlikely to be a practical technique due to the typically low NMR signal-to-noise in LC-NMR. DOSY should prove useful within DI-NMR, however, to verify compound purity in a chemical library.

Conclusions

A variety of different flow NMR techniques are currently being used in the pharmaceutical sciences, each with their own unique advantages and disadvantages. LC-NMR techniques offer separation capabilities, and are powerful, but are often slow. FIA-NMR is well suited to fast, repetitive, destructive analyses that do not need chromatographic separations, where carryover must be minimized, and which do not require much NMR time-averaging. DI-NMR is best suited to repetitive, non-destructive analyses that do not need chromatographic separations, where the sample needs to be recovered, where slight carryover is not a problem, and which may need time-averaging of the NMR data.

References

1. Dorn HC. In: DM Grant, RK Harris (Eds). Encyclopedia of Nuclear Magnetic Resonance. John Wiley & Sons Ltd.: West Sussex, England, 1996, p. 2026.
2. Keifer PA. Curr. Opin. Chem. Biol. 2003;7:388.
3. Albert K. On-Line LC-NMR and Related Techniques. Wiley: Chichester, 2002.
4. Keifer PA, Smallcombe SH, Williams EH, Salomon KE, Mendez G, Belletire JL, Moore CD. J. Comb. Chem. 2000;2:151.
5. Keifer PA. Magn. Reson. Chem. 2003;41:509.
6. Keifer PA. In: E Jucker (Ed). Progress in Drug Research, Vol. 55. Birkhauser Verlag: Basel, 2000, p. 137.
7. Schlotterbeck G, Tseng LH, Haendel H, Braumann U, Albert K. Anal. Chem. 1997;69:1421.
8. Watanabe N, Niki E. Proc. Jpn. Acad. Ser. B 1978;54:194.
9. Smallcombe SH, Patt SL, Keifer PA. J. Magn. Reson. A 1995;117:295.
10. Bendel P, Lai CM, Lauterbur PC. J. Magn. Reson. 1980;38:343.
11. Corcoran O, Spraul M. Drug Discov. Today 2003;8:624.
12. Wilson ID. J. Chromatogr. A 2000;892:315.
13. Dear GJ, Plumb RS, Sweatman BC, Ayrton J, Lindon JC, Nicholson JK, Ismail IM. J. Chromatogr. B Biomed. Sci. Appl. 2000;748:281.
14. Albert K. J. Chromatogr. A 1995;703:123.
15. Olson DL, Lacey ME, Sweedler JV. Anal. Chem. 1998;70:257A.
16. Gfrorer P, Schewitz J, Pusecker K, Bayer E. Anal. Chem. 1999;71:315A.
17. Ute K, Niimi R, Hongo SY, Hatada K. Polym. J. 1998;30:439.
18. Kramer I, Pasch H, Handel H, Albert K. Macromol. Chem. Phys. 1999;200:1734.
19. Ludlow M, Louden D, Handley A, Taylor S, Wright B, Wilson ID. J. Chromatogr. A 1999;857:89.
20. Ruckert M, Wohlfarth M, Bringmann G. J. Chromatogr. A 1999;840:131.
21. Dear GJ, Plumb RS, Sweatman BC, Parry PS, Roberts AD, Lindon JC, Nicholson JK, Ismail IM. J. Chromatogr. B Biomed. Sci. Appl. 2000;748:295.
22. Griffiths L, Horton R. Magn. Reson. Chem. 1998;36:104.
23. de Koning JA, Hogenboom AC, Lacker T, Strohschein S, Albert K, Brinkman UAT. J. Chromatogr. A 1998;813:55.
24. Bao D, Thanabal V, Pool WF. J. Pharm. Biomed. Anal. 2002;28:23.
25. Nyberg NT, Baumann H, Kenne L. Magn. Reson. Chem. 2001;39:236.
26. Tugnait M, Lenz EM, Hofmann M, Spraul M, Wilson ID, Lindon JC, Nicholson JK. J. Pharm. Biomed. Anal. 2003;30:1561.
27. Mortensen RW, Corcoran O, Cornett C, Sidelmann UG, Lindon JC, Nicholson JK, Hansen SH. Drug Metab. Dispos. 2001;29:375.
28. Jiang T, Phelan M. J. Pharm. Sci. Technol. 2003;57:287.
29. Sharman GJ, Jones IC. Magn. Reson. Chem. 2003;41:448.
30. Segmuller BE, Armstrong BL, Dunphy R, Oyler AR. J. Pharm. Biomed. Anal. 2000;23:927.
31. Fischer H, Gyllenhaal O, Vessman J, Albert K. Anal. Chem. 2003;75:622.
32. Wolfender JL, Verotta L, Belvisi L, Fuzzati N, Hostettmann K. Phytochem. Anal. 2003;14:290.
33. Stintzing FC, Conrad J, Klaiber I, Beifuss U, Carle R. Phytochemistry 2004;65:415.
34. Elipe MVS, Huskey SE, Zhu B. J. Pharm. Biomed. Anal. 2003;30:1431.
35. Gavaghan CL, Nicholson JK, Connor SC, Wilson ID, Wright B, Holmes E. Anal. Biochem. 2001;291:245.
36. Stockman BJ. Curr. Opin. Drug. Discov. Dev. 2000;3:269.
37. Cardoza LA, Williams TD, Drake B, Larive CK. ACS Symp. Ser. 2003;850:146.
38. Hostettmann K, Marston A. Phytochem. Rev. 2003;1:275.
39. Wolfender JL, Ndjoko K, Hostettmann K. J. Chromatogr. A 2003;1000:437.
40. Bobzin SC, Yang S, Kasten TP. J. Chromatogr. B Biomed. Sci. Appl. 2000;748:259.
41. Sandvoss M, Weltring A, Preiss A, Levsen K, Wuensch G. J. Chromatogr. A 2001;917:75.
42. Queiroz EF, Wolfender JL, Raoelison G, Hostettmann K. Phytochem. Anal. 2003;14:34.
43. Maiwald M, Fischer HH, Kim YK, Albert K, Hasse H. J. Magn. Reson. 2004;166:135.
44. Petritis K, Gillaizeau I, Elfakir C, Dreux M, Petit A, Bongibault N, Luijten W. J. Sep. Sci. 2002;25:593.

45. Yurek DA, Branch DL, Kuo MS. J. Comb. Chem. 2002;4:138.
46. Ruzicka J, Hansen EH. Flow Injection Analyses, John Wiley and Sons: New York, NY, 1988.
47. Keifer PA. Drug Discov. Today 1997;2:468.
48. Keifer PA. Drugs Future 1998;23:301.
49. Lenz E, Taylor S, Collins C, Wilson ID, Louden D, Handley A. J. Pharm. Biomed. Anal. 2002;27:191.
50. Louden D, Handley A, Taylor S, Lenz E, Miller S, Wilson ID, Sage A. Analyst 2000;125:927.
51. Keifer PA. In: H-Y Mei, AW Czarnik (Ed), Integrated Drug Discovery Technologies. Marcel Dekker Inc.: New York, NY, 2002, p. 485.
52. Segundo MA, Rangel AOSS. J. Flow Injection Anal. 2002;19:3.
53. Hamper BC, Snyderman DM, Owen TJ, Scates AM, Owsley DC, Kesselring AS, Chott RC. J. Comb. Chem. 1999;1:140.
54. Lewis K, Phelps D, Sefler A. Am. Pharm. Rev. 2000;3:63.
55. Combs A, Rafalski M. J. Comb. Chem. 2000;2:29.
56. Kalelkar S, Dow ER, Grimes J, Clapham M, Hu H. J. Comb. Chem. 2002;4:622.
57. Eldridge GR, Vervoort HC, Lee CM, Cremin PA, Williams CT, Hart SM, Goering MG, O'Neil-Johnson M, Zeng L. Anal. Chem. 2002;74:3963.
58. Leo GC, Krikava A, Caldwell GW. Anal. Chem. 2003;75:1954.
59. Potts BCM, Deese AJ, Stevens GJ, Reily MD, Robertson DG, Theiss J. J. Pharm. Biomed. Anal. 2001;26:463.
60. Robertson DG, Reily MD, Sigler RE, Wells DF, Paterson DA, Braden TK. Toxicol. Sci. 2000;57:326.
61. Gmeiner WH, Cui W, Konerding DE, Keifer PA, Sharma SK, Soto AM, Marky LA, Lown JW. J. Biomol. Struct. Dyn. 1999;17:507.
62. Ross A, Schlotterbeck G, Klaus W, Senn H. J. Biomol. NMR 2000;16:139.
63. Stockman BJ, Farley KA, Angwin DT. Methods Enzymol. 2001;338:230.
64. Haner RL, Llanos W, Mueller L. J. Magn. Reson. 2000;143:69.
65. Spitzer T, Sefler AM, Rutkowske R. Magn. Reson. Chem. 2001;39:539.
66. Gerritz SW, Sefler AM. J. Comb. Chem. 2000;2:39.
67. Pinciroli V, Biancardi R, Colombo N, Colombo M, Rizzo V. J. Comb. Chem. 2001;3:434.
68. Wolters AM, Jayawickrama DA, Sweedler JV. Curr. Opin. Chem. Biol. 2002;6:711.
69. Lacey ME, Subramanian R, Olson DL, Webb AG, Sweedler JV. Chem. Rev. 1999;99:3133.
70. Webb AG. Prog. Nucl. Magn. Reson. Spectrosc. 1997;31:1.
71. Li Y, Logan TM, Edison AS, Webb A. J. Magn. Reson. 2003;164:128.
72. Griffin JL, Nicholls AW, Keun HC, Mortishire-Smith RJ, Nicholson JK, Kuehn T. Analyst 2002;127:582.
73. Schlotterbeck G, Ross A, Hochstrasser R, Senn H, Kuehn T, Marek D, Schett O. Anal. Chem. 2002;74:4464.
74. Lacey ME, Tan ZJ, Webb AG, Sweedler JV. J. Chromatogr. A 2001;922:139.
75. Kautz RA, Lacey ME, Wolters AM, Foret F, Webb AG, Karger BL, Sweedler JV. J. Am. Chem. Soc. 2001;123:3159.
76. Wolters AM, Jayawickrama DA, Larive CK, Sweedler JV. Anal. Chem. 2002;74:2306.
77. Krucker M, Lienau A, Putzbach K, Grynbaum MD, Schuler P, Albert K. Anal. Chem. 2004;76:2623.
78. Lacey ME, Webb AG, Sweedler JV. Anal. Chem. 2002;74:4583.
79. Wolters AM, Jayawickrama DA, Larive CK, Sweedler JV. Anal. Chem. 2002;74:4191.
80. Lacey ME, Sweedler JV, Larive CK, Pipe AJ, Farrant RD. J. Magn. Reson. 2001;153:215.
81. Anderson WA, Brey WW, Brooke AL, Cole B, Delin KA, Fuks LF, Hill HDW, Johanson ME, Kotsubo VY, Nast R, Withers RS, Wong WH. Bull. Magn. Reson. 1995;17:98.
82. Spraul M, Freund AS, Nast RE, Withers RS, Maas WE, Corcoran O. Anal. Chem. 2003;75:1536.
83. Exarchou V, Godejohann M, Van Beek TA, Gerothanassis IP, Vervoort J. Anal. Chem. 2003;75:6288.
84. Webb AG, Sweedler JV, Raftery D. On-Line LC-NMR and Related Techniques. John Wiley & Sons Ltd.: West Sussex, England, 2002, p. 259.
85. Raftery D. Anal. Bioanal. Chem. 2004;378:1403.
86. Macnaughtan MA, Hou T, Xu J, Raftery D. Anal. Chem. 2003;75:5116.
87. Zhang X, Sweedler JV, Webb AG. J. Magn. Reson. 2001;153:254.
88. Ciobanu L, Jayawickrama DA, Zhang X, Webb AG, Sweedler JV. Angew. Chem. Int. Ed. Engl. 2003;42:4669.
89. Wolters AM, Jayawickrama DA, Webb AG, Sweedler JV. Anal. Chem. 2002;74:5550.
90. Macnaughtan MA, Hou T, Macnamara E, Santini RE, Raftery D. J. Magn. Reson. 2002;156:97.
91. Macnaughtan MA, Smith AP, Goldsbrough PB, Santini RE, Raftery D. Anal. Bioanal. Chem. 2004;378:1520.
92. Mandelshtam VA. Prog. Nucl. Magn. Reson. Spectrosc. 2001;38:159.
93. Frydman L, Scherf T, Lupulescu A. Proc. Natl. Acad. Sci. U.S.A. 2002;99:15858.
94. Shapira B, Karton A, Aronzon D, Frydman L. J. Am. Chem. Soc. 2004;126:1262.
95. Johnson CS Jr. Prog. Nucl. Magn. Reson. Spectrosc. 1999;34:203.
96. Hodge P, Monvisade P, Morris GA, Preece I. Chem. Commun. 2001;3:239.
97. Huo R, Wehrens R, Van Duynhoven J, Buydens LMC. Anal. Chim. Acta 2003;490:231.

Developments in NMR Hyphenation for Pharmaceutical Industry

Manfred Spraul

Bruker BioSpin GmbH, Silberstreifen, D-76287 Rheinstettan, Germany

Introduction

First introduced in 1978 [1], LC-NMR since the early 1990s has grown into a standard analytical method in pharmaceutical industry with an increasing number of publications every year [2–24].

This has been enabled by improved magnet design, dedicated NMR flow probes, automatic multiple solvent suppression, and full automation. The typical application areas for LC-NMR in the pharmaceutical industry are:

- screening for drug metabolites in biofluids or extracts;
- drug impurity profiling;
- drug stability profiling;
- detection of new biomarkers of toxicity or disease;
- combinatorial chemistry and parallel synthesis to investigate product mixtures;
- natural product screening.

The following operation modes have evolved and can be executed in an automated fashion on a modern hyphenated system which will be explained in more detail later and include:

- on-flow LC-NMR;
- direct stop-flow;
- loop collection [17];
- post-column solid phase extraction (SPE) [14,23];
- time slicing.

To be able to perform the stop-flow, loop collection, or post-column SPE, a non-NMR detector is needed to locate LC peaks of interest for the NMR transfer. UV detection was used exclusively in early LC-NMR experiments. Currently, DAD and MS are used for LC peak detection in most cases. Refractive index detectors are less suitable for LC-NMR due to the gradient LC methods that are required to keep the eluting peaks as sharp as possible to preserve the NMR sensitivity.

In the following sections, the operation modes mentioned above will be explained and the required hardware discussed.

On-Flow LC-NMR

In this mode, the flow from the chromatographic column is guided through a flow cell in a dedicated NMR flow probe. Repeated NMR measurements are performed in a pseudo-two-dimensional (2D) acquisition mode; however in contrast to real 2D experiments, there is no incremented evolution time and timing stays constant over the whole experiment. The time for one measurement is defined by the flow rate of the chromatographic separation and typically lies between 8 and 20 s for standard analytical columns. The flow cells used in the NMR probe head typically have an active volume of 120 or 60 μl, with the active volume defined by the detection coil dimensions. The total volume of such a flow cell is about 200 or 100 μl, respectively.

The possible accumulation time in one experiment of an on-flow run is defined not only by the flow rate, but also by the steepness of the LC gradient. Normally in reversed phase separations, water is replaced by D_2O for locking purposes. In addition to the locking advantage, the need for solvent suppression on two signals is also reduced. If water is used, the signal can be very broad, especially when methanol is used as the cosolvent through exchange with the methanol-OH. The suppression power required for such a system would be large and would also reduce desired signals in the vicinity of the water resonance. When D_2O is used, non-deuterated methanol or acetonitrile are added for the LC gradient. The dynamic range of the NMR detection system is limited by the large concentration difference between the compounds of interest and the organic solvent signal, which is orders of magnitude larger. Suppression of the solvent signals therefore is mandatory in order to achieve the optimized detection of the weak solute signals of interest.

A further complication arises through the fact that the chemical shifts of the NMR signals are strongly dependent on the composition of the solvent mixture and the pH. This means in an LC-gradient run, the relative position of the organic solvent signal to the water frequency constantly moves. If one has to suppress the solvent, one has to follow this change in frequency constantly. Within one experiment the suppression frequencies can be defined

Graham A. Webb (ed.), Modern Magnetic Resonance, 1203–1210.

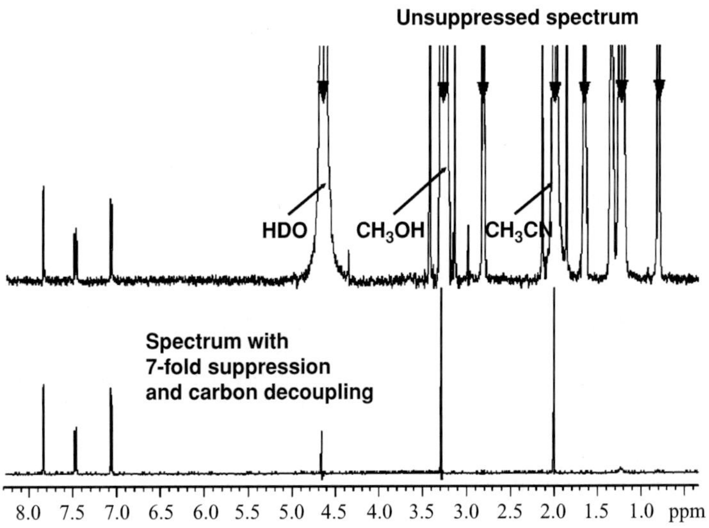

Fig. 1. The effect of automatic solvent suppression in LC-NMR on a solvent system containing acetonitrile, methanol, water, and ion-pair reagent, the signals marked with an arrow are selected for suppression automatically.

automatically by a short preparation experiment, which only takes a few seconds and delivers the correct frequencies that are then loaded for the experiment with suppression. In order not to lose suppression quality, the length of one accumulation is restricted due to the steepness of the LC gradient. Solvent signals like acetonitrile or methanol show a single resonance line, which is easy to suppress. However, at the 0.5% level of this signal, a doublet is visible. This is due to the fact that 1% of the carbon atoms are not carbon-12, but carbon-13, which are NMR active and cause the splitting of the resonance line to a doublet. These carbon satellite lines, as they are called, are still large compared to the solute signals and also increase the disturbed area in the spectrum. Therefore it is advantageous to remove the doublet by also irradiating the carbon frequencies at the same time; by this action the doublet collapses into the singlet line of the C12-based solvent, which is already suppressed. The automation control of a modern NMR spectrometer allows the suppression of as many solvent lines as needed, by using phase-shifted-shaped radio frequency pulses and carbon decoupling. However, it is obvious that the observable region in a spectrum can be diminished seriously when suppressing too many solvent or buffer lines.

Figure 1 shows the typical effect solvent suppression, comparing the non-suppressed spectrum with the result of the automatic solvent suppressed spectra. In this case, an ion-pair reagent separation was performed and the signals of acetonitrile, methanol, residual water, and octane-1-sulfonic acid were suppressed.

The lines to be suppressed are indicated by arrows in the upper spectrum. The solute in this case was 5-hydroxy-salicylic acid.

A typical on-flow experiment is shown in Figure 2, along the horizontal axis the proton-NMR chemical shift is shown, along the vertical axis the retention time of the chromatogram is increasing from bottom to top. The lines at 2 and 4.7 ppm are due to the suppressed acetonitrile and water signals. A 100 μl of urine, obtained 4 h after ingestion of 500 mg of paracetamol (male adult) was injected onto a 150 × 4.6 mm RP-18 column using a flow rate of 0.8 ml/min and a time resolution on the NMR detection of 16 s. The acetonitrile signal shows the change in frequency due to the LC gradient, it moves to higher ppm values with increasing amount of acetonitrile. Some paracetamol metabolites can be identified through the aromatic signals between 6.8 and 7.3 ppm. The anomeric protons of the glucuronides show resonances at ~5.2 ppm.

Taking a horizontal row out of the on-flow spectrum generates the ^{1}H-spectrum at a given retention time.

Due to the short residence time in the flow cell and by this the limited number of scans per row, sensitivity in this operation mode of LC-NMR is limited to the microgram range of detection. Therefore the need for stopped flow measurements emerged, especially for pharmaceutical applications, where often impurities or low concentration metabolites have to be identified.

Direct Stop-Flow

To be able to stop the flow exactly when the maximum of an LC peak of interest has reached the center of the flow cell, fast detection of the LC peaks is needed. NMR is too slow and not sensitive enough, therefore other analytical tools are needed. UV detection is most often used as a

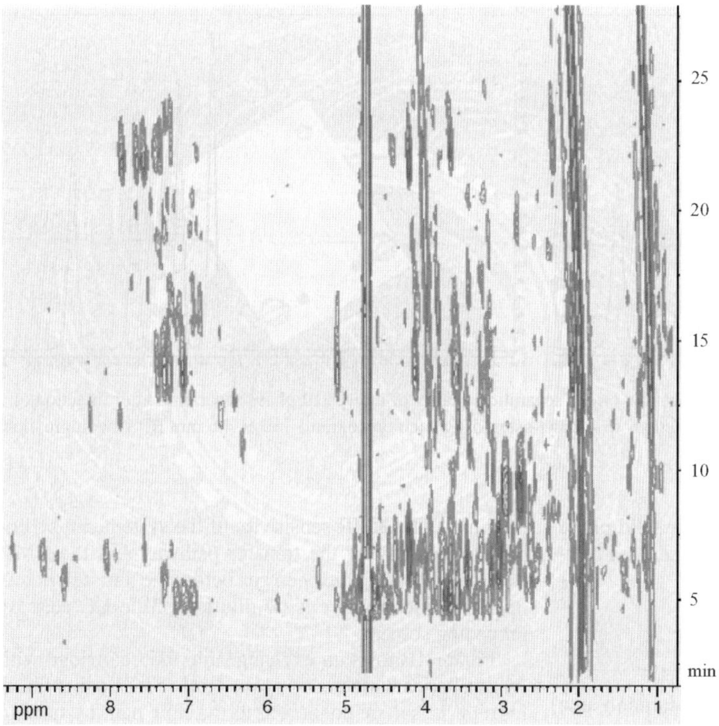

Fig. 2. On-Flow LC-NMR spectrum of a reversed phase separation after injection of a human urine obtained 4 h after ingestion of 500 mg of paracetamol (150 × 4.6 mm RP-18 column using a flow rate of 0.8 ml/min). (See also Plate 92 on page LXIV in the Color Plate Section.)

chromatographic detector and it is much more sensitive compared to NMR. Therefore a UV cell has to be positioned between the LC column outlet and the NMR flow cell. An additional valve is needed to switch the NMR flow cell off-line when an LC peak has reached the flow cell. Knowing the capillary length and diameter, as well as the flow rate, the time for an LC peak to the NMR flow cell can be calculated. In order to perform automatic stop-flow, software is needed that monitors the UV signal and stops the pump at the correct time. Since the pump flow is not stopped immediately, the valve mentioned, will turn on the stop command, to isolate the NMR flow path and prevent all movement of the LC peak inside the NMR flow cell.

In this mode of operation, long-term NMR acquisitions can be executed, also allowing fully quantitative measurements, impossible in the on-flow mode. This is also the mode to execute 2D experiments to get further structural information.

The direct stop-flow mode has limitations, if multiple stops are needed to look to all LC peaks of interest. All peaks still residing in the column or on the way to the NMR when a stop occurs will undergo diffusion. The longer the stop, the more diffusion will lead to LC peak broadening by this degrading the chromatographic resolution. After a restart, the peaks still on the column will be partly refocused; however, an effect is clearly visible. Broadening of LC peaks leads to loss of NMR sensitivity due to a smaller fraction of the peak fitting into the flow cells active region. Pump control also has to assure, that after a restart, the gradient ratio is the same as when the flow stopped.

Figure 3 shows a comparison of a chromatographic separation on an RP-18 column at 1 ml/min flow with injection of a mixture of four *p*-hydroxy benzoic acid esters. The left side shows the run with stop-flow on every peak, and the right side shows the on-flow result. The degradation of the chromatographic separation is clearly visible; even so only four short stops to acquire a one-dimensional (1D) NMR spectrum were done. The problem of the multiple stop-flow degradation can be overcome by the use of a loop collection system, which is explained in the next section.

An extension of the direct stop-flow experiment is the so-called time slicing, in this case the flow is stopped for example every 30 s, independent of LC peak positions. This allows a pseudo on-flow experiment to be obtained with reduced time resolution along the chromatographic axis and improved NMR sensitivity due to the longer acquisition time. It is also possible to do the time slicing

Techniques

Part II

New Approaches to NMR Data Acquisition, Assignment and Protein Structure Determination: Potential Impact in Drug Discovery

Antonio Pineda-Lucena

Associate Principal Scientist, Protein NMR Laboratory, AstraZeneca Pharmaceuticals, Macclesfield Cheshire England

Introduction

The majority of drugs available today were discovered either from chance observations or from the screening of synthetic or natural product libraries. The chemical modification of lead compounds, on a trial-and-error basis, typically led to compounds with improved potency, selectivity, bioavailability, and reduced toxicity. However, this approach is labor- and time-intensive, and researchers in the pharmaceutical industry are constantly developing methods with a view to increasing the efficiency of the drug discovery process. One of these relatively new approaches is the so-called rational drug design that relies heavily in the determination, either by X-ray crystallography or nuclear magnetic resonance (NMR) spesctroscopy, of the three-dimensional structures of the targets and the ligand-target complexes.

NMR spectroscopy is a biophysical technique with wide applicability in drug discovery research, particularly for the detection and characterization of molecular interactions. NMR can also be used as a tool for the determination of structure and dynamics of proteins, and protein-ligand complexes. In drug design, NMR can be used to validate or invalidate hits from high-throughput-screening (HTS) and can also generate lead compounds by a fragment-based approach. Thus, NMR is not only a technique for structure-guided drug design, but can also be integrated with HTS and combinatorial chemistry. Being so versatile, NMR can be applied to all phases of a drug discovery program, including target selection, lead generation, and lead optimization [1].

To most pharmaceutical companies structural information is an important contributor to drug discovery programs, and this information is sought and utilized whenever a target appears structurally feasible. For structure determination of large proteins, or the complex of a large protein with a tightly bound inhibitor, structure determination by X-ray crystallography is generally faster. On the contrary, smaller proteins sometimes do not readily crystallize and can be more rapidly solved by NMR. Furthermore, NMR can be applied to the structure determination of weakly bound ligands. The two methods are complementary, and each has its own strengths and weaknesses.

From a practical point of view, the main challenges NMR faces in the pharmaceutical world are the costs associated with the production of large amounts of labeled material for screening and structural studies, and, more importantly, the speed with which data are generated. Only when these issues are resolved, will NMR be able to fully exploit its potential in the drug discovery field. This chapter will review some of the methodologies recently developed to increase the efficiency of strategies for data acquisition, analysis, and protein structure calculation by NMR.

Fast Multidimensional NMR Spectroscopy

Data acquisition for multidimensional NMR spectra is not optimal in terms of speed. Systematic exploration of the indirect evolution dimensions one step at a time has the advantage that no signal frequencies are overlooked, but this is not the fastest mode of data acquisition for relatively sparse spectra. In a traditional three-dimensional NMR experiment, the evolution of the signal is followed as a function of t_1, t_2, and t_3 independently. On the contrary, a variety of methods [2] introduced in the last few years seek to simplify this process by cutting down the number of individual measurements, either in the time domain or in the frequency domain. Two of the most promising approaches are described below.

In GFT-NMR, the speed problem of multidimensional NMR is tackled by incrementing several evolution periods jointly, either exactly in step with one another or with suitable scaling factors between them. In a three-dimensional case where the evolution rates are all equal, this is equivalent to exploring the body diagonal of a cube representing the time-domain data space. This is clearly much quicker than incrementing these time dimensions independently. The chemical shifts appear in the experimental time-domain data set as linear combinations, but if several scans are carried out with the radiofrequency

Graham A. Webb (ed.), Modern Magnetic Resonance, 1221–1228.

mixing pulses systematically phase-shifted in 90° steps, a separation can be achieved by a procedure based on a G-matrix [3].

The reduction in measurement time improves exponentially as the number of locked dimensions increases, despite the increase in mandatory scans. However, the technique sacrifices sensitivity for each stage of dimensionality reduction, and higher-dimensional experiments necessarily incur increased signal loss through relaxation. Eventually a point is reached where the sensitivity, rather than the duration, becomes the limiting factor. Thus, the strength of the GFT method lies in the possibility of reducing the dimensionality of high-dimensional spectra, and the limitations of the GFT method might be expected to become evident in cases of severe overlap and in situations where sensitivity is a problem.

Another strategy aims to use projection-reconstruction (PR) techniques to recreate the full three-dimensional NMR experiments. This approach is related to that used in the X-ray tomography field [4] where it is employed to study complex systems. PR-NMR [5,6] reduces the time required to perform a three-dimensional experiment by an order of magnitude, and four-dimensional spectra by two orders of magnitude, by replacing the slow exploration of the evolution dimensions one-by-one by the acquisition of a relatively small number of plane projections.

If F_1 and F_2 are the evolution dimensions and F_3 the direct acquisition dimension of a three-dimensional NMR experiment, then the projections onto the orthogonal F_1F_3 and F_2F_3 are readily recorded by setting $t_2 = 0$ and $t_1 = 0$, respectively. These are the first planes of the three-dimensional spectrum and can be measured quite quickly. In an ideal situation where not too many peaks contribute to the intensity of a single response in the projection, and where the measured NMR intensities in both projections are reliable, the entire three-dimensional spectrum can be extracted from information on these two planes. In actual practice, ambiguities are to be expected and must be resolved if the three-dimensional spectrum is to be recovered. One solution is to measure a skew projection intermediate between the F_1F_3 and F_2F_3 planes by incrementing t_1 and t_2 jointly, with a suitable scaling factor. Skew projections are obtained by replacing the normal evolution time increments, t_1 and t_2, with $t_1\cos\alpha$ and $t_2\sin\alpha$, where α is the desired projection angle (Figure 1). The projection method relies on a Fourier transform theorem [7,8] which states that a section through the origin of a two-dimensional time domain signal $S(t_1, t_2)$, subtending an angle α with respect to the t_1 axis, transforms as the projection onto a line through the origin of the two-dimensional frequency-domain signal $S(F_1,F_2)$, subtending the same angle α with respect to the F_1 axis. By extension, a three-dimensional NMR experiment where the evolution parameters t_1 and t_2 are incremented simultaneously in the ratio $\Delta t_2/\Delta t_1 = \tan\alpha$, projects the three-dimensional NMR spectrum onto a plane tilted through an angle α with respect to the F_1F_3 plane. With the usual hypercomplex Fourier transformation, where real and imaginary signal components are acquired in both evolution dimensions [9] and suitably combined [10], projections are obtained for both $+\alpha$ and $-\alpha$, providing two independent views of the three-dimensional array of NMR peaks.

Although the sensitivity per unit time is the same as that of the conventional mode, it is reduced overall in proportion to the square root of the experimental time. Consequently, PR can realize its full theoretical speed advantage only if the inherent sensitivity of the sample is already adequate. Fortunately, modern high-field NMR spectrometers have improved in sensitivity, particularly those equipped with a cryoprobe receiver coil. As opposed to other methods of simultaneous incrementation of two evolution parameters [3,9,11], PR-NMR uses a simple and transparent method for data processing, providing a direct comparison with the traditional three-dimensional spectrum.

The inverse Radon transform can be used to reconstruct a three-dimensional NMR spectrum from a small number of different plane projections [12,13]. The Radon procedure calculates multiple line integrals of the absorption along parallel paths through the sample, and repeats the process for a set of different angles of incidence arranged around a circle. The Radon transform converts these line integrals into a related set of projections at right angles to the original beams. In practice, it is the inverse Radon transform that is of particular interest.

The reconstruction treatment aims to use the unique multidimensional array of NMR peaks compatible with all the measured plane projections. Projections involve the integration of the absorbed intensity along a series of parallel beams through the sample, and the Radon transform calculates these line integrals. The actual approach followed to find a density function compatible with all the measured projections depends on the quality of the NMR spectra. Thus, spectra with good sensitivity favor a processing scheme based on only a few projections. In this case, a three-dimensional array $S(F_1, F_2, F_3)$ is reconstructed by processing one F_1F_2 plane at a time. In a typical F_1F_2 plane, the raw experiment data for $\alpha = 0°$ and $\alpha = 90°$ are one-dimensional absorption spectra, representing the projections of NMR intensity onto the F_1 and F_2 axes, and a provisional map defining all the possible cross-peak coordinates can be derived easily. This provisional map is obtained by a two-dimensional convolution of the one-dimensional F_1 and F_2 traces. Some of these locations correspond to actual peaks, and others will later, after acquisition of additional projections at $\pm\alpha$, turn out to be unoccupied. Once all F_1F_2 planes have been processed, the three-dimensional spectrum is complete and can be displayed in any of the usual formats.

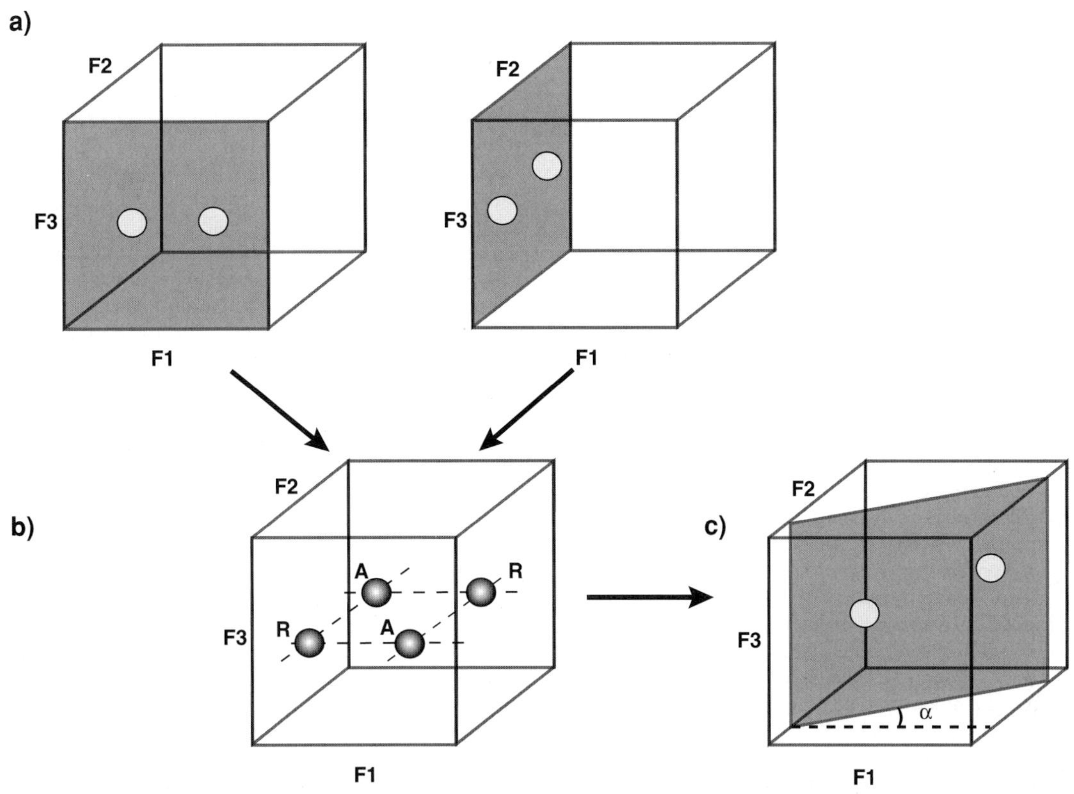

Fig. 1. Reconstruction of a three-dimensional NMR spectrum based on PR NMR techniques. First plane projections obtained for F_1F_3 and F_2F_3 at $t_2 = 0$ and $t_1 = 0$, respectively (a) cannot define unambiguously the positions of the signals in the three-dimensional spectrum (b), but this uncertainty can be resolved by measuring the projection onto a plane tilted by an angle α with respect to the F_1F_3 plane. "A" indicates artefact signals generated when the only information used for the reconstruction are first planes, and "R" indicates real signals that only become apparent when the tilted projection is measured.

For spectra with poor sensitivity, it is an advantage to record a larger number of different projections, usually spaced uniformly around a circle as in the Radon transform methodology. In this case, reconstruction is achieved by adding the contributions from the many different back-projections. There is a strong reinforcement of intensity at positions of genuine NMR peaks, enhancing the sensitivity in proportion to the number of different projections. This is an advantage as long as the number of projections remains smaller than the number of evolution increments in the conventional mode. This additive approach is preferred in situations where the intrinsic signal-to-noise is poor because each new projection enhances the final signal intensities, and the weak artefacts tend to disappear below the general noise level. Under these circumstances, it is better to enhance the signal-to-noise ratio by sampling the time-domain signals at more closely spaced angles of incidence, than to employ multiscan averaging.

In summary, provided that sensitivity is not a limiting factor, it is clear that multidimensional NMR spectroscopy can be sped up by linking the evolution periods together, rather than by varying them independently.

Speeding Up the Assignment Process

A crucial step in determining solution structures of proteins using NMR spectroscopy is the process of sequential assignment, which correlates backbone resonances to corresponding residues in the primary sequence of the protein. It has been recognized for some time that many of the interactive tasks carried out by an expert during the process of spectral analysis, could be carried out more efficiently and rapidly by computational systems. With the advent of triple-resonance and multidimensional strategies for determining resonance assignments and 3D structures, it became increasingly clear that the quality and information content of protein NMR spectra could allow largely automated analyses of assignments.

Successful automated spectral analysis is normally preceded by high-quality (manual or automatic) peak picking. Several high-quality peak-picking programs have been described over the last few years. For example, AUTOPSY [14] is a comprehensive program for automating peak picking with facilities for determining noise level, segmenting the spectra into peak-containing regions, identifying well-separated peaks, resolving spectral overlap and integrating peaks. In this context, SAMPA [15] is the only program attempting to automatically assign proteins in the absence of peak picking. As input, the module uses processed 2D and/or 3D spectra from XWIN-NMR (Bruker Biospin GmbH) and the amino acid sequence of the investigated protein. The result is a sequential assignment of the backbone atoms.

The algorithm is based on the consecutive comparison and analysis of 2D and 3D NMR-spectra whereby 2D spectra of selective experiments and 3D spectra of triple resonance experiments may be included. The program uses basic principles of pattern recognition theory and numerical recipes of mathematical analysis. The main idea is the search for characteristic patterns in different 2D and 3D spectra. A pattern is derived from the relationship of the amino acid type itself and its neighbors in the spectra.

A considerable amount of effort has been devoted in the last few years towards the automated analysis of resonance assignments [16–21] using triple-resonance NMR data. As described in Moseley and Montelione [22], most automation programs use the same general analysis scheme which includes filtering of peaks and referencing resonances from different spectra, grouping resonances into spin systems, identifying the amino acid type of the spin system (typing), linking sequential spin systems into segments, and, finally, mapping the spin-system segments onto the primary sequence of the protein. Linking and mapping can be achieved by two main methods: deterministic best-first methods [18–20,23] that require a one-time, all-to-all, full comparison between every spin system, and optimization algorithms [17,21] that use a pseudo-energy function with simulated annealing to evaluate potential connections between nearest neighbor spin systems. Best-first approaches establish criteria to select the best candidate from several potential connectivities. Unfortunately, wrong choices made early in the assignment process can lead to solutions which can be substantially incorrect [20]. On the other hand, the optimization algorithms are generally rather slow and can be susceptible to becoming trapped in local minima that correspond to incorrect assignment configurations.

Although the development of automated approaches for sequential assignment has facilitated the analysis of triple-resonance experiments, the performance of these programs is usually less satisfactory for large proteins, especially in the cases of missing connectivity or severe chemical shift degeneracy. In addition, successful algorithms must be able to address missing peaks, extra peaks from impurity or multiple conformers, uncertainty in the amino acid type determinations, and broken connectivities due to the uncertainty of resonance positions caused by the distortion or overlap of peaks.

An interesting approach to solve these problems have been proposed by Coggings and Zhou with the introduction of PACES [24], an interactive program for Protein Sequential Assignment by Computer-Assisted Exhaustive Search. The program conducts an exhaustive search of all spin systems both for establishing sequential connectivities and then for assignment. By running the program iteratively with user intervention after each cycle, ambiguities can be eliminated efficiently and backbone resonances can be assigned rapidly. PACES establishes sequential assignment based on the sequential connectivity and residue type information derived from C^{α}, C^{β}, CO, and H^{α} triple-resonance data, or any subset of these data. Additional information derived from other experiments, such as residue types or NOESY constraints, can also be introduced as PACES input. PACES anchors the resonances of a spin system to its HSQC peak, and subsequently refers to spin systems by HSQC peak numbers. The identification of residue type information is done by comparing C^{α}, C^{β}, CO shifts to the statistical chemical shift distribution of each amino acid type derived from the BioMagResBank (BMRB). If the protein under study is perdeuterated, PACES can be directed to adjust its amino acid ranges accordingly [25,26]. The mapping process involves aligning a connectivity fragment at the beginning of the protein sequence, and then moving it down the sequence, one residue at a time. When the C-terminus of the protein is encountered, spin systems are looped back to align with the N-terminus of the protein, until every spin system in the fragment has been examined at every position of the protein sequence. If at any position, the list of possible amino acid types for a spin system includes the amino acid type of the paired residue in the sequence, that pairing is considered a match. If multiple sequentially connected spin-systems match, the algorithm identifies the resulting segment as a matching fragment. Short matching fragments are filtered out unless they contain a spin system with less than five possible amino acid types. The chemical shift data and additional constraints are provided by the user on a properly formatted Excel worksheet, in a text file or from XEASY peak lists. Results are also presented on a formatted Excel worksheet, with one column for each residue. In summary, PACES is based on the methods used by NMR spectroscopists to assign proteins manually, but taking advantage of the abilities of a computer to work through all of the possibilities for connecting peaks and mapping them to the protein sequence.

Another strategy has been proposed by Reed *et al.* [27] for the semi-automated assignment of PGK (394 amino acids) based on a simulated annealing Monte Carlo technique using the following experiments as input: 2D ^{15}N-TROSY, 3D TROSY ct-HNCA, ct-HN(CO)CA, HN(CA)CB, HN(COCA)CB, HNCO, and HN(CA)CO. The program attempts to reorder the spin systems and place them in the amino acid sequence of the protein, by minimising an energy function based on both matching up the chemical shift data for adjacent residues, and on characteristic chemical shift ranges for resonances in particular residue types. Carbonyl chemical shifts are only used at the end of the process to confirm the assignments. After the initial peak picking and one run of the assignment program, a 60–80% complete assignment was obtained for this protein. The final 20–40% of the assignment took up the majority of the assignment time. Improving the level of assignment to 94% and then to 95% involved two further periods of 5 days each.

Finally, Lagnmead and Donald have published a new algorithm [28] that performs Nuclear Vector Replacement (NVR) by Expectation/Maximization (EM) to compute assignment using a combination of experimentally measured residual dipolar couplings, chemical shifts (2D H^N-N HSQC), H^N-H^N NOEs and amide exchange rates to a given *a priori* whole-protein 3D structural model. The authors proposed that this algorithm could form the basis for Molecular Replacement by NMR because of its resemblance with the technique used in X-ray crystallography to solve the phase problem using a close or homologous model. NVR adopts a minimalist approach [29], demonstrating the large amount of information available in a few key spectra. So far, NVR have been only tested on models with both high sequence and structural homology. Consequently, the present form of the algorithm may best be applied to scenarios where a crystal structure of the same protein is available, as may be the case in an SAR by NMR study. Models with significantly less homology will likely have somewhat different networks of hydrogen bonds and NOEs, as well as different amide bond-vector orientations.

Automated Protein Structure Determination

Until very recently, NMR protein structure determination has remained a laborious undertaking that occupied a trained spectroscopist over several months for each new protein structure. It has been recognized that many of the time-consuming interactive steps carried out by an expert during the process of spectral analysis could be accomplished by automated computational approaches [22]. Today automated methods for NMR structure determination are playing a more prominent role and will most likely supersede the conventional manual approaches to solving three-dimensional protein structures in solution [30].

The assignment of NOESY cross peaks requires as a prerequisite a knowledge of the chemical shifts of the spins from which nuclear Overhauser effects (NOEs) are arising. In *de novo* three-dimensional structure determination of proteins in solution by NMR spectroscopy, the key conformational data are upper distance limits derived from NOEs [31–34]. In order to extract distance constraints from a NOESY spectrum, its cross peaks have to be assigned. The NOESY assignment is based on previously determined chemical shift values that result from the chemical shift assignment. Because of the limited accuracy of chemical shift values and peak positions many NOESY cross peaks cannot be attributed to a single unique spin pair, but have an ambiguous NOE assignment comprising multiple spin pairs. Once available, a preliminary three-dimensional structure may be used to resolve ambiguous NOE assignments. Obtaining a comprehensive set of distance constraints from a NOESY spectrum is thus by no means straightforward, but becomes an iterative process in which preliminary structures, calculated from limited numbers of distance constraints, serve to reduce the ambiguity of cross peak assignments. In addition to this problem of resonance and peak overlap, considerable difficulties may arise from spectral artefacts and noise, and from the absence of expected signals because of fast relaxation.

Automated procedures follow the same general scheme but do not require manual intervention during the assignment/structure calculation cycles. Two main obstacles have to be overcome by an automated approach starting without any prior knowledge of the structure. First, the number of cross peaks with unique assignment based on chemical shifts is, in general, not sufficient to define the fold of the protein. Therefore, the automated method must have the ability to make use also of NOESY cross peaks that cannot yet be assigned unambiguously. Second, the automated program must be able to cope with the erroneously picked or inaccurately positioned peaks and with the incompleteness of the chemical shift assignment of typical experiment data sets.

In NOAH [35,36], an automated NOESY assignment routine present in the programs DIANA [37] and DYANA [38], the multiple assignment problem is treated by temporarily ignoring cross peaks with too many assignment possibilities, and instead generating independent constraints for each of the assignment possibilities of the remaining low-ambiguity cross peaks. In order to reduce the impact of these incorrect constraints on the structure, an error-tolerant target function is used [35,36]. NOAH requires a high accuracy of the input chemical shifts and peak positions, and makes use of the fact that only a set of correct assignments can form a self-consistent network.

In the initial assignment cycle with NOAH, all peaks with one or two assignment possibilities are included into the structure calculation. In view of the large number of erroneous conformational constraints that are likely to be included at this stage, it seems non-trivial that the NOAH/DIANA approach may ultimately converge to the correct structure. The explanation is related to the fact that the structure calculation algorithm attempts to satisfy a maximum number of conformational constraints simultaneously. The correctly assigned constraints form a large subset of self-consistent constraints, whereas, in contrast, the erroneously assigned constraints are randomly distributed in space, generally contradicting each other. As a consequence, erroneously assigned constraints may distort the structure, but will not lead to a distinctly different protein fold. One must keep in mind that the elimination of erroneously assigned constraints through contradiction with correct constraints will in general be less efficient in regions of low NOE density than in the well-defined protein core. Another peculiarity of the randomly distributed erroneously assigned constraints is that they are more likely to be long-range than short-range or intra-residual. This contrasts with the overall constraint distribution of a correctly assigned NOESY spectrum, where more than 50% of all cross peaks are from short-range NOEs [39].

Another widely used automated NOESY assignment procedure is ARIA [40–43] that was originally interfaced with XPLOR [44] and later with the program CNS [45]. ARIA introduced many new concepts, most importantly the use of ambiguous distance constraints [46,47] for handling ambiguities in the initial, chemical shift-based NOESY cross peak assignments. Prior to the introduction of ambiguous distance constraints, in general only unambiguously assigned NOEs could be used as distance constraints in a structure calculation. Since the majority of NOEs cannot be assigned unambiguously from chemical shift information alone, this lack of a general way to directly include ambiguous data into the structure calculation considerably hampered the performance of automatic NOESY assignment algorithms.

When using ambiguous distance constraints, each NOESY cross peak is treated as the superposition of the signals from each of its multiple assignments, using relative weights proportional to the inverse sixth power of the corresponding inter-atomic distance. ARIA starts from lists of peaks and chemical shifts in the format of the common spectral analysis programs, and proceeds in cycles of NOE assignment and structure calculation. The experimental peak lists can in practice not be assumed to be completely free of errors, especially in the early stages of a structure determination or if they originate from automatic peak picking. In addition, if the chemical shift assignment is incomplete, even the most carefully prepared peak list will contain peaks that cannot be assigned correctly, namely those involving unassigned spins, because the ARIA algorithm does not attempt to extend or modify chemical shift assignments provided by the user. When building a three-dimensional structure from NOE data, most erroneous distance constraints will be inconsistent with each other and with the correct ones. The erroneous constraints can therefore be identified by analyzing the violations of constraints with respect to the bundle of three-dimensional structures from the previous cycle of calculation. The only drawback with this program could be obtaining a correct fold at the outset of a *de novo* structure determination because the powerful structure-based filters used for the elimination of erroneous cross peak assignments are not yet operational at that stage.

The CANDID algorithm [48] in the program CYANA [49] combines features from NOAH and ARIA, such as the use of three-dimensional structure-based filters and ambiguous distance constraints, with the new concepts of network-anchoring and constraint combination that further enable an efficient and reliable search for the correct fold in the initial cycle of *de novo* NMR structure determination. Network-anchoring exploits the observation that the correctly assigned constraints form a self-consistent subset in any network of distance constraints that is sufficiently dense for the determination of a protein 3D structure. On the other hand, the concept of constraint combination aims to reduce the effect of spurious distance constraints that may arise from misinterpretation of noise and spectral artefacts. It is used in the first two CANDID cycles, and consists of generating distance constraints with combined assignments from different, in general unrelated, cross peaks. Constraint combination is applied only to the long-range peaks because in case of error their effect on the global fold of a protein is more pronounced than that of erroneous short- and medium-range constraints.

The automated CANDID method proceeds in iterative cycles of ambiguous NOE assignment followed by structure calculation with the CYANA torsion angle dynamics algorithm. Between subsequent cycles, information is transferred exclusively through the intermediary three-dimensional structures, in that the molecular structure obtained in a given cycle is used to guide the NOE assignments in the following cycle. Otherwise, the same input data are used for all cycles, that is, the amino acid sequence of the protein, one or several chemical shift lists from sequence-specific resonance assignment, and one or several lists containing the positions and volumes of cross peaks in 2D, 3D, or 4D NOESY spectra.

A CANDID cycle starts by generating for each NOESY cross peak an initial assignment list containing the hydrogen atom pairs that could, based on the chemical shift lists, contribute to the peak. Subsequently, for each cross peak these initial assignments are weighted with respect to several criteria (agreement between the values of the chemical shift list and the peak positions, and compatibility, if available, with the three-dimensional structure from the previous cycle), and initial assignments with low overall scores discarded. For each cross peak, the retained assignments are interpreted in the form of an upper distance limit derived from the cross peak volume. Thus, a conventional distance constraint is obtained for cross peaks with a single retained assignment. Otherwise an ambiguous distance constraint is generated that embodies several assignments. The structure calculations typically comprise seven cycles. Since the precision of the structure determination improves with each subsequent cycle, the criteria for accepting assignments and distance constraints are tightened in more advanced cycles of the CANDID calculation. In the final CANDID cycle, an additional filtering step ensures that all NOEs have either unique assignments to a single pair of hydrogen atoms, or are eliminated from the input for the structure calculation. This allows for the direct use of the NOE assignments in subsequent refinement and analysis programs that do not handle ambiguous distance constraints.

A very different strategy has been proposed by Kuszweski *et al.* [50] with the development of PASD, a new highly error-tolerant Probabilistic Assignment algorithm for Automated Structure Determination. The PASD algorithm which has been incorporated into the molecular structure determination package XPLOR-NIH [51], is conceptually and philosophically different from previously implemented algorithms in that it has been expressly designed with the aim of ensuring that the results from successive iteration cycles are not biased by the global fold of structures calculated in the preceding cycles. No NOE assignment is ever removed permanently, and consequently the likelihood of becoming trapped in a false minimum of NOE assignment space is significantly reduced.

Finally, although it is universally assumed that a protein structure determination by NMR requires sequence-specific resonance assignments, the chemical shift assignment by itself has no biological relevance. It is required only as an intermediate step in the interpretation of the NMR spectra. Several attempts have been made to devise a strategy for NMR protein structure determination that circumvents the tedious chemical shift assignment step. There is an analogy between these approaches and the direct phasing methods in X-ray crystallography [52].

The underlying idea of assignment-free NMR structure calculation methods is to exploit the fact that NOESY spectra provide distance information even in the absence of any chemical shift assignments. This proton-proton distance information can be exploited to calculate a spatial proton distribution. Since there is no association with the covalent structure at this point, the protons of the protein are treated as a gas of unconnected particles. Provided that the emerging proton distribution is sufficiently clear, a model can then be built into the proton density in a manner analogous to X-ray crystallography in which the structural model is constructed into the electron density. The most recent approach to NMR structure determination without chemical shift assignment is the CLOUDS protocol [53,54]. For the first time, the feasibility of the method has been demonstrated using experimental data rather than simulated data sets.

Conclusion

This review has presented some of the new tools available to improve data acquisition, analysis, and protein structure calculation by NMR. Given its wide applicability, NMR ought to play a very important role, in biomedical research and in the development of drugs for novel targets, during the post-genomic era. From the drug discovery perspective, an environment where a fast turnover of structural information is required, NMR will be competitive, if it is able to increase the speed with which assignments and protein structures are generated. This is particularly true if we consider that the sequencing of the human genome suggests that there are between 600 and 3000 genes that code for potential drug targets for human diseases [55]. About 400 of these gene products are currently in the research portfolio of pharmaceutical companies, and marketed drugs target only 120 of them. Therefore, the identification and characterization of new targets is of high interest for pharmaceutical research. NMR can contribute to this effort by determining the solution structure of a potential new target, particularly for monomeric proteins under 40 kDa or oligomeric proteins under 60 kDa [56–58], and in the characterization of proteins that do not have a defined three-dimensional structure [59]. Furthermore, NMR can also be used downstream of the drug discovery process in the identification of new compounds with biological activity against particular targets by fragment-based screening approaches [60], and in the three-dimensional characterization of the ligand-target complexes.

References

1. Widmer H, Jahnke W. Cell. Mol. Life Sci. 2004;61(5):580.
2. Freeman R, Kupce E. J. Biomol. NMR. 2003;27(2):101.
3. Kim S, Szyperski T. J. Am. Chem. Soc. 2003;125(5):1385.
4. G. Hounsfield N. Br. J. Radiol. 1973;46(552):1016.
5. Kupce E, Freeman R. J. Biomol. NMR. 2003;27(4):383.
6. Kupce E, Freeman R. J. Am. Chem. Soc. 2004;126(20): 6429.
7. Bracewell RN. Aust. J. Phys. 1956;9:198.
8. Nagayama K, Bachmann P, Wuthrich K, Ernst RR. J. Magn. Reson. 1978;31:133.
9. Brutscher B, Morelle N, Cordier F, Marion DJ. J. Magn. Reson. B. 1995;109:338.
10. Ottiger M, Delaglio F, Bax A. J. Magn. Reson. 1998;131: 373.
11. Ding K, Gronenborn A. J. Magn. Reson. 2002;156:262.

12. Dean SR. The Radon Transform and Some of Its Applications. John Wiley and Sons: New York, 1992.
13. Kupce E, Freeman R. Concepts Magn. Reson. 2004;22A(1): 4.
14. Koradi R, Billeter M, Engeli M, Güntert P, Wüthrich K. J. Magn. Reson. 1998;135:288.
15. Labudde D, Leitner D. personal communication.
16. Bartels C, Güntert P, Billeter M, Wüthrich K. J. Comput. Chem. 1997;18:139.
17. Buchler NEG, Zuiderweg ERP, Wang H, Goldstein RA. J. Magn. Reson. 1997;125:34.
18. Li K-B, Sanctuary BC. J. Chem. Inform. Comput. Sci. 1997;37:467.
19. Lukin JA,Grove AP, Talukdar SN, Ho C. J. Biomol. NMR. 1997;9:151.
20. Zimmerman DE, Kulikowski CA, Huang Y, Feng W, Tashiro M, Shimotakahara S, Chien C-Y, Powers R, Montelione GT. J. Mol. Biol., 1997;269:592.
21. Leutner M, Gschwind RM, Liermann J, Schwarz C, Gemmecker G, Kessler H. J. Biomol. NMR, 1998;11:31.
22. Moseley HN, Montelione GT. Curr. Opin. Struct. Biol. 1999;9(5):635.
23. Szyperski T, Baneck B, Braun D, Glaser RW. J. Biomol. NMR, 1998;11:387.
24. Coggins BE, Zhou P. J. Biomol. NMR. 2003;26(2):93.
25. Venters RA, Farmer 2nd BT, Fierke CA, Spicer LD. J. Mol. Biol. 1996;264(5):1101.
26. Farmer 2nd BT, Venters RA. Biol. Magn. Reson. 1999;16:75.
27. Reed MA, Hounslow AM, Sze KH, Barsukov IG, Hosszu LL, Clarke AR, Craven CJ, Waltho JP. J. Mol. Biol. 330(5):1189 (2003).
28. Langmead CJ, Donald BR. J. Biomol. NMR. 2004;29(2):111.
29. Bailey-Kellogg C, Widge A, Kelley III JJ, Berardi M, Bushweller J, Donald B. J. Comput. Biol. 2000;7: 537.
30. Guntert P. Prog. Nucl. Mag. Res. 2003;43:105.
31. Solomon I. Phys. Rev. 1955;99:559.
32. Macura S, Ernst RR. Mol. Phys. 1980;41:95.
33. Kumar A, Ernst RR, Wüthrich K. Biochem. Biophys. Res. Commun. 1980;95:1.
34. Neuhaus D, Williamson MP. The Nuclear Overhauser Effect in Structural and Conformational Analysis, VCH, 1989.
35. Mumenthaler C, Güntert P, Braun W, Wüthrich K. J. Biomol. NMR. 1997;10:351.
36. Mumenthaler C, Braun W. J. Mol. Biol. 1995;254:465.
37. Güntert P, Braun W, Wüthrich K. J. Mol. Biol. 1991;217:517.
38. Güntert P, Mumenthaler C, Wüthrich K. J. Mol. Biol. 1997;273:283.
39. Wüthrich K. NMR of Proteins and Nucleic Acids, Wiley, 1986.
40. Nilges M, Macias MJ, O'Donoghue SI, Oschkinat H. J. Mol. Biol. 1997;269:408.
41. Nilges M, O'Donoghue SI. Prog. NMR Spectrosc. 1998;32:107.
42. Linge JP, O'Donoghue SI, Nilges M. Methods Enzymol. 2001;339:71.
43. Linge JP, Habeck M, Rieping W, Nilges M. Bioinformatics 2003;19:315.
44. Brünger AT. X-PLOR Version 3.1. A system for X-ray crystallography and NMR. Yale University Press, 1993.
45. Brünger AT, Adams PD, Clore GM, DeLano WL, Gros P, Grosse-Kunstleve RW,Jiang JS, Kuszewski J, Nilges M, Pannu NS, Read RJ, Rice LM, Simonson T, Warren GL. Acta Crystallogr. 1998;54:905.
46. Nilges M. Proteins. 1993;17:297.
47. Nilges M. J. Mol. Biol. 1995;245:645.
48. Herrmann T, Güntert P, Wüthrich K. J. Mol. Biol. 2002;319: 209.
49. CYANA version 1.0, www.guentert.com.
50. Kuszewski J, Schwieters CD, Garrett DS, Byrd RA, Tjandra N, Clore GM. J. Am. Chem. Soc. 2004;126(20):6258.
51. Schwieters CD, Kuszewski JJ, Tjandra N, Marius Clore G. J. Magn. Reson. 2003;160(1):65.
52. Drenth J. Principles of Protein X-ray Crystallography, Springer, 1994.
53. Grishaev A, Llinás M. Proc. Natl Acad. Sci. U.S.A. 2002;99: 6707.
54. Grishaev A, Llinás M. Proc. Natl Acad. Sci. U.S.A. 2002;99: 6713.
55. Hopkins AL, Groom CR. Nat. Rev. Drug Discov. 2002;1(9): 727.
56. Clore GM, Gronenborn AM. Curr. Opin. Chem. Biol. 1998;2(5):564.
57. Staunton D, Owen J, Campbell ID. Acc. Chem. Res. 2003;36(3):207.
58. Montelione GT, Zheng D, Huang YJ, Gunsalus KC, Szyperski T. Nat. Struct. Biol. 2000;7:982.
59. Wright PE, Dyson HJ. J. Mol. Biol. 1999;293(2):321.
60. Huth JR, Sun C. Comb. Chem. High Throughput Screen. 2002;5:631.

Transferred Cross-Correlated Relaxation: Application to Drug/Target Complexes

T. Carlomagno and C. Griesinger
Max Planck Institut for Biophysical Chemistry, 37077 Göttingen, Germany

Introduction

The NMR spectra of biomolecules without and especially with isotopic labeling provide a wealth of information about interatomic distances and angular geometries that can be used as conformational restraints in structure determination. In this review we will focus on the application of cross-correlated relaxation rates to obtain structural information on a small-molecule/protein complex that has pharmaceutical implications. Cross-correlated relaxation complements information derived from NOEs. The chosen complex: epothilone/tubulin could not be subjected to X-ray crystallography. Electron microscopy can only be performed on tubulin sheets induced with Zn ions, which show an assembly of the tubulin dimers that is not found under physiological conditions. A structure of epothilone when bound to tubulin has been determined by electron crystallography. However, the resolution is too low to deduce the conformation of epothilone in a unique way. Thus, NMR constitutes a unique tool for the investigation of this drug/protein complex to obtain a high-resolution structure of the drug molecule in the bound state.

Cross-Correlated Relaxation for the Measurement of Projection Angles between Tensors [1]

Angular information on molecules is not only encoded in through-bond scalar or through-space dipolar couplings, but can also be derived from measurements of relaxation rates. Relaxation is caused by the correlation of two tensorial interactions. If these interactions are different, then relaxation caused by the cross-correlation of the two interactions is called cross-correlated relaxation. The cross-correlated relaxation rate depends on the strengths of the tensorial interactions as well as on the projection angle between the two tensors. To be effective the time-dependence of the tensorial interactions must be similar. Such a situation exists, e.g. between dipolar interactions of two proximate spin pairs within well-structured parts of the molecule, which display the same time-dependence due to overall tumbling. The most important tensorial interactions are the dipolar coupling and the chemical shift anisotropy. The dipole tensor between two spins I_k and I_l (I_m and I_n) is axially symmetric and the main axis is collinear with the respective internuclear bond vectors.

The cross-correlated relaxation rate observed for double- or zero-quantum coherence involving I_k or I_l and I_m or I_n for two dipolar interactions takes the following form:

$$\Gamma^c_{kl,mn} = \frac{\gamma_k\gamma_l}{(r_{kl})^3}\frac{\gamma_m\gamma_n}{(r_{mn})^3}\left(\frac{\mu_0}{4\pi}\hbar\right)^2\frac{1}{5}(3\cos^2\theta_{kl,mn}-1)\times\tau_c \tag{1}$$

The characterization of the angular dependence of the interaction of two nuclear dipoles I_kI_l and I_mI_n is therefore straightforward, namely it depends on the projection angle of the two bonds between I_k and I_l and between I_m and I_n. As can be seen from the equation for the cross-correlated relaxation rates, they depend on the correlation time of the molecule. Therefore, cross-correlated relaxation rates will increase with increasing size of the biomolecule under investigation as opposed to J coupling or dipolar coupling constants whose size does not depend on the molecular weight of the molecule. A second application lies in the fact that in analogy to the transferred NOESY experiment, in which distances are measured for ligands when bound to large target molecules, a transferred cross-correlated relaxation (trCCR) experiment allows to measure projection angles of weakly bound ligands when bound to large molecules.

Effects of Cross-Correlated Relaxation on Spectra

In two pairs of nuclei I_kI_l and I_mI_n, projection angle dependent cross-correlated relaxation rates $\Gamma^c_{kl,mn}$ due to two dipolar couplings of double- and zero-quantum coherences between nuclei I_k and I_m leads to differential relaxation of the four submultiplet lines as indicated in Figure 1. The cross-correlated relaxation rate leads to

Graham A. Webb (ed.), Modern Magnetic Resonance, 1229–1236.

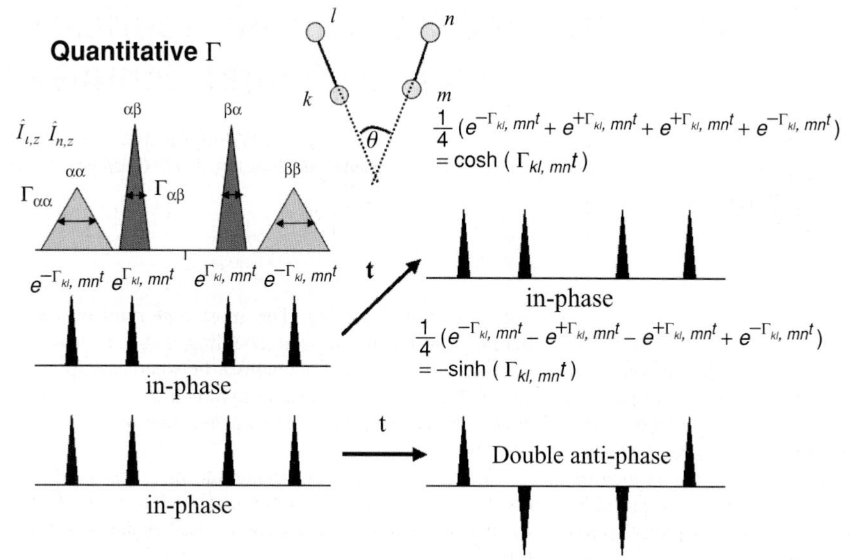

Fig. 1. Differential relaxation of the four submultiplet lines of a double-quantum coherence. The splitting originates from the J-couplings, while the line broadening originates from the cross-correlation of the two dipole–dipole couplings between k and l, as well as m and n, respectively. The differential relaxation can be expressed by the indicated product operators with evolutions that are sums and differences of exponential decays amounting to cosh and sinh dependencies.

additional line broadening of the outer lines ($\alpha\alpha$ and $\beta\beta$ for the double-quantum coherence and $\alpha\beta$ and $\beta\alpha$ for the zero-quantum coherence), while the inner lines ($\alpha\beta$ and $\alpha\beta$ for the double-quantum coherence and $\alpha\alpha$ and $\beta\beta$ for the zero-quantum coherence) narrow by the same amount. This statement is true if the cross-correlated relaxation rate is positive ($3\cos^2\theta_{kl,mn} - 1 > 0$) and thus the angle θ is between 0 and the magic angle θ^{magic} or between 180° and 180° $- \theta^{\mathrm{magic}}$. In case that θ lies in the remaining region ($3\cos^2\theta_{kl,mn} - 1 < 0$) the outer lines become narrower and the inner lines become broader. The case for positive cross-correlated relaxation rate $\Gamma^c_{kl,mn}$ is represented in Figure 1.

Often, line width measurements suffer from inaccuracies, especially since there are many contributions to the line width. Therefore, it is desirable to perform product operator transformations that create operators exclusively based on cross-correlated relaxation that otherwise could not be created. For that, we write down the product operator transformations that are effected by dipole/dipole cross-correlated relaxation. As pictorially demonstrated in Figure 1, we then find [2]:

$$I_{k,x}I_{m,y} \xrightarrow{\Gamma^c_{kl,mn}t} I_{k,x}I_{m,y}\cosh\Gamma^c_{kl,mn}t - 4I_{k,x}I_{m,y}I_{l,z}I_{n,z} \sinh\Gamma^c_{kl,mn}t \qquad (2)$$

Similarly we obtain for the evolution of an antiphase operator:

$$2I_{k,x}I_{m,y}I_{l,z} \xrightarrow{\Gamma^c_{kl,mn}t} 2I_{k,x}I_{l,z}I_{m,y} \times \cosh\Gamma^c_{kl,mn}t - 2I_{k,y}I_{m,x}I_{n,z} \times \sinh\Gamma^c_{kl,mn}t \qquad (3)$$

Obviously, operators are formed due to the evolution of cross-correlated relaxation that would not be formed without. The trick to perform now quantitative Γ experiments lies in the observation that the simultaneous evolution of the J_{kl} and J_{mn} couplings will lead to the same transformation as in Equation (3) but with known transfer amplitude provided the size of the coupling constants is known:

$$2I_{k,x}I_{m,y}I_{n,z} \xrightarrow{J_{kl},J_{mn}} 2I_{k,x}I_{m,y}I_{l,z}\sin\pi J_{kl}t\sin\pi J_{mn}t - 2I_{k,y}I_{m,x}I_{n,z}\cos\pi J_{kl}t\cos\pi J_{mn}t + \text{additional terms} \qquad (4)$$

Thus, two experiments can be performed, one that transforms the operator $2I_{k,x}I_{m,y}I_{l,z}$ into $2I_{k,y}I_{m,x}I_{n,z}$ with $\sinh\Gamma^c_{kl,mn}t$ in a so-called quantitative Γ cross experiment and a second experiment, the reference experiment, in which this transformation is effected by evolution of J

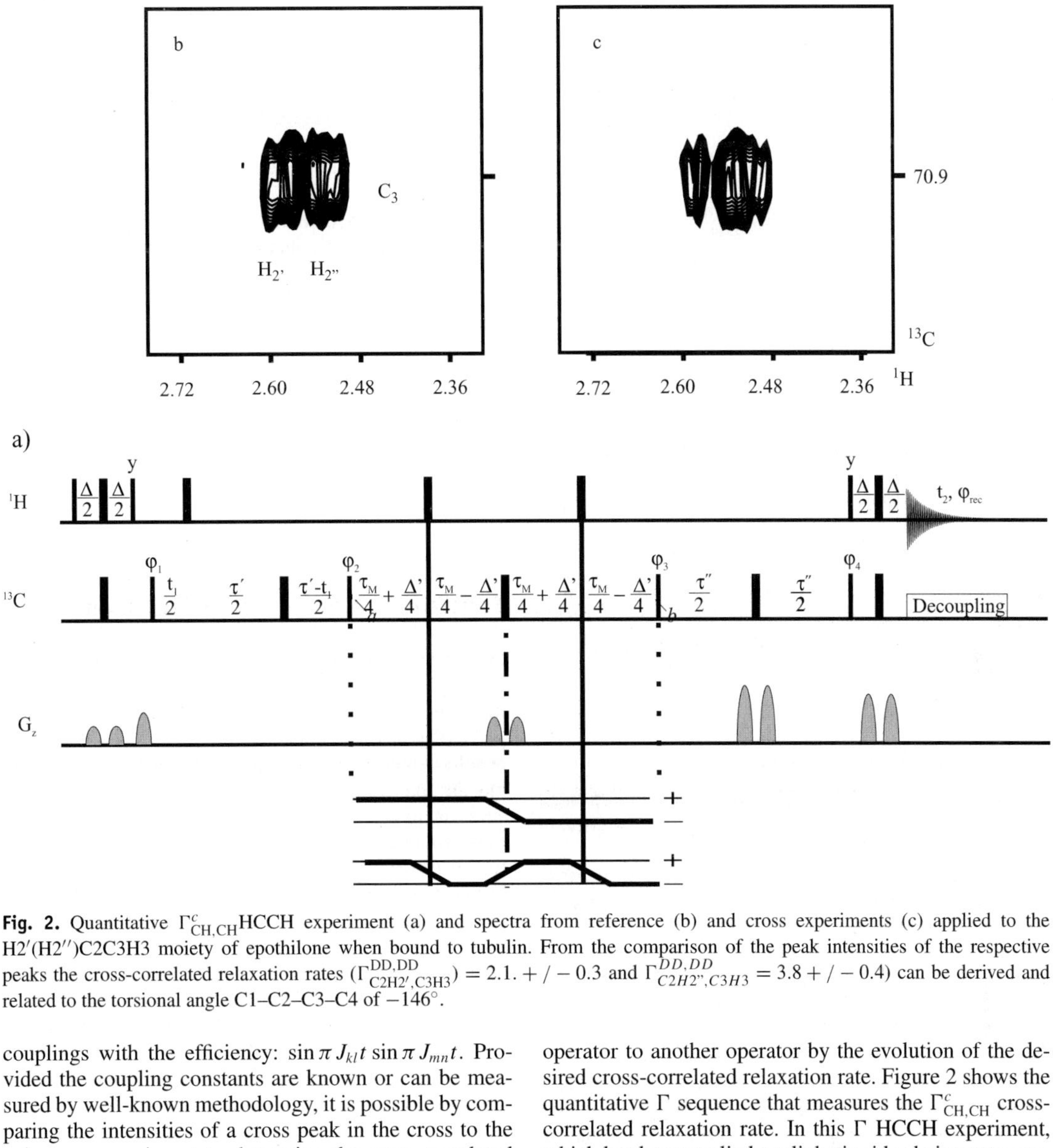

Fig. 2. Quantitative $\Gamma^{c}_{\mathrm{CH,CH}}$HCCH experiment (a) and spectra from reference (b) and cross experiments (c) applied to the H2′(H2″)C2C3H3 moiety of epothilone when bound to tubulin. From the comparison of the peak intensities of the respective peaks the cross-correlated relaxation rates ($\Gamma^{\mathrm{DD,DD}}_{\mathrm{C2H2',C3H3}}$) = 2.1. + / − 0.3 and $\Gamma^{DD,DD}_{C2H2'',C3H3}$ = 3.8 + / − 0.4) can be derived and related to the torsional angle C1–C2–C3–C4 of −146°.

couplings with the efficiency: $\sin \pi J_{kl}t \sin \pi J_{mn}t$. Provided the coupling constants are known or can be measured by well-known methodology, it is possible by comparing the intensities of a cross peak in the cross to the reference experiment to determine the cross-correlated relaxation rate.

Measurement and Application of the Quantitative Γ Method for Cross-Correlated Relaxation Rates

Quantitative Γ methods have been introduced originally by Felli *et al.* [3] and Pelupessy *et al.* [4] for dipole–dipole cross-correlated relaxation by transferring an initial operator to another operator by the evolution of the desired cross-correlated relaxation rate. Figure 2 shows the quantitative Γ sequence that measures the $\Gamma^{c}_{\mathrm{CH,CH}}$ cross-correlated relaxation rate. In this Γ HCCH experiment, which has been applied to aliphatic side chains or to sugars to measure the pseudorotation phase cross-correlated relaxation $\Gamma^{c}_{\mathrm{C}_i\mathrm{H}_i,\mathrm{C}_j\mathrm{H}_j}$ of the double- and zero-quantum coherence (DQ/ZQ) $4\mathrm{H}_{iz}\mathrm{C}_{ix}\mathrm{C}_{jy}$ generated at time point *a* creates the DQ/ZQ operator $4\mathrm{H}_{jz}\mathrm{C}_{jx}\mathrm{C}_{iy}$. In the second part of the experiment, the operator $4\mathrm{H}_{jz}\mathrm{C}_{jx}\mathrm{C}_{iy}$ is transferred via a 90° *y*-pulse applied to ^{13}C nuclei further to the proton H_j to give rise to a cross peak at $(\omega_{\mathrm{H}_j}, \omega_{\mathrm{C}_i})$. $\Gamma^{\mathrm{CSA,DD}}_{\mathrm{C}_i,\mathrm{C}_j\mathrm{H}_j}$ and $\Gamma^{\mathrm{CSA,DD}}_{\mathrm{H}_i,\mathrm{C}_j\mathrm{H}_j}$ cross-correlated relaxation are refocused by application of the 180° carbon and proton pulses during the mixing time τ_M. The following transfers are achieved in the sequence of Figure 2:

$$4H_{iz}C_{ix}C_{jy} \rightarrow 4H_{iz}C_{ix}C_{jy}\left[\cosh\left(\Gamma^c_{C_iH_i,C_jH_j}\tau_M\right)\right.$$
$$\times \cos^2\left(\pi J_{CH}\Delta'\right) - \sinh\left(\Gamma^c_{C_iH_i,C_jH_j}\tau_M\right)$$
$$\left.\times \sin^2\left(\pi J_{CH}\Delta'\right)\right] - 4H_{jz}C_{jx}C_{iy}$$
$$\times\left[\sinh\left(\Gamma^c_{C_iH_i,C_jH_j}\tau_M\right)\cos^2\left(\pi J_{CH}\Delta'\right)\right.$$
$$\left. + \cosh\left(\Gamma^c_{C_iH_i,C_jH_j}\tau_M\right)\sin^2\left(\pi J_{CH}\Delta'\right)\right] \tag{5}$$

The last term produces a cross peak at (ω_{C_i}, ω_{H_j}) due to coherence transfer between $4H_{iz}C_{ix}C_{jy}$ and $4H_{jz}C_{jx}C_{iy}$, while the first term produces a cross peak at (ω_{C_i}, ω_{H_i}) and can therefore be distinguished. In the experiment with $\Delta' = 0$, the intensity of the cross peak (I^{cross}) is proportional to $\sinh(\Gamma^c_{C_iH_i,C_jH_j}\tau_M)$ whereas for $\Delta' = 1/2J_{CH}$, the intensity of the cross peak (I^{ref}) is proportional to $\cosh(\Gamma^c_{C_iH_i,C_jH_j}\tau_M)$. By comparing the intensity of the cross peak measured in the two experiments one can determine:

$$\frac{I^{cross}}{I^{ref}} = \tanh\left(\Gamma^c_{C_iH_i,C_jH_j}\tau_M\right) \tag{6}$$

This is applied in Figure 2B and C to the torsion angle defined by C1–C2–C3–C4 of epothilone and an angle of −143° is obtained.

An extension of this method is the measurement of cross-correlated relaxation between a methyl group and a remote HC moiety in a $C^1H_3C^2C^3H$ fragment [5] which yields the torsion angle about the C^2C^3 bond. The pulse sequence used for the measurement of the $\Gamma_{CH,CH}$ rate between the C^3H vector and the C^1H_3 vectors is shown in Figure 3A. Two experiments are recorded, yielding a *cross* and a *reference* spectrum similarly as for the HCCH case except for the fact that the transfer requires one further transfer step and the double-quantum and zero-quantum coherence is excited between geminal instead of directly bound carbons. The CCR rate is extracted from the ratio of peak intensities in the *cross* and *reference* experiment. The pulse sequence is optimized for maximum sensitivity of the cross experiment: the $8H^1_zC^1_yC^2_zC^3_y$ coherence present at point *c* is transformed by $\Gamma^{DD}_{C^1H,C^3H}$ into $8C^1_xC^2_zC^3_xH^3_z$ with efficiency equal to $3\sinh(\Gamma^{DD}_{C^1H,C^3H}T_{rel})\cosh^2(\Gamma^{DD}_{C^1H,C^3H}T_{rel})$. For the reference experiment, the same operator transformation is achieved with the efficiency: $3\sin(\pi J_{C^1H}\Delta/2)\ \sin(\pi J_{C^3H}\Delta/2)$ $\cosh^2(\pi J_{C^1H}\Delta/2)\cosh^3(\Gamma^{DD}_{C^1H,C^3H}T_{rel})$. The last term stems from the fact that the cross-correlated relaxation is active also in the reference experiment. It is obvious that a comparison of the relative intensities based on the knowledge of the respective coupling constants yields the cross-correlated relaxation rate. The $8C^1_xC^2_zC^3_xH^3_z$ term is then transformed into H^3_x for detection. The delays are optimized for optimal refocusing and defocusing of the couplings. Selective π pulses on the carbons prevent loss of magnetization to other passively coupled carbons. The complete coherence transfers are given in the following:

$$\boxed{a}\,2H^1_zC^1_y \xrightarrow[\pi/2(C)]{^1J_{C^1C^2}} \boxed{b}\,4H^1_zC^1_yC^2_x \xrightarrow[\pi/2(C)]{^1J_{C^2C^3}} \boxed{c}\,8H^1_zC^1_yC^2_zC^3_y$$
$$\xrightarrow[^1J_{CH}(\text{reference})]{\text{CCR(cross)}} \boxed{d}\,8C^1_xC^2_zC^3_xH^3_z \xrightarrow{\pi/2(C)} 8C^1_zC^2_xC^3_zH^3_z$$
$$\xrightarrow[\pi/2(C)]{^1J_{C^1C^2}} \boxed{e}\,4C^2_zC^3_yH^3_z \xrightarrow{^1J_{C^2C^{23}}} \boxed{f}\,2C^3_xH^3_z \xrightarrow{\text{INEPT}} \boxed{g}\,H^3_x \tag{7}$$

Transferred Cross-Correlated Relaxation

Due to the fact that cross-correlated relaxation depends linearly on the correlation time it can be used to determine the conformation of ligands when bound to target molecules, provided the off rate is fast enough to enable detection of the cross-correlated relaxation rate via the free ligand [6, 7]. The conditions, under which such an experiment can be performed, are similar to those found for transferred NOEs [8] and for K_ds normally exceeding 10^{-6}M, ligands are in an equilibrium between a protein-bound and a free form. The period during which the molecule is in the bound conformation will contribute to relaxation to the largest extent and hence will heavily weight experiments that are based on relaxation. The population-averaged cross-correlated relaxation rate for dipole, dipole cross-correlated relaxation between two sites with fractions p_L of the free ligand and $p_{ML} = (1 - p_{ML})$ of the bound ligand is given in Equation (8):

$$\left\langle\Gamma^{DD,DD}_{C_mH_m,C_nH_n}\right\rangle = \frac{2}{5}\frac{\gamma^2_C\gamma^2_H\mu_0{}^2\hbar^2}{(4\pi)^2 r^3_{C_mH_m} r^3_{C_nH_n}}$$
$$\times\left[S^2_{mn,L}\,p_L\frac{3\cos^2\theta_{mn,L}-1}{2}\tau_{c,L}\right.$$
$$\left. + S^2_{mn,ML}\,p_{ML}\frac{3\cos^2\theta_{mn,ML}-1}{2}\tau_{c,ML}\right] \tag{8}$$

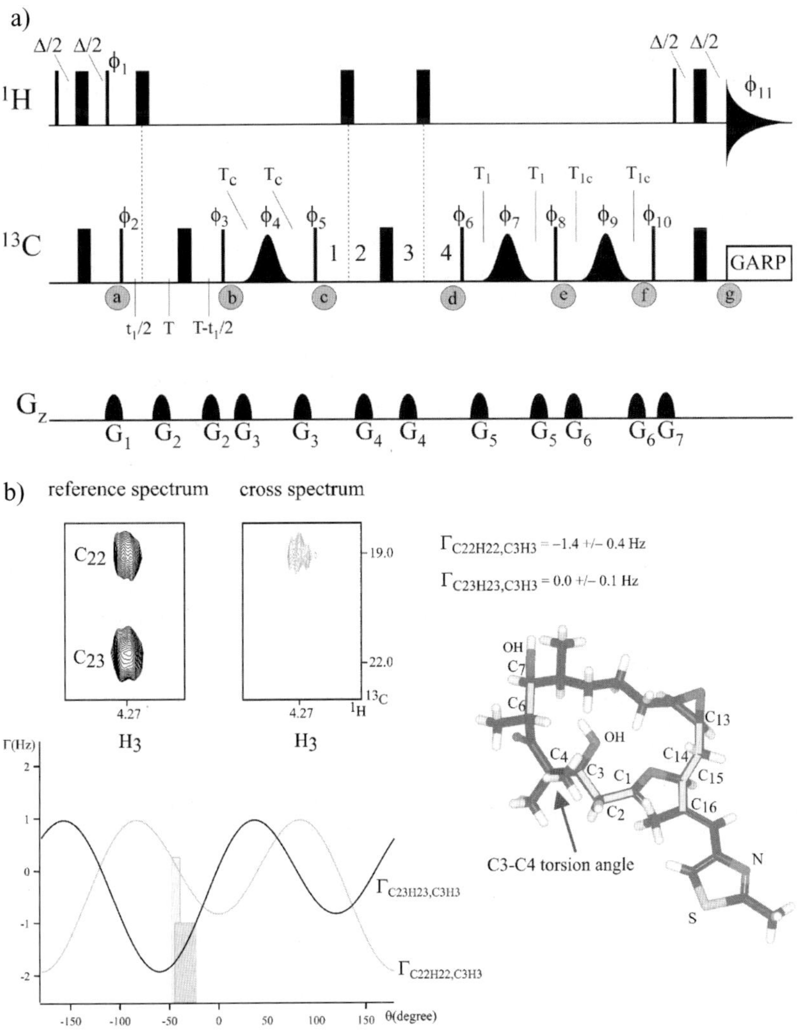

Fig. 3. (A) Pulse sequence for the measurement of dipolar–dipolar cross-correlated relaxation between the $(CH)^{22/23}_{ave.}$ vector and the $(CH)^3$ vector in epothilone. $\Delta = 1/(2^1 J_{CH})$; $T = 1/(4^1 J_{CC}) = 7$ ms; $T_c = 1/(4^1 J_{CC}) = 7$ ms; $T_1 = 1/(8^1 J_{CC}) = 3.5$ ms; $T_{1c} = T_c$. Delays 1–4 are equal to $T_{rel}/4$ ($T_{rel} = 1/^1 J_{CC} = 28$ ms) in the cross experiment; in the reference experiment delay 1 $= T_{rel}/4 + \Delta/8 =$ delay 3 and delay 2 $= T_{rel}/4 - \Delta/8 =$ delay 4. $\phi_1 = y$; $\phi_2 = x, -x$; $\phi_3 = 2(y), 2(-y)$; $\phi_5 = 4(x), 4(-x)$; $\phi_6 = 8(y), 8(-y)$; $\phi_8 = x$; $\phi_{10} = y.$; $\phi_{11} = 2(x,-x), 4(-x,x), 2(x,-x)$; $\phi_4 = \phi_7 = \phi_9 = x$. The pulses with phase ϕ_4 and ϕ_9 are Q3 pulses of 768 ms duration, centered at 63 ppm, to selectively invert the C^3 and C^4 (C^3 resonates at 70.9 ppm and C^4 at 53.0 ppm); the pulse with phase ϕ_7 is a Q3 pulse of 512 ms duration centered at 37 ppm, to selectively invert the $C^{22/23}$ and C^4 (C^{22} resonates at 19.1 ppm and C^{23} at 21.7 ppm). The proton carrier was at 4.78 ppm and the carbon carrier at 37 ppm; spectral widths were 4807.69 Hz for proton and 8333.33 Hz for carbon. Carbon decoupling in acquisition consisted of a GARP modulated pulse train at 2.38 kHz field strength. *Reference* (b) and *cross* (c) spectra acquired with the sequence in (a). The peaks visible in the reference spectrum are at the chemical shift of C^{22} (19.1 ppm) and C^{23} (21.7 ppm) in the carbon dimension and at the chemical shift of H^3 (4.27 ppm) in the proton dimension. The *cross* experiment was four times longer than the *reference* experiment. 1024 and 64 complex points were acquired in t_2 and t_1, respectively. The final matrices were 2048 × 128 points. The gray lines represent negative contours. The sample contained 0.5 mM epothilone A and 5 μM tubulin dissolved in D_2O. (D) Dependence of cross-correlated relaxation rates on the torsion angle about the C3–C4 bond. The two cross-correlated relaxation rates determine the angle to be −45°. (E) Numbering scheme of epothilone A and location of the torsion angles studied (indicated by bars drawn along the relevant bonds. (See also Plate 94 on page LXIV in the Color Plate Section.)

If the correlation time of the free ligand $\tau_{c,L}$ is much smaller than the correlation time of the bound ligand $\tau_{c,ML}$, the observed averaged cross-correlated relaxation rate $\langle \Gamma^{DD,DD}_{C_m H_m, C_n H_n} \rangle$ is dominated by the bound conformation and therefore, precise information on the conformation of a bound ligand can be obtained. This concept has been introduced on the determination of the conformation of a phosphotyrosine peptide weakly bound to STAT-6 [7] []and of a tRNA analog bound to EF-Tu· GDP complex [6]. Here we will show the power of the method on the determination of the epothilone conformation in the tubulin-bound state.

Application to the Epothilone/Tubulin Complex

In this review, we describe the application of trCCR measurements for the determination of the conformation of epothilone when bound to tubulin. Epothilones are microtubule stabilizers and inhibit cell proliferation through a mechanism of action which is analagous to that of the renowned clinical anti-cancer drug Taxol® (paclitaxel).[9] Epothilones exhibit extraordinary antiproliferative activity in vitro and they efficiently induce cell death in paclitaxel-resistant tumor cell lines at up to 5000-fold lower concentrations than paclitaxel [9–11]. In addition, they possess higher water solubility than paclitaxel [10,12] which allows delivery *in vivo* with non cremaphor-containing formulation vehicles, thus eliminating formulation-based side effects [major hypersensitivity reactions (HSR)] [13] The potential clinical utility of epothilones is supported by *in vivo* experiments with epothilone B in a variety of nude mouse human tumor models [10,14].

Fortunately, the kinetic and thermodynamic properties of the complex between epothilone A and tubulin are in the desirable range for trNOE and trCCR experiments. Indeed, the trNOE and trCCR data are found in agreement with a dissociation constant K_d in the range of 10–100 μM. Evidence of *specific* binding of epothilone A to tubulin is provided by the restoration of the NOE spectrum of free epothilone A upon addition of epothilone B to the mixture, which proves the quantitative displacement of epothilone A from tubulin by the tighter binder epothilone B (not shown). The existence of specific and transient binding of epothilone A to tubulin enables the structural analysis of the active conformation of the epothilones by NMR.

The tubulin-bound conformation of epothilone A was calculated from 46 inter-proton distance restraints and seven torsion angle restraints measured for a 0.5 mM solution of epothilone A in water in the presence of 5 μM tubulin. The distance restraints were derived from transferred NOE experiments. To filter out spin-diffusion mediated peaks, only those signals were taken into account which were present with opposite sign to the diagonal peaks in a transferred ROESY experiment. The dihedral angle restraints were obtained by measuring CH–CH dipolar–dipolar and CH–CO dipolar-CSA transferred CCR rates for 60–70% ^{13}C labeled epothilone A. The trCCR experiments [6,7] were indispensable to obtain a unique description of the bound conformation, as more than one structure of the macrolide ring is compatible with the same H–H distance set (NOE intensities). Seven trCCR rates defined the torsion angles C1–C2–C3–C4, C5–C6–C7–C8, C12–C13–C14–C15, C13–C14–C15–O11, and C14–C15–C16–C17 were all determined using the quantitative Γ HCCH. The C2–C3–C4–C5 was determined from the ratio of the cross-correlated relaxation rates between the $C^{22}H_3$ and $C^{23}H_3$, respectively, with the C^3H dipole. The O11–C1–C2–C3 was determined from CSA/DD trCCR (not described here) [15]. Figure 3B and C shows a representation of the methyl group experiment with the cross and reference experiments, respectively. In the cross spectrum the $C^{23}(\omega 1)$–$H^3(\omega 2)$ peak is missing, indicating that the dipolar–dipolar CCR between the $(CH)^{23}_{ave.}$ vector and the $(CH)^3$ vector is close to zero. The quantitative value for the two CCR rates can be extracted according to the equation:

$$I_{cross}/I_{ref} = \tanh(\Gamma T_{rel})/\sin^2(\pi J_{CH}\Delta/2)\cos^2 \times (\pi J_{CH}\Delta/2) \qquad (9)$$

where I_{cross} and I_{ref} are the intensities of the peaks in the cross and reference experiment, respectively and Γ indicates the cross-correlated relaxation rate. The measured CCR rates between the $(CH)^3$ vector and the $(CH)^{23}_{ave.}$ and $(CH)^{22}_{ave.}$ vectors are 0.1 ± 0.1 and -1.4 ± 0.4, respectively. The dihedral angle θ that complies with the ratio between the two CCR rates is $-45° \pm 5°$. We used the ratio of the two CCR rates, instead of the rates themselves, to determine the dihedral angle because this approach is independent from both the τ_c and the internal motion order parameter S^2, if equal internal reorientation of all the CH vectors is assumed. The change in the θ dihedral angle from the gauche + conformation in the free form to the gauche—conformation in the bound form is corroborated by transfer-NOE data.

The tubulin-bound conformation of epothilone A [15] is shown in green in Figure 4 and is compared with the free (unbound) conformation of epothilone A determined by X-ray crystallography, [15,16] which is shown in gray. We choose to compare the tubulin-bound conformation of epothilone with the X-ray structure and not with the solution structure available in CD_2Cl_2, because of the extensive flexibility of epothilone in solution in absence of tubulin [17]. However, the most populated conformer in

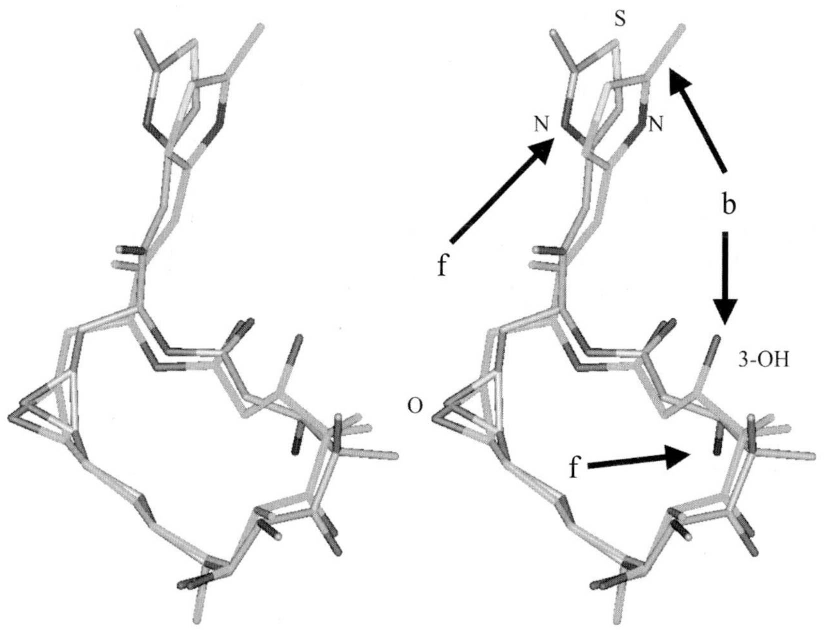

Fig. 4. Comparison of the conformations of epothilone as a stereoview. Arrows indicate free epothilone A (f: X-ray) and epothilone A bound to tubulin (B). Note the inversion of the orientation of the thiazole ring as well as the change in position of the 3-OH group. (See also Plate 95 on page LXIV in the Color Plate Section.)

solution is indeed very similar to the X-ray conformer. The position of the thiazole nitrogen (blue in Figure 4) and 3-OH (red in Figure 4), which are important for the delineation of a pharmocophore model, change significantly upon binding.

The structure of epothilone as derived from transferred NOE and cross-correlated relaxation is not identical with the structure derived from an electron crystallography study on Zn induced tubulin sheets [18]. The difference may be due to differences between the conformation of tubulin when assembled in Zn induced flat sheets as opposed to soluble tubulin or the low resolution of the electron crystallography that required extensive modeling of the epothilone conformation.

Conclusion

In conclusion, we have shown that transfer NOE and tr-CCR rates can be used for the determination of the conformation of epothilone when bound to tubulin in physiological conditions. So far, it has not been possible to obtain crystal structures of tubulin or microtubules. The structure differs from the one proposed from electron crystallography on Zn induced tubulin sheets which could be due to differences in the preparation of the samples under investigation or due to the low resolution of the electron crystallography structure.

References

1. Reif B, Hennig M, Griesinger C. Science. 1997;276:1230–3; Reif B, Diener A, Hennig M, Maurer M, Griesinger C. J. Magn. Reson. 2000;143:45–68.
2. Schwalbe H, Carlomagno T, Junker J, Hennig M, Reif B, Richter C, Griesinger C. Methods Enzymol. 2001;338:35–81.
3. Felli IC, Richter C, Griesinger C, Schwalbe H J. Am. Chem. Soc. 1999;121:56–7.
4. Pelupessy P, Chiarparin E, Ghose R, Bodenhausen G. J. Biomol. NMR. 1999;13:375–80.
5. Carlomagno T, Sánchez V, Blommers MJJ, Griesinger C. Angew. Chem. 2003;115:2619–21; Angew. Chem. Int. Ed. 2003;42:2515–7
6. Carlomagno T, Felli IC, Czech M, Fischer R, Sprinzl M, Griesinger C. J. Am. Chem. Soc. 1999;121:1945–8.
7. Blommers MJJ, Stark W, Jones CE, Head D, Owen CE, Jahnke W. J. Am. Chem. Soc. 1999;121:1949–53.
8. Clore GM, Gronenborn AM. J. Magn. Reson. 1982;48:402–17; Clore GM, Gronenborn AM. J. Magn. Reson. 1983;53:423–42; Ni F. Prog. NMR Spectrosc. 1994;26:517; Lian LY,

Barsukov IL, Sutcliffe MJ, Sze KH , Roberts GCK. Methods Enzymol. 1994;239:657–707
9. Bollag DM, McQueney PA, Zhu J, Hensens O, Koupal L, Liesch J, Goetz M, Lazarides E, Woods CM. Cancer Res. 1995, 55, 2325–33.
10. Altmann K-H, Wartmann M, O'Reilly T. BBA-Rev. Cancer 2000;1470:M79–91.
11. Kowalski J, Giannakakou P, Hamel E. J. Biol. Chem. 1997;272:2534–41.
12. Höfle G, Bedorf N, Steinmetz H, Schomburg D, Gerth K, Reichenbach H. Angew. Chem. 1996;108:1671–2; Angew. Chem. Int. Ed. 1996;35:1567–9.
13. Rowinsky EK. Ann. Rev. Med. 1997;48:353–74.
14. Altmann K-H, Bold G, Caravatti G, End N, Florsheimer A, Guagnano V, O'Reilly T, Wartmann M, Chimia 2000;54:612–21.
15. Carlomagno T, Blommers MJJ, Meiler J, Jahnke W, Schupp T, Petersen F, Schinzer D, Altmann K-H, Griesinger C. Angew. Chem. 2003;115:2615–9, Angew. Chem. Int. Ed. 2003;42:2511–5.
16. Rihs G, Walter HR. unpublished results.
17. Taylor RE, Zajicek J. J. Org. Chem. 1999;64:7224–8.
18. Nettles JH, Li H, Cornett B, Krahn JM, Snyder JP, Downing KH. Science 2004;305:866–69

Novel Uses of Paramagnets to Solve Complex Protein Structures

R. Andrew Byrd, C. Andrew Fowler, Robert L. McFeeters, and Vadim Gaponenko
Structural Biophysics Laboratory; Center for Cancer Research; National Cancer Institute; Frederick, MD 21702-1201, USA

Abbreviations: PCS, pseudocontact shift; RDC, residual dipolar coupling; pmiRDC, paramagnetic induced residual dipolar coupling; PRE, paramagnetic relaxation enhancement; PAS, paramagnetic anisotropic susceptibility.

Introduction

NMR is a critical tool in current structural biology, and plays a significant role in the understanding of pharmacologically significant proteins, as well as the development of modulators for these proteins. The general approach for structure determination has been well established; however, as the systems of interest become larger (e.g. >25 kDa) and less well behaved in solution (e.g. solubility limits of 50–200 μM), new approaches are required to enable NMR to remain a key player. The introduction of cryogenic probes has dramatically impacted our ability to study lower concentration and/or lower stability proteins. Nevertheless, there has been a need to develop new approaches to provide long-range structural restraints to solve structures of highly deuterated proteins and to determine the orientation of domains in multidomain proteins, or in multimeric complexes. This article will focus on our recent efforts in this area of research, specifically the introduction of paramagnetic metal ions into non-metalloproteins and the unique and powerful structural restraints that can be obtained from this approach.

The ability of paramagnetic centers to provide valuable structural information for proteins has been recognized for a long time [1,2]. The principal significance of this information is the long-range distance (10–40 Å) or global (distance and orientation) nature of the restraints. By augmenting traditional short-range, scalar restraints (such as NOEs) with long-range restraints, more accurate protein structures can be obtained. The relative proximity and/or orientation of protein domains or proteins within a complex can also be investigated. There are three main types of data that are obtainable: paramagnetic relaxation enhancement (PRE), pseudocontact shifts (PCSs), and residual dipolar couplings (RDCs).

Paramagnetic relaxation enhancements provide long-range distance information [3–6]; however, while relaxation enhancement is a powerful tool, we will focus our discussion here on the use of PCSs and RDCs. Paramagnetic induced residual dipolar couplings (pmiRDCs) and cross-correlation rates between dipolar and Curie relaxation pathways offer unique ways for obtaining orientational information on bond vectors in protein molecules [7–10]. PCSs provide both distance and orientational information that is very important for protein structure determination [11–14]. However, until recently, observation of these valuable structural parameters was limited to naturally metal binding proteins. Investigators have now begun to examine ways to introduce paramagnetic centers into non-metal binding proteins. We will illustrate two different approaches to achieve paramagnetic tagging and discuss new methods that enhance this general approach by providing more precise measurements of RDCs. Finally, we will demonstrate the impact of PCSs and pmiRDCs on solving difficult protein structures (>25 kDa) and on defining domain orientation in two pharmacologically significant proteins, STAT4 and hepatocyte growth factor (HGF).

Methods to Bind Paramagnets to Non-Metalloproteins

One way to introduce paramagnetic probes into non-metalloproteins is through chemical modification that introduces a metal-chelating site. A facile modification site is either a unique, natural cysteine residue or an engineered cysteine residue within the protein of interest. Two early examples of this type of protein modification with paramagnetic agents involved the attachment of nitroxide spin labels to lysozyme and bovine pancreatic trypsin inhibitor [15,16], allowing the measurement of relaxation enhancements. One limitation of this approach is evident for proteins with multiple, structurally important cysteine residues, wherein complications due to cross-linking, aggregation or refolding of the unmodified protein may occur. In this case, alternative approaches to attachment of the metal binding ligand can be utilized (*vide infra*). More recently, EDTA based ligands were successfully employed for paramagnetic labeling of macromolecules. Proteins modified with

Graham A. Webb (ed.), Modern Magnetic Resonance, 1237–1243.

S-(2-pyridylthio)-cysteaminyl-EDTA have been shown to bind Co^{2+} and allow measurement of PCSs and pmiRDCs. This approach has been demonstrated on barnase and the N-terminal domain of STAT4 [17,18]. Other paramagnetic ions, e.g. lanthanides, can be bound to the EDTA ligands and yield a wealth of PCS and pmiRDC data. It has been shown that under some conditions the EDTA-metal complex can exist in two forms, believed to be stereoisomers. The two forms result in two different paramagnetic ansisotropic susceptibility (PAS) tensors and two sets of shifted resonances, as observed for two proteins, trigger factor and ArgN [19,20]. Further inspection indicates that the multiple forms may be due to the protonation state of the EDTA moiety at the time of complexation (Gaponenko *et al.*, to be published). Additionally, a different cysteine-ligation paramagnetic tagging reagent has been proposed [21]; however, multiple isomers were observed and the reagent may prove problematic. To address the problem of multiple forms, a slightly modified and enantiometrically pure EDTA-derived tag has been proposed [19]. This new ligand is anticipated to facilitate broader use of the cysteine-ligation approach, and it is clear that further advances in ligand design will be forthcoming soon. A related approach employed cysteaminyl-EDTA-derivatized 4-thio-deoxythymidine incorporated into DNA that was subsequently used to form a protein-DNA complex. By binding Mn^{2+} to the EDTA moiety, long-range intermolecular distances were derived from PREs of protein resonances [22,23].

Since PCSs are dependent on the distance and orientation of nuclei from the paramagnet, it is likely that not all segments of a molecule will reveal PCS data using a single ligation site. However, it is straightforward to serially engineer multiple ligand attachment sites, thereby obtaining PCS data for all segments of secondary structure (Figure 1) [17]. The multiple ligand attachment sites also yield separate, distinct alignment tensors, relative to the molecular coordinate system. This is particularly attractive, since it is often difficult to reduce the degeneracy for RDC solutions with one alignment [24,25], while multiple alignments can be difficult to achieve due to protein interactions with the alignment media or correlation of the tensor orientations (C.A. Fowler *et al.*, in preparation).

A second method of achieving paramagnetic binding to proteins is through the use of chelating peptide tags. Several peptides that can be cloned at either terminus of an expressed protein have been reported to bind paramagnetic metals. ATCUN (amino terminal Cu^{2+}(Ni^{2+}) binding) motifs [26] with bound Cu^{2+} have been used to provide distance restraints by PRE in both ubiquitin [27] and several calmodulin-peptide complexes [28]. Another short peptide tag, HHP (His-His-Pro), has been cloned at the N-terminus of thioredoxin [29], allowing the measurement of PRE in the presence of Ni^{2+}. Longer peptides can also chelate metals. A 17 residue lanthanide binding peptide tag has been cloned at the N-terminus of ubiquitin and used to obtain partial orientation and observe pmiRDCs [30]. Finally, chelated metals can be bound using a combination approach. A 26 residue, calmodulin binding peptide was attached via a short linker to the C-terminus of dihydrofolate reductase (DHFR) and Tb^{3+}-saturated calmodulin was added and observed to bind to DHFR, transferring a degree of paramagnetic alignment and allowing the measurement of pmiRDCs [31].

We have proposed a simpler, common peptide tag that is used in immobilized metal affinity chromatography (IMAC) [32–35]. IMAC is a technique generally used to purify recombinant proteins expressed with a chelating peptide tag. The tag binds to partially chelated metal ions in the column resin, allowing contaminants to be washed away followed by elution of the purified protein. One of the most common tags used for IMAC is hexahistidine (the so called his-tag) [36,37], which can be added at either the N- or C-terminus of a protein. Since his-tags can bind resin-associated metal ions, it follows that they can also bind free metal ions in solution.

It has recently been shown that several his-tagged proteins are capable of binding free cobalt ions in solution. The paramagnetic anisotropy of the cobalt-his-tag complex allows measurement of PCSs and also leads to partial alignment of the sample in a magnetic field, allowing the observation of pmiRDCs. The PCSs and pmiRDCs measured from the protein-his_6-Co^{2+} species

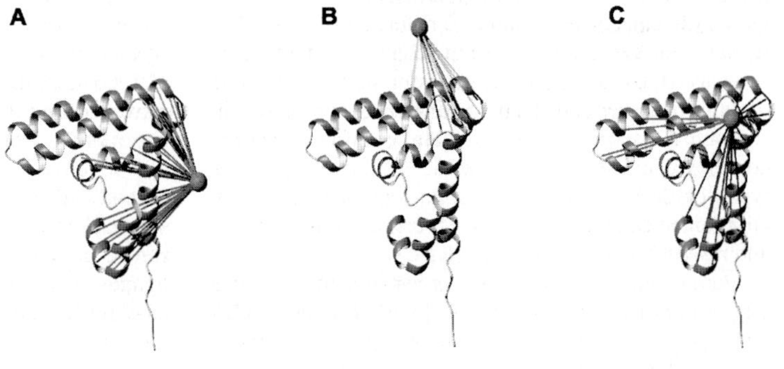

Fig. 1. Co^{2+} ligation at three distinct sites on $STAT4_{NT}$. (A) wt-C107, (B) T50C, and (C) K92C. The vectors illustrate the PCSs observed for H^N protons.

Table 1: Magnitudes of pmiRDCs observed for various protein-metal complexes

Protein and chelating tag	Metal or metal chelate	Largest magnitude ^{1}H-^{15}N RDC (Hz)
NK1-his$_6$	Co^{2+}	−3.0
Ubiquitin-his$_6$ (N-terminus)	Co^{2+}	−2.5
Ubiquitin-his$_6$ (C-terminus)	Co^{2+}	−0.7
Ubiquitin-LBP	Tb^{3+}	−7.6
Ubiquitin-LBP	Dy^{3+}	−6.6
Ubiquitin-LBP	Tm^{3+}	+4.5
DHFR-CBP	calmodulin(Tb^{3+})$_4$	−7.4
YmoA-his$_6$	Co^{2+}	+2.0
NusB-his$_6$	Co^{2+}-NTA	+2.0
STAT4-EDTA	Co^{2+}	−4.0
Trigger factor-EDTA-carbonic acid	Dy^{3+}	+8.0

LBP, lanthanide binding peptide; DHFR, dihydrofolate reductase; CBP, calmodulin binding peptide; NTA, nitrilotriacetic acid

have been shown to be relatively small; however, they are structurally significant and readily obtained (*vide infra*). Table 1 lists the magnitudes of observable pmiRDCs for several different chelating tags and conditions.

PCS Assignment and Use of PCSs and pmiRDCs as Structural Restraints

First, we will consider the PCSs. The presence of a metal ion introduces a PAS that results in the observation of PCSs for the NMR resonances. PCSs are measured as differences ($\Delta\delta_{pc}$) in chemical shift between the paramagnetic and diamagnetic species. Such differences are readily measured in the metal-free and metal-bound states. PREs can degrade the sensitivity of NMR experiments for sites close to the metal. For this reason, we have chosen to use Co^{2+}, which provides less relaxation enhancement than lanthanides or Mn^{2+}. Assignment of PCSs can be achieved using traditional NMR assignment strategies [38]. An alternative assignment strategy utilizes a preliminary structure of the macromolecule to estimate the PAS tensor orientation from several unambiguous PCSs [11], which have been assigned by inspection of HSQC spectra. The subsequent assignment of the remaining PCSs is accomplished by an iterative procedure of predicting the PCS values, using the PAS tensor orientation and the preliminary structure coordinates, and then re-optimizing the PAS based on new assignments. We have found that this procedure can assign all observable PCSs, including shifts up to 650 ppb, which were not previously assignable by inspection or other methods (Gaponenko *et al.* in preparation, 2005).

If the metal is chelated only by the EDTA or His$_6$ ligand, then there is no Fermi-contact contribution to $\Delta\delta_{pc}$, since there is no direct overlap of the molecular orbitals of the paramagnetic center with the protein atoms [2,12]. The pseudocontact shift is described by the following equation:

$$\Delta\delta_{pc} = \frac{P_{ax}}{r^3}\left(3\cos^2\theta - 1\right) + \frac{P_{rh}}{r^3}\sin^2\theta\cos(2\varphi). \quad (1)$$

where P_{ax} and P_{rh} are the axial and rhombic components of the PAS, r is the distance between the unpaired electron and the observed nucleus, and θ and ϕ are the polar angles describing the orientation of the electron-nucleus vector in the PAS tensor frame. PCSs follow the same functional form as RDCs and the structural restraints are very similar to homonuclear dipolar couplings [39]. However, the PCS restraints are more useful than homonuclear dipolar couplings because the sign information is readily available.

It is a straightforward procedure to incorporate PCS restraints into structure calculation programs, e.g. Xplor-NIH, that already have utilities for RDC refinement. The restraint is expressed in terms of the actual PCS and does not have an explicit input distance or orientation. Since many PCS restraints are included (for each tagged site) and combined with other NOE and torsion-angle restraints, a unique solution can be obtained. The distance of the atom, corresponding to the PCS, from the paramagnetic center is determined from the final structures. The corresponding distances range from 10 to 40 Å for the two cases described below.

The use of RDCs in protein structure determinations has become very significant and has been recently reviewed [40]. In the present application, partial alignment of the protein arises due to the paramagnetic susceptibility of the tagged protein. The alignment is known to be small [9]; nevertheless, the pmiRDCs are readily obtained without the need for steric alignment media, and have shown to be extremely valuable, both in terms of structure determination and domain orientation (*vide infra*). The pmiRDCs obtained from paramagnetic alignment are used in structure calculations exactly as has been described for steric alignment induced RDCs. They have also been useful for cross-validation.

New Approaches to Measurement of Small, Paramagnetically Induced RDCs

Paramagnetic alignment can induce relatively small RDCs (±2 to ±8 Hz for H^N-N), and it is important to have accurate methods to measure these couplings. Typical procedures for measuring RDCs involve the use of direct frequency measurements of resolved multiplet components, often augmented with the IPAP procedure [41]. These methods can suffer from measurement precision for small couplings. We have recently introduced a suite of experiments, based on the principal of J-modulation, that facilitate the measurement of H^N-N, H^α-C^α, C^α-C', and C^α-C^β RDCs [58]. The experiments encode the coupling of interest into peak modulation within familiar two-dimensional H^N-N (equivalent to HSQC spectra) or HN-C' correlation spectra.

Briefly, J-modulation can be incorporated into a pulse sequence in several different forms (see Figure 2), and the underlying principal remains that the coupling of interest is evolved for an amount of time, Δ, which can be varied. The observed resonance intensity is modulated by the coupling of interest in a sinusoidal manner dependent on Δ. Different types of J-modulation can be introduced into pulse sequences and result in different modulation functions of the coupling of interest. Simple, non-constant time J-modulation (Figure 2A) measures coupling for a total time 2Δ, but the intensity is a composite of J-modulation and I-spin relaxation, which complicates data analysis. Constant-time J-modulation incorporates the J-modulation delay into a constant time period, keeping relaxation effects constant and providing a means to decouple a third spin (Figure 2B). Chemical shift evolution can be reintroduced to constant time J-modulation rather easily (Figure 2C), thus expanding the repertoire of experiments that can be designed. Acquiring a series of spectra in which Δ is incremented allows the coupling of interest to be extracted from a plot of resonance intensity vs. total evolution time (Figure 2D). Examples of simple J-modulation [42–45], and constant-time J-modulation [46] have appeared, and we have extended these concepts and developed a suite of experiments to facilitate the rapid and precise measurement of a complete set of RDCs. With adequate signal-to-noise and a sufficient number of data, the precision of J-modulated couplings can exceed those of direct and E.COSY measurements. For a detailed discussion of these methods and the overall suite of experiments see McFeeters *et al.* [58].

In general, the precision of RDC measurement using the J-modulation experiments can be of the order of ±0.1 Hz and is significantly improved compared to IPAP methods or three-dimensional methods used to decrease spectral overlap. The convenience of spectral analysis for the familiar HSQC-like H^N-N 2D spectra facilitates a rapid, accurate measurement of multiple RDCs for samples with even weak alignment, such as found for paramagnetic induced alignment. These advances ensure that the maximum amount of structural data can be obtained from the paramagnet-tagged protein system.

Structural Applications of PCSs and pmiRDCs

We have addressed two difficult problems in protein structure determination: (1) improving structural accuracy for large, perdeuterated proteins, and (2) orientation of protein domains (or separate molecules) without the use of interdomain/intermolecular NOEs. Each of these problems reflects the use of NMR methods in larger, more biologically and pharmacologically significant proteins. The use of deuteration and selective methyl-protonation to enable structural studies of proteins ≥25 kDa is becoming common and has been reviewed [47].

Structural accuracy can be significantly improved by combining observed H^N (Figure 1) and methyl-^{1}H PCSs with the limited NOE-restraints available from an ILV-labeled protein to calculate the three-dimensional structure, as demonstrated for the N-terminal domain of STAT4 [17]. When PCSs were included in structure calculations the RMSD for backbone atoms in secondary structure elements between the solution structure and the crystal structure decreased from 2.9 to 2.0 Å, while the quality factor, Q [48], calculated from pmiRDCs *not used* in structure refinement decreased from 0.6 to 0.2. Furthermore, PCSs restraints for methyl groups of Ile, Leu, and Val afforded improvement in accuracy of the sidechain positions for these residues. The RMSD values for all heavy atoms in Ile, Leu, and Val sidechains relative to the crystal structure decreased from 2.97 to 2.43 Å, while the precision within the ensemble improved from 2.37 to 1.66 Å. Using the N-terminal domain of STAT4, we have also demonstrated that it is possible to break magnetic symmetry in homodimers, including the difficult case of head-to-head dimers, thus yielding unique structural information for each monomeric domain within the dimer [38] By binding subequimolar amounts of the paramagnet, only one monomer was tagged with paramagnet, and the PCSs to otherwise symmetric residues rendered the two sites magnetically inequivalent. The asymmetric system provided structural restraints, in the form of PCSs and pmiRDCs for each monomer, which enabled the

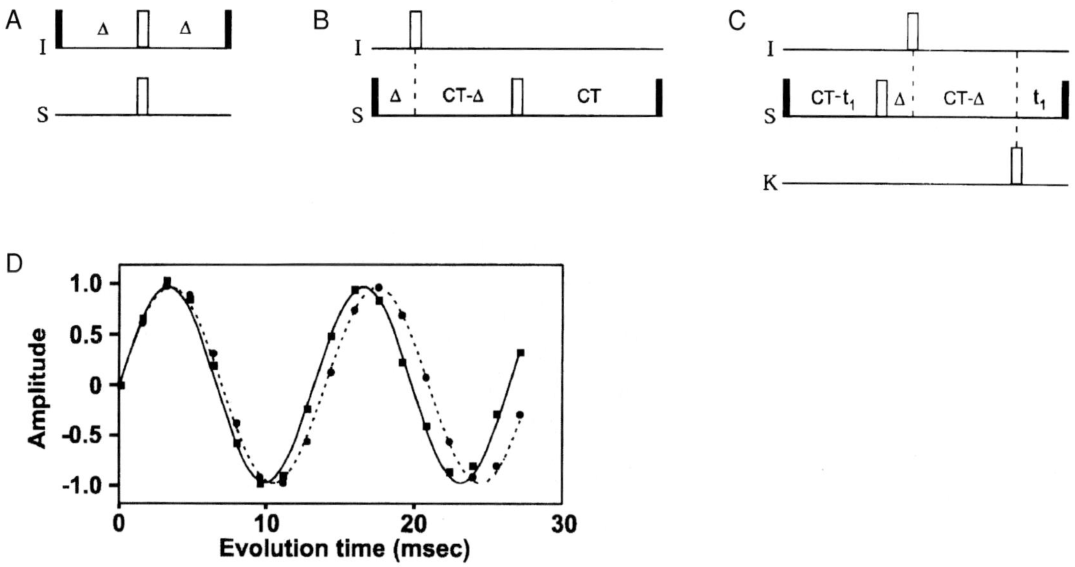

Fig. 2. Improved measurement of RDCs via J-modulation. (A) Basic pulse sequence element for non-constant time J-modulation. Intensity as a function of Δ, is a composite of both relaxation and modulation according to $\sin(2\pi J_{IS}\Delta)\bullet\exp(-R2\Delta)$. (B) Constant-time J-modulation removes relaxation damping resulting in modulation by $\sin(2\pi J_{IS}\Delta)$. (C) Frequency labeling can be reintroduced into constant-time J-modulation with no alteration in the form of modulation. (D). Typical modulation data, H^{α}-C^{α} couplings in an isotropic (solid line with squares) and aligned (dashed with circles) sample.

positioning of the two monomers with respect to one another (Gaponenko, unpublished). The improvements in structure calculations derive directly from the fact that PCSs, unlike one-bond RDCs, contain translational (distance) information in addition to orientational information. This fact heightens the value and future impact of using PCS restraints in general for protein structure determinations.

The second problem that we address is the ability to orient domains relative to one another within a multidomain protein or complex in the absence of traditional, NOE-based distance restraints, which is especially pertinent where deuteration is employed. This is a general form of the problem relating to the positioning of monomers within a homodimer structure (see above). The general approach assumes that any relatively rigid molecular fragments can be oriented relative to one another, provided enough data can be measured to determine the alignment tensor for each fragment. If the overall molecule is determined to behave as a single rigid unit, then each fragment must experience the same degree of alignment. Superimposing the alignment tensors allows the fragments to be correctly oriented with respect to one another. This technique has been successfully applied to fragments as small as a few residues [49,50], as well as to the problem of orienting entire protein domains. Recent examples of domain orientation using RDCs include the B and C domains of barley lectin [51], the domains of maltodextrin binding protein [52], and sequential domains in polyubiquitin chains [53–55]. The tensorial information may also be obtained from relaxation analyses [56,57]; however, we will focus on the use of RDCs and PCSs in this discussion.

Although useful, domain orientation using RDCs poses two significant challenges. The first of these is overcoming the well known "inversion problem" [25]. Although an alignment tensor can be determined using as few as five RDCs, it is impossible to determine which direction along each axis is positive and which is negative. This leads to four degenerate orientations, and additional independent data is required to overcome this inversion problem. A second problem is that while domains can be properly oriented, RDCs provide no translational (or distance) information and additional data must be used to position the two domains relative to one another.

We have recently combined PCSs with RDCs to orient the two domains in the NK1 fragment of HGF (C.A. Folwer *et al.*, in preparation, 2005). RDCs for the two domains were measured in phospholipid bicelles, yielding four possible domain orientations. Several additional alignment media were screened to find a second, independent alignment tensor; however, the protein either interacted with the media or the alignment tensor was

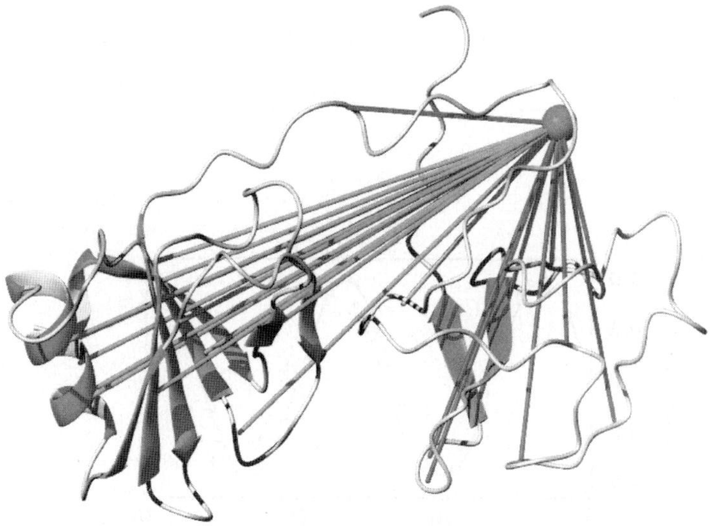

Fig. 3. Solution structure of NK1 refined against bicelle RDCs and cobalt PCSs (PCSs illustrated as orange and purple vectors to the N and K1 domains, respectively). (See also Plate 96 on page LXV in the Color Plate Section.)

not sufficiently different to solve the inversion problem. The protein was then re-cloned with a his-tag at the C-terminus, and shown to bind cobalt specifically to the his-tag. A saturating amount of cobalt was added to the sample, resulting in the observation of PCSs (±80 ppb) for amide protons in both domains. Furthermore, partial alignment of NK1 in the magnetic field, due to paramagnetic anisotropy, allowed the observation of H^N-N pmiRDCs on the order of ±3 Hz. The alignment tensor determined from the pmiRDCs was unique relative to the steric alignment tensors and was used to resolve the inversion problem, leading to a unique orientation of the two domains. The observed H^N PCSs for both domains were subsequently used as restraints and combined with RDCs measured from bicelle alignment, in a structure calculation to position the two domains of NK1 with respect to each other. The PCSs provided the requisite translational (distance) restraints to achieve a unique structural solution, and the significance is clearly seen in the long-range distances (10–40 Å) that correspond to the PCS restraints (Figure 3). The cobalt pmiRDCs were withheld from the structure calculations and subsequently used for cross-validation, which confirmed the structure. The final structure was obtained without the use of *any* intermolecular NOEs, and it is currently being analyzed to improve our understanding of the interaction of HGF with the cMet receptor.

Conclusion

We have summarized some of the exciting new applications of paramagnets in non-metalloprotein systems. The structures generated using the newly accessible restraints are of high quality and hold the potential for more accurate structures, access to larger structures, and more rapid structure determinations. While, to date, a limited number of applications of paramagnetic labeling have been described in the literature, it is already clear that this approach offers a wealth of valuable structural information. Improved tagging techniques and better software for data analysis should further enhance the advantages offered by this methodology for NMR structure determination and studies of molecular dynamics.

References

1. La Mar GN, Overkamp M, Sick H, Gersonde K. Biochem. 1978;17:325.
2. Bertini I, Luchinat C, Piccioli M. Meth. Enzymology. 2001;339:314.
3. Gillespie JR, Shortle D. J. Mol. Biol. 1997;268:170.
4. Gillespie JR, Shortle D. J. Mol. Biol. 1997;268:158.
5. Battiste JL, Wagner, G. Biochem. 2000;39:5355.
6. Gaponenko V, Howarth JW, Columbus L, Gasmi-Seabrook G, Yuan J, Hubbell WL, Rosevear PR. Prot. Sci. 2000;9:302.
7. Hus JC, Marion D, Blackledge M. J. Mol. Biol. 2000;298:927.
8. Ghose R, Prestegard JH. J. Magn. Reson. 1997;128:138.
9. Tolman JR, Flanagan JM, Kennedy MA, Prestegard JH. Proc. Natl. Acad. Sci. USA. 1995;92:9279.
10. Banci L, Bertini I, Huber JG, Luchinat C, Rosato A. J. Am. Chem. Soc. 1998;104:12903.
11. Allegrozzi M, Bertini I, Janik MBL, Lee YM, Liu G, Luchinat C. J. Am. Chem. Soc. 2000;122:4154.
12. Gochin M. J. Biomol. NMR. 1998;12:243.
13. La Mar GN, Del Gaudio J, Frye JS. Biochim. Biophys. Acta. 1977;498:422.

14. Ubbink M, Ejdeback M, Karlsson BG, Bendall DS. Structure. 1998;6:323.
15. Schmidt PG, Kuntz ID. Biochem. 1984;23:4261.
16. Kosen PA, Scheek RM, Naderi H, Basus VJ, Manogaran S, Schmidt PG, Oppenheimer NJ, Kuntz ID. Biochem. 1986;25:2356.
17. Gaponenko V, Sarma SP, Altieri AS, Horita DA, Li J, Byrd RA. J. Biomol. NMR. 2004;28:205.
18. Dvoretsky A, Gaponenko V, Rosevear PR. FEBS Lett. 2002;528:189.
19. Ikegami T, Verdier L, Sakhaii P, Grimme S, Pescatore B, Saxena K, Fiebig KM, Griesinger C. J. Biomol. NMR. 2004;29:339.
20. Pintacuda G, Moshref A, Leonchiks A, Sharipo A, Otting G. J. Biomol. NMR. 2004;29:351.
21. M. Prudencio, Rohovec J, Peters JA, Tocheva E, Boulanger MJ, Murphy ME, Hupkes HJ, Kosters W, Impagliazzo A, Ubbink M. Chemistry. 2004;10:3252.
22. Iwahara J, Schwieters CD, Clore GM. J. Am. Chem. Soc. 2004;126:5879.
23. Iwahara J, Anderson DE, Murphy EC, Clore GM. J. Am. Chem. Soc. 2003;125:6634.
24. Ramirez BE, Bax A. J. Am. Chem. Soc. 1998;120:9106.
25. Al-Hashimi HM, Valafar H, Terrell M, Zartler ER, Eidsness MK, Prestegard JH. J. Magn. Reson. 2000;143:402.
26. Harford C, Sarkar B. Acc. Chem. Res. 1997;30:123.
27. Donaldson LW, Skrynnikov NR, Choy WY, Muhandiram DR, Sarkar B, Forman-Kay JD, Kay LE. J. Am. Chem. Soc. 2001;123:9843.
28. Mal TK, Ikura M, Kay LE. J. Am. Chem. Soc. 2002;124:14002.
29. Jensen MR, Lauritzen C, Dahl SW, J. Pedersen, Led JJ. J. Biomol. NMR. 2004;29:175.
30. Wöhnert J, Franz KJ, Nitz M, Imperiali B, Schwalbe H. J. Am. Chem. Soc. 2003;125:13338.
31. Feeney J, Birdsall B, Bradbury AF, Biekofsky RR, Bayley PM. J. Biomol. NMR. 2001;21:41.
32. Ueda EKM, Gout PW, Morganti L. J. Chromatogr. A. 2003;988:1.
33. Lindner P, Guth B, Wülfing C, Krebber C, Steipe B, Müller F, Plückthun A. Methods. 1992;4:41.
34. Porath J, Carlsson J, Olsson I, Belfrage G. Nature. 1975;258:598.
35. Kågedal L. In: J-C Janson and L Rydén (Eds). *Protein Purification: Principles, High-Resolution Methods, and Applications*, 2nd ed. Wiley-VCH, Inc.: New York, 1998, p. 311.
36. Hochuli E. In: JK Setlow (Ed.). *Genetic Engineering; Vol. 12*, Plenum Press: New York, 1990, p. 87.
37. Hochuli E, Bannwarth W, Döbeli H, Gentz R, Stüber D. Biotechnology. 1998;6:1321.
38. Gaponenko V, Altieri AS, Li J, Byrd RA. J. Biomol. NMR. 2002;24:14S3.
39. Tjandra N, Marquardt J, Clore GM. J. Magn. Reson. 2000;142:393.
40. Bax A. Protein Sci. 2003;12:1.
41. Ottiger M, Delaglio F, Bax A. J. Magn. Reson. 1998;131:373.
42. Tjandra N, Grzesiek S, Bax A. J. Am. Chem. Soc. 1996;118:6264.
43. Wirmer J, Schwalbe H. J. Biomol. NMR. 2002;23:47.
44. Evenas J, Mittermaier A, Yang DW, Kay LE. J. Am. Chem. Soc. 2001;123:2858.
45. Mittermaier A, Kay LE. J. Am. Chem. Soc. 2001;123:6892.
46. Hitchens TK, McCallum SA, Rule GS. J. Magn. Reson. 1999;140:281.
47. Kay LE, Gardner KH. Curr. Opin. Struct. Biol. 1997;7:722.
48. Ottiger M, Bax A. J. Biomol. NMR. 1999;13:187.
49. Fowler CA, Tian F, Al-Hashimi HM, Prestegard JH. J. Mol. Biol. 2000;304:447.
50. Delaglio F, Kontaxis G, Bax A. J. Am. Chem. Soc. 2000;122:2142.
51. Fischer MWF, Losonczi JA, Weaver JL, Prestegard JH. Biochem. 1999;38:9013.
52. Skrynnikov NR, Goto NK, Yang D, Choy WY, Tolman JR, Mueller GA, Kay LE. J. Mol. Biol. 2000;295:1265.
53. Fushman D, Varadan R, Assfalg M, Walker O. Prog. Nucl. Magn. Reson. Spectrosc. 2004;44:189.
54. Varadan R, Walker O, Pickart C, Fushman D. J. Mol. Biol. 2002;324:637.
55. Varadan R, Assfalg M, Haririnia A, Raasi S, Pickart C, Fushman D. J. Biol. Chem. 2004;279:7055.
56. Bruschweiler R, Liao X, Wright PE. Science. 1995;268: 886.
57. Fushman D, Xu R, Cowburn D. Biochem. 1999;38:10225.
58. McFeeters RL, Fowler CA, Gaponenko VV, Byrd RA. J. Biomol. NMR. 2005;31:35.

Fast Assignments of ^{15}N-HSQC Spectra of Proteins by Paramagnetic Labeling

Guido Pintacuda[1,2], Thomas Huber[3], Max A. Keniry[1],
Ah Young Park[1], Nicholas E. Dixon[1], and Gottfried Otting[1]
[1] *Research School of Chemistry, Australian National University, Canberra ACT 0200, Australia*
[2] *Karolinska Institute, Tomtebodavägen 6, S-17177 Stockholm, Sweden*
[3] *Departments of Mathematics and Biochemistry, The University of Queensland, Brisbane QLD 4072, Australia*

Abstract

^{15}N-HSQC spectra of ^{15}N-labeled proteins are widely used to identify ligand-binding sites on protein surfaces, providing a way to assess protein–protein interactions as well as to screen for the binding of small chemical compounds. The method is particularly powerful if the cross peaks in the ^{15}N-HSQC spectrum have been sequence-specifically assigned. The present article reviews the use of paramagnetic labeling to obtain these resonance assignments quickly with the use of ^{15}N-labeled protein, provided the crystal structure of the protein is known and the protein can be labeled with a paramagnetic ion. The method also yields the anisotropy tensor of the magnetic susceptibility, which can be used to model the orientation of the ligand with respect to the protein.

Introduction

^{15}N-HSQC spectra of uniformly ^{15}N-labeled proteins play a unique role in pharmaceutical research by providing a two-dimensional fingerprint of the protein, where each amino acid residue, except proline, is represented by a peak from its backbone amide proton [1,2]. The cross peaks serve as sensitive reporters of ligand binding, as changes in the local chemical environment affect their chemical shifts and line widths, providing positive evidence of binding and information about the location of interacting surfaces. The binding of small molecules and interactions with macromolecular partners can both be assessed in this way. The method presents a tool for target validation (e.g. where do natural interaction partners bind?) but is even more attractive for the screening of chemical compound libraries (which compounds bind?) [3,4], since weak and strong binding can be detected equally well.

Assignment of the spectral changes observed in ^{15}N-HSQC spectra to specific binding sites (where does the compound bind?) requires the assignment of the cross peaks to the individual amino-acid residues in the protein. Traditional ways of assigning the ^{15}N-HSQC spectrum of a protein of molecular weight above 15 kDa require the preparation of doubly- (^{15}N and ^{13}C) or triply labeled (^{15}N, ^{13}C, and ^{2}H) samples and the recording of three-dimensional NMR spectra. These experiments are much less sensitive and more time consuming than recording of a ^{15}N-HSQC spectrum. Although uniformly labeled protein samples can, in principle, readily be prepared by expressing the protein in cell cultures grown in isotope-labeled media, labeling with isotopes other than ^{15}N proves prohibitively expensive in many cases. Therefore, resonance assignments are restricted to proteins which can be expressed in high yields and which are sufficiently soluble to record the NMR spectra necessary for resonance assignment. As a result, NMR resonance assignments seem to be out of reach for many target proteins of pharmaceutical interest, even when crystal structures and ^{15}N-HSQC spectra may be available.

The present article describes a different strategy for assigning ^{15}N-HSQC spectra of ^{15}N-labeled proteins. It requires prior knowledge of the three-dimensional protein structure and a lanthanide-binding site in the protein. The strategy has been demonstrated for the 30 kDa complex of the subunits ε (N-terminal segment with residues 2–186) and θ of *Escherichia coli* DNA polymerase III, where ε had been selectively labeled in three separate samples with ^{15}N-labeled Leu, Val, and Phe [5]. Resonance assignments are obtained from the comparison of two highly sensitive ^{15}N-HSQC spectra, recorded with and without a paramagnetic lanthanide ion bound to the complex.

ε186–θ Complex

The crystal structure of the N-terminal domain of ε (ε186) has been solved by X-ray crystallography [6]. The same fragment has been studied by NMR spectroscopy, both free and in complex with θ, using doubly and triply

Graham A. Webb (ed.), Modern Magnetic Resonance, 1245–1251.

isotope-labeled samples [7,8]. Two Mn^{2+} ions are bound to the active site of the enzyme in the crystal structure. In analogy to the related domain of DNA polymerase I, these two ions can be replaced by a single lanthanide [9].

Paramagnetic Restraints Derived from ^{15}N-HSQC Spectra of Paramagnetic and Diamagnetic ε186–θ Complexes

^{15}N-HSQC spectra of the complex between ε186 and θ with a Dy^{3+} ion bound (the paramagnetic complex) and of the complex without an ion bound or with La^{3+} bound (the diamagnetic complex) were recorded at a ^{1}H NMR frequency of 600 MHz. Spectra recorded in the presence or absence of La^{3+} were very similar; either spectrum provided an adequate diamagnetic reference for the measurement of paramagnetic restraints.

The cross peaks recorded for the paramagnetic complex are strongly shifted with respect to those of the diamagnetic spectrum (Figure 1). The chemical shift difference between the paramagnetic and diamagnetic spectrum presents the pseudocontact shift (PCS). The PCS values, measured in ppm, are almost the same in the ^{15}N and ^{1}H dimension of the spectrum. Consequently, the paramagnetic signals are shifted along approximately parallel lines, enabling the pairwise association of diamagnetic and paramagnetic cross peaks. Ambiguities resulting from situations of cross-peak overlap (Figure 1) can often be resolved by the assignment algorithm discussed below.

The ^{15}N-HSQC spectra are best recorded without decoupling during the acquisition time [5], resulting in a shorter pulse sequence and hence, reduced loss of magnetization by relaxation. More importantly, analysis of the doublet fine structure provides access to paramagnetic parameters, which would not be observable in a decoupled spectrum (Figure 2). For example, differential line broadening of the doublet components reflects the cross-correlated relaxation (CCR) between dipole–dipole (with the amide proton and amide nitrogen as the dipolar-coupled spin system) and Curie relaxation of the paramagnetic center [10–12]. In addition, the coupling constant displayed by the doublet contains the residual dipolar coupling (RDC) which arises from the partial alignment of the paramagnetic molecule with the magnetic field [13,14]. Finally, the overall broadening of the doublet components reflects the paramagnetic relaxation enhancement (PRE) by the paramagnetic center. The PRE effect is a function of the distance from the Dy^{3+} ion [15,16]. In practice, cross peaks from amide protons closer than 14 Å from the metal ion were broadened beyond detection.

PCS, RDC, CCR, and PRE effects [17] present four paramagnetic parameters arising from different structural information (Figure 3). Of these four effects, PCS data of protons are most easily measured with high accuracy. Their dependence on the anisotropy of the magnetic

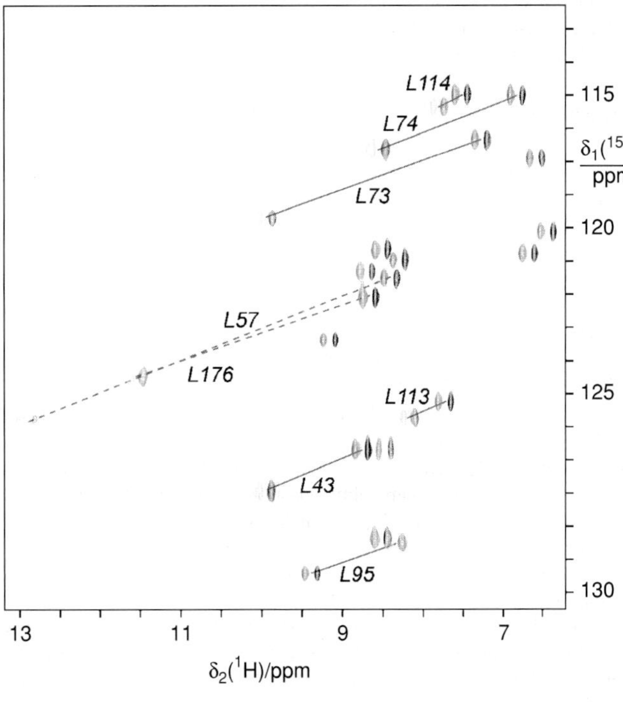

Fig. 1. ^{15}N-HSQC spectra of ^{15}N-Leu labeled ε186 complexed to unlabeled θ. Two spectra are superimposed, one recorded with paramagnetic Dy^{3+} and the other with diamagnetic La^{3+} complexed to ε186. No decoupling was applied during the acquisition time. The cross peaks appear as antiphase doublets split by the one-bond ^{1}H–^{15}N coupling constant. Positive and negative contours are plotted, respectively, with blue and yellow lines for the spectrum of the paramagnetic complex and black and red lines for the spectrum of the diamagnetic complex. Straight lines connect pairs of diamagnetic and paramagnetic cross peaks. Dashed lines identify the two most strongly shifted paramagnetic cross peaks for which the diamagnetic partners could not be identified by simple visual inspection. The spectra were recorded with 0.5 mM solutions of the complex at 25°C and pH 7.0, using total recording times of 5 and 16 h for the diamagnetic and paramagnetic samples, respectively, on a Varian Inova 600 MHz NMR spectrometer. Reproduced with permission from Ref. [5]. © 2004 Am. Chem. Soc.

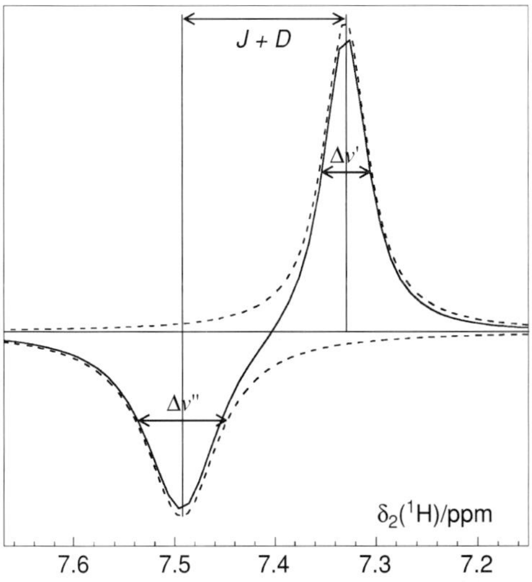

Fig. 2. Cross section through a cross peak from the spectrum of Figure 1 with two Lorentzian lineshapes fitted to the doublet components (dashed lines). The fit provides the accurate chemical shift of the doublet, the line widths of the individual doublet components, and the coupling constant of the doublet. Reproduced with permission from Ref [5]. © 2004 Am. Chem. Soc.

susceptibility tensor also provides stringent geometric restraints that are relatively insensitive with respect to local structural variations. In contrast, in our experience PRE effects yield the least accurate data, partly because line widths can vary between samples due to aggregation effects and chemical exchange phenomena [5]. The values of all four paramagnetic effects can be predicted provided that the three-dimensional structure of the protein, the position of the metal ion, and the susceptibility anisotropy tensor are known (Figure 4) [17].

PLATYPUS Algorithm for Resonance Assignments from Paramagnetic Restraints

In our assignment strategy, the resonance assignment is obtained from a comparison of the experimentally measured paramagnetic data with the data predicted from the three-dimensional structure of the protein. An exhaustive evaluation of all $n!$ possible assignments, however, is not viable when the number of cross peaks, n, is large. Consequently, a novel algorithm named PLATYPUS (paramagnetic labeling for assignment with terrific yields of proteins using their structures) was developed to derive the resonance assignments from the paramagnetic data automatically (Figure 5). Since the parameters defining the susceptibility anisotropy tensor are not known a priori, the algorithm employs a grid search with a large number of different tensors. A target function, defined as the sum of the squared deviations between experimentally measured and back-calculated paramagnetic data, is minimized to determine the best-fitting resonance assignment. Starting from each grid point, PLATYPUS uses the "Hungarian method for minimal cost assignment" [18] to arrive at resonance assignments that are continuously updated, while the tensor parameters and rotational correlation time are optimized by gradient minimization of the target function (Figure 5).

In principle, the grid search over different possible susceptibility anisotropy tensors is five-dimensional, since five parameters are required to define the shape and orientation of the tensor. In practice, the algorithm

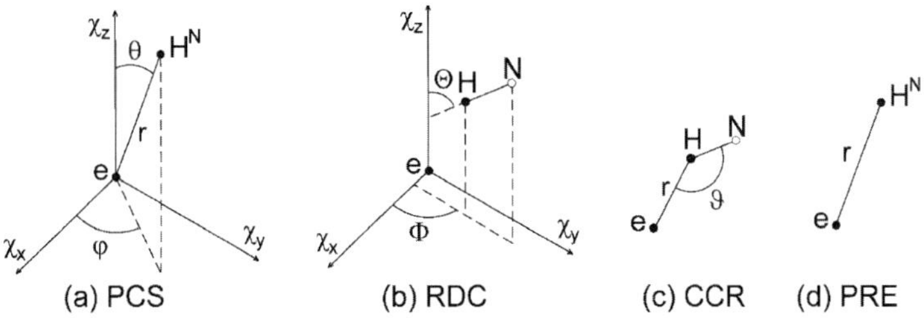

Fig. 3. Dependence of paramagnetic effects on local geometry. (a) Pseudocontact shifts (PCS) depend on the location of the nuclear spin with respect to the susceptibility anisotropy tensor. They result in chemical shift changes of the NMR signal. (b) Residual dipolar couplings (RDC) depend on the angle formed between the vector connecting the coupled spins and the axes of the alignment tensor resulting from alignment of the paramagnetic protein in the magnetic field. They are manifested in altered multiplet splittings. (c) Cross-correlated relaxation (CCR) between the Curie spin of the paramagnetic protein and the dipole–dipole interaction between two nuclear spins depends on the angle between the internuclear vector and the vector between the nuclear and electronic spins. This relaxation mechanism results in different line widths of the two doublet components. (d) Paramagnetic relaxation enhancements (PRE) depend on the distance from the paramagnetic center. The relaxation enhancement results in equal line broadening of both doublet components.

[30–34]. Lanthanide-binding peptide motifs avoid the generation of different diastereomers, but restrict the choice of attachment sites to the terminal ends of the protein. Furthermore, many of the paramagnetic effects are reduced when the linker to the lanthanide-binding peptide is flexible.

A paramagnetic tag attached to a selected site on the protein surface offers a powerful tool for the detection and structural characterization of ligand binding. The attachment of TEMPO spin labels has been shown to enhance the sensitivity of NMR methods for the screening of compound libraries by several orders of magnitude [35]. TEMPO spin labels have also been shown to provide long-range structural information in large biomolecules as well as access to the study of transient interactions in partially or completely unfolded proteins. The relaxation enhancement effect of TEMPO spin labels is, however, governed by a $1/r^6$ distance dependence, whereas PCSs induced by lanthanide ions follow a $1/r^3$ distance law, resulting in effects observable over longer distances. Furthermore, the isotropic nature of the magnetic susceptibility tensor in TEMPO prohibits the observation of the PCS, CCR, and RDC effects that contain the most valuable structural information (Figure 3). Therefore, more information can be obtained from site-specifically attached lanthanides. In addition, lanthanide ions could not only detect the presence of ligand binding, but also provide a detailed picture of the relative orientation of the ligand with respect to the magnetic susceptibility anisotropy tensor and hence, with respect to the protein. This information could be obtained with a minimal number of resonance assignments of the protein.

The present version of PLATYPUS was designed for the assignment of selectively ^{15}N-labeled proteins. For some proteins, cell-free protein expression systems offer a cost-effecitve and fast route to protein samples with selectively labeled amino acids [36–38]. Selective labeling of proteins becomes expensive when the proteins cannot be expressed in high yields. For those situations, a modified strategy would be desirable, which provides reliable cross-peak assignments for uniformly ^{15}N-labeled proteins. Work towards this goal is in progress.

Acknowledgments

Complete resonance assignments of ε186 or ε186 in complex with θ were not available at the time we published the ^{15}N-HSQC assignments of Leu, Val, and Phe labeled ε186 in complex with θ. We thank Drs. Eugene F. DeRose and Robert E. London for making their resonance assignments available to us afterwards to confirm the correctness of the assignments. G.P. thanks the EU for a postdoctoral fellowship within the Research Training Network on Cross Correlation (HPRN-CT-2000-00092). G.O. thanks the Australian Research Council for a Federation Fellowship. Financial support by the Australian Research Council is gratefully acknowledged.

References

1. Pellecchia M, Sem DS, Wüthrich K. Nat. Rev. Drug Discov. 2002;1:211.
2. van Dongen M, Weigelt J, Uppenberg J, Schultz J, Wikström M. Drug Discov. Today 2002;7:471.
3. Shuker SB, Hajduk PJ, Meadows RP, Fesik SW. Science 1996;274:1531.
4. Hajduk PJ, Meadows RP, Fesik SW. Q. Rev. Biophys. 1999;32:211.
5. Pintacuda G, Keniry MA, Huber T, Park AY, Dixon NE, Otting G. J. Am. Chem. Soc. 2004;126:2963.
6. Hamdan S, Carr PD, Brown SE, Ollis DL, Dixon NE. Structure 2002;10:535.
7. DeRose EF, Li D, Darden T, Harvey S, Perrino FW, Schaaper RM, London RE. Biochemistry 2002;41:94.
8. DeRose EF, Darden T, Harvey S, Gabel S, Perrino FW, Schaaper RM, London RE. Biochemistry 2003;42:3635.
9. Brautigam CA, Aschheim K, Steitz TA. Chem. Biol. 1999; 6:901.
10. Boisbouvier J, Gans P, Blackledge M, Brutscher B, Marion D. J. Am. Chem. Soc. 1999;121:7700.
11. Pintacuda G, Hohenthanner K, Otting G, Müller N. J. Biomol. NMR 2003;27:115.
12. Ghose R, Prestegard JH. J. Magn. Reson. 1997;128:138.
13. Tolman JR, Flanagan JM, Kennedy MA, Prestegard JH. Proc. Natl. Acad. Sci. USA 1995;92:9279.
14. Barbieri R, Bertini I, Cavallaro G, Lee YM, Luchinat C, Rosato A. J. Am. Chem. Soc. 2002;124:5581.
15. Guéron M. J. Magn. Reson. 1975;19:58.
16. Vega AJ, Fiat D. Mol. Phys. 1976;31:347.
17. Bertini I, Luchinat C, Parigi G. Solution NMR of Paramagnetic Molecules: Applications to Metallobiomolecules and Models. Elsevier: Amsterdam, 2001.
18. Kuhn HW. Nav. Res. Logist. Q. 1955;2:83.
19. Hus JC, Prompers JJ, Brüschweiler R. J. Magn. Reson. 2002;157:119.
20. Zweckstetter M. J. Biomol. NMR 2003;27:41.
21. Allegrozzi M, Bertini I, Janik MBL, Lee YM, Lin GH, Luchinat C. J. Am. Chem. Soc. 2000;122:4154.
22. Liu S, Edwards DS. Bioconjugate Chem. 2001;12:7–34.
23. Caravan P, Ellison JJ, McMurry TJ, Lauffer RB. Chem. Rev. 1999;99:2293.
24. Selvin PR. Annu. Rev. Biophys. Biomol. Struct. 2002;31: 275.
25. Voss J, Wu J, Hubbell WL, Jacques V, Meares CF, Kaback HR. Biochemistry. 2001;40:3184.
26. Purdy MD, Ge P, Chen J, Selvin PR, Wiener MC. Acta Crystallogr. D. 2002;58:1111.
27. Dvoretsky A, Gaponenko V, Rosevear PR. FEBS Lett. 2002;528:189.
28. Ikegami T, Verdier L, Sakhaii P, Grimme S, Pescatore B, Saxena K, Fiebig KM, Griesinger C. J. Biomol. NMR, 2004;29:339.

29. Pintacuda G, Moshref A, Leonchiks A, Sharipo A, Otting G. J. Biomol. NMR, 2004;29:351.
30. Ma C, Opella SJ. J. Magn. Reson. 2000;146:381.
31. Biekofsky RR, Muskett FW, Schmidt JM, Martin SR, Browne JP, Bayley PM, Feeney J. FEBS Lett. 1999;460:519.
32. Feeny J, Birdsall B, Bradbury AF, Biekofsky RR, Bayley PM. J. Biomol. NMR 2001;21:41.
33. Wohnert J, Franz KJ, Nitz M, Imperiali B, Schwalbe H. J. Am. Chem. Soc. 2003;125:13338.
34. Caravan P, Greenwood JM, Welch JT, Franklin SJ., Chem. Commun. 2003;20:2574.
35. Jahnke W, Rudisser S, Zurini M. J. Am. Chem. Soc. 2001; 123:3149.
36. Kigawa T, Muto Y, Yokoyama S. J. Biomol. NMR 1995;6: 129.
37. Guignard L, Ozawa K, Pursglove SE, Otting G, Dixon NE. FEBS Lett. 2002;524:159.
38. Yabuki T, Kigawa T, Dohmae N, Takio K, Terada T, Ito Y, Laue ED, Cooper JA, Kainosho M, Yokoyama S. J. Biomol. NMR 1998;11:295.
39. Koradi R, Billeter M, Wüthrich K. J. Mol. Graph. 1996;14: 51.

Phospholipid Bicelle Membrane Systems for Studying Drug Molecules

Jianxin Guo[1], Xiaoyu Tian[1], Spiro Pavlopoulos[2], and Alexandros Makriyannis[1]
[1]*Center for Drug Discovery, Bouve College of Health Sciences, Northeastern University, 360 Huntington Avenue, Boston, MA 02115, USA; and*
[2]*Department of Pharmaceutical Sciences, University of Connecticut, U-3092, 69 North Eagleville Road, CT 06269, USA*

Abbreviations: NMR, nuclear magnetic resonance; DSC, differential scanning calorimetry; DPPC, 1,2-dipalmitoyl-*sn*-glycero-3-phosphocholine; DMPC, 1,2-dimyristoyl-*sn*-glycero-3-phosphocholine; DHPC, 1,2-dihexanoic-*sn*-glycero-3-phosphocholine; CHAPSO, 3-(cholamidopropyl) dimethylammonio-2-hydroxy-1-propanesulfonate; SDS, sodium dodecyl sulfate; DPC, dodecylphosphocholine; Δ^8-THC, $(-)$-Δ^8-tetrahydrocannabinol; Me-Δ^8-THC, $(-)$-O-methyl-Δ^8-tetrahydrocannabinol

Introduction

The design of drugs that modulate membrane bound receptors has always been of prime importance to the pharmaceutical industry. It is increasingly being recognized that the cellular membrane plays an important role in drug action and that understanding the manner in which drug molecules interact with lipid bilayers can enhance our abilities to design and develop improved medications. It has been proposed that lipophilic drugs reach their sites of action through cell membrane penetration and lateral diffusion within the membrane leaflet [1–4]. The protein supporting lipid membranes may be capable of a high degree of structural discrimination by positioning the ligand in a proper location, orientation, and active conformation for a productive interaction with the receptor. Therefore, acquiring detailed knowledge of drug and ligand physical properties that underlie their interaction profiles with the cellular membrane is of great interest in the discovery, design, and delivery of novel therapeutic agents. Information related to drug–membrane interactions, however, can rarely be derived directly from experiments with complex and labile natural membranes. Thus, various membrane models with different levels of complexity have been developed in order to exploit all the biophysical techniques available [5–7]. Whether it is in the form of liposomes, micelles, or other membrane mimetics, the interplay between the ligand and membranes is central to many aspects in pharmaceutical research.

Liposomes (in most cases, multilamellar vesicles) have been extensively utilized to study the manner in which drug molecules perturb cellular membranes as well as the location, conformation, and preferred orientation of a drug molecule within the membrane bilayer using differential scanning calorimetry (DSC), nuclear magnetic resonance (NMR) spectroscopy, and small angle X-ray diffraction [3,8–15]. Among these techniques, NMR spectroscopy holds a prominent place by being particularly sensitive to the fine structural and dynamic details of drug molecules within the bilayer system. However, liposomes are not suitable model membrane systems for exploring the conformational properties of embedded ligands since even the smallest lipid vesicles are too large to yield a quality high-resolution NMR spectrum [16]. Alternatively, fast tumbling micelles consisting of various detergents have been widely used to mimic the membrane environment in high-resolution NMR conformational analysis [5,17–19], even though it has long been a concern that micelles may not accurately represent cellular membrane bilayers [20,21]. Recently, phospholipid bilayered-micelles (often referred to as bicelles) have received much attention as promising membrane-mimetic systems [22–27]. Work in our laboratory has demonstrated that bicelle systems can serve as excellent bilayer membrane models for studying drug–membrane interactions using high-resolution NMR. We have developed an isotropic bicelle system for studying the conformational properties of small- and medium-sized ligands without the need of isotopic labeling [21]. We also used magnetically oriented bicelles to determine the preferred orientations of several lipophilic drugs from deuterium NMR experiments [28].

In this chapter, we briefly review the commonly used model membrane systems for studying drug–membrane interactions using NMR with an emphasis on phospholipid bicelles. To highlight the advantages of using bicelle systems for such work, we have sought to determine the conformational properties as well as the preferred orientation of two lipophilic ligands in different membrane-mimetic systems. A comparison of the ligand spectra in

Graham A. Webb (ed.), Modern Magnetic Resonance, 1253–1259.

the different media clearly demonstrates the value of phospholipid bicelles in such studies.

Membrane Systems for NMR Studies

Cell membranes consist of a variety of glycerolphospholipids, sphingolipids, and cholesterol with local heterogeneities and segregated microdomains. Although much work has been dedicated to understand their structural and dynamic properties, many questions still remain unanswered [29]. Opportunities for studying the drug–membrane interactions with native membranes are presently confined to the use of solid-state NMR methods. Only few published examples can be found of the interactions between small- or medium-sized molecules with native membranes where the ligand is isotopically labeled in specific positions for effective NMR observation [30,31]. Simplified model membranes composed of certain synthetic or native lipids have become the most commonly used systems for such work.

It can be argued that notwithstanding their great functional diversity, membranes share the common features of an anisotropic bilayer system consisting of a hydrophobic interior and a hydrophilic headgroup region. This fundamental structure defines many of the key structural and functional properties of biological membranes. For this reason, it is possible to use simplified lipid systems as legitimate mimics of the cellular membrane.

Liposomes

Liposomes are the most used and best-characterized model membrane system. They can be produced in a variety of sizes and lipid compositions including multilamellar vesicles and small or large unilamellar vesicles [32,33]. Liposomes are generally prepared from glycerolphospholipids, mainly phosphatidycholines such as DPPC, DMPC, and POPC, while cholesterol and charged lipids can also be added. Because multilamellar vesicles are relatively easy to prepare and multilayering makes it possible to incorporate larger amounts of drug, they have been extensively employed for studying drug–membrane interactions using various solid-state NMR techniques, such as stationary ^{2}H, ^{31}P, and ^{13}C NMR as well as the cross polarization magic angle spinning (CP/MAS) experiments [11,34,35]. Liposomes in the L_{α} liquid crystalline phase have been used to determine the preferred orientations of native and xenobiotic molecules and the effects of these ligands on the membrane system [8,9,11,12,14,36–39]. Liposomes are usually not suitable for studying the conformation of a ligand using solution NMR because of the relatively slow tumbling rates resulting from their large size. For such applications, other model membrane systems consisting of much smaller aggregates must be used.

Organic Solvent Mixtures

The simplest membrane-mimetic environment resembling the membrane interior and interfacial regions can be obtained by carefully mixing two organic solvents of low and intermediate polarity [40]. However, such systems are overly simplified membrane models and are of limited value.

Micelles

The most commonly used spherical micelle preparations can be obtained from SDS, DPC, or DHPC. Because of effective motional averaging, the small and uniformly sized micelles can yield satisfactory high-resolution NMR spectra of an enclosed ligand. Micelles have been extensively utilized as membrane-mimetic media for studying the conformational properties of small molecules and peptides [18,19]. However, as micelles do not consist of natural cell membrane lipids and have a highly curved surface, their value as a reliable model membrane system is somewhat reduced [21,26].

Phospholipid Bicelles

An optimal membrane-mimetic system for solution NMR experiments should consist of a relatively smaller lipid aggregate with a bilayer structure. With a short correlation time, such a membrane preparation enables more detailed structural studies of drugs in a bilayer environment. In order to form such a lipid aggregate, the mean chemical potential μ_N^0 must initially decrease and reach its minimum at a certain aggregation number N. If there were no free energy contributions, such small lipid aggregates would be entropically favored. However, because of the large cohesive binding energies of lipid molecules, discrete planar bilayer systems have never been observed when only one type of lipids is present. For a hypothetical planar bilayer disc, the free energy for molecules sequestered along the rim is expected to be much higher than that of molecules in the planar center. The mean free energy per molecule in such a planar bilayer disc is [41]:

$$\mu_N^0 = \mu_\infty^0 + \alpha kT/N^{1/2}$$

where μ_∞^0 is the mean free energy in an aggregate of infinite size, k is the Boltzmann constant and T is the temperature. As μ_N^0 decreases monotonically with N, the expectation is that an infinite number of lipid molecules will aggregate spontaneously at critical micelle concentrations while single component lipid bilayers with exposed edges are not favored.

However, discrete planar bilayer disc assemblies can be obtained if detergents are added to stabilize the energetically unfavorable edges [23,26]. Such systems,

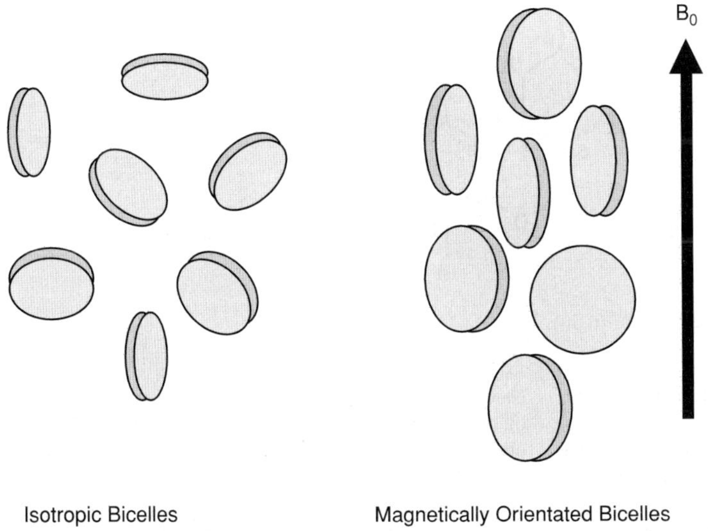

Fig. 1. Isotropic (left) and magnetically oriented bicelles (right).

commonly referred to as phospholipid bicelles, are typically composed of DMPC with either DHPC or a bile-salt derivative CHAPSO in 0.1 M KCl solution [26]. Above the phase transition temperature (T_m), the DMPC-rich domain resembles the liquid crystalline L_α phase multilamellar membrane bilayers [22,42]. Phospholipid bicelles combine the attractive structural and dynamic properties of both micelles and lipid vesicles and maintain most of the key bilayer properties that are absent in the micellar systems. Additionally, compared to the liposome preparations, bicelles are relatively small, noncompartmentalized, effectively monodisperse and ideally suited for high-resolution NMR studies.

Phospholipid bicelles (Figure 1) are a versatile model membrane system whose morphology and dynamic properties can be adjusted over wide ranges of lipid to detergent ratio (q), total lipid concentration, buffer pH, and temperature [26]. With appropriate manipulation, bicelles can provide isotropic membrane systems suitable for studying the conformational properties of embedded ligands using high-resolution NMR [21,43]. Alternatively, bicelles can serve as anisotropic membrane preparations capable of orienting in an applied magnetic field [23,25].

Optimizing Isotropic Bicelles for Drug Conformational Studies

At a DMPC to DHPC molar ratio (q) of 0.5 and an overall lipid concentration of 15% (w/v), the bicelle system has been shown to be at the isotropic state and the bicelles are small enough for high-resolution experiments NMR [43]. This is promising in that without isotopic labeling, such a preparation can be used to explore the conformational properties of incorporated drugs. However, such bicelles may not serve as an ideal membrane environment for studying our lipophilic drugs. As the diameter of the bicelle is estimated to be approximately 80 Å [44], the embedded drugs may have much higher chances of interacting with the stabilizing detergents rather than with the lipids. In addition, small bicelles are incapable of incorporating adequate quantities of lipophilic drugs and precipitation has been frequently observed in our research.

For these reasons, we have sought to develop larger isotropic bicelles that would better represent cellular membrane bilayer systems. Our approach to maximizing bicelle size while maintaining isotropic conditions was to systematically adjust the lipid to detergent ratio (q) as well as the overall lipid concentration. Additionally, preparations free of KCl allow us to use higher values of q and the stability of bicelles can be significantly improved by adding trace amount of amphiphiles. We found that a preparation of a much higher DMPC/DHPC ratio ($q = 2.0$) and an overall lipid concentration of 8% (w/v) is optimal for not only allowing effective drug incorporation, but also providing an isotropic system suitable for high-resolution NMR studies. In such a preparation, no precipitation of either the drug or lipid was observed over the course of several days. Well-resolved proton resonances reflect the isotropic nature of the optimized bicelle membrane media within the NMR time scale. This can be attributed to the ample spacing between the individual bicelles, which allow for faster motional averaging [25]. Although a certain degree of broadening in the ^{1}H resonances is still observed, the spectral resolution is adequate for standard 2D-NMR experiments such as DQF-COSY

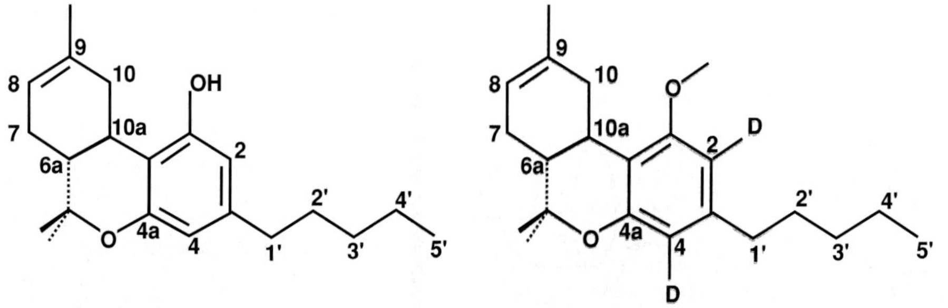

Fig. 2. The structure of Δ^8-THC (left) and Me-Δ^8-THC (right).

and NOESY from which ligand conformational properties can be obtained.

To evaluate the experimental benefits of our isotropic bicelle preparation, we compared the ^{1}H spectra of Δ^8-THC (Figure 2) in chloroform, SDS micelles, and two different isotropic bicelle preparations with $q = 0.5$ and $q = 2.0$, respectively. Figure 3 shows the NOE patterns observed between the aromatic ring protons and protons of the pentyl tail. In chloroform, only two NOEs were observed between the H2, H4 aromatic protons, and the pentyl tail H1′ protons. Additional NOEs were observed between the aromatic H4 resonance and resonances due to protons beyond H1′ of the pentyl sidechain with their relative intensities varying depending on the membrane

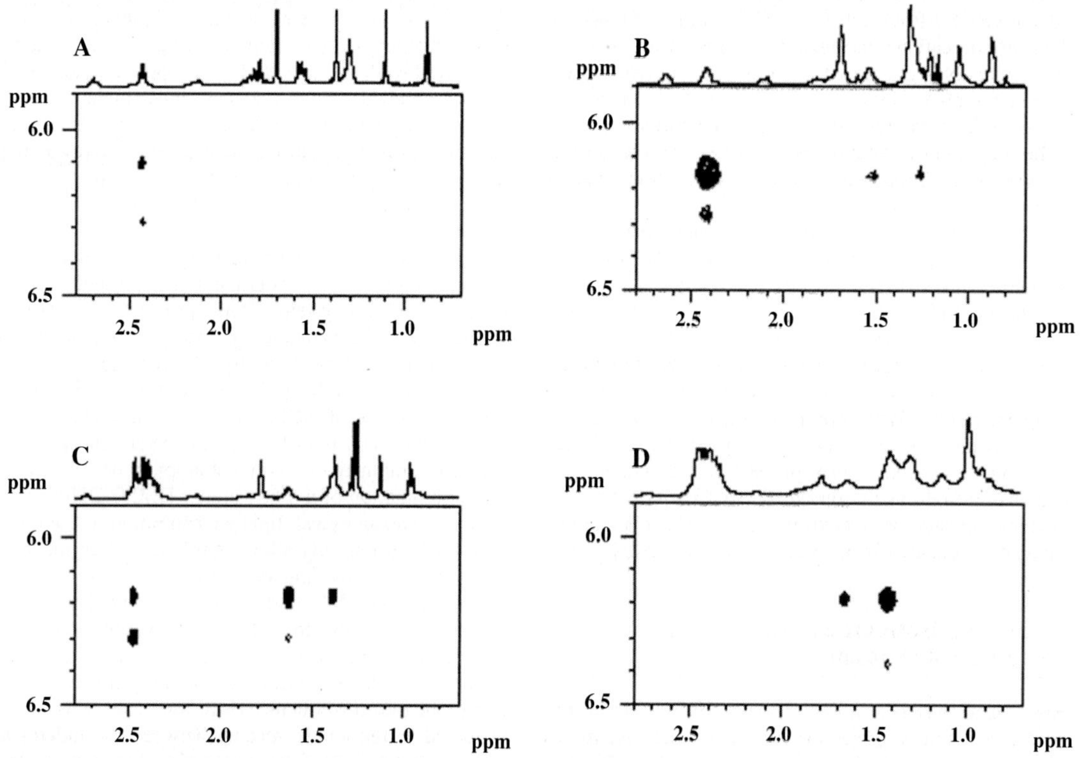

Fig. 3. Expansions of NOESY spectra showing NOEs between the aromatic ring and the pentyl-tail of Δ^8-THC in (A) $CDCl_3$; (B) SDS micelles; (C) $q = 0.5$ bicelle, and (D) $q = 2.0$ bicelle.

preparations used. In SDS micelles, the NOESY spectrum from Δ^8-THC is comparable to that observed in chloroform where strong NOEs were observed between the H2, H4 aromatic protons, and the H1′ protons in the pentyl group. Two weak NOEs were also observed between H4 and H2′, and H3′, H4′ overlapping protons, suggesting a conformational preference for orientation of the pentyl chain towards the H4 side of the tricyclic ring system. In the $q = 0.5$ bicelle solution, the H4–H2′ and H4–H3′/H4′ NOEs are similar in intensity to the H4–H1′, H2–H1′ NOEs; while in the $q = 2.0$ solutions, the former set of NOEs are significantly stronger than the latter. This observation shows that higher proportions of DMPC within the bicelle lead to an increased preference for an "amphipathic" conformation of the pentyl sidechain towards the H4 side of the tricyclic ring system.

The above results indicate that the choice of model membrane systems can greatly influence the structural and dynamic properties of the incorporated drugs. It is also worth noting that in both the SDS micelle and the $q = 0.5$ bicelle preparations, Δ^8-THC was observed to precipitate over 3–6 h whereas it was well solubilized in the $q = 2.0$ bicelle solution. Our results obtained from the $q = 2.0$ bicelle preparation are congruent with previous studies using DPPC multilamellar membrane bilayers [11,14,15], which we attribute to the more bilayer-like morphology of the $q = 2.0$ lipid bicelles compared to the SDS micelle and $q = 0.5$ bicelle solutions. Therefore, the $q = 2.0$ bicelles represent a satisfactory membrane bilayer system suitable for high-resolution NMR conformational studies.

Magnetically Aligned Bicelles for Studying Drug Orientation

The formation of magnetically oriented phospholipid bicelles under various conditions and the corresponding properties of the bicelles have been reviewed [22,45]. By using a series of titration studies, it has been found that the liquid crystalline L_α DMPC/DHPC bicelles with q ranging from 2.0 to 3.5 are well aligned in the applied magnetic field as indicated by their ^{31}P and ^2H NMR spectra, while addition of KCl can drastically improve the degree of orientation.

The orientation may be explained by the anisotropy of the diamagnetic susceptibility tensor χ of the lipid molecule. If we assume an axis of rotational symmetry, the diamagnetic anisotropy $\Delta\chi$ will be the difference between the susceptibilities parallel and perpendicular to this axis. For the bicelle forming long acyl chain lipid DMPC in the L_α phase, $\Delta\chi$ was found to be $(-1.0 \pm 0.1) \times 10^8$ erg cm^{-3} G^{-2} [46]. Since the magnetic potential energy ΔU can be expressed by:

$$\Delta U = -N \Delta\chi B_0^2$$

where N is the number of long acyl chain DMPC lipids within one bicelle disc, and B_0 is the external magnetic field, when $\Delta\chi$ is less than zero, the energically favorable orientation of a bicelle aggregate will be with its principal axis perpendicular to the applied magnetic field.

In earlier work, we have described how the preferred orientation of a drug molecule within the membrane bilayer system can be obtained from the quadrupolar splittings ($\Delta\nu_Q$) of ^{2}H-labels introduced in specific sites of the molecule [9,11]. We have, thus, sought to explore whether such data could be obtained in a magnetically oriented bicellar system. Figure 4 compares three deuterium spectra of 2,4-D_2-Me-Δ^8-THC (Figure 2) obtained in DHPC micelles, DPPC multilamellar vesicles, and magnetically oriented DMPC/DHPC bicelle preparations, respectively. The spectrum acquired in a DHPC micelle solution displays a single isotropic deuterium peak, while that from the DPPC multilamellar vesicles corresponds to two pairs of overlapping Pake patterns from which the quadrupolar splittings can be extracted. Conversely, in magnetically oriented bicelles, the spectrum is reduced to two pairs of doublets representing the two and four deuterium atoms in Me-Δ^8-THC. The sharp deuterium resonances indicate that the bicelles are completely aligned in the magnetic field. Due to the fact that the bicelles with the incorporated drug molecules are magnetically oriented with their

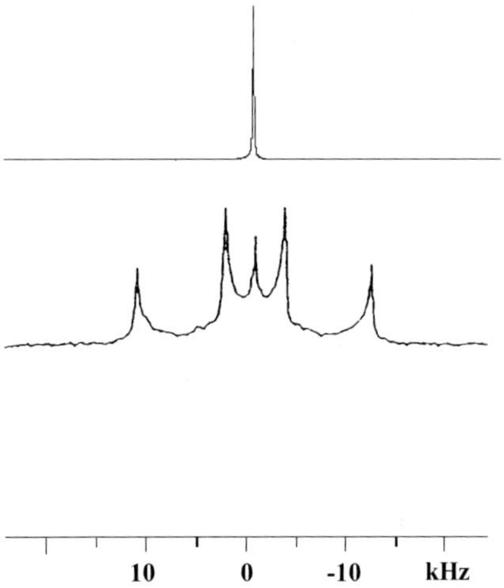

Fig. 4. A comparison of deuterium spectra of 10% 2,4-D_2-Me-Δ^8-THC in DHPC micelles (top), DPPC multilamellar bilayers (middle), and magnetically oriented DMPC/DHPC bicelles (bottom).

Part II

bilayer normal perpendicular to the external magnetic field, the observed splittings in the ^{2}H spectrum correspond to the 90° edges of the respective Pake patterns from the semi-solid hydrated DPPC sample. However, the $\Delta\nu_Q$ values are much reduced in the bicelle system, reflecting a lower order parameter (S_{mol}) when compared to that from the multilamellar preparations. The calculated orientation of Me-Δ^8-THC (to be reported elsewhere) closely matches our result from the multilamellar vesicles, thus attesting to the reliability of the bicelle system as a model membrane.

In contrast to micelle preparations where quadrupolar couplings are completely averaged out due to fast isotropic tumbling, magnetically oriented bicelles can serve as an alternative to multilamellar membrane preparations for studying the orientation of a ligand in an anisotropic membrane system. In addition, the experiment described above was carried out under solution NMR conditions and offers the advantages of higher sensitivity and more accurate direct measurement of the $\Delta\nu_Q$ values from the sharp ^{2}H doublets.

Conclusion

Bicelles can serve as satisfactory membrane systems for isotropic and anisotropic NMR experiments aimed at studying the conformational properties and preferred orientation of lipophilic drug molecules in the cell membrane. Using the lipophilic Δ^8-THC as an example, we have demonstrated that high-resolution NOESY spectra of this ligand in optimized isotropic bicelles can be obtained without the need for isotopic labeling. Our results clearly showed that isotropic bicelle systems are better membrane bilayer models than the commonly used micelles. Arguably, the specific "amphipathic" conformation of Δ^8-THC obtained in the optimized bicelles may be necessary for a productive interaction with its receptor. We also demonstrated that the preferred orientation of the embedded cannabinoids can be determined by adjusting bicelle conditions to produce larger bicelle aggregates capable of aligning in the magnetic field.

Phospholipid bicelles offer many opportunities for use as successful membrane mimetics. The preparation may be further improved by using lipids with certain negative charges and some degree of unsaturation, and by incorporating appropriate amounts of cholesterol to more closely resemble native cell membranes. A most recent application of bicelles is for determining the relative stereochemistry of multiple stereocenters [47]. Also, crystallization of membrane proteins has been achieved from bicelle-forming lipid/detergent mixtures [48], an experiment which may be extended to study the drug/membrane–protein complex.

References

1. Herbette LG, Chester DW, Rhodes DG. Biophys. J. 1986;49: 91–3.
2. Mason RP, Rhodes DG, Herbette LG. J. Med. Chem. 1991;34: 869–77.
3. Makriyannis A. In: RG Pertwee (Ed). Cannabinoid Receptors. Academic Press: London, UK, 1995, pp 87–115.
4. Xie XQ, Melvin LS, Makriyannis A. J. Biol. Chem. 1996;271: 10640–7.
5. Henry GD, Sykes BD. Bull. Can. Biochem. Soc. 1987;24:21–6.
6. Scrimin P, Tecilla P. Curr. Opin. Chem. Biol. 1999;3:730–5.
7. Milhaud J. Biochem. Biophys. Acta 2004;1663:19–51.
8. Makriyannis A, Banijamali A, Van Der Schyf C, Jarrell H. NIDA Res. Monogr. 1987;79:123–133.
9. Makriyannis A, Banijamali A, Jarrell HC, Yang DP. Biochem. Biophys. Acta 1989;986:141–5.
10. Mavromoustakos T, Yang DP, Charalambous A, Herbette LG, Makriyannis A. Biochim. Biophys. Acta 1990;1024:336–44.
11. Yang DP, Banijamali A, Charalambous A, Marciniak G, Makriyannis A. Pharmacol. Biochem. Behav. 1991;40:553–7.
12. Makriyannis A, Yang DP, Mavromoustakos T. NIDA Res. Monogr. 1991;112:106–28.
13. Yang DP, Mavromoustakos T, Makriyannis A. Life Sci. 1993;53:117–22.
14. Makriyannis A, Yang DP. In: BN Dhawan, RC Sirmal, R Raghubir, RS Rapaka (Ed). Recent Advances in the Study of Neurotransmitter Receptors. Central Drug Research Institute: Lucknow, India, 1994, pp 329–48.
15. Mavromoustakos T, Yang DP, Makriyannis A. Biochim. Biophys. Acta 1995;1237:183–8.
16. Henry GD, Sykes BD. Methods Enzymol 1994;239:515–35.
17. Opella SJ, Kim Y, McDonnell P. Meth Enzymol 1994;239:536–60.
18. Mavromoustakos T, Theodoropoulou E, Yang DP, Lin SY, Koufaki M, Makriyannis A. Chem. Phys. Lipids 1996;84:21–34.
19. Naito A, Nishimura K. Curr. Top. Med. Chem. 2004;4:135–45.
20. Sanders CR, Oxenoid K. Biochem. Biophys. Acta 2000;1508: 129–45.
21. Guo J, Pavlopoulos S, Tian X, Lu D, Nikas SP, Yang DP, Makriyannis A. J. Med. Chem. 2003;46:4838–46.
22. Sanders CR, Schwonek JP. Biochemistry 1992;31:8898–905.
23. Sanders CR, Hare BJ, Howard KP, Prestegard JH. Prog. Nucl. Magn. Reson. Spectrosc. 1994;26:421–44.
24. Howard KP, Prestegard JH. J. Am. Chem. Soc. 1996;118: 3345–53.
25. Tjandra N, Bax A. Science 1997;278:1111–4.
26. Sanders CR, Prosser RS. Structure 1998;6:1227–34.
27. Glover KJ, Whiles JA, Wu G, Yu N-J, Deems R, Struppe JO, Stark RE, Komives EA, Vold RR. Biophys. J. 2001;81:2163–71.
28. Guo J. Ph.D. Thesis. University of Connecticut, Storrs, 2004.

29. Dowhan W, Bogdanov M. In: DE Vance, JE Vance (Ed). Biochemistry of Lipids, Lipoproteins and Membranes. Elsevier Press: Amsterdam, 2002, pp 1–35.
30. Watts A, Ulrich AS, Middleton DA. Mol. Membr. Biol. 1995;12:233–46.
31. Kisselev OG, Kao J, Ponder JW, Fann YC, Gautam N, Marshall GR. Proc. Natl. Acad. Sci. USA 1998;95:4270–5.
32. Bloom M, Evans E, Mouritsen OG. Q. Rev. Biophys. 1991;24:293–397.
33. Bangham AD. Chem. Phys. Lipids 1993;64:275–85.
34. Tang P, Yan B, Xu Y. Biophys. J 1997;72:1676–82.
35. Gutierrez ME, Garcia AF, Africa de Madariaga M, Sagrista ML, Casado FJ, Mora M. Life Sci. 2003;72:2337–60.
36. Davis J. Biochim. Biophys. Acta 1983;737:117–71.
37. Makriyannis A, Rapaka RS. Life Sci. 1990;47:2173–84.
38. Bechinger B. Biochem. Biophys. Acta 1999;1462:157–83.
39. Yamaguchi S, Huster D, Waring A, Lehrer RI, Kearney W, Tack BF, Hong M. Biophys. J. 2001;81:2203–14.
40. Girvin ME, Rastogi VK, Abildgaard F, Markley JL, Fillingame RH. Biochemistry 1998;37:8817–24.
41. Israelachvilli JN, Mitchell DJ, Ninham BW. J. Chem. Soc. Faraday Trans. II 1976;72:1525–68.
42. Sanders C. Biophys. J. 1993;64:171–81.
43. Vold RR, Prosser RS, Deese AJ. J. Biomol. NMR 1997;9:329–35.
44. Vold RR, Prosser RS. J. Magn. Reson. B 1996;113:267–71.
45. Sternin E, Nizza D, Gawrisch K. Langmuir 2001;17:2610–6.
46. Scholz F, Boroske E, Helfrich W. Biophys. J. 1984;45:589–92.
47. Yan J, Delaglio F, Kaerner A, Kline AD, Mo H, Shapiro MJ, Smitka TA, Stephenson GA, Zartler ER. J. Am. Chem. Soc. 2004;126:5008–17.
48. Faham S, Bowie JU. J. Mol. Biol. 2002;316:1–6.

Partial Alignment for Structure Determination of Organic Molecules

Burkhard Luy and Horst Kessler
TU München, Department Chemie, Lehrstuhl für Organische Chemie II, Lichtenbergstrasse 4, D-85747 Garching, Germany

Introduction

NMR spectroscopy is the most important method for structure elucidation in solution. Once the atomic connectivity is established (constitution of the molecule) the proof of stereochemistry is the next step. NMR as an achiral method provides only relative stereochemical information unless chiral environments (solvents, interaction with shift reagents, chemical modification via chiral auxiliaries) are used. For determining the relative stereochemistry, the classical parameters are NOE effects and J coupling constants. Only recently, two different new sources of stereochemical information have been introduced for biomolecules: cross-correlated relaxation [1] and residual dipolar couplings (RDCs) [2,3]. We will here consider the application of RDCs for small molecules such as drugs or drug-like molecules.

Residual Dipolar Couplings

Magnetically active nuclei in molecules interact through space, as they are magnetic dipoles. These couplings, which depend on the relative orientation of the dipoles and their distance, are on the order of several kHz and cause immense numbers of splittings in solid-state NMR. In solution, the rapid isotropic molecular tumbling averages out all dipolar couplings, resulting in the narrow line width of high-resolution NMR spectroscopy. With the removal of all anisotropic interactions, however, there is also a significant loss in potential structural information. Therefore, an intermediate state between solid and liquid is desirable in which the narrow line width is maintained, but anisotropic interactions are re-established and measurable. In general, there are two ways to reach this intermediate state: if the sample of interest is available in a microcrystalline phase, high-speed magic angle spinning can effectively reduce the dipolar interactions to give resolved resonances similar to the liquid state; in contrast, a sample in solution can be transferred into a directional molecular lattice, which then partially aligns the molecule to have measurable anisotropic interactions. For example, if the molecules are oriented to a low percentage, let us say on the order of 0.1%, the dipolar coupling is also reduced to ~0.1% of its original size. A molecule containing a CH fragment, which would normally exhibit a $^1J_{CH}$ coupling of about 130 Hz in isotropic solution will give rise to an additional RDC that adds or subtracts to this coupling depending on the tiny preferred orientation. If one measures the $^1J_{CH}$ coupling constant with and without partial orientation, the difference directly yields the RDC.

The Alignment Tensor

In the following section, a very brief and handwavy introduction to the alignment tensors is given and includes only basic equations. For a more detailed and exact derivation of key equations, we refer to [4].

In the static case, the dipolar coupling constant is given by [5]

$$D = \frac{\kappa}{R^3}\left(\cos^2\theta - \frac{1}{3}\right), \tag{1}$$

where θ is the angle between an internuclear vector and the magnetic field B_0. The term

$$\kappa = -\frac{3}{8\pi^2}\gamma_I\gamma_S\mu_0\hbar \tag{2}$$

only depends on physical constants [5]: the gyromagnetic ratios γ_I and γ_S of the two vector-spanning spins I and S, respectively, the Planck constant $\hbar = h/2\pi$ and the permeability of vacuum μ_0. For example, for ^{1}H–^{1}H and ^{13}C–^{1}H spin pairs, $\kappa = -360.3$ and -90.6 kHz Å^3, respectively. However, in the case of a tumbling molecule, the dipolar coupling is further averaged in the way that

$$\bar{D} = \frac{\kappa}{R^3}\left(\overline{\cos^2\theta} - \frac{1}{3}\right) \tag{3}$$

represents the so-called RDC constant between the two spins, which depends on the average alignment of the molecule.

In the case of isotropic tumbling as in conventional liquid-state NMR, $\overline{\cos^2\theta} = 1/3$ in all three dimensions

Graham A. Webb (ed.), Modern Magnetic Resonance, 1261–1267.

and the dipolar interaction is quenched. In a partially oriented molecule instead, the term $\overline{\cos^2\theta}$ generally is a three-dimensional tensor. This tensor is equivalent to the probability tensor **P** to find the axis of the magnetic field $B_0(t)$ along the internuclear vector **r** expressed in a reference frame, which is fixed to the molecule. This (3×3) probability tensor is symmetric with respect to the diagonal and the diagonal elements must fulfill the condition $P_{xx} + P_{yy} + P_{zz} = 1$, which implicitly leads to five independent tensor components. In reverse, this implies that in an arbitrary case a minimum of five non-identical RDCs have to be measured to unambiguously identify a probability tensor of a given sample.

Although the concept of the probability tensor **P** is very intuitive and sufficient to completely describe RDCs in rigid molecules, it is more common in the NMR literature to express the partial orientation of a molecule by the so-called alignment tensor **A**. This tensor is related to the probability tensor **P** via

$$\mathbf{A} = \mathbf{P} - \frac{1}{3}\mathbf{1} \qquad (4)$$

and is therefore proportional to the anisotropy of the magnetic field distribution in the molecular frame. The eigenvalues of **A** are usually denoted as A_{xx}, A_{yy}, and A_{zz} and fulfill the condition $A_{xx} + A_{yy} + A_{zz} = 0$. For further definitions like the axial and rhombic components of an alignment tensor of the Saupe matrix, we refer to the appendix of [4].

Alignment Media

So far, we have not discussed how to partially orient the molecules. When a molecule has considerable magnetic susceptibility anisotropy, the molecule orients partially under the strong magnetic field. But this effect—although already observed in the eighties [6]—is normally too small to be observed. On the other hand, orientation can also be induced by an oriented molecular lattice, which then partially aligns the molecule of interest via steric or electrostatic interactions. All known alignment media can basically be assigned to one of two classes: liquid crystalline phases and stretched polymer gels.

Liquid Crystalline Phases

In 1963, the first spectrum of benzene partially oriented in the nematic mesophase of 4,4′-di-*n*-hexyloxyazoxybenzene was reported, a spectrum with at least 30 resonances and a width of the multiplet pattern of approximately 2500 Hz [7]. Following this result, a large number of liquid crystals for partially aligning small molecules was found in the sixties and seventies (see, for example, reviews [8–10]), but it turned out that the orientation introduced by liquid crystalline phases generally is very strong, yielding numerous large splittings by RDCs that can hardly be interpreted for more complex organic molecules. Poly-γ-benzyl-L-glutamate, dissolved in solvents like dichloromethane, chloroform, or dimethylformamide, was introduced by Panar and Phillips [11]. It was one of the least orienting liquid crystals known at that time and was used for the first successful measurements of RDCs in an organic solvent to obtain structural information of a small molecule [12–14]. Currently, new liquid crystalline phases for organic solvents have been developed that can achieve lower degrees of anisotropies. 4-*n*-Pentyl-4′-cyanobiphenyl [15] and poly-γ-ethyl-L-glutamate [16] seem to be two promising candidates for the measurement of RDCs in small molecules. The existence of liquid crystalline phases with very low induced anisotropies was proven in the last 7 years in the field of biomolecular NMR. Several lipid/detergent mixtures [2,3], filamentous phage [17], and other liquid crystalline phases [18,19] were used to successfully measure RDCs for structure refinement of proteins and nucleic acids. These alignment media are, of course, also applicable to small molecules in aqueous solutions [20–26].

Stretched Polymer Gels

Deloche and Samulski showed in their pioneering work in 1981 that partial alignment can also be achieved by stretching polymer gels [27], usually mechanically. This technique now is a standard approach in polymer NMR to obtain information about polymer properties. Not until 2000 was the use of these stretched polymer gels for aligning molecules dissolved inside the gel to obtain RDCs for structural investigations realized. The new approach is called "strain-induced alignment in a gel" and is demonstrated on stretched polyacrylamide [28–31] and polyacrylamide/acrylate [32] copolymers. Several ways of aligning by using a shigemi plunger [28], teflon funnels [33], or glass capillaries [32] have been developed and successfully applied to proteins as well as small molecules.

Recently, we were able to show that stretched polymers can also be used to partially align small molecules in organic solvents [34]. Cross-linked polystyrene (PS) sticks were simply swollen in NMR tubes, where they are automatically stretched by the boundaries of the glass walls (cf. Figure 1). In the meantime, the approach was successfully applied to a number of other cross-linked polymers like polymethylmethacrylate (with no polymer signals in the aromatic region, unpublished results) and polydimethylsiloxane (with strongly improved spectral quality [35]). Very recently, the gap of alignment media for polar

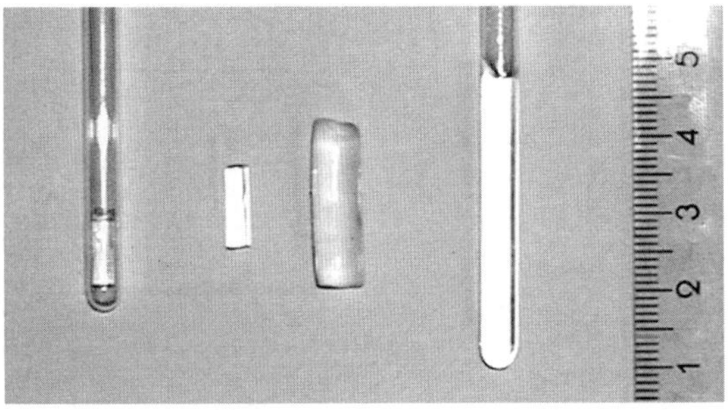

Fig. 1. Photographs of a cross-linked PS stick designed for partial alignment in different states of swelling. From left to right: unswollen polymer stick in standard 5-mm NMR tube, polymer stick directly after polymerization, free polymer stick completely swollen, and polymer stick swollen in the NMR tube.

solvents was closed by stretched gels that swell in DMSO [36,37] and other solvents like methanol, acetonitrile, or acetone [36]. The main advantage of stretched polymer gels compared to liquid crystalline phases is that no lower limit of alignment is imposed because an unstretched gel principally does not show any alignment. The induced anisotropy is generally field independent, but as in liquid crystals, a significant temperature dependence is observed (35,72). Despite the microheterogeneity of the gels, line widths below 1 Hz can be achieved. Very importantly, the strength of alignment can be adjusted in many ways, like changing the diameter of the polymer sticks, the amount of cross-linking agent, or the polymerization conditions [34–36]. Of course, swelling properties and thus the strength of alignment are different in different solvents. It is also interesting to note that the orientation and overall form of the alignment tensor not only depends on the polymer chosen but also on the solvent used [36, 72].

RDC Measurement

NMR experiments on partially aligned samples are special with respect to some details in the setup. The lock, for example, is usually split by the quadrupolar coupling of the deuterium nuclei. As a consequence, the spectrometer might lock on one or the other half of the doublet and built-in spectrometer frequency references that are based on the defined lock signal typically lead to wrongly referenced chemical shifts. Also, shimming is affected: since the lock signal is split, it cannot be used for shimming purposes. Instead, it is necessary to shim on the integral of the FID. Having learned how to deal with these specialties, the acquisition of NMR experiments is straightforward. Some general considerations concerning the practicability of measuring certain RDCs should, however, be taken into account. We therefore have a quick look at the relative size of different types of RDCs before we give an overview of a number of available experiments.

Relative Size of RDCs

Equations (1) and (2) contain all the information necessary to estimate the relative size of different types of RDCs. The dipolar coupling is proportional to the gyromagnetic ratios γ_I and γ_S and to the inverse cube of the internuclear distance r of an arbitrary pair of spins. Since the RDCs also depend on the angle θ, their size generally cannot be predicted without knowing the alignment tensor. Nevertheless, the maximum possible RDC values can be estimated and we will concentrate on these maximum values in the following text. In Figure 2, the relative size of maximum ^{1}H-X-RDCs is shown for different nuclei with respect to the internuclear distance r. It turns out that typical D_{HH} RDCs over two or three bonds and one bond D_{CH} RDCs show the strongest couplings, while long-range D_{CH} or direct D_{CC} RDCs (one-fourth of the D_{CH} couplings for a given distance r) are one order of magnitude smaller in size. Because of potential line broadening and to avoid strong coupling or second order effects in NMR spectra, the largest occurring RDCs should be ideally limited to be $\leq\sim$30 Hz by adjusting the alignment medium. In this range, one bond D_{CH} and D_{HH} couplings can be accurately measured and provide valuable structural information. All other RDCs to carbon nuclei, however, are very small in size and require highly accurate techniques to obtain structural information. Therefore, the goal must be to measure as many strong RDCs as possible and only in cases where the easily obtained information is not sufficient, more advanced techniques should be used to measure the weaker RDCs, which generally have a larger relative error. In addition to proton and carbon bound RDCs, one bond D_{NH} couplings should provide useful structural information with RDCs of intermediate size. Whenever present, D_{HF} and D_{FF} RDCs behave very similar to D_{HH} RDCs. Although having a high gyromagnetic ratio and 100% natural abundance, RDCs involving ^{31}P nuclei are usually small because of the large internuclear distance to the next NMR-active nucleus.

Experimental Methods

After having the assignment, RDCs are in principle measurable in 1D experiments. Proton-coupled ^{13}C spectra provide useful information in this regard. In fact, if the alignment is so strong that RDCs are on the order of $^1J_{CH}$ coupling constants, one-dimensional experiments provide the only viable way to measure RDCs [12]. However, if the largest RDCs are in the desired range of $\pm$30 Hz

Fig. 2. Relative size of maximum dipolar couplings for protons for various NMR-active nuclei with respect to the internuclear distance r [see Equations (1) and (2)]. ^{13}C–^{13}C dipolar couplings are one-fourth of the ^{1}H–^{13}C dipolar couplings.

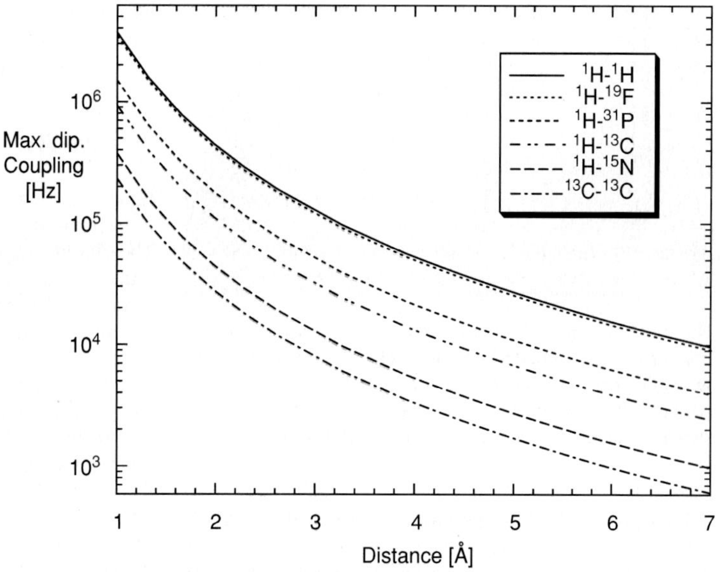

more advanced NMR experiments with high sensitivity and better resolution can be applied. The most easily measurable RDCs are one bond D_{CH} couplings, which can be measured using standard coupled HSQC or HMQC experiments [12–14,25,26,31,34–36] (cf. Figure 3, for example, spectra of strychnine in PS/$CDCl_3$ gel). Because of the variation in ($^{1}J_{CH} + D_{CH}$)-effective couplings, a procedure in which every multiplet component is phased separately should be applied to get reliable coupling constants [25].

If the resolution of standard HSQC/HMQC experiments is not sufficient, a large number of specialized techniques developed in the field of biomolecular NMR spectroscopy are available to accurately measure one bond D_{CH} and D_{NH} couplings. Spin-state selective excitation [38,39] and coherence transfer [40], IPAP [41],

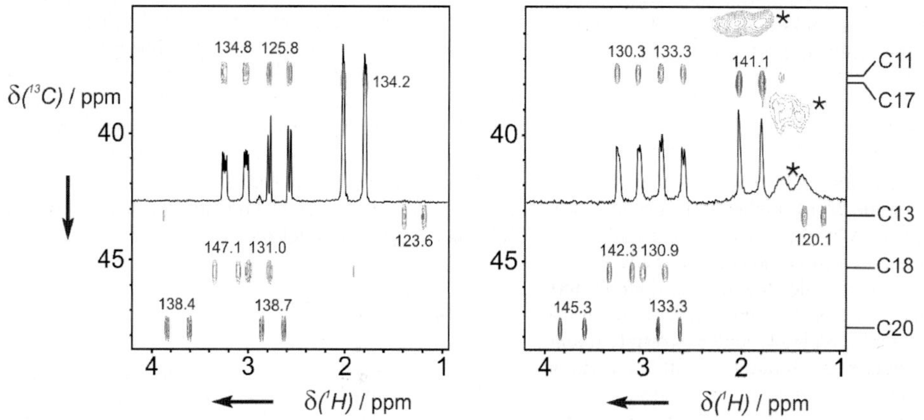

Fig. 3. Region of the ^{1}H,^{13}C-correlation spectra (HSQC) acquired on ~50 mg strychnine in $CDCl_3$ (left) and in a PS gel swollen in $CDCl_3$ (right) at 300 K. $^{1}J_{HC}$ and $^{1}J_{HC} + D_{HC}$ couplings, respectively, are given next to the corresponding cross peaks. As can be seen from the 1D slice at 37.8 ppm D_{CH} RDCs are easily measured. The broad signals of the PS polymer are indicated with an asterisk (*).

J modulation [42], JE-TROSY [43], and quantitative J [44] experiments are some useful examples. For methylene groups, the so-called SPITZE-HSQC provides a separation of multiplet components in four spectra that allow RDC determination in more crowded regions [45]. It should also be mentioned that in almost all experiments, the use of BIRD filtering techniques can further improve spectra by selectively decoupling long-range C–H couplings [46–48]. The experiments designed for measuring D_{CH} RDCs should also be easily adaptable to the D_{NH} case. Due to the fast averaging D_{CH} RDCs of methyl groups cannot be used directly as structural information. However, they can be converted to D_{CC} RDCs to the attached carbon via the formula $D_{CC} = D_{CH_3}(-3\gamma_C/\gamma_H)(r^3_{CH}/r^3_{CC})$ [13].

The measurement of D_{HH} couplings is more difficult: in simple cases $(J + D)$ splittings can be measured directly in a one-dimensional experiment. However, since the sign of the splitting is not known and J and D are of comparable size, two possible RDC values must be considered (e.g. $J = 5$ Hz, $|J + D| = 7$ Hz leads to either $D = 2$ or -12 Hz). A sign-sensitive measurement of coupling constants is therefore highly desirable, which can be achieved in principle by E-COSY [49–51] and S^{3}E-E-COSY [39,52] methods. Unfortunately, the applicability of at least the homonuclear E-COSY methods might be strongly reduced because of the broadened lines of partially aligned samples. A more sophisticated homonuclear approach to accurately measure RDCs is the SIAM-TOCSY [53,54], which uses in-phase and anti-phase components in TOCSY-type experiments in combination with the Keeler–Titman fitting procedure [55]. Integration-based techniques like the CT-COSY [56,57] or even the integration of conventional COSY cross peaks [58] offer an alternative for measuring the size of RDCs. The sign of couplings might then be determined with additional pulse sequences that use recently developed Hartmann–Hahn like multiple pulse elements [59]. The sign of the dipolar coupling might also be measurable by slightly changing the strength of alignment in the same alignment medium. The difference in measured RDCs should then lead to the sign of the dipolar coupling relative to the scalar coupling.

Long range D_{CH} couplings can basically obtained in two ways: if the carbon atom has a proton attached, S^{3}E-E-COSY type methods [52,60] provide an efficient, sign-sensitive procedure; in the case that the carbon of interest does not have a proton attached, recently developed HMBC-based methods can be applied that use a sophisticated fitting procedure to obtain the size of the coupling [13,61,62].

Methods for measuring D_{CC} couplings at natural abundance are the well-known INADEQUATE [63,64] and ADEQUATE [65,66] experiments, which inherently have very low sensitivity. RDCs to ^{31}P nuclei have been measured in nucleic acids [67,68]. Fluorine with its high gyromagnetic ratio and 100% natural abundance is perfectly suited for RDC measurements. D_{CF} couplings should be measurable the same way as D_{CH} RDCs. In the case when a rather strong J_{HF} coupling is present, specially designed E-COSY and S^{3}E-E-COSY type experiments allow the very sensitive measurement of D_{HF}, D_{HH}, and D_{FF} couplings [69,70].

Applications

The first applications to obtain structural information from RDCs for sugars were reported in 2000 and for medium-sized molecules in organic solvents in 2003. Considering this short time-span, the demonstrated uses of RDCs are quite remarkable.

Maybe the most elegant application of RDCs is found in six-membered chair-like rings, where RDCs can be used directly to distinguish axial and equatorial protons without having the need to derive an alignment tensor [25]. The method uses the fact that all axial C–H vectors are oriented in the same direction and therefore must have virtually identical D_{CH} couplings. The assignment then simply can be achieved by looking at the occurrence of identical RDCs: all protons with very similar D_{CH} RDCs are axial while all others are equatorial (cf. Figure 4).

But even for arbitrary molecules the prochiral assignment of methylene groups can be achieved in most cases without having to rely on NOE connectivities as long as the alignment tensor is known [12,13]. This is especially important for isolated CH_2 groups, which have few neighbors for NOE identification.

A major application can also be found in the determination of Z/E configurations [14] and the determination of the relative chirality of stereochemical centers [26,35] even in different parts of a molecule that might be spatially distant. For this technique, structural models of all possible configurations have to be built first. The measured RDCs are then compared with RDCs backcalculated from the different structures. The quality of the fits expressed in

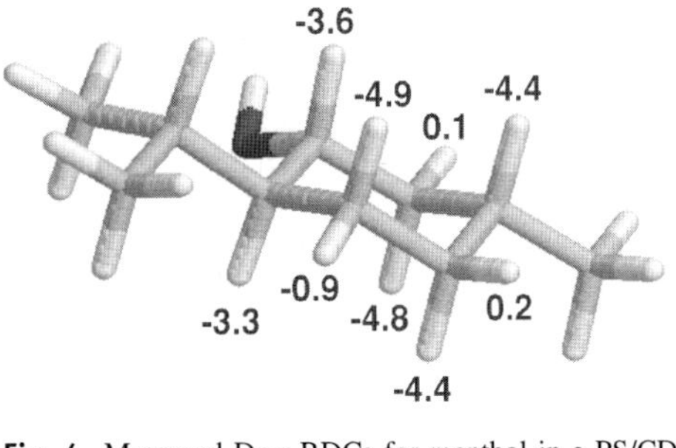

Fig. 4. Measured D_{CH} RDCs for menthol in a PS/$CDCl_3$ gel. Axial and equatorial protons can easily be distinguished without knowing the alignment tensor.

χ^2 or the correlation factor R^2 leads to the best structural model.

A challenging application is the determination of the sugar pucker with RDCs. Freedberg [24] has shown that the RDCs measured for sucrose cannot be explained by a single conformation. On the other hand, a large number of potential sugar puckers can be ruled out, giving more insight into the population of structures and the dynamic behavior of sugars.

Dynamic regions in molecules are generally difficult to handle and there is no exception in the case of RDCs. RDCs in dynamic regions are averaged over all populated states and might lead to misinterpretations if the populated structures differ significantly. However, the use of several different alignment media should make it possible to unambiguously identify flexible regions in a molecule and together with the improved prediction of alignment tensors [71], the determined RDCs might be fitted to a small ensemble of structures in the future.

A potential application of RDCs as upper distance restraints surprisingly has not yet been reported. RDCs over distances larger than 7Å have been measured. For a given measured RDC, an upper distance restraint can be immediately specified, as long as the axial component of the alignment tensor is known.

Conclusion

The development of partial alignment media over the last years allows for the first time the broad use of anisotropic parameters like RDCs or residual chemical shift anisotropy in high-resolution NMR spectroscopy. Liquid crystalline phases and stretched polymer gels with scalable alignment properties for practically all useful NMR solvents are available. The RDCs of these samples can be measured by simply calculating the difference of couplings measured in isotropic solution and couplings measured in aligned samples. So far mostly solutes in aqueous solutions have been studied. Recently developed alignment techniques in stretched polymer gels will open the measurement of RDCs to structural studies in all NMR relevant solutions. Quite a number of applications with small organic molecules have been demonstrated, including the assignment of axial/equatorial or prochiral methylene protons, and the determination of relative stereochemistry and configuration. The method significantly widens the potential of modern liquid-state NMR spectroscopy and it will be interesting to see new developments in this promising young field.

References

1. Reif B, Hennig M, Griesinger C. Science. 1997;276:1230.
2. Tjandra N, Bax A. Science. 1997;278:1111.
3. Prestegard JH. Nat. Struct. Biol. 1998;5:517.
4. Kramer F, Deshmukh MV, Kessler H, Glaser SJ. Concepts Magn. Reson. Part A 2004;21A:10.
5. Ernst RR, Bodenhausen G, Wokaun A. Principles of Nuclear Magnetic Resonance in One and Two Dimensions. Oxford University Press: New York, 1987.
6. Bothnerby AA, Domaille PJ, Gayathri C. J. Am. Chem. Soc. 1981;103:5602.
7. Saupe A, Englert G. Phys. Rev. Lett. 1963;11:462.
8. Emsley JW, Lindon JC. NMR Spectroscopy Using Liquid Crystal Solvents. Pergamon Press: Oxford, 1975.
9. Diehl P, Henrichs PM. Specialist Report on NMR Spectroscopy. Chemical Society, Vol. 1, 1972.
10. Diehl P, Niederberger W. Specialist Report on NMR Spectroscopy. Chemical Society, Vol. 3, 1974.
11. Panar M, Phillips WD. J. Am. Chem. Soc. 1968;90:3880.
12. Thiele CM, Berger S. Org. Lett. 2003;5:705.
13. Verdier L, Sakhaii P, Zweckstetter M, Griesinger C. J. Magn. Reson. 2003;163:353.
14. Aroulanda C, Boucard V, Guibe F, Courtieu J, Merlet D. Chem. Eur. J. 2003;9:4536.
15. Bendiak B. J. Am. Chem. Soc. 2002;124:14862.
16. Thiele CM. J. Org. Chem. 2004;69:7403.
17. Hansen MR, Mueller L, Pardi A. Nat. Struct. Biol. 1998; 5:1065.
18. Rückert M, Otting G. J. Am. Chem. Soc. 2000;122:7793.
19. Fleming K, Gray D, Prasannan S, Matthews S. J. Am. Chem. Soc. 2000;122:5224.
20. Martin-Pastor M, Bush CA. Abstracts of Papers of the Am. Chem. Soc. 2000;220:U117.
21. Martin-Pastor M, Bush CA. J. Biomol. NMR 2001;19: 125.
22. Neubauer H, Meiler J, Peti W, Griesinger C. Helvetica Chim. Acta. 2001;84:243.
23. Azurmendi HF, Bush CA. Carbohydr. Res. 2002;337:905.
24. Freedberg DI. J. Am. Chem. Soc. 2002;124:2358.
25. Yan JL, Kline AD, Mo HP, Shapiro MJ, Zartler ER. J. Org. Chem. 2003;68:1786.
26. Yan JL, Delaglio F, Kaerner A, Kline AD, Mo HP, Shapiro MJ, Smitka TA, Stephenson GA, Zartler ER. J. Am. Chem. Soc. 2004;126:5008.
27. Deloche B, Samulski ET. Macromolecules. 1981;14:575.
28. Tycko R, Blanco FJ, Ishii Y. J. Am. Chem. Soc. 2000; 122:9340.
29. Ishii Y, Markus MA, Tycko R. J. Biomol. NMR 2001;21:141.
30. Sass HJ, Musco G, Stahl SJ, Wingfield PT, Grzesiek S. J. Biomol. NMR 2000;18:303.
31. Mangoni A, Esposito V, Randazzo A. Chem. Commun. 2003; 154.
32. Meier S, Haussinger D, Grzesiek S. J. Biomol. NMR 2002;24: 351.
33. Chou JJ, Gaemers S, Howder B, Louis JM, Bax A. J. Biomol. NMR 2001;21:377.
34. Luy B, Kobzar K, Kessler H. Angew. Chem. Int. Ed. 2004; 43:1092.
35. Freudenberger C, Spitteler P, Bauer R, Kessler H, Luy B. J. Amer. Chem. Soc. 2004;126:14690.
36. Freudenberger C, Kobzar K, S. Knör, Heckmann D, Paululat T, Kessler H, Luy B. Angew. Chem. Int. Ed. 2005;44:423.
37. Haberz P, Farjon J, Griesinger C. Angew. Chem. Int. Ed. 2005;44:427.

38. Meissner A, Duus JO, Sørensen OW. J. Biomol. NMR 1997; 10:89.
39. Meissner A, Duus JO, Sørensen OW. J. Magn. Reson. 1997; 128:92.
40. Sørensen MD, Meissner A, Sørensen OW. J. Biomol. NMR 1997;10:181.
41. Ottiger M, Delaglio F, Bax A. J. Magn. Reson. 1998;131: 373.
42. Tjandra N, Bax A. J. Magn. Reson. 1997;124:512.
43. Luy B, Marino JP. J. Magn. Reson. 2003;163:92.
44. Vuister GW, Bax A. J. Am. Chem. Soc. 1993;115:7772.
45. Carlomagno T, Peti W, Griesinger C. J. Biomol. NMR 2000;17:99.
46. Garbow JR, Weitekamp DP, Pines A. Chem. Phys. Lett. 1982; 93:504.
47. Uhrin D, Liptaj T, Kover KE. J. Magn. Reson. Ser. A 1993; 101:41.
48. Feher K, Berger S, Kover KE. J. Magn. Reson. 2003;163:340.
49. Griesinger C, Sørensen OW, Ernst RR. J. Am. Chem. Soc. 1985;107:6394.
50. Griesinger C, Sørensen OW, Ernst RR. J. Chem. Phys. 1986; 85:6837.
51. Griesinger C, Sørensen OW, Ernst RR. J. Magn. Reson. 1987; 75:474.
52. Meissner A, Sørensen OW. Magn. Reson. Chem. 2001;39:49.
53. Prasch T, Gröschke P, Glaser SJ. Angew. Chem. Int. Ed. 1998;37:802.
54. Möglich A, Wenzler M, Kramer F, Glaser SJ, Brunner E. J. Biomol. NMR 2002;23:211.
55. Titman JJ, Keeler J. J. Magn. Reson. 1990;89:640.
56. Tian F, Bolon PJ, Prestegard JH. J. Am. Chem. Soc. 1999;121:7712.
57. Wu ZR, Bax A. J. Magn. Reson. 2001;151:242.
58. Delaglio F, Wu ZR, Bax A. J. Magn. Reson. 2001;149:276.
59. Kramer F, Jung A, Brunner E, Glaser SJ. J. Magn. Reson. 2004;169:49.
60. Meissner A, Sørensen OW. Chem. Phys. Lett. 1997;276:97.
61. Edden RAE, Keeler J. J. Magn. Reson. 2004;166:53.
62. Schulte-Herbrüggen T, Meissner A, Papanikos A, Meldal M, Sørensen OW. J. Magn. Reson. 2002;156:282.
63. Bax A, Freeman R, Kempsell SP. J. Am. Chem. Soc. 1980;102:4849.
64. Meissner A, Sørensen OW. Concepts Magn. Reson. 2002;14: 141.
65. Reif B, Köck M, Kerssebaum R, Kang H, Fenical W, Griesinger C. J. Magn. Reson. Ser. A 1996;118:282.
66. Reif B, Køck M, Kerssebaum R, Schleucher J, Griesinger C. J. Magn. Reson. Ser. B 1996;112:295.
67. Wu ZR, Tjandra N, Bax A. J. Biomol. NMR 2001;19:367.
68. Carlomagno T, Hennig M, Williamson JR. J. Biomol. NMR 2002;22:65.
69. Luy B, Marino JP. J. Biomol. NMR 2001;20:39.
70. Luy B, Barchi JJ, Marino JP. J. Magn. Reson. 2001;152:179.
71. Zweckstetter M, Bax A. J. Am. Chem. Soc. 2000;122:3791.

Part II

Measurement of Residual Dipolar Couplings and Applications in Protein NMR

Keyang Ding and Angela M. Gronenborn

Laboratory of Chemical Physics, National Institute of Diabetes and Digestive and Kidney Diseases, National Institutes of Health, Bethesda, MD 20892

Abstract

Residual dipolar couplings (RDCs) provide important constraints for the determination and refinement of protein NMR structures. Based on echo–anti-echo manipulation or IPAP principle, a suite of sensitivity enhanced experiments are described for measuring backbone $^1H^N$–^{15}N, ^{15}N–$^{13}C'$, $^1H^N$–$^{13}C'$, $^{13}C'$–$^{13}C^\alpha$, $^{13}C^\alpha$–$^1H^\alpha$, $^{15}N(i)$–$^{13}C^\alpha(i)$, $^1H^N(i)$–$^{13}C^\alpha(i)$, $^{15}N(i)$–$^{13}C^\alpha(i-1)$, and $^1H^N(i)$–$^{13}C^\alpha(i-1)$ dipolar couplings in proteins. The accuracy of the measured couplings can be assessed by comparing the experimentally obtained values with those predicted based on high resolution structures. Even for very small RDCs, such as the $^{15}N(i)$–$^{13}C^\alpha(i-1)$ couplings that are smaller than 0.3 Hz, a correlation coefficient of 0.83 is obtained, attesting to the accuracy of couplings obtained with these sensitivity-enhanced IPAP experiments. We also present a novel application for the use of RDCs. Under certain conditions, the folded state of a protein comprises detectable, conformational sub-states. Such sub-states at local sites, so-called melting hot spots, are characterized by re-orienting bond vectors. Determination of RDCs allows for efficient and easy detection of such hot spots.

Introduction

The importance of being ordered has been highly recognized in the field of protein NMR spectroscopy [1]. When a protein is placed into a magnetic field, it will experience a small degree of alignment. This causes any pair of spins within the protein to exhibit a dipolar interaction in addition to the scalar interaction that is manifested in the J coupling. It was noted earlier [2,3] that these small dipolar couplings are a direct reflection of the protein structure, since they provide a measure of the distance and the orientation of the vector connecting the two nuclei. Most proteins lack sufficient intrinsic magnetic susceptibility anisotropies for practical alignment purposes, however, alignment of any biological macromolecule can be achieved using liquid crystalline media. Generally, there are two major aspects that need to be addressed in the residual dipolar coupling (RDC)-related techniques: selection and optimization of the alignment media to yield measurable dipolar couplings and the development of accurate methods for measuring these small couplings.

Any alignment medium should be inert to the solutes (proteins) under investigation and needs to be stable over reasonably wide pH and temperature ranges. In addition, it should not cause any adverse effects on the NMR line widths, so that the NMR spectra in the alignment media remain at comparable resolution to those in the isotropic aqueous solution. The first alignment medium ever used was a lyotropic liquid crystal consisting of binary mixtures of dihexanoyl phosphatidylcholine and dimyristoyl phosphatidylcholine [3]. At present, a large variety of different alignment media are in use, including suspensions of charged, rod-shaped viruses, such as tobacco mosaic virus and filamentous bacteriophages fd/M13 and Pf1, quasi-ternary systems of surfactant/salt/alcohol or aqueous alkyl-poly(ethylene glycol)/alcohol mixtures, known to form stacked lamellar phases, and vertical or radial strained polyacrylamide gels [4–10]. Such gels are generally believed to be the most inert media for proteins. The degree of alignment can be tuned by adjusting the concentration of the liquid crystalline materials or the degree of strain/compression in case of the polyacrylamide gels. A suitable degree of alignment results in RDC values for $^1D_{NH}$ between 15 and 25 Hz.

The most commonly investigated dipolar couplings for proteins are the five backbone one-bond ^{15}N–$^1H^N$, ^{15}N–$^{13}C'$, $^{13}C'$–$^{13}C^\alpha$, and $^1H^\alpha$–$^{13}C^\alpha$ and two-bond $^1H^N$–$^{13}C'$ RDCs [11]. In addition, one-bond $^{13}C^\alpha$–$^{13}C^\beta$ and $^1H^\beta$–$^{13}C^\beta$, and one-bond ^{13}C–^{13}C and ^1H–^{13}C couplings in methyl groups are also of interest [12,13]. Long-range ^1H–^1H RDCs of protons close in space or connected via multiple bonds may yield important information with respect to quantitative NOE connectivity or local conformation [14].

Accurate measurement of these small dipolar couplings is still a challenge in protein NMR spectroscopy. Generally, all presently available methods for RDC determination can be classified into two major categories.

Graham A. Webb (ed.), Modern Magnetic Resonance, 1269–1273.

Fig. 1. Schematic representation of the echo–anti-echo manipulation. The upper left box displays the normal area of a 2D ^{15}N–$^{1}H^{N}$ HSQC spectrum. Only one ^{15}N–$^{1}H^{N}$ cross peak is shown, split by the coupling J_{NX} and J_{HX} along the ^{15}N and ^{1}H dimensions, respectively. Using echo–anti-echo manipulation, a spectrum such as shown in the right box is obtained. One peak of the pair remains at its original position, while the other one is flipped around the zero ^{15}N frequency line. In this manner, coupled ^{15}N–$^{1}H^{N}$ cross peaks are separated. The splitting J_{NX} and J_{HX} can be the active coupling J_{NH}, or the passive coupling from a third spin X.

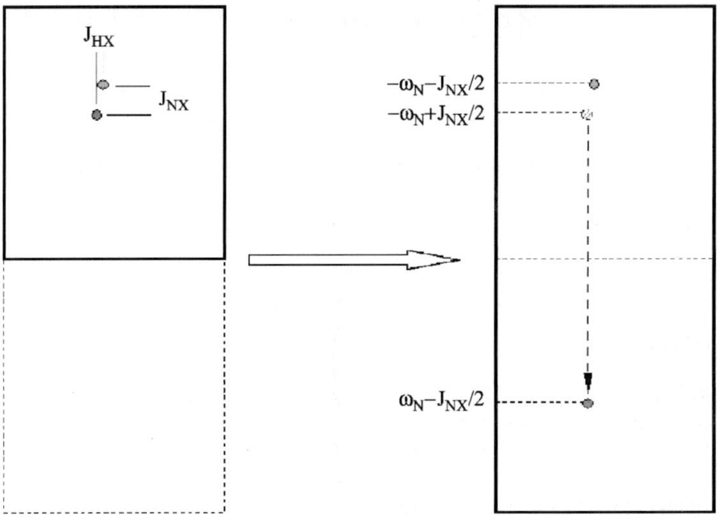

One methodology is called the quantitative J correlation and is based on intensity analysis [15]. Here, the coupling constant is often estimated from two points in the period during which this coupling evolves. If the measurement is based on a series of points, the resulting experiment is equivalent to a three-dimensional (3D) experiment. The accuracy of quantitative J correlation experiments relies critically on the uniqueness of the selected magnetization transfer pathways and the model employed to de-convolute the coupling from all other complex relaxation processes. The second methodology for the determination of dipolar couplings involves measuring the corresponding splitting directly from coupled multidimensional spectra. In these spectra the number of peaks is doubled due to coupling splitting, resulting in a reduction of resolution in these coupled spectra compared to their decoupled counterparts. In some cases, the splitting may actually be too small to be resolved with respect to the digital resolution of the spectra. It therefore is highly desirable to separate the two peaks within a split doublet.

In the following section, we describe techniques for separating the two peaks within a coupled doublet. They result in a suite of sensitivity-enhanced experiments for accurately and efficiently measuring backbone RDCs in proteins [16–20]. In addition, we describe some of the applications with respect to using RDCs to study protein folding and unfolding [21].

Measurement of Backbone Residual Dipolar Couplings in Proteins

Generally, there are two methods to separate the couplings within a split pair of resonances in a coupled spectrum: echo–anti-echo manipulation [16–18] and spin-state-selective technique [22–24]. The echo–anti-echo manipulation is schematically represented in Figure 1. The two peaks within a doublet are manipulated into P-type and N-type peaks [18], respectively. In a series of FIDs, the t_1-modulation alternates between echo and anti-echo for one peak and between anti-echo and echo for the other peak within the doublet. In this way, sensitivity enhancement is achieved naturally and directly. The loss in resolution of such coupled spectra, when compared to the decoupled spectra, can be easily compensated for by setting the spectral width in the ^{15}N dimension to twice the chemical shift range and by implementing TPPI [25,26] to shift the carrier frequency to the edge of the ^{15}N chemical shift range. In special cases, either J_{NX} or J_{HX} may be zero. Based on the echo–anti-echo manipulation, four sensitivity-enhanced experiments [16,17] were devised for measuring one-bond $^{1}H^{N}$–^{15}N dipolar couplings and one sensitivity-enhanced experiment [18] for simultaneously measuring one-bond ^{15}N–$^{13}C'$ and two-bond $^{1}H^{N}$–$^{13}C'$ dipolar couplings. We demonstrated the accuracy of the developed experiments using the protein GB1 aligned in liquid crystalline Pf1 (15 mg/ml) in 95% H_2O/5% D_2O at pH 7 and 25 °C and measured a large set of RDCs. Comparing the experimentally obtained RDCs with those calculated from a previously obtained high resolution model structure [27] (PDB code: 3GB1), excellent agreement between these values can be noted [16–18]. For one-bond ^{15}N–$^{1}H^{N}$, ^{15}N–$^{13}C'$ and two-bond $^{1}H^{N}$–$^{13}C'$ RDCs the correlation coefficients are 0.998, 0.98, and 0.96, respectively. Interestingly, the presence of cross-correlated relaxation between ^{15}N–$^{1}H^{N}$ and $^{1}H^{N}$–$^{1}H^{\alpha}$ dipolar interactions causes one-bond ^{15}N–$^{1}H^{N}$ couplings measured in the ^{15}N dimension to be different from those measured in

Table 1: Modulation of the coupling J_{MX} evolution and ^{15}N chemical shift frequency ω_N to the raw and manipulated FIDs in the 2D series for values of n from 1 to $TD_1/2$

2D series number	Real part	Imaginary part
Raw dataset		
$4n-3$	$\cos(\pi kJ_{MX}t_1)\cos(\omega_H t_2 - \omega_N t_1)$	$-\cos(\pi kJ_{MX}t_1)\sin(\omega_H t_2 - \omega_N t_1)$
$4n-2$	$\sin(\pi kJ_{MX}t_1)\sin(\omega_H t_2 - \omega_N t_1)$	$\sin(\pi kJ_{MX}t_1)\cos(\omega_H t_2 - \omega_N t_1)$
$4n-1$	$\cos(\pi kJ_{MX}t_1)\cos(\omega_H t_2 + \omega_N t_1)$	$-\cos(\pi kJ_{MX}t_1)\sin(\omega_H t_2 + \omega_N t_1)$
$4n$	$-\sin(\pi kJ_{MX}t_1)\sin(\omega_H t_2 + \omega_N t_1)$	$\sin(\pi kJ_{MX}t_1)\cos(\omega_H t_2 + \omega_N t_1)$
New dataset 1		
$(2n-1) = (4n-3) + (4n-2)$	$\cos[\omega_H t_2 - (\omega_N + \pi kJ_{MX})t_1]$	$-\sin[\omega_H t_2 - (\omega_N + \pi kJ_{MX})t_1]$
$2n = (4n-1) + (4n)$	$\cos[\omega_H t_2 + (\omega_N + \pi kJ_{MX})t_1]$	$-\sin[\omega_H t_2 + (\omega_N + \pi kJ_{MX})t_1]$
New dataset 2		
$(2n-1) = (4n-3) - (4n-2)$	$\cos[\omega_H t_2 - (\omega_N - \pi kJ_{MX})t_1]$	$-\sin[\omega_H t_2 - (\omega_N - \pi kJ_{MX})t_1]$
$2n = (4n-1) - (4n)$	$\cos[\omega_H t_2 + (\omega_N - \pi kJ_{MX})t_1]$	$-\sin[\omega_H t_2 + (\omega_N - \pi kJ_{MX})t_1]$

the ^1H dimension [17]. Therefore, it is in general preferable to measure the amide RDCs in the ^{15}N dimension in these spectra, although the orientational information is identical in both dimensions.

The spin-state-selective technique combined with sensitivity enhancement and interleaved data acquisition is summarized in Table 1 [19]. This approach is especially useful for the cases where the coupling is to be measured and the ^{15}N chemical shift evolve during different time periods. In these cases, it is often necessary to record a 3D experiment, which in turn is reduced in dimensionality from 3 to 2 [28–30] using the accordion principle [31,32]. Generally, a scaling factor of k can be introduced to scale the J_{MX} coupling evolution. By carefully setting the t_1-modulation according to Table 1, sensitivity-enhanced experiments were implemented for measuring one-bond $^{13}C'$–$^{13}C^\alpha$ and $^1H^\alpha$–$^{13}C^\alpha$ dipolar couplings [19]. In this way, two sub-spectra with resolution comparable to the decoupled spectra are obtained and the corresponding peak displacement along the ^{15}N dimension provides a direct measure of the one-bond $^{13}C'$–$^{13}C^\alpha$ and $^1H^\alpha$–$^{13}C^\alpha$ dipolar couplings. Again, the applicability of these experiments was demonstrated for protein GB1 aligned in liquid crystalline Pf1 (15 mg/ml) in 95% H_2O/5% D_2O at pH 7 and 25 °C. Excellent correspondence between measured and calculated one-bond $^{13}C'$–$^{13}C^\alpha$ and $^1H^\alpha$–$^{13}C^\alpha$ dipolar couplings was found [19], with correlation coefficients of 0.96 and 0.96, respectively.

In a further application of the spin-state-selective techniques, sensitivity-enhanced experiments for measuring $^{15}N(i)$–$^{13}C^\alpha(i)$, $^1H^N(i)$–$^{13}C^\alpha(i)$, $^{15}N(i)$–$^{13}C^\alpha(i-1)$, and $^1H^N(i)$–$^{13}C^\alpha(i-1)$ dipolar couplings were also developed [20]. Since the $^{15}N(i)$–$^{13}C^\alpha(i)$ and $^{15}N(i)$–$^{13}C^\alpha(i-1)$ coupling evolution and the ^{15}N chemical shift evolution are in a common time period, there is no need for the accordion principle. Every cross peak in the two sub-spectra exhibits E.COSY splittings from the intra-residue $^{13}C^\alpha$ spin. Because one-bond ^{15}N–$^{13}C^\alpha$ J couplings are small, ranging from 8.5 to 13.5 Hz, very high digital resolution in ^{15}N dimension is necessary for resolving the E.COSY doublet. Consequently, the application of this type of experiment is limited to small proteins with an apparent ^{15}N line width ≤ 8.5 Hz. The comparison between measured and calculated one-bond and two-bond ^{15}N–$^{13}C^\alpha$, as well as two-bond and three-bond $^1H^N$–$^{13}C^\alpha$ RDCs for the protein GB1 aligned in liquid crystalline Pf1 (15 mg/ml) in 95% H_2O/5% D_2O at pH 7 and 25 °C, yields correlation coefficients of 0.983, 0.833, 0.976, and 0.944, respectively. We wish to point out that the two-bond ^{15}N–$^{13}C^\alpha$ RDCs are extremely small, namely ≤ 0.3 Hz, still yielding a remarkable correlation coefficient of 0.833 [20]. This implies that the experimentally measured values have to be highly accurate and that the quality of the model structure is also very good.

Applications of Residual Dipolar Couplings in Proteins

The most direct application of RDCs in protein NMR is their use as orientational constraints for 3D structure determination. For a rigid molecule, the orientations of all the two-spin vectors are fixed with respect to the molecular frame. Therefore, the corresponding RDC is proportional to the scalar product between a 3×3 traceless order tensor, global to any pair of spins, and the second-order tensor of the two-spin vector [33]. It is this relationship between the order tensors that is the basis for using RDCs in NMR structure determination. If, however, the directions of some two-spin vectors reorient with respect to the molecular frame, the proportionality to the scalar product is broken and the corresponding RDCs cannot

Fig. 2. Characterization of melting hot spots in a destabilized mutant of GB1 (S14) by RDC measurement. The measured D_{NH} for S14 against the predicted values are shown in (A) and (B), and data points associated with local melting hot spots are displayed as filled triangles (T11, L12, K13) and filled circles (A24, T25). (C) Measured D_{NH} of S14 in the molten state vs. those predicted from the folded structure. (D) Correlation between measured D_{NH} of folded S14 at 25 °C in a stretched gel at pH 5.5 and at pH 3.0. (Data points for the residues comprising the melting hot spots are excluded.)

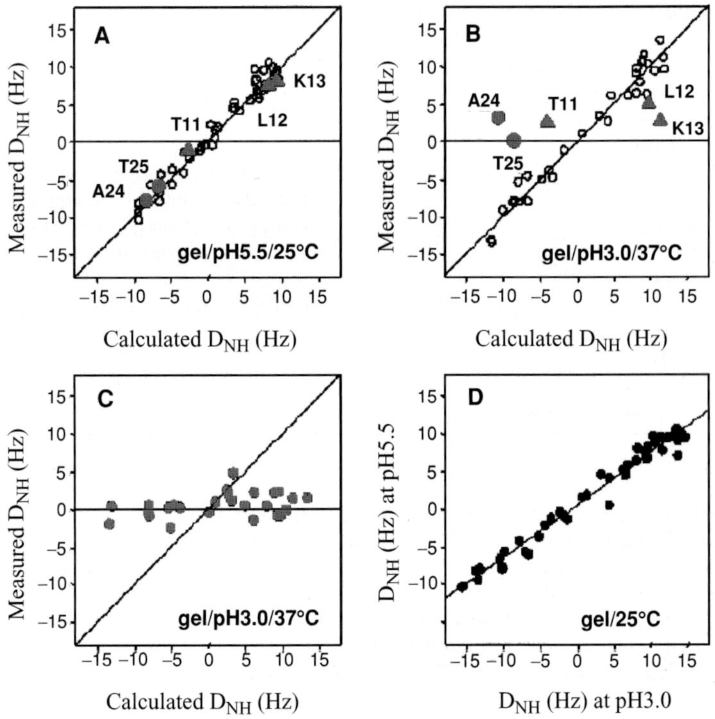

provide meaningful structural constraints. They can, on the other hand, reveal important information about motions, either relating to inter-domain motions [34] or local melting events [21]. Local melting hot spots are easily identified since local order tensor parameters will be different from the global tensor, caused by the local reorientation motion at this site. For example, we examined the structure and dynamics of a series of core mutants of the protein GB1. For one of these, S14, local melting was discovered under destabilizing conditions based on RDC measurements. The identification of the melting hot spots is illustrated in Figure 2 [21]. RDCs were measured in a stretched polyacrylamide gel at temperatures between 25 and 37 °C and at pH 5.5 they correlate well (Pearson correlation coefficient = 0.99) with calculated values based on the refined NMR structure of wild-type GB1 (PDB code: 3GB1). However, at pH 3.0 and 37 °C, the data points for residues T11, L12, K13, A24, and T25 deviate from the diagonal correlation line and are scattering along a horizontal line (Figure 2B). Under those experimental conditions S14 exists in a folded and unfolded state, as evidenced by a second set of resonances, not belonging to native state. For the unfolded protein, no correlation with the native, folded structure is preserved and all experimental RDCs scatter along the same horizontal line (Figure 2C) as noted before for the outliers of the folded state. We also established that the changes in RDCs are not caused by a pH change on the gel, since the couplings of the folded GB1 at pH 3.0 and 5.5 are in excellent agreement (Figure 2D).

In conclusion, our application of RDC measurements to the GB1 folding landscape clearly demonstrates their power for identifying melting hot spots in proteins.

Discussion

There is now ample data available (see above) that indicate the importance of RDC derived constraints for NMR structure determination of proteins as well as for identifying local areas of mobility. Proteins are also rich in ^1H–^{13}C spin pairs in the side chains. Therefore long-range ^1H–^1H RDCs will probably attract more and more attention since they will be able to provide quantitative distance constraints. Once accurate and efficient methods are developed for measuring these side-chain ^1H–^{13}C and the long-range ^1H–^1H dipolar couplings, NMR structural biology will move into a new area again.

Acknowledgments

This work was supported in part by the Intramural AIDS Targeted Antiviral Program of the Office of the Director of the National Institutes of Health to AMG.

References

1. Gronenborn AM. C. R. Biol. 2002;325:957–66.
2. Tolman JR, Flanagan JM, Kennedy MA, Prestegard JH. Proc. Natl. Acad. Sci. U.S.A. 1995;92:9279–83.
3. Tjandra N, Bax A. Science. 1997;278:1111–4.
4. Clore GM, Starich MR, Gronenborn AM. J. Am. Chem. Soc. 1998;120:10571–2.
5. Hansen MR, Mueller L, Pardi A. Nat. Struct. Biol. 1998;5: 1065–74.
6. Prosser RS, Losonczi JA, Shiyanovskaya IV. J. Am. Chem. Soc. 1998;120:11010–1.
7. Barrientos LG, Dolan C, Gronenborn AM. J. Biomol. NMR. 2000;16:329–37.
8. Ruckert M, Otting GJ. Am. Chem. Soc. 2000;122:7793–7.
9. Tycko R, Blanco FJ, Ishii Y. J. Am. Chem. Soc. 2000;122: 9340–1.
10. Sass HJ, Musco G, Stahl SJ, Wingfield PT, Grzesiek SJ. Biomol. NMR. 1998;18:303–9.
11. deAlba E, Tjandra N. Prog. Nucl. Magn. Reson. Spectrosc. 2002;40:175–97.
12. Ottiger M, Bax A. J. Am. Chem. Soc. 1999;121:4690–5.
13. Chou JJ, Bax A. J. Am. Chem. Soc. 2001;123;3844–5.
14. Wu Z, Bax A. J. Am. Chem. Soc. 2002;124:9672–3.
15. Bax A, Vuister GW, Grzesiek S, Delaglio F, Wang AC, Tschudin R, Zhu G. Methods Enzymol. 1994;239:79–105.
16. Ding K, Gronenborn AM. J. Magn. Reson. 2002;158:173–7.
17. Ding K, Gronenborn AM. J. Magn. Reson. 2003;163:208–14.
18. Ding K, Gronenborn AM. J. Am. Chem. Soc. 2003;125: 11504–5.
19. Ding K, Gronenborn AM. J. Magn. Reson. 2004;167:253–8.
20. Ding K, Gronenborn AM. J. Am. Chem. Soc. 2004;126: 6232–3.
21. Ding K, Louis JM, Gronenborn AM. J. Mol. Biol. 2004;335: 1299–307.
22. Ottiger M, Delaglio F, Bax A. J. Magn. Reson. 1998;131: 373–8.
23. Cordier F, Dingley AJ, Grzesiek S. J. Biomol. NMR. 1999;13: 175–80.
24. Lerche MH, Meissner A, Poulsen FM, Sorensen OW. J. Magn. Reson. 1999;140:259–63.
25. Drobny G, Pines A, Sinton S, Weitekamp D, Wemmer D. Faraday Div. Chem. Soc. Symp. 1979;13:49.
26. Bodenhausen G, Vold RL, Vold RR. J. Magn. Reson. 1980; 37:93–106.
27. Kuszewski J, Gronenborn AM, Clore GM. J. Am. Chem. Soc. 1999;121:2337–8.
28. Szyperski T, Yeh DC, Sukumaran DK, Moseley HNB, Montelione GT. Proc. Natl. Acad. Sci. U.S.A. 2002;99:8009–14.
29. Ding K, Gronenborn AM. J. Magn. Reson. 2002;156:262–8.
30. Kozminski W, Zhukov I. J. Biomol. NMR. 2003;26:157–66.
31. Bodenhausen G, Ernst RR. J. Magn. Reson. 1981;45:367–73.
32. Bodenhausen G, Ernst RR. J. Am. Chem. Soc. 1982;104: 1304–9.
33. Tolman JR, Al-Hashimi HM, Kay LE, Prestegard JH. J. Am. Chem. Soc. 2001;123;1416–24.
34. Braddock DT, Cai ML, Baber JL, Huang Y, Clore GM. J. Am. Chem. Soc. 2001;123:8634–5.

Using Chemical Shift Perturbations to Validate and Refine the Docking of Novel IgE Antagonists to the High-Affinity IgE Receptor

Melissa A. Starovasnik and Wayne J. Fairbrother
Department of Protein Engineering, Genentech Inc., South San Francisco, CA 94080, USA

The binding of IgE to its high-affinity receptor, FcεRI, is fundamental to allergic disease. Molecules that block this interaction could therefore act as useful therapeutics for the treatment of asthma, allergic rhinitis, and other forms of atopy. To this end, binding selections using the extracellular portion of the α-chain of FcεRI (FcεRIα) and polyvalent peptide-phage libraries have yielded two distinct classes of peptide ligands that antagonize IgE binding to its receptor and prevent downstream IgE-mediated signaling events in basophils [1,2]. NMR spectroscopy has been used to characterize the structures of these peptide antagonists and their modes of binding to FcεRIα.

Structure determination of the ~32-kDa peptide/receptor complexes by NMR methods was not feasible for these systems because the heavily glycosylated receptor protein could not be bacterially expressed and thus could not be isotopically labeled (at least not readily). The high carbohydrate content of the receptor also makes these systems challenging for structure determination using X-ray crystallography. As described below, however, large ^{1}H chemical shift perturbations observed for the isotopically labeled peptide ligands upon binding to the receptor protein in solution could be used qualitatively to validate hypothesis-driven models of ligand binding, and quantitatively to refine these models. The resulting structural models will be useful in the development of novel IgE antagonists.

Hairpin Peptide Structure

The first family of peptides identified by phage display have the general form $X_4CX_2GPX_4CX_4$ and bind FcεRIα with affinities in the 1–10 μM range [1]. Analysis of one of these peptides, IGE06, by NMR spectroscopy revealed that the peptide has a well-defined three-dimensional (3D) structure in solution comprising two β-strands (residues 5–8 and 11–14) connected by a type I β-turn centered at residues Pro9 and Trp10 (Figure 1A). The three N-terminal residues are not as well defined by the NMR data and appear to be more flexible in solution than residues Cys5-Cys14. Residues required for structure and/or function were identified by analysis of peptide variants using both NMR and activity measurements. Importantly, NMR analysis of peptides having alanine substitutions for Gly8, Pro9, and Trp10 indicate that these substitutions do not disrupt the hairpin structure even though they have a significant impact on FcεRIα-binding affinity (Figure 1A).

Zeta Peptide Structure

Using an expanded set of peptide-phage libraries, a second family of peptides with a simpler motif, X_2CPX_2CYX, was identified [2]. When synthetic peptides were assayed for inhibition of IgE binding to cell-surface FcεRI, however, IC_{50} values of 250 μM or greater were obtained. Surprisingly, this activity improved over time and stabilized at 64 μM when assayed similarly 10 days after solubilization. One-dimensional (1D) and two-dimensional (2D) NMR analysis of a freshly prepared nine-residue monomeric disulfide-bonded peptide demonstrated at least four conformationally distinct forms, with the major form showing no evidence of stable structure (Figure 2). Analysis of 1D NMR spectra over time showed changes in the relative populations of the different forms (Figure 2), and after 7 days at room temperature ~90% of the peptide had converted to a single new form that also showed significantly longer retention time than the original sample in an analytical HPLC run. This form had a stable structure in solution as evidenced by significant chemical shift dispersion relative to "random coil," and extreme values for backbone and side chain ^{3}J coupling constants. Mass spectra of the new peptide form showed it to be a covalent disulfide-linked dimer. Comparison of analytical HPLC chromatograms and NMR spectra of synthetic parallel or antiparallel dimeric peptides confirmed that the active state of the phage-derived peptide was an antiparallel dimer with two intermolecular disulfide bonds. Subsequent linking of the monomeric peptides with a three-residue linker sequence to make a 21-residue "single-chain dimer" species, and further optimization on phage, resulted in a 32 nM-affinity peptide antagonist (e131).

Graham A. Webb (ed.), Modern Magnetic Resonance, 1275–1280.

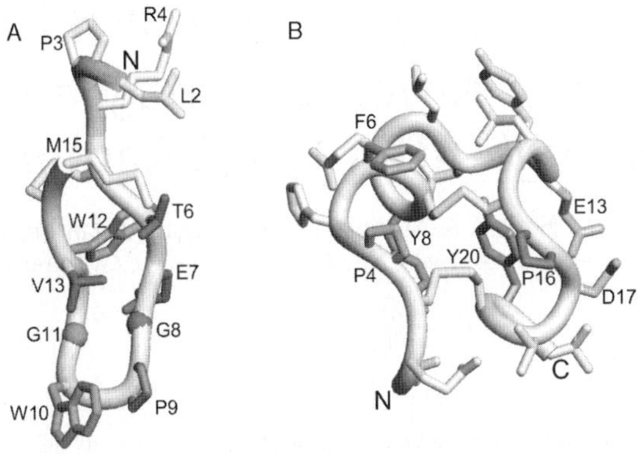

Fig. 1. Solution structures of representative (A) hairpin and (B) zeta peptides. Shown are ribbon diagrams of the first structure from the ensembles in PDB files 1JBF (hairpin) [1] and 1KCO (zeta) [2]. Side chains are colored to indicate those residues that are important for maintaining the structure of the peptide (green; disulfide bonds in yellow), and those that are important for receptor binding, but are not important for maintaining a stable structure of the peptide (from most important to least, red > orange > blue). The backbone of residues 2–4 of the hairpin peptide is colored to indicate that truncation of this region results in significant loss in receptor-binding affinity. (See also Plate 98 on page LXV in the Color Plate Section.)

NMR structural analysis of e131 showed that the peptide adopts a stable structure comprised of two small antiparallel 3_{10} helices connected by two disulfide bonds and a flexible linker region (Figure 1B). The helical backbone conformation is stabilized by packing of two Tyr side chains (Tyr8 and Tyr20) "below" the disulfide bonds, with each Tyr hydroxyl donating a hydrogen bond across the "dimer interface" to the carbonyl oxygen of a Cys residue. The backbone structure is pseudo-symmetric with residues 1–9 adopting essentially the same conformation as residues 13–21. Peptides of this class were designated "zeta" peptides due to a resemblance between the structurally conserved regions and the Greek letter zeta (ζ). As with the hairpin class of peptides, residues in the zeta peptide required for structure and/or function were identified from analysis of peptide analogs both by NMR and activity measurements. Again, the most important residue for high-affinity receptor binding was found to be a proline (Pro16) (Figure 1B).

Receptor Binding

Despite their significantly different structures, the two-peptide classes were shown to compete for binding at the same site on the receptor. Alanine-scanning mutagenesis of FcεRIα demonstrated further that the peptides bind to a region of the receptor that had been shown previously to contact IgE. This aromatic-rich IgE interaction site is centered around two tryptophan residues on the receptor (Trp87 and Trp110), between which Pro426 of bound IgE is buried [3]. In order to characterize further the receptor-binding modes of the different peptide classes, $^{13}C/^{15}N$-labeled peptides were obtained by expression of protein A Z-domain fusion proteins [4] in *Escherichia coli* using a previously described alkaline phosphatase induction system [5], followed by cleavage (using trypsin or CNBr for the zeta or hairpin peptide, respectively) and purification of the peptides [6]. Isotopic labeling of FcεRIα was not practical because the heavily glycosylated extracellular domain (average molecular mass ~30 kDa) could be produced only via baculoviral infection of insect cells [6].

NMR Analysis of Zeta Peptide/Receptor Complex

The 23-residue zeta peptide selected for NMR structural studies (e117) was shown previously to bind FcεRI and block IgE binding with an IC_{50} of 80 nM [2]. The resonances of the free zeta peptide were assigned readily by standard analysis of 2D $^1H/^{15}N$-HSQC, 3D $^1H/^{15}N$-TOCSY-HSQC, 2D $^1H/^{13}C$-HSQC, 2D CBCA(CO)NH, and 2D $^1H/^{13}C$-HCCH-TOCSY spectra [7]. Resonance assignments of the receptor-bound zeta peptide were complicated, however, by the overall reduced sensitivity of J-correlated spectra for the ~32 kDa complex, and by large variations in line widths, with those residues at the peptide termini and in the GGH "linker" being sharper and more intense than those from the rest of the peptide. Nevertheless, unambiguous assignments were obtained for a majority of the zeta peptide resonances based on analysis of 3D ^{15}N-edited and ^{13}C-edited NOESY spectra.

Comparison of the resonance assignments of the free and receptor-bound peptide showed that many of the peaks from residues within the "disulfide-bonded core" of the peptide experienced significant chemical shift perturbations; the most dramatic chemical shift changes upon receptor binding were seen for the ^1H resonances of Pro16 (Hα, Hβ, Hγ, and Hδ all shifted upfield by 1.7–3.1 ppm; Figure 3A). Shifts of this magnitude observed for all proton resonances within the proline side chain can be explained if the proline is packed between aromatic rings in the protein/peptide complex, in much the same way as Pro426 of IgE is found "sandwiched" between two tryptophan residues in the crystal structure of the Fc fragment of IgE in complex with FcεRIα [3]. By contrast, the sharper peaks from the bound state of the zeta peptide underwent only minimal changes in chemical shift relative to the free state (Figure 3A), consistent with this group of residues being of minimal importance for receptor binding [2].

In addition to the chemical shift perturbations indicating that the zeta peptide binds proximal to aromatic residues on the receptor, ^{13}C-edited/^{12}C-filtered NOESY

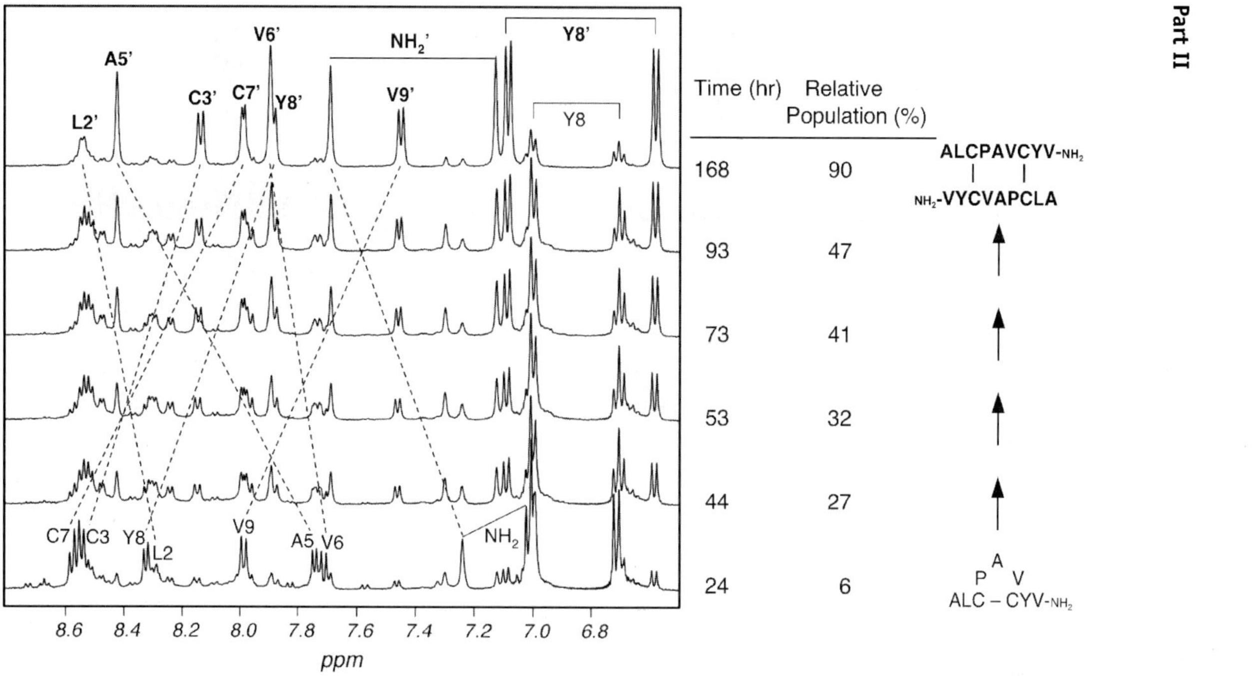

Fig. 2. Spontaneous rearrangement of a monomeric 9-mer peptide into a covalent disulfide-linked dimer monitored by 1D NMR spectroscopy. Amide and aromatic region of a series of 1D ^{1}H NMR spectra acquired at 500 MHz, 25 °C, on a sample of peptide e101 (ALCPAVCYV-NH_2), originally synthesized and purified as a monomeric peptide with an intramolecular disulfide bond. Lyophilized peptide was resuspended in water and spectra were acquired intermittently over the course of a week, with the time after resuspension indicated. The population of well-folded dimeric peptide present over time is estimated based on the relative intensities of the two most upfield peaks shown, corresponding to the Tyr8 $^1H^{\varepsilon}$ resonance in the ordered dimeric state and in the monomeric or intermediate states. Note that if the peptide existed in a single conformation in solution, that no more than 11 peaks would be expected in this region of the spectrum.

spectra [8] showed NOE correlations between at least two distinct aromatic groups on the receptor and peptide residues Val1, Leu14, and Val18. Peptide residues Val1, Gln2, and Pro4 also had NOE correlations with a pair of peaks resonating at 0.75 and 0.73 ppm that were assigned tentatively to a pair of methyl groups on the receptor based on resonance intensity and frequency. No NOE correlations were observed from Pro16 of the peptide, despite the large chemical shift changes observed for this residue, due to the increased line widths in the receptor-bound state. The resonance broadening is most likely due to the large chemical shift gradient in which the proline is located; even small changes in the orientation of Pro16 with respect to the aromatic residues on the surface of the receptor would cause chemical shift changes resulting in the observed resonance broadening.

Receptor contacts involving aromatic and methyl-containing side chains are consistent with the zeta peptide binding to the IgE-binding site as identified by receptor mutagenesis. The peptide was therefore docked manually into the IgE-binding site of FcεRIα using coordinates of the free peptide and the previously published crystal structure of the IgE/FcεRIα complex [3]. The initial peptide placement was based on the assumption that the functionally important proline residue, Pro16, which experienced the most significant ^{1}H chemical shift perturbations upon receptor binding (Figure 3A), would bind between Trp87 and Trp110 of the receptor, in approximately the same way as observed for Pro426 of IgE. The resulting peptide orientation appeared consistent with the observed intermolecular NOE correlations; in particular, intermolecular NOE correlations between zeta peptide residues Val1, Gln2, and Pro4 and the unidentified pair of receptor methyl groups could be assigned tentatively to Leu158 of the receptor because these are the only methyl groups in this region of the receptor.

The aliphatic ^{1}H chemical shifts of the manually docked peptide were calculated using the program SHIFTS 4.1 (X. Xu and D.A. Case, The Scripps Research Institute, La Jolla, CA; http://www.scripps.edu/case/

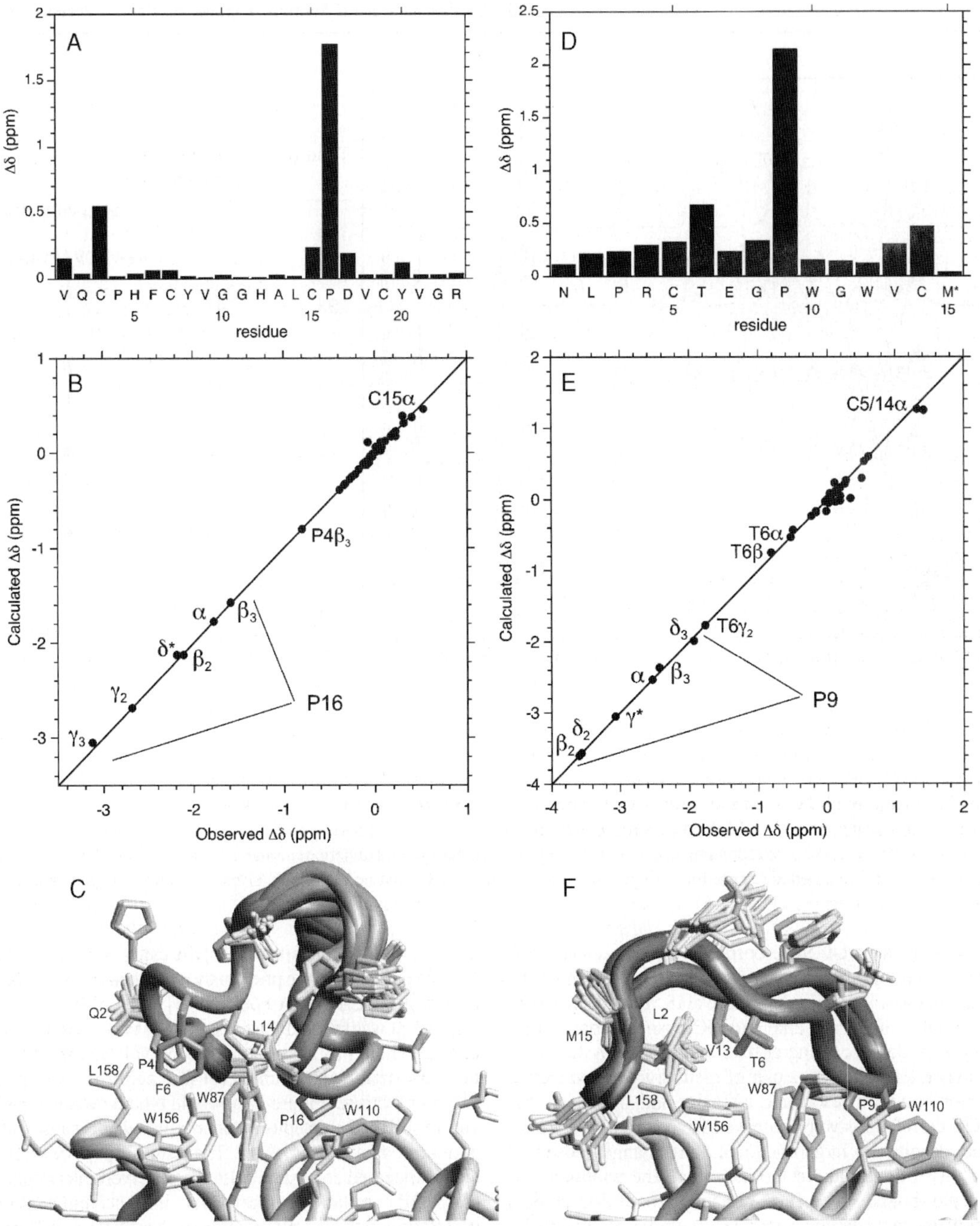

Fig. 3. (A) Plot of the net change in zeta peptide ${}^{1}H^{\alpha}$ chemical shifts upon binding FcεRIα ($\Delta\delta = [\Delta\delta^2_{{}^1H^\alpha}]^{1/2}$). (B) Correlation between observed and calculated chemical shifts ($\Delta\delta = \delta_{\text{total}} - \delta_{\text{random coil}}$) following refinement of the zeta peptide/FcεRIα complex structure using AMBER 6.0 (rms error = 0.038 ppm, $R^2 = 0.9978$). (C) Superposition of 20 models of the zeta peptide/FcεRIα complex structure obtained by docking and refinement against the ${}^{1}H$ chemical shifts of the peptide (average rms error = 0.041 ± 0.005 ppm). (D) Plot of the net change in hairpin peptide ${}^{1}H^{\alpha}$ chemical shifts upon binding FcεRIα. (E) Correlation between observed and calculated chemical shifts following refinement of the hairpin peptide/FcεRIα complex structure using AMBER 6.0 (rms error = 0.090 ppm, $R^2 = 0.9954$). (F) Superposition of 20 models of the hairpin peptide/FcεRIα complex structure obtained by docking and refinement against the ${}^{1}H$ chemical shifts of the peptide (average rms error = 0.12 ± 0.02 ppm). (See also Plate 99 on page LXVI in the Color Plate Section.)

qshifts/qshifts.htm), in which the "structural" effects on chemical shifts due to contributions of ring currents of aromatic groups [9,10], magnetic anisotropy from peptide groups [11], and electrostatic effects arising from charges and dipoles [12], are calculated using empirical functions [13–15]. The total chemical shift of a given resonance is defined as the sum of the "random coil" value, as observed in short disordered linear peptides [16,17], and the conformation-dependent contributions ($\Delta\delta$). The conformation-dependent contribution to the total chemical shift, as calculated by SHIFTS 4.1, is thus defined as:

$$\Delta\delta = \delta_{\text{total}} - \delta_{\text{random coil}} \approx \sum \delta_{\text{rc}} + \sum \delta_{\text{m}} + \sum \delta_{\text{el}} + \delta_{\text{const}} \quad (1)$$

where rc, m, and el, refer to ring current, magnetic anisotropy, and electrostatic contributions, respectively. The δ_{const} term is an empirical "fitting" parameter that includes contributions from δ_{m} and δ_{el} that are present in the random coil state [14]. Good general agreement was found between the calculated and experimentally observed chemical shifts, although differences suggested that the manually docked model could be refined further.

The docked model was therefore refined by optimizing the agreement between the calculated and observed ^{1}H chemical shifts of the receptor-bound peptide using functionally equivalent code within the molecular modeling package AMBER 6.0 (http://amber.scripps.edu/doc6/). All the unambiguously assigned aliphatic ^{1}H chemical shifts were incorporated as restraints (force constant = 25 kcal/·mol/·ppm) in a simple simulated annealing protocol. For initial rounds of refinement the chemical shifts of prochiral methylene and methyl protons were averaged. The chemical shifts of all peptide protons were subsequently back-calculated using SHIFTS 4.1 and compared to the experimentally determined chemical shifts. In some cases the chemical shifts of prochiral proton pairs were sufficiently different, and the agreement between the calculated and experimentally observed values was sufficiently close, to allow for stereospecific assignments that could be incorporated into subsequent rounds of refinement. In this way stereospecific assignments were possible for Pro4 ^{1}H$^{\beta}$, Leu14 ^{1}H$^{\beta}$, and Pro16 ^{1}H$^{\beta}$ and ^{1}H$^{\gamma}$.

In addition to the chemical shift restraints, distance and dihedral angle restraints derived from those used to calculate the solution structure of the free peptide were incorporated into the refinement to ensure that the peptide structure was maintained. The coordinates of the receptor were restrained to be nearly fixed during the simulated annealing and subsequent restrained minimization calculations using a harmonic potential with a force constant of 5 kcal/·mol. A weak harmonic potential (force constant = 0.5 kcal/·mol) was also used to restrain the disulfide-bonded core (residues Cys3-Tyr8 and Ala13-Tyr20) of the zeta peptide. Finally, tentatively assigned intermolecular NOE distance restraints were incorporated iteratively into the refinement calculations as upper-bound distance restraints of 5.8 Å, only if they were clearly satisfied in the current models; in total, 17 intermolecular distance restraints were included in the final simulated annealing calculation. Importantly, the intermolecular distance restraints were not required for convergence of the refined models. Indeed, trial calculations in which the starting zeta peptide and receptor molecules were separated by ~20 Å still converged to essentially the same peptide/receptor orientation as those starting from manually docked models, even in the absence of intermolecular distance restraints (W.J. Fairbrother, unpublished results).

The resultant agreement between the calculated and experimentally observed ^{1}H chemical shifts was excellent (Figure 3B), suggesting that the refined model of the zeta peptide/receptor complex (Figure 3C) represents a good approximation of the structure. In this model Pro16 packs between Trp87 and Trp110 on the receptor, and the N-terminal region of the peptide makes additional hydrophobic contacts near Leu158. A subsequent 3 Å-resolution X-ray crystal structure of the complex between a closely related zeta peptide (e131) and FcεRIα was in very close agreement with the NMR-based model [6], confirming the validity of the approach.

NMR Analysis of Hairpin Peptide/Receptor Complex

Resonances of the free 15-residue hairpin peptide, IGE32, were assigned by analysis of 2D ^{1}H/^{13}C-HSQC, 2D CBCA(CO)NH, and 2D ^{1}H/^{13}C-HCCH-TOCSY spectra [7]. Assignments of the receptor-bound state were obtained by analysis of 2D ^{1}H/^{13}C-HSQC and 3D ^{13}C-edited NOESY spectra. However, as observed for Pro16 of the zeta peptide, the resonances of the functionally important Pro9 were broad and did not show observable cross peaks in the ^{13}C-edited NOESY spectrum; assignment of these resonances was thus confirmed by analysis of a 2D ^{1}H/^{13}C-HSQC spectrum of a sample prepared with ^{13}C-Pro9 specifically labeled hairpin peptide generated by chemical synthesis.

Similar dramatic chemical shift changes to those found for the zeta peptide were observed when the hairpin peptide was added to FcεRIα (Figure 3D). In particular the ^{1}H resonances of residues Thr6 and Pro9 were shifted upfield by 1.7–3.2 ppm upon binding the receptor. As for the zeta peptide, these large chemical shift changes imply that the hairpin peptide packs closely with aromatic rings in the complex. In this case, a total of 40 intermolecular NOE correlations involving peptide residues Asn1, Leu2, Pro3, Thr6, Val13, and Met15, were identified. These included NOE correlations between peptide residues Val13 and Met15 and a pair of methyl groups on the receptor; as with the zeta peptide complex, the receptor methyl resonances could be assigned tentatively to Leu158. This assignment allowed for manual docking of the hairpin peptide so that the C-terminal residues were proximal to Leu158 on the receptor while, at the same time, Pro9 packs between Trp87 and Trp10 in the IgE-binding site. The resulting docked model was refined in a similar fashion to the refinement of the zeta peptide complex, as outlined above. In the case of the hairpin peptide calculations, however, the weak harmonic potential (force constant = 0.25 kcal/·mol) used to restrain the peptide structure was applied only during the initial iterations, and was never applied to peptide residues Asn1, Leu2, Thr6, Trp10, or Trp12. Conformational freedom of these residues was found to be necessary in order to satisfy the ^{1}H chemical shift restraints; in particular the side chain conformation of Trp10 changed significantly during the refinement. Stereospecific ^{1}H resonance assignments were possible for Pro9 $^1H^{\beta}$ and $^1H^{\delta}$, and Val13 $^1H^{\gamma}$, and using the same criteria as for the zeta peptide complex a total of 22 intermolecular distance restraints could be included in the final simulated annealing calculations.

The final correspondence between the calculated and observed ^{1}H chemical shifts was again excellent (Figure 3E) indicating that the resulting model (Figure 3F) is a good approximation of the structure of the hairpin peptide/receptor complex. As was found for the zeta peptide complex, and consistent with original assumptions, the functionally important Pro9 is sandwiched between Trp87 and Trp110 on the receptor. Thr6 is proximal to Trp156 on the receptor, which accounts predominantly for the upfield shifts observed for resonances of this residue.

Comparison of the zeta and hairpin peptide/FcεRIα complex models reveals that even though the phage-derived peptides are structurally distinct they make remarkably similar contacts with the receptor. For instance, in addition to the zeta peptide Pro16 and the hairpin peptide Pro9 being "sandwiched" between Trp87 and Trp110 of the receptor, Pro4 and Phe6 of the zeta peptide occupy essentially the same space on the receptor as Val13 and Thr6, respectively, of the hairpin peptide (Figure 3C and F). Interestingly, the latter interactions are not utilized by the natural ligand, IgE, and thus, together with the proline pocket defined by Trp87 and Trp110, represent a novel binding epitope for IgE antagonists.

Conclusion

In the examples presented above the readily obtained ^{1}H chemical shifts of peptide ligands were used to validate docked models of the peptide/receptor complexes that were based originally on assumption and limited experimental data, such as mutational analysis. More extensive data, such as receptor chemical shift assignments, and consequently assigned intermolecular NOE correlations, were not available. The ^{1}H chemical shifts of the bound peptides, and in particular the large upfield shifts of ^{1}H resonances observed for some of the peptide residues, could be accounted for quantitatively by directly refining the models against the observed chemical shifts.

Some caveats of the approach are worth mentioning. First, this is not a *de novo* docking method, but rather a docking validation method, and thus requires some degree of independent data in order to localize the binding site. Starting models could come from *de novo* docking algorithms, or could be based on experimental data such as obtained from mutational analysis, as used here. Secondly, reasonably accurate structural models of the ligand and receptor are necessary and conformational changes upon binding should be minimal (although as noted in the hairpin peptide example some degree of conformational flexibility can be accommodated and might be necessary in order to obtain a good fit). Finally, the most sensitive restraints of ligand orientation are the ring current contributions of aromatic groups. Thus, the presence of aromatic groups that cause significant perturbations in the ligand chemical shifts is likely a prerequisite for successful application of this approach.

References

1. Nakamura GR, Starovasnik MA, Reynolds ME, Lowman HB. Biochemistry. 2001;40:9828.
2. Nakamura GR, Reynolds ME, Chen YM, Starovasnik MA, Lowman HB. Proc. Natl. Acad. Sci. U.S.A. 2002;99:1303.
3. Garman SC, Wurzburg BA, Tarchevskaya SS, Kinet JP, Jardetzky TS. Nature. 2000;406:259.
4. Dennis MS, Roberge M, Quan C, Lazarus RA. Biochemistry. 2001;40:9513.
5. Reilly D, Fairbrother WJ. J. Biomol. NMR. 1994;4:459.
6. Stamos J, Eigenbrot C, Nakamura GR, Reynolds ME, Yin J, Lowman HB, Fairbrother WJ, Starovasnik MA. Structure. 2004;12:1289.
7. Cavanagh J, Fairbrother WJ, Palmer AG III, Skelton NJ. Protein NMR Spectroscopy: Principles and Practice. Academic Press: San Diego, 1995.
8. Zwahlen C, Sébastien PL, Vincent JF, Greenblatt J, Konrat R, Kay LE. J. Am. Chem. Soc. 1997;119:6711.
9. Johnson CE, Bovey FA. J. Chem. Phys. 1958;29:1012.
10. Haigh CW, Mallion RB. Prog. NMR Spectrosc. 1980;13:303.
11. McConnell HM. J. Chem. Phys. 1957;27:226.
12. Buckingham AD. Can. J. Chem. 1960;38:300.
13. Ösapay K, Case DA. J. Am. Chem. Soc. 1991;113:9436.
14. Ösapay K, Case DA. J. Biomol. NMR. 1994;4:215.
15. Dejaegere AP, Bryce RA, Case DA. In: JC Facelli, AC de Dios (Eds). Modeling NMR Chemical Shifts. American Chemical Society: Washington, 1999, p 194.
16. Bundi A, Wüthrich K. Biopolymers. 1979;18:285.
17. Merutka G, Dyson HJ, Wright PE. J. Biomol. NMR. 1995;5:14.

Dual-Region Hadamard-Encoding to Improve Resolution and Save Time

Ronald Crouch
Varian NMR Systems, NC 27513, USA

The advent of indirect detection has revolutionized the world of NMR spectroscopy in all areas of chemistry. For the purpose of assembling fragments of moieties as defined by NMR resonances into meaningful structural fragments or even complete structures, HMBC [1] is one of the most powerful tools available to the modern NMR spectroscopist. In situations of low sensitivity brought about most commonly by the availability of vanishingly small amounts of sample to analyze, direct observation of ^{13}C resonances can be difficult or impossible. Accordingly, the sensitivity afforded by indirect detection with HMBC and variants provides a tool that has the potential to access the key and difficult quaternary ^{13}C atoms while at the same time assembling the puzzle pieces into a coherent structure.

The strength of ^{1}H detection, thereby casting the ^{13}C chemical shift information into the indirect dimension, is also the weakness of HMBC. This is especially true in the more difficult situations where there are a large number of ^{13}C resonances spread out over a large chemical shift range. At a ^{1}H observe frequency of 500 MHz, 225 ppm of ^{13}C chemical shift represents nearly 28,000 Hz of frequency space. Resolving two ^{13}C resonances separated by10 Hz would require ~2800 t1 increments (or ~1000 t1 increments with high enough signal to noise to allow reliable linear prediction) in F1! Region selection in the ^{13}C dimension with HMBC is a very old idea [2–3]. Utilization of excitation sculpting [4–5] added considerable robustness to the method. Combination of excitation sculpting with a simple Hadamard-encoding step [6] provides a simple and reliable methodology to greatly reduce the number of t1 increments required for a needed resolution. This encoding methodology affords a mean to either record lower resolution HMBC information in less time or much higher resolved spectra in equivalent time relative to traditional sampling methods.

Consider the simplest two-element Hadamard matrix $+, +$ and $+, -$. Addition of the two elements leads to one enhanced signal to noise in one region, whereas subtraction results in cancelation of the first region and enhanced signal to noise in the second region. Bear in mind that with modern NMR spectrometers, given an actual ^{13}C spectrum from which to build the needed shaped ^{13}C pulses, a Hadamard matrix can automatically be created to record and construct a complete 2D HMBC spectrum in n increments, where n represents the number of ^{13}C resonances [7–10]. A major advantage of using a simple two-element matrix is that the regions can be quite wide and general and not require a precise understanding of carbon chemical shifts. It is quite reasonable to simply divide the carbon regions by half, and immediately obtain a time saving of a factor of two, because by employing Hadamard-encoding, the number of t1 increments is cut in half to maintain exactly the same digital resolution as in a full single-region HMBC. The magic is in the encoding pulses and it is quite practical to create whatever is needed on the fly to obtain the two required pulses for a simple two-element matrix. Alternatively, a pair of pulses appropriate for a routine experiment with any desired frequency bounds can be created, and the pulse width and power parameters remembered for future use. A pulse sequence optimized for long-range ^{1}H–X correlation spectroscopy with an option for single or multiple X-band selection is presented in Figure 1 [11].

As an introduction to the illustration of the method, consider the gHMBC spectrum acquired with a sample of cyclosporine-A shown in Figure 2. The F1 window required to observe all ^{13}C resonances is not particularly large. A ^{13}C chemical shift range from 0 to 190 ppm is more than enough. Inspection of the data in Figure 2 reveals that is quite practical to consider splitting the F1 window into two ~80 ppm regions; 0–80 and 105–185 ppm. Simply acquiring a simultaneous dual F1 region-selected long-range $^{1}H/^{13}C$ chemical shift correlation experiment with the pulse sequence shown in Figure 1, immediately affords a time saving of a factor of two over the full single-region data set, simply because the number of t1 increments for equivalent digital resolution in the ^{13}C dimension can be cut in half.

A direct comparison of a single-region gHMBC and the simultaneous dual F1 region-selected gHMBC variant is presented in Figure 3, panels A–D. The two data sets

Graham A. Webb (ed.), Modern Magnetic Resonance, 1281–1286.

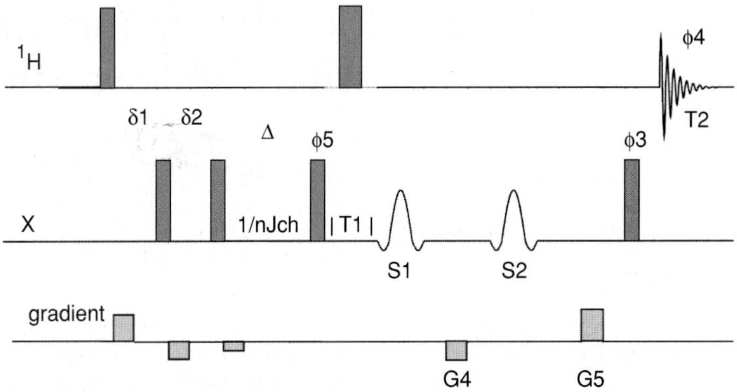

Fig. 1. Gradient coherence-selected pulse sequence for HMBC with adiabatic swept refocusing. Narrow bars denote 90° pulses. The wide bar and two selective pulses (S1 and S2) are 180° pulses with the t1 evolution time split by the ^{1}H 180. All of the data presented in this note were acquired with a constant adiabaticity cosine^2 (WURST2i) 180° pulse [12–13]. Gradients G4 and G5 are for coherence selection and can be in ratios of −3/5 or 5/−3 for N or P type selection of $^{13}CH_n$ resonances. The $\delta 1$ and $\delta 2$ delays in the low pass filter are optimized for $^1J_{max}$ of 165 Hz and $^1J_{min}$ of 130 Hz, as described by Sorensen and co-workers [14]. The delay for long-range XH coherence selection (Δ) can be set as desired. The gradients in the low pass filter are in the ratios of 9, −6, and −3. The basic phase cycling is as follows: $\phi 3 = x, x, -x, -x$; $\phi 5 = x, -x$; $\phi 4 = x, -x, -x, x$. Pulses S1 and S1 are adiabatic shaped pulses with downfield to upfield inversion sweep.

were obtained in precisely the same time with 128 t1 increments in each. The dual-region data (panels B and D) were acquired with two scans per t1 increment (with the two required band-selected shapes) and the single-region data (panels A and C) used four scans. Clearly the digital resolution and overall quality of data are improved for the dual-region Hadamard-encoded data presented in Figure 3, panels B and D.

A simple single-region band-selected experiment can be performed with the pulse sequence shown in Figure 1 to obtain extremely high digital resolution with efficient selection of the desired F1 region and rejection of other responses. Figure 4 shows a very high ^{13}C resolution long-range ^{1}H–^{13}C correlation experiment wherein the correlations between the protons and only the carbonyl ^{13}C resonances have been selected. The time to acquire the data was once again ~20 min. Comparison with the data in Figure 2 (or Figure 3, panel C) reveals the much higher digital resolution afforded by confining the F1 sampling solely to the carbonyl region of the ^{13}C spectrum.

At this point, it is useful to address the details of the experiment in terms of the requirements imposed by the band-selected pulses. A good means for illustration would be to graphically compare the pulses required for the acquisition of the single-region data presented in Figure 4, with those required to acquire the simultaneous dual-region data presented in the Figure 3 panels. This comparison is shown in Figure 5. Application of a very narrow band ^{13}C inversion pulse pair, as illustrated in Figure 5A, affords the possibility of quite high ^{13}C digital resolution with relatively few t1 increments precisely because only the portion of the ^{13}C frequency space refocused by the narrow band pulse is detected.

For simultaneous dual-region selection, it is necessary to interleave data acquisition using the two pulses graphically presented in Figure 5B and C. After data acquisition, 2DFT sum of the two interleaved data sets would result in full signal to noise for the upfield region of the ^{13}C frequency space, whereas 2DFT difference would give full signal to noise for the downfield ^{13}C frequency space. It is important to understand that the upfield and downfield regions are the result of two separate data processing steps that are graphically represented in Figure 5D and E.

The method presented here is both robust and simple to implement. It is the author's feeling that the full Hadamard-encoded techniques outlined in Refs. [7–10] may well represent the beginnings of a major revolution in the general approach to multi-dimensional NMR spectroscopy typically used in support of the elucidation of chemical structure in the pharmaceutical industry. It is hoped that the presentation of this simple two-step procedure serves to illustrate the general concepts.

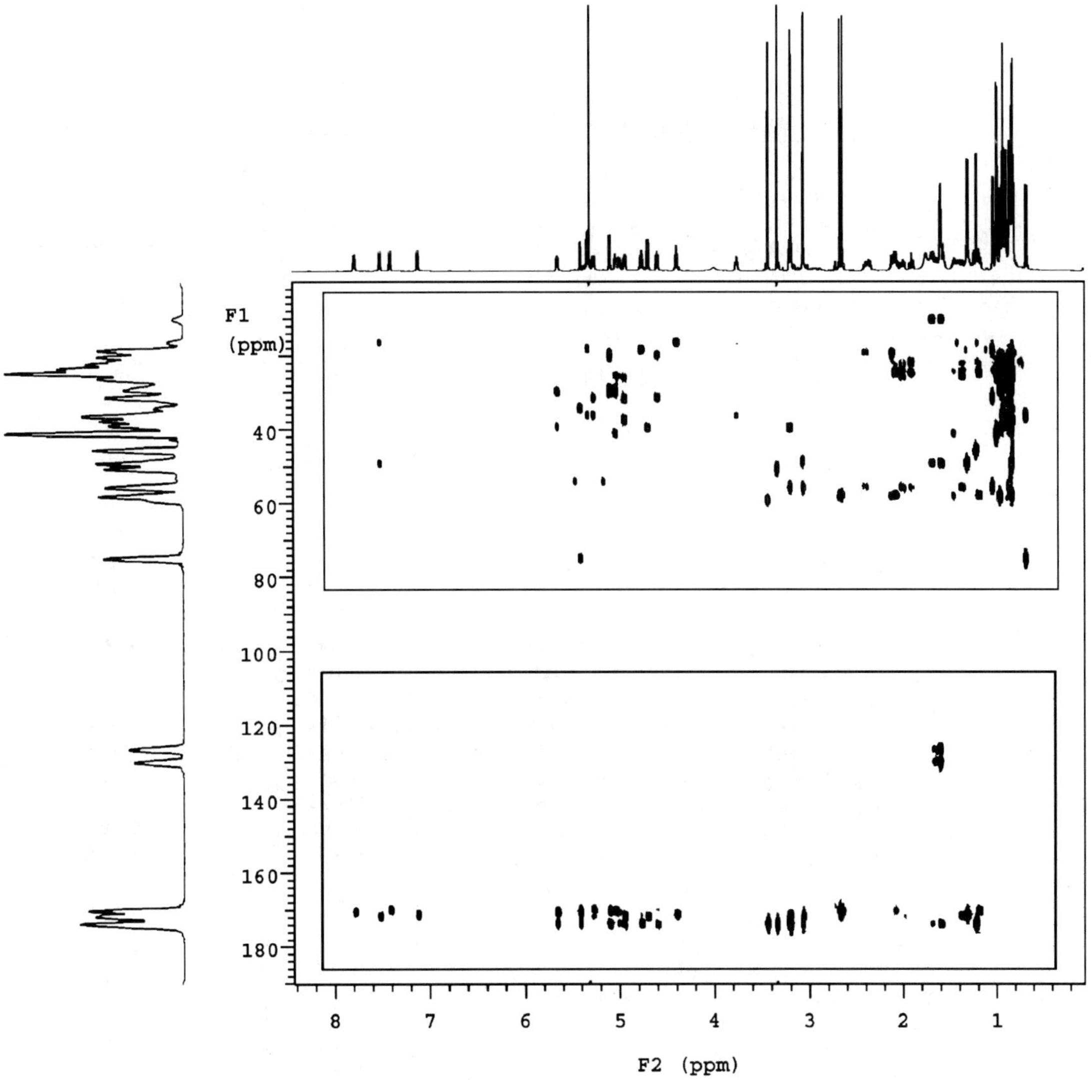

Fig. 2. Plot of the full single-region gHMBC acquired with a 2 mg sample of cyclosporine-A using an INOVA 600 with HCN Cold Probe. Data were acquired with four scans in each of 128 t1 increments in 22 min. A normal ^{1}H spectrum is on top and a projection showing the detected ^{13}C resonances is shown on the side. The boxed areas denote the regions to be selected in subsequent dual Hadamard-encoded experiments.

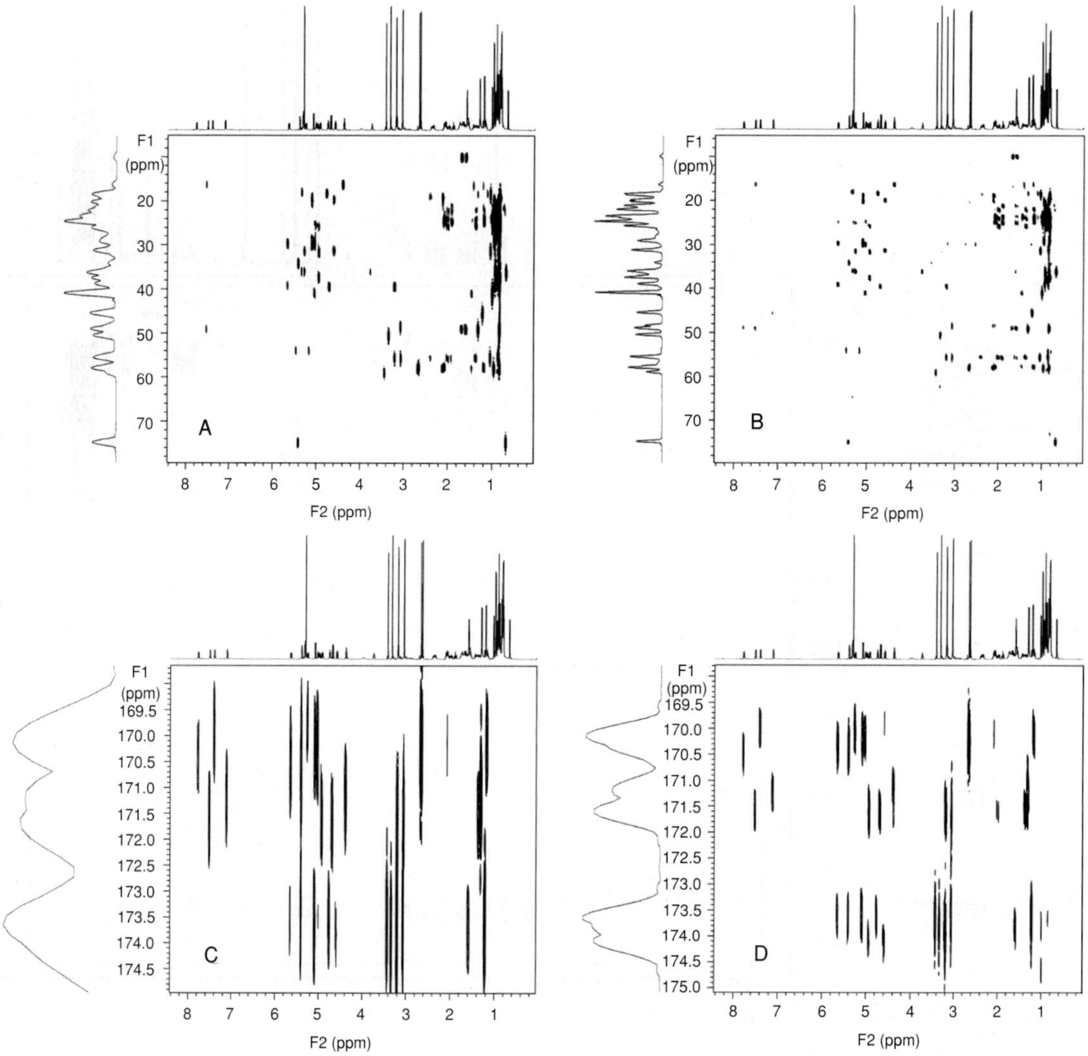

Fig. 3. Panels A–D compare regions from both traditional gHMBC and simultaneous dual-region Hadamard-encoding using the pulse sequence described in Figure 1. Panels A and C are simply expansions from the normal gHMBC depicted in Figure 2, whereas panels B and D are the result of 2DFT sum or difference of Hadamard-encoded data acquired in a single experiment in exactly the same experimental time as the gHMBC data shown in panels A and C. The data in panels A and C (gHMBC) were acquired with four scans/t1 increment. The Hadamard-encoded data in panels B and D were acquired with two scans per increment for both of the region-selected encoding ^{13}C pulses.

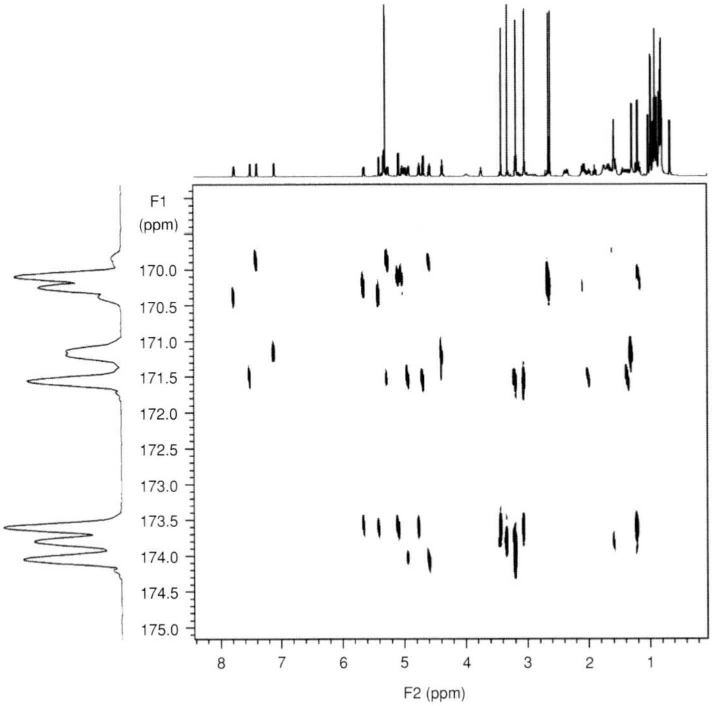

Fig. 4. Single C=O region-selected data acquired with the 2 mg sample of cyclosporine-A in 20 min. Data were acquired with two scans for each of the 256 t1 increments. The data were acquired and processed as a single phase with sine-bell window functions.

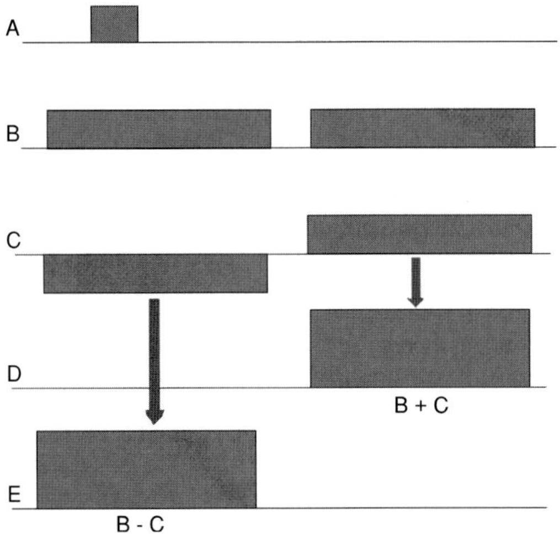

Fig. 5. Graphical representation of band-selection pulses required to acquire data in a two-step Hadamard matrix to generate the results shown in Figure 3. (A) For the single-region band-selection of carbonyl ^{13}C resonances described in Figure 4 only a very narrow refocusing of the ^{13}C carbonyl resonances is required. Figure 5B and C illustrate the pair of band-selected inversion pulses required for Hadamard-encoding. The anti-echo refocusing shown in the inverted portion of Figure 5C is achieved by application of a 90° phase shift for the band-selection of that π pulse portion of the ^{13}C frequency window. In Figure 5D and E the effect of processing the two encoded data sets, respectively, as sum and difference is illustrated. Figure 5D depicts 5B plus 5C, whereas Figure 5E depicts 5B minus 5C.

frequencies. A different approach to multiplexing has been developed by Frydman and colleagues [11], which utilizes orthogonal magnetic field gradients to partition the sample into "subsamples," each reporting a different part of the spin response. This approach is not practical when sensitivity is already a limiting factor for measuring the response of the entire sample, but as advances in probe technology continue to improve sensitivity and gradient strength (which determines the resolution), it will likely find increasing application in biomolecular NMR. Parametric methods that model the NMR signal as a sum of exponentially decaying sinusoids, such as LP singular value decomposition [12], Hankel SVD [13], filter diagonalization method (FDM) [14], maximum likelihood [15], and Bayesian [16] methods are in principle capable of producing high-resolution spectra from very short data records. These methods often exhibit "spontaneous splitting," resulting in multiple sinusoids for a single resonance when applied to noisy or nonideal (non-Lorentzian) data, complicating the parametric interpretation. Indeed the regularized resolvent transform [17], in which an FDM-like approach is used to extrapolate the signal, rather than compute a parametric decomposition, was developed to avoid this problem.

Another class of methods utilizes symmetries or redundancy in multidimensional spectra to avoid acquiring all combinations of the evolution times for the different dimensions. An approach inspired by the methods used in medicine to reconstruct images in computerized axial tomography exploits the relationship between line projections in the frequency domain and linear cross sections in the time domain. Called back-projection reconstruction [18], this approach can, in favorable circumstances recover a complete two-dimensional spectrum from a small number of cross sections through the full time domain data matrix. However, it has been pointed out that in the general case n cross sections are needed to unambiguously resolve n peaks in multiple dimensions [19], and for complex biomolecules, this may not result in significant time savings. Another approach, called three-way decomposition [20], exploits the fact that line shapes in multiple dimensional spectra can be decomposed into products of one-dimensional line shapes.

Back-projection reconstruction and three-way decomposition are special cases of nonuniform sampling in the time domain. In principle, any method that approaches spectrum reconstruction as an inverse problem, that is, computes spectra and inverts the spectrum to obtain "mock" data that are then compared with the empirical data for consistency, can be used with nonuniformly sampled time domain data. A very general method that makes no assumptions regarding the nature of the signals, and imposes no particular constraints on the subset of the time domain data matrix that must be collected, is maximum entropy (MaxEnt) reconstruction [21]. It has been shown recently to be particularly robust, yielding fewer false positives and more accurate frequencies than LP extrapolation [7]. It is also versatile: it can be used to construct spectra from the same data utilized by back-projection reconstruction, or can be used in combination with other methods, such as GFT (Jim Sun and Gerhard Wagner, personal communication). In principle, it can be used with any pulse sequence, and the computed spectra are compatible with conventional analysis tools developed for Fourier spectra. In this chapter, we describe the use of nonuniform sampling and MaxEnt reconstruction in multidimensional NMR experiments and its application to biomolecules. In practice, factors of two to three savings in data acquisition time per indirect time dimension are readily achieved. The robustness and versatility of MaxEnt reconstruction make it a powerful tool for improving sensitivity and resolution, and saving time in biomolecular NMR experiments.

MaxEnt Reconstruction

The MaxEnt reconstruction of the spectrum of a complex-valued time series **d** is the spectrum **f** which maximizes the entropy $S(\mathbf{f})$, subject to the constraint that the mock data **m**, given by the inverse DFT of the spectrum, is consistent with the time series **d**. Consistency is defined by the condition

$$C(\mathbf{f}, \mathbf{d}) \leq C_0 \tag{1}$$

where $C(\mathbf{f}, \mathbf{d})$ is the unweighted chi-squared statistic,

$$C(\mathbf{f}, \mathbf{d}) = \sum_{i=0}^{M-1} |m_i - d_i|^2 = \sum_{i=0}^{M-1} |\mathrm{IDFT}(\mathbf{f})_i - d_i|^2, \tag{2}$$

and C_0 is an estimate of the noise level; IDFT is the inverse DFT. A definition of the entropy $S(\mathbf{f})$ applicable to complex-valued spectra can be derived from consideration of the entropy of an ensemble of spin-1/2 particles [22] or by extending the Shannon information formula to the complex domain along with suitable differentiability conditions [23]. The two approaches yield equivalent formulations for the entropy of a complex spectrum

$$S(\mathbf{f}) = -\sum_{n=0}^{N-1} \frac{|f_n|}{def} \log\left(\frac{|f_n|/def + \sqrt{4 + |f_n|^2/def^2}}{2}\right) - \sqrt{4 + |f_n|^2/def^2}, \tag{3}$$

where def is a scale factor. In principle, the quantum-mechanical derivation prescribes the value of def (it depends on the sensitivity of the spectrometer and the number of spins in the sample), but it is more convenient to treat *def* as an adjustable parameter. Essentially, it determines the scale at which the nonlinearity of MaxEnt becomes pronounced. The MaxEnt solution is found by maximizing the objective function

$$O = S(\mathbf{f}) - \lambda C(\mathbf{f}, \mathbf{d}), \qquad (4)$$

where the value of the Lagrange multiplier λ is adjusted to obtain $C = C_0$. The value of λ depends on the values of the parameters *def* and C_0, and on the data. Practical guidelines for choosing the values of *def* and C_0 are described elsewhere [24].

The formulas for entropy and constraint readily generalize to multidimensional data. However, because MaxEnt reconstruction is nonlinear, MaxEnt must be the last time-to-frequency transformation applied. So if MaxEnt is to be used in t_1, then the t_2 dimension must be transformed to the frequency domain first. And if MaxEnt is to be used for both t_1 and t_2, then it must be applied to both dimensions simultaneously, rather than one at a time. The IDFT operation appearing in Equation (2), and the corresponding sums in Equations (2) and (3), can be applied along as many dimensions as one likes.

Applying MaxEnt to reconstruct separately each t_1 row, or each t_1–t_2 plane, of a three-dimensional spectrum raises another problem. The extent of the nonlinearity of the reconstructions can vary from row to row (or from plane to plane), resulting in distorted peak shapes. One way to avoid this is to process the entire data set as a single unit, extending the sums in Equations (2) and (3) to cover the entire spectrum, rather than working on a single row or plane at a time. Of course, such an approach will entail drastically increased data storage requirements. A simpler solution is to use a fixed value of λ rather than iterating to attain $C = C_0$ [25]. The correct value for λ can be determined by choosing a representative row, then computing the normal MaxEnt reconstruction with an appropriate value for C_0. In quantitative applications, it is important to use one of these methods—full data set MaxEnt reconstruction or constant λ reconstruction—to assure that the nonlinearity is uniform across the spectrum, and thus it can be calibrated in a reliable way.

The nature of the nonlinearity of MaxEnt reconstructions is fairly well understood—in general, MaxEnt reconstruction tends to scale intensities down, compared to the DFT, with small amplitudes scaled down more than large amplitudes [26]. In addition to requiring calibration when used for quantitative applications, such as measuring nuclear Overhauser effects, the nonlinearity of MaxEnt reconstruction heightens the distinction between sensitivity, which is the ability to distinguish peaks from noise, and signal-to-noise ratio (S/N). For MaxEnt reconstruction, S/N is not a reliable indicator of sensitivity.

Nonuniform Sampling

The matched filter is a function that follows the decay envelope of the FID; multiplying the FID by the matched filter results in optimal S/N in the DFT spectrum [27]. In essence, the matched filter places more emphasis on the portion of the data where the S/N (in the time domain) is highest. The same idea can be applied to sampling in the time domain: instead of sampling at uniform intervals, the signal can be sampled more frequently when it is strong, and less frequently when it is weak [28–32]; we call this nonuniform sampling. Nonuniform sampling prevents the use of LP or the DFT, since they require data sampled at uniform intervals, but MaxEnt handles such data because the chi-squared statistic in Equation (2) can be computed from just the collected samples. Nonuniform sampling provides additional flexibility in balancing the trade-offs between acquisition time, resolution, and sensitivity. The potential for time saving accrues in the indirect time dimensions of multidimensional experiments, since the relaxation delay between transients renders the total acquisition time insensitive to the number of samples acquired in the direct time dimension. In contrast, the total experiment time increases in direct proportion to the number of indirect time samples.

A list of evolution times for which data are acquired is called a sampling schedule. Examples of nonuniform sampling schedules are depicted in Figure 1. A uniform grid is used to signify evenly spaced points in the various time dimensions. A large dot indicates a sampled time interval and small dots indicate times not sampled. Panel A illustrates an exponential sampling schedule in one indirect dimension (t_1), and panel B illustrates an exponential sampling schedule in two indirect dimensions. Panel C also illustrates a schedule employing randomly selected points chosen from a two-dimensional exponential distribution; this helps to suppress coherent artifacts that can be introduced by schedules that use exactly the same time values for successive rows or columns (such as the schedule in panel B). Panel D illustrates a schedule appropriate for reduced sampling of constant-time dimensions, where there is no decay. For COSY-type spectra, an optimal sampling schedule can be constructed from a sine-modulated exponential distribution. In principle, it is possible to construct more complex sampling schedules that make explicit consideration of the dimensionality of the experiment or the distribution of expected frequencies, for example, along the lines of the samples used for back-projection reconstruction.

Example Applications

Comparisons of spectra obtained using conventional linear sampling and nonuniform sampling applied to two

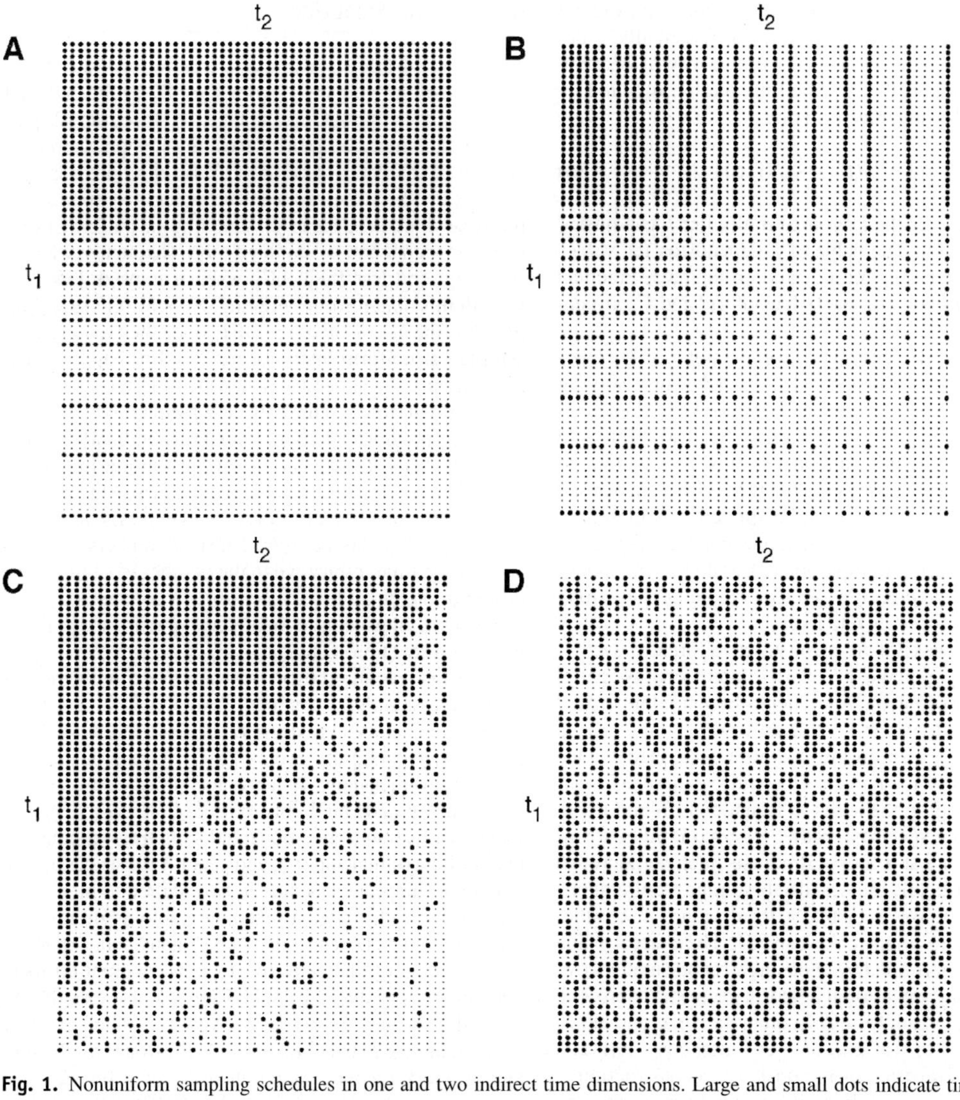

Fig. 1. Nonuniform sampling schedules in one and two indirect time dimensions. Large and small dots indicate times for which samples are collected and not collected, respectively. The schedules depicted are (A) exponential in t_1 and linear in t_2, (B) exponential in both t_1 and t_2, (C) randomly sampled from an exponential distribution in t_1 and t_2, and (D) random in t_1 and t_2.

experiments important for pharmaceutical applications of protein NMR are shown in Figures 2 and 3. Figure 2 shows ^{15}N–^{1}H HSQC spectra of a 42-residue atracotoxin from the Australian funnel-web spider *Hadronyche versuta*. Panel A is the spectrum computed using LP extrapolation, shifted sine-bell windowing, and DFT to produce a 1024-point spectrum in f_1 from 512 uniformly spaced samples in t_1. Panel B was computed in the same fashion and to the same size in f_1, using 64 samples in t_1. The spectrum in panel C was computed using MaxEnt reconstruction, using 64 t_1 samples with exponentially increasing time between samples. Panels D, E, and F show an expanded region of panels A, B, and C, respectively. The resolution obtained from 64 t_1 samples using MaxEnt reconstruction and nonuniform sampling is comparable to that obtained using uniform sampling and conventional processing; however, data acquisition required a factor of eight less time. This time saving can be important for high-throughput applications such as screening compound libraries for molecules that bind to a protein target.

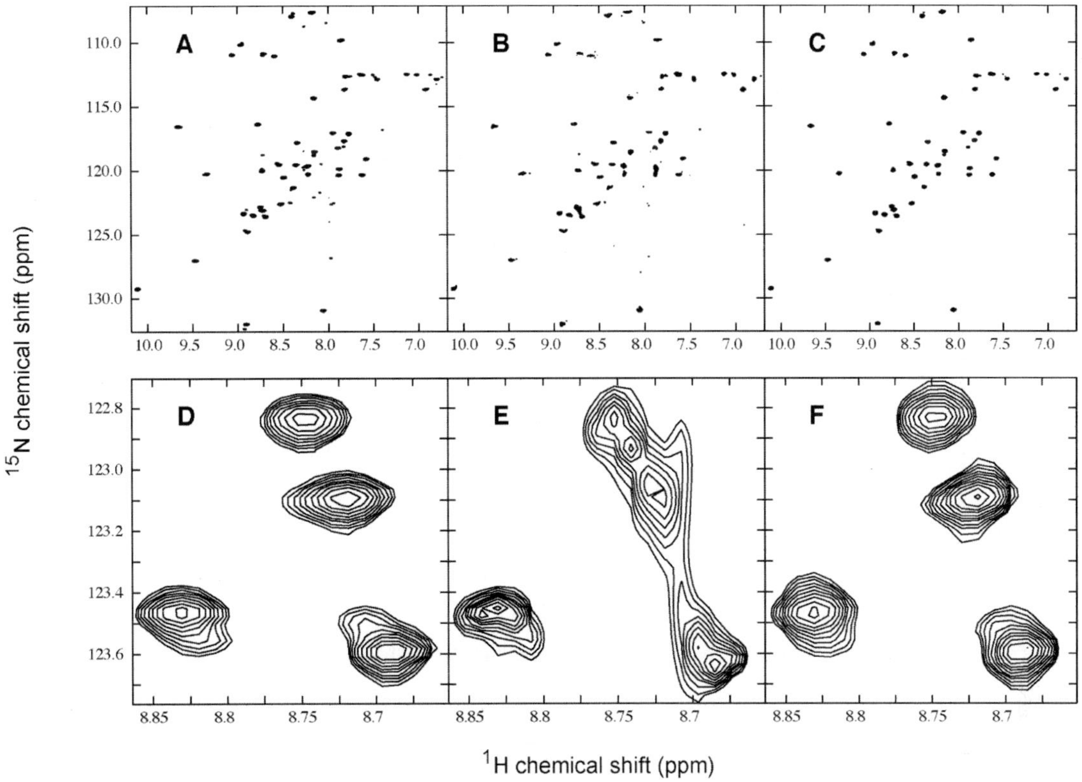

Fig. 2. ^{15}N–^{1}H HSQC spectra for a 42-residue atracotoxin. (A) Computed using LP extrapolation and DFT from 256 linearly spaced samples in t_1, (B) using LP extrapolation and DFT from 64 linearly spaced samples in t_1, and (C) using MaxEnt reconstruction from 64 exponentially spaced samples in t_1. Panels D, E, and F depict an enlarged region from panels A, B, and C, respectively.

Extension of nonuniform sampling and MaxEnt reconstruction to two indirect time dimensions is straightforward, enabling the acquisition of triple resonance spectra for sequential assignment of proteins in 1 day [33]. Experiments in which the signals of interest span a wide range of intensities (such as those involving nuclear Overhauser effects) remain a challenge. The very nature of the nonlinearity of MaxEnt reconstruction, which is to scale down weak signals more than strong signals, makes it more difficult to choose parameters of the reconstruction (*def* and C_0) that enable weak signals to be distinguished from noise. Nonuniform sampling can introduce additional nonlinearities. An example is shown in Figure 3, in which exponential nonuniform sampling and MaxEnt reconstruction have been applied to the two indirect dimensions of a 3D ^{15}N–^{1}H HSQC-NOESY experiment. Panels A and B shows f_1–f_3 cross sections at the f_2 frequency 120.27 ppm, close to the amide ^{15}N chemical shift of residue Cys 21 (120.24 ppm); A is from the linearly sampled DFT spectrum and B is computed using MaxEnt reconstruction from data containing one-eighth fewer samples. Remarkably, the MaxEnt spectrum exhibits resolution superior to the spectrum computed from the larger linearly sampled data set. This is evidenced by the disappearance of resonances at the f_3 frequencies of 8.69 and 8.22 ppm, which correspond to residues with amide ^{15}N chemical shifts of 120.53 and 120.30 ppm, respectively, which have maxima in adjacent f_2 planes. However, the nonlinearity (which depends on the values chosen for *def* and C_0, as well as the sampling schedule) in this instance makes it difficult to observe several weak peaks. Panels C and D are linear cross sections from panels A and B at the f_3 frequency of the amide proton of Cys 21, scaled vertically to the height of the maximum in each panel. While the noise level is greatly reduced, so are the weak peaks, making clear the difficulty caused by high dynamic range.

While careful choice of the nonuniform sampling schedule and parameters of the MaxEnt reconstruction can in principle permit observation of the missing weak

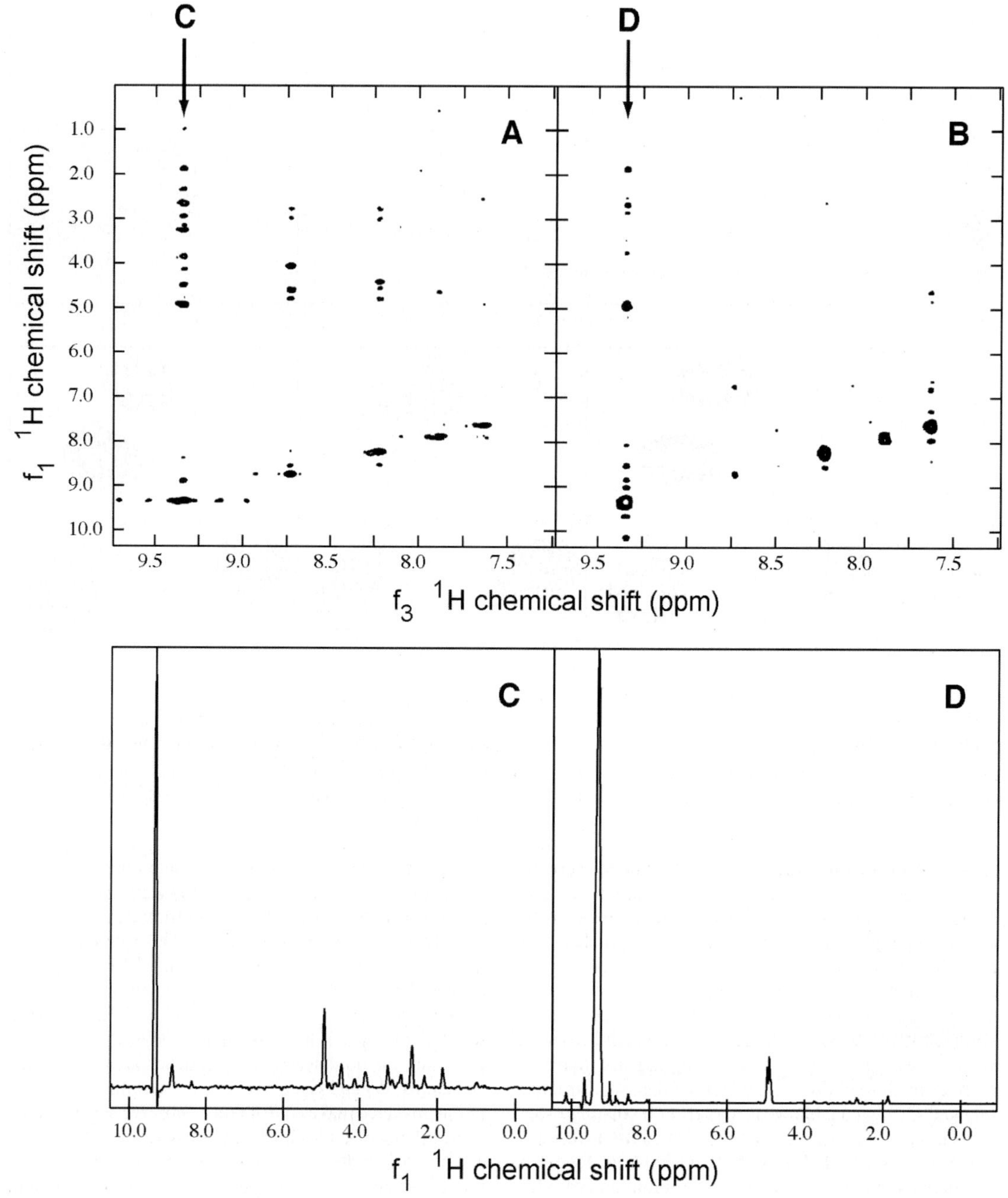

Fig. 3. Two-dimensional cross sections at $f_2 = 120.27$ ppm from 3D ^{15}N–^{1}H HSQC-NOESY spectra of a 42-residue atracotoxin. (A) Computed using LP extrapolation and DFT, (B) using nonuniform sampling in t_1 and t_2 and MaxEnt reconstruction. (C) and (D) are linear cross sections from panels A and B at the f_3 frequency of the amide proton of Cys 21 (9.34 ppm).

peaks (for example, using a small value for C_0 decreases the nonlinearity of the reconstruction), determination of the appropriate values is by no means straightforward, and remains the subject of active research [34]. Regardless of the application, the nonlinearities make error analysis an important adjunct to nonuniform sampling and MaxEnt reconstruction.

Concluding Remarks

MaxEnt reconstruction is a particularly versatile method of spectrum analysis, capable of providing high-resolution spectral estimates from short data records. Its ability to determine spectra from data sampled at arbitrary times enables the use of nonuniform sampling to improve resolution or reduce data acquisition time. Model-based methods, such as Bayesian analysis or FDM, offer many of the same capabilities but are not suited to the computation of spectra containing non-Lorentzian lines. The availability of efficient algorithms and powerful computers has removed what were once perceived as significant barriers to the widespread application of MaxEnt reconstruction in NMR. The ability to collect high-resolution spectra in substantially less time not only enables more efficient use of expensive spectrometers, but also enables investigation of biomolecular systems that are fleetingly stable. These capabilities are particularly well suited to the demands of applications of NMR in drug discovery, including high-throughput screening and rapid assignment of protein spectra.

Acknowledgments

We thank Drs. Peter Schmieder, David Rovnyak, and Gerhard Wagner for sage advice. Financial support was provided by the US National Institutes of Health (GM47467), the US National Science Foundation (MCB 9316938 and MCB 0234638), and The Rowland Institute at Harvard.

References

1. Ernst RR, Anderson WA. Rev. Sci. Instrum. 1966;37:93-102.
2. Jeener J. Oral Presentation, Ampere International Summer School: Yugoslavia, 1971.
3. Aue WP, Bartholdi E, Ernst RR. J. Chem. Phys. 1976; 64: 2229-46.
4. Oschkinat H, Griesinger C, Kraulis PJ, Sorensen OW, Ernst RR, Gronenborn AM, Clore GM. Nature. 1996;332:374-6.
5. Zeng Y, Tang J, Bush CA, Norris JA. J. Magn. Reson. 1989;83:473-83.
6. Ni F, Scheraga HA. J. Magn. Reson. 1986;70:506-11.
7. Stern AS, Li K-B, Hoch JC. J. Am. Chem. Soc. 2002;124: 1982-93.
8. Kim S, Szyperski T. J. Am. Chem. Soc. 2003;125:1385-93.
9. Ding K, Gronenborn AM. J. Magn. Reson. 2002;156:262-8.
10. Kupce E, Freeman R. J. Magn. Reson. 2003;162:300-10.
11. Frydman L, Scherf T, Lupulescu A. Proc. Natl. Acad. Sci. USA 2002;99:15858-62.
12. Barkhuijsen H, de Beer R, Bovee WM, Creyghton JH, van Ormondt D. Magn. Reson. Med. 1985;2:86-9.
13. de Beer R, van den Boogaart A, van Ormondt D, Pijnappel WW, den Hollander JA, Marien AJ, Luyten PR. NMR Biomed. 1992;5:171-8.
14. Mandelshtam VA, Taylor HS, Shaka AJ. J. Magn. Reson. 1998;133:304-12.
15. Chylla RA, Markley JL. J. Biomol. NMR 1995;5:245-58.
16. Kotyk JJ, Hoffman NG, Hutton WC, Bretthorst GL, Ackerman JJH. J. Magn. Reson. 1995;A116:1-9.
17. Chen J, Shaka AJ, Mandelshtam VA. J. Magn. Reson. 2000; 147:129-37.
18. Kupce E, Freeman R. J. Am. Chem. Soc. 2003;125:13958-9.
19. Coggins BE, Venters RA, Zhou P. J. Am. Chem. Soc. 2004; 126:1000-1.
20. Orekhov VY, Ibraghimov I, Billeter M. J. Biomol. NMR 2003; 27:165-73.
21. Wernecke SJ, D' Addario LR. IEEE Trans. Comp. 1977;26: 351-64.
22. Daniell GJ, Hore PJ. J. Magn. Reson. 1989;84:515-36.
23. Hoch JC, Stern AS, Donoho DL, Johnstone IM. J. Magn. Reson. 1990;86:236-46.
24. Hoch JC, Stern AS. Methods Enzym. 2001;338:159-78.
25. Schmieder P, Stern AS, Wagner G, Hoch JC. J. Magn. Reson. 1997;125:332-9.
26. Donoho DL, Johnstone IM, Stern AS, Hoch JC. Proc. Natl. Acad. Sci. USA 1990;87:5066-8.
27. Ernst RR. Adv. Magn. Reson. 1966;2:1-135.
28. Barna JCJ, Laue ED. J. Magn. Reson. 1987;75:384-9.
29. Barna JCJ, Laue ED, Mayger MR, Skilling J, Worrall SJP. J. Magn. Reson. 1987;73:69-77.
30. Schmieder P, Stern AS, Wagner G, Hoch JC. J. Biomol. NMR 1993;3:569-76.
31. Schmieder P, Stern AS, Wagner G, Hoch JC. J. Biomol. NMR 1994;4:483-90.
32. Robin M, Delsuc MA, Guittet E, Lallemand JY. J. Magn. Reson. 1991;92: 645-50.
33. Rovnyak D, Frueh DP, Sastry M, Sun Z-Y, Stern AS, Hoch JC, Wagner G. J. Magn. Reson., 2004;170:15-21.
34. Rovnyak D, Hoch JC, Stern AS, Wagner G. J. Biomol. NMR. 2004;30:1-10.

Applications

Structural Characterization of Antimicrobial Peptides by NMR Spectroscopy

Leonard T. Nguyen, Elmar J. Prenner, and Hans J. Vogel
Structural Biology Research Group, Department of Biological Sciences, University of Calgary Alberta, Canada T2N 1N4

Introduction

Antimicrobial peptides are ubiquitous in nature and they aid in host defenses against bacterial infections. While initial discoveries of bacterial peptides go as far back as the 1950s, the 1990s saw an explosion in research to isolate novel peptides from all kingdoms of life. Out of the more than 800 eukaryotic peptides reported in the Antimicrobial Sequences Database (found at www.bbcm.units.it/~tossi/pag1.htm), more than 700 entries were reported since 1990. The rapid growth of this field is fueled by the potential of these compounds to be used in a clinical setting [1,2] and replace commonly used antibiotics. The latter are losing potency against disease-causing bacterial strains that are increasingly becoming resistant. The unique qualities of the various antimicrobial peptides make them an incredibly diverse family in terms of sequence, structure and by extension, mechanism of action [3].

Typically shorter than 50 amino acid residues, antimicrobial peptides are prime candidates for structural study by NMR spectroscopy. Cyclic and disulfide cross-linked peptides usually have a stable structure in aqueous solution. However, many of the linear peptides do not form defined structures in aqueous solution and do so only in a membrane mimetic environment [4]; this finding supports the idea that peptide-lipid interactions are critical for antimicrobial action and/or insertion [5]. In addition to solution structures calculated mostly from standard two-dimensional TOCSY and NOESY ^{1}H experiments [6], much progress has been made to characterize the insertion and orientation of the peptides in membrane bilayers through solid-state NMR measurements [7]. This chapter will examine the progress in structural studies of antimicrobial peptides by NMR. Recent reviews have described the solution structures of antimicrobial peptides in an attempt to elucidate general themes and mechanisms [8,9]. Therefore to avoid overlap in this chapter we will highlight the versatility of NMR as a tool to characterize the solution structures of antimicrobial peptides which possess some unique features. In addition, we will describe some solid-state NMR studies which focus on several well known peptides.

Solution Structures of Antimicrobial Peptides

To date, more than 100 unique three-dimensional structures of antimicrobial peptides have been solved by NMR. The complete list consists of more than 80 parent peptides and roughly 20 synthetic peptide derivatives, resulting in great variability. Table 1 summarizes the information for selected solution structures of peptides that have been studied because of their potential as bactericidal agents. In order to be useful, such peptides cannot have high toxicity toward mammalian cells. Some representative structures are shown in Figures 1 and 2. Well-characterized antimicrobial peptides include magainin 2 from the skin of the African clawed frog (Figure 1A) [10], protegrin-1 from mice (Figure 1B) [11], and several members of the defensin family (Figure 2). Some rather distinct antimicrobial peptides have been included in Table 1 as well: These are human hepcidin (Figure 1C) [12], which is better known as a peptide involved in the regulation of iron absorption [13], three gramicidin S derivatives [14], which are listed in spite of the fact that the parent gramicidin S compound, which is secreted from a Bacillus species [15], possesses significant hemolytic activity, and a completely synthetic peptide, combi-1 [16], that was obtained by screening from a library of hexa-peptides produced by combinatorial chemistry.

Various structural classification schemes have been used in the past ([8] for example). However, given the wide diversity of the peptides, a satisfactory organization scheme can misrepresent the properties of certain peptides. Nevertheless, the eukaryotic cationic peptides are often classified into broad groups according to the secondary structure of the parent peptide. As such researchers often talk about helical, β-sheet, turn or extended peptides. This nomenclature can be somewhat misleading at times as some of the larger peptides should really be considered as mini-proteins with various types of secondary structure. For example, insect defensin A [17], as a defensin, is grouped as a β-sheet peptide even though there is an α-helix present in the structure in addition to several β-strands. It should also be noted that sometimes the structure of a synthetic derivative deviates from that of the parent peptide. This is seen for a shortened derivative of

Graham A. Webb (ed.), Modern Magnetic Resonance, 1297–1305.

Table 1: Examples of NMR derived high-resolution solution structures of antimicrobial peptides

Peptide	Class	Membrane Mimetic[a]	Source	Ref.
Cecropin A	α-helix	HFIP	*Hyalophora cecropia*	[87]
Cecropin A-Magainin 2 hybrid (CA-MA) (4)[b]	α-helix	DPC, None	Synthetic	[88]([89])c
Cecropin A-Melittin hybrid (CA-ME)	α-helix	DPC	Synthetic	[88]
Magainin 2 (1)	α-helix	DPC, SDS,TFE	*Xenopus laevis*	[10]([90])
Ovispirin-1 (2)	α-helix	TFE	Synthetic (from *Ovis aries*)	[91]([91])
PGLa	α-helix	DPC	*Xenopus laevis*	[92]
Sheep myeloid antimicrobial peptide (SMAP-29)	α-helix	TFE	*Ovis aries*	[93]
Bovine lactoferricin (1)	β-sheet	None	*Bos taurus*	[19]([18])
Drosomysin	β-sheet	None	*Drosophila melanogaster*	[41]
γ-1-H thionin	β-sheet	None	*Hordeum vulgare*	[42]
Hepcidin-20 and -25	β-sheet	None	*Homo sapiens*	[12]
Human β defensin-3 (HBD-3)	β-sheet	None	*Homo sapiens*	[44]
Human neutrophil protein-1 (HNP-1)	β-sheet	None	*Homo sapiens*	[39]
Insect defensin A/Phormicin	β-sheet	None	*Protophormia terraenovae*	[17]
NaD1 floral defensin	β-sheet	None	*Nicotiana alata*	[94]
Protegrin-1 (2)	β-sheet	None, DMSO	*Sus scrofa*	[11]([95,96])
Rabbit kidney defensin 1 (RK-1)	β-sheet	None	*Oryctolagus cuniculus*	[97]
Rhesus θ defensin-1 (RTD-1)	β-sheet	Acetonitrile	*Macaca mulatta*	[47]
Tachyplesin I (3)	β-sheet	None	*Tachypleus tridentatus*	[98]([98])
Tachystatin A	β-sheet	None	*Tachypleus tridentatus*	[99]
Ac-RRWWRF-NH2 (Combi-1)	Extended	SDS	Synthetic	[16]
Indolicidin (1)	Extended	SDS, DPC	*Bos taurus*	[21]([20])
Tritrpticin	Turn	SDS	Synthetic (from *Sus scrofa*)	[100]
NK-lysin	Loop	None	*Homo sapiens*	[101]
Thanatin	Loop	None	*Podisus maculiventris*	[102]
Bacteriocins				
Carnobacteriocin B2	α-helix	TFE	*Carnobacterium piscicola*	[103]
Gramicidin S derivatives (3)	Loop	None	Synthetic (from *Bacillus brevis*)	([14])
Microcin J25	Loop	None	*Escherichia coli*	[36,61,62]
Nisin A	Turn	DPC, SDS	*Lactococcus lactis*	[30]

[a] Abbreviations used for membrane mimetic present are as follows: TFE, trifluoroethanol; DPC, dodecylphosphocholine micelles; HFIP, hexafluoroisopropanol; SDS, sodium dodecyl sulfate micelles, and DMSO, dimethyl sulfoxide.
[b] Number in parentheses denotes the number of derivatives from a parent peptide for which structures are available.
[c] Number in parentheses in the reference column refers to the structures of the derivatives.

bovine lactoferricin [18] which forms an extended peptide structure even though intact bovine lactoferricin itself [19] adopts a β-sheet conformation. Also, a cyclized derivative [20] of indolicidin [21] has quite a different structure than its parent compound.

The prevalent theme among all eukaryotic cationic peptides is a structure with an amphipathic character. Thus the vast majority of the peptides have a preponderance of basic and hydrophobic amino acids, yet they have no sequence homology. The cationic side of the structure is responsible for electrostatic interactions with the bacterial membranes, which have a high abundance of negatively charged phosphatidylglycerol phospholipids head groups. The charge-charge interactions are thought to be an important factor in providing long-range attraction between the peptide and bacterial membranes [3]. This mechanism allows for broad specificity in target cells while retaining low toxicity to eukaryotic membranes, which mostly have

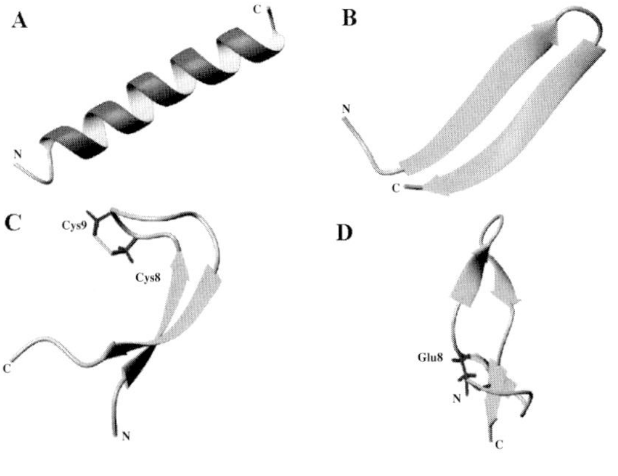

Fig. 1. Representative solution structures of antimicrobial peptides presented as ribbon diagrams: (A) magainin 2 (PDB ID 2MAG) in TFE solution, (B) protegrin-1 (PDB ID 1PG1), (C) hepcidin-20 showing the vicinal disulfide link between Cys8 and Cys9 (PDB ID 1M4E), and (D) microcin J25 showing the amide link between the sidechain of Glu8 and the N-terminal amino group (PDB ID 1Q71). These figures were prepared using MOLMOL [104]. (See also Plate 100 on page LXVI in the Color Plate Section.)

zwitterionic phosphatidylcholine head groups. However, a few anionic peptides from eukaryotic sources have been identified, for example in human lungs where Zn^{2+} chelation is required for activity [22], or from the toad *Bombina maxima* where the activity does not require the presence of cations [23]. The hydrophobic portion of the antimicrobial peptides allows them to interact with the acyl chain

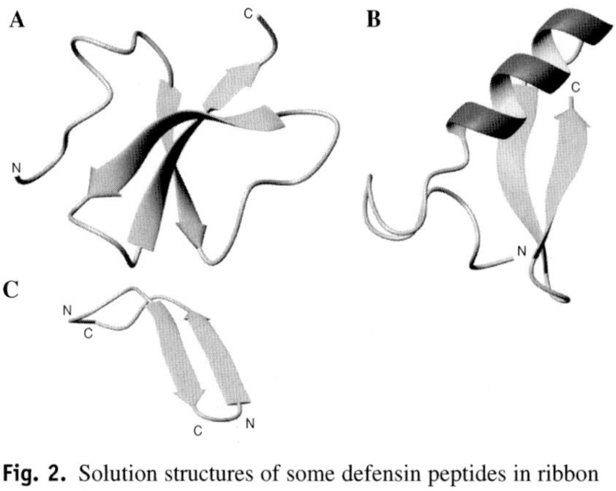

Fig. 2. Solution structures of some defensin peptides in ribbon diagram: (A) bovine neutrophil β-defensin-12 (BNBD-12) (PDB ID 1BNB), (B) insect defensin A (PDB ID 1ICA), and (C) rhesus θ-defensin (RTD-1) (PDB ID 1HVZ). These figures were prepared using MOLMOL [104]. (See also Plate 101 on page LXVI in the Color Plate Section.)

region of membrane bilayers. The amphipathicity that balances both facets of the peptides plays a key role in target selectivity, as peptides that have greater hydrophobicity usually act as general toxins, such as the bee venom melittin [24].

Almost all α-helical linear peptide structures were solved using either a micelle suspension or an organic co-solvent environment because the vast majority does not form a defined structure in aqueous solution. Highly represented in this category are the peptides obtained from the skins of frog or toad species. β-sheet peptides meanwhile, which encompass the mini-protein defensin family, come from a variety of plants and animals and are known for their characteristic disulfide linkages between cysteine residues [25,26]. Highly disulfide bridged peptides quite frequently act as antimicrobial peptides because the crosslinking would reduce their flexibility and promote stability against proteolytic enzymes. In contrast to the α-helical peptides, structures of the β-sheet peptides have mostly been solved in an aqueous environment and in fact, binding of protegrin-1 to lipid bilayers resulted in aggregation, which prevented detailed structural analysis [27,28]. Highly oligomerized peptides on membrane-mimetic surfaces, while possibly more representative of *in vivo* behavior, unfortunately pose difficulties in structure elucidation.

Also included in Table 1 are a few antimicrobial peptides from bacterial origin. Bacteria use these to protect themselves from unrelated microbes in their ecological niche. These "biological warfare" peptides contain many unusual amino acids which are either the result of extensive post-translational modifications of ribosomally synthesized peptides (e.g. lantibiotics and microcins) or they are synthesized non-ribosomally on large peptide synthesis complexes (e.g. peptaibols). These peptides, often referred to as bacteriocins [29], have been the focus of many studies and nisin A [30] from *Lactococcus lactis* particularly has received a lot of attention to date, because this peptide is currently widely used as a food preservative [32]. Moreover, it is active at lower concentrations than the cationic antimicrobial peptides. The more potent activity could be explained when it was discovered that nisin, and related lantibiotics, form a specific complex with phospholipid II, a precursor for bacterial cell wall biosynthesis [33,34]. Very recently, an NMR paper was published showing the complex structure of nisin with lipid II (Figure 3) [35]. This is a unique contribution as lipid/peptide interactions are usually too transient to experimentally capture their complex structures. The reader is encouraged to read this intriguing paper describing the "pyrophosphate cage" structure for the nisin/lipid II complex. The solution structure of microcin J25 (Figure 1D) [36], another bacterial peptide with uncommon biochemical properties, will be discussed later in this chapter.

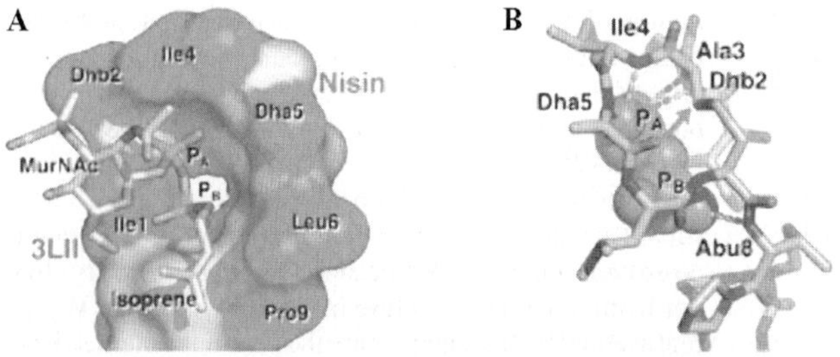

Fig. 3. The contact region between nisin and 3LII, a lipid II variant, as determined from the NMR solution structure of the complex in dimethyl sulfoxide. (A) The N-terminal part of nisin (shown in van der Waals surface) encages the two phosphate groups (P_A and P_B) of 3LII, which also contains the amino sugar N-acetylmuramic acid (MurNAc) and an isoprene tail. The side chains of nisin are labeled, with the dehydrated residues dehydroalanine, Dha, and dehydrobutyrine, Dhb. (B) The intermolecular hydrogen bond network between the nisin backbone and the 3LII pyrophosphate group (in spheres). Hydrogen bonds with high occurrence in the ensemble of structures are indicated by dashed lines, with corresponding residues labeled (with Abu representing α-aminobutyric acid). An arrow indicates the position of Ala3-Cβ, at which bioactivity is disrupted by the addition of a methyl group. The figures were taken with permission from Hsu *et al.* [35]. (See also Plate 102 on page LXVII in the Color Plate Section.)

Defensins; Structure Determination and Dimerization Properties

The defensins (Figure 2) are cationic antimicrobial mini-proteins that play a key role in the innate immune defense of organisms. Major families are found in plants, insects, and vertebrates including humans. They are distinguished by the position and cross-linking patterns of six disulfide-bonded cysteine residues. A three-stranded β-sheet structure is found in most of the defensins. Plant and insect defensins have their own distinct disulfide connectivity patterns while the disulfides of the mammalian defensins [26] are separated into three major classes: the α- and β-defensins and the more recently discovered θ-defensins. Along with the cathelicidins [37], the defensins make up the major family of antimicrobial peptides produced in humans. They are of great biological importance as they can also play a key role in stimulating the adaptive immune response, substituting for chemokines [28]. They are found on the skin and other epithelial linings and in phagocytic neutrophils [26].

Defensins are of a small enough size (30–50 amino acids) that their proton NMR spectra can be fully assigned by regular TOCSY and NOESY NMR spectroscopy [6]. The first structures solved were those of rabbit neutrophil defensin-5 (NP-5) [38], NP-2 [39], and human neutrophil protein-1 (HNP-1) [38]. These early studies revealed a typical monomeric defensin structure with an amphipathic triple stranded antiparallel β-sheet. The structure was further characterized by hairpin formation and several other tight turns which appear to be stabilized by the disulfide bonds. The structure of the bovine neutrophil β-defensin-12 (Figure 2A) was also shown to be similar to the previously studied α-defensins [40]. While the structure of insect defensin A (Figure 2B) has similar features to the mammalian defensins, it contains an additional "cysteine-stabilized α-helix" (CSH) motif where the α-helix is stabilized by disulfide bridging to the nearby antiparallel β-sheet [17]. This CSH motif is also found in other insect defensins such as in drosomycin from *Drosophila melanogaster* [41] as well as in a few plant defensins such as the γ-1-H and γ-1-P thionins from *Hordeum vulgare* and *Triticum turgidum*, respectively [42].

While the majority of the structures of the defensins show monomeric molecules, a few human peptides are thought to dimerize at low concentrations in aqueous solution. The rationale for this is unclear but it may be a first-step toward oligomerization which is required for pore formation in membranes. For example, the crystal structure of human neutrophil protein-3 (HNP-3) [43] showed an amphipathic symmetrical dimer of peptides. Diffusion NMR and dynamic light scattering, which were used to estimate the approximate size of human β-defensin-3 (HBD-3) in solution, also suggest the presence of a dimeric complex for this peptide, but not for other β-defensins [44]. Distinct chemical shifts for a region within the first β-strand of HBD-3 indicate that it is likely involved in intermolecular β-sheet formation. However, unambiguous NOE crosspeaks in the spectra could not be found to identify the exact dimerization contacts between the two β-defensin molecules in the dimer. Diffusion NMR, using the PG-SLED approach [45], is a facile technique to study the dimerization and multimerization properties of peptides and small proteins in solution [12,44,46].

A recently solved structure of interest is that of the cyclic θ defensin-1 (RTD-1) [47] from rhesus macaque leukocytes. The biosynthesis of RTD-1 involves the translation of two α-defensin-related nonapeptides which are combined by head-to-tail backbone ligation of the peptides and the formation of three disulfide bridges between these two molecules [48]. The resulting NMR-derived structure (Figure 2C), determined in 10% acetonitrile, shows an antiparallel β-sheet with very well defined turns but an overall flexible structure as judged from the poor superposition of the final 10 structures. The lack of a well-defined structure was attributed to bending motions at the center of the cyclic peptide [47].

Hepcidin; Identification of Disulfide Cross-Linking

The structure determination of the liver-synthesized peptide hepcidin presented a challenge in establishing the correct disulfide bond connectivities. Hepcidin was originally discovered in human urine [49] and plasma ultrafiltrate [50] in two forms of 20 and 25 amino acid residues long. The peptide plays an important role in the control and regulation of iron uptake; in addition, hepcidin has also been found to have antimicrobial and antifungal properties [50]. As observed with several antimicrobial peptides, there is a high proportion of Cys in hepcidin. All eight of its Cys residues are involved in disulfide bonds [50], but the specific connectivities between the residues could not be determined by conventional proteolysis methods [51]. Therefore, the solution structure determination of the two peptides by 2D ^{1}H NMR was used to establish the connectivities [12]. As done previously for the 51-residue pheromone Er-23, where the disulfide bond pattern could also not be determined by chemical methods [52], initial structure calculations incorporated only NOE restraints and bond angle constraints, without the disulfide bonds or the hydrogen bonds being included. The initial NOE patterns of the hepcidins themselves only clarified two out of the four unambiguous disulfide links. Hydrogen exchange experiments in D_2O helped to establish the position of some hydrogen bonds, and once these were introduced all the disulfide cross-links could be assigned in further rounds of the structure calculations. This approach resulted in a highly unusual vicinal cysteine bridge between the adjacent Cys residues 8 and 9. Support for this unexpected connectivity was obtained from other NMR parameters, such as a lack of amide proton correlations in the fingerprint regions for these two residues in the TOCSY and NOESY spectra, as well as broad peaks for their α and β-protons, indicating a conformational exchange process on the NMR timescale. The role of this unique structural feature in the hepcidins is still unclear. Vicinal disulfides have been observed in a few enzymes where they were found to be critical for enzyme function [53]. They have not yet been found in any other antimicrobial peptide. The overall structure of the hepcidins is a distorted two-stranded β-sheet where the vicinal disulfide bridge is found at the turn of the hairpin. In a similar manner, disulfide connectivities have been determined on the basis of NMR experiments alone for other peptides such as the antifungal Alo-3 [54] and the insecticidal pea albumin 1, subunit b [55]. These authors used ambiguous disulfide restraints and reduced charges on the sulfur groups in the early stages of the structure calculation, with the ARIA software [56] to establish the correct connectivities.

Microcin J25; Characterization of Unusual Linkages

The bacteriocin microcin J25 is secreted by certain strains of *Escherichia coli*; its antibacterial mechanism of action has been suggested to involve cell membrane disruption [57]. However, several bacterial strains with specific mutations in the β-subunit of RNA polymerase have been found to be resistant to the peptide [58], suggesting another mode of action. The original solution structure determined for the peptide [59] revealed a distorted, highly compact β-sheet structure with the main chain N- and C-termini joined together by an amide bond.Unexpectedly, microcins with their circular backbone cleaved by thermolysin retained their antimicrobial activity as well as the core structure [60]. Subsequent NMR analysis of chemically synthesized linear and head-to-tail cyclic peptides revealed that the spectra of these synthetic peptides were different from those of the natural compound [36]. A reanalysis of the original NMR data for native microcin J25 revealed that the expected sequential NOEs linking the glycines of both termini were absent. Nevertheless, mass spectrometry of the native peptide showed that there was one extra link present in the peptide, which had been formed through a dehydration reaction. A careful NMR analysis of microcin J25 revealed that the compact peptide did not have a head-to-tail cyclized backbone, but rather a novel side chain-to-backbone ring structure (Figure 1D). This linkage involves an amide link between the amino terminus of the peptide and the carboxyl side chain of Glu 8, with an overall knot structure where the C-terminal end is threaded through the ring in a rigid conformation. In fact, this noose-shaped structure was simultaneously reported in two other independent NMR studies [61,62]. The internal cross-link found in microcin J25 has been found in other bacterial proteins, such as small peptide inhibitors [63], where it is believed to contribute to a greater resistance to proteolytic degradation.

Solid-State NMR Experiments: Peptide Orientation in Bilayers

Magainin 2

As a classical example of amphipathic α-helical antimicrobial peptides, magainin, and its natural and synthetic analogues have been intensively studied [64]. Membrane pore formation is thought to be at the center of its mechanism of action, given that magainin has been shown to disrupt the electrochemical gradient across cell membranes [65]. Hence, recent NMR studies have focused on characterizing its interactions with membrane bilayers. Residue-specific ^{15}N labeling of a series of synthetic magainin 2 peptides has been used in solid-state NMR measurements of oriented membrane samples. This work conclusively showed that the helix formed by magainin, and the related PGLa peptide, lie in an orientation that is parallel to the membrane surface of oriented bilayer samples [66], in agreement with fluorescence quenching experiments [67]. The cationic side of the magainin helix is exposed to the aqueous environment and interacts with the surrounding polar lipid head groups, while the hydrophobic half of the peptide is embedded into the membrane and interacts with the fatty acyl chains in the bilayer. Additionally, a recent hydrogen/deuterium exchange NMR experiment confirms this parallel arrangement in SDS micelles [68], with lower exchange factors associated with the hydrophobic residues and higher exchange rates seen for the polar residues.

These findings have helped in explaining the mechanism of pore formation for these peptides by excluding the "barrel-stave" model of pore formation in favor of the "toroidal" pore model [5] (shown in a schematic in Figure 4). In the barrel-stave model, lytic peptides oligomerize at very low peptide:lipid ratios on the membrane and then form peptide-lined pores by orienting perpendicularly to the membrane plane. A good example of such a barrel-stave pore is provided by alamethicin [9], a peptide that is inhibitory of prokaryotic as well as eukaryotic cells. In the toroidal pore model, after reaching a high peptide:lipid threshold, the peptides associate loosely with each other to form transmembrane pores that are lined with peptides separated by lipid head groups. This lipid-peptide rearrangement has been supported in the case of magainin 2 through fluorescence experiments showing rapid flip-flop of the lipids from one side of the bilayer leaflet to the other upon addition of the peptide [70]. Solid-state ^{31}P NMR studies of lipid bilayers showed that upon addition of MSI-78, a synthetic analogue of magainin, the temperature-dependent transition from the fluid lamellar to the hexagonal phase was inhibited [71]. This means that the peptide induced a positive curvature strain in the membrane which is consistant with toroidal pore formation.

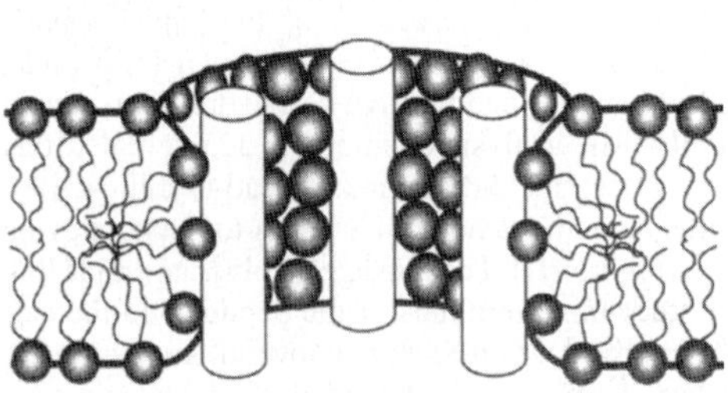

Fig. 4. The toroidal pore model as originally proposed for magainin 2. After a threshold concentration is reached, the peptides change their conformation from a parallel orientation on the bilayer plane. The peptides transiently cross the membrane to adopt a conformation perpendicular to the bilayer plane. Contacts with the phospholipid head groups remain because the pore is lined with peptides intercalated with lipids.

Recently, TRNOE experiments [72] of magainin 2 transiently bound to dilaurylphosphatidylcholine (DLPC) vesicles suggested that the peptide formed an antiparallel dimer structure; this interpretation was based on numerous long-range NOE crosspeaks that were not compatible with the peptide being in a monomeric state [73]. This study contests that solid-state NMR studies of magainin 2 missed this dimer formation because the ^{15}N labeling only provides information for the main chain conformation. Meanwhile, side chain conformations and therefore oligomerization contacts are harder to determine. The impact of this dimer formation on the currently accepted model of pore formation by magainin is presently unclear.

Protegrin-1

Solid-state NMR measurements with protegrin-1, a β-sheet peptide, showed different results compared to magainin 2. Chemical shifts of ^{13}CO and ^{15}N labels on Val-16 of the peptide were measured, and the two-stranded hairpin was calculated to be rotated by $48 \pm 5^\circ$ from the bilayer normal of the DLPC bilayers [74]. This orientation and the length of the peptide still allow for the positively charged arginine side chains to protrude out of the hydrophobic core of the membrane and interact with the lipid head groups. Three of the six Arg's are near the termini of the peptide and the other three are adjacent to each other near the hairpin turn, while the residues on the strands are hydrophobic to allow for favorable interactions with the acyl portion of the membrane. In a related solid-state NMR experiment, relatively low concentrations of paramagnetic Mn^{2+} were used to induce distance-dependent dipolar relaxation to investigate the depth of insertion of specific residues of protegrin-1 bound to DLPC bilayers [75]. This was accomplished through specific labeling of four selected residues with

^{13}C, and the order of penetration of the residues was in agreement with the tilted model for protegrin-1 insertion. While these experiments shed an interesting light on the arrangement of protegrin-1 in bilayers, it should be recognized that the DLPC lipids used have shorter acyl chains by 4–6 carbons than physiologically relevant phospholipids.

Solid-state NMR studies making use of the magic-angle-spinning (MAS) technique to obtain higher resolution have also been reported. These allowed the study of protegrin-1 bound to palmitoyloleoylphosphatidylcholine (POPC) bilayers, a more relevant membrane system [27]. However, the ^{13}C signals of the isotopically labeled protegrin-1 indicated a high level of aggregation in these bilayers; therefore proper characterization of the bound peptides could not be achieved. Be that as it may, information about the lipid arrangement could be obtained by studying ^{2}H NMR spectra of deuterium-labeled POPC with increasing concentrations of protegrin-1. The results clearly showed a greater proportion of unordered lipids in the lamellar phase [74]. These findings support the toroidal pore mechanism in the case of the protegrin-1 peptide where the membrane is under positive curvature stress and the lipids can move between the inner and outer leaflets.

Conclusions and Future Directions

Standard 2D proton NMR spectroscopy remains the primary method to obtain reliable and detailed structural information about antimicrobial peptides in solution or bound to micelles. As we have discussed, when performing structure calculations based on NMR data alone, one must carefully consider at what stage of the process it is justified to include hydrogen bonding, disulfide bridging or special features such as the knot motif in microcin J25 [36,61,62] as a constraint in the calculations. As the population of easily solvable peptide structures decreases, new NMR strategies will be necessary to fully characterize antimicrobial peptides in a more realistic membrane environment, where many are thought to exert their antimicrobial effects [3]. To this end solid-state NMR has already started to provide a snapshot of such peptides in a membrane bilayer. In the future, it should be possible to use fully isotope-labeled and deuterated peptides and TROSY NMR approaches to study these antimicrobial peptide in small vesicles, as is already being done for some membrane proteins [76,77,78]. Vesicles are a better membrane mimetic than the frequently used micelles.

The development of lipid bicelles, bilayers composed of lipids with different chain lengths forming small discoidal shapes [79], could also prove useful in determining more accurate membrane-bound solution structures. Multimerization of the peptides presents difficulties in data analysis and yet, the oligomerization states could be more reflective of biologically relevant conditions. Also, the structural data coming from solution NMR studies should be supplemented by dynamic information to gain further insights. Novel approaches using homonuclear NMR are now available [80].

Additionally, a multitude of other aspects of the mechanism of action of antimicrobial peptides remain to be characterized. In particular, studies of their interactions with the lipopolysaccharides of the outer membrane of Gram negative bacteria need to be done. Although many peptides exert their cytotoxicity through membrane disruption, in the case of peptides that do not act on the membrane, interaction studies with their intracellular targets will need to be pursued.

In addition to being bactericidal, other useful properties are starting to emerge for many of these peptides [81]. A fair number of them also selectively exhibit toxicity toward tumor cells [82], fungi [83], and viruses [84]. Interaction studies with relevant receptors can lead to an understanding of a peptide's role in the anti-inflammatory response [85], cell signaling pathways [86], or its capacity to stimulate the adaptive immune response [28]. These alternate properties present further directions in which research in this field can go.

Great strides have been made in the characterization of the solution structures of antimicrobial peptides by high-resolution NMR. In addition, solid-state NMR is beginning to provide valuable insights into the positioning of these peptides in membrane environments. Such information will be crucial for the design and development of novel classes of powerful pharmaceuticals to combat bacterial infections.

Acknowledgments

The authors would like to thank Drs. David Schibli, Howard Hunter, and Weiguo Jing for many insightful discussions. This research is supported by the Canadian Institute for Health Research and the Alberta Heritage Foundation for Medical Research.

References

1. Hancock RE. Expert. Opin. Investig. Drugs. 2000;9:1723.
2. Boman HG. J. Intern. Med. 2003;254:197.
3. Epand RM, Vogel HJ. Biochim. Biophys. Acta. 1999;1462: 11.
4. Marassi FM, Opella SJ. Curr. Opin. Struct. Biol. 1998;8: 640.
5. Matsuzaki K. Biochim. Biophys. Acta. 1999;1462:1.
6. Wuthrich K (Ed). NMR of Proteins and Nucleic Acids, John Wiley and Sons: New York, 1986.
7. Bechinger B. Biochim. Biophys. Acta. 1999;1462:157.
8. Hwang PM, Vogel HJ. Biochem. Cell Biol. 1998;76:235.

9. Powers JP, Hancock RE. Peptides. 2003;24:1681.
10. Gesell J, Zasloff M, Opella SJ. J. Biomol. NMR 1997;9:127.
11. Fahrner RL, Dieckmann T, Harwig SS, Lehrer RI, Eisenberg D, Feigon J. Chem. Biol. 1996;3:543.
12. Hunter HN, Fulton DB, Ganz T, Vogel HJ. J. Biol. Chem. 2002;277:37597.
13. Leong WI, Lonnerdal B. J. Nutr. 2004;134:1.
14. Lee DL, Hodges RS. Biopolymers. 2003;71:28.
15. Ando S, Nishikawa H, Takiguchi H, Lee S, Sugihara G. Biochim. Biophys. Acta. 1993;1147:42.
16. Jing W, Hunter HN, Hagel J, Vogel HJ. J. Pept. Res. 2003;61:219.
17. Cornet B, Bonmatin JM, Hetru C, Hoffmann JA, Ptak M, Vovelle F. Structure. 1995;3:435.
18. Schibli DJ, Hwang PM, Vogel HJ. FEBS Lett. 1999;446:213.
19. Hwang PM, Zhou N, Shan X, Arrowsmith CH, Vogel HJ. Biochemistry. 1998;37:4288.
20. Rozek A, Powers JP, Friedrich CL, Hancock RE. Biochemistry. 2003;42:14130.
21. Rozek A, Friedrich CL, Hancock RE. Biochemistry. 2000;39:15765.
22. Fales-Williams AJ, Gallup JM, Ramirez-Romero R, Brogden KA, Ackermann MR. Clin. Diagn. Lab. Immunol. 2002; 9:28.
23. Lai R, Liu H, Hui LW, Zhang Y. Biochem. Biophys. Res. Commun. 2002;295:796.
24. Matsuzaki K, Yoneyama S, Miyajima K. Biophys. J. 1997;73:831.
25. White SH, Wimley WC, Selsted ME. Curr. Opin. Struct. Biol. 1995;5:521.
26. Ganz T. Science 2002;298:977.
27. Buffy JJ, Waring AJ, Lehrer RI, Hong M. Biochemistry 2003;42:13725.
28. Roumestand C, Louis V, Aumelas A, Grassy G, Calas B, Chavanieu A. FEBS Lett. 1998;421:263.
29. Riley MA, Wertz JE. Annu. Rev. Microbiol. 2002;56:117.
30. Van Den Hooven HW, Doeland CC, Van De KM, Konings RN, Hilbers CW, Van De Ven FJ. Eur. J. Biochem. 1996;235: 382.
31. Breukink E, de Kruijff B. Biochim. Biophys. Acta. 1999; 1462:223.
32. Cleveland J, Montville TJ, Nes IF, Chikindas ML. Int. J. Food Microbiol. 2001;71:1.
33. Twomey D, Ross RP, Ryan M, Meaney B, Hill C, Antonie Van Leewenhoek 2002;82:165.
34. Breukink E, Wiedemann I, van Kraaij C, Kuipers OP, Sahl H, Kruijff B. Science 1999;286:2361.
35. Hsu ST, Breukink E, Tischenko E, Lutters MA, de Kruijff B, Kaptein R, Bonvin AM, van Nuland NM. Nat. Struct. Mol. Biol. 2004;11:963.
36. Rosengren KJ, Clark RJ, Daly NL, Goransson U, Jones A, Craik DJ. J. Am. Chem. Soc. 2003;125:12464.
37. Zanetti M, Gennaro R, Scocchi M, Skerlavaj B. Adv. Exp. Med. Biol. 2000;479:203.
38. Pardi A, Zhang XL, Selsted ME, Skalicky JJ, Yip PF. Biochemistry 1992;31:11357.
39. Pardi A, Hare DR, Selsted ME, Morrison RD, Bassolino DA, Bach AC. J. Mol. Biol. 1988;201:625.
40. Zimmermann GR, Legault P, Selsted ME, Pardi A. Biochemistry 1995;34:13663.
41. Landon C, Sodano P, Hetru C, Hoffmann J, Ptak M. Protein Sci. 1997;6:1878.
42. Bruix M, Gonzalez C, Santoro J, Soriano F, Rocher A, Mendez E, Rico M. Biopolymers. 1995;36:751.
43. Hill CP, Yee J, Selsted ME, Eisenberg D. Science 1991;251:1481.
44. Schibli DJ, Hunter HN, Aseyev V, Starner TD, Wiencek JM, McCray PB Jr, Tack BF, Vogel HJ. J. Biol. Chem. 2002;277:8279.
45. Wilkins DK, Grimshaw SB, Receveur V, Dobson CM, Jones JA, Smith LJ. Biochemistry. 1999;38:16424.
46. Weljie AM, Yamniuk AP, Yoshino H, Izumi Y, Vogel HJ. Prot. Sci. 2003;12:228.
47. Trabi M, Schirra HJ, Craik DJ. Biochemistry. 2001;40:4211.
48. Tang YQ, Yuan J, Osapay G, Osapay K, Tran D, Miller CJ, Ouellette AJ, Selsted ME. Science. 1999;286:498.
49. Park CH, Valore EV, Waring AJ, Ganz T. J. Biol. Chem. 2001;276:7806.
50. Krause A, Neitz S, Magert HJ, Schulz A, Forssmann WG, Schulz-Knappe P, Adermann K. FEBS Lett. 2000;480:147.
51. Creighton TE. In: TE Creighton (Ed). Protein Structure, A Practical Approach. IRL Press Ensyham: Oxford, 1989, p 155.
52. Zahn R, Damberger F, Ortenzi C, Luporini P, Wuthrich K. J. Mol. Biol. 2001;313:923–931.
53. Wang X, Connor M, Smith R, Maciejewski MW, Howden ME, Nicholson GM, Christie MJ, King GF. Nat. Struct. Biol. 2000;7:505.
54. Barbault F, Landon C, Guenneugues M, Meyer JP, Schott V, Dimarcq JL, Vovelle F. Biochemistry. 2003;42:14434.
55. Jouvensal L, Quillien L, Ferrasson E, Rahbe Y, Gueguen J, Vovelle F. Biochemistry. 2003;42:11915.
56. Nilges M, O'Donoghue SI. Prog. NMR Spec. 1992;32:107.
57. Rintoul MR, de Arcuri BF, Salomon RA, Farias RN, Morero RD. FEMS Microbiol. Lett. 2001;204:265.
58. Delgado MA, Rintoul MR, Farias RN, Salomon RA. J. Bacteriol. 2001;183:4543.
59. Blond A, Cheminant M, Segalas-Milazzo I, Peduzzi J, Barthelemy M, Goulard C, Salomon R, Moreno F, Farias R, Rebuffat S. Eur. J. Biochem. 2001;268:2124.
60. Blond A, Cheminant M, Destoumieux-Garzon D, Segalas-Milazzo I, Peduzzi J, Goulard C, Rebuffat S. Eur. J. Biochem. 2002;269:6212.
61. Wilson KA, Kalkum M, Ottesen J, Yuzenkova J, Chait BT, Landick R, Muir T, Severinov K, Darst SA. J. Am. Chem. Soc. 2003;125:12475.
62. Bayro MJ, Mukhopadhyay J, Swapna GV, Huang JY, Ma LC, Sineva E, Dawson PE, Montelione GT, Ebright RH. J. Am. Chem. Soc. 2003;125:12382.
63. Katahira R, Uosaki Y, Ogawa H, Yamashita Y, Nakano H, Yoshida M, J. Antibiot. (Tokyo) 1998;51:267.
64. Matsuzaki K. Biochim. Biophys. Acta. 1998;1376:391.
65. Westerhoff HV, Juretic D, Hendler RW, Zasloff M. Proc. Natl. Acad. Sci. U.S.A. 1989;86:6597.
66. Bechinger B, Zasloff M, Opella SJ. Protein Sci. 1993;2: 2077.
67. Matsuzaki K, Murase O, Tokuda H, Funakoshi S, Fujii N, Miyajima K. Biochemistry 1994;33:3342.
68. Veglia G, Zeri AC, Ma C, Opella SJ. Biophys. J. 2002;82:2176.

69. Laver DR. Biophys. J. 1994;66:355.
70. Matsuzaki K, Murase O, Fujii N, Miyajima K. Biochemistry. 1996;35:11361.
71. Hallock KJ, Lee DK, Ramamoorthy A. Biophys. J. 2003; 84:3052.
72. Post CB. Curr. Opin. Struct. Biol. 2003;13:581.
73. Wakamatsu K, Takeda A, Tachi T, Matsuzaki K. Biopolymers 2002;64:314.
74. Yamaguchi S, Hong T, Waring A, Lehrer RI, Hong M. Biochemistry 2002;41:9852.
75. Buffy JJ, Hong T, Yamaguchi S, Waring AJ, Lehrer RI, Hong M. Biophys. J. 2003;85:2363.
76. Fernandez C, Hilty C, Wider H, Guntert P, Wuthrich K. J. Mol. Biol. 2004;336:1211.
77. Hwang PM, Choy WY, Lo EI, Chen L, Forman-Kay JD, Raetz CR, Prive GG, Bishop RE, Kay LE. Proc. Natl. Acad. Sci. U.S.A. 2002;99:13560.
78. Arora A, Abildgaard J, Bushweller JH, Tamm LK. Nat. Struct. Biol. 2001;8:334.
79. Glover KJ, Whiles JA, Wu G, Yu N, Deems R, Struppe JO, Stark RE, Komives EA, Vold RR. Biophys. J. 2001;81:2163.
80. Schleucher J, Wijmenga SS. J. Am. Chem. Soc. 2002;124: 5881.
81. Kamysz W, Okroj M, Lukasiak J. Acta. Biochim. Pol. 2003;50:461.
82. Lichtenstein A, Ganz T, Selsted ME, Lehrer RI, Blood. 1986;68:1407.
83. Lupetti A, Danesi R, van 't Wout JW, van Dissel JT, Senesi S, Nibbering PH. Expert. Opin. Investig. Drugs. 2002; 11:309.
84. Cole AM, Lehrer RI. Curr. Pharm. Des. 2003;9:1463.
85. Murphy CJ, Foster BA, Mannis MJ, Selsted ME, Reid TW. J. Cell Physiol. 1993;155:408.
86. Charp PA, Rice WG, Raynor RL, Reimund E, Kinkade JM Jr, Ganz T, Selsted ME, Lehrer RI, Kuo JF. Biochem. Pharmacol. 1988;37:951.
87. Holak TA, Engstrom A, Kraulis PJ, Lindeberg G, Bennich H, Jones TA, Gronenborn AM, Clore GM. Biochemistry 1988;27:7620.
88. Shin SY, Kang JH, Hahm KS. J. Pept. Res. 1999;53: 82.
89. Oh D, Shin SY, Kang JH, Hahm KS, Kim KL, Kim Y. J. Pept. Res. 1999;53:578.
90. Hicks RP, Mones E, Kim H, Koser BW, Nichols DA, Bhattacharjee AK. Biopolymers 2003;68:459.
91. Sawai MV, Waring AJ, Kearney WR, McCray PB Jr, Forsyth WR, Lehrer RI, Tack BF. Prot. Eng. 2002;15:225.
92. Bechinger B, Zasloff M, Opella SJ. Biophys. J. 1998; 74:981.
93. Tack BF, Sawai MV, Kearney WR, Robertson AD, Sherman MA, Wang W, Hong T, Boo LM, Wu H, Waring AJ, Lehrer RI. Eur. J. Biochem. 2002;269:1181.
94. Lay FT, Schirra HJ, Scanlon MJ, Anderson MA, Craik DJ. J. Mol. Biol. 2003;325:175.
95. Lai JR, Huck BR, Weisblum B, Gellman SH. Biochemistry 2002;41:12835.
96. Shankaramma SC, Athanassiou Z, Zerbe O, Moehle K, Mouton C, Bernardini F, Vrijbloed JW, Obrecht D, Robinson JA. Chembiochem. 2002;3:1126.
97. McManus AM, Dawson NF, Wade JD, Carrington LE, Winzor DJ, Craik DJ. Biochemistry. 2000;39:15757.
98. Laederach A, Andreotti AH, Fulton DB. Biochemistry. 2002;41:12359.
99. Fujitani N, Kawabata S, Osaki T, Kumaki Y, Demura M, Nitta K, Kawano K. J. Biol. Chem. 2002;277:23651.
100. Schibli DJ, Hwang PM, Vogel HJ. Biochemistry. 1999; 38:16749.
101. Andersson M, Gunne H, Agerberth B, Boman A, Bergman T, Olsson B, Dagerlind A, Wigzell H, Boman HG, Gudmundsson GH. Vet. Immunol. Immunopathol. 1996;54:123.
102. Mandard N, Sodano P, Labbe H, Bonmatin JM, Bulet P, Hetru C, Ptak M, Vovelle F. Eur. J. Biochem. 1998;256: 404.
103. Wang Y, Henz ME, Gallagher NL, Chai S, Gibbs AC, Yan LZ, Stiles ME, Wishart DS, Vederas JC. Biochemistry. 1999;38:15438.
104. Koradi R, Billeter M, Wuthrich K. J. Mol. Graph. 1996; 14:29.

Pharmaceutical Applications of Ion Channel Blockers: Use of NMR to Determine the Structure of Scorpion Toxins

Muriel Delepierre[1] and Lourival D. Possani[2]
[1] *Unité de Résonance Magnétique Nucléaire des Biomolécules, CNRS URA 2185, 75 724 Paris Cedex 15, France*
[2] *Department of Molecular Medicine, Institute of Biotechnology, National Autonomous University of Mexico, Cuernavaca 62210, Mexico*

Introduction

Animal venoms, such as those from snakes, scorpions, cone snails, sea anemones, spiders and honeybees, are rich sources of different classes of ligands that affect the function of eukaryotic cells and tissues by blocking ion channels [1]. These ligands recognize membrane bound proteins with different specificities and affinities, thus providing scientists with unique tools to investigate diverse areas of neuroscience, protein chemistry and evolution, including the ionic channels of excitable membranes, the phylogeny of proteins and structure–function relationships of proteins [2,3]. It is unclear why toxins with similar structures display such different functions, for example why do they show specific affinities for ion channels from a particular animal? Specific recognition of a wide range of ion channel types allows the use of their pharmacological properties.

A given toxin can act on several different channels, of which only one may be of therapeutic interest [4], rendering its use as a therapeutic agent dangerous [5]. Nevertheless, successful approaches have been developed for the production of immunosuppressive agents from toxins and structural analogs [4] for the treatment of chronic pain [6] and for the treatment of various human disorders such as multiple sclerosis [7], cancer, diabetes, cardiovascular, and neurological diseases [1,8]. The fact that animal toxins are built around limited disulfide bonded frameworks, makes them resistant to degradation and permissive to mutations, meaning that they can be used as templates for protein engineering [9,10]. Before the detailed structure of a potassium channel was solved by X-ray crystallography [11], animal toxins were mainly used to map the channel-binding site and to identify the structural attributes that make a channel unique [12]. With present knowledge, the use of these toxins can be switched to the design or identification of new compounds that compete or interfere with their action upon ion channels. Unfortunately, for Na^+ channels no high-resolution structural model is yet available [13]. Therefore, further structure–activity relationship studies are required on animal toxins to determine the molecular basis of the toxin–ion channel interaction in order to engineer molecules with novel biological functions and/or therapeutic potential [1].

This chapter will focus on toxins extracted from scorpion venoms that are active on sodium and potassium channels. Sodium channel toxins are much more abundant in scorpion venom than potassium channel toxins, which account for less than 0.1% of total venom. In addition, toxins affecting potassium channels usually make only a minor contribution to the effects of envenomation on humans. It is therefore not surprising that the first toxin structure to be solved was that of a sodium channel-specific toxin, the *Centruroides sculpturatus* variant-3 toxin [14].

Use of NMR to Determine the Structure of Rare Components

It is essential to solve the three-dimensional structure of a toxin to determine how it interacts with its receptors. NMR and X-ray crystallography are the two most powerful tools available for structural determination and have been extensively used to obtain structural information on toxins. X-ray crystallography is limited by the necessity to obtain crystals, whereas NMR is limited by the size of the molecule and the low sensitivity of the method, making these two techniques complementary. NMR, which is particularly useful for determining the structure of small molecules, can also be used to obtain information on the dynamics of macromolecules over a wide timescale range and to study molecular interactions [15,16]. However, NMR is inherently less sensitive than almost all other analytical methods as a result of the small energy gap between ground and excited states. Indeed, the minimal quantity that can be analyzed by NMR is usually in the

Graham A. Webb (ed.), Modern Magnetic Resonance, 1307–1312.

nanomolar range or higher, precluding its routine use in trace analysis.

Nevertheless, the sensitivity of NMR has been improved considerably in the last decade by increasing the field strength—900 MHz is now available—and by increasing probe performance [17]. The use of cryogenic probes increases the coil quality factor, improving sensitivity by four- or fivefold compared to conventional 5 mm probes [18]. Sensitivity can be also improved by optimizing the sample volume as a function of the specific availability or solubility of a given sample.

Three types of small sample volume probes have been developed:

(i) Microprobes that use small diameter (3 mm) vertical sample tubes with volumes ranging from 120 to 150 μl. These probes, designed to couple HPLC with NMR, use saddle-type radio frequency (RF) coils. They were initially used in the pharmaceutical industry and are now becoming more widely used in academic laboratories for the characterization of small volumes of natural products and other rare samples [19]. The sample volume can be further reduced to 70–80 μl by using 3 mm Shigemi tubes. In this case, the bottom of the tube that is not in the receiver coil is filled with glass with the same dielectric properties as the solvent used. The amount of material necessary and/or the experimental time can be decreased further by using a 3-mm cryogenic probe [20]. Finally, 1.7 mm submicro inverse-detection gradient NMR probes first described in 1998 are compatible with sample volumes of just 20–30 μl [21].

(ii) The nano-NMR probe was first developed to characterize molecules covalently attached to solid-phase synthesis beads [22]. It uses a 40 μl observable volume that does not need to be filled without compromising the line widths [23] and identical concentrations to standard probes. High detection efficiency is achieved by placing the entire sample in the receiver coil. Magnetic susceptibility contributions to the line width around or within the sample are removed by spinning samples very rapidly (1–4 kHz) at the magic angle. This produces a very narrow water lineshape even at the base of the water signal (Figure 1). However, the spinning of samples in the presence of both RF and magnetic field inhomogeneities modulates the effective field, dramatically reducing the performance of isotropic mixing in TOCSY experiments when conventional composite-pulse mixing sequences are used. Adiabatic mixing sequences are less susceptible to such modulations [24] and perform considerably better in TOCSY experiments at the magic angle [25].

(iii) Microcoil NMR probes were first described in 1994 [26]. Their capillaries can contain nanoliter to microliter volumes and the probes are built with a solenoid coil of about 1 mm. They were developed to couple capillary HPLC or electrophoresis to NMR [27]. Although impressive results have been obtained [28], most of these microprobes are still in the prototype phase.

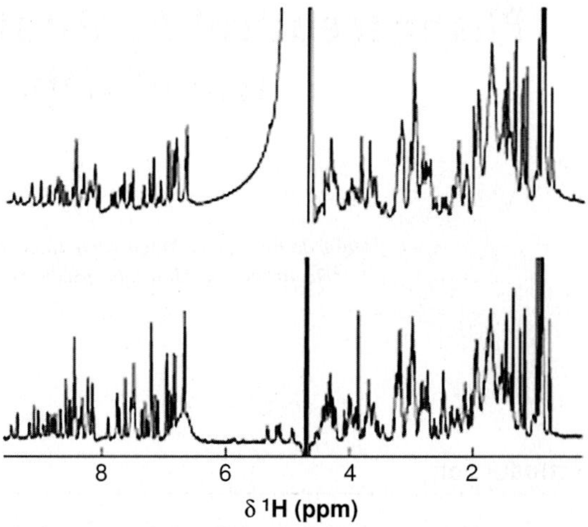

Fig. 1. One-dimensional spectra of the Pi4 toxin [56]. The lower spectrum corresponds to natural ($CONH_2$) Pi4 toxin at 380 μM 15% D_2O obtained at 500 MHz with a nano-NMR probe (Varian INNOVA) while top spectrum corresponds to synthetic (COO^-) Pi4 toxin at 100 μM 15% D_2O, 3 mm Shigemi tube obtained at 600 MHz with a cryoprobe (Varian INNOVA). In both cases 80 μg of protein was dissolved in 5 mM pH 4.0 CD_3COONa buffer. The water was suppressed with presaturation (courtesy of Inaki Guijarro).

The small volume probes make it possible to collect spectra with excellent resolution, sensitivity, and lineshape with small amounts of product, making this technique a promising option when sample size is a real limitation. The combined use of small volume probes—3 mm and below—cryogenic technology, and Shigemi tubes makes it possible to determine chemical structures at unprecedented low concentrations. The use of these technological improvements has allowed the reduction of sample amounts to less than 2% of the initial quantity required. This is evidenced by the fact that for the first scorpion toxin solved by NMR, the sodium channel-specific insect toxin AaHIT, about 12 mg was needed [29], whereas for the first potassium channel-specific scorpion toxin, charybdotoxin (ChTX) 8 mg was required [30]. Nowadays, as little as 200 μg is necessary to obtain structure of potassium channel-specific toxin with a reasonable precision, i.e. with an average of 8–14 constraints per residue [31,32].

NMR Structures of Toxins Active on Sodium Channels

Sodium channel toxins typically contain 60–70 amino acid residues. Two classes of toxin were initially described based on their pharmacological properties and their ability to bind to two different sites on the extracellular surface of the sodium channel [33,34]. Alpha scorpion toxins (α-ScTxs) slow down sodium channel inactivation by binding to receptor site 3 in a voltage-dependent manner. These have been mostly found in the venom of Asian and African scorpions. Beta scorpion toxins (β-ScTxs) bind to receptor site 4 of sodium channels and alter the mechanism of activation of the channel. They were initially found in the New World scorpion (North and South America). This rule is not absolute, because α-ScTxs have been found in the New World and conversely β-ScTxs in the Old World. Displacement experiments on excitable tissues have shown that the two binding sites are not related.

In addition to channel specificity, the sodium channel toxins also display species specificity, for example some α-like toxins are active on insects and mammals [33]. Based on toxicity tests, binding experiments and electrophysiological experiments, sodium channel toxins can be divided into peptides that are specifically toxic to either one or combinations of groups of animals. In this context, the term specific does not mean "exclusively effective" but refers to low toxicity or affinity in one group and/or higher toxicity or affinity in another.

The three-dimensional structures of several sodium channel toxins have been determined by NMR or X-ray crystallography. All toxins (α, β, and α-like) have the same overall structure based on a conserved scaffold with different insertions, deletions, and mutations. Inserted fragments are named loops J, M, B, and F and correspond, respectively, to loop 1 between β strand 1 and the α-helix, loop 2 between the α-helix and the β strand 2, loop 3 between β strands 2 and 3 and finally the C-terminal end. The α-ScTxs have a short J loop and a long B loop, while the β-ScTxs have a long J loop and a short B loop. Excitatory toxins have short J and B loops. An example of such insertion is the excitatory insect toxin from *Buthotus judaicus* (BjxtrIT) in which a five-amino acid insertion occurs just before the α-helix in loop 1 and a short α-helix is inserted in the C-terminal part of the molecule [35]. The α-ScTxs contain an insertion at around residue 42 of the B loop that is believed to play a key role in the modulation of toxic activity [36].

Based on experimental evidence, three criteria appear to be important for sodium channel specificity, namely: (i) the presence of a positively charged residue at the N-terminus (usually lysine and sometimes arginine), immediately followed by a negatively charged residue such as glutamic acid or aspartic acid at position 2; (ii) a conserved aromatic cluster; and (iii) a positively charged group (e.g. lysine or arginine) at position 13 in β-type toxins or position 58 in α-type toxins.

Most of the scorpion toxin structures that have been reported so far have been assigned to the α-type group or are weakly toxic to mammals. In North and South America, two of the most dangerous scorpions for mammals are *Centruroides noxius* and *Tityus serrulatus*, respectively. In each case, the most toxic components for mammals are β-type toxins active on sodium channels, respectively Ts1 and Cn2. Ts1 is a β-ScTxs and is the major component of the venom of *T. serrulatus*. The structure of Ts1, also known as Ts-γ or toxin VII, has been determined by X-ray diffraction to a resolution of 1.7 Å [37] while the structure of Cn2 has been determined by NMR [38]. Cn2 is composed of one α-helix and three β strands. It contains 10 aromatic residues, several of which are located on the same face of the molecule, face A, as is generally observed for sodium channel toxins. This face was thought to be the determinant for the sodium channel specificity. The orientation of the C-terminal region differs between α- and β-toxins as do the sizes of the J loop and B loop regions. The unique orientation of the Cn2 C-terminal region stems from a proline in a *cis* conformation. Interestingly Ts1 C-terminal orientation is similar to that of Cn2 but here the *cis*-proline is replaced by a helix turn. Whereas binding of β toxins to a common site on the sodium channel may be explained by the same recognition pattern on face A (i.e. the face containing the hydrophobic cluster), the difference in their toxicities may be related to the distribution of residues on the opposite face (face B). In Cn2 and Ts1, this face, formed by loop 1 and the C-terminal end, displays a large number of basic amino acid residues, whereas in weakly active toxins these residues are replaced by neutral amino acids.

The fact that the first crustacean-specific toxin for which a structure was solved came from a scorpion of the same genus as Cn2 (*Centruroides limpidus limpidus*) revealed structural features that might be responsible for the species specificity: mammals versus crustaceans. Whether a particular toxin of the genus *Centruroides* is specific towards mammals or arthropods depends on the nature of a few residues clustered in two well-defined regions (mainly loops between secondary structure elements). Mutational analyses in which the two regions are exchanged or point mutations introduced at these residues are required to corroborate these results.

NMR Structures of Toxins Active on Potassium Channels

Potassium channel toxins also vary considerably in terms of sequence and specificity, and can act on a wide range of potassium channels. Although potassium channel toxins from scorpions of the family Buthidae are mostly present as minor components of the venom, the increasing interest in venom constituents has led to efforts to improve the separation and identification of all venom compounds. So far, more than 120 different K^+ channel toxins have been found in scorpion venoms [39,40]. They have been divided into three subfamilies based on their primary structures and functions: α-, β- and γ-KTxs. The α-KTx family is the largest one, containing more than 80 members in 18 subfamilies active on Kv, BK_{Ca}, or SK_{Ca} channels. They are short-chain toxins, consisting of 23–41 amino acids and three or four disulfide bridges [41]. The four members of the β-Ktxs family identified so far are long-chain toxins composed of 60–64 amino acids and containing only three disulfide bridges [42]. Twenty-six members of the γ-Ktxs family have been identified. Members of this family are HERG channel blockers consisting of 36–47 amino acids and three or four disulfide bridges [43].

With the exception of the κ-Hefutoxins [44], all the K^+ channel-specific toxins isolated from scorpion venoms are structurally related. They are characterized by the cysteine-stabilized α/β motif (CS-αβ) in which Ci–Cj and $Ci+4$–$Cj+2$ disulfide bridges link the α-helix to the second strand of the β-sheet. κ-Hefutoxin 1 adopts a unique three-dimensional fold, consisting of two parallel helices linked by two disulfide bridges and without any β-sheets [44].

The first potassium channel toxin to be identified was noxiustoxin in 1982 [45]. Its three-dimensional structure was solved in 1995 [46]. The first potassium channel for which the structure was solved was charybdotoxin [47] discovered in 1985 [48]. The structures of several potassium channel toxins have been elucidated in the last 15 years either by X-ray or by NMR. This was made possible by following developments including: (i) chemical synthesis associated with efficient refolding, (ii) recombinant methods allowing uniform ^{15}N and ^{13}C labeling of these toxins, (iii) advances in methods for the extraction and purification, and (iv) recent developments in highly sensitive NMR methodology allowing structure determination with just a few hundred micrograms of compound [49, and references therein].

Among the 120 K^+ channel toxins the α-Ktx6 subfamily, which now comprises five different members, is characterized by the presence of an additional disulfide bridge that, in most members, fixes the C-terminus to the turn following the α-helix. Two different arrangements of disulfide bonds have been described for members of this subfamily. Pi1 [50], Pi4 [51], Pi7 [32,51] and HsTX1 [52] show half-cystine pairing of C1–C5, C2–C6, C3–C7, C4–C8, whereas MTX displays a different S–S bond topology C1–C5, C2–C6, C3–C4, C7–C8 [53]. The different S–S bonds in MTX link contiguous regions in the sequence; however MTX displays the characteristic structure of a small scorpion toxin [54]. Interestingly, no target has yet been found for toxin Pi7 despite it having a similar overall fold to all potassium channel toxins [32]. Another toxin containing four disulfide bridges has also been purified from *T. serrulatus*, TXT α-KTx17, but the additional disulfide is on the N-terminal end and the structure of this toxin has not yet been solved [55].

The three-dimensional structures of the three *Pandinus imperator* toxins have been determined using nanomolar amounts of the natural compounds and the nano-NMR probe approach and/or 3 mm Shigemi tubes [31–32,56]. The structures of HsTX1 [57] and MTX [54] were determined using synthetic molecules. The structure of Pi4 was determined using both the natural and synthetic toxins, showing that both have the same fold and disulfide pairing [56].

Disulfide bridges can be identified from the spatial interactions observed between Hα and Hβ and between Hβ and Hβ of cysteines involved in the disulfide bridge. However, in cysteine-rich proteins, such as scorpion venom toxins, it is not always possible to obtain conclusive results, as connections between Hβ protons can involve more than two cysteines. S–S bonds have been successfully assigned by NMR and modeled using different methodologies for several proteins [57,58]. Structures are calculated using NMR data only, with no disulfide bridges, to propose topologies compatible with NMR restraints for all possible combinations of disulfide bridges. Given that the bond between the sulfur atoms can only be formed if the distance between the two sulfur atoms is 2.03 Å, the structures with lowest experimental energies are analyzed in terms of the statistical distribution of distances between cystine sulfur atoms [32]. Even when other experimental data are available, NMR and structure calculations offer the possibility of testing the disulfide bridge pattern determined by other techniques and can be routinely used to locate the S–S bonds of proteins of unknown structures.

The structures of only two members of the γ-KTx family have been solved: BeKm-1 [59] and CnErg1 [60]. They have different disulfide bridging patterns and do not share any significant homology in primary structure, but both block the turret region of the HERG channel [61,62]. Their structure was solved using the recombinant toxin [59], the synthetic toxin [63] and the native toxin in the case of CnErg1 [64]. They adopt the α-KTx CS-αβmotif; however the turn of helix observed in the N-terminal part of CnErg1 is unique to the family of potassium channel toxins [63,64]. BeKm-1 and CnErg1 share similar interaction surfaces with the HERG channel. Evolutionary [65] and structural analyses demonstrated the presence of two important functional residues that are mainly located in two patches—one hydrophobic and the other hydrophilic—located at the two opposite heads of the toxin molecules [40,54,61,62]. Therefore, scorpion γ-KTx toxins seem to share the same interaction mode regardless of their primary structures.

Conclusion

The structures determined so far for scorpion toxin acting on ion channels seem to converge to a unique global fold CS-$\alpha\beta$motif even though they exert a wide range of biological functions. It is therefore reasonable to expect that molecular modeling will be enough to establish reasonable structural models. However, it is probably the unique fine details of structure (i.e. subtle changes in secondary structure elements and/or the relative orientation of side chains and/or the dynamic properties of the molecule) that are important for channel recognition and species specificity. Two recent findings emphasize how fine structure can help to elucidate the structure–function relationships of scorpion toxins. The first one concerns the finding that a α-KTx toxin, BmTx3, isolated from the venom of the scorpion *Buthus martensi* Karsch, can display dual activity, blocking HERG activity and A-type potassium currents [66]. The second is the finding of the inhibitor cystine knot (ICK) [67] fold among scorpion toxins [68]. Furthermore, preliminary results on venom proteomics [39] seem to suggest that a large number of toxins from the venom of various animal species remain to be structurally characterized, thus it is reasonable to expect that new folds and new potent molecules will be discovered.

References

1. Lewis RJ, Garcia ML. Nat. Rev. Drug Discov. 2003:2;790.
2. Menez A. Toxicon. 1998:36;1557.
3. Mouhat S, Jouirou B, Mosbah A, De Waard M, Sabatier J-M. Biochem. J. 2004:378;717.
4. Kem W, Pennington MW, Norton RS. Drug Discov. Des. 1999:15/16;111.
5. Kalman K, Pennington MW, Lanigan MD, Nguyen A, Rauer H, Mahnir V, Paschetto K, Kem WK, Grissmer S, Gutrman GA, Christian EP, Cahalan MD, Norton RS, Chandy KG. J. Biol. Chem. 1998:273;32697.
6. Nielsen K, Schroeder T, Lewis RJ. J. Mol. Recognit. 2000:13;55.
7. Beeton C, Wulff H, Barbaria J, Clot-Faybesse O, Pennington M, Bernard D, Cahalan MD, Chandy KG, Béraud E. Proc. Natl. Acad. Sci. U.S.A. 2001:98;13942.
8. Chandy KG, Wulff H, Beeton C, Pennington M, Gutman GA, Chalan MD. Trends Pharmacol. Sci. 2004:25;280.
9. Vita C, Roumestand C, Toma F, Menez A. Proc. Natl. Acad. Sci. U.S.A. 1995:92;6404.
10. Mer G, Kellenberger E, Lefevre JF. J. Mol. Biol. 1998:281;235.
11. Doyle DA, Morais Cabral JM, Pfuetzner RA, Kuo A, Gulbis JM, Cohen SL, Chait BT, MacKinnon J. Science. 1998:280;69.
12. Aiyar J, Withka JM, Rizzi JP, Singleton DH, Andrews GC, Lin W, Boyd J, Hanson D, Simon M, Dethlefs B, Lee CL, Hall JE, Gutman GA, Chandy G. Neuron. 1995:15;1169.
13. Blumenthal KM, Seibert AL. Cell Biochem. Biophys. 2003:38;1.
14. Fontecilla Camp JC, Almassy RJ, Suddath FL, Watt DD, Bugg CE. Proc. Natl. Acad. Sci. U.S.A. 1980:77;6496.
15. Wolff N, Guenneugues M, Gilquin B, Drakopoulou E, Vita C, Ménez A, Zinn-Justin S. Eur. J. Biochem. 2000:267;6519.
16. Cui M, Shen J, Briggs JM, Fei W, Wu J, Zhang Y, Luo X, Chi Z, Ji R, Jiang H, Chen K. J. Mol. Biol. 2002:318;417.
17. Hoult D, Richards RE. J. Magn. Reson. 1976:24;71.
18. Styles P, Soffe NF, Scott CA, Cragg DA, Row F, White DJ, White PCJ. J. Magn. Reson. 1984:60;397.
19. Crouch RC, Martin GE. J. Nat. Prod. 1992:55;1343.
20. Russel DJ, Hadden CE, Martin GE, Gibson AA, Zen AP, Cavolan JL. J. Nat. Prod. 2000:63;1047.
21. Subramanian R, Lam MM, Web AG. J. Magn. Reson. 1998:133;227.
22. Keifer PA. Drug Discov. Today 1997:2;468.
23. Delepierre M, Roux P, Chaffotte A-F, Goldberg ME. Magn. Reson. Chem. 1998:36;645.
24. Kupce E, Freeman R. J. Magn. Reson. A. 1995:115;273.
25. Kupce E, Keifer PA, Delepierre M. J. Magn. Reson. 2001:148;115.
26. Wu N, Peck TL, Webb AG, Magin RL, Sweedler JV. Anal. Chem. 1994:66;3849.
27. Olson DL, Peck TL, Webb AG, Magin RL, Sweedler JV. Science. 1995:270;1967.
28. Martin GE, Guido JE, Robins RH, Sharaf MHM, Tackie AN, Schiff PL Jr. J. Nat. Prod. 1998:61;555.
29. Darbon H, Weber C, Braun W. Biochemistry. 1991:30;1836.
30. Bontems F, Roumestand C, Boyot P, Gilquin B, Doljansky Y, Menez A, Toma F. Eur. J. Biochem. 1991:196;19.
31. Delepierre M, Prochnicka-Chalufour A, Possani LD. Biochemistry. 1997:36;2649.
32. Delepierre M, Prochnicka-Chalufour A, Boisbouvier J, Possani LD. Biochemistry. 1999:38;16756.
33. Gordon D, Savarin P, Gurevitz M, Zinn-Justin S. J. Toxicol. Toxin Rev. 1998:17;131.
34. Possani LD, Becerril B, Delepierre M, Tytgat J. Eur. J. Biochem. 1999:264;287.
35. Oren DA, Froy O, Amit E, Kleinberger-Doron V, Gurevitz M, Shaanan B. Structure. 1998:6;10995.
36. Fontecilla-Camps JC, Habersetzer-Rochat C, Rochat H. Proc. Natl. Acad. Sci. U.S.A 1988:85;7443.
37. Polikarpov I, Matilde MS Jr, Marangoni S, Toyama MH, Teplyakov A. J. Mol. Biol. 1999:290;175.
38. Pintar A, Possani LD, Delepierre M. J. Mol. Biol. 1999:287;359.
39. Tytgat J, Chandy KG, Garcia ML, Gutman GA, Martin-Eauclaire M-F, van der Walt JJ, Possani LD. Trends Pharmacol. Sci. 1999:20;444.
40. Rodriguez de la Vega RC, Merino E, Becerril B, Posan LD. Trends Pharmacol. Sci. 2003:24;222.
41. Batista CVF, del Pozo L, Zamudio FZ, Contreras S, Becerril B, Wanke E, Possani LD. J. Chromatogr. B. 2004:803; 55.
42. Legros C, Ceard B, Bougis PE, Martin-Eauclaire MF. FEBS Lett. 1998:431;375.
43. Corona M, Zurita M, Possani LD, Becerril B. Toxicon. 1996:34;251.
44. Srinivasan KN, Sivaraja V, Huys I, Sasaki T, Cheng B, Kumar TKS, Sato K, Tytgat J, Yu C, Cheng San BC, Ranganathan

S, Bowie HJ, Kini RM, Gopalakrishnakone P. J. Biol. Chem. 2002:277;30040.
45. Possani LD, Martin BM, Svendsen I. Carlsberg Res. Commun. 1982:47;285.
46. Dauplais M, Gilquin B, Possani LD, Gurrola-Briones G, Roumestand C, Menez A. Biochemistry. 1995:34; 16563.
47. Bontems F, Roumestand C, Gilquin B, Menez A, Toma F. Science. 1991:254;1521.
48. Miller C, Moczydlowski E, Latorre R, Philipps M. Nature. 1985:313;316.
49. Possani LD, Becerrill B, Tytgat J, Delepierre M. In: Lopatin A, Nichols CG (Eds). Ion Channel Localization Methods and Protocols. Humana Press Inc., Totowa, NJ, 2000, pp 145–66.
50. Olamendi-Portugal T, Gomez-Lagunas F, Gurrola GB, Possani LD. Biochem. J. 1996:315;977.
51. Olamendi-Portugal T, Gomez-Lagunas F, Gurrola GB, Possani LD. Toxicon. 1998:36;759.
52. Lebrun B, Romi-Lebrun R, Martin-Eauclaire M-F, Yasuda A, Ishiguro M, Oyama Y, Pongs O, Nakajima T. Biochem. J. 1997:328;321.
53. Kharrat R, Mabrouk KP, Crest M, Darbon H, Oughideni R, Martin-Eauclaire M-F, Jacquet G, El Ayeb M, Van Rietschoten J, Rochat H, Sampieri F. Eur. J. Biochem. 1996:242;491.
54. Blanc E, Sabatier JM, Kharrat R, Meunier S, El Ayeb M, Van Rietschoten J, Darbon H. Proteins Struct. Funct. Genet. 1997:29;321.
55. Becerril B, Marangoni S, Possani LD. Toxicon. 1997:35; 821.
56. Guijarro IJ, M'Barrek S, Olamendi-Portugal T, Gomez-Lagunas F, Garnier D, Rochat H, Sabatier J-M, Possani LD, Delepierre M. Protein Sci. 2003:12;1844.
57. Savarin P, Romi-Lebrun R, Zinn-Justin S, Lebrun B, Nakajima T, Gilquin B, Ménez A. Protein Sci. 1999:8;2672.
58. Nilges M, Macias MJ, O'Donoghue SI, Oschkinat H. J. Mol. Biol. 1997:269;408.
59. Boisbouvier J, Albrand J-P, Blackledge M, Jaquinod M, Schweitz H, Lazdunski M, Marion D. J. Mol. Biol. 1998:283;205.
60. Korolkova YV, Bocharov EV, Angelo K, Maslennikov IV, Grinenko OV, Lipkin AV, Nosyreva ED, Pluzhnikov KA, Olesen SP, Arseniev AS, Grishin EV. J. Biol. Chem. 2002:277;43104.
61. Gurrola GB, Rosati B, Rocchetti M, Pimienta G, Zaza A, Arcangeli A, Olivotto M, Possani LD, Wanke E. FASEB J. 1999:13;953.
62. Zhang M, Korolkova YV, Liu J, Jiang M, Grishin EV, Tseng GN. Biophys. J. 2003:84;3022.
63. Pardo-Lopez L, Zhang M, Liu J, Jiang M, Possani LD, Tseng GN. J. Biol. Chem. 2002:277;16403.
64. Torres AM, Bansal P, Alewood PF, Bursill JA, Kuchel PW, Vandenberg JI. FEBS Lett. 2003:539;138.
65. Frenal K, Xu C-Q, Wolff N, Wecker K, Gurrola G, Zhu S-Y, Chi C-W, Possani LD, Tytgat J, Delepierre M. Proteins Struct. Funct. Bioinform. 2004:56(2);367–75.
66. Huys I, Xu C-Q, Wang C-Z, Vacher H, Martin-Eauclaire M-F. Biochem. J. 2004:378;745.
67. Craik DJ. Toxicon. 2001:39;1809.
68. Zhu S, Darbon H, Dyason K, Verdonk F, Tytgat J. Faseb J. 2003:17;1765.

Structure and Dynamics of Inhibitor and Metal Binding to Metallo-β-Lactamases

Christian Damblon and Gordon C.K. Roberts
Biological NMR Centre, Department of Biochemistry, University of Leicester, Leicester LE1 7RH, UK

Introduction

The β-lactam antibiotics are among the most useful antibacterial chemotherapeutic agents, for both human and animal use, but their efficiency is continuously being challenged by the emergence of resistant strains of pathogenic bacteria. Production of β-lactamases, which inactivate these antibiotics by hydrolyzing their endocyclic amide bond, is the most important resistance mechanism [1]. β-Lactamases have been divided into four classes on the basis of their amino acid sequences and catalytic mechanisms [2]. Class B enzymes, the metallo-β-lactamases (MBLs), are ~30 kDa metallo-proteins which require one or two zinc ion(s) for their activity [3]. In the last 20 years, MBL-mediated resistance has appeared in several pathogenic strains including *Bacteroides fragilis*, *Aeromonas hydrophila, Stenotrophomonas maltophilia*, and *Serratia marcescens* [4], and some MBL genes are being disseminated rapidly by horizontal transfer, involving both plasmid and integron-borne genetic elements [5–9]. The spread of MBL-mediated bacterial resistance to β-lactams is thus a matter of concern, particularly in view of the very wide substrate profile displayed by many MBLs, which can hydrolyze almost all the known β-lactam antibiotics, including the carbapenems such as imipenem [10].

The MBLs share a small number of conserved motifs, but otherwise show significant sequence diversity. Crystal structures have been determined for the MBLs from *B. cereus* [11,12], *B. fragilis* [13–15], *Chryseobacterium meningosepticum* [16] and *S. maltophilia* [17], and for the IMP-1 enzyme from *Pseudomonas aeruginosa* [18]. These structures reveal a unique "αββα sandwich" characteristic of this family of enzymes and distinct from other zinc-dependent amide hydrolases such as thermolysin and carboxypeptidase A. It has subsequently become apparent that the metallo-β-lactamase fold is a widespread one [19–21]. Two zinc cations are present in the active site of the MBLs for which crystal structures are available. For example, in site I of the *B. cereus* enzyme, the zinc is co-ordinated by the imidazole rings of three histidine residues, H86, H88, and H149 (H116, H118, and H196 according to the standard numbering of MBLs [22]) and one water molecule. This water (or hydroxide) bridges to the zinc in site II, which is also co-ordinated by a histidine (H210 in *B. cereus*), a cysteine, an aspartate and a second water (or a carbonate ion). The zinc ligands are not absolutely conserved among MBLs, some enzymes having an Asn in place of a His in position 116 (site I ligand) and others having an Asp in place of the Cys in site II. These differences may contribute to some of the observed differences in substrate profiles and zinc affinities between different MBLs.

The precise role of the two metals in catalysis is still unclear (e.g. [3,23–26]); mechanisms have been proposed in which only the zinc in site I is involved in catalysis [23], and in which both zinc cations play essential catalytic roles [3,26]. In fact, it remains unclear whether the mononuclear or binuclear enzyme is the more physiologically relevant; recent measurements suggest that at physiological metal ion concentrations most MBLs may exist as the apoenzyme, and that the presence of substrate leads to enhanced zinc binding in the mononuclear form of the enzyme [27]. It is notable that, while the presence of the second zinc does seem to be necessary for obtaining the maximum catalytic efficiency in MBLs, most of them can catalyze substrate hydrolysis with only one zinc cation bound [28–30]. Indeed, catalytic activity of the *A. hydrophila* MBL is actually inhibited by the binding of a second zinc cation [30]. It has been proposed that the observed conservation of the ligating residues at site II may be explained if a translocation of the metal ion between the two metal sites is involved in catalysis, and that this might explain the observation of catalytic activity with only one zinc per molecule of enzyme.

The range of active site architectures for the MBLs makes the discovery of useful broad-spectrum inhibitors a challenging task, although compounds of several different chemical classes which inhibit the enzyme *in vitro* have been described [15,18,23,31–44]. Structural information, from crystallography or NMR, is available for the binding of a biphenyl tetrazole, which binds to the zinc in site II [31], and for a number of mercaptocarboxylates [18,41,44,45]. We recently demonstrated that a simple mercaptocarboxylate, thiomandelic acid

Graham A. Webb (ed.), Modern Magnetic Resonance, 1313–1319.

(β-mercapto phenylacetic acid), is a reasonably potent (sub-micromolar) inhibitor for most MBLs, except for the enzyme from *A. hydrophila* [45]. The kinetic data for thiomandelic acid fitted a competitive pattern of inhibition, and no evidence was obtained for irreversible binding of thiomandelic acid, at least to the *B. cereus* enzyme. Such a wide spectrum of activity against MBLs is unprecedented in previously published data concerning MBL inhibitors. For instance, thiomandelic acid is only about 25-fold less potent as an inhibitor of the enzyme from *B. fragilis* than of the IMP-1 enzyme from *P. aeruginosa*. This is in marked contrast to the reported thioester inhibitors, which are very much poorer inhibitors of the *B. fragilis* enzyme [37,38,43].

To obtain structural information to guide the design of improved broad-spectrum MBL inhibitors, we have used NMR to study the interaction of thiomandelic acid with the *B. cereus* MBL, BcII, particularly with its metal ions. The approaches we have used are described below.

Effect of Inhibitor Binding on the Backbone Amide Resonances

A two-dimensional proton–nitrogen correlation spectrum (^{1}H-^{15}N HSQC) of ^{15}N-labelled BcII allows the rapid observation of most of the backbone amide ^{1}H and ^{15}N resonances. We have assigned these resonances to individual residues in the free protein [45], and monitoring changes in these resonances thus provides a convenient method for identifying the regions of the enzyme affected by inhibitor binding—the "chemical shift mapping" approach has become a widely used method for the initial characterization of protein–ligand and protein–protein interactions.

Figure 1 shows a comparison of the ^{1}H-^{15}N HSQC spectra of the free enzyme and of the enzyme in the presence of *R*-thiomandelic acid (*R*-TM). Most of the spectrum remains essentially unaffected, but some significant changes are observed, generally involving residues which are close to the zinc binding sites. Only very few residues in the secondary structure elements of the core of the protein are affected by the inhibitor; some 90 residues in this part of the structure which give well-resolved resonance signals are unaffected by the addition of inhibitors. While the resonance assignment for the free enzyme is complete, the resonances in the complex are not yet assigned, so that the magnitude of the shifts of individual resonances cannot be accurately determined. We have thus used the "minimum chemical shift" approach [46–48], in which the chemical shift difference from a given cross-peak in the free protein to the closest cross-peak in the complex is calculated, to identify the residues most affected by inhibitor binding; these are indicated on the structure in Figure 1. As would be expected for a competitive inhibitor, many of the residues showing significant changes in chemical shift are in the active-site region of the structure, near the zinc ions. Only two or three residues in the structural core of the enzyme show significant changes.

Substantial amide chemical shift perturbations are, however, observed for residues in the "flexible-flap" region (the β_3–β_4 loop, residues 30–40), which in the absence of inhibitor is some distance from the active site (Figure 1). Shift changes are seen for E30, L31, F34, and V39 in this loop. (A significant shift perturbation is also observed for W59, which is located at the beginning of a loop on the top of the cavity; since the side-chain of W59 extends towards the β_3–β_4 loop, this could result from a movement of the loop.) There is good evidence that this loop plays a role in ligand binding and catalytic activity [18,41,44,45,49,50]. Loop deletion and loop swapping experiments with the *B. fragilis, B. cereus* and IMP-1 MBLs [49,50] have shown that the loop influences not only the K_i of inhibitors but also both the K_M and k_{cat} of substrates, suggesting that it plays a role in the catalytic mechanism. In the structure of the complex of IMP-1 MBL with a large mercaptocarboxylate inhibitor [18], there are clear hydrophobic contacts between the inhibitor and residues of the β3–β4 loop. These interactions might stabilize the loop, which is flexible in the absence of the inhibitor [44], in a "closed" position. However, our own studies with the *B. cereus* enzyme have shown that the residues of this loop are affected by the binding of much smaller inhibitors, not only thiomandelic acid [45] but even the simple mercaptoacetic acid [Kumar, S. unpublished work]. This clearly shows that hydrophobic interactions with the bound ligand cannot be the origin of the loop movement; there must be a "trigger" mechanism related to the coordination of the metal ions.

Effect of Inhibitor Binding on the Imidazole Resonances of the Metal Ligands

In examining the effect of inhibitor binding on the active site of the *B. cereus* enzyme, the imidazole resonances of the metal-binding histidine residues (H86, H88, and H149 in site I, H210 in site II) provide valuable probes. The imidazole N*H* resonances appear between 12 and 15 ppm (Figure 2). They have been assigned to individual residues by using ^{1}H-^{15}N HMQC spectra optimized for observation of the long-range ^{1}H-^{15}N couplings in the imidazole ring [29], which allow one to connect the N*H* resonances to those of the two nitrogens and the two C*H* protons in each imidazole. At pH 6.4, 298K, in the absence of inhibitors, relatively sharp resonances are observed for H86 and H88, and a broad resonance for H210, but no signal is apparent for H149. At lower pH or temperature, the H210 resonance sharpens and a signal for H149 appears. These differences in linewidth reflect the

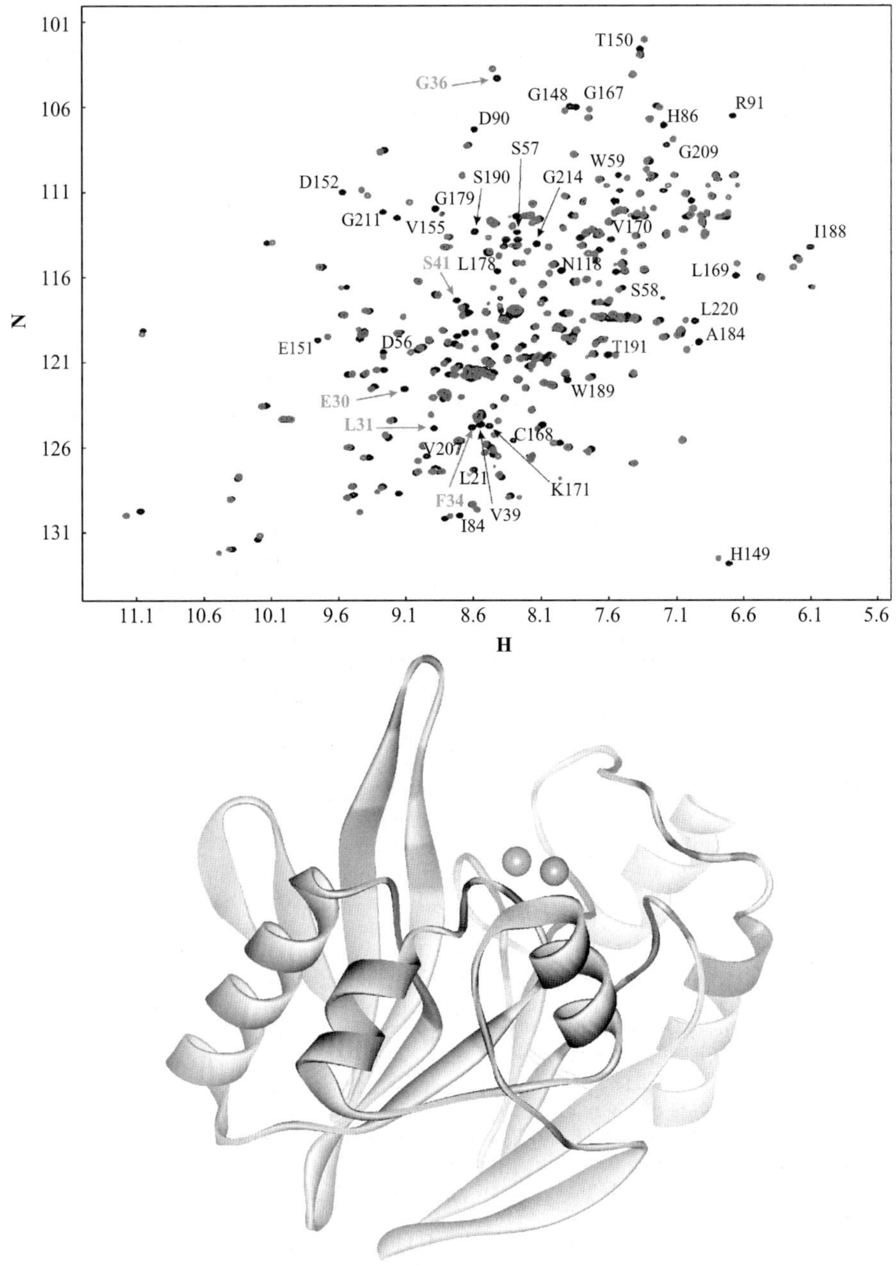

Fig. 1. ^{1}H-^{15}N HSQC spectra of *B. cereus* MBL BcII alone in black, and in its complex with *R*-TM in red. Assignments of residues affected by ligand binding are labelled on the HSQC spectra and are displayed on the X-ray structure of the free BcII. Residues affected in the flexible β3-β4 loop are labelled in blue on the HSQC and on the structure. Other residues affected by the presence of the inhibitor are coloured in green on the structure. (See also Plate 103 on page LXVII in the Color Plate Section.)

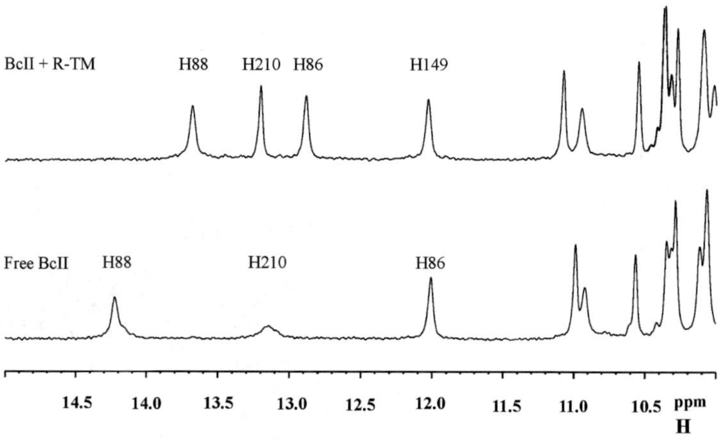

Fig. 2. Part of the ^1H NMR spectrum of *B. cereus* MBL BcII in the presence and absence of *R*-TM, showing the imidazole N*H* resonances of the metal-binding histidine residues.

differences in accessibility of the histidine imidazole N*H*s for exchange with the solvent. Gradual addition of *R*-thiomandelic acid to the enzyme results in a progressive decrease in the intensity of the imidazole N*H* signals from the free enzyme, and a progressive increase in a new set of signals attributable to the enzyme-inhibitor complex. The changes are complete at a 1:1 ratio of inhibitor to enzyme, and the spectrum of the fully formed complex is shown in Figure 2. It is clear that the imidazoles in both metal binding sites are markedly affected by inhibitor binding. In site I, both H86 and H88 show large (>0.5 ppm) changes in N*H* chemical shift on inhibitor binding. In addition, inhibitor binding clearly tends to decrease the rate of exchange with water of the imidazole N*H*s of the metal-binding histidines, thus sharpening their resonances, although the magnitude of this effect varies significantly from one residue to another. Thus, the resonances of H86 and H88 are reasonably sharp in the spectrum of the free enzyme, and their linewidths are little affected by inhibitor binding, while the resonance of H210 is broad in the spectrum of the free enzyme and is markedly sharpened by the binding of *R*-thiomandelic acid so that in the complex it has approximately the same linewidth as H86 and H88. The N*H* resonance of H149, on the other hand, is too broad to see in the spectrum of the free enzyme; only the binding of the inhibitor decreases the imidazole N*H* exchange rate of this residue sufficiently to yield an observable resonance—in fact one as sharp as those of the other three active site histidines in this complex. The binding of *R*-thiomandelic acid thus changes not only the time-average environment of the zinc-ligating residues, as reflected in their chemical shifts, but also their dynamics, as reflected in the rate of exchange of their N*H* protons with water.

Direct Observation of the Active-Site Metals

In the majority of MBLs, zinc can be exchanged with cadmium to yield catalytically active enzymes [51], and in the case of CcrA, the structure of the cadmium-substituted enzyme has been shown to be essentially identical to that of the zinc enzyme [52]. Isotopes of cadmium provide very convenient spectroscopic probes, allowing direct studies of the co-ordination and dynamics of the metal ion. We have used NMR and, in collaboration with Prof. R. Bauer (Copenhagen), perturbed angular correlation of γ-rays (PAC) spectroscopy [29,45,53,54] to make detailed studies of metal binding to the *B. cereus* MBL. The enzyme with two cadmium ions bound gives a ^{113}Cd NMR spectrum containing two distinct resonances at ~140 and ~260 ppm [29]. We have assigned these two signals to the two metal binding sites of the enzyme by using a variety of multinuclear NMR experiments, particularly ^{113}Cd-edited ^1H and ^1H-^{15}N HSQC experiments, which allow us to connect the ^{113}Cd signals to assigned ^1H and ^{15}N resonances from the metal ligands (see Figure 3; [29,45,53,54]).

Analysis of the ^1H-^{15}N HSQC spectra of the *B. cereus* MBL enzyme with only one cadmium ion bound demonstrates the existence of a mononuclear cadmium enzyme, which is distinct from the apoenzyme, and from the species with two cadmiums bound. This mononuclear enzyme shows one single broad ^{113}Cd NMR signal at ~175 ppm, suggesting at first sight that cadmium occupies only one of the two sites under these conditions; however, two well defined cadmium PAC signals are observed [53]. The only explanation for these observations comes from considering the dynamics of metal binding and the different time regimes monitored by the two methods.

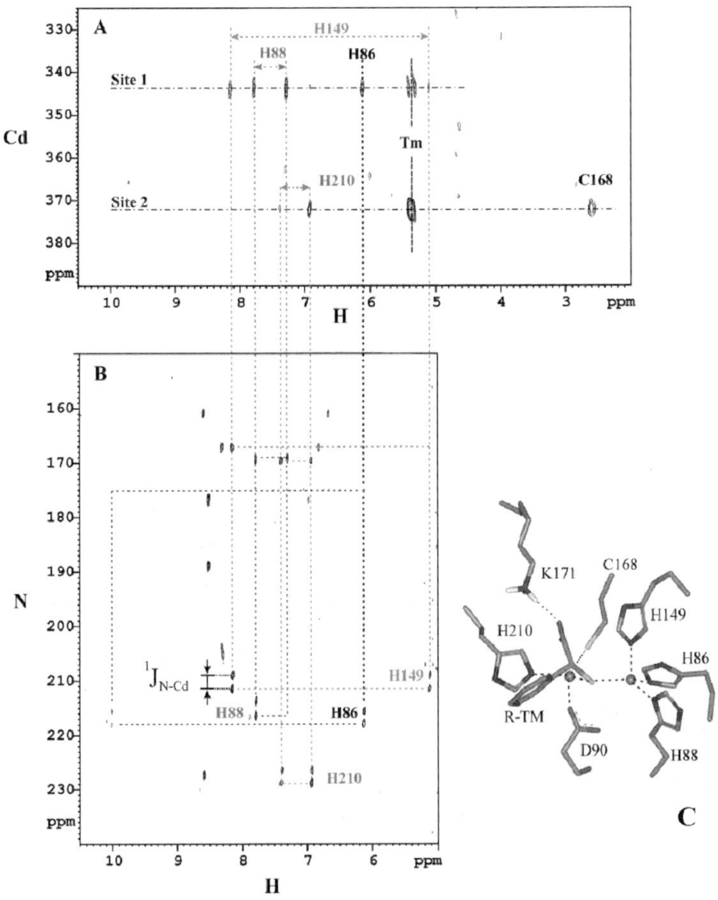

Fig. 3. NMR spectra of di-cadmium BcII in the presence of one equivalent of *R*-TM. (A) ^{1}H-^{113}Cd HMQC spectrum; (B) ^{1}H-^{15}N HMQC spectrum of the imidazole resonances, allowing assignment of the imidazole cross-peaks in A. Signal splittings in the nitrogen dimension due to the N-Cd coupling are observed for the four active site histidines. (C), Model of the binding of *R*-TM to the active site of BcII [49].

The timescales to which the two methods are sensitive are quite different: ^{111m}Cd-PAC experiments can monitor dynamics in a time regime from about 0.1 ns to 100 ns, whereas chemical exchange effects in ^{113}Cd NMR, with large chemical shift differences arising from differences in coordination, can typically monitor dynamics from 0.01 to 10 ms. The most straightforward explanation consistent with both the NMR data and the PAC data at [Cd(II)]/[E] ratios ≤ 1, is that a single cadmium bound to the enzyme jumps between the two binding sites on a time scale between 100 ns and 0.01 ms. Thus, the long-held assumption that the two different macroscopic dissociation constants for metal ion binding to the enzyme, reflecting a higher affinity of the metal for site I was shown to be wrong; they in fact reflect negative co-operativity in metal binding [53]. Recent studies of the wild-type and mutant enzyme by optical spectroscopy and EXAFS experiments indicated that there is a similar negative co-operativity in the binding of zinc and cobalt [55].

Effects of Thiomandelate Binding on the ^{113}Cd Spectrum

The BcII gives a ^{113}Cd NMR spectrum containing two distinct resonances at ~140 and ~260 ppm [29,53]. Addition of one molar equivalent of *R*-thiomandelic acid to the *B. cereus* MBL with two cadmium ions bound leads to a substantial downfield shift of both the ^{113}Cd resonances, to ~343 and ~372 ppm (Figure 3). Addition of 0.5 molar equivalent of thiomandelate leads to a spectrum containing both resonances corresponding to the free enzyme, and those corresponding to the complex, with approximately equal intensities, indicating that thiomandelate binding is in slow exchange on the ^{113}Cd NMR timescale. Further addition of the inhibitor from one to two molar equivalents does not change the NMR spectra. The direction and magnitude of the inhibitor-induced change in chemical shift is consistent with that expected for the co-ordination of a sulphur atom to each of the cadmium ions, and hence with the idea that the inhibitor binds directly to both metal sites.

Direct evidence for this, and the assignment of the two ^{113}Cd resonances in the inhibitor complexes, can be obtained from the combination of ^{1}H-^{113}Cd HMQC spectra and ^{1}H-^{15}N HSQC spectrum optimized for observation of imidazole resonances [29], shown in Figure 3. In the ^{1}H-^{113}Cd HMQC spectrum, cross-peaks are observed from the 343 ppm ^{113}Cd resonance for three imidazoles, His88, His86, and His149. The His88 can easily be assigned because it is the only metal-bound histidine which is in the N_{ε}H tautomeric form, and thus readily recognized in the ^{1}H-^{15}N HSQC spectrum[1]. The 343 ppm ^{113}Cd resonance can thus be unambiguously assigned to the cadmium in site I. There is also a strong cross-peak to a signal at 5.36 ppm, which can be assigned to the CαH (benzylic proton) of the bound *R*-thiomandelate (*R*-TM). This inhibitor resonance also gives a strong cross-peak to the low-field ^{113}Cd resonance (Figure 3), which in turn shows cross-peaks to the imidazole protons of His210 and the β-proton(s) of Cys168, and can thus be assigned to the metal in site II. The observation that, in a complex containing one molar equivalent of inhibitor, the *R*-TM resonance at 5.36 ppm gives a cross-peak in the ^{1}H-^{113}Cd HMQC spectrum to both ^{113}Cd resonances, is clear evidence that *R*-TM co-ordinates through its sulphur atom to both cadmium ions.

As described above, the ^{113}Cd spectrum of the enzyme having one cadmium equivalent bound contains a single resonance, reflecting a rapid intramolecular exchange of the single metal between the two sites. Addition of one molar equivalent of *R*-TM to the mono-cadmium enzyme led to a ^{113}Cd spectrum having two resonances at the same chemical shifts as those observed for the complex of the di-cadmium enzyme. Furthermore, the ^{113}Cd-edited ^{1}H spectra of the *R*-TM complex demonstrated that the thiomandelate CαH resonance at 5.35 ppm showed ^{1}H-^{113}Cd scalar coupling to two cadmium nuclei, just as observed for the di-cadmium enzyme, demonstrating that the thiomandelate-enzyme complex contains two cadmium ions. These observations clearly demonstrate that addition of thiomandelate to a sample of enzyme containing cadmium at a [Cd]/[E] ratio of one, leads to the formation of the *R*-TM complex of the di-cadmium enzyme together with a corresponding amount of apoenzyme [54]. This was confirmed by analysis of the backbone amide ^{1}H-^{15}N HSQC spectrum. The presence of the inhibitor, which binds to both metal ions, thus induces positive co-operativity in metal binding, in marked contrast to the negative co-operativity in cadmium binding observed in the absence of inhibitor [29].

[1] The other two histidines have not been individually assigned in the cadmium-substituted enzyme, but can be tentatively assigned by comparison with the assigned resonances in the zinc enzyme [45].

Conclusion

The power of NMR in determining the three-dimensional structure of proteins and their complexes in solution is well-established. Full structure determination of course provides the most detailed picture of a drug–protein complex, and in the present case the determination of the structure of the complex of *R*-thiomandelic acid with the *B. cereus* MBL is in progress. However the experiments described here demonstrate some of the range of NMR experiments which can provide very valuable structural and dynamic information on ligand binding short of a full structure; others are reviewed elsewhere [56,57]. Experiments of this kind, which can be carried out rapidly, have considerable value in understanding drug–protein interactions and in comparing the binding of related compounds to a target, and promise to have a continuing place in structure-based drug design.

References

1. Frere JM, Mol. Microbiol. 1995;16(3):385.
2. Matagne A, Dubus A, Galleni M, Frère JM. Nat. Prod. Rep. 16,1 (1999).
3. Wang Z, Fast W, Valentine AM, Benkovic SJ. Curr. Opin. Chem. Biol. 1999;3(5):614.
4. Payne DJ. Med J. Microbiol. 1993;39(2):93.
5. Laraki N, Franceschini N, Rossolini GM, Santucci P, Meunier C, de Pauw E, Amicosante G, Frere JM, Galleni M. Antimicrob. Agents Chemother. 1999;43(4):902.
6. Riccio ML, Franceschini N, Boschi L, Caravelli B, Cornaglia G, Fontana R, Amicosante G, Rossolini GM. Antimicrob. Agents Chemother. 2000;44(5):1229.
7. Iyobe S, Kusadokoro H, Ozaki J, Matsumura N, Minami S, Haruta S, Sawai T, O'Hara K. Antimicrob. Agents Chemother. 2000;44(8):2023.
8. Chu YW, Afzal-Shah M, Houang ET, Palepou MF, Lyon DJ, Woodford N, Livermore DM. Antimicrob. Agents Chemother. 2001;45(3):710.
9. Yano H, Kuga A, Okamoto R, Kitasato H, Kobayashi T, Inoue M. Antimicrobial. Agents and Chemother. 2001;45(5): 1343.
10. Livermore DM, Woodford N. Curr. Opin. Microbiol. 2000; 3(5):489.
11. Fabiane SM, Sohi MK, Wan T, Payne DJ, Bateson JH, Mitchell T, Sutton BJ, Biochemistry 1998;37(36):12404.

12. Carfi A, Du Eée, Galleni M, Frère JM and Dideberg O. Acta Cryst. 1998;D54:313.
13. Carfi A, Duee E, Paul-Soto R, Galleni M, Frere JM, Dideberg O, Acta Crystallogr. D. Biol. Crystallogr. 1998;54(Pt 1):45.
14. Concha NO, Rasmussen BA, Bush K, Herzberg O. Structure 1996;4(7):823.
15. Fitzgerald PM, Wu JK, Toney JH. Biochemistry 1998; 37(19):6791.
16. Garcia-Saez I, Hopkins J, Papamicael C, Franceschini N, Amicosante G, Rossolini GM, Galleni M, Frere JM, Dideberg O. J. Biol. Chem. 2003;278(26):23868.
17. Ullah JH, Walsh TR, Taylor IA, Emery DC, Verma CS, Gamblin SJ, Spencer J. Mol. J. Biol. 1998;284(1):125.
18. Concha NO, Janson CA, Rowling P, Pearson S, Cheever CA, Clarke BP, Lewis C, Galleni M, Frere JM, Payne DJ, Bateson JH, Abdel-Meguid SS, Biochemistry. 2000;39(15):4288.
19. Aravind L. In Silico Biol. 1999;1(2):69.
20. Daiyasu H, Osaka K, Ishino Y, Toh H. FEBS. Lett. 2001; 503(1):1.
21. Melino S, Capo C, Dragani B, Aceto A, Petruzzelli R. Trends Biochem. Sci. 1998;23(10):381.
22. Galleni M, Lamotte-Brasseur J, Rossolini GM, Spencer J, Dideberg O, Frere JM. Antimicrob. Agents Chemother. 2001; 45(3):660.
23. Bounaga S, Laws AP, Galleni M, Page MI. Biochem. J. 1998; 331(Pt 3):703.
24. Rasia RM, Vila AJ. Biochemistry. 2002;41(6):1853.
25. Wang ZG, Fast W, Benkovic SJ. Biochemistry. 1999;38(31): 10013.
26. Cricco JA, Orellano EG, Rasia RM, Ceccarelli EA, Vila AJ. Coordination Chem. Rev. 192,519 (1999).
27. Wommer S, Rival S, Heinz U, Galleni M, Frere JM, Franceschini N, Amicosante G, Rasmussen B, Bauer R, Adolph HW. J. Biol. Chem. 2002;277(27):24142.
28. Paul-Soto R, Hernadez-Valladares M, Galleni M, Bauer R, Zeppezauer M, Frère JM, Adolph HW. FEBS Lett. 1998;438: 137.
29. Damblon C, Prosperi C, Lian LY, Barsukov I, Paul-Soto R, Galleni M, Frère JM, Roberts GCK. J. Am. Chem. Soc. 1999;121:11575.
30. Hernandez Valladares M, Felici A, Weber G, Adolph HW, Zeppezauer M, Rossolini GM, Amicosante G, Frere JM, Galleni M. Biochemistry. 1997;36(38):11534.
31. Toney JH, Fitzgerald PM, Grover-Sharma N, Olson SH, May WJ, Sundelof JG, Vanderwall DE, Cleary KA, Grant SK, Wu JK, Kozarich JW, Pompliano DL, Hammond GG. Chem. Biol. 1998;5(4):185.
32. Nagano R, Adachi Y, Imamura H, Yamada K, Hashizume T, Morishima H, Antimicrob. Agents Chemother. 1999;43(10): 2497.
33. Walter MW, Felici A, Galleni M, Paul-Soto R, Adlington RM, Baldwin JE, Frère JM, Gololobov M, Schofield CJ. Bioorg. Med. Chem. Lett. 1996;6(20):2455.
34. Walter MW, Hernandez Valladres M, Adlington RM, Amicosante G, Baldwin JE, Frère JM, Galleni M, Rossolini GM, Schofield CJ. Bioorg. Chem. 1999;27:35.
35. Toney JH, Hammond GG, Fitzgerald PMD, Sharma N, Balkovec JM, Rouen GP, Olson SH, Hammond ML, Greenlee ML, Gao YD, J. Biol. Chem. 2001;276(34):31913.
36. Goto M, Takahashi T, Yamashita F, Koreeda A, Mori H, Ohta M, Arakawa Y. Biol. Pharm. Bull. 1997;20(11):1136.
37. Payne DJ, Bateson JH, Gasson BC, Proctor D, Khushi T, Farmer TH, Tolson DA, Bell D, Skett PW, Marshall AC, Reid R, Ghosez L, Combret Y, Marchand-Brynaert J. Antimicrob. Agents Chemother. 1997;41(1):135.
38. Payne DJ, Bateson JH, Gasson BC, Khushi T, Proctor D, Pearson SC, Reid R, Fems. Microbiol. Lett. 1997;157: 171.
39. Payne DJ, Bateson JH, Gasson BC, Khushi T, Proctor D, Pearson SC, Reid R. FEMS Microbiol. Lett. 1997;157(1):171.
40. Page MI, Laws AP. Chem. Commun. 1998;16:1609.
41. Scrofani SDB, Chung J, Huntley JJA, Benkovic SJ, Wright PE, Dyson HJ. Biochemistry. 1999;38(44):14507.
42. Greenlee ML, Laub JB, Balkovec JM, Hammond ML, Hammond GG, Pompliano DL, Epstein_Toney JH. Bioorgan. Med. Chem. Lett. 1999;9(17):2549.
43. Hammond GG, Huber JL, Greenlee ML, Laub JB, Young K, Silver LL, Balkovec JM, Pryor KD, Wu JK, Leiting B, Pompliano DL, Toney JH. FEMS Microbiol. Lett. 1999;179(2):289.
44. Huntley JJA, Scrofani SDB, Osborne MJ, Wright PE, Dyson HJ. Biochemistry. 2000;39(44):13356.
45. Mollard C, Moali C, Papamicael C, Damblon C, Vessilier S, Amicosante G, Schofield CJ, Galleni M, Frere JM, Roberts GCK. J. Biol. Chem. 2001;276(48):45015.
46. Lian LY, Barsukov I, Golovanov AP, Hawkins DI, Badii R, Sze KH, Keep NH, Bokoch GM, Roberts GCK. Structure 2000;8(1):47.
47. Farmer BT. Nat. Struct. Biol. 1996;3(12):995.
48. Williamson RA, Carr MD, Frenkiel TA, Feeney J, Freedman RB. Biochemistry. 1997;36(45):13882.
49. Yang Y, Keeney D, Tang X, Canfield N, Rasmussen BA, J. Biol. Chem. 1999;274(22):15706.
50. Moali C, Anne C, Lamotte-Brasseur J, Groslambert S, Devreese B, Van Beeumen J, Galleni M, Frere JM. Chem. & Biol. 2003;10(4):319.
51. Paul-Soto R, Bauer R, Frere JM, Galleni M, Meyer-Klaucke W, Nolting H, Rossolini GM, de Seny D, Hernandez-Valladares M, Zeppezauer M, Adolph HW. Biol. J. Chem. 1999;274(19):13242.
52. Concha NO, Rasmussen BA, Bush K, Herzberg O, Protein Sci. 1997;6(12):2671.
53. Hemmingsen L, Damblon C, Antony J, Jensen M, Adolph HW, Wommer S, Roberts GCK, Bauer R, Amer. J. Chem. Soc. 2001;123:10329.
54. Damblon C, Jensen M, Ababou A, Barsukov I, Papamicael C, Schofield CJ, Olsen L, Bauer R, Roberts GCK, J. Biol. Chem. 2003;278(31):29240.
55. de Seny D, Heinz U, Wommer S, Kiefer M, Meyer-Klaucke W, Galleni M, Frere JM, Bauer R, Adolph HW. Biol. J. Chem. 2001;276(48):45065.
56. Roberts GCK, Curr. Opin. Biotech. 1999;10(1):42.
57. Roberts GCK. Drug Discov. Today. 2000;5(6):230.

Part II

NMR Spectroscopy in the Analysis of Protein–Protein Interactions

David A. Gell and Joel P. Mackay
School of Molecular and Microbial Biosciences, University of Sydney, Sydney, NSW 2006, Australia

Introduction

Specific interactions between pairs or larger sets of proteins are central to all processes that take place within an organism, and many of these interactions may prove to be suitable targets for therapeutic intervention. In order to target such complexes effectively, detailed structural information is highly desirable, and NMR spectroscopy is one of the methods of choice for providing this information.

Recent methodological advances have enabled full structure determination for increasingly large proteins and protein complexes. These methods include TROSY pulse sequences, sample deuteration, and the use of additional long-range structural restraints derived from residual dipolar couplings (RDCs) and paramagnetic effects to complement traditional NOE-based methods. For protein–protein complexes, filtered/edited NOE experiments with differentially labeled samples greatly reduce spectral complexity through selecting only intra- or intermolecular NOEs.

While NMR cannot currently compete with X-ray crystallography in the determination of the structures of huge macromolecular machines such as the ribosome, it has many advantages for the study of more transient protein complexes, for which crystals are less likely to form. Relatively weak or transient interactions (high μM–mM dissociation constants) allow cells to respond dynamically to rapidly changing conditions; for example, many cell surface receptor interactions [1] and electron transport processes [2] rely on transient interactions. Likewise, transcription factor complexes are often formed from combinations of low-affinity interactions [3], facilitating exchange of factors that alter the transcriptional activity of the complex [4]. For weak protein complexes full structure determination by NMR may be prevented by exchange processes, but protein–protein interfaces and association constants may still be determined using chemical shift perturbation experiments. In such cases, shift perturbation data or long-range conformational restraints can be used to drive the docking of a complex, based on known structures of subunit components.

This chapter will outline the approaches mentioned above, together with experimental difficulties that are likely to be encountered. Examples both from our own work and from that of others will be used for illustration.

Tackling the Size Issue for Larger Protein Complexes

In general, the study of protein complexes involves dealing with larger species. Unfortunately, the slower tumbling (measured by τ_c, the molecular correlation time) of larger species leads to increased transverse relaxation rates. Faster transverse relaxation reduces the efficiency with which magnetization can be transferred through scalar couplings. As a consequence, the experiments used to make resonance assignments perform more poorly as τ_c increases. This has effectively placed a rough upper limit of ~30 kDa on the size of a system that can be solved using standard triple resonance methods [5]. However, several recently developed approaches have opened the door to the analysis of much larger proteins and protein complexes.

Deuteration

The first strategy involves the preparation of protein samples in which some percentage of protons is substituted by deuterons. This can be achieved by protein over-expression in cells grown on deuterated media [6]. Two advantages are conferred by the incorporation of ^{2}H. Firstly, the relaxation of ^{13}C nuclei is brought about primarily through the dipolar interaction with their attached proton(s). Because the magnitude of the dipolar interaction is directly proportional to the gyromagnetic ratios of the nuclei involved, substitution of ^{1}H for ^{2}H ($\gamma_H/\gamma_D = 6.5$) reduces the magnitude of this interaction significantly. With the addition of ^{2}H decoupling that removes the line broadening effects that arise from scalar coupling to a quadrupolar nucleus, ^{13}C line widths are dramatically reduced. Secondly, the general reduction in

Graham A. Webb (ed.), Modern Magnetic Resonance, 1321–1328.

proton density reduces spin diffusion significantly, resulting in narrower ^{1}H line widths.

Growth medium for protein deuteration is made up in $^{2}H_2O$, and cells often require acclimatization on this media to grow well [7]. To achieve complete deuteration (perdeuteration), labeled carbon sources are required in addition to $^{2}H_2O$ (e.g. ^{2}H, ^{13}C-glucose or acetate). The extent of deuteration aimed for will depend on the goal of the study. Perdeuteration, followed by back exchange of labile backbone and side chain amide deuterons for protons, as will generally occur during protein purification in $^{1}H_2O$, will yield the greatest improvements in sensitivity and resolution for backbone triple resonance experiments and chemical shift perturbation experiments (see below). However, lower levels of deuteration, down to ~50% [8], are required if side chain NOEs are to be detected. Alternatively, several groups have pioneered the use of biosynthetically directed labeling strategies that can be used to produce, for example, ^{2}H-, ^{13}C-, and ^{15}N-labeled proteins that bear protonated methyl groups on Leu, Val, and the $\delta 1$ methyl of Ile [9]. Such protein samples allow NOEs to be detected between methyl groups, which often lie in the protein core and provide a substantial number of structural constraints. Indeed, H^N–H^N, H^N–methyl, and methyl–methyl NOEs alone may be sufficient to define the global fold of a large protein [6].

TROSY

The other recent breakthrough that is revolutionizing the analysis of large systems is transverse relaxation optimized spectroscopy (TROSY) [10]. Transverse relaxation of amide ^{1}H and ^{15}N spins at high field strengths is dominated by two main mechanisms: dipole–dipole (DD) interactions and chemical shift anisotropy (CSA). Amide protons additionally experience significant relaxation (~40% of T_2 [11]) from remote ^{1}H spins, but this latter term can be largely eliminated by perdeuteration, as described above. Interference between DD and CSA relaxation (cross-correlation) has the effect of enhancing the relaxation of one component of the H^N doublet while slowing the relaxation of the other; the two doublet components thus display different line widths. The same is true for the ^{15}N doublet, and an undecoupled 2×2 multiplet in an HSQC has one component that is considerably narrower in both dimensions. The degree of relaxation interference changes with field strength, because CSA but not DD relaxation is field dependent, and at ~1 GHz the two relaxation mechanisms essentially cancel each other in the narrow component. The line width of this component consequently becomes relatively insensitive to the rotational correlation time of the protein, and the TROSY pulse sequence element (a simple modification of the HSQC pulse sequence) selects this signal. TROSY promises a great deal for the analysis of large systems and high-quality TROSY-HSQC spectra have already been recorded of systems up to ~900 kDa (a GroEL–GroES complex [12]). The TROSY element has been incorporated into a number of triple resonance experiments, and in combination with deuteration these sequences have been used to make massive strides in the size of systems that are amenable to solution NMR analysis. Backbone assignments of malate synthase, a 723-residue monomer [13], and aldolase, a 110-kDa octamer [14], have been reported recently. Perhaps even more excitingly, the structures of several integral membrane proteins solubilized in lipid micelles have been determined [15–17]. These latter studies hold substantial promise for the analysis of interactions between receptors and effector proteins.

The benefits of TROSY and deuteration have already been applied to map the interaction surfaces of large protein complexes using chemical shift perturbation experiments. Complexes formed between Ras and Byr2 [18], calreticulin and Erp57 [19], p53 and Hsp90 [20], and FimC and FimH [21] have all been characterized in this fashion.

Reducing Complexity: Differential Isotope Labeling

An additional issue that arises in NMR studies of protein complexes is the increase in the number of resonances. A simple method to alleviate this problem involves preparing complexes in which individual subunits carry different isotope labeling patterns. Most commonly, one protein will be ^{15}N- and ^{13}C-labeled, while its partner will be unlabeled. This arrangement has two advantages. First, HSQC spectra show signals from only the labeled partner, simplifying chemical shift titration experiments. Second, heteronuclear filters can be incorporated into pulse programs to either select or filter out magnetization from protons attached to specific nuclides, and hence select specific subsets of inter- or intramolecular resonances. The double half-filtered NOESY spectrum comprises four sub-spectra that are created by the linear combination of four data sets recorded with different phase cycles [22]. The most useful of these sub-spectra contain only NOEs between protons attached to different nuclides. For example, the $^{13}C(\omega_1)$-filtered/$^{13}C(\omega_2)$-selected sub-spectrum of a $^{13}C(\omega_1, \omega_2)$ double half-filtered NOESY will display NOEs between protons not bound to ^{13}C in ω_1 and protons exclusively bound to ^{13}C in ω_2. Figure 1 shows such a sub-spectrum from a data set recorded on a complex of two zinc finger domains [23] in our laboratory, and many other studies have utilized this technique to gather NOE constraints for structure determination.

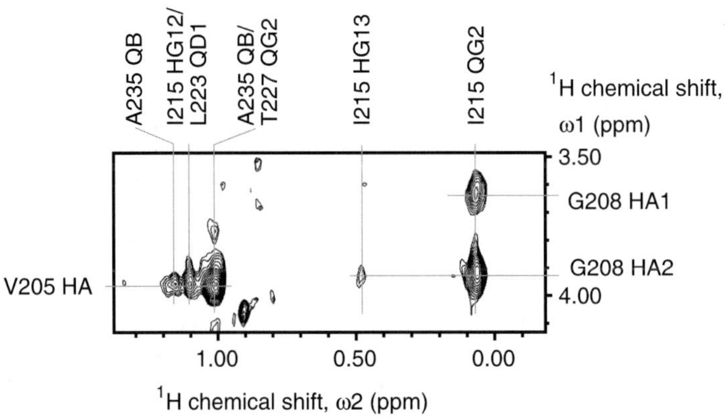

Fig. 1. Heteronuclear filtered/selected NOESY data. A portion of a two-dimensional ^{13}C-filtered, ^{13}C-edited NOESY spectrum of a 1:1 complex formed from ^{15}N and ^{13}C-labeled GATA-1 N-finger and unlabeled U-shaped finger 1, highlighting a number of intermolecular NOEs (Liew *et al.* (2005) PNAS 102, 583–8). This sub-spectrum contains only NOEs between protons attached to ^{12}C ($\omega 2$) and protons attached to ^{13}C ($\omega 1$).

Obtaining Long-Range Structural Information

A recent innovation has been the use of long-range structural information to supplement the conventional NOE data. Two methods for deriving long-range structure calculation restraints are described below; these methods have particular significance for protein complexes as they provide information about the relative orientation of subunits.

Line 'em Up—Residual Dipolar Couplings

In NMR spectra of species that do not undergo rotational motion, resonances are split not only by scalar couplings (J), but also by dipolar couplings (D) involving nuclei that are nearby in space. These dipolar couplings can be extremely large (thousands of Hz), but average to zero if the molecule tumbles rapidly in solution. Because the magnitude of a dipolar coupling is dependent on the orientation of each internuclear vector relative to the static field, couplings between different pairs of nuclei reveal the orientation of those vectors relative to a common frame of reference (and therefore relative to each other). This represents valuable structural information, not least because it reports on the relative orientations of distant parts of a molecule or complex, in contrast to the more usual NOEs and scalar coupling constants [24].

The trick in creating a useful structural tool from this concept has been to create a physical environment where a protein exhibits a small degree of preferential alignment. Too much alignment results in the observation of many dipolar couplings that render spectra un-interpretable, but a small amount yields RDCs of up to ~10–20 Hz that are observed only for spin pairs that are strongly coupled: generally one-bond interactions. Solutions containing low concentrations of a highly aligned and asymmetric component provide suitable environments for these experiments. Bacteriophage, disc-shaped lipid bicelles (both of which align completely in a magnetic field) and compressed polyacrylamide gels have all been used [24], and in such media, the reorientation of a protein becomes slightly non-isotropic. The magnitudes of the RDCs can be tuned by altering the concentration of the alignment medium.

These splittings (e.g. $^{1}D_{\mathrm{HNN}}$, $^{1}D_{\mathrm{H}\alpha\mathrm{C}\alpha}$) can be measured directly in simple 2D and 3D NMR experiments such as undecoupled HSQC and HNCO spectra, and methodologies have been developed to incorporate these data into standard structure calculation programs [25]. It should be noted that there is some degeneracy in the derived data, with more than one possible relative orientation of a vector leading to the same RDC. This fact, together with issues with the energy landscapes generated by dipolar couplings in simulated annealing protocols, have somewhat hindered the widespread use of such data to date [24].

Dipolar coupling data have been demonstrated to be beneficial in a number of scenarios, including validation of NMR structures [26] and DNA structure determination [27]. One of their most powerful uses may prove to be in defining the relative orientations of domains in a multi-domain proteins [28–30] or of protein subunits in a complex [31,32].

Paramagnetic Nuclei

Paramagnetic species cause a multitude of effects on the relaxation and chemical shifts of neighboring nuclei [33]. Paramagnetics induce both T_1 and T_2 relaxation in nearby nuclei through a number of mechanisms, and traditionally cause problems in the structure determination of metalloproteins such as copper(II) plastocyanin. However, these effects show predictable distance and angle dependence, and can therefore provide useful structural information [33,34]. Pseudocontact shifts (PCSs) are chemical shift changes induced through the interaction of nuclei with the anisotropic susceptibility tensor χ of the paramagnetic center. These changes are dependent on the position of the nucleus relative to the χ tensor and fall off with the third power of distance, thereby providing useful long-range structural information [35]. Furthermore, interaction between the χ tensor and the strong static magnetic field can lead to preferential orientation of the protein, generating RDCs.

Because PCSs are long-range effects, chemical shift changes can be induced across protein–protein interfaces and the information can therefore be used to orient protein complexes. This approach has been successful for complexes containing cytochromes [36–38] as well as for a DNA polymerase III complex [34]. Many proteins contain metal sites that can accommodate paramagnetic ions. For example many Ca^{2+} binding sites can bind paramagnetic lanthanide ions, which due to their highly anisotropic paramagnetic susceptibilities are good choices for PCS and RDC studies [39]. In the absence of natural metal binding sites, spin-labels, such as gadolinium-EDTA [40], cobalt-EDTA [35], and TEMPO (which contains a stable nitroxide radical [41]), can be attached to proteins by chemical cross-linking.

Mapping Protein–Protein Interfaces

Chemical Shift Perturbation Experiments

The chemical shift of a nuclear spin is extremely sensitive to the physical and chemical environment of the spin, and has been used extensively to map the amino acids in a protein that directly contact a binding partner. Figure 2A shows ^{1}H, ^{15}N HSQC data from the titration of a solution of unlabeled α-hemoglobin into uniformly ^{15}N-labeled alpha hemoglobin stabilizing protein (AHSP [42]). ^{1}H, ^{15}N HSQC spectra are generally used in such titrations because of their high sensitivity and excellent resonance dispersion. It can be seen that a subset of the signals shift significantly upon the addition of α-hemoglobin, and it can be inferred that the amino acids that correspond to these signals (Figure 2B) either reside on the α-hemoglobin binding surface of AHSP, or are involved in a conformational change upon α-hemoglobin binding Feng *et al.* (2004) CELL 119, 629–40. Because of this latter possibility, care must be taken in interpreting the results of chemical shift titrations; complementary data from mutagenesis studies are often extremely useful [23,43]. Identification of binding surfaces is potentially an

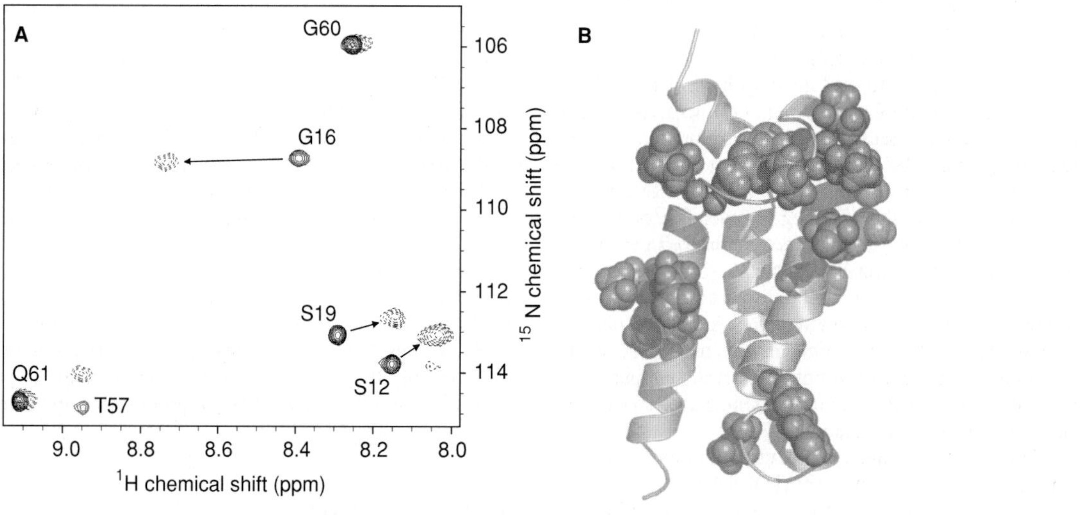

Fig. 2. Chemical shift perturbation experiment. (A) Sections of ^{1}H, ^{15}N HSQC spectra of ^{15}N-labeled AHSP are shown in the absence (solid lines) and the presence (dashed line) of 1.2 molar equivalents of unlabeled α-hemoglobin. Resonance assignments are indicated for several residues that undergo significant chemical shift changes (indicated by arrows), as well as several that remain unperturbed. (B) A ribbon representation of the AHSP NMR structure, highlighting the residues that undergo significant chemical shift changes upon titration with α-hemoglobin (space-fill). These residues constitute the major portion of the α-hemoglobin binding surface of AHSP (Feng *et al.* (2004) CELL 119, 629–40).

important early step in the design of targeted therapeutics, and chemical shift titrations have been used extensively in the screening of small molecule libraries to find lead compounds that bind a target protein [44]. Chemical shift titration data can also be used to determine equilibrium constants (for $K_A \sim 10^3$–10^6 M^{-1}).

Cross-saturation

Recently, a new method for mapping protein interfaces, called cross-saturation [45], has been employed to map interaction interfaces of a number of protein complexes [46–48]. In this approach, the protein whose interaction surface is to be mapped is ^{2}H- and ^{15}N-labeled and mixed with its unlabeled partner. Saturation of high-density ^{1}H nuclei of the unlabeled protein spreads to the back-exchanged amide protons on the interaction surface of the double-labeled protein by spin diffusion and can be detected through quenching of signals in a ^{1}H, ^{15}N TROSY-HSQC. Because this technique relies on through-space proximity of the interaction surfaces, conformational changes would not be expected to affect the results as they do with chemical shift mapping studies.

Protein–Protein Interactions and Chemical Exchange

The phenomenon of chemical exchange is observed in many different guises in NMR spectra. In the examination of protein–protein interactions, the rate of exchange between free and bound protein forms can lead to poor NMR spectra if the timescale of the binding process compared to the chemical shift timescale is inappropriate.

As is well known, slow and fast exchange regimes exist [49]. Consider a single proton at a protein–protein interface. In slow exchange, the lifetimes of both the bound and free states are sufficiently long ($>\sim 1$ s) that a separate signal is observed for each. In fast exchange (~μs or shorter lifetimes), a single signal at an averaged chemical shift (determined by the populations of the bound and free states) is observed. At the extremes of these regimes, no signal broadening occurs. However, in between lies the intermediate exchange regime, where substantial line broadening and non-Lorentzian lineshapes can occur.

In the situation where two species with equal populations interconvert in a unimolecular reaction (A ↔ B), the maximum amount of line broadening occurs when the rate of exchange (k_{ex}) is given by:

$$k_{ex} = \frac{\pi \Delta \nu}{\sqrt{2}} \quad (1)$$

where $\Delta\nu$ is the frequency difference for a spin between two states. For bimolecular reactions (A + B ↔ C), the mathematics is more complex, but analogous expressions, albeit approximate, can be derived. In cases of intermediate exchange, the degree of line broadening depends on several parameters, namely the off-rate for the complex (k_d), the equilibrium constant (K_A), the line widths of the free and bound species, and the concentrations of each component. For typical frequency changes that might be observed in the formation of a protein complex (~10–600 Hz for ^{1}H), and assuming diffusion controlled on-rates for complex formation, equilibrium constants in the range 10^3–10^6 M^{-1} are most likely to give rise to NMR spectra exhibiting intermediate exchange. As noted above, regulatory interactions of this strength are common in cells.

Chemical shift titration data can be used to determine interaction affinities. For systems in fast exchange, non-linear least squares fitting of the population averaged chemical shift to a simple hyperbolic binding function can be carried out. For slow exchange, disappearance and/or reappearance of signals from the free or bound protein states can be used to generate a binding curve. In the case of intermediate exchange, where significant line broadening occurs, the analysis is more complex, but binding constants may still be calculated [50] (Figure 3).

Line widths often increase substantially during the course of such a titration as a result of exchange broadening, but then decrease toward the end as the fraction of labeled protein in the complexed state nears one. If the

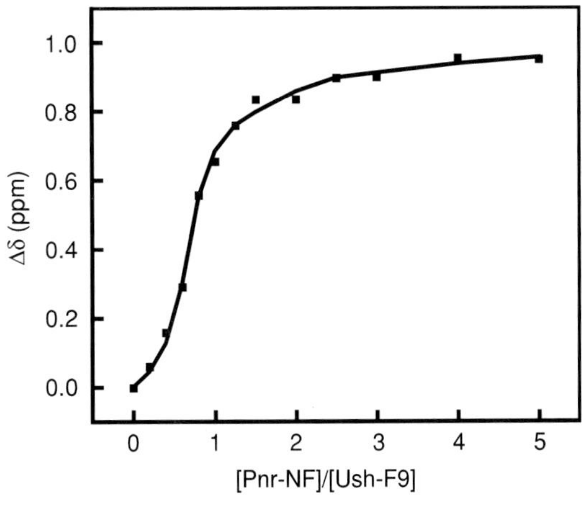

Fig. 3. Deriving binding affinity from NMR data. Plot of the change of ^{15}N chemical shift ($\Delta\delta$) for residue V21 of U-Shaped finger 9 (Ush-F9) upon binding to the N-finger of Pannier (Pnr-NF). The analysis [62] was based using a model of intermediate exchange [50] and reveals an association constant of 2.5×10^4 M^{-1}.

10. Pervushin K, Riek R, Wider G, Wüthrich K. Proc. Natl. Acad. Sci. U.S.A. 1998;94:12366.
11. Volkman BF, Prehoda KE, Scott JA, Peterson FC, Lim WA. Cell. 2002;111(4):565.
12. Fiaux J, Bertelsen EB, Horwich AL, Wuthrich K. Nature. 2002;418(6894):207.
13. Tugarinov V, Muhandiram R, Ayed A, Kay LE. J. Am. Chem. Soc. 2002;124(34):10025.
14. Salzmann M, Pervushin K, Wider G, Senn H, Wuthrich K. J. Am. Chem. Soc. 2000;122:7543.
15. Fernandez C, Adeishvili K, Wuthrich K. Proc. Natl. Acad. Sci. U.S.A. 2001;98(5):2358.
16. Arora A, Abildgaard F, Bushweller JH, Tamm LK. Nat. Struct. Biol. 2001;8(4):334.
17. Hwang PM, Choy WY, Lo EI, Chen L, Forman-Kay JD, Raetz CR, Prive GG, Bishop RE, Kay LE. Proc. Natl. Acad. Sci. U.S.A. 2002;99(21):13560.
18. Gronwald W, Huber F, Grunewald P, Sporner M, Wohlgemuth S, Herrmann C, Kalbitzer HR. Structure. 2001;9(11):1029.
19. Frickel EM, Riek R, Jelesarov I, Helenius A, Wuthrich K, Ellgaard L. Proc. Natl. Acad. Sci. U.S.A. 2002;99(4):1954.
20. Rudiger S, Freund SM, Veprintsev DB, Fersht AR. Proc. Natl. Acad. Sci. U.S.A. 2002;99(17):11085.
21. Pellecchia M, Sebbel P, Hermanns U, Wuthrich K, Glockshuber R. Nat. Struct. Biol. 1999;6(4):336.
22. Otting G, Wüthrich K. J. Magn. Reson. 1989;85:586.
23. Liew CK, Kowalski K, Fox AH, Newton A, Sharpe BK, Crossley M, Mackay JP. Structure. 2000;8:1157.
24. Bax A, Kontaxis G, Tjandra N. Methods Enzymol. 2001; 339:127.
25. Tjandra N, Omichinski JG, Gronenborn AM, Clore GM, Bax A. Nat. Struct. Biol. 1997;4(9):732.
26. Clore GM, Garrett D. J. Am. Chem. Soc. 1999;121:9008.
27. Ramos A, Grunert S, Adams J, Micklem DR, Proctor MR, Freund S, Bycroft M, St Johnston D, Varani G. EMBO J. 2000;19(5):997.
28. Fisher M, Losonczi J, Prestegard J. Biochemistry. 1999; 38:9013.
29. Walters KJ, Lech PJ, Goh AM, Wang Q, Howley PM. Proc. Natl. Acad. Sci. U.S.A. 2003;100(22):12694.
30. Tsui V, Zhu L, Huang TH, Wright PE, Case DA. J. Biomol. NMR. 2000;16(1):9.
31. Garrett DS, Seok YJ, Peterkofsky A, Gronenborn AM, Clore GM. Nat. Struct. Biol. 1999;6(2):166.
32. Wang G, Louis JM, Sondej M, Seok YJ, Peterkofsky A, Clore GM. EMBO J. 2000;19(21):5635.
33. Ubbink M, Worrall JA, Canters GW, Groenen EJ, Huber M. Annu. Rev. Biophys. Biomol. Struct. 2002;31:393.
34. Pintacuda G, Keniry MA, Huber T, Park AY, Dixon NE, Otting G. J. Am. Chem. Soc. 2004;126(9):2963.
35. Gaponenko V, Sarma SP, Altieri AS, Horita DA, Li J, Byrd RA. J. Biomol. NMR. 2004;28(3):205.
36. Ubbink M, Ejdeback M, Karlsson BG, Bendall DS. Structure. 1998;6(3):323.
37. Guiles RD, Sarma S, DiGate RJ, Banville D, Basus VJ, Kuntz ID, Waskell L. Nat. Struct. Biol. 1996;3(4):333.
38. Worrall JA, Liu Y, Crowley PB, Nocek JM, Hoffman BM, Ubbink M. Biochemistry. 2002;41(39):11721.
39. Barbieri R, Bertini I, Cavallaro G, Lee YM, Luchinat C, Rosato A. J. Am. Chem. Soc. 2002;124(19):5581.
40. Arumugam S, Hemme CL, Yoshida N, Suzuki K, Nagase H, Berjanskii M, Wu B, Van Doren SR. Biochemistry. 1998;37(27):9650.
41. Jain NU, Venot A, Umemoto K, Leffler H, Prestegard JH. Protein Sci. 2001;10(11):2393.
42. Kihm AJ, Kong Y, Hong W, Russell JE, Rouda S, Adachi K, Simon MC, Blobel GA, Weiss MJ. Nature. 2002; 417(6890):758.
43. Schmiedeskamp M, Rajagopal P, Klevit RE. Protein Sci. 1997;6(9):1835.
44. Shuker SB, Hajduk PJ, Meadows RP, Fesik SW. Science. 1996;274(5292):1531.
45. Takahashi H, Nakanishi T, Kami K, Arata Y, Shimada I. Nat. Struct. Biol. 2000;7(3):220.
46. Lane AN, Kelly G, Ramos A, Frenkiel TA. J. Biomol. NMR. 2001;21(2):127.
47. Nishida N, Sumikawa H, Sakakura M, Shimba N, Takahashi H, Terasawa H, Suzuki EI, Shimada I. Nat. Struct. Biol. 2003;10(1):53.
48. Shao W, Im SC, Zuiderweg ER, Waskell L. Biochemistry. 2003;42(50):14774.
49. Sanders J, Hunter B. Modern NMR Spectroscopy. Oxford University Press: Oxford, 1987.
50. Feeney J, Batchelor J, Albrand J, Roberts G. J. Magn. Reson. 1979;33:519.
51. Demarest SJ, Martinez-Yamout M, Chung J, Chen H, Xu W, Dyson HJ, Evans RM, Wright PE. Nature. 2002;415(6871): 549.
52. Deane JE, Sum E, Mackay JP, Lindeman GJ, Visvader JE, Matthews JM. Protein Eng. 2001;14:493.
53. Deane J, Mackay J, Kwan A, Sum E, Visvader J, Matthews J. EMBO J. 2003;22:2224.
54. Hofmann RM, Muir TW. Curr. Opin. Biotechnol. 2002;13(4): 297.
55. Clore GM. Proc. Natl. Acad. Sci. U.S.A. 2000;97(16):9021.
56. Dosset P, Hus JC, Marion D, Blackledge M. J. Biomol. NMR. 2001;20(3):223.
57. Dobrodumov A, Gronenborn AM. Proteins. 2003;53(1):18.
58. Dominguez C, Boelens R, Bonvin AM. J. Am. Chem. Soc. 2003;125(7):1731.
59. Fahmy A, Wagner G. J. Am. Chem. Soc. 2002;124(7):1241.
60. Brunger AT, Adams PD, Clore GM, DeLano WL, Gros P, Grosse-Kunstleve RW, Jiang JS, Kuszewski J, Nilges M, Pannu NS, Read RJ, Rice LM, Simonson T, Warren GL. Acta. Crystallogr. D. Biol. Crystallogr. 1998;54(Pt 5):905.
61. Nilges M. J. Mol. Biol. 1995;245(5):645.
62. Kowalski K, Liew CK, Matthews JM, Gell DA, Crossley M, Mackay JP. J. Biol. Chem. 2002;277:35720.

Identification and Characterization of Ternary Complexes Using NMR Spectroscopy

Robert E. London

Laboratory of Structural Biology, National Institute of Environmental Health Sciences, Research Triangle Park, NC 27709, USA

Introduction

The complex macromolecules that have evolved to serve a broad range of biological functions present surfaces with highly variable topological, electrostatic, and hydrogen bonding properties. These surfaces provide the basis for interactions that play critical roles in catalysis and signal transduction, and offer a limitless landscape of potential targets for chemical agents designed to interfere with these functions. A broad range of analytical techniques has been developed to exploit this topological information when it is available, but it is clear that flexible ligands can interact in unanticipated ways with their theoretical targets, and often with unintended targets as well. Substitution of an amino group for a single oxygen substituent to convert folate to aminopterin results in a potent antifolate drug which, unexpectedly, adopts a dramatically different conformation in the complex with the target enzyme, dihydrofolate reductase [1,2]. Studies of the inhibition of alpha-chymotrypsin by a series of 1-acetamidoboronic acids unexpectedly revealed that when the side chain of the inhibitor was altered from a *p*-chlorophenyl to a 1-naphthyl group, the enzyme reversed its usual preference for L-enantiomers, which more closely resemble the natural substrates, and bound more tightly to the D-enantiomer [3].

The sensitivity of NMR spectroscopy to ligand–macromolecule interactions makes it of considerable value for the process of ligand identification and modification, and a broad range of approaches have been described that involve direct observations of either the target macromolecule [4] or the ligand [5]. The use of any particular approach is generally strongly dependent on the particular problem under study, however direct NMR studies of biologically important macromolecules are limited by the size and complexity of the macromolecule, its solubility, the general requirement for isotopic labeling, and the more extensive analysis which may be required. Alternatively, observations of ligands have been demonstrated to be useful for selecting lead compounds from complex mixtures [6] and for obtaining insight into the conformation of bound ligands [7], and are often more straightforward to interpret.

One approach for the analysis of ligand–macromolecule interactions involves the identification of ternary complexes formed from pairs of ligands in the presence of a macromolecule. In some cases, biologically interesting complexes have formed from adventitious inorganic ions such as aluminum or beryllium fluorides [8]. The presence of adventitious zinc ions led to the identification of a trypsin-Zn^{2+}-bis(5-amidino-2-benzimidazolyl)methane ternary complex, which formed a basis for the further development of serine protease inhibitors [9]. Arsenate and borate complexes have been identified due to the use of cacodylate and borate buffers [10,11]. Vanadate complexes have been identified in toxicological and bioinorganic studies [12–14]. These complexes have been detected based on their biochemical effects, or by structural methods including X-ray crystallography and NMR spectroscopy. Arsenate complexes often mimic phosphate complexes, both structural and biochemically, while vanadate complexes often form pentavalent structures that mimic the transition state involved in nuclease and phosphatase biochemistry [12–15]. Vanadate, and aluminum and beryllium fluorides are all capable of mimicking the terminal phosphate group of nucleotides, e.g. turning ADP into the ADP-VO_4 analog of ATP [16–18]. Boronate and borate complexes most frequently mimic the tetrahedral transition state of serine proteases and mechanistically related enzymes [19–21]. Alternatively, enzymatic catalysis frequently involves the formation of ternary complexes that include two, and occasionally more, organic ligands. For example, many enzyme-catalyzed redox reactions involve ternary enzyme-substrate–cofactor complexes. Formation of transient complexes involving macromolecules and organic ligands can be studied using transferred interligand Overhauser effects (ILOEs) [22–27]. A brief outline of NMR approaches used to study ternary complexes is described in the present article.

Borate Complexes and Their Study by NMR Spectroscopy

Studies of ternary and higher order complexes formed from inorganic ions have provided insights into the

Graham A. Webb (ed.), *Modern Magnetic Resonance*, 1329–1338.

toxicological properties of these ions, valuable structural information, insight into catalytic mechanisms, as well as fundamental information of use for the design of inhibitors. In some cases, these ions interact with active site residues of enzymes in ways that mimic transition state structures. A comprehensive review of this area is beyond the scope of the present article, but our recent studies of borate complexes illustrate many of the features of these studies.

Boric acid (pK = 9.0) is a convenient buffer for high pH assays of enzymes or other biochemical phenomena. The use of borate buffers has led to a number of unanticipated observations that resulted from the presence of borate–ligand complexes. For example, borate has been found to inhibit pyridine nucleotide and flavin-requiring enzymes as a result of complex formation with these cofactors [28,29]. A more dramatic effect is observed in studies of the membrane protein γ-glutamyl transpeptidase (γ-GT), an enzyme important for glutathione metabolism that transfers γ-linked glutamyl groups between peptides, amino acids, or other acceptors. For this enzyme, serine and borate, which individually exert little effect on the enzyme, exhibit a dramatic inhibitory synergy [11]. This inhibitory synergy apparently results from the stabilization of a ternary γ-GT–serine–borate complex, in which the borate adopts a tetrahedral, anionic form that is esterified to both an enzyme threonine hydroxyl group, as well as to the hydroxyl group of the serine ligand [20,21]. This ternary complex thus mimics a putative transition state in which the threonine hydroxyl group of the active site binds to the γ-carbon of the glutamyl group in glutathione or another substrate, creating a transition state with tetrahedral geometry.

Although direct NMR observations of membrane associated enzymes such as γ-GT are difficult, the general features of such ternary complexes have been observed for the serine protease trypsin [30–32]. In these studies, several cationic alcohols with presumed affinity for the trypsin S1 specificity binding pocket were studied. The molecules were selected so that the alcohol function would be appropriately positioned to interact with a borate bound to the active site serine-195. Boron-11 spectra of samples containing borate, trypsin, and 4-aminobutanol (4AB) are characterized by a resonance with a chemical shift that corresponds to the boric acid–borate equilibrium at the pH used for the study, as well as an upfield shifted resonance corresponding to a tetrahedral borate which is now in slow exchange with the free boric acid/borate species (Figure 1). Binary mixtures of borate–trypsin or borate–4AB, or a ternary mixture of borate, trypsin, and 3-aminobutanol do not show the upfield shifted resonance. The resonance observed at −17.3 ppm can therefore be assigned to a ternary trypsin–borate–4AB complex.

Analogous complexes have now been observed in the crystalline state [31,32]. Interestingly, the crystallographic studies demonstrate that the nature of the borate complex formed at the active site is significantly influenced by lattice contacts, so that a ternary trypsin–borate–4AB complex analogous to that observed in solution could be observed in $P3_121$ crystals [32], while only quaternary trypsin–borate–4AB–ethylene glycol complexes were observed in $P2_12_12_1$ crystals [31]. In the $P2_12_12_1$ crystals, formation of the active site borate complex was found to be correlated with a 4% expansion of the crystal along the b-axis. These results were interpreted to indicate that the additional stability of the quaternary complex was required to overcome crystal packing effects. This study therefore illustrates the significant effect that lattice contacts can have on crystallographic screening for enzyme ligands.

Most of the inorganic complexes which have been studied by NMR, involve nuclei with spin $>1/2$, and hence are subject to quadrupolar relaxation. For quadrupolar nuclei bound to very large molecules such that $\omega\tau \gg 1$, where ω is the Larmor frequency of the nucleus and τ its rotational correlation time, the relaxation behavior is multiexponential. For nuclei with half-integral spin in this limit, the central $1/2 \leftrightarrow -1/2$ transition can give rise to a relatively narrow signal, while the remaining transitions are very broad and generally unobserved [33–36]. Subject to the above assumptions, the linewidth of the central transition is then inversely related to $\omega^2\tau$, i.e. the linewidth actually narrows as the rotational correlation time of the complex τ increases:

$$\Delta\nu_{1/2} = \frac{1}{\pi T_2} = \frac{2s+3}{\pi s^2(2s-1)}\left(2J_1^Q(\omega_S) + ((s+3/2) \times (s-1/2)-1)\left(J_2^Q(2\omega_S)\right)\right) \quad (1)$$

where

$$J^Q(\omega) = \frac{3\chi^2}{160}\left(\frac{\tau}{1+\omega^2\tau^2}\right) \quad (2)$$

χ is the quadrupole coupling constant, nucleus S has spin s and Larmor frequency ω_S, τ is the rotational correlation time, and isotropic motion is assumed. In the limit $\omega\tau \gg 1$, we obtain:

$$\Delta\nu_{1/2} = \frac{1}{\pi T_2} \approx \left(\frac{2(2s+3)}{\pi s^2(2s-1)}\right)\left(1+\frac{1}{8}((s+3/2) \times (s-1/2)-1)\right)\frac{3\chi^2}{160}\frac{1}{\omega^2\tau} \quad (3)$$

This inverse dependence on τ contrasts with the linewidth predictions based on other relaxation mechanisms such as the dipolar interaction and chemical shift anisotropy, which predict $\Delta\nu_{1/2} \sim J(0) \sim \tau_c$. Exchange

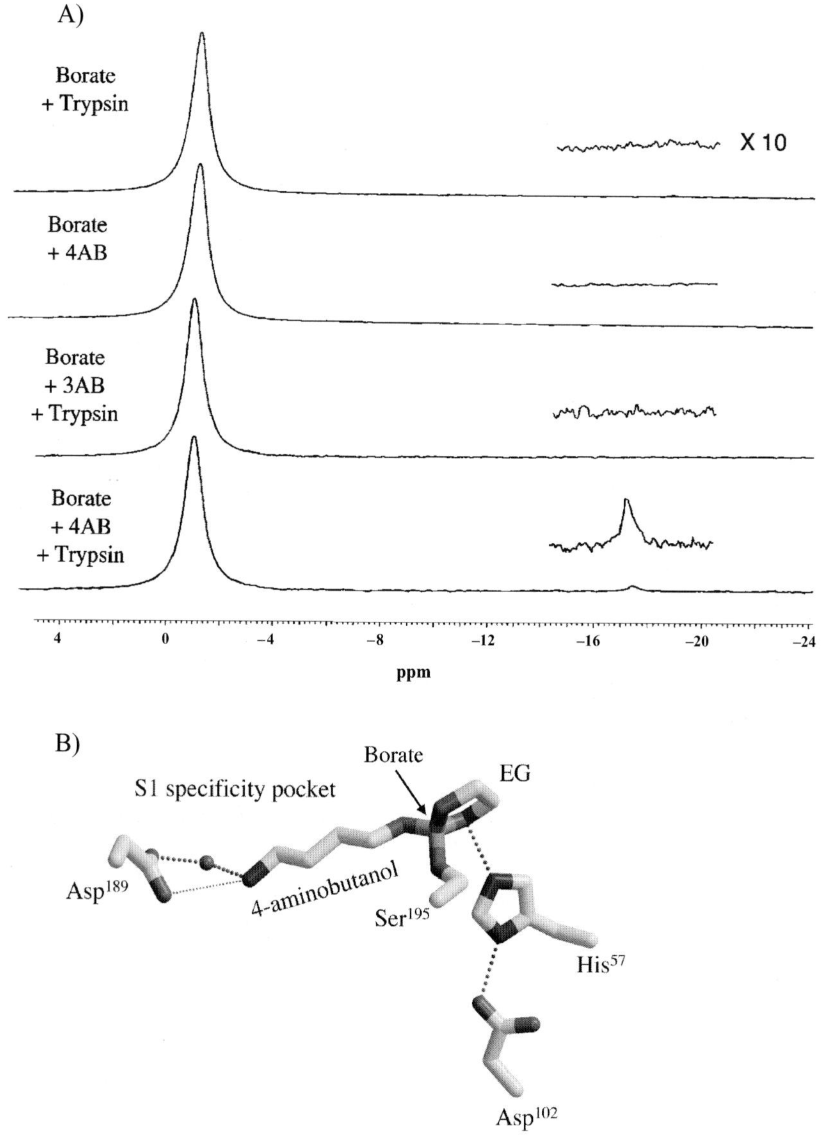

Fig. 1. (A) Boron-11 NMR spectra obtained at 160.6 MHz, 25°C on samples containing (top to bottom) 5 mM borate and 2 mM porcine pancreatic trypsin; 25 mM borate and 250 mM 4-aminobutanol; 5.0 mM borate, 50 mM 3-aminopropanol, and 2 mM trypsin; 5.0 mM borate, 50 mM 4-aminobutanol, and 2 mM trypsin. Insets show a 10-fold vertical expansion. All samples contained 20 mM HEPES, pH 8.0 (based on Figure 1 of Ref. [30]). (B) Active site of trypsin showing complex formed with borate, 4-aminobutanol, and ethylene glycol (EG) (based on Figure 1F of Ref. [31]). A similar ternary complex without the EG has also been observed for crystals in the $P3_121$ space group [32].

broadening due to the chemical exchange between uncomplexed and variously complexed forms of the ion also can make significant contributions. The observation of narrow resonances for quadrupolar nuclei in macromolecules is illustrated by the ^{11}B NMR spectrum of 4-fluorobenzeneboronic acid (FBBA) in the presence of a small ligand, sorbitol, and the enzyme subtilisin Carlsberg (MW = 27.3 kD), shown in Figure 2. As

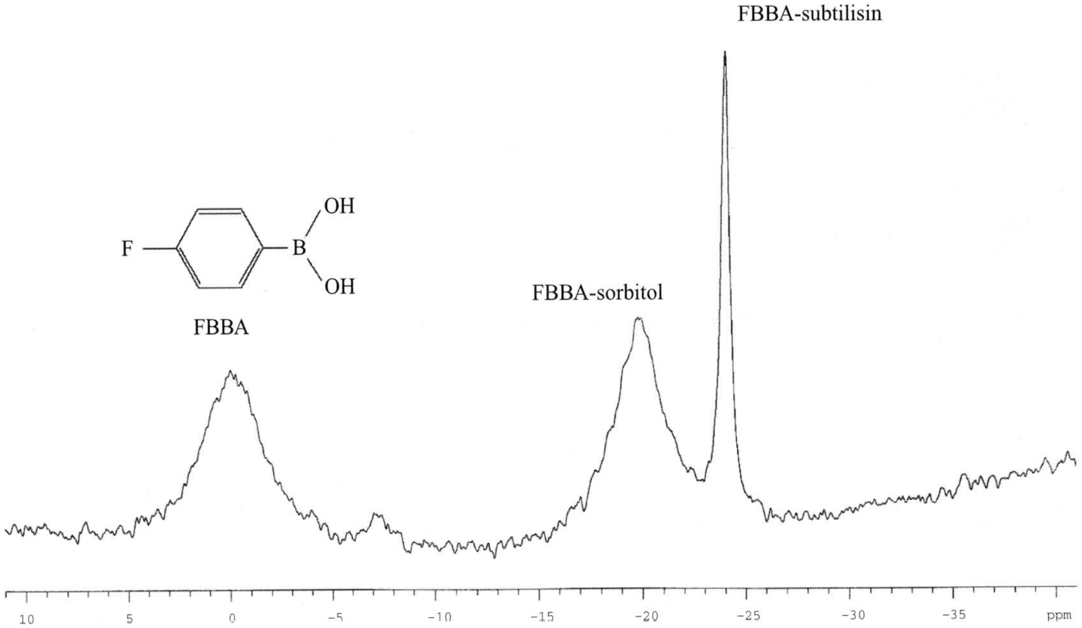

Fig. 2. Boron-11 NMR spectrum of 5 mM 4-fluorobenzeneboronic acid, 1.5 mM sorbitol, 2 mM subtilisin Carlsberg (MW = 27.3 kD) in 50 mM HEPES, pH 7.5 obtained at 11.75 T (160.6 MHz ^{11}B frequency), 5°C.

predicted above, the resonance arising from the boronate–subtilisin complex is significantly narrower than the resonances for either free boronate or for the boronate–sorbitol complex. The ^{11}B linewidth of the uncomplexed FBBA is determined by quadrupolar relaxation of the ^{11}B nucleus (spin = 3/2) and by boronic acid–boronate exchange, while the linewidths of the FBBA–sorbitol and FBBA–subtilisin complexes are probably determined primarily by quadrupolar ^{11}B relaxation. In addition to this unintuitive dependence of linewidth on rotational correlation time, the quadrupolar nuclei are also subject to a second-order dynamic frequency shift which is also inversely correlated with ω (i.e. B_0) [35,36]:

$$\delta = \frac{8s(s+1)-6}{s^2(2s-1)^2}\left(L_1^Q(\omega_S) - L_2^Q(2\omega_S)\right) \quad (4)$$

where

$$L^Q(\omega) = \frac{3\chi^2}{160}\left(\frac{\omega\tau^2}{1+\omega^2\tau^2}\right) \quad (5)$$

so that

$$\delta \approx \frac{4s(s+1)-3}{s^2(2s-1)^2}\left(\frac{3\chi^2}{160}\right)\left(\frac{1}{\omega}\right) \quad \text{for } \omega\tau \gg 1 \quad (6)$$

Expression of the shift in ppm results in a ω^{-2} dependence. Observations on the relaxation behavior of various metal ion complexes with transferrin have verified the predicted linewidth and shift behavior [37–39].

In summary, ^{11}B NMR studies provide a useful way of characterizing transient complexes that may form involving borate esters. The exchange behavior among various complexes is often sufficiently slow on the NMR time scale to allow observation of separate resonances. Similar complexes can be observed in the crystalline state, however in this case, lattice contacts can significantly alter the observation of these weak complexes. The observation of such complexes can provide useful insight into the development of boronated as well as non-boronated ligands.

Ternary Complexes Involving Organic Molecules

Although ternary complexes formed from inorganic ions can provide a useful basis for the development of high affinity ligands, it is desirable to generalize this approach by removing the restriction to inorganic ions. The observation of ILOEs provides one basis for obtaining information on the spatial proximity and relative orientation of two (or potentially more) ligands that form a relatively weak complex with a macromolecular binding site. Nuclear Overhauser effect (NOE) interactions involving

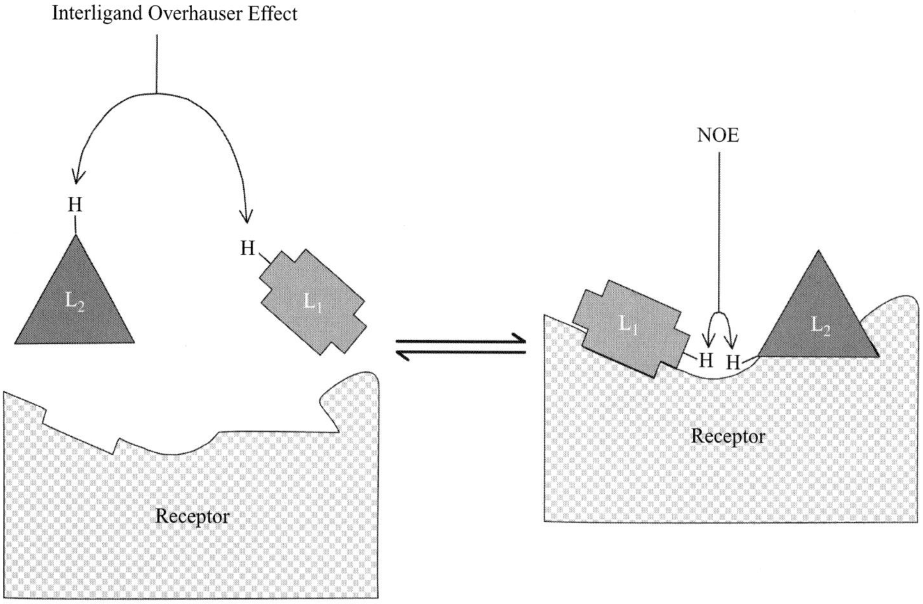

Fig. 3. Schematic illustration of the interaction of two ligands, L_1 and L_2, with a macromolecule. NOE interactions between the protons on the two different ligands can be transferred to the free ligands due to dissociation of the complex.

protons on the two ligands are transferred to the pair of uncomplexed ligands much as the intramolecular NOE information is transferred to the free ligand in a transferred NOE experiment (Figure 3). As in the transferred NOE experiment, these experiments are typically performed using a significant excess of ligand concentration, i.e. $[L_1]$, $[L_2] \gg [E]$, where L_1 and L_2 are the two ligands and E is the enzyme or other macromolecule forming the complex. Since this approach involves direct observation of the ligand resonances, but not the resonances of the macromolecule, it does not require complex resonance assignment strategies. Further, as with the transferred NOE experiment, it works well on macromolecules that are too large for direct structural determination using currently available NMR methodology. However, in contrast with the transferred NOE experiment, this experiment is considerably less subject to misinterpretation resulting from non-specific binding, ligand aggregation, or ligand misbinding. The latter effect arises because ligand binding is essentially an exploratory process in which a ligand may become transiently lodged in an enzyme cleft in a non-productive conformation such that the structural constraints will not allow it to readily adopt a productive conformation. It then becomes most probable for the ligand to dissociate, and subsequently it may reassociate in a conformation which can lead to a productive complex. If the misbound ligand remains associated with the macromolecule for a time that is on the order of the rotational correlation time, it is then able to develop an NOE characteristic of the bound state, and this NOE information characteristic of the misbound complex will be transferred to the free ligand, along with the NOE information from the correctly bound ligand.

For both transferred NOE and interligand NOE studies to provide useful information, the exchange rates for both ligands must be reasonably fast compared with the T_1 time scale of the protons that are under observation. This allows NOE interactions characteristic of the bound ligand(s) to be observed regardless of whether the exchange is fast or slow on the chemical shift time scale, although in general, slow exchange on the shift time scale is less likely to satisfy the T_1 time scale constraint noted above, and leads to non-linear NOE build-up curves. The long T_1 values for the uncomplexed ligand allow it to be used as a "magnetic storage device" for NOE information that is formed in the macromolecular complex. The kinetic behavior of such a system is described using eight rate constants, seven of which are independent, as shown in Scheme 1. The fractional populations of the uncomplexed enzyme/macromolecule, pE, the two binary complexes pEL_1 and pEL_2, and the ternary complex pEL_1L_2, can be expressed as a function of the concentrations and rate constants, as described previously [24]. From a practical standpoint, we have generally found that

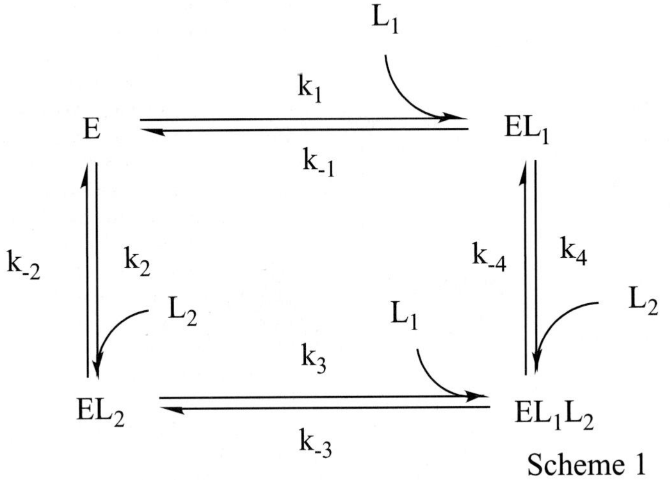

Scheme 1

association rate constants of $\sim 10^8$/M/s and dissociation rate constants of 10^3/s that correspond to equilibrium dissociation constants of ~ 10 μM produce strong effects that are readily observed under typical sample conditions.

An illustrative 2D NOESY spectrum for a system with two spins is shown in Figure 4. The 1D spectrum for this system consists of two sets of three resonances: for spin 1 on L_1, the shift will generally be dependent on whether the ligand is free, in a binary complex with E, or in a ternary complex with E and L_2. The diagonal peaks are shown as open circles. The exchange cross peaks which connect specific spins interconverting among the free, binary, and ternary complexed forms are in black. The NOE cross peaks connecting spins 1 and 2 on the ternary complex are shown in light gray. Of course, since the spins are postulated to be on different ligands, the values for the binary complexes and the free ligands will be zero in the absence of exchange. Finally, the dark gray cross peaks are exchange-mediated NOE interactions which become non-zero when both chemical exchange and NOE interactions are present. For example, such cross peaks can connect spin 1 in a ternary complex with spin 2 in the free state. When the exchange is sufficiently rapid compared with the relevant spin lattice relaxation rates, the sets of nine cross peaks in each of the four boxes shown in the figure exhibit the same time dependence, differing only in a magnitude. When this condition is not met, the elements in each box can exhibit different dependences on the mixing time. In a typical experiment of this type, one observes only a single resonance for each spin due to the fact that the concentrations of the binary and ternary complexes are much lower than the free ligand concentrations, or, more typically, due to fast exchange on the chemical shift time scale. If the latter condition is met, it is appropriate to sum the NOE build-up curves for the nine peaks in each of the indicated boxes. This represents the optimal condition for the transferred NOE and interligand NOE observations. In this case, the sum of the nine NOE cross peaks, A_{ij}, can be represented by another matrix, C_{ij} [24].

Simulations of the dependence of the interligand NOE (ILOE) on mixing time have been performed for a range of geometric structures and kinetic rate constants [24]. In general, the ILOE curves exhibit a dependence on mixing time which is similar to the transferred NOE curves, characterized by an initial build up and a subsequent decay as relaxation effects become dominant. Calculations corresponding to the intensities of the NOESY cross peaks, A_{ij}, and to the sums of elements, C_{ij}, have been performed for a model system that contains two ligands with three spins arranged linearly and separated by 2.5 Å. A ternary complex was modeled in which all six nuclei of both ligands are also arranged in a line, with nucleus 3 on ligand 1 and nucleus 4 on ligand 2 also separated by 2.5 Å, as shown below:

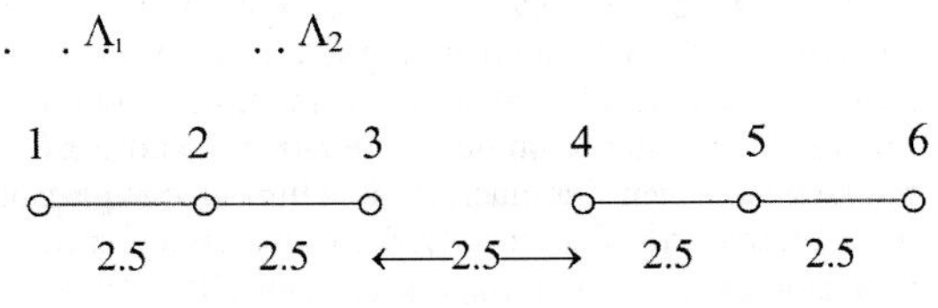

Using parameters: [E] = 0.4 mM, $[L_1] = [L_2] = 5$ mM, $\tau_{1F} = \tau_{2F} = 10^{-10}$ s; $\tau_B = 10^{-7}$ s, $k_1 = k_2 = k_3 = k_4 = 10^8$/M/s, $k_{-1} = k_{-2} = k_{-3} = k_{-4} = 10^{-3}$/s, and other parameters as given in Ref. [24], curves showing

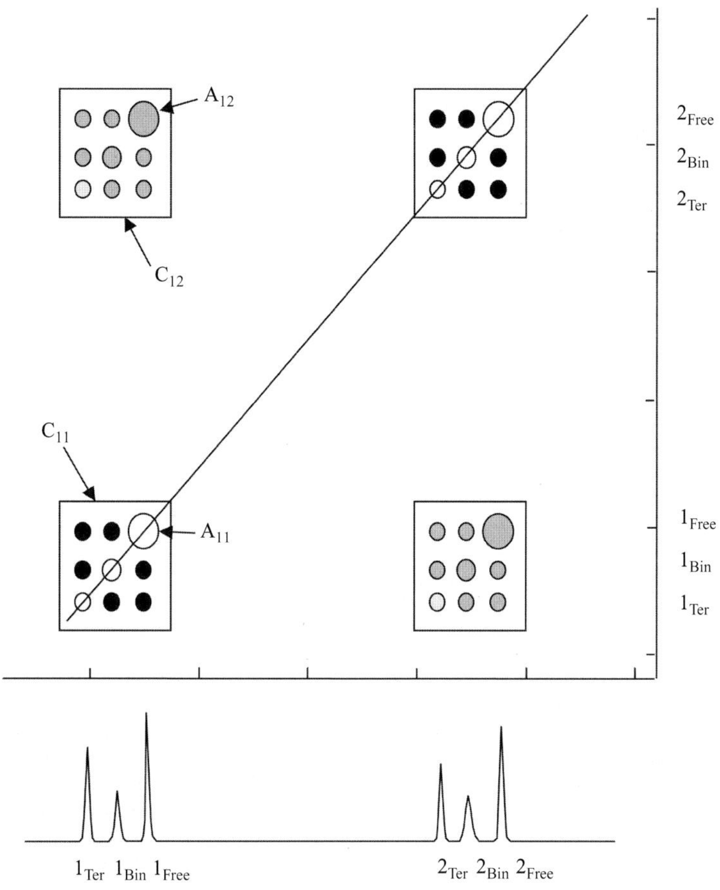

Fig. 4. Schematic representation of a 2D NOESY spectrum for a system containing a macromolecule, E, and two ligands, L_1 and L_2, which can form binary or ternary complexes with E. Nucleus 1 is assumed to be positioned on L_1, and nucleus 2 on L_2. The observation of separate resonances for each spin in the free, binary, and ternary complexes would only be possible under slow exchange conditions on the chemical shift time scale, but is useful for illustrative purposes. The exchange peaks which interconvert nucleus 1 (or 2) between free, binary, and ternary complexes are shown in black. Cross peaks arising from NOE interactions between nuclei 1 and 2 in the ternary complex are shown in light gray. Note that if spins 1 and 2 are on different ligands, there will be no cross peaks for the uncomplexed ligands and for either binary complex. Cross peaks that arise due to a combination of exchange and NOE interactions are shown in dark gray. Under conditions of fast exchange, only the sums of the peaks in each square, i.e. the C_{ij} matrix elements, are observed (based on Figure 1 of Ref. [24]).

the dependence of C_{12}, C_{56}, and C_{34} on the mixing time are shown in Figure 5A. Note that the initial slopes of the C_{12}, C_{56}, and C_{34} curves are identical, consistent with the equal r_{12}, r_{56}, and r_{34} internuclear distances in the model ternary complex which, based on the above parameters, corresponds to 0.996 of the total enzyme concentration available. Most of the difference that can be observed between C_{12} and C_{34} arises due to the faster decay of the C_{34} NOESY interaction, which arises primarily from the presence of two nearest neighbor spins in the ternary complex, compared with one nearest neighbor spin for nucleus 1. Thus, a comparison of the C_{23} with C_{34} curves shows greater similarity (not shown).

In a typical transferred NOE experiment, the internuclear distances of the complexed ligand can be calibrated by comparing the initial slope of the NOE build-up curves with that obtained for a pair of nuclei that have a fixed internuclear distance. For example, in a study involving NADPH, the calibration curve could correspond to the H5–H6 protons on the nicotinamide ring. For the interligand NOE, the analysis becomes more complex if the two ligands have significantly different kinetic parameters. If the fraction of the ternary complex pEL_1L_2 becomes significantly less than one and the fraction of one binary complex, e.g. pEL_1 becomes significant, ligand 1 will be subject to greater transferred NOE effect than ligand 2. In general, the ILOE will scale with the weaker binding ligand. As shown in Figure 5B, increasing the off rates for L_2 to 10^5/s results in a relatively modest increase in pEL_1 and decreased initial slopes for the C_{56} and C_{34} build-up curves. Further increasing the off rate of L_2 to 10^6/s lowers the fraction of E in the ternary complex to 1/3, resulting in a significantly greater initial slope for C_{12} than for C_{56} or C_{34}, which exhibit similar values (Figure 5C). Increasing $k_{-2} = k_{-4}$ to 10^7/s largely eliminates the binding of ligand 2, giving $pEL_1 = 0.95$ (Figure 5D). Interestingly, the initial slope calculated

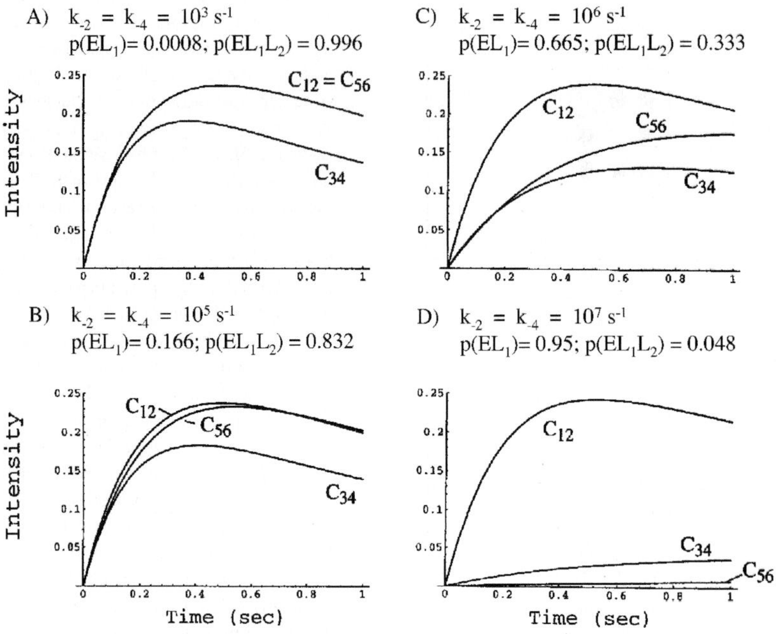

Fig. 5. Sensitivity of interligand Overhauser effects to weaker binding. The set of simulations shown above illustrates the effect of increasing the dissociation rate constant of L_2, while leaving that of L_1 fixed, with $k_{-1} = k_{-3} = 10^3$/s. All association rate constants are set at 10^5/M/s. The calculated fractions of E involved in the ternary (pEL_1L_2) and binary (pEL_1) complexes are shown for the corresponding dissociation rate constants above each figure. (A) $k_{-1} = k_{-2} = k_{-3} = k_{-4} = 10^3$/s; (B) $k_{-2} = k_{-4} = 10^5$/s; (C) $k_{-2} = k_{-4} = 10^6$/s; (D) $k_{-2} = k_{-4} = 10^7$/s. Other parameters as given in Ref. [24] (based on Figure 6 in Ref. [24]).

for C_{34} is greater than that calculated for C_{56}, and no significant transferred NOEs are predicted for ligand 2. This results because for these parameters, L_2 is largely free, and free L_2 experiences an NOE of opposite sign to the bound ligand. Hence, the presence of free L_2 reduces the observed build-up. However, no such subtraction effect arises for the interaction between nuclei 3 and 4, since these do not interact when L_2 is not bound to E.

ILOE Observations—Type II Dihydrofolate Reductase

Interligand Overhauser effects have been observed in several systems of interest [22,23,25–27]. In general, such effects can be observed in non-productive ternary enzyme complexes that involve inhibitors or other molecules that are not substrates for the enzyme under study. Ternary complexes involving enzyme substrates can be observed if the enzyme has been inactivated by mutagenesis. Alternatively, enzymes involved in redox chemistry can be studied if the pair of ligands are both in the oxidized or reduced state, rather than the reduced–oxidized combination that functions as a substrate–cofactor pair. Of course, binding to other types of macromolecules can also be studied, and in this case, there are no concerns about ligand tur over.

Dihydrofolate reductase catalyzes the NADPH-dependent reduction of dihydrofolate to tetrahydrofolate. Type II dihydrofolate reductase is a plasmid-encoded enzyme that is structurally unrelated to the extensively studied type I enzyme, and confers antifolate drug resistance on bacteria containing the plasmid. The enzyme is a symmetric tetramer that binds the substrate and cofactor in a central pore [27]. Figure 6 illustrates portions of a NOESY experiment for a sample containing 0.1 mM R67 DHFR, 5 mM folate, and 5 mM NADP [25]. The region of the spectrum including the CH_2 protons of folate (at the F9 position) are shown in Figure 6. Transferred NOE cross peaks to the folate F7, proton, and to the folate F12/16 and F13/15 protons on the *p*-aminobenzoyl group are readily observed. ILOE peaks connecting the folate F9 protons to the nicotinamide NH-4 and NH-5 protons on the NADP also can be observed in the spectra. All of these peaks exhibit time-dependent build-up curves expected for the

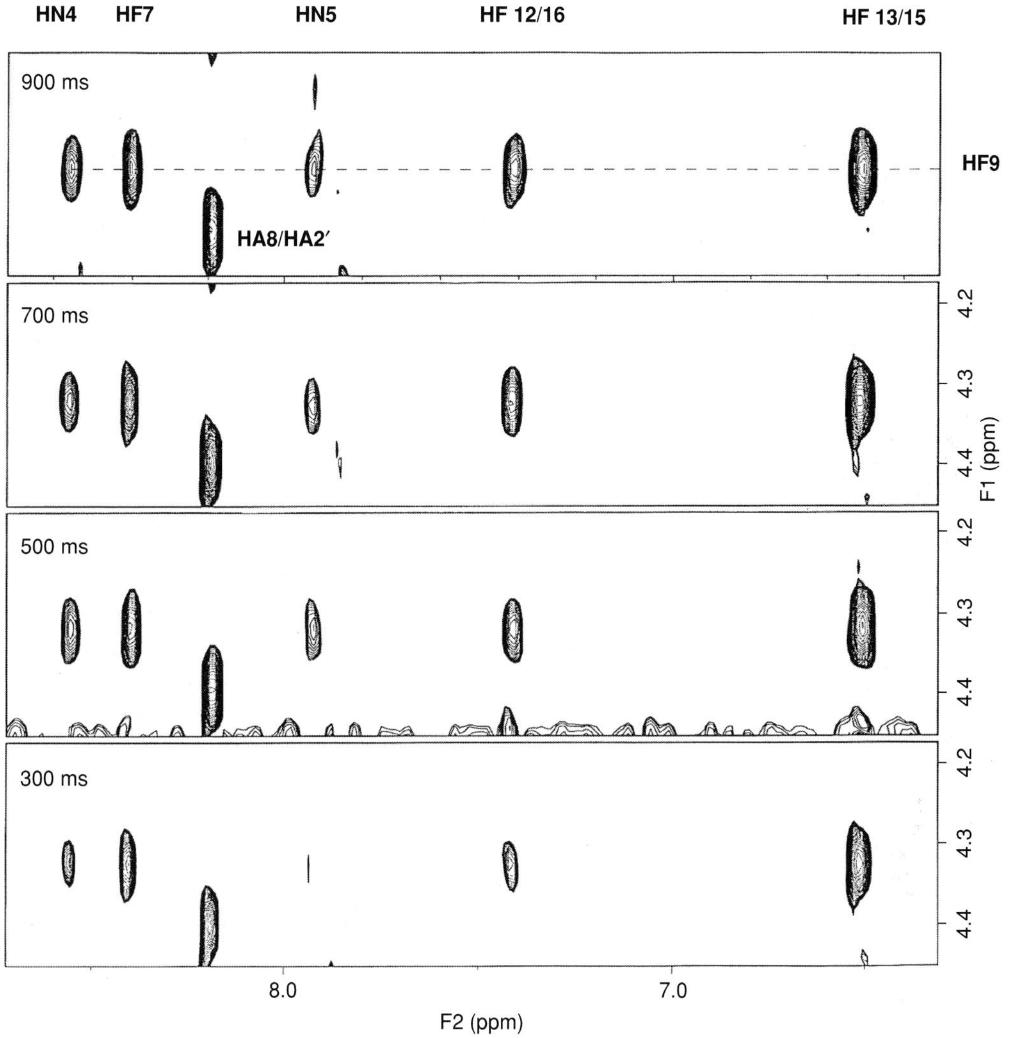

Fig. 6. A portion of the NOESY spectrum illustrating the cross peaks connecting the folate H-9 protons with the aromatic protons of folate and $NADP^+$, obtained on a sample containing 0.1 mM R67 DHFR, 5 mM folate, and 5 mM $NADP^+$ in 100 mM Tris-d11, pH 8.0 in D_2O. The spectra correspond to mixing times of 300, 500, 700, and 900 ms, as indicated (based on Figure 2 of Ref. [25]).

NOE experiment. The results obtained in this study are intuitive since the ILOE peaks are observed near the positions (NH-4 on NADPH, C-6 on folate) at which redox chemistry occurs. The very weak ILOE peaks involving the folate F-7 were also useful for determining the relative positions of the two ligands.

The potential application of ILOE studies for ligand discovery has been demonstrated in experiments on alcohol dehydrogenase, in which a sample contained NADH and an inhibitor, *m*-methylbenzamide, as well as a reducing agent, sodium cyanoborohydride, included in order to maintain the NADH in the reduced state [26]. In this study, unanticipated cross peaks were observed between the BH_3 protons on the cyanoborohydride and the methylbenzamide protons. It was concluded that the methylbenzamide binds primarily to a hydrophobic pocket in the active site that is involved in binding larger substrates, such as benzyl alcohol, while the cyanoborohydride formed a transient complex with the active site Zn^{2+} ion [26].

Summary

The identification and characterization of ternary complexes involving organic molecules and inorganic ions should provide a useful basis for the further development of targeted ligands. Although the example of borate discussed above involved the serine protease trypsin, recent structural data indicate that borate complexes can form outside of the active site, supporting the possibility of targeting protein–protein interactions in a more general way. Comparisons with crystallographic analyses of the same complexes indicate that crystal screening approaches, while generally very useful, are subject to potentially significant influences that result from lattice contacts. ILOEs can provide useful insight into the relative binding and orientation of ligands that reversibly form ternary complexes. The presence of unanticipated complexes has been observed in some studies, supporting the potential application of this approach to the design of novel ligands.

References

1. Charlton PQ, Young DW, Birdsall B, Feeney J, Roberts GCK. J. Chem. Soc. Chem. Commun. 1979;922.
2. Bystroff C, Oatley SJ, Kraut J. Biochemistry 1990;29: 3263.
3. Stoll VS, Eger BT, Hynes RC, Martichonok V, Jones JB, Pai EF. Biochemistry 1998;37:451.
4. Shuker SB, Hajduk PJ, Meadows RP, Fesik SW. Science 1996;274:1531.
5. Meyer B, Peters T. Angew. Chem. Int. Ed. 2003;42:864.
6. Mayer M, Meyer B. Angew. Chem. Int. Ed. 1999;38:1784.
7. Post CB. Curr. Opin. Struct. Biol. 2003;13:581.
8. Sternweis PC, Gilman AG. Proc. Natl. Acad. Sci. U.S.A. 1982;79:4888.
9. Katz BA, Clark JM, Finer-Moore JS, Jenkins TE, Johnson CR, Ross MJ, Luong C, Moore WR, Stroud RM. Nature 1998;391:608–12.
10. Boyle FA, Cook ND, Peters TJ. Clin. Chim. Acta 1988;172:291–6.
11. Revel JP, Ball EG. J. Biol. Chem. 1959;254:577.
12. Lindquist RN, Lynn JL Jr, Lienhard GE. J. Am. Chem. Soc. 1973;95:8762.
13. Borah B, Chen C-W, Egan W, Miller M, Wlodawer A, Cohen JS. Biochemistry 1985;24:2058.
14. Crans DC, Smee JJ, Gaidamauskas E, Yang L. Chem. Rev. 2004;104:849.
15. Davies DR, Interthal H, Champoux JJ, Hol WGJ. J. Med. Chem. 2004;47:829.
16. Goodno CC. Proc. Natl. Acad. Sci. U.S.A. 1979;76:2620.
17. Maruta S, Henry GD, Sykes BD, Ikebe M. J. Biol. Chem. 1993;268:7093.
18. Fisher AJ, Smith CA, Thoden JB, Smith R, Sutoh K, Holden HM, Rayment I. Biochemistry 1995;34:8960.
19. Walker B, Lynas JF. Cell. Mol. Life Sci. 2001;58:596.
20. Tate SS, Meiser A. Proc. Natl. Acad. Sci. U.S.A. 1978;75: 4806.
21. London RE, Gabel SA. Arch. Biochem. Biophys. 2001;385: 250.
22. Barsukov IL, Lian LY, Ellis J, Sze KH, Shaw WV, Roberts GCK. J. Mol. Biol. 1996;262:543.
23. Li D, DeRose EF, London RE. J. Biomol. NMR 1999;15:71.
24. London RE. J. Magn. Reson. 1999;141:301.
25. Li D, Levy LA, Gabel SA, Lebetkin MS, DeRose EF, Wall MJ, Howell EE, London RE. Biochemistry 2001;40:4242–52.
26. Li D, London RE. Biotechnol. Lett. 2002;24:623–9.
27. Pitcher WH III, DeRose EF, Mueller GA, Howell EE, London RE. Biochemistry 2003;42:11150.
28. Roush A, Norris ER. Arch. Biochem. 1950;29:345.
29. Smith KW, Johnson SL. Biochemistry 1976;15:560.
30. London RE, Gabel SA. Biochemistry 2002;41:5963.
31. Transue TR, Krahn JM, Gabel SA, DeRose EF, London RE. Biochemistry 2004;43:2829.
32. Babine RE, Rynkiewicz MJ, Jin L, Abdel-Maguid SS. Lett. Drug Des. Discov. 2004;1:35.
33. Bull TE, Forsen S, Turner DL. J. Chem. Phys. 1979;70:3106.
34. Werbelow LG. J. Chem. Phys. 1979;70:5381.
35. Werbelow LG, Pouzard G. J. Phys. Chem. 1981;85:3887.
36. Werbelow LG. In: DM Grant (Ed). Encyclopedia of Nuclear Magnetic Resonance, Vol. 6. John Wiley & Sons: New York, 1995, p 4092.
37. Aramini JM, Vogel HJ. J. Am. Chem. Soc. 1993;115:245.
38. Aramini JM, Germann MW, Vogel HJ. J. Am. Chem. Soc. 1993;115:9750.
39. Aramini JM, McIntyre DD, Vogel HJ. J. Am. Chem. Soc. 1994;116:11506.

Part II

The Transferred NOE

Mike P Williamson
Department of Molecular Biology and Biotechnology, University of Sheffield, Firth Court, Western Bank, Sheffield S10 2TN, UK

Abstract

The transferred NOE allows NOE information to be obtained on a bound ligand, but appearing on the resonances of the free ligand. It is thus a very useful method for gaining information on the bound ligand. Its major limitation is that it only works well in a range of dissociation constants between 1 mM and 10 nM. A key negative control experiment is to demonstrate the absence of NOEs arising from nonspecific binding, by addition of competitor ligands. A second is to measure NOEs in the absence of protein. For quantitative work, it is also necessary to minimize spin diffusion, by using relatively short mixing times. Recent work has shown that spin diffusion in TR-NOE is no worse than spin diffusion in standard NOE spectra of proteins. With care, TRNOE can thus provide powerful quantitative information on bound conformation.

Affinities and Timescales

The appearance and information content of an NMR spectrum of a protein/ligand mixture depend strongly on the rates of binding and tumbling of the components. This is why it is important to think about these rates, and understand their implications.

The binding of a ligand to a protein can be written

$$L + P \underset{k_{off}}{\overset{k_{on}}{\rightleftharpoons}} L \cdot P \tag{1}$$

The overall on-rate is k_{on}[L][P], and the off-rate is k_{off}[L.P]. At equilibrium, these are by definition equal. The dissociation constant K_d is [L][P]/[LP], which is therefore also equal to k_{off}/k_{on}. The association constant K_a is simply $1/K_d$.

For most ligand/protein interactions, binding happens as fast as ligand and protein can come together: in other words, the on-rate is diffusion controlled. The on-rate (k_{on}) is usually assumed to have a value of around 10^8 $M^{-1}s^{-1}$, and will only be significantly slower if the protein binding site is somehow buried or hidden (which is unusual, particularly for inhibitors as compared to substrates). We can therefore get a good estimate of the off-rate if we know the dissociation constant, since $k_{off} \approx 10^8 K_d$. A good drug often has a K_d in the low nM range, and therefore has a k_{off} of around 1 s^{-1} or less. By contrast, initial targets derived from screening programes will typically have K_d values in the high μM or low mM range, and will therefore have off-rates of around 10^5 s^{-1}. This leads to very different NMR behavior.

We also need to consider that the concentrations of samples in NMR are comparable to some of these numbers, and therefore the fraction of ligand bound is strongly dependent on K_d. It is easily shown that for the simple binding equilibrium shown above, the concentration of free ligand is given by

$$[L] = 0.5([L_0] - [P_0] - K_d + \{([L_0] - [P_0] - K_d)^2 + 4[L_0]K_d\}^{1/2}) \tag{2}$$

where $[L_0]$ and $[P_0]$ are the total concentrations of ligand and protein respectively. So for example, if the total protein concentration is 50 μM, K_d is 10 nM, and there is a 1:1 ratio of ligand to protein, then almost all the ligand is bound (only ca. 1.4% is free); if there is a 10-fold excess of ligand over protein, then to a good approximation one equivalent is bound and the other 9 are free. By contrast, with the same protein concentration, a 10-fold ligand excess and a much weaker K_d of 1 mM, the concentration of bound ligand is only 16 μM (i.e. only one-third of the protein is bound), and 97% of the ligand is free. Thus, the amount of bound complex depends strongly on K_d. This will be important when we come to think about what the NMR spectrum looks like.

The off-rate is important for another reason, namely because it defines whether exchange processes are in "fast" or "slow" exchange [1]. Confusingly, there are different definitions of what is meant by fast and slow, depending on what we are interested in. The most obvious is the chemical shift timescale: whether we see one signal or two. This depends on $\Delta\nu$, the difference in resonance frequency between the two signals, expressed in Hz. Fast exchange is when $k_{off} \gg \Delta\nu$. Thus if for example we are considering a ^{1}H spectrum at 500 MHz, in which free and bound signals are 0.3 ppm different, $\Delta\nu = 150$ Hz, so k_{off} must be 1000 s^{-1} or greater for fast exchange (corresponding to a K_d of 10 μM or weaker). Slow exchange requires k_{off} <30 s^{-1} and K_d 0.3 μM or

Graham A. Webb (ed.), Modern Magnetic Resonance, 1339–1344.

stronger. Thus, almost all initial leads and screening hits are in or close to fast exchange, while strong ligands are in slow exchange. Intermediate off-rate values give broadened signals. It is however worth noting that intermediate exchange looks different when the amounts of the two species are unequal, which would happen if for example there were an excess of free over bound ligand. In this situation, the appearance of the free signal is almost unaffected, while the bound signal is broadened so much that it is frequently unobservable [2]. It is therefore not easy to tell whether the ligand is in fast exchange on the chemical shift timescale simply by looking at the appearance of the signal.

A ligand undergoing exchange can of course be in slow exchange for some signals and fast exchange for others, depending on the chemical shift differences involved. Going to higher magnetic field pushes the system toward slower exchange, as does cooling the sample. A typical chemical shift change could be 0.3 ppm or 150 Hz in ^{1}H, as illustrated above, but 0.5 ppm or 25 Hz in ^{15}N and 0.3 ppm or 38 Hz in ^{13}C. Thus, a spectrum in slow exchange for ^{1}H could be in fast or intermediate exchange for ^{15}N or ^{13}C. Since the biggest difficulty for carrying out TRNOE experiments on many "interesting" (i.e. tight binding) ligands is that their off-rate is too slow, it is worth noting that better results can often be obtained by increasing the temperature and sometimes by going to lower field.

The other exchange rates that affect the spectra obtained are relaxation rates, and in particular the spin-lattice relaxation rate and the cross-relaxation rate. The spin-lattice relaxation rate R_1 is the inverse of the more familiar spin-lattice relaxation time T_1, and is the ultimate entropic sink: if the off-rate is much slower than R_1 (and in particular, R_1 of the bound state) then there is no way that any information on the bound state (such as an NOE) can ever be transferred over into the free state, because it is lost to relaxation before it can be seen. R_1 is typically around 10 s^{-1}, and therefore off-rates slower than 1 s^{-1} (i.e. K_d values stronger than 10 nM) usually mean that no information can be obtained about the bound state by observation of the free state. We will see an example of this below. The cross-relaxation rate is the rate responsible for the NOE (and a similar relaxation rate is responsible for cross-correlated relaxation), and it is typically slower than k_{off}, being around 1–10 s^{-1}. The important consequence of this is that NOEs can build up in one state, for example the bound state, and be transferred across to the free state by chemical exchange without significant loss of magnetization from spin-lattice relaxation: hence its name of transferred NOE or TRNOE. (Some authors have suggested that the abbreviation TRNOE can be confused with the transient NOE, and prefer other terms, such as exchange-transferred NOE (et-NOE) [3].) The TRNOE has recently been reviewed [3]. However, before we can consider the transferred NOE, we need to clarify a few things about the NOE.

The NOE

The NOE is a very useful technique for measuring distance in NMR, because the rate of buildup of the NOE (the cross-relaxation rate) is proportional to the inverse sixth power of the internuclear distance. In a homonuclear two-spin system, the cross-relaxation rate is given by [1]

$$\sigma = \left(\frac{\mu_0}{4\pi}\right)^2 \frac{\hbar^2\gamma^4}{10}\left(\frac{6\tau_c}{1+4\omega^2\tau_c^2} - \tau_c\right) \tag{3}$$

where $\mu_0/4\pi$ is a conversion factor, γ is the gyromagnetic ratio, τ_c is the correlation time of the internuclear vector, and ω is the spectrometer frequency (in rad s^{-1}, i.e. 2π times larger than the more conventional value in Hz). The correlation time is the time it takes for the relevant internuclear vector to rotate by approximately one radian. The crucial factor here is the term involving τ_c, which in practice means that the cross-relaxation rate scales with the correlation time, except for a hiatus around $\omega\tau_c = 1$, where the cross-relaxation rate passes from positive to negative (Figure 1). The important point is that the cross-relaxation rate is much faster for large τ_c, i.e. for large proteins.

In principle, the relationship between NOE intensity and distance makes the NOE a very precise measure of distance. However, there are several factors that in practice make it much less precise than it should be, of which the most important are spin diffusion and internal mobility.

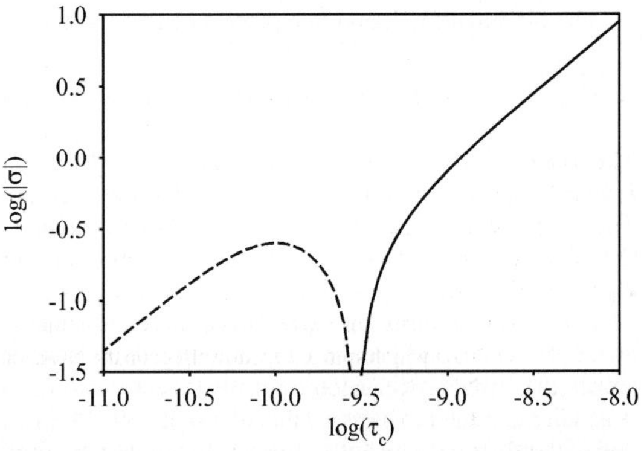

Fig. 1. Variation of the cross-relaxation rate with correlation time, calculated for two protons 2 Å apart for a 600 MHz spectrometer. For $\omega\tau_c < 1.12$, σ is positive (dashed line), and for $\omega\tau_c > 1.12$, it is negative (solid line).

Spin Diffusion

This is a process analogous to heat diffusion, whereby an NOE from one proton to another can subsequently diffuse to other protons throughout the protein. There are thus several pathways for NOE transfer, and therefore any spin diffusion immediately renders distance information less useful. This can be illustrated with a simple three-spin example (Figure 2), which shows that a third proton that bridges between the two protons being measured can reduce the apparent distance between them if it is significantly closer to both of them than they are to each other. Clearly, this is much more of a problem when trying to measure long internuclear distances. It is hard to detect the presence of spin diffusion (except using ROESY experiments, as described below), and therefore spin diffusion is normally handled by using rather wide ranges for NOE distance calibration and short mixing times. This clearly makes structure calculation less precise.

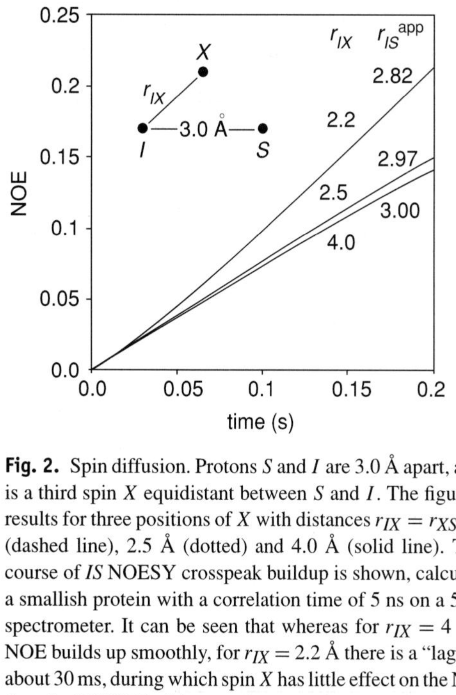

Fig. 2. Spin diffusion. Protons S and I are 3.0 Å apart, and there is a third spin X equidistant between S and I. The figure shows results for three positions of X with distances $r_{IX} = r_{XS}$ of 2.2 Å (dashed line), 2.5 Å (dotted) and 4.0 Å (solid line). The time course of *IS* NOESY crosspeak buildup is shown, calculated for a smallish protein with a correlation time of 5 ns on a 500 MHz spectrometer. It can be seen that whereas for $r_{IX} = 4$ Å the *IS* NOE builds up smoothly, for $r_{IX} = 2.2$ Å there is a "lag time" of about 30 ms, during which spin X has little effect on the NOE, but then the *IS* NOE buildup rate increases because of spin diffusion via X. The initial buildup over 200 ms was fitted by hand and converted to an apparent distance by taking the sixth root of the buildup rate. Spin diffusion reduces the apparent distance r_{IS} from 3.0 to 2.82 Å when proton X is close to both I and S. In most cases, such a reduction in apparent distance would have little effect on the calculated structure.

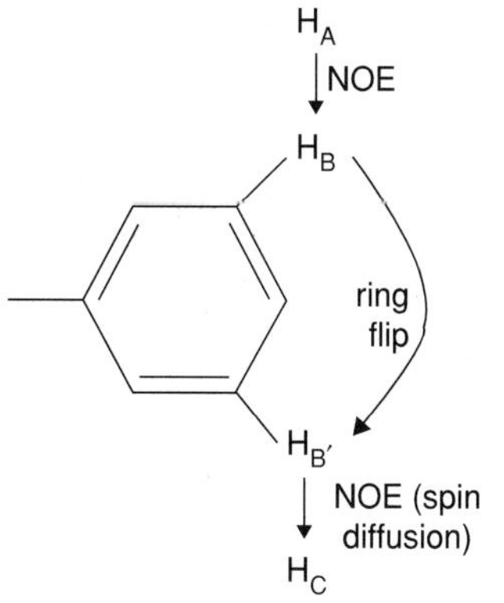

Fig. 3. The effect of ring flipping on NOEs. Proton A generates an NOE at B. If the phenylalanine ring flips 180°, then B ends up at B', and can pass an NOE on to C by spin diffusion. This will be significant as long as the flip rate is greater than the NOE cross-relaxation rate, which is almost always the case. The buildup rate of NOE at proton C suggests it is close to A, when in fact it can be easily 8–10 Å away.

Internal Mobility

Internal mobility can affect the NOE in several ways. The main result, as for spin diffusion, is to make the relationship between NOE and distance less precise. Any mobility will cause the protons to have a range of distance separations. Because of the r^{-6} averaging, this almost always results in an apparent shortening of the distance. Very close protons cannot be brought any closer together, and so the distance averaging effect is more significant with more distant protons. In practice, this often makes the r^{-6} effect end up looking more like a r^{-5} or even r^{-4} effect. Potentially the most severe effect of motion is when there is an NOE to an aromatic proton. Flipping of the ring can lead to the NOE then being passed on by spin diffusion to protons on the other side of the ring (Figure 3). This will have a marked effect on apparent distances in the vicinity of the ring.

The Transferred NOE

The transferred NOE makes use of a particularly useful phenomenon, namely that the rate of buildup of the NOE is approximately proportional to the correlation time (Figure 1). The average cross-relaxation rate in a ligand exchanging between free and bound states is given by

$$<\sigma> = N_F\sigma_F + N_B\sigma_B \qquad (4)$$

where N_F and N_B are the fractions of free and bound ligand. The cross-relaxation rate of bound ligand, σ_B, is much greater than the cross-relaxation rate of free ligand, σ_F. It is therefore easy to produce a situation where the observed NOE is dominated by cross-relaxation in the bound state. Depending on the exact values of σ_B and σ_F (which depend on the tumbling rate of bound and free ligand), ligand:protein ratios of 10–40 can be and are often used, and give useful quantitative results. A further helpful aspect of this high ligand:protein ratio is that one can use protein concentrations in the low μM range and still measure NOEs. An NOE measurement (almost always run as a NOESY experiment) then produces NOEs characteristic of the bound state. In the usual situation in which the off-rate is fast on the chemical shift timescale, the only signals observed are sharp and virtually identical to those of free ligand, and consequently are easily assigned. We therefore observe NOEs characteristic of the bound state, but seen on the signals of the free state. The experiment has a characteristic unusual for NMR spectra of proteins, that the experiment actually gets better as the protein gets bigger, because σ_B increases, and therefore the NOE is dominated more and more by the cross-relaxation rate in the bound state.

For many ligands, σ_F is so small as to be insignificant. It can (and should) always be checked by running a control NOE experiment in the absence of protein. For intermediate sized ligands, such as peptides, there is a particularly advantageous situation, in that the cross-relaxation rate goes through zero at $\omega\tau_c \approx 1$; as it happens, this is where many peptides come. This greatly reduces the need to worry about NOEs in the free peptide.

There have been a number of detailed simulations carried out on the TRNOE [4–7]. Figure 4 shows simulation results of a closely similar experiment, the STD experiment (discussed later), which show a complex dependence on K_d [8]. At K_d values weaker than 1 mM (right-hand side of Figure 4), the TRNOE falls off to zero because there is not enough ligand bound to give rise to an NOE. Conversely, at K_d values stronger than about 10 nM (left-hand side of Figure 4), the off-rate is so slow that bound ligand hardly ever dissociates from the complex, and the TRNOE again goes to zero. There is thus in this simulation a "window" of K_d of between 10 nM and 100 μM where the TRNOEs are not greatly affected by exchange rates and can be related simply to distance (see however the discussion in the figure legend). For really quantitative results (i.e. an NOE that is related *only* to the distance in the complex and is not affected by exchange), it is necessary to have k_{off} much greater than the cross-relaxation rate. An approximate guide to estimating cross-relaxation rates in proteins has been suggested: [3] $\sigma \approx -24 \times \text{MWt (kDa)}/r^6$ (Å). Protons cannot get much closer than 2 Å apart, which means that the largest possible cross-relaxation rate is approximately $0.375 \times$ MWt, or 19 s^{-1} for a "typical" 50 kDa protein. This implies that k_{off} should be at least 100 s^{-1}, or alternatively K_d should

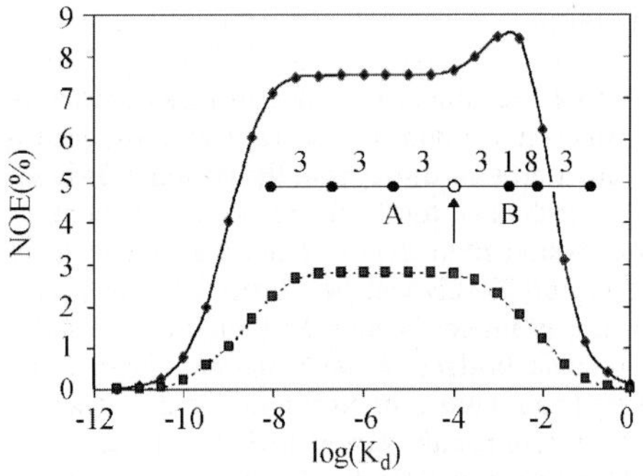

Fig. 4. The effect of irradiating a protein resonance (open circle) on ligand signal intensity, shown for two ligand signals, A (diamonds, solid line) and B (squares, dashed). The effect calculated is STD, which is closely similar to what would be measured in a TRNOE experiment. All distances are 3 Å except for the distance from ligand proton B to its nearest neighbor which is 1.8 Å. This near neighbor causes a rapid loss of NOE from B by spin diffusion: this loss (and the consequent reduction in the NOE measured at B) is not as severe in the case of TRNOE. A 10:1 excess of ligand over protein is used in the calculation, with $k_{on} = 10^9$ $s^{-1}M^{-1}$, $\tau_c = 10^{-7}$ s, a 600 MHz spectrometer and a uniform leakage relaxation rate of 0.3 s^{-1} on all protons. The free ligand is assumed to have zero NOE. Using data from Jayalakshmi and Krishna [8].

be 1 μM or weaker. Of course, for protons at more realistic distances, this limit can be relaxed somewhat. As the molecular weight of the protein increases, the limitations on K_d become more stringent, such that for example for a 500 kDa protein, K_d should be 10 μM or weaker. This implies that for properly quantitative results, the window of K_d is very small, being only between 1 μM and 100 μM. Fortunately this is a range that includes a good proportion of ligands of interest, and one can go at least a factor of 10 to either side without losing much quantitative information. It should also be added that TRNOEs have been observed and used even with much more tightly bound ligands. Useful TRNOEs were for example observed for a ligand with a K_d of 30 nM [9].

Some of the early work on TRNOEs was overly precise and optimistic in the power of TRNOE to produce structural information. The pendulum therefore later swung in the opposite direction, with dire warnings about the dangers of over-interpreting TRNOE spectra [1]. More recent work has brought the pendulum back towards its proper position, and has shown that with proper controls, TRNOEs can give distance information that is very nearly as accurate as from conventional structure calculations. Apart from the problems discussed below, the main limitation is simply that ligands often bind in fairly extended geometries on protein surfaces, which means that most of the distances measurable by NOE are already close in terms of the covalent structure, and thus there are relatively few informative NOEs that can be measured. It is worth looking at the two most important pitfalls, as listed below:

1. Spin Diffusion. This has already been described above, but in the TRNOE there is a further problem, that there could be spin diffusion from one ligand proton to another, going *via* a protein proton [10]. Such a transfer would be very hard to detect, and would seriously affect measured distances. Recent simulations have shown that in practice, this very seldom happens, basically because most ligands bind on the exposed surfaces of proteins and therefore are not close enough to the protein for this to happen (cf. Figure 2) [11,12]. In particular, if the ligand has a reasonably high density of protons (for example, a peptide) it would be uncommon for ligand-protein-ligand spin diffusion to affect a distance measurement. The spin diffusion problem is less severe if a short presaturation time is used [12]. However, the large excess of free ligand effectively dilutes the buildup of NOE on the bound ligand by exchanging with it, with the consequence that one can use longer mixing time for TRNOE experiments than for conventional NOE experiments on proteins: mixing times of 100–200 ms are common, even with very large proteins. (It should also be added that the large number of TRNOEs that can be measured also means that one inaccurate TRNOE is less likely to perturb the structure.) This does not remove the problem of spin diffusion within the ligand, which is as bad as it is in a conventional NOE measurement, and still implies a need for caution in interpreting NOEs as distances.

It has been noted that spin diffusion (of any kind) can be detected using a transferred ROESY experiment, in which spin diffusion gives rise to smaller NOEs of opposite sign to direct NOEs. Bax and co-workers have provided an elegant example of a TRNOE that disappeared on using a transferred ROESY experiment, and was thereby shown to be due to spin diffusion [13]. However, this experiment typically requires a low peptide:protein ratio (and more careful control experiments in the absence of protein), because the ratio of σ_B to σ_F is not as great as for NOESY. It is also more prone to artifacts, and has not proved as popular as one might expect. Where used, it tends to be used as a check that measured NOEs are genuine, rather than as a stand-alone method.

2. Nonspecific Binding. Usually, one uses something like a 30-fold excess of ligand over protein. Under these circumstances, a ligand that binds specifically, but in addition binds nonspecifically but 100-fold more weakly, will be fully bound at the tight binding site and almost fully bound at the weaker site, Equation (2). It will thus give equally strong TRNOEs from its strongly and weakly bound conformations (indeed, if it binds very strongly at the more specific site, these TRNOEs will be small for reasons shown in Figure 4, and thus almost all the TRNOE will in fact come from nonspecific binding!). This gives rise to averaged TRNOEs, which, if not spotted, will produce misleading distance information. This is best detected using a control experiment, in which a strong competitive ligand is added [7]. This will displace the ligand from the binding site, and any remaining TRNOEs must therefore arise from nonspecific binding. Nonspecific binding appears to be very common, and therefore a control experiment is extremely valuable: indeed, for quantitative results such an experiment is crucial.

Practical Implementation

One of the beauties of TRNOE is that it is simple to do: one just needs to carry out a conventional NOESY experiment (in conjunction with a control using a tight inhibitor, to rule out nonspecific binding). It is common

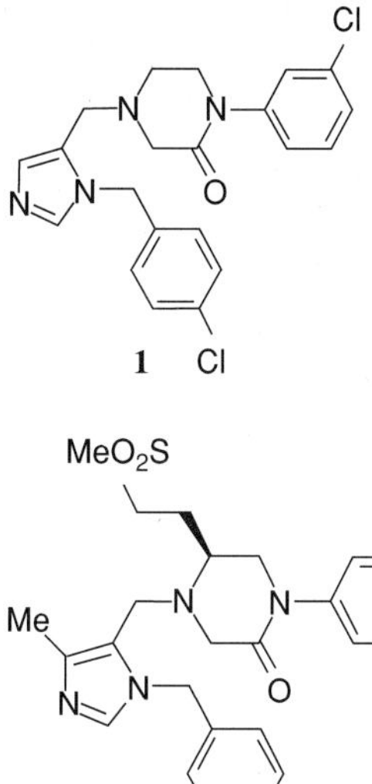

1

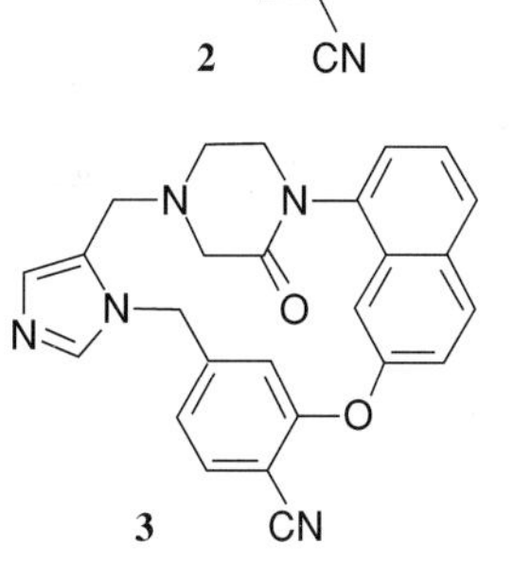

2

3

to add a relaxation filter to the experiment, often in the form of a T_2 or $T_{1\rho}$ filter, to reduce broad signals from the protein and therefore make the baseline flatter and reduce integration problems. As an example, compound **1** is a lead compound as an inhibitor of farnesyltransferase (FTase), and therefore a potential antitumor agent. It binds to FTase with an IC_{50} of 2 nM, and its off-rate is therefore too slow to give TRNOEs. However, an analogue **2** has an IC_{50} of only 475 nM and is therefore suitable for TRNOE experiments. Measurements were carried out using 64 μM enzyme and 2 mM **2** (ca. 1:30), a typical concentration for such experiments. TRNOE experiments were carried out using several mixing times, going up to 300 ms. Secondary binding was identified by addition of an excess of a competitive ligand. Measurement of TRNOEs revealed an NOE between the cyanophenyl ring and the piperazinone ring, which prompted the authors to design a cyclised ligand **3** that resembled the bound conformation and has an IC_{50} of 0.1 nM.

Related Experiments

The TRNOE is closely related to two other experiments described in this volume, namely saturation transfer difference (STD) and transferred cross-correlated relaxation. The only difference between TRNOE and STD is that in STD one saturates the protein and transfers magnetization from protein to bound ligand, whereas in TRNOE one uses intramolecular NOEs in the bound ligand. (It is also true that STD is normally run as a steady-state NOE experiment, with irradiation of the protein resonance for several seconds, while TRNOE is normally run as a 2D transient experiment.) The dependence on off-rate and molecular size is the same, and it is therefore applicable in the same circumstances. This of course means that STD does not produce a signal for very tightly bound ligands. A neat method has been devised to get around this limitation, using competition experiments—for example, an STD seen to a relatively weakly binding ligand disappears on addition of a much tighter binding competitor, thus allowing the tight binding competitors to be identified in high-throughput screens. The same would hold with the TRNOE also, but it would of course be a less useful result. A combination of STD and TRNOE experiment is particularly valuable in screening strategies [14,15].

References

1. Neuhaus D, Williamson MP. The Nuclear Overhauser Effect in Structural and Conformational Analysis, 2nd ed. Wiley-VCH: New York, 2000.
2. Sandström J. Dynamic NMR Spectroscopy. Academic Press: Paris, 1982.
3. Post CB. Curr. Op. Struct. Biol. 2003;13:581–588.
4. Clore GM, Gronenborn AM. J. Magn. Reson. 1982;48:402–417.
5. Campbell AP, Sykes BD. Annu. Rev. Biophys. Biomolec. Struct. 1993;22:99–122.
6. Lippens GM. Cerf C, Hallenga K. J. Magn. Reson. 1992;99:268–281.
7. Ni F. Prog. Nucl. Magn. Reson. Spectrosc. 1994;26:517–606.
8. Jayalakshmi V, Krishna NR. J. Magn. Reson. 2002;155:106–118.
9. Weimar T, Petersen BO, Svensson B, Pinto BM. Carbohydr. Res. 2000;326:50–55.
10. Jackson PL. Moseley HNB, Krishna NR. J. Magn. Reson. Ser. B 1995;107:289–292.
11. Eisenmesser EZ. Zabell APR, Post CB. J. Biomol. NMR, 2000;17:17–32.
12. Zabell APR, Post CB. J. Biomol. NMR, 2002;22:303–315.
13. Arepalli SR. Glaudemans CPJ. Daves GD. Kovac P, Bax A. J. Magn. Reson. Ser. B 1995;106:195–198.
14. H. Kogelberg, Solis D, Jimenez-Barbero J. Curr. Op. Struct. Biol. 2003;13:646–653.
15. Stockman BJ, Dalvit C. Prog. Nucl. Magn. Reson. Spectrosc. 2002;41:187–231.

NMR Kinetic Measurements in DNA Folding and Drug Binding

Mark S. Searle[1], Graham Balkwill[1], Huw E.L. Williams[1], and Evripidis Gavathiotis[2]
[1]*Centre for Biomolecular Sciences, School of Chemistry, University Park, Nottingham NG7 2RD, UK; and*
[2]*Laboratory of Molecular Biophysics, The Rockefeller University, New York, NY 10021, USA*

Drug–Quadruplex Interactions Studied by NMR

DNA quadruplex structures have been of considerable interest as supramolecular self-assembling systems [1] and because of their potential biological importance in the regulation of a variety of processes within the cell cycle including replication, transcription, and recombination [2]. The observation that the telomeric repeats at the ends of chromosomes (5′-TTAGGG in humans) are able to assemble in the presence of monovalent cations into folded structures containing stacked G-tetrads has bought them into focus as a drug target, notably to interfere with telomere maintenance in immortalized human tumor-derived cell lines by inhibiting the enzyme telomerase [1–6]. The novel fluorinated polycyclic quinoacridinium cation RHPS4 (Figure 1a) shows enhanced binding to higher-ordered DNA structures (triplex/quadruplex) and is a potent inhibitor of telomerase function [7,8]. To investigate the drug–quadruplex interaction, as part of a rational ligand design approach, we have studied by NMR the structure (Figure 1b) and dynamics of the RHPS4 complex with the intermolecular parallel-stranded quadruplex d(TTAGGGT)$_4$, formed from the human telomeric repeat [9,10].

Exchange Rates for Drug Binding to Quadruplex DNA

Fluorescence quenching studies have established that RHPS4 interacts with d(TTAGGGT)$_4$ with a binding affinity of 2.2×10^5 M [7]. NMR lineshape analysis during drug titration studies shows that the drug is in fast exchange between free and bound states with the quadruplex accommodating drug molecules at both the 5′-AG and 5′-GT steps at the ends of the core G-quadruplex structure [10]. Drug complexation was monitored using the downfield guanine imino proton resonances between 10 and 12 ppm. These initially shift upfield due to ring current effects from the drug, broadening significantly at a bound drug:DNA ratio of ~0.3, but subsequently sharpen as the fully bound state is reached.

For a ligand-binding interaction where free and bound states are in fast exchange [11,12], the observed chemical shift δ_{obs} is given by,

$$\delta_{obs} = \delta_f p_f + \delta_b p_b \qquad (1)$$

The parameters δ_f and δ_b are the chemical shifts for the free DNA and the fully bound DNA:ligand complex, and p_f and p_b are the fractional populations of these species ($p_f + p_b = 1$), enabling the chemical shift change to be used to determine p_b as a function of drug concentration. The observed line width (LW$_{obs}$) is related to the populations of free and bound states with an additional exchange contribution given by the expression,

$$\mathrm{LW}_{obs} = p_f \mathrm{LW}_f + p_b \mathrm{LW}_b + [p_b(1-p_b)^2 4\pi(\Delta\nu)^2]/k_{off} \qquad (2)$$

where k_{off} is the off-rate (per second) and $\Delta\nu$ the chemical shift difference (Hz) between the *free* and fully *bound* states, and LW$_f$ and LW$_b$ the line widths in the free and bound states. The exchange contribution, and hence the off-rate, can be determined by monitoring LW$_{obs}$ as a function of the fraction bound (p_b) at each site. Selected 1D NMR spectra from the RHPS4 titration study at 303 K are shown in Figure 2a, with line width changes as a function of the p_b shown in Figure 2b. Data for the guanine imino proton resonances G5 and G6 were fitted to Equation (2) using values of $\Delta\nu$ of 205 and 415 Hz, giving estimated values for k_{off} of 1814 (±170)/s and 3812 (±250)/s. Given the uncertainty in line width measurements at values of p_b close to ~0.3 (Figure 2), the factor of ~2 difference in exchange rates measured using the two different probes probably represents a more realistic estimate of possible errors, at least in this context, and hence we report a mean value for k_{off} of 2800 (±1000)/s.

Thus, despite the fact that RHPS4 is a potent inhibitor of telomerase its interaction with quadruplex DNA is highly dynamic. Structural analysis reveals that the drug is not able to intercalate between the G-tetrads since the core structure is highly stabilized by K^+ ions which are

Graham A. Webb (ed.), Modern Magnetic Resonance, 1345–1349.

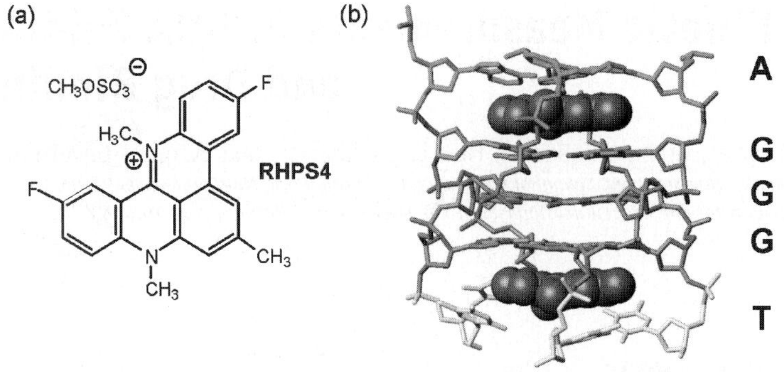

Fig. 1. (a) Chemical structure of the telomerase inhibitor RHPS4; (b) NMR structure of the 2:1 complex of RHPS4 bound at the 5′-AG and 5′-GT sites in the intermolecular parallel-stranded quadruplex formed from the human telomeric repeat d(TTAGGGT)$_4$. Only the core AGGGT structure is shown (5′-terminal thymines are relatively disordered). From Ref. [10]. (See also Plate 104 on page LXVIII in the Color Plate Section.)

octahedrally co-ordinated to the O6 carbonyl of guanine bases in adjacent tetrads. Displacement of these ions is energetically costly. Consequently, quadruplex-binding ligands appear to associate with the terminal G-tetrads, binding at the 5′-AG and 5′-GT steps within the sequence 5′-TTAGGGT (Figure 1b), where the adjacent nucleotides are less well structured [9,10]. As we have shown, small energy barriers to binding and dissociation permit fast kinetics at these sites.

DNA Hairpin Folding and Slow Exchange Equilibria

The mechanisms and pathways by which biomolecules fold to their native structures is well developed in the protein folding arena; however, the folding of DNA and large RNA assemblies is still poorly understood [13]. The process is likely to be just as complex, involving multiple pathways and kinetic traps originating from misfolded structures and non-native interactions [13]. DNA and RNA hairpins, consisting of a double stranded stem region and connecting loop, occur naturally in single stranded nucleotide sequences and serve a range of important biological functions. A number of DNA mini-hairpin sequences have been identified with extraordinary stability containing GNNA and GNA loop sequences (N = any nucleotide) where the purine bases fold back to form a G–A wobble base pair. In the most stable hairpin structures, the G–A pair preferentially stacks on an adjacent C–G base pair in the adjoining double stranded stem region [14,15]. We have exploited the 5′-C*GNA*G mini-hairpin motif (with a 5′-GAA loop) in the design of hairpins with mismatched or extrahelical bases for studies of drug recognition and binding [16]. We recently identified a 13-mer sequence (5′-GCTACGTAGTCGC) that folds in solution to form a hairpin with C–T mismatch (Figure 3a). In the presence of 100 mM NaCl (10 mM sodium phosphate, pH 7.0) the folded state is in slow exchange with the unstructured single strand with two sets of resonances assignable in 2D NMR spectra. The observation of two distinct species in slow exchange is unusual and indicative of a large activation energy barrier between folded and unfolded conformations. Significant chemical shift differences are observed between resonances in the two conformers (up to ∼0.3 ppm), making this an ideal system for characterizing the folding kinetics via magnetization transfer experiments.

Slow Exchange Between Two Conformers

In the general case, two conformers A and B are in slow exchange when the first order rate constants k_A and k_B for interconversion are $\ll \Delta\nu_{AB}$, where $\Delta\nu_{AB}$ is the chemical shift difference in Hz between a nucleus X in the two structural environments A and B. Selective perturbation of one signal can result in magnetization transfer to the other due to the exchange between environments. Assuming that the longitudinal relaxation rates in the two conformers are not considerably faster than the exchange rate, the time-dependent change in intensity after the perturbation can enable the rate of exchange to be measured [11,17]. Selective inversion of nucleus X in conformation A results in the recovery of the magnetization of A via a double exponential process,

$$M_A(t) = C_1 e^{\lambda 1t} + C_2 e^{\lambda 2t} + M_A^{\infty} \tag{3}$$

Similarly, inversion of A leads to transfer of magnetization to B with recovery of the magnetization of B also

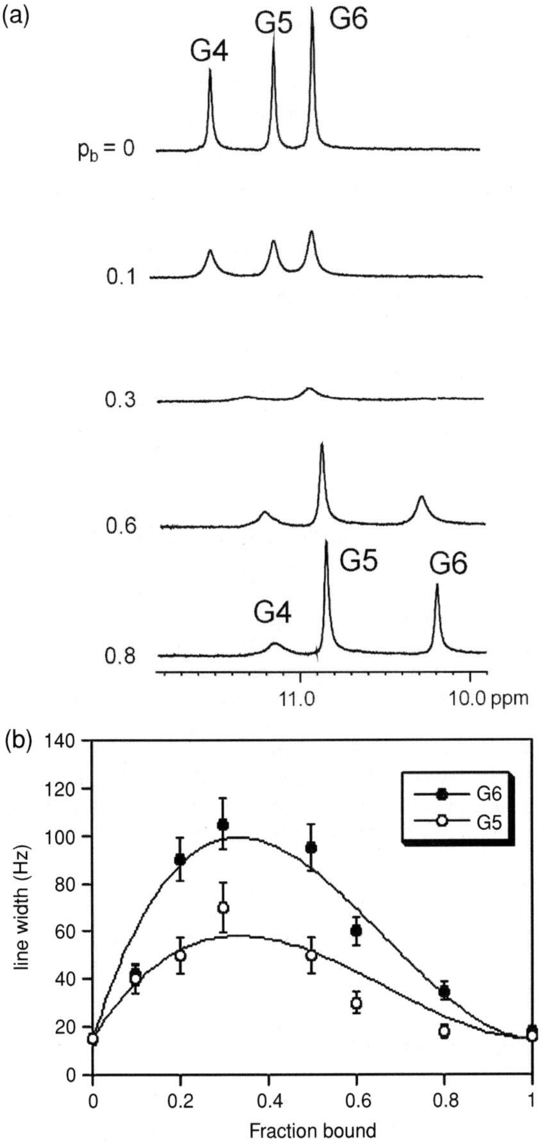

Fig. 2. (a) Selected 600 MHz 1D NMR spectra at 303 K showing the guanine imino proton resonances of the G-quadruplex structure d(TTAGGGT)$_4$ as a function of the fraction of drug bound (p_b); resonances are shifted upfield by ring current effects from the bound drug and initially broaden (maximum line width at $p_b \sim 0.3$) before again sharpening as the binding sites are filled. From Ref. [10]. (b) Measured line width (LW_{obs}) as a function of the fraction of drug bound showing the best fit to Equation (2).

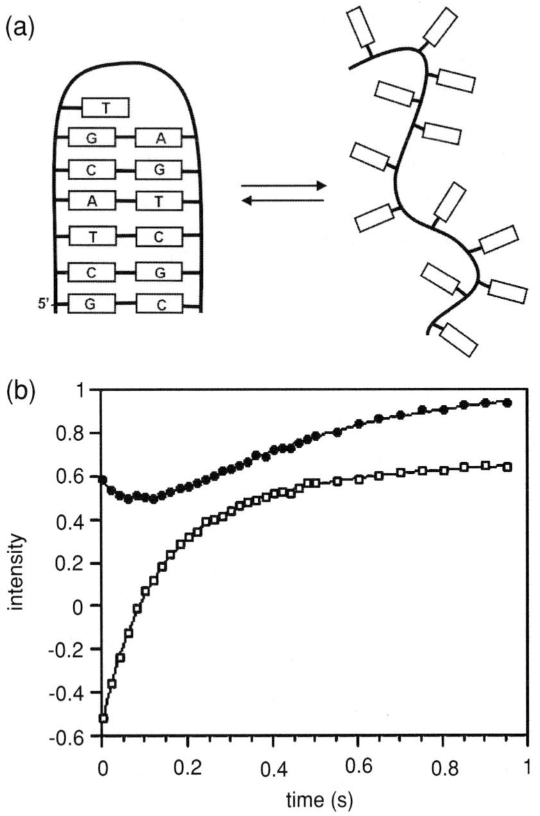

Fig. 3. (a) Proposed equilibrium between the hairpin conformation of the 13-mer d(GCTACGTAGTCGC) and the disordered single strand with interconversion rates k_A and k_B. (b) Magnetization transfer data showing the recovery of magnetization after selective inversion of the A4 H8 signal for the single stranded conformation at 8.23 ppm (open squares), and the change in intensity of A4 H8 (hairpin conformation) due to magnetization transfer (black dots); 600 MHz data collected at 308 K. Both data sets were analyzed using a double exponential fit [see Equations (3) and (4)]. The experimental parameters obtained are as follows (open squares): $C_1 = 0.458$, $\lambda_1 = 2.787$, $C_2 = 0.765$, $\lambda_2 = 9.892$, $M_A^\infty = 0.677$, and $M_A^0 = 0.521$; (black dots) $C_3 = 0.895$, $\lambda_1 = 2.572$, $C_4 = 0.455$, $\lambda_2 = 10.014$, $M_B^\infty = 1.022$, and $M_B^0 = 0.582$.

following a double exponential decay,

$$M_B(t) = C_3 e^{\lambda 1t} + C_4 e^{\lambda 2t} + M_B^\infty \qquad (4)$$

where M_A^∞ and M_B^∞ are the equilibrium longitudinal magnetization of X in the two conformations A and B. The various parameters have the following relationships as solutions of McConnell's equations [18]:

$$\lambda_1 = \{-(k_{1A} + k_{1B}) + [(k_{1A} - k_{1B})^2 + 4k_A k_B]^{1/2}\}/2 \quad (5)$$

$$\lambda_2 = \{-(k_{1A} + k_{1B}) - [(k_{1A} - k_{1B})^2 + 4k_A k_B]^{1/2}\}/2 \quad (6)$$

$$C_1 = [(\lambda_2 + k_{1A})(M_A^{\infty} - M_A^0) - k_B(M_B^{\infty} - M_B^0)]/(\lambda_1 - \lambda_2) \quad (7)$$

$$C_2 = [-(\lambda_1 + k_{1A})(M_A^{\infty} - M_A^0) + k_B(M_B^{\infty} - M_B^0)]/(\lambda_1 - \lambda_2) \quad (8)$$

$$C_3 = [-k_A(M_A^{\infty} - M_A^0) - (\lambda_1 + k_{1A})(M_B^{\infty} - M_B^0)]/(\lambda_1 - \lambda_2) \quad (9)$$

$$C_4 = [k_A(M_A^{\infty} - M_A^0) + (\lambda_2 + k_{1A})(M_B^{\infty} - M_B^0)]/(\lambda_1 - \lambda_2) \quad (10)$$

where k_{1A} and k_A, and k_{1B} and k_B are related by

$$k_{1A} = k_A + R_{1A} \quad (11)$$

$$k_{1B} = k_B + R_{1B} \quad (12)$$

where k_A and k_B are the required rate constants (Figure 3), R_{1A} and R_{1B} the longitudinal relaxation rates, and M_A^0 and M_B^0 are proportional to the initial intensities immediately after the perturbation to the magnetization of A. Since conformers A and B are in equilibrium, the following relationship must also hold:

$$k_A M_A^{\infty} = k_B M_B^{\infty} \quad (13)$$

The double exponential nature of the time-dependence of the magnetization means that λ_1 and λ_2 can only be determined independently with any certainty if they differ by more than a factor of 3, otherwise only the combination $C_1\lambda_1 + C_2\lambda_2$ can be determined, as discussed in detail elsewhere [11,17]. This is remedied by repeating the experiment by selectively inverting signal B to provide another pair of independent combinations.

DNA Hairpin Folding Kinetics by Magnetization Transfer

Inversion transfer experiments were used to characterize the hairpin to single strand equilibrium at 308 K where the two conformers are significantly populated ($M_A^{\infty}/M_B^{\infty} = k_B/k_A = 0.662$). Selective inversion was achieved using a 180° Gaussian-shaped pulse with the excitation bandwidth optimized to give the desired selectivity. Perturbation of the A4 H8 signal at 8.23 ppm in the single strand conformation results in transfer of magnetization to the resonance of the folded hairpin at 8.28 ppm. The recovery curves for both resonances are illustrated in Figure 3b. A double exponential fit resolves two rate constants which differ by a factor >3. Making use of Equation (13) to substitute into Equations (11) and (12), and subsequent substitutions into Equations (5)–(10) using experimentally determined values for λ_1, λ_2, and C_1–C_4, enables all of the unknown rate constants in Equations (11) and (12) to be determined, from which we deduce that $k_A = 3.3$/s and $k_B = 2.1$/s at 308 K. Rate constants were similarly determined from the reverse experiment (inversion of the hairpin signal at 8.28 ppm), and from inversion of other resonance pairs, with all the data in very close agreement.

We have chosen experimental conditions under which the two conformations are of similar stability (at 308 K, $K_{eq} = 1.6$) with the slow rate of interconversion indicating that the two states are separated by a large activation barrier. Recent thermodynamic investigations involving mutation of residues in various mini-loop sequences, including 5′-C*GNA*G, suggest that stacking interactions are highly co-operative with base substitutions or functional group mutations leading to large changes in stability [19,20]. Thus, a significant kinetic barrier may originate from the requirement for a highly structured loop region in the transition state which may have many of the interactions of the fully folded state. Our investigations have shown that the introduction of the destabilizing C–T mismatch has enabled the equilibrium to be sufficiently perturbed to be able to detect both folded and disordered species simultaneously in slow exchange. Magnetization transfer methods are ideally suited to measuring the kinetics of such systems, and have similarly been employed in characterizing coiled-coil transitions in the folding of GCN4 leucine zipper peptides [21].

Acknowledgments

We acknowledge our collaborators Professor Malcolm Stevens and Dr. Robert Heald on the drug–quadruplex studies, and the support from the EPSRC of the UK and the School of Chemistry, University of Nottingham.

References

1. Neidle S, Parkinson G. Nat. Rev. Drug Discov. 2002;1:383.
2. Wheelhouse RT, Sun D, Han H, Han FX, Hurley LH. J. Am. Chem. Soc. 1998;120:3261.
3. Mergny J-L, Mailliet P, Lavelle F, Riou J-F, Laoui A, Helene C. Anticancer Drug Des. 1999;14:327.
4. Davies JT. Angew. Chem. Int. Ed. 2004;43:668.

5. Haider SM, Parkinson GN, Neidle S. J. Mol. Biol. 2003;326:117.
6. Schouten JA, Ladame S, Mason SJ, Cooper MA, Balasubramanian S. J. Am. Chem. Soc. 2003;125:5594.
7. Heald RA, Modi C, Cookson JC, Hutchinson I, Laughton CA, Gowan SM, Kelland LR, Stevens MFG. J. Med. Chem. 2002;45:590.
8. Gowan SM, Heald RA, Stevens MFG, Kelland LR. Mol. Pharmacol. 2001;60:981.
9. Gavathiotis E, Heald RA, Stevens MFG, Searle MS. Angew. Chem. Int. Ed. 2001;40:4749.
10. Gavathiotis E, Heald RA, Stevens MFG, Searle MS. J. Mol. Biol. 2003;334:25.
11. Lian L-Y, Roberts GCK. In: GCK Roberts (Ed). NMR of Macromolecules. IRL Press, Oxford, 1993, p 152.
12. Craik DJ, Pavlopoulos S, Wickham G. NMR in Drug Design. CRC Press Inc., New York, 1996, p 423.
13. Pan J, Thirumalai D, Woodson SA. J. Mol. Biol. 1997;273:7.
14. Hirao I, Kawai G, Yoshizawa S, Nishimura Y, Ishido Y, Watanabe K, Miura K. Nucleic Acids Res. 1994;22:576.
15. Yoshizawa S, Kawai G, Watanabe K, Miura K, Hirao I. Biochemistry. 1997;36:4761.
16. Colgrave ML, Williams HEL, Searle MS. Angew. Chem. Int. Ed. 2002;41:4754.
17. Led JJ, Gesmar H, Abildgaard F. Methods Enzymol. 1989;176:311.
18. McConnell HM. J. Chem. Phys. 1958;28:430.
19. Moody EM, Bevilacqua PC. J. Am. Chem. Soc. 2003;125:2032.
20. Moody EM, Bevilacqua PC. J. Am. Chem. Soc. 2003;125:16285.
21. d' Avignon DA, Bretthorst GL, Holtzer ME, Holtzer A. Biophys. J. 1998;74:3190.

The Use of NMR in the Studies of Highly Flexible States of Proteins: Relation to Protein Function and Stability

Søren M. Kristensen[1], Marina R. Kasimova[2], and Jens J. Led[1]
[1]*Department of Chemistry, University of Copenhagen, DK-2100 Copenhagen, Denmark; and*
[2]*Optics and Fluid Dynamics Department, Risø National Laboratory, DK-4000 Roskilde, Denmark*

Introduction

The power of NMR as a method for characterizing the solution state of proteins stems from the highly specific way that NMR reports on protein structure and dynamics. This is due to the fact that NMR parameters, such as chemical shifts, relaxation rates and exchange-rate parameters report specifically about the local environment around the investigated nuclei. This obviously requires that the relevant NMR resonances have been assigned to the individual nuclei of the protein structures, which with present day techniques for isotope labeling and multidimensional NMR methods is possible for proteins with molecular weights higher than 30 kDa. After the resonances have been assigned, the overall three-dimensional structure of a protein can be determined from distance and angular constraints derived from additional spectral information.

However, a detailed NMR determination of the three-dimensional solution structure of proteins may be hampered by the dynamics of the protein molecules. At the same time, a detailed characterization of the flexible regions in the protein is highly interesting due to the functional importance of these regions, as shown by several NMR studies in recent years [1–4]. To solve this immediate conflict, that is, to alleviate the impact of the dynamics on the NMR structure determination of proteins, while at the same time obtaining important information on their dynamics and flexibility, several approaches can be used. Here we illustrate some of the approaches that have been used in the studies of the structure and function of two flexible insulin mutants and a flexible state of human growth hormone (hGH).

One of the approaches concerns the determination of the relatively fast exchanging amide protons involved in weak intra- and intermolecular hydrogen bonds. This approach is useful in cases where the exchange rates of the amide protons are relatively fast while separate amide proton signals can still be observed, that is, the exchange takes place within an hour or less. This approach is illustrated using a monomeric insulin mutant that, although flexible, still forms a well-defined structure in water.

Another approach focuses on the use of the helix-promoting solvent 2,2,2-trifluoroethanol (TFE) to stabilize helices inherently encoded in the amino acid residue sequence of the protein. This approach is illustrated in the study of a highly flexible insulin mutant without a well-defined structure in water. TFE strengthens helical structures in peptides and proteins and induces helical structures in segments which have the propensity for forming helices [5,6]. It has also been found that TFE stabilizes β-turns and β-hairpins in proteins [7,8]. On the other hand, TFE disrupts the quaternary structure of proteins and reduces interactions between non-polar residues [9,10].

However, a complete assignment of the resonances may not always be possible. In particular, resonances in highly flexible regions in proteins may be broadened beyond detection and thereby preventing an assignment of these resonances. Still, NMR can provide valuable information about the structure and dynamics of such highly flexible proteins, despite an incomplete assignment. Here, this is exemplified by the study of hGH. A partial assignment can be used to track structural changes caused by variations in the solution conditions. Changes in pH, for instance, may result in alterations in the charge distribution on the protein surface, which may modulate both intra- and intermolecular coulombic interactions and hydrogen bond interactions. This, in turn, may give rise to changes in NMR parameters such as chemical shifts, relaxation rates and exchange rates of labile protons. Similarly, changes in the salt concentration and the buffer composition can modulate structure and stability of specific regions in proteins, as well as intermolecular interactions, while changes in temperature are likely to influence the stability of the structure of a protein and to affect exchange phenomena. Hence, by measuring the NMR parameters for a protein at different solution conditions, it is possible to obtain information about non-native and destabilized structural states, even if only a partial assignment is available.

Graham A. Webb (ed.), Modern Magnetic Resonance, 1351–1358.

Insulin Flexibility and Activity

Insulin is a 5.8-kDa peptide hormone with two chains, a 30-residue long A-chain and a 21-residue long B-chain, and with a three-dimensional structure as shown in Figure 1. Native insulin forms well-defined dimers and higher aggregates thereof in water solution, and in the presence of Zn^{2+} it forms a well-defined hexamer complexed with two zinc ions. However, the active form of insulin is the monomer. Therefore, the structure of several monomeric insulin mutants have been determined by NMR, and their *in vivo* activities have been compared, in order to elucidate the function of insulin and the mechanism of its interaction with the receptor, through the correlation between its structure and flexibility and its activity. However, active monomeric mutants are often highly flexible, making it necessary to use special NMR techniques to extract the dynamical and structural information. Here we describe the NMR investigations of two monomeric mutants, the des-[Phe(B25)] mutant, that is, an insulin mutant in which Phe in position B25 has been removed, and the T(B27)P, P(B28)T mutant (PT insulin), where Thr(B27) and Pro(B28) of the native insulin have been interchanged. In both cases, a hydrophobic patch that is surface exposed in native insulin is turned toward the inside of the molecule, resulting in intramolecular hydrophobic interactions that eliminate the aggregation propensity. Both mutants have a biological activity that is higher than that of native insulin. However, they also have flexible structures. In the case of des-[Phe(B25)]insulin this results in increased exchange rates of the H-bonded amide protons involved in the stabilization of the α-helices, while in the case of PT insulin the structure is too unstable in water solution to be determined by NMR.

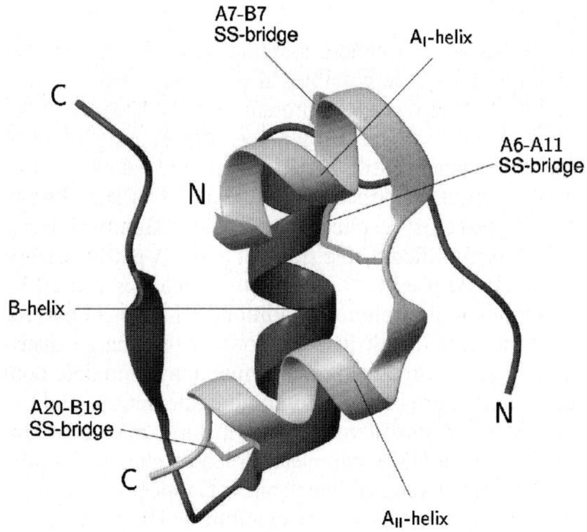

Fig. 1. The native insulin monomer consists of a 21-residue long A-chain (yellow) and a 30-residue long B-chain (red). Two disulfide bridges (SS bridge), CysA7-CysB7 and CysA20-CysB19, tie the two chains together, while a third disulfide bridge connects CysA6 and CysA11. The N- and C-terminal ends of the A-chain form α-helices (A_I: A2–A8 and A_{II}: A13–A20) connected by a loop. The central part of the B-chain forms a third helix (B10–B19). The C-terminal end of the B-chain forms an extended structure, while the N-terminal end is unstructured. (See also Plate 105 on page LXVIII in the Color Plate Section.)

Des-[Phe(B25)] Insulin: Quantitative Exchange Rates of Weakly H-bonded Amide Protons in Proteins from a Single 2D Experiment

To get more detailed information about the structure of the des-[Phe(B25)] mutant, the relatively fast exchange rates of the H-bonded amide protons were determined quantitatively using the two-dimensional (2D) NMR method for measuring exchange rates of the order of reciprocal hours (h^{-1}). This method is based on a single two-dimensional NOESY or TOCSY spectrum of the protein. It exploits the line broadening of the amide cross-peaks in the indirect dimension, caused by the hydrogen/deuterium exchange of the amide proton during the indirect evolution period when the protein is dissolved in D_2O. The method is described in detail in Ref. [11]. It allows the determination of amide proton exchange rates that are considerably faster than those, which can be measured by the conventional 2D approach based on series of NOESY spectra [12]. Furthermore, it is far less time consuming than the conventional 2D approach, since it is based on a single 2D experiment. At the same time, the accuracy of the rates determined by the method is high because data points from all the slices in the indirect dimension of the 2D experiment contribute to the determination of the rates. The accuracy can be further improved if the method is combined with a spectral analysis base on linear prediction. Moreover, the range of exchange rates covered by the method can be further expanded using the linear prediction model method described previously [13].

Using this method together with standard methodologies for NMR structure determination of proteins, it was found [14] that, although the des-[Phe(B25)] mutant has the normal insulin structure in water (Figure 1), the exchange rates of the H-bonded amide protons are relatively fast. Thus, the rates are about an order of magnitude faster than those of an insulin mutant where the hydrophobic patch is surface exposed, namely the B9(Asp) or S(B10)D mutant where Ser in position B9 has been substituted by Asp [15,16]. This clearly indicates that the structure of the des-[Phe(B25)] mutant is considerably less stable than that of the B9(Asp) mutant. Most importantly, the absence in the des-[Phe(B25)] insulin of

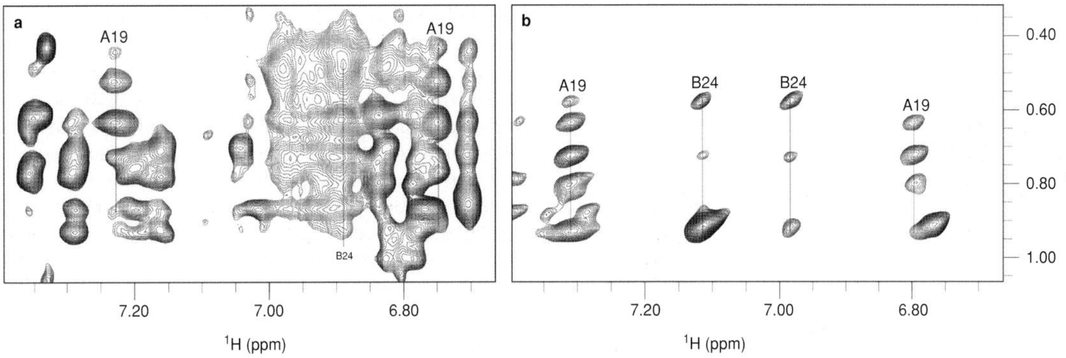

Fig. 2. The amide/aromatic region of an 800 MHz ^{1}H–^{1}H NOESY spectra of PT insulin showing the correlations between the aromatic side chain protons and aliphatic side chain protons (a) in pure water and (b) in 35% TFE-d_3. The broadening of the signals in the spectrum recorded in water reflects the high flexibility of the side chains. (Reproduced with permission from Ref. [17].)

slow exchanging amide protons, corresponding to the intermolecular hydrogen bonds associated with the dimer formation, show unambiguously that the mutant does not form a dimer in water solution. This is in contrast to the B9(Asp) mutant that forms a well-defined dimer in water, as reflected by the slow exchange of amide protons involved in intermolecular hydrogen bonds that stabilize the dimer.

PT Insulin: Stabilizing the Structure of an Unstructured Protein by 2,2,2-trifluoroethanol

In the case of proteins that are sufficiently flexible in water to escape any NMR structure determination, secondary or tertiary, the helix-promoting solvent TFE may stabilize the structure [10], allowing the determination of a sufficient number of NOE constraints for a regular NMR structure determination. This was demonstrated in a study of the structure and folding propensity of the highly flexible PT insulin mutant [17].

PT insulin is biologically active with an activity that is 50% higher than that of native insulin [18]. It can, therefore, adapt the conformation necessary for its binding to the insulin receptor (IR). However, it is also unusually flexible with a poorly defined structure in water. This is indicated by the broad NOESY cross-peaks between the side chain protons observed for the mutant dissolved in water (Figure 2a). Moreover, only a few slowly exchanging amide protons can be detected in the protein dissolved in water, and only at low temperature (15 °C). The slowly exchanging amide protons are all involved in weak hydrogen bonds associated with very loose secondary or tertiary structure elements. The strength of a hydrogen bond can be measured by the protection factor, P, of the amide proton involved in the H bond [19]. A more detailed description of the protection factor is given below together with the studies of hGH. Suffice it to mention here that a normal hydrogen bond corresponds to a protection factor of ≥ 25 [20], while the protection of the observed slow exchanging amide protons in PT insulin in water solution are in the range from 2.1 to 18.7.

When dissolved in a mixture of 35% TFE and water, PT insulin assumes a more stable structure, as shown by the relatively sharp and well-defined NOESY cross-peaks between the side chain protons observed for the mutant dissolved in this solvent (Figure 2b). Thus, a sufficient number of NOE-derived distance constraints and slowly exchanging amide protons could be identified to allow a regular NMR structure determination using the program X-PLOR [21]. Moreover, the structure that PT insulin assumes is identical to the structure of native insulin shown in Figure 1. That is, even though PT insulin does not have a well-defined structure in water, it has the propensity to form the native fold.

Model for the Insulin–Receptor Interaction

It is well established that the flexibility of the insulin molecule is important for its function. Thus, it was found [22] that an insulin mutant, where the N-terminal end of the A-chain was tied to the C-terminal end of the B-chain by a short peptide link, was biologically inactive even though it has a three-dimensional structure similar to that of native insulin (Figure 1). Also NMR studies of insulin mutants in which the C-terminal end of the B-chain is turned away from the rest of the molecule [23,24] indicate that such a structural change is required for insulin to interact with its receptor.

In the light of these findings, the concomitant high flexibility and biological activity of PT insulin is interesting.

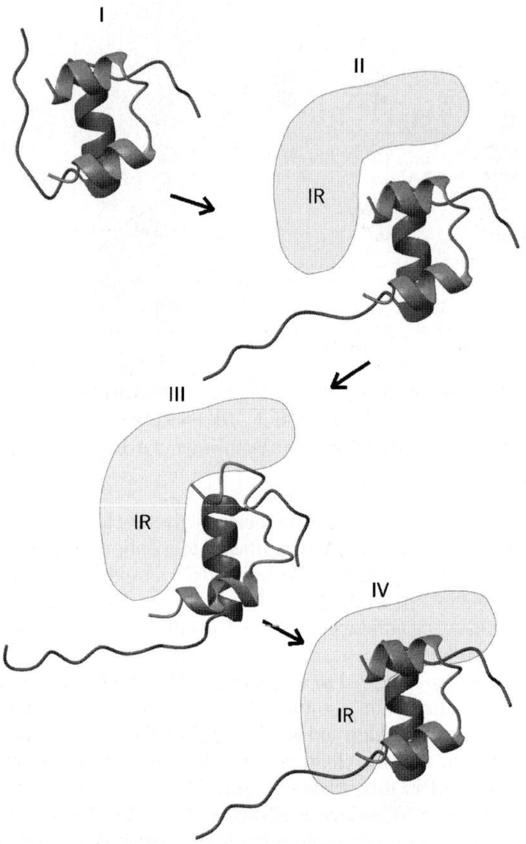

Fig. 3. Model of the interaction of native insulin with the insulin receptor (IR). I, free insulin in solution; II, detachment of the C-terminal part of the B-chain (vertical helix) from the rest of the molecule; III, unfolding of the A_I-helix; IV, refolding of the A_I-helix after navigating into the binding site of receptor (gray) and binding to the receptor. The enhanced biological activity of the PT insulin mutant could result from the fact that the A_I-helix of PT insulin is very loose or non-existing in water, yet it has the propensity to form the helix (see text). Therefore, PT insulin does not need to go through a genuine unfolding before navigating into the active site of the receptor. (Reproduced with permission from Ref. [17].) (See also Plate 106 on page LXVIII in the Color Plate Section.)

Together with the observation that the mutant has the propensity to form the native fold, these characteristics led to the suggestion of the mechanism for the binding of insulin to its receptor (IR) shown in Figure 3. This mechanism also includes a detachment of the C-terminal end of the B-chain from the rest of the molecule before the interaction with the receptor can take place, as suggested in previous studies [23,24].

The Acid State of Human Growth Hormone

In order to rationalize a rather puzzling study of hGH, the so-called acid state that occurs at pH 2.7, we shall compare this state with the native state of hGH at pH 7.0. hGH is a 191-residue four-helix bundle protein with a classical up-up-down-down cytokine topology [25]. The native state is characterized by a high thermal and chemical stability, similar to most native states of globular proteins. The acid state of hGH is more difficult to categorize, because it has characteristics of both the molten globule and the native state. Similar to the native state, the acid state has a high stability toward both chemical [26] and thermal denaturation and shows a cooperative unfolding behavior [27]. These features are contradictory to a classic definition of molten globules as defined by Ptitsyn [28]. They also differ from the behavior of other four-helix bundle proteins at similar solution conditions. Thus, the two cytokines, interleukin-2 and interleukin-4, were reported to adopt a classical molten globule state [29] and a "highly ordered molten globule" state [30], respectively, at low pH. In accordance with the stability data, far-UV CD measurements indicate that the structural properties of the acid state of hGH are very similar to those of its native state, suggesting that the helical content of the two states are similar. However, the near-UV CD spectrum of the acid state deviates from the spectrum of the native state and is more similar to the spectrum of denatured hGH [26]. This indicates that the aromatic side chains in the acid state are disordered, just like in the denatured state or a molten globule state. It is puzzling how it is possible to have a structural state with both a high stability toward denaturation and, at the same time, a disordered and flexible protein core.

Using standard three-dimensional triple-resonance techniques combined with uniform ^{13}C- and ^{15}N-labeling, only partial assignments of the hGH molecule were obtained [31]. In total, 60% and 75% of the backbone resonances were assigned for the native and acid states, respectively. Complete resonance assignment was not possible because of severe resonance overlap and line broadening effects [31].

However, even without access to the solution structure of hGH, NMR investigations of the protein allow a detailed comparison of the native and the acid states. The investigation includes analyses of the backbone chemical shifts, the amide and indole hydrogen/deuterium exchange rates, and backbone amide ^{15}N relaxation rates. Backbone C^{α} and H^{α} chemical shifts contain information about the secondary structure, and provide means for accessing the structural integrity of a protein under a particular set of experimental conditions. This information is derived by calculation of secondary chemical shifts and a subsequent comparison with reference values for helix,

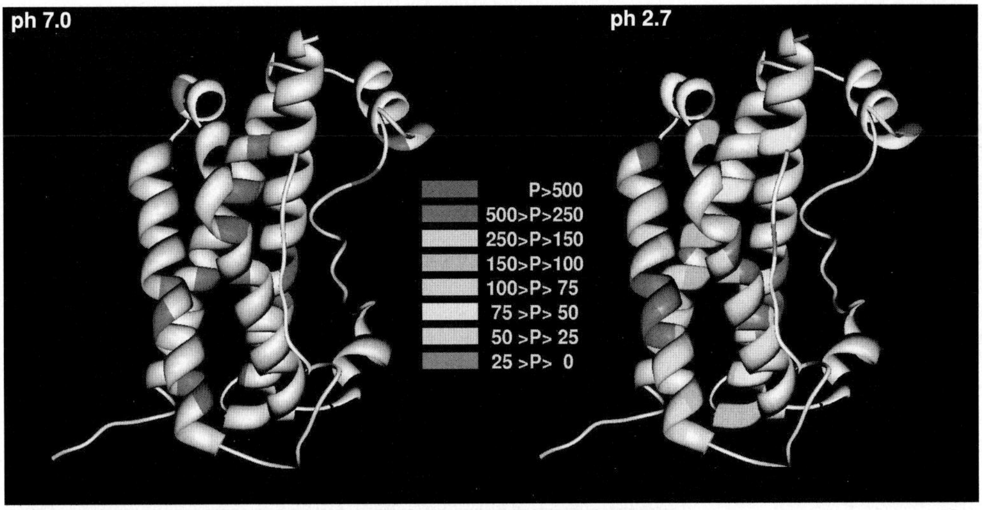

Fig. 4. Distribution of measured protection factors (P) for the native state (pH 7) and the acid state (pH 2.7) of human growth hormone. The protection factors are color mapped onto the three-dimensional structure as indicated in the figure. Structural parts colored in gray correspond to regions where the amide proton exchange rates could not be quantified. The figure was created with UCSF chimera [40]. (See also Plate 107 on page LXIX in the Color Plate Section.)

sheet, and extended structure [32]. Exchange-rate constants of labile protons report directly on solvent accessibility and in particular on the protection of labile protons by hydrogen bonds [33]. Sequence-related differences in intrinsic exchange-rate constants, k_{intr}, and their dependencies upon pH, are taken into account by calculation of protection factors $P = k_{\text{ex}}/k_{\text{intr}}$ [19], where k_{ex} is the experimentally derived exchange-rate constant. Measurement of ^{15}N longitudinal (R_1) and transverse (R_2) relaxation rates, and [^{1}H]–^{15}N heteronuclear NOE for each assigned amide ^{1}H–^{15}N pair, and analysis within the framework of the Lipari–Szabo model-free formalism [34,35]

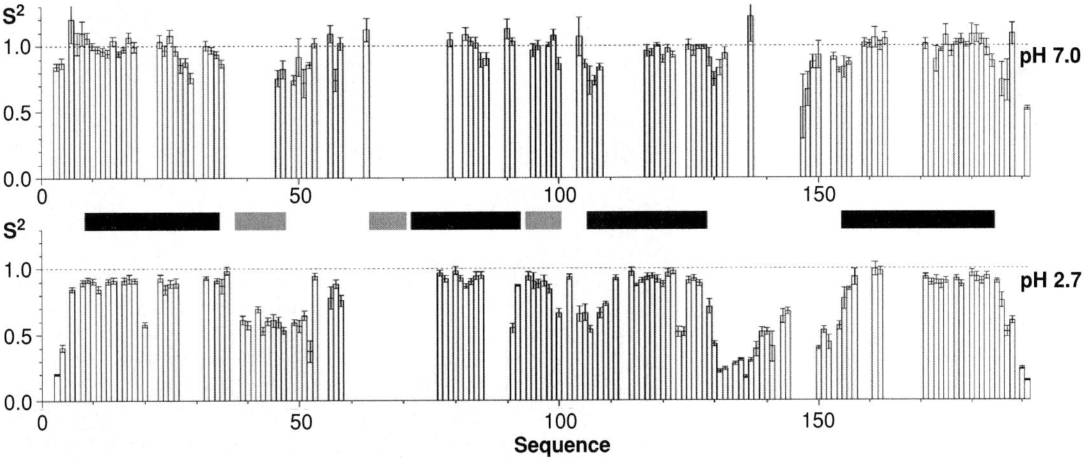

Fig. 5. Squared generalized order parameters (S^2) for the native state (pH 7.0) and the acid state (pH 2.7) of human growth hormone mapped as a function of the amino acid sequence. The black horizontal bars indicate the position of the four central helices A, B, C, and D forming the helix bundle (residues 9–34, 72–97, 106–128, and 155–184, respectively) and the gray horizontal bars indicate the position of the minor helices observed in the crystal structure of the hormone–receptor complex (PDB entry 3HHR).

provide a structural mapping of the so-called generalized order parameters, which reflect the restriction of the N–H bond vector on the pico- to nanosecond dynamics on a scale from zero to one. The ^{15}N relaxation measurements also provide information about exchange processes on the micro- to millisecond timescale as reflected in line broadening effects, and finally, information about the overall molecular tumbling rate and rotational diffusion anisotropy [36,37].

Backbone C^{α} and H^{α} assignments for the two states form the basis for a chemical-shift-based prediction of secondary structure [32]. This provides secondary structural data for the protein at the level of the individual residues without *a priori* knowledge of the overall structure. A comparison with the crystal structures of the free hormone (PDB entry 1HGU) and of the hormone bound to the extracellular domain of the receptor (PDB entry 3HHR) shows a high degree of similarity between the solution-state predictions at high and low pH and the two crystal structures [31]. The general conclusion that can be drawn from the NMR data is that the helix bundle scaffold is preserved at both high and low pH. However, subtle differences exist, including evidence for a helical propensity from residue 102 to 107 in the native solution state at neutral pH, which is not seen in the acid state. Also, the secondary chemical shifts indicate that the mini-helix from residue 94 to 100, observed in the crystal structure of the hGH–receptor complex but not in the crystal structure of free hGH, is present in the solution states.

The partial assignment of hGH gives access not only to structural information. NH exchange and backbone ^{15}N relaxation report on the flexibility of a molecule, and hence can help in the understanding of molecular stability.

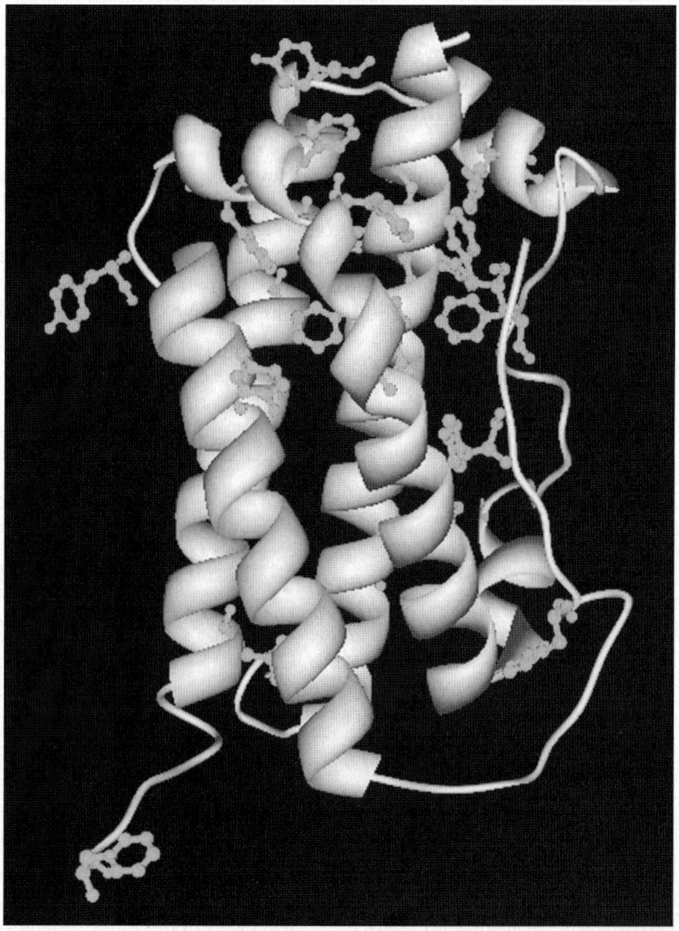

Fig. 6. Mapping of the aromatic residues in ball-and-stick representation on the ribbon diagram of human growth hormone shows a high density of aromatic groups in the north end of the molecule and a low density of aromatic groups in the south end of the molecule. The figure was created with UCSF chimera [40]. (See also Plate 108 on page LXIX in the Color Plate Section.)

For the native state of hGH, protection factors were determined for 22 backbone amide proton and despite the scarcity of the data, it is apparent that the protection factors are high throughout the helix bundle with values in the range 6×10^3 to 1×10^6. Furthermore, the indole NH of tryptophan 86 shows a high degree of protection ($P \sim 1 \times 10^6$) consistent with the hydrogen bond to $O^{\delta 1}$ of Asp169 first suggested by Bewley and Li [38] and later verified in the crystal structures. In the acid state, the backbone protection factors are orders of magnitude smaller than in the native state, and are in the range from 9×10^3 to 2×10^3, with the highest protection factors in the south end of the molecule. Figure 4 graphically illustrates the distribution of protection factors in the two solution states. Also, the Trp86 indole NH is exhibiting fast exchange with the solvent consistent with absence of the hydrogen bond to $O^{\delta 1}$ of Asp169 observed in the native state.

The backbone ^{15}N order parameters, S^2, for the native state are uniformly high throughout the sequence, corresponding to a highly rigid globular protein. This is in agreement with the high protection factors observed for this state. The acid state, on the other hand, displays increased flexibility throughout the sequence, but most pronounced in the long loop regions between the four main helices, as shown in Figure 5. In particular, the loop between helices C and D display order parameters as low as ~0.2, which is comparable to what is expected for an unfolded state [39]. Thus, the acid state of hGH is characterized by a preserved helix bundle structure with slightly increased overall flexibility as compared to the native state, but with highly mobile inter-helical loop regions.

The notion of the acid state as a well-defined, though more flexible, helix bundle with flexible loop regions is consistent with the high stability of the acid state toward denaturation. Also the native-like far-UV CD spectrum is indicative of a helix bundle structure with largely preserved inter-helical stabilizing interactions. On the other hand, the similarity of the near-UV CD spectrum with that of the denatured state of the protein suggests a certain degree of disorder of the aromatic groups, and seems to contradict the picture of a native-like helix bundle with preserved inter-helix stabilizing interactions. As shown in Figure 6, the majority of the aromatic side chains are located in the north end of the molecule. Thus, a unified picture is emerging, where the north end of the helix bundle is opening up resulting in a disorder of aromatic groups and an increased solvent access to amide groups—in agreement with near-UV CD and NH exchange data. At the same time, inter-helix interactions in the south end of the molecule are preserved, which can account for the high stability toward denaturation.

All taken together, the NMR investigation of hGH shows that important and detailed information about the structure and stability of proteins can be obtained from a multitude of available NMR data, despite an incomplete assignment of the NMR spectra.

Acknowledgement

The 800 MHz spectra were acquired at The Danish Instrument Center for NMR Spectroscopy of Biological Macromolecules.

References

1. Feher VA, Cavanagh J. Nature. 1999;400:289.
2. Eisenmesser EZ, Bosco DA, Akke M, Kern D. Science. 2002; 295:1520.
3. Ma LX, Hass MAS, Vierick N, Kristensen SM, Ulstrup J, Led JJ. Biochemistry. 2003;42:320.
4. Hansen DF, Hass MAS, Christensen HM, Ulstrup J, Led JJ. J. Am. Chem. Soc. 2003;125:6858.
5. Buck M, Radford SE, Dobson CM. Biochemistry. 1993;32: 669.
6. Hamada D, Kuroda Y, Tanaka T, Goto Y. J. Mol. Biol. 1995; 254:737.
7. Blanco FJ, Ortiz AR, Serrano L. Fold. Des. 1997;2:123.
8. Mabrouk K, Vanrietschoten J, Rochat H, Loret EP. Biochemistry. 1995;34:8294.
9. Albert JS, Hamilton AD. Biochemistry. 1995;34:984.
10. Buck M. Q. Rev. Biophys. 1998;31:297.
11. Olsen HB, Gesmar H, Led JJ. J. Am. Chem. Soc. 1993;115: 1456.
12. Wagner G, Wüthrich K. J. Mol. Biol. 1982;160:343.
13. Moss R, Gesmar H, Led JJ. J. Am. Chem. Soc. 1994;116: 747.
14. Jørgensen AMM, Olsen HB, Balschmidt P, Led JJ. J. Mol. Biol. 1996;257:684.
15. Kristensen SM, Jørgensen AMM, Led JJ, Balschmidt P, Hansen FB. J. Mol. Biol. 1991;218:221.
16. Jørgensen AMM, Kristensen SM, Led JJ, Balschmidt P. J. Mol. Biol. 1992;227:1146.
17. Keller D, Clausen R, Josefsen K, Led JJ. Biochemistry. 2001;40:10732.
18. Clausen R, Jørgensen TG, Jørgensen KH, Johnsen AH, Led JJ, Josefsen K. Eur. J. Endocrinol. 2002;147:227.
19. Bai YW, Milne JS, Mayne L, Englander SW, Proteins Struct. Funct. Genet. 1993;17:75.
20. Mori S, Abeygunawardana C, Berg JM, Vanzijl PCM. J. Am. Chem. Soc. 1997;119:6844.
21. Brunger AT. X-PLOR: A system for X-ray Crystallography and NMR. Yale University Press: New Haven, Connecticut, 1992.
22. Derewenda U, Derewenda Z, Dodson EJ, Dodson GG, Bing X, Markussen J. J. Mol. Biol. 1991;220:425.
23. Hua QX, Shoelson SE, Kochoyan M, Weiss MA. Nature. 1991; 354:238.
24. Ludvigsen S, Olsen HB, Kaarsholm NC. J. Mol. Biol. 1998;279:1.
25. de Vos AM, Ultsch M, Kossiakoff AA. Science. 1992;255: 306.

26. DeFelippis MR, Kilcomons MA, Lents MP, Youngman KM, Havel HA. Biochim. Biophys. Acta-Prot. Struct. Mol. Enzymol. 1995;1247:35.
27. Kasimova MR, Milstein SJ, Freire E. J. Mol. Biol. 1998; 277:409.
28. Ptitsyn OB. J. Protein Chem. 1987;6:273.
29. Dryden D, Weir MP. Biochim. Biophys. Acta. 1991;1078:94.
30. Redfield C, Smith RAG, Dobson CM, Nat. Struct. Biol. 1994;1:23.
31. Kasimova MR, Kristensen SM, Howe PWA, Christensen T, Matthiesen F, Petersen J, Sørensen HH, Led JJ. J. Mol. Biol. 2002;318:679.
32. Wishart DS, Sykes BD, Methods Enzymol. 1994;239:363.
33. Englander WS, Downer NW, Teitelba H. Annu. Rev. Biochem. 1972;41:903.
34. Lipari G, Szabo A. J. Am. Chem. Soc. 1982;104:4546.
35. Lipari G, Szabo A. J. Am. Chem. Soc. 1982;104:4559.
36. Tjandra N, Feller SE, Pastor RW, Bax A. J. Am. Chem. Soc. 1995;117:12562.
37. Tjandra N, Wingfield P, Stahl S, Bax A. J. Biomol. NMR. 1996;8:273.
38. Bewley TA, Li CH. Arch. Biochem. Biophys. 1984;233:219.
39. Farrow NA, Zhang OW, Formankay JD, Kay LE. Biochemistry. 1997;36:2390.
40. Huang CC, Couch GS, Pettersen EF, Ferrin TE, Pac. Symp. Biocomput. 1996;1:724.

NMR-based Metabonomics Techniques and Applications

John C. Lindon, Elaine Holmes, and Jeremy K. Nicholson
Biomedical Sciences Division, Imperial College London, Sir Alexander Fleming Building, South Kensington, London SW7 2AZ, UK

Introduction

Since the decoding of the human genome, there has been much interest in using changes in gene expression to discover the basis of disease and for the identification of new drug targets. However to date, the promise of the approach is yet to be comprehensively realized and there remains a difficulty in relating such changes to real conventional end points used in diagnosis and pharmaceutical development. The simultaneous measurement of many gene expression changes is termed transriptomics, and is usually carried out in an automatic fashion using so-called gene microarrays. The relevance of such gene changes is not always clear, which has led subsequently to efforts focused on the consequent protein level changes (a subject termed proteomics), but again it is not always possible to relate such changes directly to pathological events.

On the other hand, metabonomics, a whole systems approach for evaluating metabolite changes and pathway analysis through study of biofluids and tissues [1] offers a process whereby real end points can be obtained. In complex organisms, these three levels of biomolecular organization and control are highly interdependent, but they can have very different time scales of change. All of the technologies, which rely on analytical chemistry methods, result in complex multivariate data sets that require a variety of chemometric and bioinformatic tools for interpretation. The aim of such procedures is to extract biochemical information that is of diagnostic or prognostic value, and that reflects actual biological events. The areas where metabonomics impacts pharmaceutical R&D are:

- validation of animal models of disease, including genetically modified animals;
- preclinical evaluation of drug safety and ranking of compounds;
- assessment of safety in clinical trials and after product launch;
- quantitation, or ranking, of the beneficial effects of pharmaceuticals both in development and clinically;
- improved understanding of idiosyncratic toxicity;
- improved, differential diagnosis and prognosis of clinical diseases;
- better understanding of environmental population effects through epidemiological studies;
- patient stratification (pharmaco-metabonomics);
- nutrition, interactions between drugs, and between drug and diet;
- studies in environmental science.

Importantly, metabonomics also allows time-dependent patterns of change in response to stimuli to be measured. In multi-cellular organisms, there are a broad variety of time scales, varying widely according to gene, protein, pathway, and tissue. One important potential role for metabonomics therefore is to direct the timing of proteomic and genomic analyses in order to maximize the probability of observing "omic" biological changes that are relevant to functional outcomes.

Metabonomics, formally defined as the quantitative measurement of the dynamic multi-parametric metabolic response of living systems to pathophysiological stimuli or genetic modification, thus provides an approach which leads to real world end points. Metabolites can be identified and quantified, and changes can be related to health and disease, and are changeable by therapeutic intervention. The subject of metabonomics has been reviewed recently [2–4].

Metabonomics Analytical Technologies

The two principal methods that can produce metabolic profiles of biomaterials, comprise ^{1}H NMR spectroscopy [2] and mass spectrometry (MS) [5], the latter usually including a separation stage such as GC or LC. NMR has the advantages of being non-destructive, also applicable to intact tissues using magic angle spinning (MAS) methods [6], and provides detailed information on molecular structure, especially in complex mixture analysis. In addition, NMR can also be used to probe molecular dynamics as well as concentrations. MS is inherently considerably more sensitive than NMR, but it is necessary generally to employ different separation techniques for different classes of substances. Quantitation in MS can also be impaired by variable ionization effects, such that derivatization may be necessary. Most published mammalian studies have used NMR spectroscopy, but LC–MS is increasing in usage.

Graham A. Webb (ed.), Modern Magnetic Resonance, 1359–1367.

Typically metabonomics is carried out on biofluids that provide an integrated view of the whole systems biology. The biochemical profiles of the main diagnostic fluids, blood plasma, cerebrospinal fluid (CSF), and urine, reflect both normal variation and the impact of drug effects or disease on single or multiple organ systems [7]. Urine and plasma are obtained in a non- or minimally invasive fashion, and hence are appropriate for clinical trials monitoring and disease diagnosis. Different biofluids have characteristically different metabolic profiles (Figure 1). Although urine and plasma are the main diagnostic fluids, others have also been used in special cases. These include CSF, seminal fluids, digestive fluids, pathological fluids such as cyst fluid, and administered fluids such as dialysis fluids. A standard ^{1}H NMR spectrum of urine typically contains thousands of sharp lines from predominantly low molecular weight metabolites. Plasma contains low and high molecular weight components, which give a wide range of signal line widths. Protein and lipoprotein signals dominate ^{1}H NMR spectra of plasma, with small molecule fingerprints superimposed on them. Standard editing experiments based on T_1, $T_{1\rho}$, T_2, or diffusion coefficients can be used to select only the contributions from proteins, and other macromolecules and micelles, or alternatively to select only the signals from the small molecule metabolites. Each biofluid yields a characteristic ^{1}H NMR

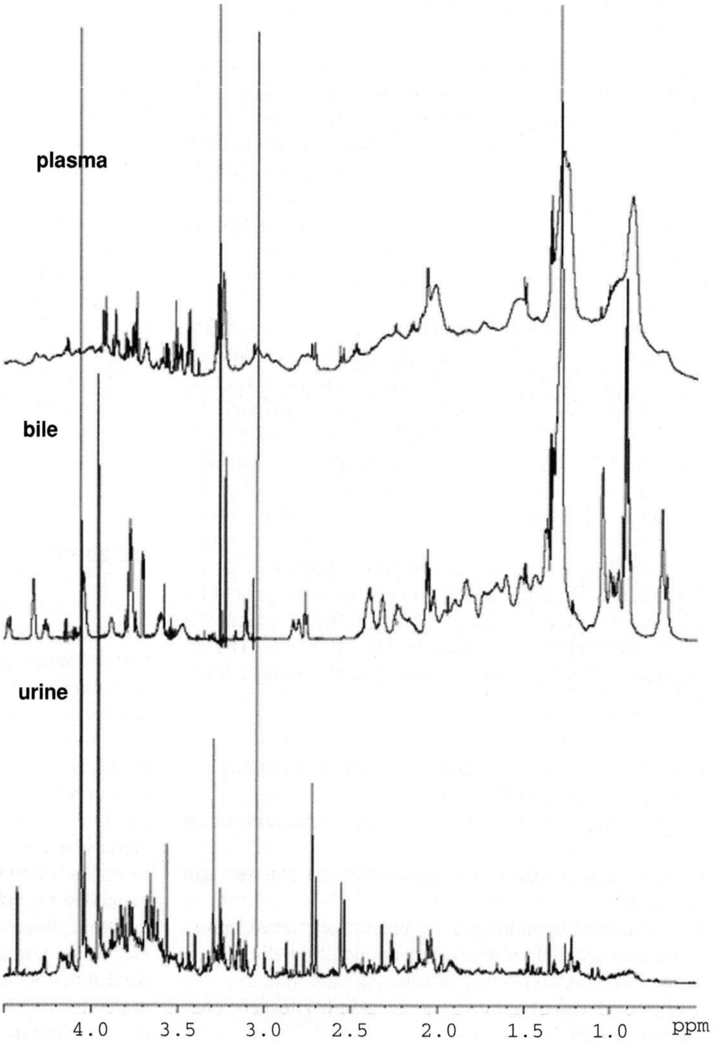

Fig. 1. 800 MHz ^{1}H NMR spectra with water peak suppression of control human biofluids, urine, gall bladder bile, and blood plasma showing characteristically different biochemical profiles.

spectroscopic fingerprint in which the spectral intensity distribution is determined by the relative concentrations of solutes, and in some cases by their intermolecular interactions. Standard spectra typically take only a few minutes to acquire using robotic flow injection methods. These entail robotic sample preparation involving buffering, addition of a chemical shift and quantitation standard such as TSP, and addition of D_2O as a magnetic field lock signal for the spectrometer. The large interfering NMR signal arising from water in all biofluids is easily eliminated by use of appropriate NMR solvent suppression methods such as NOESY presaturation or WET. Using NMR flow probes, the capacity for NMR analysis has increased enormously and now up to 200–300 samples per day can be measured.

Identification of the biomarkers (i.e. metabolites that change in level as a consequence of the pathology) detected in a biofluid NMR spectrum can involve the application of a number of other NMR techniques including two-dimensional (2D) experiments [2]. Although ^{1}H NMR spectra of urine and other biofluids are very complex, many resonances can be assigned directly based on their chemical shifts, signal multiplicities, and by adding authentic material and remeasuring the spectrum. 2D NMR spectroscopy can also be useful for spreading the signals out and for working out the connectivities between signals, thereby enhancing the information content and helping to identify biochemical substances. These include the ^{1}H–^{1}H J-resolved experiment, which reduces the contribution of macromolecules and yields information on the multiplicity and coupling patterns of resonances. Other 2D experiments such as COSY and TOCSY provide ^{1}H–^{1}H spin–spin coupling connectivities. Use of other nuclei can be important to help assign NMR peaks and here inverse-detected heteronuclear correlations, usually ^{1}H–^{13}C, can also be obtained by use of sequences such as HSQC or HMBC.

A major improvement in the scope of metabonomics has been made possible by the commercialization of miniaturized NMR probes. Now it is possible to study metabolic profiles by NMR using as little as 2–20 μl of sample, and examples have been published using CSF and blood plasma [8,9].

Cryogenic NMR probe technology, whereby the NMR detector coil and preamplifier are cooled to about 20 K, is now commercially available and provides an improvement in spectral signal–noise ratios of up to 500%. This improvement permits the routine use of natural abundance ^{13}C NMR spectroscopy of biofluids, such as urine or plasma, with acquisition times that enable a high throughput of samples. Information-rich ^{13}C NMR spectra of urine can be obtained using appropriately short acquisition times suitable for biochemical samples when using a cryogenic probe [10].

For metabolite identification, directly coupled chromatography–NMR spectroscopy methods can be used. The most powerful of these "hyphenated" approaches being HPLC–NMR–MS, [11] which can provide the full array of NMR and MS-based molecular identification tools. These include MS–MS for fragment ions and Fourier transform (FT)-MS or time-of-flight (TOF)-MS for accurate mass measurement and hence derivation of molecular empirical formulae.

Within the last few years, the development of high-resolution ^{1}H MAS NMR spectroscopy has had a substantial impact on the ability to analyze intact tissues [12]. Rapid spinning of the sample (~4–6 kHz typically) at an angle of 54.7° relative to the applied magnetic field serves to reduce line broadening effects caused by sample heterogeneity, residual dipolar couplings, and residual chemical shift anisotropy. Thus, it is possible to obtain very high quality NMR spectra of whole tissue samples with no sample pretreatment. Such experiments indicate that diseased or toxin-affected tissues have substantially different metabolic profiles to those taken from healthy organs [13]. In addition, MAS NMR spectroscopy can be used to access information regarding the compartmentalization of metabolites within cellular environments. Such MAS NMR-based metabonomics can also be applied to *in vitro* systems such as tissue extracts [14], Caco-2 cells [15], or spheroids [16].

The NMR spectrum of a sample can be thought of as an object in a multi-dimensional set of metabolic coordinates, the values of which could be the spectral intensity at every data point. Similarity or differences between samples can then be evaluated using multivariate statistical methods or other pattern recognition approaches. One simple approach that has been widely used is to effect a dimension reduction from the typically 32K or 64K NMR intensity points describing each spectrum, down to a few dimensions to visualize similarities and differences between samples. This can be accomplished using principal components analysis (PCA). This constructs latent variables from linear combinations of the original descriptors to describe as much variation as possible in the data set, with first principal component (PC) explaining the maximum variation and successive components explaining decreasing amounts of variance with the constraint that all PCs are orthogonal to each other. The approach gives rise to two matrices based on the data, a scores and a loadings matrix. A plot of the PC scores shows the relationships between the samples (each point on the plot comprises one sample) and the PC loadings provide information on those variables, which contribute to the position of the samples in the scores plot. This is illustrated in Figure 2 for rat urine NMR spectra. The scores plot shows three clusters of samples, those of NMR spectra of control urine and those of NMR spectra of urine from animals dosed with two different liver toxins, hydrazine and a substance known as ANIT. These cause different biochemical changes and produce the separate clusters (Figure 2a). Examination of

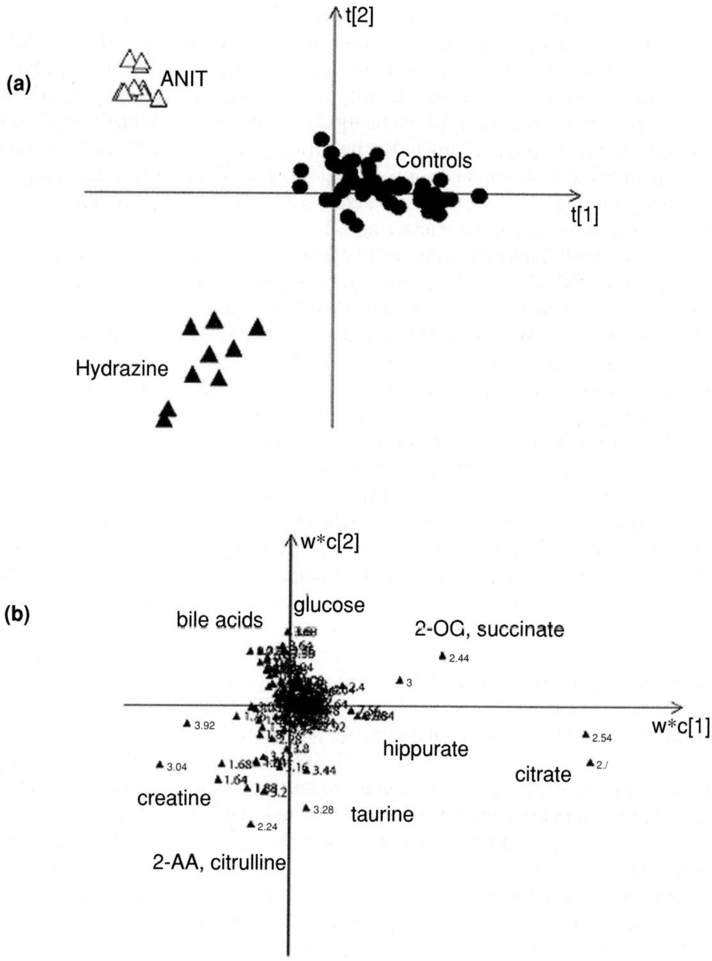

Fig. 2. Classification of rat urine samples based on their ^{1}H NMR spectra using principal components (PC) analysis. Animals were dosed with hydrazine or ANIT, two model liver toxins operating by different biochemical mechanisms, or with dosing vehicle alone (controls). (a) PC scores plot where each spectrum is reduced to a single point and samples with similar metabolic profiles appear close to each other. (b) PC loadings plot where each point corresponds to the signal intensity a single NMR spectral region of width 0.04 ppm and the position of a spectral region point determines which regions are responsible for the corresponding clustering in the scores plot [17].

the loadings plot, Figure 2b, gives information on which NMR spectral regions are responsible for the clustering in the scores plot, i.e. a cluster with a low PC1 and a low PC2 score (such as hydrazine) will appear in the bottom left hand quadrant and the corresponding loadings or altered NMR spectral regions will be in the same quadrant.

In the real world, biochemical changes caused by disease, nutrition, lifestyle, drug therapy, or drug toxicity develop and recover in real time and there can be complex, time-related changes in NMR spectra of biofluids. Hence, in order to develop automatic classification methods, it has proved efficient to use advanced chemometrics and bioinformatics approaches, known as supervised methods, that also take time into account. In general, these methods allow the quantitative description of the multivariate boundaries that characterize and separate each class of sample in terms of their metabolic profiles, some providing a level of probability for the classification. By using a variety of chemometric methods, it is possible to use such models to provide classification probabilities or even quantitative response factors for a wide range of sample types. It is important to build and test such chemometric models using independent training data and validation data sets.

Subtle biochemical changes in ^{1}H NMR spectroscopic profiles of biofluids can be obscured in such analyses by interfering factors such as variations in pH, which can cause changes in NMR chemical shifts because of differences in ionization state of some molecules. In NMR spectroscopy of urine, one means of limiting the effects of pH on the chemical shift of sensitive moieties is to add a standard amount of buffer to the sample prior to NMR spectroscopic analysis. Alternatively, mathematical algorithms can be used to realign the chemical shifts of resonances from protons near ionizable groups displaced by pH effects [18]. Some spectral regions, such as those containing water or urea, are very variable due to water NMR peak suppression effects. In addition, many drug compounds or their metabolites are excreted in biofluids and these can obscure significant changes in the concentration of endogenous components. Therefore, it is usual to remove these redundant spectral regions prior to chemometric analysis.

One example of a robust automatic data reduction method that has been widely used is the division of the NMR spectrum into regions of equal chemical shift ranges followed by signal integration within those ranges [19]. Automatic data reduction of 2D NMR spectra can be performed using a similar procedure, in which the spectrum is divided by a grid containing squares or rectangles of equal size, and the spectral integral in each volume element is calculated.

Selected Applications of Metabonomics

Preclinical Drug Candidate Safety Assessment

There is evidence through the recurring need to withdraw drugs from the market place that drug safety assessment approaches used in the pharmaceutical industry can still fail. There is a need for methodologies that can pick up potential problems earlier, faster, more cheaply, and more reliably. A recent survey of market withdrawals during the period 1960–1999 has identified hepatotoxicity as the most common reason for withdrawal [20]. Minimizing attrition caused by drug adverse effects is therefore one of the most important aims of pharmaceutical R&D, and metabonomics has been used extensively in evaluating the adverse effects of candidate drugs.

In this application, NMR-based metabonomics can be used for (i) definition of the metabolic hyperspace occupied by normal samples, (ii) rapid and simple classification of the sample as normal or abnormal (this enables spectrometer automation for data acquisition), (iii) classification of target organ toxicity, (iv) site and mechanism of action within the organ, (v) identification of biomarkers of toxic effect, and (vi) evaluation of the time course of effect, e.g. the onset, evolution, and regression of toxicity. The role that metabonomics has in the evaluation of xenobiotic toxicity has been comprehensively defined by the Consortium for Metabonomic Toxicology (COMET) formed between five pharmaceutical companies and Imperial College, London, UK [21]. The aim of this project is to define methodologies and to apply metabonomic data generated using ^{1}H NMR spectroscopy of urine and blood serum for preclinical toxicological screening of candidate drugs. This has been achieved by generating databases of spectral and conventional results for a wide range of model toxins (147 in total) that serve as the raw material for computer-based expert systems for toxicity prediction. The project goals of the generation of comprehensive metabonomic databases (now around 35,000 NMR spectra) and multivariate statistical models for prediction of toxicity (initially for liver and kidney toxicity in the rat and mouse) have completely been achieved.

A study operated to the same detailed protocol and using the same model toxin was carried out over seven sites in the companies and contract research organizations. This was used to probe both the analytical and biological variation that could both arise through the use of metabonomics. The inter-site NMR analytical reproducibility revealed a high degree of robustness where split samples were analyzed both at Imperial College and at various company sites, giving a coefficient of variation of about 1.6%. The biological variability was evaluated by a detailed comparison of the ability of the companies to provide consistent urine and serum samples with all samples measured at Imperial College. There was a high degree of consistency between samples from the various companies, and the differences between samples were small compared to the biochemical effects of the toxin, where dose-related effects could be distinguished [22]. Following this successful start, metabonomic models have been constructed for urine from control rats and mice, enabling identification of outlier samples and the metabolic reasons for the deviation. Building on this, and with the completion of all planned studies, and the development of new chemometrics methodology to meet the challenges thrown up by this project, workable expert systems for prediction of liver and kidney toxicity have been generated.

To achieve this goal, a new approach to the classification of a large data set of COMET samples has been developed, termed—Classification Of Unknowns by Density Superposition (CLOUDS)—a novel non-neural implementation of a classification technique developed from probabilistic neural networks [23]. Modeling the urinary NMR data according to organ of effect (control, liver, kidney, or other organ), using a model training set of 50% of the samples and predicting the other 50%, over 90% of the test samples were classified as belonging to the correct group with only a 2% misclassification rate between the classes. This work showed that it is possible to construct predictive and informative models of metabonomic data, delineating the whole time course of toxicity—the ultimate goal of the COMET project.

Genetic Differences and other Physiological Effects by Metabonomics

In order to determine therapeutic effects, it is necessary to understand any underlying physiological variation and to this end, metabonomics can be used to separate classes of experimental animals such as mice and rats according to a number of inherent and external factors based on the endogenous metabolite patterns in their biofluids [24]. Such differences may help explain differential toxicity of drugs between strains and inter-animal variation within a study.

Metabonomics is also being used for the phenotyping of mutant or transgenic animals and the investigation of the consequences of transgenesis such as the transfection process itself [25]. This suggests that the method may be appropriate for following treatment regimes such as gene therapy. It is important to differentiate often-seen unintended consequences of the genetic engineering process from the intended result, because pharmaceutical companies are developing genetically engineered animal models of disease using transfection procedures. Metabonomic approaches can give insight into the metabolic similarities or differences between mutant or transgenic animals and the human disease processes that they are intended to simulate and their appropriateness for monitoring the efficacy of novel therapeutic agents.

The importance of gut microfloral populations on urine composition has been highlighted by a study in which axenic (germ-free) rats were allowed to acclimatize in normal laboratory conditions and their urine biochemical composition was monitored for 21 days [26].

Many other effects can be distinguished using metabonomics, including male/female differences, age-related changes, estrus cycle effects in females, diet, diurnal effects, and interspecies differences and similarities [24].

Integrated Metabonomic Studies

The value of obtaining multiple NMR data sets from various biofluid samples and tissues of the same animals collected at different time points has been demonstrated. This procedure has been termed "integrated metabonomics," [13,14] and can be used to describe the changes in metabolic chemistry in different body compartments affected by exposure to toxic drugs. Such timed profiles in multiple compartments are themselves characteristic of particular types and mechanisms of pathology and can be used to give a more complete description of the biochemical consequences than can be obtained from one fluid or tissue alone. An example is given in Figure 3 that shows ^{1}H NMR spectra from intact liver tissue, tissue extracts, and blood plasma from a mouse after administration of a toxic dose of paracetamol. Building on this, it has also been possible to integrate data from transcriptomics and metabonomics to find common metabolic pathways implicated by both gene expression changes and changes in metabolism [27].

Disease Diagnosis

Many examples exist in the literature on the use of NMR-based metabolic profiling to aid human disease diagnosis, including the use of plasma to study diabetes, CSF for investigating Alzheimer's disease, synovial fluid for osteoarthritis, seminal fluid for male infertility and urine in the investigation of drug overdose, renal transplantation, and various renal diseases. Most of the earlier studies have been reviewed [7].

Some studies have been undertaken in the area of cancer diagnosis using perchloric acid extracts of various types of human brain tumor tissue.[28] The spectra were classified using neural network software giving ~85% correct classification. Tissues themselves can be studied by metabonomics using the MAS technique and published examples include prostate cancer [29] and renal cell carcinoma [30]. Other recent studies include an NMR-based urinary metabonomic study of multiple sclerosis in humans and non-human primates [31].

Currently, the only reliable diagnostic method for coronary heart disease (CHD) is the injection of X-ray opaque dye into the blood stream and visualization of the coronary arteries using X-ray angiography. This is both expensive and invasive with an associated 0.1% mortality and 1–3% of patients experiencing adverse effects. Recently metabonomics has been applied to provide a method for diagnosis of CHD non-invasively through analysis of a blood serum sample using NMR spectroscopy [32]. Patients were classified into two groups, those with normal coronary arteries and those with triple coronary vessel disease, as based on an angiographic examination. Around 80% of the NMR spectra were used as a training set to provide a two-class model after appropriate data filtering techniques had been applied and the samples from the two classes were easily distinguished. The remaining 20% of the samples were used as test set and their class was then predicted based on the derived model with a sensitivity of 92% and a specificity of 93% based on a 99% confidence limit for class membership.

It was also possible to diagnose the severity of the CHD that was present by employing serum samples from patients with stenosis of one, two, or three of the coronary arteries. Although this is a simplistic indicator of disease severity, separation of the three sample classes was evident even though none of the wide range of conventional clinical risk factors that had been measured was significantly different between the classes.

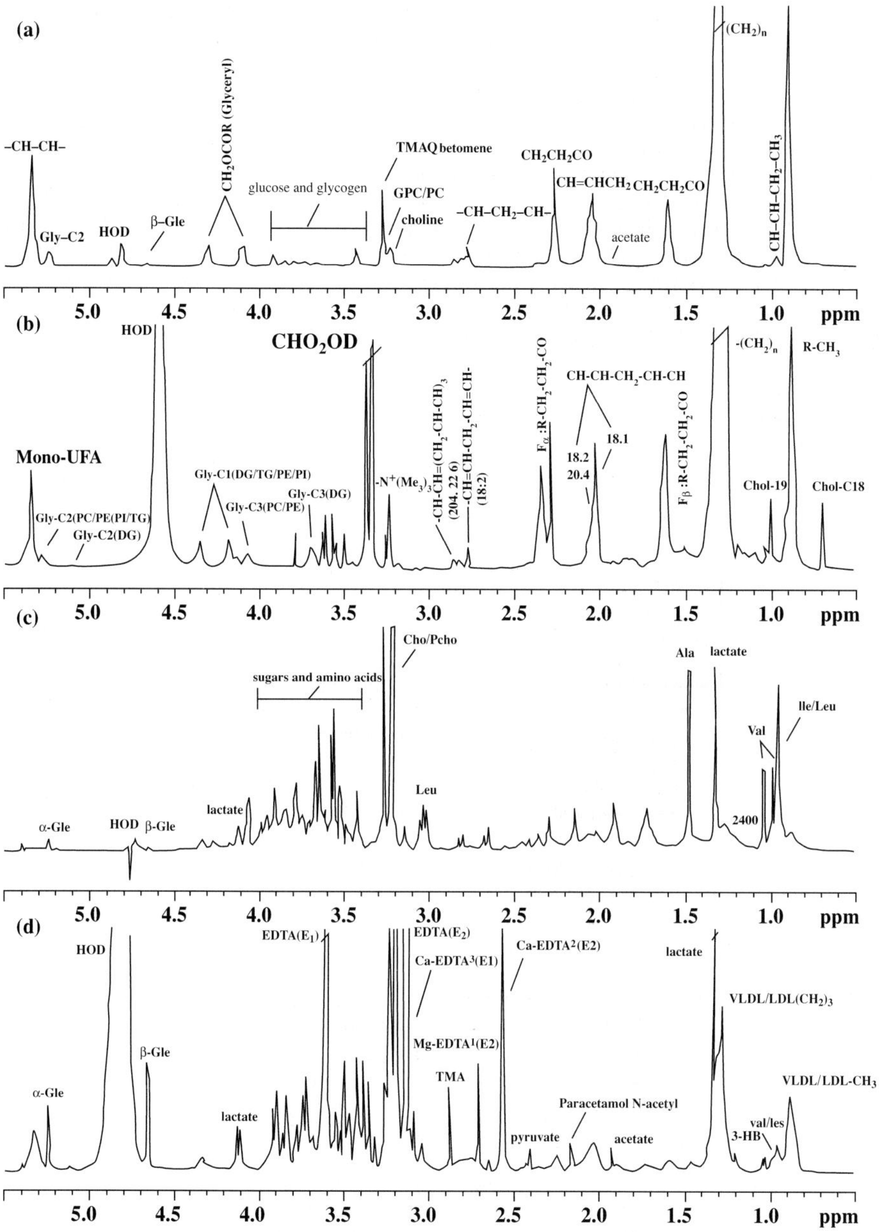

Fig. 3. (a) 600 MHz ^{1}H MAS NMR CPMG spin-echo spectrum of intact liver tissue; (b) 600 MHz standard ^{1}H NMR spectrum of a lipid-soluble liver tissue extract; (c) 600 MHz solvent suppressed ^{1}H NMR spectrum of an aqueous-soluble liver tissue extract; (d) 500 MHz ^{1}H NMR CPMG spectrum of blood plasma. All spectra are from animals treated with a toxic dose of paracetamol (500 mg/kg) and sacrificed at 240 min after dosing. Key: 3HB, 3-D-hydroxybutyrate; Cho, choline; Chol, cholesterol; Glu, glucose; GPC, glycerophosphorylcholine; Gly, glycerol; LDL, low-density lipoprotein; P Cho, phosphocholine; TMAO, trimethylamine-*N*-oxide; VLDL, very low-density lipoprotein.

Conclusions

Although there continues to be a need for advances in metabonomic analytical technologies both in NMR and MS, NMR is likely to remain the method of choice for a broad impartial survey of metabolic profiles, especially given recent gains in sensitivity through the use of cryoprobe detectors. MS coupled to a separation stage is always likely to yield better detection limits for specific classes of metabolite, but is inherently less general.

NMR-based metabonomics is now recognized as an independent and widely used technique for evaluating the toxicity of drug candidate compounds, and it has been adopted by a number of pharmaceutical companies into their drug development protocols. For drug safety studies, it is possible to identify the target organ of toxicity, derive the biochemical mechanism of the toxicity, and determine the combination of biochemical biomarkers for the onset, progression and regression of the lesion. Additionally, the technique has been shown to be able to provide a metabolic fingerprint of an organism (metabotyping) as an adjunct to functional genomics, and hence has applications in the design of drug clinical trials and for evaluation of genetically modified animals as disease models. The potential and real impact of metabonomics on all stages of pharmaceutical R&D is encapsulated in Figure 4.

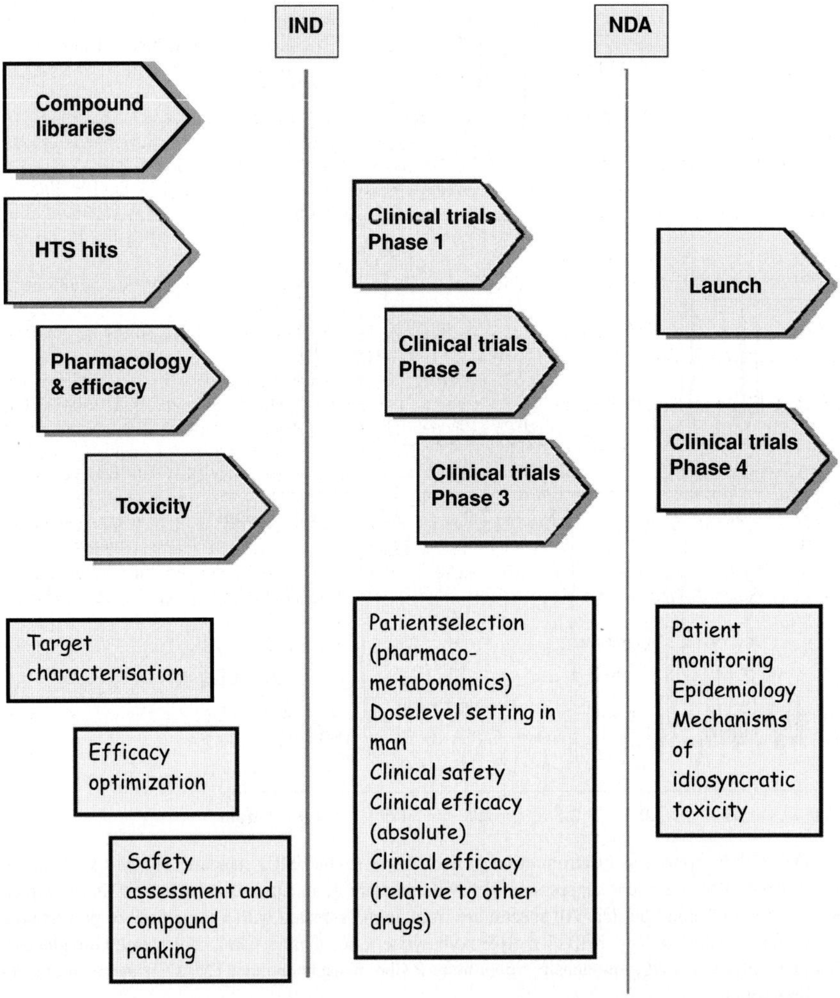

Fig. 4. The role of metabonomics in the various stages of pharmaceutical R&D. IND—Investigational new drug application—required before administration to man is permitted. NDA—new drug application—required before a new product launch.

Using metabonomics, it has proved possible to derive new biochemically based assays for disease diagnosis and to identify combination biomarkers for disease, which can then be used to monitor the efficacy of drugs in clinical trials. Thus, based on differences observed in metabonomic databases from control animals and from animal models of disease, diagnostic methods and biomarker combinations might be derivable in a preclinical setting. Similarly, the use of databases to derive predictive expert systems for human disease diagnosis and the effects of therapy, require compilations from both normal human populations and patients before, during, and after therapy. In human studies, metabonomics also has the potential for disease diagnosis and possibly even prognosis.

References

1. Nicholson JK, Lindon JC, Holmes E. Xenobiota. 1999;29: 1181.
2. Lindon JC, Nicholson JK, Holmes E, JR Everett. Concepts Magn. Reson. 2000;12:289.
3. Nicholson JK, Connelly J, Lindon JC Holmes E. Nat. Rev. Drug Discov. 2002;1:153.
4. Lindon JC, Holmes E Nicholson JK. Anal. Chem. 2003;75: 384A.
5. Plumb RS, Stumpf CL, Gorenstein MV, Castro-Perez JM, Dear GJ, Anthony M, Sweatman BC, Connor SC Haselden JN. Rapid Commun. Mass Spectrom. 2002;16:1991.
6. Moka D, Vorreuther R, Schicha H, Spraul M, Humpfer E, Lipinski M, Foxall PJD, Nicholson JK Lindon JC. Anal. Commun. 1997;34:107.
7. Lindon JC, Nicholson JK Everett JR. Annu. Rep. NMR Spectrosc. 1999;38:1.
8. Khandelwal P, Beyer CE, Lin Q, McGonigle P, Schechter LE, Bach C. J. Neurosci. Methods. 2004;133:181.
9. Griffin JL, Nicholls AW, Keun HC, Mortishire-Smith RJ, Nicholson JK, Kuehn T. Analyst. 2002;127:582.
10. Keun HC, Beckonert O, Griffin JL, Richter C, Moskau D, Lindon JC, Nicholson JK. Anal. Chem. 2002;74:4588.
11. Lindon JC, Nicholson JK, Wilson ID. J. Chromatogr. B. 2000;748:233.
12. Garrod SL, Humpfer E, Spraul M, Connor SC, Polley S, Connelly J, Lindon JC, Nicholson JK, Holmes E. Magn. Reson. Med. 1999;41:1108.
13. Waters NJ, Holmes E, Williams A, Waterfield CJ, Farrant RD, Nicholson JK. Chem. Res. Toxicol. 2001;14:1401.
14. Coen M, Lenz EM, Nicholson JK, Wilson ID, Pognan F, Lindon JC. Chem. Res. Toxicol. 2003;16:295.
15. Lamers RJAN, Wessels ECHH, van der Sandt JJM, Venema K, Schaafsma G, van der Greef J, van Nesselrooij JHJ. J. Nutr. 2003;133:3080.
16. Bollard ME, Xu JS, Purcell W, Griffin JL, Quirk C, Holmes E Nicholson JK. Chem. Res. Toxicol. 2002;15:1351.
17. Lindon JC, Holmes E, Nicholson JK. Prog. NMR Spectrosc. 2001;39:1.
18. Brown TR, Stoyanova R. J. Magn. Reson. 1996;112:32.
19. Farrant RD, Lindon JC, Rahr E, Sweatman BC. J. Pharm. Biomed. Anal. 1992;10:141.
20. Fung M, Thornton A, Mybeck K, Wu JH-H, Hornbuckle K, Muniz E. Drug Inf. J. 2001;35:293.
21. Lindon JC, Nicholson JK, Holmes E, Antti H, Bollard ME, Keun H, Beckonert O, Ebbels TM, Reily MD, Robertson D, Stevens GJ, Luke P, Breau AP, Cantor GH, Bible RH, Niederhauser U, Senn H, Schlotterbeck G, Sidelmann UG, Laursen SM, Tymiak A, Car BD, Lehman-McKeeman L, Colet JM, Loukaci A, Thomas C. Toxicol. Appl. Pharmacol. 2003;187:137.
22. Keun HC, Ebbels TMD, Antti H, Bollard M, Beckonert O, Schlotterbeck G, Senn H, Niederhauser U, Holme E, Lindon JC, Nicholson JK. Chem. Res. Toxicol. 2002;15:1380.
23. Ebbels T, Keun H, Beckonert O, Antti H, Bollard M, Holmes E, Lindon J, Nicholson J. Anal. Chim. Acta. 2003;490:109.
24. Lindon JC, Holmes E, Bollard ME, Stanley EG, Nicholson JK. Biomarkers. 2004;9:1–31.
25. Griffin JL, Sang E, Evens T, Davies K, Clarke K. FEBS Lett. 2002;530:109.
26. Nicholls AW, Mortishire-Smith RJ, Nicholson JK. Chem. Res. Toxicol. 2003;16:1395.
27. Coen M, Nicholson JK, Lindon JC, Lenz EM, Wilson ID, Ruepp SU, Pognan F. J. Pharm. Biomed. Anal. 2004;35:93.
28. Maxwell RJ, Martinez-Perez I, Cerdan S, Cabanas ME, Arus C, Moreno A, Capdevila A, Ferrer E, Bartomeus F, Aparicio A, Conesa G, Roda JM, Carceller F, Pascual JM, Howells SL, Mazucco R, Griffiths JR. Magn. Reson. Med. 1998;39:869.
29. Tomlins A, Foxall PJD, Lindon JC, Lynch MJ, Spraul M, Everett JR, Nicholson JK. Anal. Commun. 1998;35:113.
30. Moka D, Vorreuther R, Schicha H, Spraul M, Humpfer E, Lipinski M, Foxall PJD, Nicholson JK, Lindon JC. J. Pharm. Biomed. Anal. 1998;17:125.
31. 't Hart BA, Vogels JTWE, Spijksma G, Brok HPM, Polman C, van der Greef J. J. Neuro. Sci. 2003;212:21.
32. Brindle JT, Antti H, Holmes E, Tranter G, Nicholson JK, Bethell HWL, Clarke S, Schofield PM, McKilligin E, Mosedale DE, Grainger DJ. Nat. Med. 2002;8:1439.

Protein Misfolding Disease: Overview of Liquid and Solid-State High Resolution NMR Studies

Harald Schwalbe and Julia Wirmer

Institute for Organic Chemistry and Chemical Biology, Center for Biomolecular Magnetic Resonance, Johann Wolfgang Goethe-Universität, Marie-Curie-Strasse 11, D-60439 Frankfurt/M, Germany

Protein Misfolding Diseases

The failure of proteins to fold into their functional forms can occasionally lead to "protein misfolding" or "protein conformational" diseases. Many among the most common and debilitating of these diseases are associated with the formation of protein amyloid, an insoluble material that is deposited as fibrils or plaques in different tissues and organs of the body. Amyloid formation is known to be accelerated by a variety of cellular factors, including metal ions, such as copper and zinc, and interactions with other species, such as lipids and RNA. It is implicated in many medical conditions including Alzheimer's disease and the transmissible prion disorders. It is becoming increasingly recognized that the switch from a normal to a diseased state of the cell in protein misfolding diseases is induced by a shift in the equilibrium between different conformational and aggregation states of a polypeptide chain that are present under normal conditions. The native state [N] of a protein can be investigated in great detail using X-ray crystallography and NMR-spectroscopy. Our understanding of states other then the native one, however, is only now emerging. Liquid and solid-state high-resolution NMR spectroscopy are the major structural techniques to determine structure and dynamics of the polypeptide chain under a variety of different conditions. They describe proteins in their native state [N], in unfolded states [U], in transient intermediates [I], and in their fibrillar and prefibrillar states (Figure 1).

Natively Unfolded Proteins Involved in Protein Misfolding Diseases

Non-native and unfolded states of proteins have also come into interest recently from the observation that an axiomatic linking of the function of a protein to a persistent fold might not be general, because a number of proteins have been identified that lack intrinsic globular structure in their normal functional form [1–3]. The expressions "intrinsically unstructured" and "natively unfolded" are being used synonymously, the latter being coined by Schweers *et al.* in 1994 in the context of structural studies of the protein tau [4]. Intrinsically unstructured proteins are extremely flexible, non-compact, and reveal little if any secondary structure under physiological conditions. In 2000, the list of natively unfolded protein comprised 100 entries [5]. Natively unfolded proteins are implied in the development of a number of neurodegenerative diseases including Alzheimer's disease (deposition of amyloid-β, τ-protein, α-synuclein), Down's syndrome and Parkinson disease to name a few [5]. They are predicted to be ubiquituous in the proteome [6,7] and algorithms available as a web-program (http://dis.embl.de/) have been developed to predict protein disorder [8]. According to the predictions, 35–51% of eucaryotic proteins have at least one long (>50 residues) disordered region and 11% of proteins in Swiss-Prot and between 6 and 17% of proteins encoded by various genomes are probably fully disordered [6]. Proteins predicted to be intrinsically unstructured show low compositional complexity. These regions sometimes correspond to repetitive structural units in fibrillar proteins. Therefore, it does not seem unlikely that lack of structure of the polypeptide chains in some states of a protein plays an important role in the development of fibrillar states and this further supports the importance for detailed structural and dynamic investigations of non-native states of proteins. It has also been noted by Gerstein that the average genomially encoded protein is significantly different in terms of size and amino acid composition from folded proteins in the PDB [9]. This difference would indicate that the structures deposited in the PDB are not random and in turn that they cannot be taken as representative for the entire structural diversity of polypeptide chains.

Brief Background in NMR Parameters

NMR is able to provide both dynamic and structural information about proteins in a variety of different states at atomic resolution. NMR has the potential for probing residual structure, the size of aggregating molecules, and variation in the internal dynamical properties on the basis of diffusion-weighted NMR spectroscopy, heteronuclear relaxation measurements, paramagnetic enhancement of

Graham A. Webb (ed.), Modern Magnetic Resonance, 1369–1373.

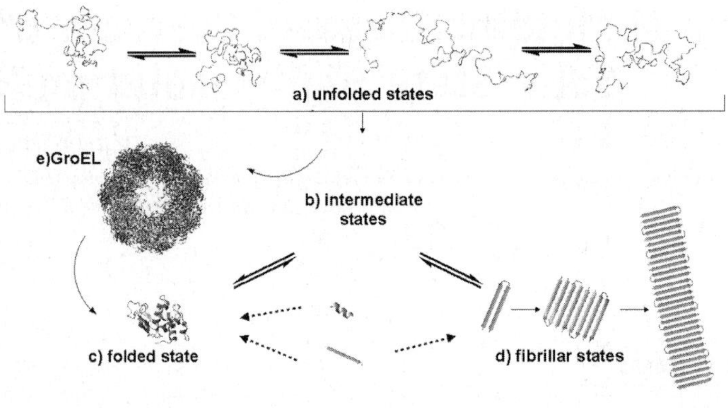

Fig. 1. The characteristics of states accessible to a protein differ widely in their structural and kinetic properties. The different conformations adopted by a polypeptide chain can range from the native, often monomeric state (c), in which a single conformation exists and which is built up from secondary structure elements and their specific arrangements, to the ensemble of conformers representing the random coil state of a protein (a). The individual members of this ensemble have widely different compaction, dynamics, local, and non-local conformations. Protein folding preceeds via formation of folding intermediates (b) whose structure and dynamics may be modulated by protein–protein interactions with molecular chaperones such as GroEL (scaled down by a factor of 2) (e) shown in the figure. At the other extreme of conformational states, proteins can aggregate and form oligomeric states called fibrils (d). Liquid and solid-state NMR spectroscopy can provide detailed information on structure and dynamics in all of these states.

relaxation induced by paramagnetic spin labels, and residual dipolar couplings (summarized in Table 1). More recently, NMR techniques have been developed to characterize the structural transitions from unfolded states of protein, via low molecular oligomers to fibrils. Such studies include both static and time-resolved experiments.

Proteins Involved in Misfolding Diseases Studied by NMR

Two proteins intimately linked to neurodegenerative illnesses seem to have a possible function in copper homeostasis: the amyloid precursor protein (APP) of Alzheimer's disease, and the prion protein (PrP), which causes transmissible spongiform encephalopathies (TSE). APP gives rise to the amyloid Aβ peptide and the prion protein PrP^{c} can convert to the variant isoform PrP^{Sc} of Creutzfeldt-Jakob's disease (CJD). The protein Cu,Zn dismutase (SOD) is a metalloprotein for which familial mutants are linked to familial amyotrophic lateral sclerosis (FALS). The three proteins are known to interact: The prion protein specifically binds copper and native PrP charged with copper has superoxide dismutase (SOD1) activity. In turn, overexpression of APP alters brain copper content and attenuates SOD1. On the other hand,

Table 1: Overview of NMR parameters and their conformational dependence

NMR parameter	Conformational dependence
Chemical shift δ (ppm)	Multiple torsion angles: ϕ, ψ, ω, χ_1
Scalar couplings ^{n}J (Hz)	Single torsion angles via Karplus equations
Homonuclear NOEs (a.u.)	Distances, dependence on correlation time, and motional properties
Heteronuclear relaxation (Hz)	Motional properties, dependence on τ_c, S^2, τ_e [a]
Residual dipolar couplings RDC (Hz)	Overall shape, dynamics, S [b]
H/D exchange	Exchangeable H^N
Diffusion	Radius of hydration (R_h)
Photo CIDNP	Accessible Trp, Tyr, His

[a] τ_c correlation time for overall rotational tumbling, S^2 order parameter of local dynamics, τ_e correlation time for local dynamics
[b] S order parameter for local dynamics

for the other systems discussed here, α-synuclein and transthyretin, no metal induced misfolding is currently being discussed in the literature.

Amyloid Precursor Protein

Cleavage of APP by β- and γ-secretases results in amyloid forming peptides containing 39–43 residues referred to as the Amyloid β-peptide (Aβ). The Aβ is the major constituent of the plaques in the brains of Alzheimer patients. The presence of copper inhibits Aβ production and stimulates the non-amyloidogenic pathway of APP in cells and in transgenic animals overexpressing APP. The soluble and the fibrillar form of various peptides of Aβ have been studied using liquid and solid-state NMR spectroscopy: conformations in the soluble form range from α-helical conformations via random coil to β-sheet conformations, depending on solvent condititions, pH values and peptide concentration. In membrane mimicking environments such as SDS, the N-terminus is unstructured (residues 1–14) while α-helical conformation is found from residue 15–36 with a kink from residues 25–27 [10–13]. A high propensity for aggregation is found in solution and the conformation in solution is often found to be in an equilibrium between random coil conformation [14] and β-sheet conformers [15] with increased β-sheet propensity at high concentration, high temperature and high salt conditions. However, at very low concentrations and at 0 °C, a conformational equilibrium between a left handed 3(10) helix and random coil conformations is found [16]. Solid-state NMR studies revealed the structure of the amyloid fibrils formed by Aβ (1–40): approximately the first 10 residues of the peptide are disordered in the fibril, residues 12–24 and 30–40 adopt β-strand conformations and form parallel β-sheets through intermolecular hydrogen bonding. The two β-sheets within a single peptide unit form an antiparallel sheet [17]. The structural model found for the Aβ(1–40) is very similar to structural models of smaller (11–25, 16–22) and larger (1–42) peptides [18,19].

Prion Protein

Prion protein (PrP^c) is a glycosyl-phosphatidyl-inositol (GPI) anchored membrane protein of 35 kD. Its conformational altered scrapie form (PrP^{Sc}) is associated with neurodegenerative diseases. Cellular PrP^c is synthesized in three topologic forms, secreted ^{sec}PrP and the single-spanning membrane proteins of opposite orientations ^{Ntm}PrP, and ^{Ctm}PrP. Increases in ^{Ctm}PrP are associated with the development of neurodegenerative disease. The conformational properties of a protein are generally thought to be determined by specific elements in the sequence; for the prion protein these are the N-terminal sequence, a hydrophobic stretch around residue 100, a putative transmembrane domain, and the C-terminus for GPI anchor addition. The signal sequence plays a decisive role in determining the membrane orientation of the PrP. ^{Ctm}PrP contains an uncleaved N-terminal signal peptide.

NMR solution structures have been reported for PrP^c from several species [20–31]. Interestingly, it was shown that only few of the disease related mutations lead to reduced stability of N [26]. Residues 23–121 comprise a flexibly disordered tail in contrast to the well-ordered 125–228 core part. Cu^{2+} binding to histidines located in the flexible part of the protein (high affinity octarepeat regions are found between residues 51–90 and low affinity binding sites are formed by His96 and His111) induces a shift towards β-sheet conformation as shown by liquid-state NMR [32–36]. In particular, the conformational shift induced by binding of His96 and His111 is remarkable, as these residues are part of the infectious region (residues 90–231) [24]. β-sheet conformation has also been observed by liquid and solid-state NMR studies on fibrillar states: a hydrogen exchange study of the fibrillar form of the 106–121 peptide revealed 50% β-sheet structure, which is located in the center portion of the peptide [37]. Complete β-sheet structure was found by solid-state NMR investigation on the fibrillar state of a peptide corresponding to the infectious P101L mutant of residues 89–143 [38].

α-Synuclein

α-synuclein is associated with Parkinson's disease, a neurodegenerative disorder affecting some five million people worldwide. This disease is characterized by the formation of proteinaceous inclusions called Lewy bodies that are known to be composed primarily of aggregated α-synuclein. There is mounting evidence that the oligomeric species that are formed during the process of aggregation of the monomeric α-synuclein are at the origin of the pathology. In solution, α-synuclein can be either found in a monomeric, substantially unfolded conformation, or it can form several morphologically different types of aggregates, including oligomers, amorphous aggregates, and amyloid-like fibrils. Its normal function is poorly understood, but it is believed that it includes transient or reversible lipid interactions [39]. NMR studies of the lipid bound form of α-synuclein revealed that the C-terminus of the protein remains free in solution, while the N-terminus of the protein binds to the lipids in a helical conformation [40–43]. This helical conformation in the lipid bound state remains nearly unchanged in the early onset mutations (A30P, A53T) [43]. More interesting results were found by NMR studies on the natively unfolded monomeric form free in solution: the protein is more compact than expected for a highly denatured protein [44] and residual helical structure is present in the N-terminal part of α-synuclein (residues 18–31) [40,45,46]. This residual helical structure is diminished in one of the early onset mutants A30P, while enhanced β-sheet propensity around the mutation site in the other early onset mutant A53T is observed [45]. Changes in these parts of the sequence (residues 22–93) were also observed upon binding of aggregation promoting polyamines at the C-terminal part (residues 109–140) [47]. Taking all NMR investigations on the monomeric form in solution together, clearly the N-terminal part of the protein is the aggregation nucleus.

Cu-Zn-Superoxide Dismutase

Mutations in the gene for cytosolic SOD are linked with familial FALS. The active site of each monomer of SOD (153 residues) contains one copper and one zinc ion. The enzyme catalyzes the dismutation of superoxide to dioxygen and hydrogen peroxide. FALS-related mutations are concentrated either in the metal-binding loop or in loops III (38–40) and V (90–93). In this protein, metal ions have a fundamental role for protein stabilization, correct folding, and in increasing protein rigidity. Binding of the metal ions is of fundamental importance to stabilize its secondary structure and to define the tertiary interactions relevant for the relative orientation of the two sheets forming as can be shown by a comparison of the NMR strucutures in the presence and the absence of metal ions [48,49]. A comparison of WT SOD and the FALS-related mutant G93A based on NMR dynamics revealed that indeed a in the β-barrel loops III and V are a lot more dynamic and thus less stable in the mutant than in the wild type proteins [50].

Transthyretin

Transthyretin (TTR) is a homotetrameric protein that is involved in the transport of thyroid hormones and retinol in human serum. A large number of mutations (more than 80 different ones) have been found that lead to misfolded forms of the protein. These misfolded forms are implicated in amyloid diseases such as familial amyloidotic polyneuropathy and senile systemic amyloidosis. A combination of hydrogen exchange experiments coupled with NMR detection have compared WT TTR under native conditions and under conditions that are known to promote amyloid formation: These studies revealed that amyloidogenic behavior is linked to destabilization of one half of the β-sandwich structure of TTR [51]. These findings were further confirmed by additional NMR studies of some of the amyloidogenic mutants [52,53]. Liquid-state NMR studies of amyloidogenic peptides (TTR10–20 and TTR105–115) showed that the peptides are unstructured in solution [54,55]. In contrast, the TTR105–115 peptide in its fibrillar state is in an extended β-sheet conformation, with its backbone and side chain torsion angles close to their optimal values for this secondary structure element. In addition, long-range order could be detected that is generally associated with crystalline materials by solid-state NMR [56,57].

References

1. Dyson HJ, Wright PE. Curr. Opin. Struct. Biol. 2002;12:54.
2. Wright PE. Dyson HJ. J. Mol. Biol. 1999;293:321.
3. Uversky VN. Eur. J. Biochem. 2002;269:2.
4. Schweers O, Schonbrunn-Hanebeck E, Marx A, Mandelkow E. J. Biol. Chem. 1994;269:24290.
5. Uversky VN, Gillespie JR, Fink AL. Proteins. 2000;41:415.
6. Tompa P. Trends Biochem. Sci. 2002;27:527.
7. Uversky VN. Protein Sci. 2002;11:739.
8. Linding R, Jensen LJ, Diella F, Bork P, Gibson TJ, Russell RB. Structure (Camb.) 2003;11:1453.
9. Gerstein M. Fold Des. 1998;3:497.
10. Coles M, Bicknell W, Watson AA, Fairlie DP, Craik DJ. Biochemistry 1998;37:11064.
11. Shao H, Jao S, Ma K, Zagorski MG. J. Mol. Biol. 1999;285:755.
12. D'Ursi AM, Armenante MR, Guerrini R, Salvadori S, Sorrentino G, Picone D. J. Med. Chem. 2004;47:4231.
13. Crescenzi O, Tomaselli S, Guerrini R, Salvadori S, D'Ursi AM, Temussi PA, Picone D. Eur. J. Biochem. 2002;269:5642.
14. Riek R, Guntert P, Dobeli H, Wipf B, Wuthrich K. Eur. J. Biochem. 2001;268:5930.
15. Jarvet J, Damberg P, Bodell K, Eriksson LEG, Graslund A. J. Amer. Chem. Soc. 2000;122:4261.
16. Jarvet J, Damberg P, Danielsson J, Johansson I, Eriksson LE, Graslund A. FEBS Lett. 2003;555:371.
17. Petkova AT, Ishii Y, Balbach JJ, Antzutkin ON, Leapman RD, Delaglio F, Tycko R. Proc. Natl. Acad. Sci. U.S.A. 2002;99:16742.
18. Petkova AT, Buntkowsky G, Dyda F, Leapman RD, Yau WM, Tycko R. J. Mol. Biol. 2004;335:247.
19. Antzutkin ON, Leapman RD, Balbach JJ, Tycko R. Biochemistry. 2002;41:15436.
20. Riek R, Hornemann S, Wider G, Billeter M, Glockshuber R, Wuthrich K. Nature, 1996;382:180.
21. Glockshuber R, Hornemann S, Riek R, Wider G, Billeter M, Wuthrich K. Trends Biochem. Sci. 1997;22:241.
22. Billeter M, Riek R, Wider G, Hornemann S, Glockshuber R, Wuthrich K. Proc. Natl. Acad. Sci. U.S.A. 1997;94:7281.
23. Riek R, Hornemann S, Wider G, Glockshuber R, Wuthrich K. FEBS Lett. 1997;413:282.
24. James TL, Liu H, Ulyanov NB, Farr-Jones S, Zhang H, Donne DG, Kaneko K, Groth D, Mehlhorn I, Prusiner SB, Cohen FE. Proc. Natl. Acad. Sci. U.S.A. 1997;94:10086.
25. Donne DG, Viles JH, Groth D, Mehlhorn I, James TL, Cohen FE, Prusiner SB, Wright PE, Dyson HJ. Proc. Natl. Acad. Sci. U.S.A. 1997;94:13452.
26. Riek R, Wider G, Billeter M, Hornemann S, Glockshuber R, Wuthrich K. Proc. Natl. Acad. Sci. U.S.A. 1998;95:11667.

27. Liu H, Farr-Jones S, Ulyanov NB, Llinas M, Marqusee S, Groth D, Cohen FE, Prusiner SB, James TL. Biochemistry. 1999;38:5362.
28. Zahn R, Liu A, Luhrs T, Riek R, von Schroetter C, Lopez Garcia F, Billeter M, Calzolai L, Wider G, Wuthrich K. Proc. Natl. Acad. Sci. U.S.A. 2000;97:145.
29. Lopez Garcia F, Zahn R, Riek R, Wuthrich K. Proc. Natl. Acad. Sci. U.S.A. 2000;97:8334.
30. Calzolai L, Lysek DA, Guntert P, von Schroetter C, Riek R, Zahn R, Wuthrich K. Proc. Natl. Acad. Sci. U.S.A. 2000;97: 8340.
31. Zahn R, Guntert P, von Schroetter C, Wuthrich K. J. Mol. Biol. 2003;326:225.
32. Stockel J, Safar J, Wallace AC, Cohen FE, Prusiner SB. Biochemistry. 1998;37:7185.
33. Viles JH, Cohen FE, Prusiner SB, Goodin DB, Wright PE, Dyson HJ. Proc. Natl. Acad. Sci. U.S.A. 1999;96:2042.
34. Jackson GS, Murray I, Hosszu LL, Gibbs N, Waltho JP, Clarke AR, Collinge J. Proc. Natl. Acad. Sci. U.S.A. 2001;98:8531.
35. Jones CE, Abdelraheim SR, Brown DR, Viles JH. J. Biol. Chem. 2004;279:32018.
36. Belosi B, Gaggelli E, Guerrini R, Kozlowski H, Luczkowski M, Mancini FM, Remelli M, Valensin D, Valensin G. Chembiochem. 2004;5:349.
37. Kuwata K, Matumoto T, Cheng H, Nagayama K, James TL, Roder H. Proc. Natl. Acad. Sci. U.S.A. 2003;100:14790.
38. Laws DD, Bitter HM, Liu K, Ball HL, Kaneko K, Wille H, Cohen FE, Prusiner SB, Pines A, Wemmer DE. Proc. Natl. Acad. Sci. U.S.A. 2001;98:11686.
39. Clayton DF, George JM. Trends Neurosci. 1998;21:249.
40. Eliezer D, Kutluay E, Bussell R Jr, Browne G. J. Mol. Biol. 2001;307:1061.
41. Bussell R Jr, Eliezer D. J. Mol. Biol. 2003;329:763.
42. Chandra S, Chen X, Rizo J, Jahn R, Sudhof TC. J. Biol. Chem. 2003;278:15313.
43. Bussell R Jr, Eliezer D. Biochemistry. 2004;43:4810.
44. Morar AS, Olteanu A, Young GB, Pielak GJ. Protein Sci. 2001;10:2195.
45. Bussell R Jr, Eliezer D. J. Biol. Chem. 2001;276:45996.
46. Yao J, Chung J, Eliezer D, Wright PE, Dyson HJ. Biochemistry. 2001;40:3561.
47. Fernandez CO, Hoyer W, Zweckstetter M, Jares-Erijman EA, Subramaniam V, Griesinger C, Jovin TM. Embo. J. 2004; 23:2039.
48. Assfalg M, Banci L, Bertini I, Turano P, Vasos PR. J. Mol. Biol. 2003;330:145.
49. Banci L, Bertini I, Cramaro F, Del Conte R, Viezzoli MS. Biochemistry. 2003;42:9543.
50. Shipp EL, Cantini F, Bertini I, Valentine JS, Banci L. Biochemistry. 2003;42:1890.
51. Liu K, Cho HS, Lashuel HA, Kelly JW, Wemmer DE. Nat. Struct. Biol. 2000;7:754.
52. Liu K, Kelly JW, Wemmer DE. J. Mol. Biol. 2002;320: 821.
53. Niraula TN, Haraoka K, Ando Y, Li H, Yamada H, Akasaka K. J. Mol. Biol. 2002;320:333.
54. Jarvis JA, Craik DJ. J. Magn. Reson. B. 1995;107:95.
55. Jarvis JA, Kirkpatrick A, Craik DJ. Int. J. Pept. Protein. Res. 1994;44:388.
56. Jaroniec CP, MacPhee CE, Astrof NS, Dobson CM, Griffin RG. Proc. Natl. Acad. Sci. U.S.A. 2002;99:16748.
57. Jaroniec CP, MacPhee CE, Bajaj VS, McMahon MT, Dobson CM, Griffin RG. Proc. Natl. Acad. Sci. U.S.A. 2004;101: 711.

^{19}F NMR Spectroscopy for Functional and Binding High-Throughput Screening

Marina Veronesi and Claudio Dalvit
Chemistry Department, Nerviano Medical Sciences, 20014 Nerviano, Milan, Italy

NMR screening has emerged as a powerful and reliable approach for identification of potential drug candidates [1]. The technique is now recognized for its impact on the drug discovery process and has become an important tool for lead identification, lead validation, and lead optimization in many pharmaceutical companies and universities [2–17]. A plethora of different NMR experiments has been proposed in the literature for performing these tasks. Two of these approaches, recently introduced, use fluorine NMR spectroscopy. FAXS (fluorine chemical shift anisotropy and exchange for screening) [18,19] and 3-FABS (three fluorine atoms for biochemical screening) [20] allow one to perform binding and functional high-throughput screening (HTS), respectively, and determine the dissociation binding constant (K_D) and the 50% mean inhibition concentration (IC_{50}) of the identified binders and inhibitors, respectively. This chapter provides an insight into the theory and practical aspects of these two experiments and presents applications to the screening of different biomolecular targets.

FAXS

Competition ligand-based NMR screening experiments were introduced for overcoming all the limitations associated with ligand-based NMR screening experiments [21–23]. The screening of chemical mixtures or single compounds against the biomolecular target of interest is performed in the presence of a weak- to medium-affinity ligand of known binding constant referred to in its role as the spy or reporter. Changes in the transverse or selective longitudinal relaxation rates of the spy resonances are monitored. Signals from the molecules screened are not utilized. Often the screening is performed in the presence of an additional molecule that does not interact with the receptor and referred to in its role as the control molecule. This compound represents an internal reference. Competition NMR-based screening, originally proposed with proton detection experiments, was subsequently extended to fluorine detection experiments [18,19]. For these experiments it is sufficient to have a spy molecule containing either a CF or a CF_3 moiety. This approach offers some unique advantages. (i) Absence of overlap permits the screening of large chemical mixtures and automated analysis of the spectra. (ii) Protonated solvents, buffers, or detergent do not interfere with the measurements thus allowing, for example, the screening against membrane proteins. (iii) The ^{19}F transverse relaxation rate R_2, given by Equation (1), is a sensitive parameter of binding events since it contains spectral densities calculated at zero frequency for both the heteronuclear ^{19}F–^{1}H dipolar interactions and the ^{19}F chemical shift anisotropy (CSA) interaction [24]:

$$
\begin{aligned}
R_2^{F} = {} & \frac{\gamma_F^2 \gamma_H^2 \hbar^2 \tau_c}{20} \sum_{H_i} \frac{1}{r_{FH_i}^6} \left\{ 4 + \frac{1}{1+(\omega_F-\omega_H)^2\tau_c^2} \right. \\
& + \frac{3}{1+\omega_F^2\tau_c^2} + \frac{6}{1+\omega_H^2\tau_c^2} \\
& \left. + \frac{6}{1+(\omega_F+\omega_H)^2\tau_c^2} \right\} \\
& + \frac{2}{15}\Delta\sigma^2 \left(1 + \frac{\eta_{CSA}^2}{3}\right) B_0^2 \gamma_F^2 \tau_c \\
& \times \left\{ \frac{2}{3} + \frac{1}{2(1+\omega_F^2\tau_c^2)} \right\} \qquad (1)
\end{aligned}
$$

The H_i correspond to all the protons of the spy compound and of the protein close in space to the fluorine atom and r_{FH_i} is the internuclear distance between proton H_i and the fluorine atom of the spy molecule. $\Delta\sigma$ is the CSA of the ^{19}F atom and is given by $\Delta\sigma = \sigma_{zz} - (\sigma_{xx} + \sigma_{yy})/2$ where the different σs are the components of the chemical shift tensor. The asymmetry parameter η_{CSA} is given by $\eta_{CSA} = (3/2)(\sigma_{xx} - \sigma_{yy})/\Delta\sigma$, B_0 is the strength of the magnetic field, γ_H and γ_F are the proton and fluorine gyromagnetic ratios, respectively, ω_H and ω_F are the proton and fluorine Larmor frequencies, respectively, and τ_c is the correlation time.

Owing to the large CSA of ^{19}F (as much as few hundreds ppm) it will contribute significantly, according to Equation (1), to the transverse relaxation of the fraction of bound spy molecule [18,25–28]. CSA contribution to

Graham A. Webb (ed.), Modern Magnetic Resonance, 1375–1381.

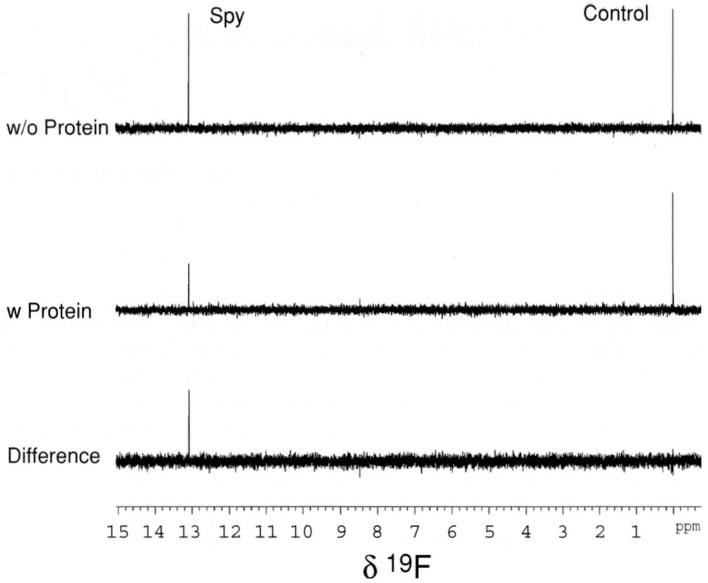

Fig. 1. One-dimensional ^{19}F NMR spin-echo ($n2\tau = 100$ ms) spectra recorded for the identification of a spy and control molecule required for FAXS. The two molecules are the weak-affinity ligand PHA-739917 and the non-interacting trifluoroacetic acid (TFA) molecule and the protein is the p21 activated kinase. The chemical shifts are referenced to TFA. The concentration of PHA-739917 and TFA was 50 and 15 μM, respectively. The spectra were recorded in the absence (top) and the presence of 1.5 μM of the protein (middle). In the difference spectrum (bottom) only the signal of the spy molecule is visible. Reprinted with permission from Dalvit *et al.* [18]. © 2002 Bentham Science Publishers.

transverse relaxation is directly proportional to the square of the magnetic field thus resulting in a more pronounced effect at stronger magnetic fields.

For a weak-affinity spy molecule the observed transverse relaxation $R_{2,\mathrm{obs}}$ is given by [29,30]:

$$R_{2,\mathrm{obs}} = \frac{[\mathrm{EL}]}{[\mathrm{L_{TOT}}]} R_{2,\mathrm{bound}} + \left(1 - \frac{[\mathrm{EL}]}{[\mathrm{L_{TOT}}]}\right) R_{2,\mathrm{free}} + \frac{[\mathrm{EL}]}{[\mathrm{L_{TOT}}]}\left(1 - \frac{[\mathrm{EL}]}{[\mathrm{L_{TOT}}]}\right)^2 \times \frac{4\pi^2(\delta_{\mathrm{free}} - \delta_{\mathrm{bound}})^2}{K_{-1}} \quad (2)$$

where $[\mathrm{EL}]/[\mathrm{L_{TOT}}]$ and $(1 - [\mathrm{EL}]/[\mathrm{L_{TOT}}])$ are the fraction of bound and free ligand, respectively, $R_{2,\mathrm{bound}}$ and $R_{2,\mathrm{free}}$ are the transverse relaxation rate constants for the ligand in the bound and free states, respectively. The last term in Equation (2) is the exchange term where δ_{bound} and δ_{free} are the isotropic chemical shifts of the fluorine resonance of the spy molecule in the bound and free states, respectively, and $1/K_{-1}$ is the residence time τ_{res} of the spy molecule bound to the protein.

Typically, a Carr–Purcell–Meibom–Gill (CPMG) spin-echo scheme [31,32] with a long 2τ interval between the train of 180° pulses (where $2\tau > 5\tau_{\mathrm{res}}$) [33,34] is used in these experiments before the acquisition time. This is possible because the evolution with the heteronuclear ^{1}H–^{19}F scalar couplings is refocused at the end of the scheme. However, the 2τ period should not be very long in order to minimize signal attenuation originating from the spatial diffusion of the spy molecule.

The steps required for the screening with FAXS are described below:

1. A library of ^{19}F (CF or CF_3) containing molecules well characterized, chemically stable, and with high aqueous solubility is tested in mixtures against the receptor of interest for the identification of the potential spy and control molecules. STD [35], WaterLOGSY [36], or spin-echo ^{19}F experiments [18,19,37], as the example of Figure 1, are used for the identification of the two molecules.
2. The K_D of the identified spy molecules are determined with either ITC or fluorescence spectroscopy. The spy molecule is then selected on the basis of its K_D and the presence of only one binding site.
3. After the selection of the spy and control molecules, titration experiments as a function of the protein concentration are recorded and the intensity ratio of the two fluorine signals is plotted as a function of the fraction of protein-bound spy molecule [21], as shown in the example of Figure 2. The fraction of bound compound is calculated by using its K_D value and the equation:

$$\frac{[\mathrm{EL}]}{[\mathrm{L_{TOT}}]} = \frac{[\mathrm{E_{TOT}}] + [\mathrm{L_{TOT}}] + K_D}{2[\mathrm{L_{TOT}}]} - \frac{\sqrt{([\mathrm{E_{TOT}}] + [\mathrm{L_{TOT}}] + K_D)^2 - 4[\mathrm{E_{TOT}}][\mathrm{L_{TOT}}]}}{2[\mathrm{L_{TOT}}]} \quad (3)$$

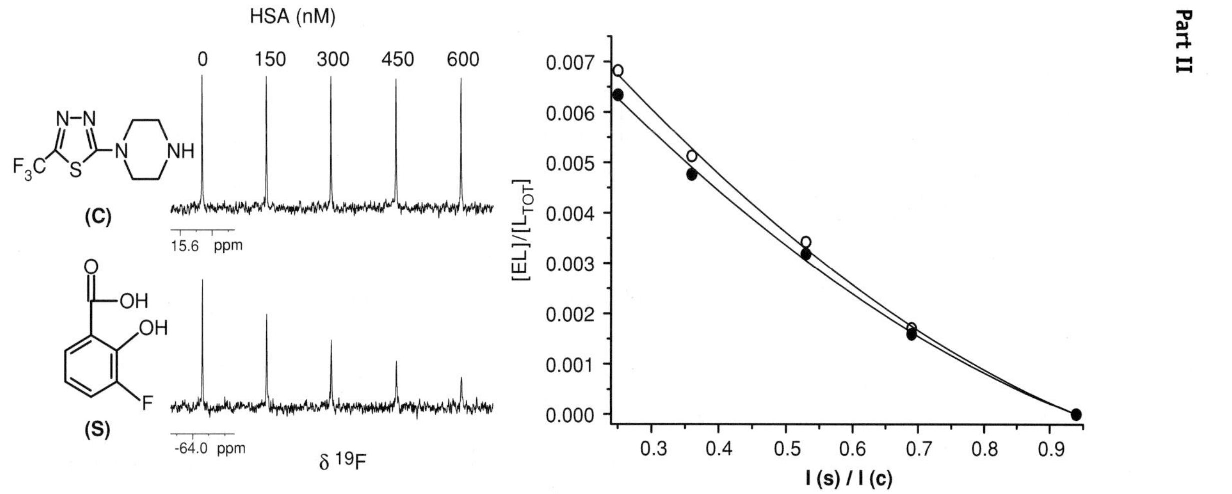

Fig. 2. (Left) One-dimensional ^{19}F NMR spin-echo spectra ($n2\tau = 80$ ms) recorded for the spy molecule 2-hydroxy 3-fluorobenzoic acid **(s)** and the control molecule 1-[5-(trifluoromethyl)1,3,4-thiadiazol-2-yl]piperazine**(c)** as a function of HSA concentration. The concentration of **(s)** and **(c)** was 50 and 25 μM, respectively. (Right) Plot of the signal intensity ratio (x-axis) of the two ^{19}F signals of **(s)** and **(c)** as a function of the fraction of bound spy molecule ([EL]/[L_{TOT}]) (y-axis). The last point on the right corresponds to the value in the absence of the protein. Two ratios ([EL]/[L_{TOT}]) were calculated using the limits of the ITC-derived K_D value of 41 ± 3.3 μM for **(s)** and using Equation (3). Values indicated by open circles were calculated with a K_D of 44.3 μM, values indicated by filled circles were calculated with a K_D of 37.7 μM. The curves represent the best fits of the experimental points. Reprinted with permission from Dalvit *et al.* [19]. © 2003 American Chemical Society.

where [E_{TOT}] and [L_{TOT}] are the total concentration of enzyme and ligand, respectively, and [EL] is the concentration of bound ligand.

4. The experimental conditions for the FAXS are then selected according to the graph of Figure 2. Compounds are tested in mixtures or as a single compound in the presence of the two molecules. Displacement of the spy molecule results in a shift of the R_2 parameter toward the intrinsic of the free state. When mixtures are used, deconvolution is then performed for the identification of the active compound. The screening with FAXS against human serum albumin (HSA) is shown in Figure 3. For screening, a total spin-echo period ($2n\tau$) is selected for which the signal of the spy molecule is approaching zero. The presence in the mixture of 5-CH_3 D, L Trp, and sucrose, known as non-binders, do not alter the spectrum of the spy molecule. In contrast, the presence in the mixture of the warfarin derivative 4-hydroxy-3-[1-(p-iodophenyl)-3-oxobutyl] coumarin (PNU-24009) results in the reappearance of the signal of the spy molecule thus identifying the NMR-hit.
5. The extent of displacement of the spy molecule is then used to calculate the binding constant K_I of the NMR-hit. Since [L_{TOT}] in the NMR screening experiments is known and fixed, the concentration of [EL] in the presence of the competing molecule can be calculated from the titration curve of Figure 2. The knowledge of [L_{TOT}], [EL], and [E_{TOT}] permits determination of the apparent dissociation binding constant K_D^{app} of the spy molecule in the presence of the competing molecule according to the equation:

$$K_D^{app} = \frac{[E_{TOT}][L_{TOT}] - [E_{TOT}][EL] + [EL]^2 - [L_{TOT}][EL]}{[EL]} \quad (4)$$

In the assumption of a simple competitive mechanism, the K_D^{app} is then used to extract the binding constant K_I of the NMR-hit according to the equation:

$$K_I = \frac{[I]K_D}{K_D^{app} - K_D} \quad (5)$$

where [I] is the concentration of the NMR-hit. The NMR-derived K_I values obtained with a single experimental point compare favorably with the values derived from full titration fluorescence or ITC measurements [18,21,38]. A major advantage of NMR applied to these measurements is the direct determination of the concentration of the NMR-hits thus providing reliable K_I values.

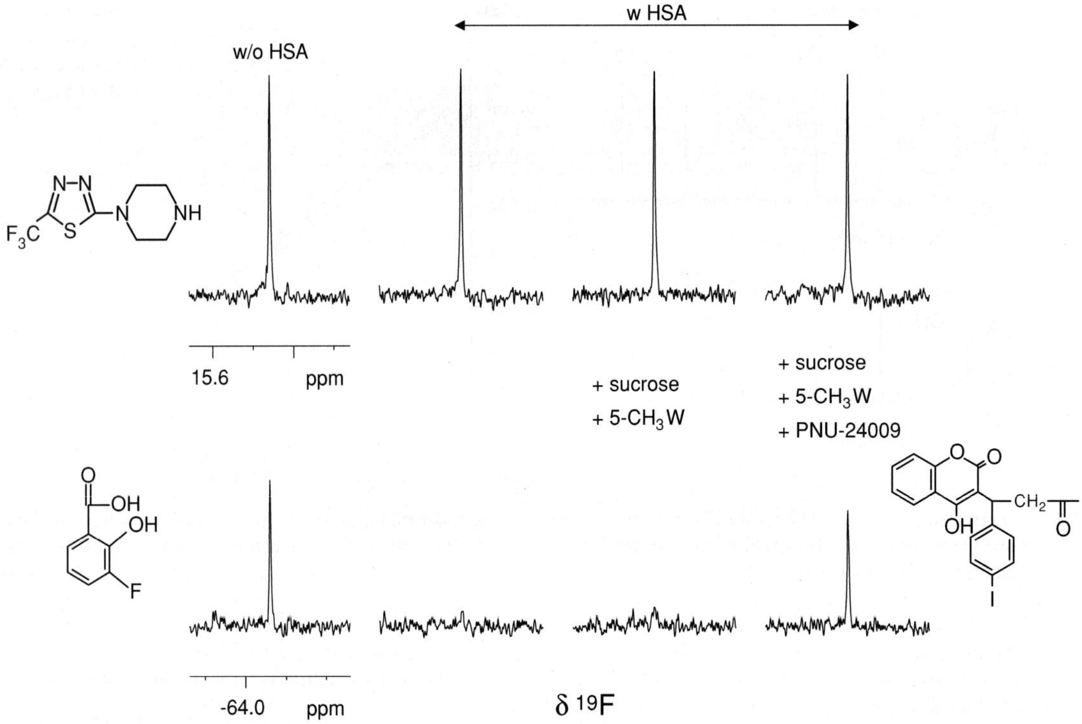

Fig. 3. FAXS performed with **(c)** (top) and **(s)** (bottom). The spectra were recorded with a total spin-echo period of 160 ms with an interval between the 180° pulses (2τ) of 40 ms. The concentration of **(s)** and **(c)** was 50 and 25 μM, respectively. The spectra on the left were recorded in the absence of protein while all the other spectra were recorded in the presence of 600 nM HSA. Reprinted with permission from Dalvit *et al.* [19]. © 2003 American Chemical Society.

3-FABS

NMR has been extensively used for characterizing the product or products of an enzymatic reaction and for gaining insight into the kinetics of the reaction (see, for example, Refs. [39,40]). A high substrate concentration was necessary for these studies due to the low sensitivity of the NMR technique. The required high concentration represents a major hurdle for the utilization of this approach to functional screening purposes, since only the very strong inhibitors would be detected. The goal in a primary screening is the identification also of weak- and medium-affinity inhibitors derived from a diversity of chemical classes.

A way to overcome the sensitivity limitation is to tag the substrate with a CF_3 moiety and use ^{19}F NMR (with proton decoupling) as the method of detection. The principle of this approach named 3-FABS [20] is described in Figure 4. The high receptivity of ^{19}F NMR spectroscopy, the 100% natural abundance of the isotope ^{19}F and the presence of three fluorine atoms result in ^{19}F NMR signals of high intensity. Modification of the substrate through the enzymatic reaction results in changes of the fluorine chemical shift tensor components even when the CF_3 moiety is distant from the reaction site. Therefore, distinct ^{19}F signals for the substrate and product (or products) are observed. The process involved with 3-FABS requires the following steps [20]:

1. Determination of the linear range (first-order region) of the enzymatic reaction. This is achieved by monitoring the course of the reaction within the NMR tube. ^{19}F spectra at different intervals are recorded and the ^{19}F signal integral of the substrate or product (products) is plotted as a function of the time elapsed from the beginning of the reaction. Screening and K_M measurements are then performed in an end point format. The reaction is quenched with either a strong inhibitor, a denaturating or a chelating molecule after a delay for which the reaction is still linear and sufficient product is formed.

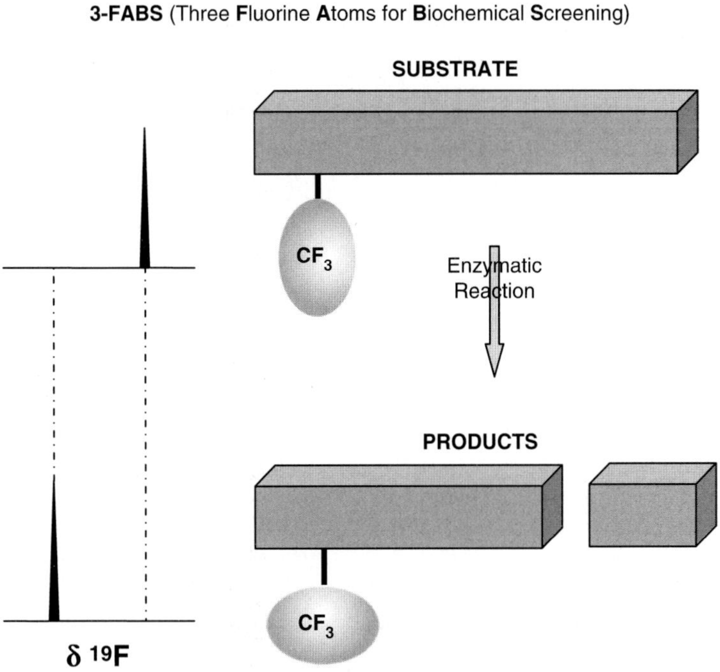

Fig. 4. Schematic diagram of the 3-FABS method. The enzyme in the example is a protease that cleaves the peptide bond with the formation of two products.

2. Determination of the K_M of the substrate and cosubstrate. This is achieved by performing the reaction at different substrate or cosubstrate concentrations. The K_M is measured using the following equation [41]:

$$V = \frac{[S]}{K_M + [S]} V_{max} \quad (6)$$

where V is the initial speed of the reaction and V_{max} is the maximum speed that would be observed if all the enzyme $[E_{TOT}]$ is complexed with the substrate ($V_{max} = K_{cat}*[E_{TOT}]$, where K_{cat} is the catalytic rate constant) and [S] is the substrate concentration. V is obtained experimentally by measuring the integral of the ^{19}F NMR signal of the product divided by the incubation time of the reaction. A plot of these values as a function of [S] permits the determination of K_M and V_{max}.

3. Screening, as shown in Figure 5, is performed in an end point format using either chemical mixtures or single compound at a substrate concentration in the range 2–5 K_M. In every screening run a sample without test molecules is recorded. This sample represents the reference with 0% inhibition. Deconvolution of the active mixtures is then performed for the identification of the inhibitors.

4. The IC_{50} value of the detected inhibitor is obtained by simply recording experiments at different inhibitor concentration and by monitoring the integral of the product or substrate ^{19}F NMR signal as shown in Figure 5. A plot of these values as a function of the inhibitor concentration allows the determination of IC_{50} according to the equations:

a) monitoring the substrate signal

$$[S_w] = \frac{[S_{w/o}] - [S_{TOT}]}{1 + ([I]/IC_{50})^n} + [S_{TOT}] \quad (7)$$

b) monitoring the product signal

$$[P_w] = \frac{[P_{w/o}]}{1 + ([I]/IC_{50})^n} \quad (8)$$

where $[S_w]$ and $[S_{w/o}]$ are given by the integrals of the substrate signal in the presence and absence of the inhibitor, respectively, $[P_w]$ and $[P_{w/o}]$ are given by the integrals of the product (or products) signal in the presence and absence of the inhibitor, respectively, and $[S_{TOT}] = [P_{w/o}] + [S_{w/o}] = [P_w] + [S_w]$. $[S_{TOT}]$ is the substrate concentration used for the experiments and it represents an internal reference, [I] is the concentration of the inhibitor, IC_{50} the concentration of the inhibitor

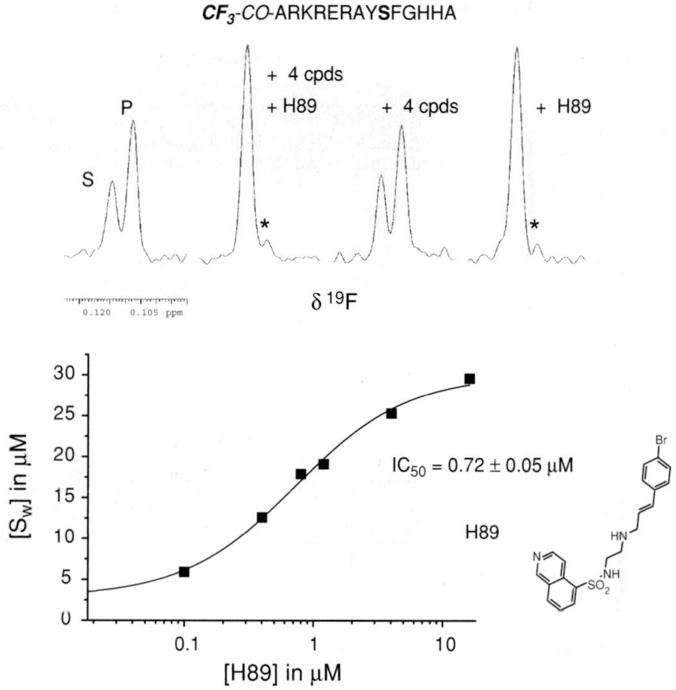

Fig. 5. Screening and deconvolution (top) and IC_{50} measurement (bottom) for compound H89 [43] performed with 3-FABS. The substrate is the N-terminal trifluoracetylated AKTide [44], the enzyme is the Ser/Thr kinase AKT1. The five compounds are (2-amino-6 methylquinazolin-4 ol, ethyl 2-quinoxalinecarboxylate, 5-methylbenzimidazole, methyl isoquinoline-3-carboxylate, and N-(2-{[(2E)-3-(4-bromophenyl)prop-2-enyl]amino}ethyl, known as H89). The activated enzyme AKT1, peptide, ATP, and compound concentrations were 25 nM, 30 μM, 131 μM (∼2 K_M), and 10 μM, respectively. S and P represent the CF_3 signal of the peptide and the phosphorylated peptide, respectively. IC_{50} was measured with Equation (7), but the same value was obtained with Equation (8). The asterisks indicate the tiny amount of phosphorylated peptide in the presence of H89. Reprinted with permission from Dalvit *et al.* [20]. © 2003 American Chemical Society.

at which 50% inhibition is observed and n is the cooperativity factor (known also as Hill slope). In the absence of allosteric effects (i.e. $n = 1$) a meaningful value for IC_{50} can be derived with a single experimental point. This is possible because the values for both plateaus are known. These are $[S_{w/o}]$ and $[S_{TOT}]$ if Equation (7) is used and $[P_{w/o}]$ and the 0 value if Equation (8) is used.

Typically, enzyme concentration in the low nanomolar is used, but if the reaction is fast (substrates with large K_{cat}/K_M) and the enzyme is stable at room temperature for few hours the concentration can be further reduced [42]. This can be appreciated in the example of Figure 6 where the screening against the protease trypsin is performed with only 50 pM enzyme concentration. With cryoprobe technology optimized to fluorine NMR detection it will be possible, in some fortunate cases, to perform 3-FABS at enzyme concentration in the femtomolar range.

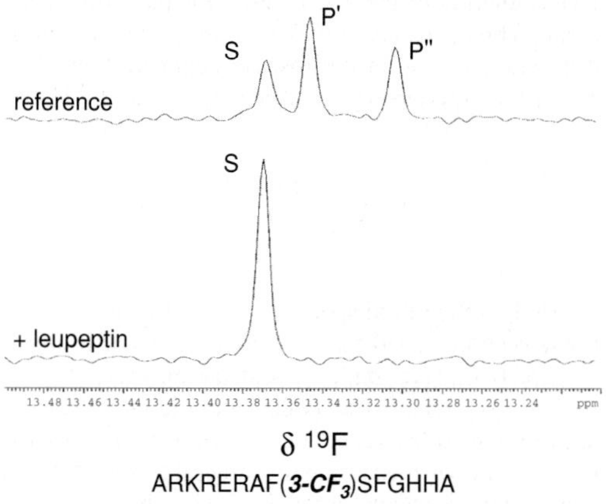

Fig. 6. 3-FABS performed with the protease trypsin (Roche Molecular Biochemical Cat. No. 1418475) at an enzyme concentration of only 50 pM. The substrate concentration was 30 μM. The reaction was performed in 50 mM Tris pH 7.5 and 0.03% Triton X-100 (Sigma X-100) at 20 °C and quenched after 6:40 hours with 0.5 mM phenylmethylsulfonyl fluoride (PMSF). S is the original peptide and P′, P″ are two different products of the enzymic hydrolysis. In the presence of 20 μM leupeptin no product formation is observed. Reprinted with permission from Dalvit *et al.* [42]. © 2004 Elsevier Science Ltd.

Conclusion

The fluorine NMR approaches FAXS and 3-FABS represent powerful tools for performing HTS and determining the K_D and IC_{50} values of the identified hits. The low protein consumption required by these methodologies compares favorably with the concentration used with the other techniques used in HTS. In addition, the two NMR techniques have some important advantages. Their simplicity together with the possibility of directly monitoring the real concentration, purity, stability, and solubility of the screened compounds results in reliable lead molecule detection and precise quantification of their strength. It is envisioned that the speed and easy set-up of FAXS and 3-FABS together with their broad range of applications will play a major role in the drug discovery process for discovering potent, bioavailable, and safe clinical candidates.

References

1. Shuker SB, Hajduk PJ, Meadows RP, Fesik SW. Science. 1996;274:1531.
2. Moore JM. Biopolymers (Peptide Science). 1999;51:221.
3. Hajduk PJ, Meadows RP, Fesik SW. Q. Rev. Biophys. 1999;32:211.
4. Roberts GCK. Drug Discov. Today. 2000;5:230.
5. Ross A, Senn H. Drug Discov. Today. 2001;11:583.
6. Peng JW, Lepre CA, Fejzo J, Abdul-Manan N, Moore JM. Methods Enzymol. 2001;338:202.
7. Diercks T, Coles M, Kessler H. Curr. Opin. Chem. Biol. 2001;5:285.
8. Pellecchia M, Sem DS, Wüthrich K. Nat. Rev. Drug Discov. 2002;1:211.
9. Van Dongen M, Weigelt J, Uppenberg J, Schultz J, Wikström M. Drug Discov. Today. 2002;7:471.
10. Wyss D, McCoy MA, Senior MM. Curr. Opin. Drug Discov. Dev. 2002;5:630.
11. Stockman BJ, Dalvit C. Prog. NMR Spectrosc. 2002;41: 187.
12. Zartler ER, Yan J, Mo H, Kline AD, Shapiro MJ. Curr. Top. Med. Chem. 2003;3:25.
13. Fielding L. Curr. Top. Med. Chem. 2003;3:39.
14. Meyer B, Peters T. Angew. Chem. Int. Ed. 2003;42:864.
15. Coles M, Heller M, Kessler H. Drug Discov. Today. 2003;8:803.
16. Salvatella X, Giralt E. Chem. Soc. Rev. 2003;32:365.
17. Jahnke W, Widmer H. Cell. Mol. Life Sci. 2004;61: 580.
18. Dalvit C, Flocco M, Veronesi M, Stockman BJ. Comb. Chem. HTS. 2002;5:605.
19. Dalvit C, Fagerness PE, Hadden DTA, Sarver RW, Stockman BJ. J. Am. Chem. Soc. 2003;125:7696.
20. Dalvit C, Ardini E, Flocco M, Fogliatto GP, Mongelli N, Veronesi M. J. Am. Chem. Soc. 2003;125:14620.
21. Dalvit C, Flocco M, Knapp S, Mostardini M, Perego R, Stockman BJ, Veronesi M, Varasi M. J. Am. Chem. Soc. 2002;124:7702.
22. Jahnke W, Floersheim P, Ostermeier C, Zhang X, Hemmig R, Hurth K, Uzunov DP. Angew. Chem. Int. Ed. 2002;41: 3420.
23. Siriwardena AH, Tian F, Noble S, Prestegard JH. Angew. Chem. Int. Ed. 2002;41:3454.
24. Canet D. Nuclear Magnetic Resonance Concepts and Methods. John Wiley & Sons: Chichester, 1996.
25. Hull WE, Sykes BD. J. Mol. Biol. 1975;98:121.
26. Gerig JT. Methods Enzymol. 1989;177:3.
27. Gerig JT. Prog. NMR Spectrosc. 1994;26:293.
28. London RE, Gabel SA. J. Am. Chem. Soc. 1994;116:2570.
29. Lian LY, Barsukov IL, Sutcliffe MJ, Sze KH, Roberts GCK. Methods Enzymol. 1994;239:657.
30. Craik DJ, Wilce JA. In: DG Reid (Ed). Protein NMR Techniques. Humana Press Inc.: New Jersey, 1997, pp 195–232.
31. Carr HY, Purcell EM. Phys. Rev. 1954;94:630.
32. Meiboom S, Gill D. Rev. Sci. Instrum. 1958;29:688.
33. Luz Z, Meiboom S. J. Chem. Phys. 1963;39:366.
34. Allerhand A, Gutowsky HS. J. Chem. Phys. 1964;41: 2115.
35. Mayer M, Meyer B. Angew. Chem. Int. Ed. 1999;38:1784.
36. Dalvit C, Pevarello P, Tatò M, Veronesi M, Vulpetti A, Sundström M. J. Biomol. NMR. 2000;18:65.
37. Tengel T, Fex T, Emtenas H, Almqvist F, Sethson I, Kihlberg J. Org. Biomol. Chem. 2004;2:725.
38. Doerr AJ, Case MA, Pelczer I, McLendon GL. J. Am. Chem. Soc. 2004;126:4192.
39. Percival MD, Withers SG. Biochemistry. 1992;31:505.
40. Evans JNS. Biomolecular NMR Spectroscopy. Oxford University press, New York, USA, 1995, pp 237–340.
41. Segel IH. Biochemical Calculations. John Wiley & Sons, New York, USA, 1976.
42. Dalvit C, Ardini E, Fogliatto GP, Mongelli N, Veronesi M. Drug Discov. Today. 2004;9:595.
43. Reuveni H, Livnah N, Geiger T, Klein S, Ohne O, Cohen I, Benhar M, Gellerman G, Levitzki A. Biochemistry. 2002;41:10304.
44. Obata T, Yaffe MB, Leparc GG, Piro ET, Maegawa H, Kashiwagi A, Kikkawa R, Cantley LC. J. Biol. Chem. 2000;46:36108.

Applications of Receptor-Based NMR Screening in Drug Discovery

Philip J. Hajduk
Global Pharmaceutical Research and Development, Abbott Laboratories, Abbott Park, IL 60064, USA

Introduction

High-throughput screening (HTS) of large corporate compound libraries has become the primary strategy for lead generation in the pharmaceutical industry. However, despite the advances in screening and chemistry technologies over the last decade, there has been little increase in the rate of discovery of quality drug leads [1,2]. One of the primary reasons for this problem is the nature of the hits that come from traditional HTS campaigns. Most hits tend to look "drug-like," with molecular weights <500 and $C\log P$ values <5, but the optimization process tends to increase both size and hydrophobicity—leading to compounds with significantly poorer drug-like properties [3]. An additional problem is the large numbers of false positives that potentially come from HTS campaigns, confounding lead triage and the identification of quality leads that act via the intended mechanism of action [4].

Recently, NMR-based screening has become a powerful complement to traditional HTS technologies to identify lead compounds that have high potential for further optimization. Unlike conventional HTS, most NMR-based screening applications focus on the identification of low molecular weight, low-affinity compounds from which high-affinity drug candidates can be constructed. In addition, NMR screening has been particularly successful when closely integrated with other drug discovery technologies, including HTS, X-ray crystallography, and high-throughput organic synthesis. There have been many excellent reviews on the applications as well as the theoretical and experimental aspects of NMR-based screening [5–14]. In this review, the rationale, advantages, and applications of fragment-based screening will be presented, with particular emphasis on receptor-based methods for NMR screening.

Fragment-Based Screening: Identifying "Hot Spots" on Protein Surfaces

Over the last decade, it has become widely recognized that the interaction energy of a protein with its natural ligand or receptor is not necessarily evenly distributed over the binding surface, but instead can be highly localized to specific regions of the protein surface called "hot spots." This was first recognized in mutational analyses of protein–protein interactions [15], but has also been extended to the study of protein–ligand interactions [16]. These studies imply that there are relatively small regions of the protein surface that impart the majority of the free energy of binding with the ligand, while additional interaction surface primarily serves to modulate specificity. This has significant implications for drug design, in that the rapid identification and optimal utilization of the energetic focal point of the binding site can accelerate the discovery of potent molecules with good lead-like or drug-like properties [2]. Unfortunately, the large compounds that come from typical HTS campaigns tend to interact sub-optimally with the hot spot, while gaining potency by interacting with peripheral sites in the binding pocket that have little potential for further optimization (Figure 1a). This can confound subsequent optimization, as the region of the compound that interacts with the hot spot (which has the greatest potential for increasing binding affinity) is initially unknown and must be discovered through several rounds of blind chemical modification. Unfortunately, even after this "core" piece is discovered, it is often difficult to modify in the context of the lead molecule without incurring unacceptable losses in potency.

As an alternative to pursuing relatively high molecular weight, sub-optimal leads, fragment-based screening is becoming a powerful tool in the drug discovery arena. In fragment-based screens, low molecular weight (typically 100–300 Da) compounds that interact optimally with the hot spot of the active site are initially pursued. By targeting only a small region of the binding site, the chances of identifying a compound that interacts optimally in the pocket are increased dramatically [17]. In addition, because of the reduced complexity of the protein–ligand interface, the number of compounds that need to be screened is drastically reduced. Typical fragment libraries reported in the literature tend to be on the order of 10^3–10^4 compounds, compared to the 10^6 compounds typically evaluated in HTS assays. There are a variety of ways to utilize fragment leads in drug design efforts, as will be described

Graham A. Webb (ed.), Modern Magnetic Resonance, 1383–1389.

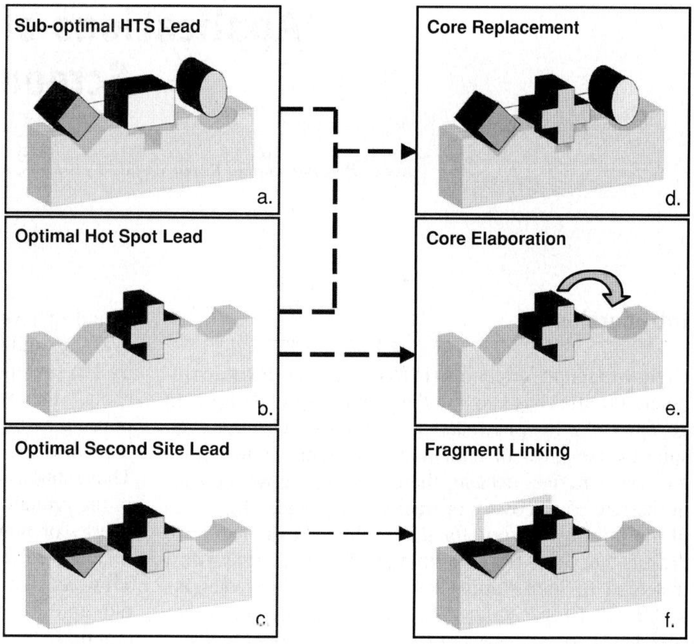

Fig. 1. Fragment-based strategies for the design of novel drug leads. (a) Representation of a high-throughput screening lead (connected shapes) binding sub-optimally to the protein target (gray surface). Sub-optimal binding is depicted by poor packing and shape complementarity. (b) Identification of a fragment lead (cross) that interacts optimally with the protein, indicated by close packing and good shape complementarity. (c) Identification of two ligands (cross and triangles) that bind to proximal subsites on a protein surface. (d) Core replacement strategy for lead design, in which the fragment identified in (b) is incorporated into the lead compound depicted in (a). (e) Core elaboration strategy, in which the fragment identified in (b) is directionally modified via high-throughput synthetic means to produce large libraries of compounds that can access neighboring subsites. (f) Fragment-linking strategy, in which the two fragments identified in (c) are linked to produce a high-affinity ligand.

below. However, these leads first need to be reliably identified. While attractive starting points for design, these fragment leads tend to bind with low affinity to the target (e.g. K_D values of 10–1000 μM)—which can complicate biochemical means of detection. As a result, a host of new technologies have been developed or elaborated for the express purpose of enabling fragment-based drug design. Such methodologies include NMR-based screening (both target-based and ligand-based) [10,11], high-throughput X-ray crystallography (e.g. CrystaLEAD) [2,18], mass spectroscopy [19], surface plasmon resonance [20], fragment tethering [21], and dynamic combinatorial chemistry [22]. This review will focus on receptor-based methods for fragment-based screening. The reader is referred to other sections in this handbook for more thorough discussions of other aspects on NMR-based screening in drug discovery and design.

Receptor-Based NMR Screening

Receptor-Based NMR screening monitors the perturbations in the protein signals upon addition of a test compound. The target is isotopically labeled and two-dimensional $^{15}N/^{1}H$- or $^{13}C/^{1}H$-correlation spectra are collected. Screening is typically performed against mixtures of 10–30 compounds, and the mixtures that produce the largest chemical shift perturbations are deconvoluted to identify the leads. NMR screening based on chemical shift perturbations of the resonances of the protein target is a profoundly reliable method for detecting protein–ligand interactions, and has significant advantages over other screening methods. The observation of discrete, stable chemical shift perturbations is strong evidence for a specific, well defined binding event. As a result, target-based NMR screening is essentially immune to false positives that can arise from non-specific binding or compound aggregation [23]. In addition, the spectral editing essentially eliminates the ligand signals, enabling the experiments to be performed even at very high ligand concentrations. Finally, the unique chemical shift perturbation "fingerprint" induced by ligand binding can be used not only to detect binding, but also to identify the ligand-binding site. This is shown in Figure 2 for $^{13}CH_3$-labeled protein kinase A (PKA) (mutant form). Addition of adenosine gives rise to characteristic chemical shift perturbations, and all screening hits that shift these same peaks can be classified as ATP-site ligands (Figure 2A). However, ligands can be identified that bind to alternative pockets on the protein (Figure 2B). These pockets can potentially be independently exploited for the development of therapeutics or ligands for these sites can be reliably classified and safely ignored if it is determined that the site is therapeutically irrelevant. In either case, this unique feature of receptor-based NMR screening makes it a powerful tool in the discovery and characterization of ligands for protein targets.

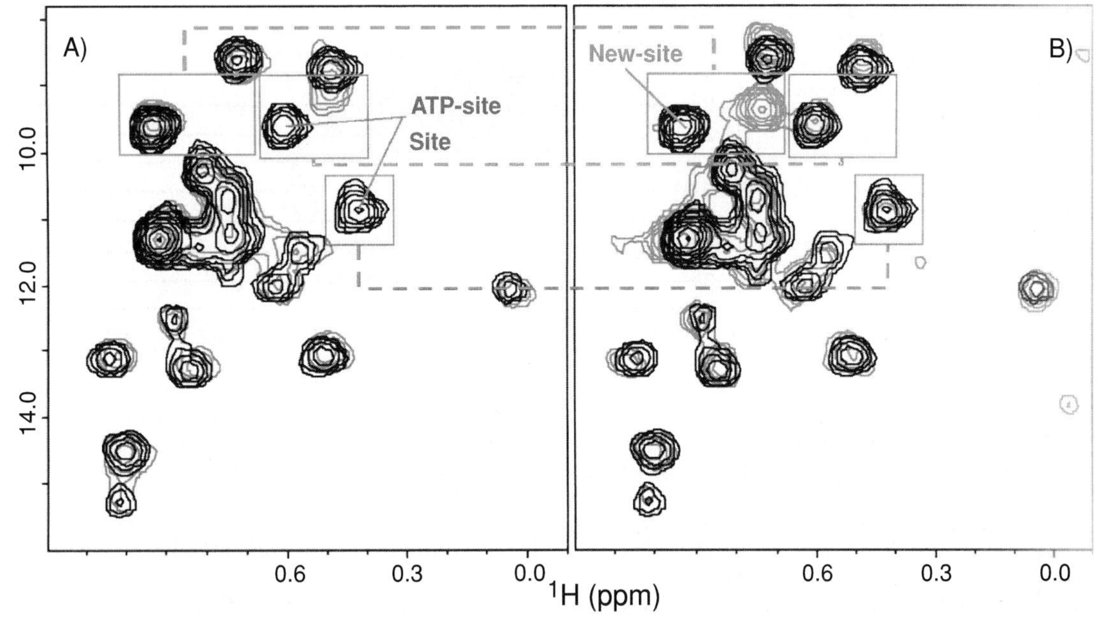

Fig. 2. Expanded section of $^1H/^{13}C$-HSQC spectra (showing isoleucine-δ signals) acquired on a mutant form of protein kinase A (PKA) in the absence (black contours) and presence (red contours) of (A) ATP-γS and (B) a test compound that binds to a site on PKA other than the ATP-binding site. The unique chemical shift fingerprint for each binding site can be used to classify new compounds as ATP- or non-ATP-site ligands. (See also Plate 109 on page LXX in the Color Plate Section.)

With the advance of cryogenic probes [24] and cost-effective ^{13}C-methyl labeling [25], target-based NMR screening utilizing ^{15}N- or ^{13}C-labeling can now routinely be applied to targets with MW in excess of 40 kDa. For larger targets, extensive deuteration and the use of TROSY pulse sequences can significantly increase this MW limit [25,26], although the high cost associated with extensive deuteration usually limits these applications to lead validation and characterization. Another limitation of receptor-based screening is that cost-effective and efficient labeling can currently only be achieved in bacterial systems. While isotopic labeling in insect cell systems has recently been reported [27], such methods are still in their infancy. Thus, for high MW targets or those that can only be produced in baculovirus or mammalian cells, the ligand-based NMR methods described elsewhere in this volume should be explored.

Utilization of Fragment Leads in Drug Design

Receptor-based NMR screening has been quite successful at its desired objective: the identification of low MW leads that bind to hot spots on protein surfaces and that would be difficult or impossible to detect using any other method. Figure 3 shows the molecular weight and pK_D values for more than 750 NMR leads that we have derived from receptor-based screens against more than 20 protein targets. By intentional library design, the leads are small (average MW of 234 ± 56 Da), and this results in leads that are quite weak (average pK_D of 2.7, which translates to a K_D value of 500 μM). The key question is how to utilize these hits in drug design. Three strategies for utilizing NMR-derived fragments in drug design will be briefly described: (1) core replacement, (2) high-throughput core elaboration, and (3) fragment linking.

Core Replacement

As mentioned above, most leads from HTS interact sub-optimally with the hot spot of the binding site. In contrast, as shown in Figure 1b, fragment-based screening attempts to identify a low molecular weight fragment that takes full advantage of the available binding interactions. One approach to exploiting these fragments is to incorporate the optimized core back into the lead molecule. This strategy is called core replacement and is shown schematically in Figure 1, where the optimized core (Figure 1b) is synthetically incorporated into the lead molecule (Figure 1a) to produce a compound with improved interactions with the target (Figure 1d). This approach can be especially

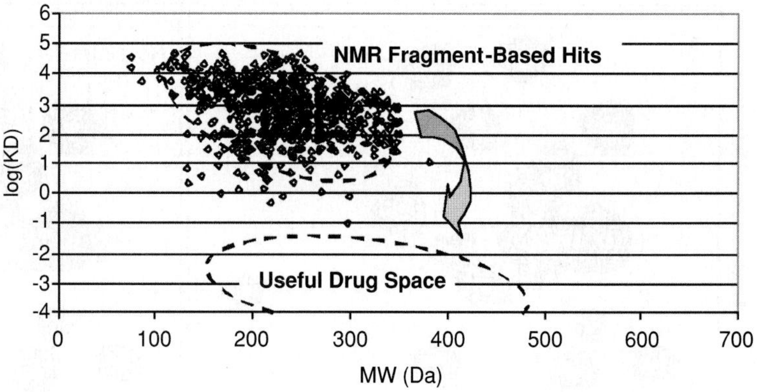

Fig. 3. Plot of molecular weight (in Da) vs. affinity [expressed in terms of $\log(K_D)$, where K_D is in μM units] for more than 750 leads derived from receptor-based NMR screens against more than 20 protein targets.

useful in breaking out of existing structural classes to address unanticipated safety, pharmacological, or intellectual property issues.

Core replacement strategies have been effectively used to address non-drug-like properties of several leads (Table 1). For example, reported inhibitors of both the SH2 domain of Lck and protein tyrosine phosphatase 1B (PTP-1B) contain highly charged phosphotyrosine (pTyr) mimetics that can severely limit cellular permeability and oral bioavailability. Receptor-based screening was able to identify singly charged moieties that could serve as superior pTyr replacements for these targets [28,29]. Significantly, in the case of PTP-1B, a class of isoxazole leads were incorporated back into the lead series, resulting in potent compounds that exhibited increased Caco-2 permeability and dose-dependent cellular activity [29]. A core replacement strategy was also implemented in the design of inhibitors of urokinase, in which the highly basic phenylguanidine group was successfully replaced with a more neutral aminobenzimidazole moiety [30].

High-Throughput Core Elaboration

While the leads from NMR-based screening are only of low molecular weight and affinity, they can be viewed as small molecule scaffolds that can be rapidly elaborated through combinatorial chemistry or high-throughput parallel synthesis. While many scaffolds have been utilized in the design of combinatorial libraries [31,32], it is typically unknown whether any scaffold possesses inherent binding affinity for any given target. In this sense, it is significant to note that an NMR-derived lead with a K_D value of only 1 mM represents more than 4 kcal/mol of binding energy. Thus, such leads are ideal for use in target-directed library design. The design of such libraries is

Table 1: Published examples of receptor-based fragment approaches in the design of novel drug leads

Target	Fragment approach	Result	References
Lck-SH2	Core replacement	Novel pTyr mimetics	[28]
UK	Core replacement	Novel, sub-μM inhibitors	[30]
PTP-1B	Core replacement	Novel, cell permeable inhibitors	[29]
Erm-AM	Core elaboration	Novel, μM inhibitors	[34]
PTP-1B	Core elaboration	Novel, selective, nM inhibitors	[35]
Oxidoreductases	Core elaboration	Novel, selective, nM inhibitors	[36]
FKBP	Linked fragment	Novel, nM ligands	[39]
Stromelysin	Linked fragment	Novel, nM ligands	[40]
Stromelysin	Linked fragment	Improved bioavailability	[41]
AK	Linked fragment	Structural diversity	[42]
LFA	Linked fragment	Improved solubility	[43]
PTP-1B	Linked fragment	Novel, nM inhibitors	[44]
DHPR	Linked fragment	Novel, nM inhibitors	[36]
HCV protease	Linked fragment	Non-peptidic, μM inhibitors	[50]

significantly enhanced when structural information on the protein–ligand complex can be obtained either from NMR or X-ray crystallography. The process of core elaboration is shown schematically in Figure 1, where the optimized fragment lead (Figure 1B) is synthetically modified to produce libraries of compounds in the expectation of accessing a neighboring pocket to increase both binding affinity and target specificity. A wide range of chemistries is available for the elaboration of suitable scaffolds. In particular, polymer assisted solution phase (PASP) synthesis is rapidly emerging as a method of choice for library preparation via parallel synthesis [33].

High-throughput core elaboration has been effectively utilized in the design of targeted libraries for ErmAM [34], PTP-1B [35], and various oxidoreductases (Table 1) [36]. In the case of ErmAM, a millimolar lead derived from receptor-based screening was rapidly elaborated at multiple positions, resulting in a more than 200-fold gain in potency [34]. For PTP-1B, a library of 76 compounds was prepared around an NMR-based lead resulting in compounds with K_I values <100 nM—representing a 30-fold gain over the parent compound [35]. Most significantly, members of this library exhibited moderate selectivity over highly homologous phosphatases, suggesting that such selectivity could be achieved with a small molecule inhibitor. NMR-SOLVE [37] was implemented to target the oxidoreductase family by first identifying a low molecular weight ligand that inhibited multiple members of the family, and then rapidly elaborating this core to increase affinity and specificity. From a library of 300 compounds, nM inhibitors of three oxidoreductases (LDH, DHPR, and DOXPR) were identified. Significantly, while the initial core was essentially non-selective with respect to these oxidoreductases, the resulting inhibitors exhibited 10- to 100-fold selectivity.

Fragment Linking

By far the most powerful utilization of fragment-based screening in drug design is the linked fragment approach. As depicted in Figure 1, a linked fragment strategy involves identifying not simply one (Figure 1B) but two fragments that bind simultaneously to the protein at neighboring sites (Figure 1C). The two fragments are then linked together (maintaining their spatial geometry) to produce high-affinity, highly optimized leads (Figure 1F). One of the most significant advantages of the linked fragment approach is that the amount of chemical space that is sampled increases exponentially with the size of the fragment library [38]. For instance, an NMR screen in which 10,000 compounds are tested for independent binding to two subsites on a protein covers as much chemical space as screening ~100 million single compounds. The linked fragment strategy has been utilized in two related approaches: (1) the *de novo* construction of high-affinity ligands and (2) the optimization of existing leads by identification of new substituents. The construction of high-affinity ligands from pieces that have been identified via NMR has been described for FKBP [39], stromelysin [40,41], adenosine kinase [42], LFA [43], PTP-1B [44], and DHPR [36]. In all of these cases, screens were performed to identify ligands for peripheral sites that bound simultaneously with a first-site ligand, and nM ligands were produced after linking. The strategies for *de novo* construction vs. lead optimization differ only in the sense of the desired end: generation of a novel structural series vs. improved physicochemical properties of an existing series.

Receptor-Based Methods for Lead Validation and Characterization

In addition to their use as screening tools, receptor-based NMR Methods are being increasingly utilized as lead validation tools. Often, HTS yields tens or even hundreds of compounds that modulate the biochemical response with apparent IC_{50} values on the order of 10 μM or less. As many assays utilized by HTS can be complex, multi-component systems, some number of the hits will actually disrupt or inhibit an assay component other than the target of interest. In addition, false positives can result from the method of detection (e.g. interference from fluorescent or colored compounds), or mechanisms such as non-specific binding, compound aggregation [23], or compound reactivity [45]. Thus, it is extremely valuable to identify false positives as early as possible in the hit evaluation process in order to focus resources on those hits that have the highest potential to become bonafide leads.

Typically, complex, target-specific secondary assays are used to eliminate false positives from the original primary screening hits. However, NMR is a universal tool that has been successfully applied in the evaluation of leads from a variety of screening formats, including activity assays, ligand competition assays, whole-cell screens, affinity-based screens, and virtual ligand screening (VLS) [4,46,47]. It has also been demonstrated that, using specific labeling strategies, multiple protein targets can be simultaneously evaluated for binding to test compounds—opening the door for monitoring compound selectivity in a single experiment [48]. Table 2 shows results on published targets for which NMR follow-up of HTS or VLS hits was employed. In all of the cases listed, ligands were confirmed that could be prioritized for further follow-up. However, the majority of compounds derived even from HTS or affinity-based screens did not form a stable complex with the target of interest. In fact, it is our experience that, on average, <20%

Table 2: Examples of receptor-based NMR methods for the validation of leads derived from HTS, affinity screening, and virtual ligand screening campaigns

Target	Discovery technology	No. of hits tested	No. of hits confirmed	Ultimate result	References
LFA	HTS	2	2	Clinical candidate	[43]
Bcl-xL	HTS/affinity	55	23	Novel structural series	[53]
MurF	Affinity	11	2	Novel, nM lead series	[51]
RGS4	Affinity	50	1	Validated lead series	[46]
FBP	VLS	100	1	Novel structural series	[54]
IGFPB-1	VLS	3	1	Novel structural series	[52]
DNA gyrase	VLS	14	7	Novel structural series	[20]

of all leads from HTS campaigns are suitable for further evaluation [4]. Thus, a combination of the inherent strengths of both HTS and NMR represents a powerful strategy for the identification of lead compounds that bind with high affinity to and specificity for the protein target. Such compounds have high potential for further optimization through medicinal chemistry and/or structure-based design approaches.

Summary

In summary, receptor-based NMR methods for fragment lead identification have been used in a number of design strategies. Significantly, these design strategies can be used throughout the entire course of pre-clinical drug discovery—including initial lead identification, potency optimization, and even modulation of physicochemical properties to aid in candidate selection. Even more significant, however, is the fact that compounds from several programs employing receptor-based NMR screening and fragment design have already gone beyond the pre-clinical setting into the clinic (e.g. matrix metalloproteinase inhibitors [49] and inhibitors of the LFA–ICAM interaction [43]). This is ample validation that the methods described in this review can be used to produce ligands not only with high affinity, but also with the appropriate drug-like properties for use in therapeutic intervention in man.

References

1. Campbell SJ. Clin. Sci. 2000;99:255–60.
2. Carr R, Jhoti H. Drug Discov. Today. 2002;7:522–7.
3. Oprea TI, Davis AM, Teague SJ, Leeson PD. J. Chem. Inf. Comput. Sci. 2001;41:1308–15.
4. Hajduk PJ, Burns DJ. Comb. Chem. High Throughput Screen. 2002;5:613–22.
5. Hajduk PJ, Meadows RP, Fesik SW. Q. Rev. Biophys. 1999; 32:211–40.
6. Peng JW, Lepre CA, Fejzo J, Abdul-Manan N, Moore JM. Methods Enzymol. 2001;338:202–30.
7. Stockman BJ, Farley KA, Angwin DT. Methods Enzymol. 2001;338:231–46.
8. Huth JR, Sun C. Comb. Chem. High Throughput Screen. 2002;5:631–44.
9. Lepre CA, Peng JW, Fejzo J, Abdul-Manan N, Pocas J, Jacobs M, Xie X, Moore JM. Comb. Chem. High Throughput Screen. 2002;5:583–90.
10. Stockman BJ, Dalvit C. Prog. NMR Spectrosc. 2002;41:187 231.
11. Meyer B, Peters T. Angew. Chem. Int. Ed. 2003;42:864–90.
12. Coles M, Heller M, Kessler H. Drug Discov. Today. 2003;8:803–10.
13. Jahnke W, Florsheimer A, Blommers MJJ, Paris CG, Heim J, Nalin CM, Perez LB. Curr. Top. Med. Chem. 2003;3:69–80.
14. Jahnke W, Widmer H. Cell. Mol. Life Sci. 2004;61:580–99.
15. DeLano WL. Curr. Opin. Struct. Biol. 2002;12:14–20.
16. Kuntz ID, Chen K, Sharp KA, Kollman PA. Proc. Natl. Acad. Sci. 1999;96:9997–10002.
17. Hann MM, Leach AR, Harper G. J. Chem. Inf. Comput. Sci. 2001;41:856–64.
18. Nienaber VL, Richardson PL, Klinghofer V, Bouska JJ, Giranda VL, Greer J. Nat. Biotechnol. 2000;18:1105–8.
19. Swayze EE, Jefferson EA, Sannes-Lowery KA, Blyn LB, Risen LM, Arakawa S, Osgood SA, Hofstadler SA, Griggey RH. J. Med. Chem. 2002;45:3816–9.
20. Boehm H-J, Boehringer M, Bur D, Gmuender H, Huber W, Klaus W, Kostrewa D, Kuehne H, Luebbers T, Meunier-Keller N, Mueller F. J. Med. Chem. 2000;43:2664–74.
21. Erlanson DA, Braisted AC, Raphael DR, Randal M, Stroud RM, Gordon EM, Wells JA. Proc. Natl. Acad. Sci. 2000;97:9367–72.
22. Ramstrom O, Lehn J-M. Nat. Rev. Drug Discov. 2002;1:26–36.
23. McGovern SL, Caselli E, Grigorieff N, Shoichet BK. J. Med. Chem. 2002;45:1712–22.
24. Hajduk PJ, Gerfin T, Boehlen J-M, Häberli M, Marek D, Fesik SW. J. Med. Chem. 1999;42:2315–7.
25. Hajduk PJ, Augeri DA, Mack J, Mendoza R, Yang J, Betz SF, Fesik SW. J. Med. Chem. 2000;122:7898–904.
26. Fernandez C, Wider G. Curr. Opin. Struct. Biol. 2003;13:570–80.

27. Strauss A, Bitsch F, Cutting B, Fendrich G, Graff P, Liebetanz J, Zurini M, Jahnke W. J. Biomol. NMR. 2003;26:367–72.
28. Hajduk PJ, Zhou M-M, Fesik SW. Bioorg. Med. Chem. Lett. 1999;9:2403–6.
29. Liu G, Xin Z, Pei Z, Zhao H, Hajduk PJ, Abad-Zapatero C, Hutchins CW, Lubben TH, Ballaron SJ, Haasch DL, Kaszubska W, Rondinone CM, Trevillyan JM, Jirousek MR. J. Med. Chem. 2003;46:4232–5.
30. Hajduk PJ, Boyd S, Nettesheim D, Nienaber V, Severin J, Smith R, Davidson D, Rockway T, Fesik SW. J. Med. Chem. 2000;43:3862–6.
31. Fecik RA, Frank KE, Gentry EJ, Menon SR, Mitscher LA, Telikepalli H. Res. Rev. 1998;18:149–85.
32. Fauchere JL, Boutin JA, Henlin JM, Kucharczyk N, Ortuno JC. Chemom. Intell. Lab. Sys. 1998;43:43–68.
33. Hinzen B. Methods Princ. Med. Chem. 2000;9:209–37.
34. Hajduk PJ, Dinges J, Schkeryantz JM, Janowick D, Kaminski M, Tufano M, Augeri DJ, Petros A, Nienaber V, Zhong P, Hammond R, Coen M, Beutel B, Katz L, Fesik SW. J. Med. Chem. 1999;42:3852–9.
35. Xin A, Oost TK, Abad-Zapatero C, Hajduk PJ, Pei Z, Szczepankiewicz BG, Hutchins CW, Ballaron SJ, Stashko MA, Lubben T, Trevillyan JM, Jirousek MR, Liu G. Biorg. Med. Chem. Lett. 2003;13:1887–90.
36. Sem DS, Bertolaet B, Baker B, Chang E, Costache AD, Coutts S, Dong Q, Hansen M, Hong V, Huang X, Jack RM, Kho R, Lang H, Ma C-T, Meininger D, Pellecchia M, Pierre F, Villar H, Yu L. Chem. Biol. 2004;11:185–94.
37. Sem DS, Yu L, Coutts SM, Jack R. J. Cell. Biochem. 2001;S37:99–105.
38. Hajduk PJ, Meadows RP, Fesik SW. Science. 1997;278:497–9.
39. Shuker SB, Hajduk PJ, Meadows RP, Fesik SW. Science. 1996;274:1531–4.
40. Hajduk PJ, Sheppard G, Nettesheim DG, Olejniczak ET, Shuker SB, Meadows RP, Steinman DH, Carrera GM, Marcotte PA, Severin J, Walter K, Smith H, Gubbins E, Simmer R, Holzman TF, Morgan DW, Davidsen SK, Fesik SW. J. Am. Chem. Soc. 1997;119:5818–27.
41. Hajduk PJ, Shuker SB, Nettesheim DG, Xu L, Augeri DJ, Betebenner D, Craig R, Albert DH, Guo Y, Meadows RP, Michaelides M, Davidsen SK, Fesik SW. J. Med. Chem. 2002; 45:5628–39.
42. Hajduk PJ, Gomtsyan A, Didomenico S, Cowart M, Bayburt EK, Solomon L, Severin J, Smith R, Walter K, Holzman TF, Stewart A, McGaraughty S, Jarvis MF, Kowaluk ES, Fesik SW. J. Med. Chem. 2000;43:4781–6.
43. Liu G, Huth JR, Olejniczak ET, Mendoza R, DeVries P, Leitza S, Reilly EB, Okasinski GF, Fesik SW, von Geldern TW. J. Med. Chem. 2001;44:1202–10.
44. Szczepankiewicz BG, Liu G, Hajduk PJ, Abad-Zapatero C, Pei Z, Xin Z, Lubben T, Trevillyan JM, Stashko MA, Ballaron SJ, Liang H, Huang F, Hutchins CW, Fesik SW, Jirousek MR. J. Am. Chem. Soc. 2003;125:4087–96.
45. Rishton GM. Drug Discov. Today. 1997;2:382–4.
46. Moy FJ, Haraki K, Mobilio D, Walker G, Powers R, Tabei K, Tong H, Siegal MM. Anal. Chem. 2001;73:571–81.
47. Rudisser S, Jahnke W. Comb. Chem. High Throughput Screen. 2002;5:591–603.
48. Zartler ER, Hanson J, Jones BE, Kline AD, Martin G, Mo H, Shapiro MJ, Wang R, Wu H, Yan J. J. Am. Chem. Soc. 2003;125:10941–6.
49. Wada CK, Holms JH, Curtin ML, Dai Y, Florjancic AS, Garland RB, Guo Y, Heyman HR, Stacey JR, Steinman DH, Albert DH, Bouska JJ, Elmore IN, Goodfellow CL, Marcotte PA, Tapang P, Morgan DW, Michaelides MR, Davidsen SK. J. Med. Chem. 2002;45:219–32.
50. Wyss DF, Arasappan A, Senior M, Wang Y-S, Beyer BM, Njoroge FG, McCoy MA. J. Med. Chem. 2004;47:2486–98.
51. Gu Y-G, Florjancic AS, Clark RF, Zhang T, Cooper CS, Anderson DD, Lerner CG, McCall O, Cai Y, Black-Schaefer CB, Stamper GF, Hajduk PJ, Beutel BA. Bioorg. Med. Chem. Lett. 2004;14:267–70.
52. Kamionka M, Rehm T, Beisel H, Lang K, Engh R, Holak T. J. Med. Chem. 2002;45:5655–60.
53. Qian J, Voorbach MJ, Huth JR, Coen ML, Zhang H, Ng S-C, Comess KM, Petros AM, Rosenberg SH, Warrior U, Burns DJ. Anal. Biochem. 2004;328:131–8.
54. Huth JR, Yu L, Collins I, Mack J, Mendoza R, Isaac B, Braddock DT, Muchmore SW, Comess KM, Fesik SW, Clore GM, Levens D, Hajduk PJ. J. Med. Chem. 2004;47: 4851–7.

NMR SHAPES Screening

Christopher A. Lepre and Jonathan M. Moore
Vertex Pharmaceuticals, Inc., Cambridge, MA 02139-4242, UK

Introduction

The SHAPES strategy is a process that employs nuclear magnetic resonance (NMR) spectroscopy to screen small, drug-like molecules in a directed search for novel (or improved) medicinal chemistry leads. This strategy typically integrates NMR screening experiments with other enabling methods in drug discovery such as enzymatic assays, high-throughput screening (HTS), X-ray crystallography, molecular modeling, and combinatorial chemistry. There are three components of any NMR-based lead discovery strategy: (1) a carefully selected library of screening compounds, (2) a NMR screening method that is well-suited to the target of interest, and (3) effective use of the information obtained through screening to generate or optimize drug leads for that target. Each of these aspects will be discussed in this chapter to provide an overview of how NMR SHAPES screening may be used in drug discovery. For additional reading, we have recently published more detailed treatments of each of these components, i.e. NMR screening library design [1,2], experimental approaches [3–5], and examples of SHAPES applications [6,7].

Principles of SHAPES Screening

The SHAPES screening strategy was developed based on two assumptions. The first assumption is that information about molecules that bind to a target with relatively low affinities may be used to direct the design of more potent inhibitors of that target. Any biophysical method that is capable of detecting weak interactions between ligands and macromolecules can be used, however, NMR methods are particularly well suited for this purpose. In many cases, NMR methods are capable of identifying inhibitors of a target that are beyond the limits of detection by traditional enzyme assays. The second assumption inherent in the SHAPES strategy is that compounds containing molecular fragments found in successful drugs are more likely to possess desirable, drug-like properties (such as good oral bioavailability, low toxicity, metabolic stability, and good synthetic accessibility) than random molecules. Thus, the SHAPES method uses simple NMR techniques to detect binding of a relatively small but diverse library of low molecular weight, soluble compounds to a drug target. The resulting SHAPES screening hits are then used to guide molecular modeling efforts, synthesize combinatorial libraries, and bias the choice of compounds selected to undergo high throughput enzymatic screening (Figure 1). Integration of the SHAPES strategy with iterative ligand structure determination has proven to be very useful for deriving pharmacophore models and rapidly achieving dramatic improvements in inhibitor potency with minimal synthetic effort.

Design of the SHAPES Compound Library

The design of an NMR screening library based on molecules derived from known drugs evolved from a modeling study carried out at Vertex Pharmaceuticals [8]. In this study, 5120 drugs covering a spectrum of therapeutic indications were selected from the Comprehensive Medicinal Chemistry database and computationally reduced to graph frameworks (i.e. unadorned rings and linkers). The authors found that only 32 simple frameworks described approximately half of the drugs in the database. When atom type and bond order were included in the analysis, only 41 "complex" frameworks, or scaffolds, described about a quarter of the known drugs. These findings implied that a relatively small number of molecular "shapes" could be used in a screening library for a wide variety of therapeutic targets.

Although many of the factors responsible for the failure of clinical candidates are not well understood, it was believed that commonly recurring drug scaffolds are predisposed to having favorable clinical properties. Therefore, the original SHAPES library was built with compounds containing the most common scaffolds found in known drugs, combined with commonly occurring drug side chains (identified in a later study [9] (Figure 2). Subsequently, the library has been expanded to incorporate drug-like and "lead-like" [10] compounds that contain linkages amenable to combinatorial chemistry, are targeted against targets from a particular gene family, or exhibit diverse pharmacophores [1,2].

The SHAPES library has been designed to satisfy stringent criteria for compound solubility, drug-likeness (or lead-likeness), and synthetic accessibility. Details of the compound selection process and overall library design have been described previously [1,2]. More generally, starting from a database of over 1 million commercially

Graham A. Webb (ed.), Modern Magnetic Resonance, 1391–1399.

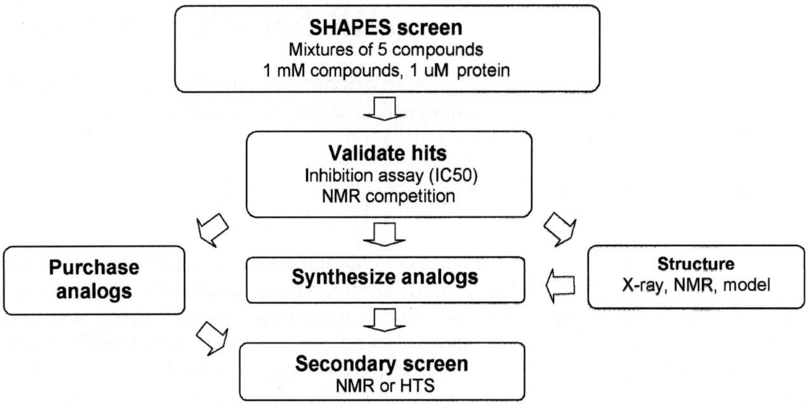

Fig. 1. Flow diagram of the SHAPES process.

available compounds, computational filters were applied to remove compounds predicted to have poor solubilities or other physicochemical properties (using "Rule of 5" [11] and other criteria [1]), as well as to eliminate compounds possessing undesirable functional groups [12]. Candidates surviving the filtering process were then clustered using the Jarvis–Patrick method [13], and individual compounds were picked manually from the centroids of the clusters. Mixtures of compounds were then chosen with NMR considerations in mind, i.e. to minimize NMR resonance overlap and avoid interactions between the components. Prior to screening, the mixtures were checked for aggregation by recording 2D NOESY experiments.

NMR Methods for Screening Compound Libraries

Because the molecules in the SHAPES screening library are typically small (<350 MW) and unelaborated, they are expected to have lower average affinities than most compounds screened by HTS. To enable the broadest possible range of affinities for detection, we use well-known NMR-based methods capable of identifying molecules that bind with dissociation constants ranging from low micromolar (and lower for some methods [14,15]) to millimolar in a single experiment.

We have recently reviewed and compared these methods in detail [4,5]. Although protein-based detection methods [16,17] are very useful for mapping the binding sites of hits on protein targets, ligand-detected methods for primary screening consume less reagents, do not require isotopic labeling, and may be used for many types of targets with no limit to molecular mass.

Many approaches have been proposed that monitor differences in ligand NMR parameters in the free vs. bound states. These approaches have used transverse relaxation rates, selective longitudinal relaxation rates, exchange-transferred NOE, and translation diffusion coefficients as probes of interactions between small molecules and a macromolecular target. In these experiments, a small molecule binder will transiently assume the much slower rotational and translational mobility of the much larger receptor, exhibiting changes in the aforementioned relaxation and diffusion parameters.

A second class of experiments does not rely on differences between ligand free and bound state parameters, but rather takes advantage of ^{1}H magnetization transfer from the target to the ligand. These experiments have distinct advantages, because binders to the target experience the magnetization transfer, while non-binders do not. This phenomenon results in significant simplification of spectra and identification and deconvolution of hits. The two most commonly used of this latter class of experiments are the Saturation Transfer Difference (STD) [18] and WaterLOGSY [19,20] methods. In the STD experiment, the target is saturated via a train of radiofrequency pulses. Upon saturation, spin diffusion occurs throughout the protein matrix to the binding site, and is transferred to ligands during the ligand residence time. WaterLOGSY relies similarly on initial saturation of the receptor, however the mechanism for saturation differs, in that either saturation or inversion of solvent water molecules is the initial step, followed by transfer of that magnetization to sites on the target via multiple pathways, and subsequently to the binding ligand. A more rigorous physical treatment of magnetization transfer processes in both the STD and WaterLOGSY classes of experiments may be found elsewhere [5].

Our preferred method for detection of ligand binding to protein targets is saturation transfer difference spectroscopy [18], which offers high sensitivity, low

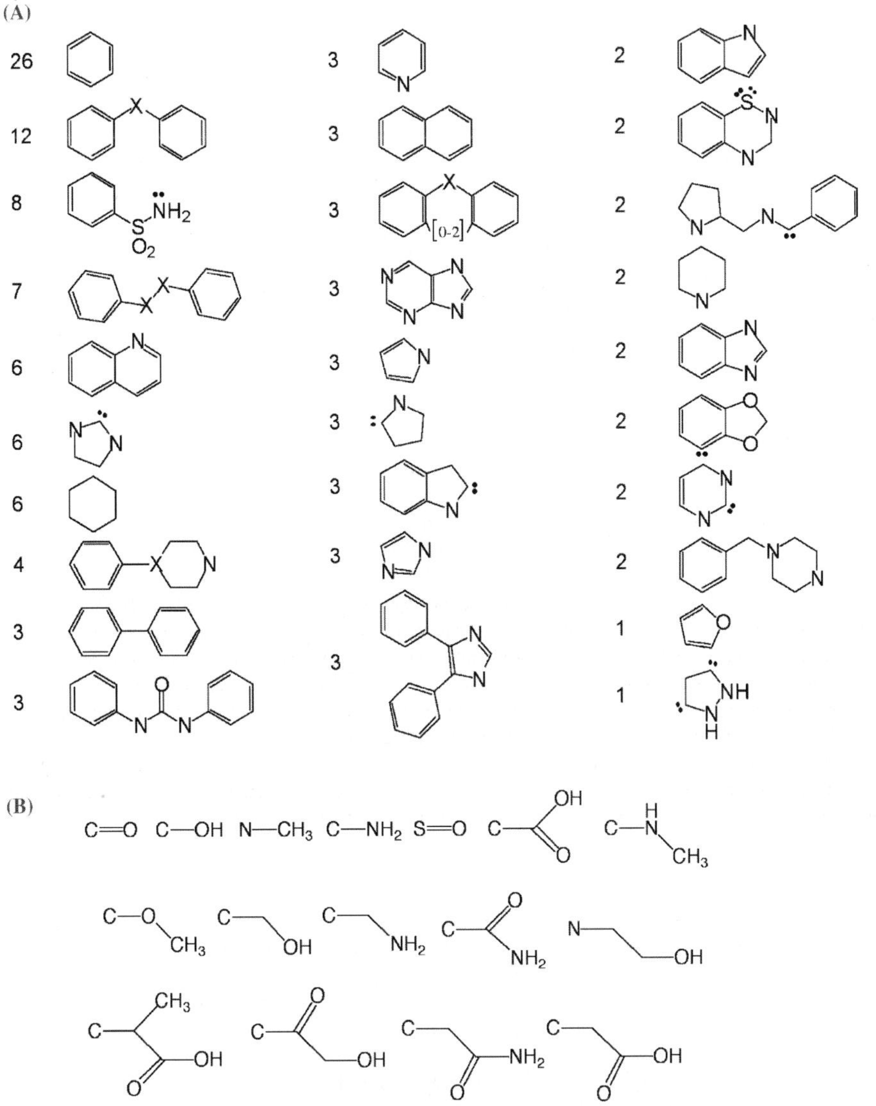

Fig. 2. (A) Molecular frameworks used in the SHAPES library. The numbers denote the frequency of occurrence of each framework. Side chain positions are indicated by lone pairs. "X" denotes a C, N, O, or S atom. Frameworks representing less common classes of drug scaffolds are not shown. (B) Common drug side chains used in the SHAPES library. The left-most atom indicates the point of attachment to the framework. *Source:* Reproduced with permission from ref. [3].

protein consumption, and the ability to tune the detection range of the experiment [3,5]. SHAPES screening hits are validated by detection of inhibition (e.g. by enzymatic assay), competition with a known inhibitor (e.g. by NMR displacement experiment), or localization to the active site (e.g. by NMR chemical shift mapping experiments or crystallographic soaking studies). An example of STD spectra for a mixture of ligands is shown in Figure 3. For nucleic acid targets, WaterLOGSY is the method of choice, as the multiple sites for interaction of water molecules with the nucleic acid molecules compensate for the reduced proton

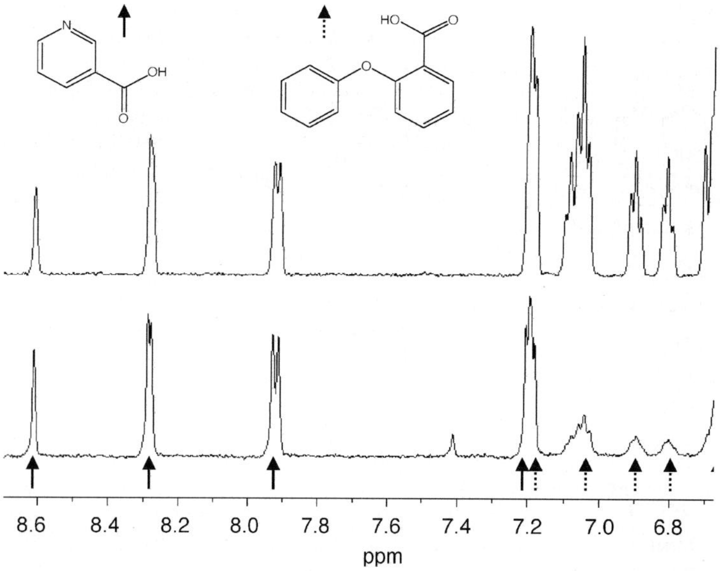

Fig. 3. Comparison of (top) the 1D ^{1}H NMR spectrum of two potential small molecule ligands in the absence of target and (bottom) the STD [18] spectrum of the same mixture in the presence of p38 MAP kinase. Resonances from nicotinic acid and 2-phenoxy benzoic acid are marked with asterisks and diamonds, respectively. Only peaks from the ligand that binds to p38, 2-phenoxybenzoic acid, appear in the STD spectrum. The sample contained 35 μM p38 MAP kinase, 1 mM ligands, 100 mM deutero-Tris, 10% deutero-glycerol, 20 mM deutero-DTT at pD* = 8.4. Experiments were carried out at 278 K. The 1D spectrum was collected using a standard NOESY pulse sequence with 16K data points, 128 transients and a relaxation delay of 3 s. 2K data points were collected with 256 transients each for the on and off resonance spectra. Internal subtraction was achieved via phase cycling. On resonance irradiation was at $\delta = 0.74$ and off-resonance at $\delta = -20$ ppm. A pulse train of 60 Gaussian selective pulses of 50 ms duration separated by a 1 ms delay was used to saturate the protein. *Source:* Reproduced with permission from ref. [3].

density in this class of macromolecule vs. protein targets [6,21].

Implementation of SHAPES Screening

SHAPES screening has been implemented at Vertex at several stages of drug discovery projects: very early in the life cycle, before a conventional enzymatic screen is run to find leads (pre-HTS); in the mid-life of a discovery project, in the event that HTS has not produced viable leads (post-HTS); and late in a project, when mature leads are undergoing final optimization to become preclinical candidates (lead optimization). Effective use of information from SHAPES screens is essential for successful implementation of the method. SHAPES hits are followed up using one or more of three strategies [2,5]: (1) primary hits are combined or linked together into larger, more potent molecules; (2) the scaffolds from primary hits are systematically elaborated by adding functional groups to make more complex molecules; and (3) portions of primary hits or a lead molecule are systematically varied. We present here examples that illustrate these three follow-up strategies applied to each of the three principal implementations of the SHAPES method.

Pre-HTS Screening

SHAPES screening is most commonly applied at the pre-HTS stage of a project, when a new protein first becomes available and the HTS screen is still under development. Because NMR screening requires no assay development and only a few milligrams of protein, it is possible to find SHAPES hits, then purchase analogs or synthesize a combinatorial library around these hits quickly enough to be included in an initial HTS run, thereby enriching the HTS library with scaffolds known to bind to the protein.

Validated hits from pre-HTS SHAPES screens are most often followed up by purchasing more elaborate analogs, as has been described elsewhere [1,7]. Retrospective analysis of HTS screening results has shown that compounds containing SHAPES scaffolds (those shown to bind by NMR) typically have hit rates about tenfold higher than general screening compounds that do not contain those scaffolds [6]. Higher enhancements are possible by designing combinatorial libraries around SHAPES scaffolds. For example, SHAPES screening of regulatory erythroid kinase (REDK) identified a novel class of ATP-competitive inhibitors with IC50s of 100–200 μM. A 64 compound combinatorial library was synthesized around this scaffold, 14 of which (22%) had IC50s better than 5 μM (including several that were sub-micromolar) when subsequently included in the HTS screen. This small, focused combinatorial follow-up library thus improved upon the potency of the primary SHAPES hits by two orders of magnitude and enhanced the HTS hit rate by around 170-fold. The observation of such enhancements validates the assumption that binding information from compounds with weak affinities can lead to discovery of significantly (e.g. several orders of magnitude) more potent inhibitors.

SHAPES screening has also been applied to targets for which high throughput enzymatic assays are unavailable, and can identify reagents for use in developing such assays. These targets include RNA [6,21] and non-enzymatic proteins, such as molecular transport proteins and receptors. For example, SHAPES screening was used as a primary screen to find leads for human adipocyte lipid binding protein (ALBP), a putative therapeutic target for treating type II diabetes [22–26] for which a HTS assay was unavailable. Using a prototype 100-compound SHAPES library, thirteen ligands were found with affinities ranging from 0.3 μM to 800 μM. A fluorimetric assay and chemical shift (^{1}H and ^{15}N) perturbation studies confirmed that these ligands bind in the lipid-binding pocket. Several fluorescent analogs of the SHAPES hits bound avidly in the ALBP lipid-binding pocket, including 1-anilinonapthlalene-8-sulfonic acid (1,8-ANS), which exhibits an apparent dissociation constant of 0.4 μM. These analogs were later used to develop a high-throughput fluorescence-based competition assay.

In order to identify more potent binders, two primary SHAPES hits were soaked into crystals of apo-ALBP and the structures of the complexes solved by crystallography. This structural information, along with molecular models of other hits, was used to identify likely protein interaction sites. Twelve commercially available analogs were screened in order to explore variations at positions predicted to make favorable protein contacts. All of the analogs inhibited lipid binding to ALBP and the bound structures of several were subsequently determined by crystallography. This information was used to design a second-generation follow-up library of 134 commercially available compounds that was calorimetrically screened, yielding nine leads with sub-micromolar affinities, including a compound class with Kd values of 80–100 nM.

Fatty acid binding proteins, unlike most enzymes, have considerable space in the lipid binding pocket that can accomodate a variety of fatty acid substrates as well as bound water molecules. Understanding the binding modes of different ligands by solving multiple crystal structures allows one to derive a pharmocophore model describing key interactions in the binding pocket that may be accessed to achieve higher affinity and selectivity. A pharmacophore model for ALBP based on eight different crystal structures is shown in Figure 4.

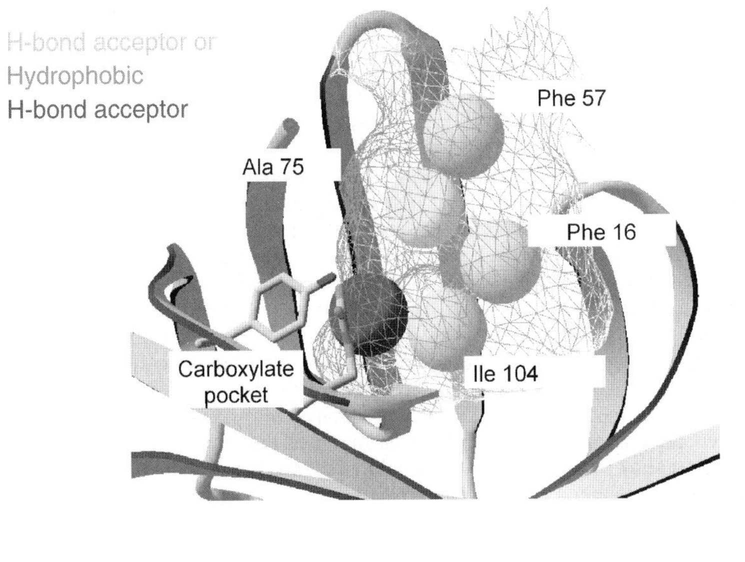

Fig. 4. Pharmacophore model for ALBP derived from ligand binding and crystallographic studies. (See also Plate 110 on page LXX in the Color Plate Section.)

This example demonstrates that by combining NMR-based screening, computational methods, and X-ray crystallography, it is possible to rapidly converge upon multiple sub-micromolar lead classes (in this case, by testing only ~250 compounds), using modest resources and without requiring chemical synthesis or a HTS assay. The information derived from this work serves as an excellent starting point for a structure-based drug design program for this unique class of drug target.

Post-HTS Screening

We have also applied SHAPES screening to targets for which conventional HTS and medicinal chemistry have failed to produce viable leads. For example, the c-Jun N-terminal kinase Jnk3, a potential target for the treatment of stroke and Parkinson's disease [27,28], was one of the first targets screened by the SHAPES method, after extensive medicinal chemistry efforts to adequately improve the potency of the HTS hits. NMR screening of a 100 compound prototype SHAPES library yielded 17 hits with very weak (approximately millimolar) affinities, 13 of which were confirmed to target the ATP site [7].

Attempts to soak the SHAPES hits into crystals of apo-Jnk3 for crystallographic studies were unsuccessful (presumably due to their low affinity). In the absence of structural information about the bound hits, two approaches were taken to design follow-up libraries. First, approximately 200 analogs containing combinations of binding fragments were chosen by using substructure-matching and similarity based searches of a database of commercially available compounds.

Second, compounds were selected using a virtual screening procedure. The SHAPES screening hits were computationally docked into the ATP site of the crystal structure, and four produced distinct, energetically reasonable bound orientations. A substructure search for commercially available analogs was then carried out using the four preferred molecules, followed by filtering for drug-like character [12], to give a starting set of 1647 compounds. These candidates were then docked into the ATP site and the best 171 conformers from the consensus of four scoring functions [29] were then manually reduced to 94 top picks.

These procedures selected approximately 300 compounds that were then screened in an enzymatic assay, yielding 8 inhibitors with potency better than 20 μM, a hit rate tenfold higher than was observed in the original HTS screen of Jnk3 using a random compound library. Interestingly, none of the SHAPES-derived inhibitor classes were found in the original HTS screen of Jnk3, even though compounds containing those scaffolds were present in the random HTS library. The compounds representing those scaffolds in the HTS library are more highly elaborated, which presumably reduces the likelihood of binding. The unelaborated SHAPES scaffolds are more likely to bind, albeit with affinities too weak to detect reliably by enzymatic methods. These results support the idea that NMR screens of weakly binding scaffolds can be used to direct HTS to novel classes of compounds that might otherwise be missed.

The SHAPES leads were followed up by medicinal chemistry after efforts to develop the leads from the original HTS screen were unsuccessful. Several hundred compounds were prepared (mostly in combinatorial libraries) based on four SHAPES-derived leads, producing three sub-micromolar classes of compounds, shown in Figure 5. X-ray structures of examples from each class of compounds complexed to Jnk3 suggested routes for improving the potency of these inhibitors. By incorporating substituents to access all available binding pockets in the active site, a number of compounds were synthesized which were able to inhibit Jnk3 activity with potencies better than 10 nM.

Lead Optimization

NMR screening of molecular fragments has been used by other laboratories to selectively optimize portions of existing lead molecules for the purpose of improving potency, novelty, or ADME properties [30,31]. A recent example from our own laboratory combines virtual screening and NMR screening to selectively optimize a portion of an ATP site inhibitor of a protein kinase involved in inflammation. The original lead molecule consisted of three fragments (depicted schematically in Figure 6): a central scaffold, a group that hydrogen-bonds to the protein backbone of the inter-domain hinge and a group that forms an unusual salt bridge with nearby charged side chains. The latter group was of particular interest because optimizing this novel interaction offered a potential means to improve the selectivity for this kinase vs. other kinases.

An NMR screening library of fragments targeting the salt bridging site was selected using virtual screening. The ATP-site conformation derived from the crystal structure of the lead compound bound to an analogous kinase was used as a template for computationally docking 20,000 small, commercially available compounds, which were pre-selected for their putative salt bridging ability. Twenty-four compounds were then manually chosen from the best-scoring 0.8% of docked compounds; these were purchased and screened by NMR. Fourteen of these compounds (58%) bound to the kinase with potencies in the millimolar range. The hit rate with this virtual screening-derived library was approximately fivefold higher than that measured for a comparable NMR screen that used the diverse SHAPES library. The binding site of the NMR hits was determined by using a double

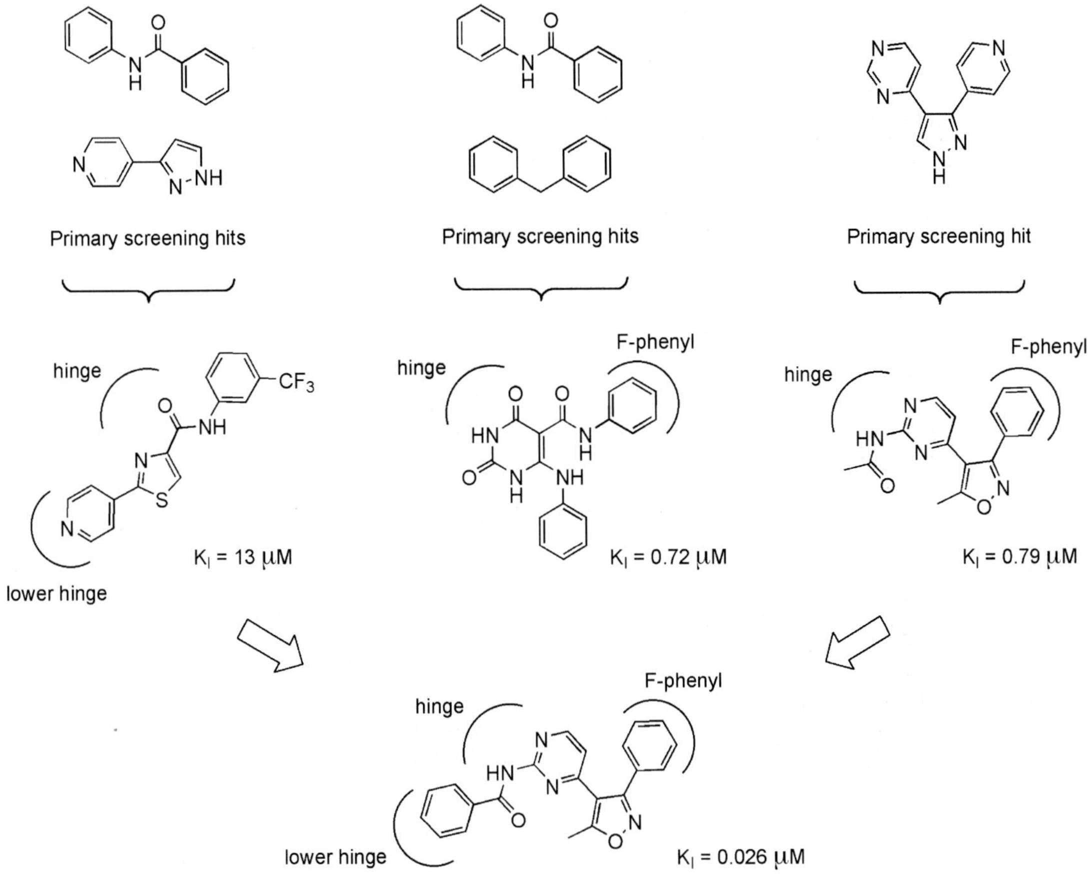

Fig. 5. Evolution of Jnk-3 SHAPES screening hits into potent leads. The SHAPES hits (top row) were followed up by screening commercially available compounds containing random combinations (left and center columns) and variations (right column) of scaffolds found in the primary hits. Crystal structures of the three lead classes (center row) revealed three ATP site subsites (hinge, fluorophenyl pocket, and lower hinge pocket) that were occupied by functional groups on the lead molecules. Each lead molecule contacts two of the three subsites, leading to synthesis of a third generation molecule (bottom row) that contacts all three. *Source*: Reprinted with permission form ref. [5].

knockoff filter (Figure 6), a two-step competition binding experiment that first uses a moderate-affinity inhibitor that binds only to the core and hinge sites, followed by a high-affinity inhibitor that binds to all three sites. Ten of the 14 hits (71%) behaved as expected in the double knockoff filter, indicating that they bound at the salt bridge site.

This example demonstrates that the combination of virtual and NMR screening can be used not only to enrich a library with active compounds, but also to preferentially discover compounds that bind to a desired site. Intermolecular NOEs or chemical shift perturbation data can then be used to determine the relative proximity and orientation of the bound fragments, enabling the design of more potent linked compounds. In this manner, thousands of putative replacement fragments can be efficiently screened prior to carrying out chemical synthesis.

Conclusion

NMR SHAPES screening can be an extremely useful tool for the identification of drug leads. Even in the absence of HTS hits, the SHAPES screening process is capable of identifying chemically accessible, drug-like fragments that can be rapidly translated into potent lead compounds. As illustrated in the examples, the unique ligand-binding data derived from NMR SHAPES screening and subsequent integration with other enabling technologies (such

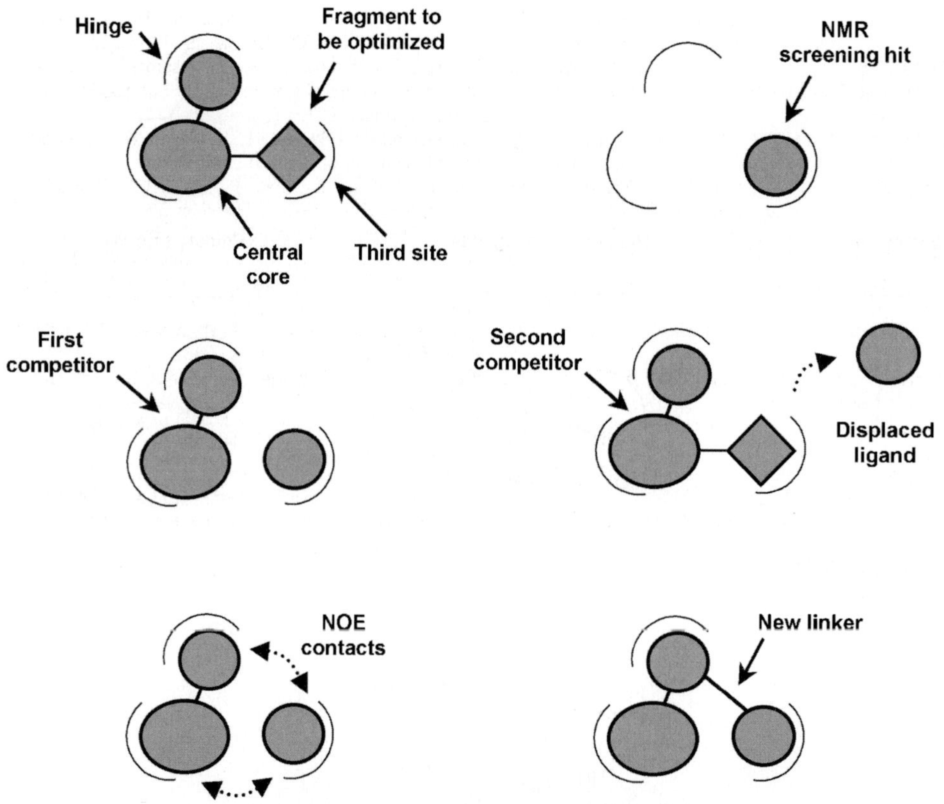

Fig. 6. Fragment optimization process. The original lead molecule (top left) consists of a hinge-binding motif, central scaffold, and a functional group (diamond) to be optimized in the third subsite. A library of fragments is selected by virtual screening and the binding fragments are identified by NMR screening (top right). The binding site of a NMR screening hit with low (e.g. millimolar) affinity is determined using a double knockoff competition experiment. The binding (or displacement) of the NMR hit at each step is measured using the same NMR method as was used for the screen. First, the sample is spiked with a moderate (e.g. micromolar) affinity competitor compound that occupies only the hinge and central sites (center left); if the NMR hit is in the third site it will not be displaced. The sample is then spiked (center right) with a second competitor compound (i.e. the original lead molecule) that occupies all three sites and has an affinity, high enough (nanomolar) to displace the first competitor; if the NMR hit is in the third s site it will be displaced as well. A new, higher concentration sample containing the first competitor and the NMR is then used to determine the relative orientations of the two bound molecules by means of intermolecular NOEs (lower left). Using this information, a linked compound is designed, possibly even incorporating a novel linkage site (lower right).

as X-ray crystallography, computational chemistry, and other biochemical and biophysical techniques) can greatly accelerate the discovery of lead molecules, as well as aid in optimization of existing leads.

References

1. Lepre CA. Drug. Discov. Today 2001;6:133.
2. Lepre CA. In: O Zerbe (Ed). Strategies for NMR Screening and Library Design. Wiley-VCH: Weinheim, 2003, p 1349.
3. Peng JW, Lepre CA, Fejzo J, Abdul-Manan N, Moore JM. Methods Enzymol. 2001;338:202.
4. Peng JW. Prog. Nucl. Mag. Reson. Spectrosc. 2004;44: 225.
5. Lepre CA, Moore JM, Peng JW. Chem. Rev. 2004;104: 3641.
6. Lepre CA, Peng J, Fejzo J, Abdul-Manan N, Pocas J, Jacobs M, Xie X, Moore JM. Comb. Chem. High Throughput Screen. 2002;5:583.
7. Fejzo J, Lepre, C, Xie, X. Curr. Top. Med. Chem. 2002;2: 1349.
8. Bemis GW, Murcko MA. J. Med. Chem. 1996;39:2887.

9. Bemis GW, Murcko MA. J. Med. Chem. 1999;42:5095.
10. Teague SJ, Davis AM, Leeson PD, Oprea T. Angew. Chem. Int. Ed. 1999;38:3743.
11. Lipinski CA, Lombardo F, Dominy BW, Feeny PJ. Adv. Drug Deliv. Rev. 1997;23:3.
12. Walters WP, Stahl MT, Murcko MA. Drug. Discov. Today. 1998;3:160.
13. Jarvis RA, Patrick EA. IEEE Trans. Comput. 1973;C-22:1025.
14. Dalvit C, Fasolini M, Flocco M, Knapp S, Pevarello P, Veronesi M. J. Med. Chem. 2002;45:2610.
15. Dalvit C, Flocco M, Knapp S, Mostardini M, Perego R, Stockman BJ, Veronesi M, Varasi M. J. Am. Chem. Soc. 2002;124:7702.
16. Shuker SB, Hajduk PJ, Meadows RP, Fesik SW. Science 1996;274:1531.
17. Pellecchia M, Meininger D, Dong Q, Chang E, Jack R, Sem DS. J. Biomol. NMR 2002;22:165.
18. Mayer M, Meyer B, Angew. Chem. Int. Ed. 1999;38: 1784.
19. Dalvit C, Pevarello P, Tato M, Veronesi M, Vulpetti A, Sundstr Möm. J. Biomol. NMR 2000;18:65.
20. Dalvit C, Fogliatto G, Stewart A, Veronesi M, Stockman B. J. Biomol. NMR. 2001;21:349.
21. Johnson EC, Feher VA, Peng JW, Moore JM, Williamson JR. J. Am. Chem. Soc. 2003;125:15724.
22. Hotamisligil GS, Johnson RS, Distel RJ, Ellis R, Papaioannou VE, Spiegelman BM. Science. 1996;274:1377.
23. Scheja L, Makowski L, Uysal KT, Wiesbrock SM, Shimshek DR, Meyers DS, Morgan M, Parker RA, Hotamisligil GS. Diabetes 1999;48:1987.
24. Uysal KT, Scheja L, Wiesbrock SM, Bonner-Weir S, Hotamisligil GS. Endocrinology 2000;141:3388.
25. Layne MD, Patel A, Chen YH, Rebel VI, Carvajal IM, Pellacani A, Ith B, Zhao D, Schreiber BM, Yet SF, Lee ME, Storch J, Perrella MA. FASEB J. 2001;15:2733.
26. Makowski L, Boord JB, Maeda K, Babaev VR, Uysal KT, Morgan MA, Parker RA, Suttles J, Fazio S, Hotamisligil GS, Linton MF. Nat. Med. 2001;7:699.
27. Bruckner SR, Tammariello SP, Kuan CY, Flavell RA, Rakic P, Estus S. J. Neurochem. 2001;78:298.
28. Xia XG, Harding T, Weller M, Bieneman A, Uney JB, Schulz JB, Proc. Natl. Acad. Sci. U.S.A. 2001;98:10433.
29. Charifson PS, Corkery JJ, Murcko MA, Walters WP. J. Med. Chem. 1999;42:5100.
30. Hajduk PJ, Gomtsyan A, Didomenico S, Cowart M, Bayburt EK, Solomon L, Severin J, Smith R, Walter K, Holzman TF, Stewart A, McGaraughty S, Jarvis MF, Kowaluk EA, Fesik SJ. J. Med. Chem. 2000;43:4781.
31. Hajduk PJ, Burns DJ, Comb. Chem. High Throughput Screen. 2002;5:613.

Part II

NMR-Based Screening Applied to Drug Discovery Targets

Jennifer J. Gesell, Mark A. McCoy, Mary M. Senior, Yu-Sen Wang, and Daniel F. Wyss
Schering-Plough Research Institute, Department of Structural Chemistry, Kenilworth, NJ 07033, USA

Abstract

While conventional bioassay-based high-throughput screening (HTS) remains a mainstream approach for lead discovery, its limitations have driven the development of alternative and complementary tools. In this regard, novel NMR-based approaches that have emerged over the last few years show great promise. We have used NMR-based screening approaches for a variety of drug targets to identify low molecular weight (MW) small molecule hits from customized libraries, which subsequently could be optimized into leads through focused, structure-guided chemistry. Focus was placed on targets for which HTS failed to identify suitable leads. This report discusses different NMR-based screening techniques and follow-up strategies for lead discovery and illustrates their application to the NS3 protease and NS3 helicase domains of the hepatitis C virus (HCV).

NMR for Lead Discovery

NMR methods are solidly established for studying molecular interactions [1]. In the last few years various NMR approaches have been developed to aid in lead discovery [2–13]. They involve different NMR screening methods [14–34] to identify initial compounds which often bind only weakly, in the μM–mM range, to the drug target. Intelligent and focused follow-up strategies enable their development into potent, sub-μM inhibitors for use as leads in drug discovery projects [35–46]. An interesting NMR-based biochemical screening method was also recently introduced [47]. NMR-based screening has several advantages over more traditional screening methods which are important for lead discovery: (1) it can reliably detect site-specific ligand binding even for compounds with very weak affinities which are typically missed in HTS (dissociation constants (K_D) up to $\sim 10^{-2}$–10^{-1} M); (2) it is a universal screening technique that unlike bioassays, requires no prior knowledge of a protein's function and therefore, does not require the development of specific assays for each target; (3) it allows the concomitant determination of K_D (or K_I) values over a large affinity range and thus, provides structure-activity relationship (SAR); information (4) it can provide crucial structural information for both the target and the ligand with atomic resolution for the subsequent optimization of weak initial hits into strongly binding lead candidates; (5) it can report on the quality of either the target or the ligand during the NMR screen which reduces the rate of false positives; (6) it is less prone to false positives due to trace contaminants; (7) the final lead does not need to be present in the initial screening library and because affinity is built up progressively, it is derived from a large virtual library. Despite these advantages, once higher affinity ligands are produced, it is always prudent to verify that they show the desired activity in a functional bioassay. Although NMR screening is, in principle, slower than conventional HTS, it can provide good, reliable starting points for focused chemistry to produce leads.

NMR-Based Screening Techniques

Many NMR observables can be used to monitor small molecule-protein interactions as a consequence of either global effects or local effects which occur upon complex formation [6,7,20]. Because the small molecule ligand adopts motional properties in the bound state that are similar to those of the drug target (slow motion), but very different from those in its unbound state (fast motion), observation of ligand binding based on global effects is restricted to the small molecule ligand where the effects are most pronounced. On the other hand, local NMR effects of complex formation such as intermolecular magnetization transfer or chemical shift perturbations (CSP) may be observed on the protein target and/or the small molecule ligand. In addition, structural information may be available as effects are mostly localized at the intermolecular binding sites.

NMR methods for detecting ligand-binding fall into two main categories that are based on whether signals from the protein target or the ligand are monitored (see also the reports by P.J. Hajduk and J.M. Moore in this volume). Methods within each category have inherent limitations, such as sample availability, solubility, stability, and molecular size, which dictate the choice of the NMR

Graham A. Webb (ed.), Modern Magnetic Resonance, 1401–1410.

screening approach taken to discover hits for a particular drug target.

Target-Detected 2D Heteronuclear NMR-Based Screening

We have used 2D heteronuclear NMR-based screening, in which the protein target is isotopically labeled with ^{15}N or ^{13}C, to assist lead discovery in several programs. HSQC-based screening detects ligand-binding based on chemical shift changes of protein signals which occur near the ligand-binding site. HSQC-based screening is simple (a single tube experiment with only the target and compound), robust (isotopic editing ensures that only the protein target is observed), and information-rich. The NMR correlation spectra not only give detailed information about whether the compound binds, with virtually no limit on the K_D value, but also yield a distinct pattern of CSP that serves as a fingerprint for binding to a specific region of the protein target. Thus, compounds that bind at the active site can be readily distinguished from those that bind outside the active site, or that, for one reason or another, do not bind to the target of interest at all. If CSP of the protein target are analyzed in a more quantitative manner, often the exact ligand-binding site can be determined rapidly [7,48]. In addition, the detailed binding orientation of the ligand with respect to the protein surface may be available, especially if good quality CSP data of closely related ligands are available [7,49] (Figure 1 and see below).

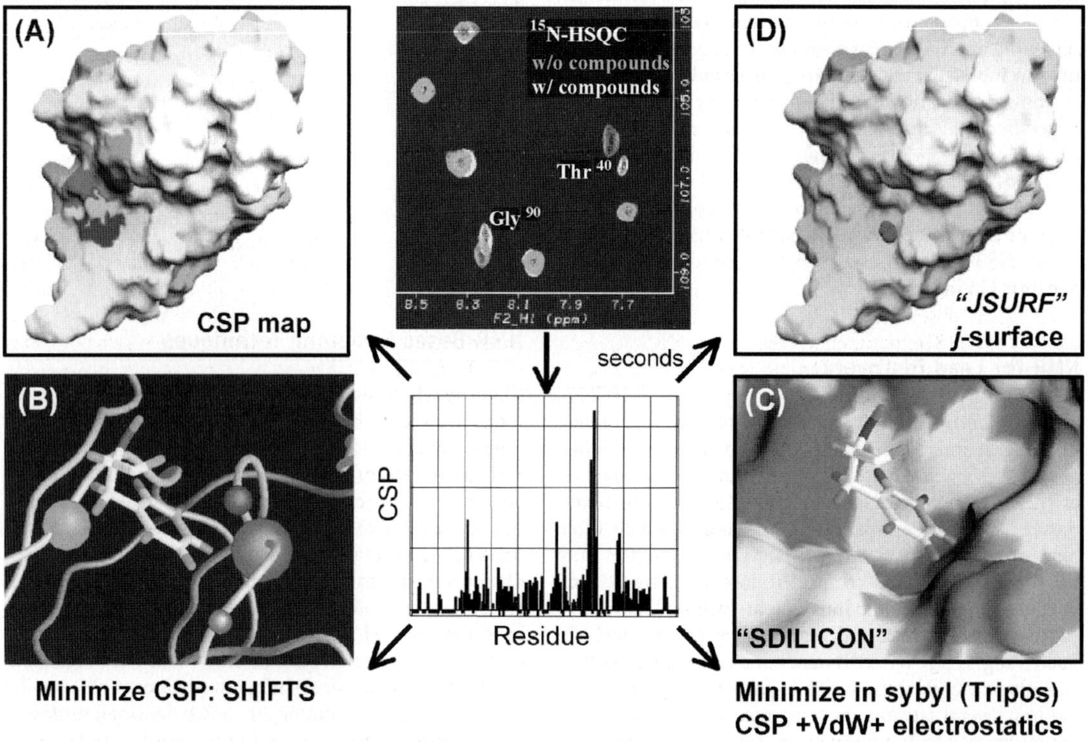

Fig. 1. New tools to determine structures of protein-small molecule ligand complexes using NMR chemical shift perturbation. (A) A CSP map is used to approximate the interaction site of an aromatic ligand binding to the HCV NS3 protease domain [7]. This map was created by coloring protein residues for which significant CSP (center) were observed in ^{15}N-^{1}H HSQC spectra; each color represents a different residue. The resulting map is diffuse; it suggests that the ligand binds the unprimed side of the substrate channel but no particular site is suggested by this data. (B) Ring location and orientation is determined on an atomic level by minimizing the difference between experimental and simulated CSP [49]. (C) Ring location and orientation when energy minimized against van der Waals and CSP [7]. (D) A ligand *j*-surface (magenta), created from the same data used in Figure 1A localizes the ligand-binding site to the S_1-pocket [48]. Recently, these procedures have been implemented into two programs, "SDILICON" and "JSURF" (http://tonga.usip.edu/gmoyna/sdilicon/index.html). (See also Plate 111 on page LXX in the Color Plate Section.)

HSQC spectra can also reveal if the protein target was folded during the screen or if addition of the compound mixture caused the drug target to aggregate, unfold or precipitate. The verification of the chemical structure of NMR-detected hits by NMR and mass spectrometry and the determination of their aqueous solubility under the screening conditions using NMR, make this type of screen very reliable. Moreover, HSQC-based NMR titrations can be used to determine accurate K_Ds for weakly interacting ligands which are in fast exchange on the NMR time scale between their bound and free states (typically $K_D > \sim 10 \mu M$).

However, efficient application of HSQC-based screening requires large amounts of isotope-labeled protein and is typically limited to smaller proteins (MW $> \sim 40$–60 kDa). This restricts its application to a subset of proteins which do not aggregate at concentrations around 100 μM in an aqueous buffer, and to those that can be overproduced in high isotope-labeled yields (mg/l cell culture). Therefore, several ligand-detected 1D NMR screening methods have emerged over the last few years [7,8,15,17,18,20,30–33] which require no isotope labeling of the target, use significantly less target (micrograms vs. milligrams), and are for proteins of almost any MW.

Ligand-Setected 1D NMR-Based Screening

The saturation transfer difference (STD) [15] and water-ligand observed via gradient spectroscopy (WaterLOGSY) [17,18] experiments are the most sensitive methods in this category. The STD method relies on transfer of saturation to detect ligand binding; only STD signals from compounds in a mixture which directly interact with the drug target are observed. STD NMR is based on the intermolecular transfer of saturation from the protein to the bound small molecule ligands which in turn, by fast chemical exchange between their bound and unbound states, is moved into solution where it is detected. For high-affinity ligands, the saturation cannot effectively be transferred further to the free state of the small molecule ligands due to slow chemical exchange resulting in no observable STD effect, that is, ligand peaks are not observed in the difference spectrum. WaterLOGSY is a related experiment that is based on the transfer of magnetization from bulk water to the protein binding site and onto a bound ligand. The concentrations of both the protein and the compound mixture can be adjusted to adjust the sensitivity of such screens.

Because the ligand-detected methods usually require no isotope labeling of the protein, use less protein ($\sim$10–1000$\times$ less) than the target-detected approaches, can be applied to protein targets of almost any MW (even perform best for large proteins), and are useful when no 3D structure of the drug target is available, they are widely used in the pharmaceutical industry. However, these methods do not reveal the ligand-binding site on the drug target and any structural information available is limited to the binding epitope of the small molecule ligand. In general, these methods are only for weaker ligands ($K_D \sim 10^{-7} - 10^{-3}$ M) and compounds must be soluble in an aqueous buffer solution at concentrations of at least $\sim$50–200 μM, which prevents the detection of ligands with high affinity and/or low solubility. In our experience, these screening methods are less robust and produce more false positives than 2D heteronuclear NMR screening for reasons such as non-specific binding and compound aggregation. However, many of these drawbacks can be overcome by combining ligand-observed NMR screening methods with competition binding experiments [24–28]. Therefore, we recently implemented a competition STD NMR method which can be used to identify ligands that bind to the same site as, for example, a natural substrate, a cofactor or a known inhibitor of a protein drug target [50].

STD NMR in Combination with Competition Experiments

A site-specific affinity marker is used in this experiment which binds weakly to an interesting subsite of the drug target (K_D typically in the tens-to-hundreds of μM range) and produces strong STD signals. Reduction or disappearance of the STD signal of the site-specific affinity marker upon addition of a compound mixture suggests the presence of at least one ligand in the mixture which competes with the marker for binding to the active site (Figure 2A). We have used this approach to screen a kinase-directed library of novel, chemically accessible cores against a variety of kinases. An example is shown in Figure 2B, in which we used ATP as a site-specific affinity marker in competition experiments (c-STD[ATP]). ATP binds this particular kinase with a K_D of $\sim$30 μM and produces an STD signal for the H-2 proton of ATP at 8.30 ppm when 200 μM ATP are added to 5 μM of this kinase target. Addition of a sub-μM ligand which competes for binding to the ATP site completely removes the STD signal of ATP and this high-affinity binder does not produce an STD epitope itself. Addition of a ligand in the μM range still effectively competes with ATP, but now generates an STD epitope due to increased exchange between its bound and unbound states. A weaker binder in the tens of μM range only partially competes with ATP and produces an even stronger STD epitope due to increased exchange. This example illustrates that c-STD[ATP] can be used to (1) rapidly rank-order the affinity of ATP-competitive binders and (2) detect high-affinity ligands which would be missed in the simple STD NMR experiment.

This method can also be used to determine K_I's of high-affinity ligands which are in slow chemical exchange

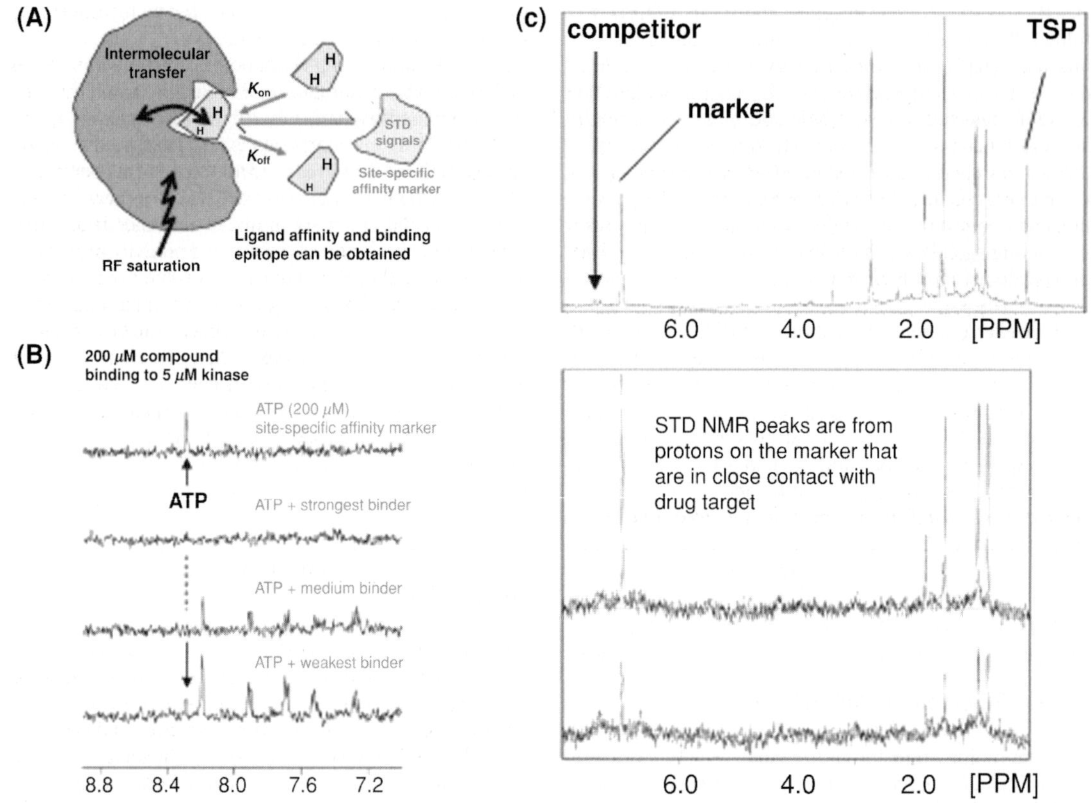

Fig. 2. 1D competition STD NMR [50].(A) A ligand which binds weakly to an interesting subsite of the drug target weakly (K_D typically in the tens-to-hundreds of μM range) and produces strong STD signals is used as a site-specific affinity marker. Reduction or disappearance of the STD signal of the site-specific affinity marker upon addition of a compound mixture suggests the presence of at least one ligand in the mixture which competes with the marker for binding to the same subsite. (B) c-STD[ATP] NMR spectra of a kinase (from top to bottom) in the absence of ligand and in the presence of a strong, medium and weak ATP-competitive binder at a concentration of 200 μM each (see text), illustrating its use to rapidly rank-order the affinity of ATP-competitive binders including high-affinity ligands which do not produce STD signals. (C) K_I's of high-affinity ligands which are in slow chemical exchange on the NMR time scale (K_D typically sub-μM) can also be determined with this method (see text). Addition of 140 μM of an active site-specific affinity marker ($K_D[^{15}N\text{-HSQC}] = 60$ μM) to 5.6 μM kinase produces strong STD signals in a reference spectrum (middle). Addition of 6.1 μM high-affinity inhibitor reduces the STD signals of the site-specific affinity marker (bottom). Based on this reduction and the experimentally determined ligand concentration relative to an internal standard TSP (top) yielded a K_I of 330 nM for the high-affinity ligand.

on the NMR time scale; this cannot be achieved with HSQC-based NMR titration experiments. Recently, we used this method in a drug discovery program to determine the K_I's of sub-μM NMR-derived lead inhibitors. An active site-specific affinity marker with a $K_D[^{15}N\text{-HSQC NMR}]$ of 60 μM was used which produced strong STD signals in a reference spectrum (Figure 2C). Addition of small amounts of the high-affinity inhibitor then reduced the STD signals of the site-specific affinity marker. Based on this reduction and the experimentally determined ligand concentration relative to an internal standard (TSP) yielded a K_I of 330 nM for the high-affinity ligand. A second experiment with a different site-specific affinity marker (K_D of 120 μM) gave a similar K_I of 360 nM for the high-affinity ligand (data not shown), which was in good agreement with the IC_{50} of 500 nM from a homogenous time-resolved fluorescence (HTRF)-based enzymatic assay.

NMR–Based Screening Applied to Drug Targets

We have used NMR to assist lead generation in a number of difficult drug discovery programs in which HTS of large chemical libraries identified either no, or only a few, suitable leads for chemical optimization (Table 1) [51]. NMR-based screening (previous section) of customized libraries containing several hundred to several thousand small molecules identified multiple weak hits to active-site regions of these drug targets in the μM–mM range. Some of these initial hits were followed-up through the evaluation of analogs by a combination of NMR, X-ray crystallography and bioassays to derive SAR and 3D structural data. Interesting scaffolds were subsequently either linked, or further optimized by the parallel synthesis of a large number of analogs, or used as templates for structure-based inhibitor design to develop novel lead series against a variety of these targets.

A critical step in these fragment-based methods [2,3,10], which take a modular approach to drug design, is the optimization of weak NMR hits ($K_D \sim \mu$M–mM range) into potent leads ($K_D < \mu$M) using chemistry guided by 3D structural data. However, such information cannot always be obtained easily. We have, therefore, developed novel tools, which utilize protein CSP data, generated in HSQC-based NMR screens, to rapidly provide an accurate structural representation of protein-ligand complexes under conditions unfavorable for traditional structural work using X-ray crystallography and/or solution NMR [48,49]. These tools use the CSP data in a more quantitative manner than the traditional "chemical shift perturbation mapping" widely used to qualitatively map

Table 1: NMR-based screening applied to drug discovery targets

Drug targets	MW (kDa)	# HTS leads	# compounds screened	# NMR hits to active site(s) [affinity range]	# NMR hit-to-lead [affinity/activity range]	# sub-sites	X-ray structure with NMR hit	Chemistry approach [lead activity]
CNS	46	0	10,100 HSQC	10 [30 μM–mM]	6 series [15–100 μM]	3	yes	• structure-based • directed libraries on-going [<0.03 μM]
Cancer	13	0	6750 HSQC	62 [<700 μM]	3 series [1–100μM]	4	no	• link • directed libraries [<0.05 μM]
Cancer	37–40	1	Hundreds (in progress) HSQC/STD	>20 [5–500 μM]	4 series (in progress)	1	One	• Directed libraries on-going [<0.6 μM]
HCV NS3 protease—NS4A	20	0	3639 HSQC	16 [100 μM–10 mM]	5 series [100 μM–mM]	3–5	No	• Directed libraries • Link [0.8 μM]
HCV NS3 helicase	15/30 /49	0	6006 HSQC	19 [mM range]	5 series [100 μM–mM]	3	Yes	• Structure-based • Directed libraries [3.0 μM]
Anti-viral	64	0	Hundreds STD	Few [<500 μM]	1 series [400 μM]	1	Yes	• Directed libraries [20 μM]
Cancer	42	Few	Hundreds (in progress) c-STD [ATP]	>60 [0.4–500 μM]	7 series (in progress)	>2	One	Being initiated
Cancer	48	Few	Hundreds (in progress) c-STD [ATP]	>60 [<0.7–500μM]	Not initiated	>2	No	Not initiated
Anti-infective	41	0	3647 HSQC	13 [~mM]	3 series [50 μM–mM]	3	Yes	Not initiated

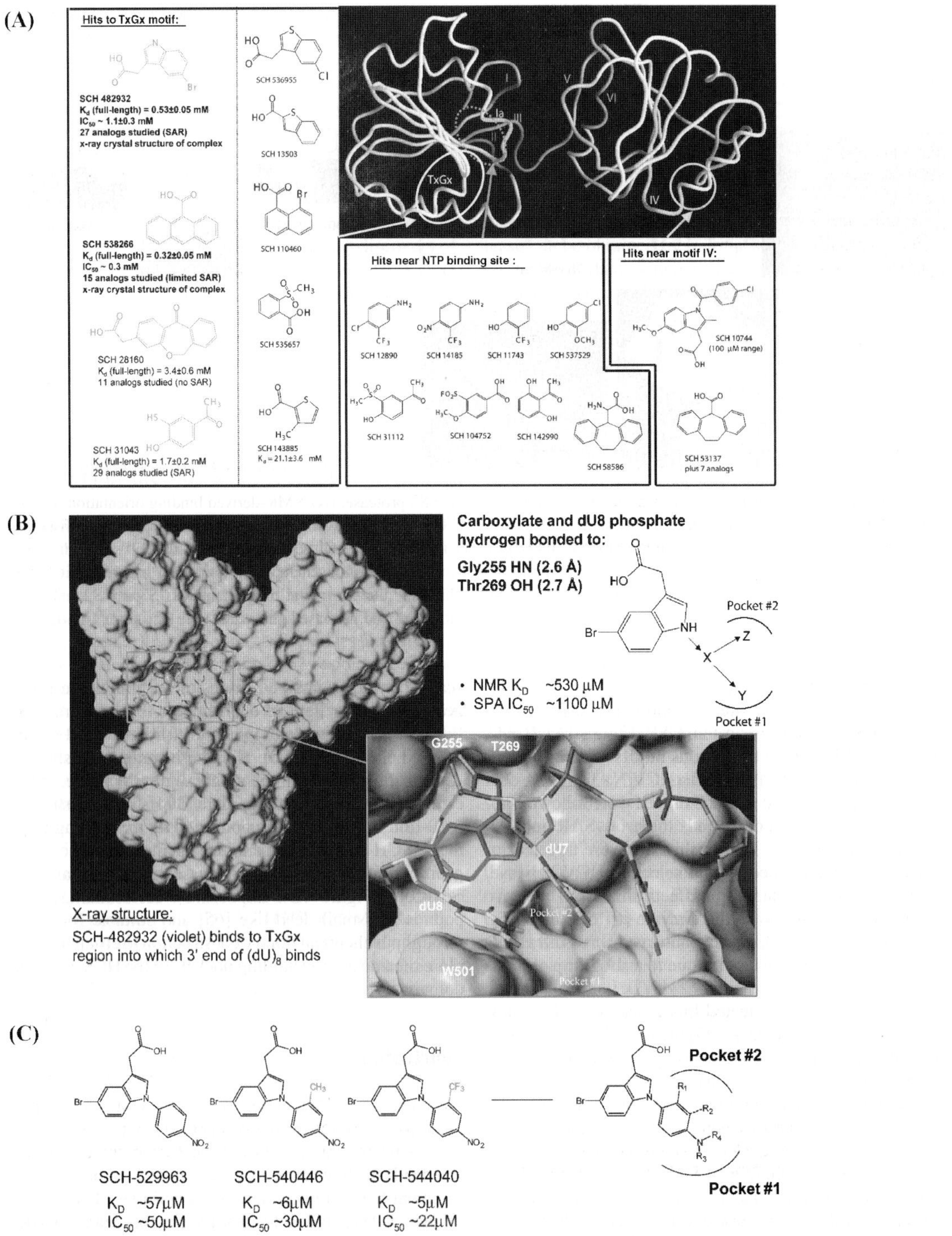

Fig. 4. NMR-derived lead inhibitors of the HCV NS3 helicase. (A) Validated active-site directed NMR hits identified by 2D ^{15}N-^{1}H HSQC-based NMR screening of 6006 compounds from a customized library against $d_{1,2\Delta}$-HCVh. Four NMR hits (red) to the TxGx motif were followed-up by evaluating related compounds to derive SAR and 3D structural information. (B) Binding orientation of SCH-482932 (in-house crystal structure) relative to bound ssDNA (PDB accession code 1A1V) illustrating that this NMR hit binds to the same subsite as the 3′ end of $(dU)_8$. (C) Simple phenyl extensions at the indole nitrogen of SCH-482932 yields mM HCV helicase lead inhibitors that are amenable to high throughput chemical synthesis. (See also Plate 113 on page LXXI in the Color Plate Section.)

modular approach to drug discovery, is the optimization of the often weak NMR hits ($K_D \sim \mu$M–mM) into potent leads ($K_D < \mu$M) using chemistry guided by 3D structural data and SAR. We have implemented a number of NMR-based tools which are designed to facilitate this process.

In addition, we have used the NMR methods described in this report in several programs to validate hits from other sources and rapidly assist the project team chemists in selecting candidate leads for synthetic efforts. In this respect, NMR-derived information often helps to eliminate false positives hits early in the drug discovery process.

References

1. Zuiderweg ERP. Biochemistry. 2002;41:1.
2. Shuker SB, Hajduk PJ, Meadows RP, Fesik SW. Science 1996;274:1531.
3. Hajduk PJ, Meadows RP, Fesik SW. Q. Rev. Biophys. 1999;32:211.
4. Fejzo J, Lepre CA, Peng JW, Bemis GW, Ajay Murcko MA, Moore JM, Chem. Biol. 1999;6:755.
5. Peng JW, Lepre CA, Fejzo J, Abdul-Manan N, Moore JM. Methods Enzymol. 2001;338:202.
6. Pellecchia M, Sem DS, Wüthrich K. Nat. Rev. Drug Discov. 2002;1:211.
7. Wyss DF, McCoy MA, Senior MM. Curr. Opin. Drug Discov. Devel. 2002;5:630.
8. Stockman BJ, Dalvit C. Prog. NMR Spectrosc. 2002;41:187.
9. Hajduk PJ, Burns DJ. Comb. Chem. High Throughput Screen. 2002;5:613.
10. Huth JR, Sun C. Comb. Chem. High Throughput Screen. 2002;5:631.
11. Jahnke W, Flörsheimer A, Blommers MJJ, Paris CG, Heim J, Nalin CM, Perez LB. Curr. Top. Med. Chem. 2003;3:69.
12. Fejzo J, Lepre C, Xie X. Curr. Top. Med. Chem. 2003;3:81.
13. Jahnke W, Widmer H. Cell. Mol. Life Sci. 2004;61:580.
14. Hajduk PJ, Gerfin T, Boehlen JM, Häberli M, Marek D, Fesik SW. J. Med. Chem. 1999;42:2315.
15. Mayer M, Meyer B, Angew. Chem. Int. Ed. Engl. 1999;38:1784.
16. Hajduk PJ, Augeri DJ, Mack J, Mendoza R, Yang J, Betz SF, Fesik SW. J. Am. Chem. Soc. 2000;122:7898.
17. Dalvit C, Pevarello P, Tatò M, Veronesi M, Vulpetti A, Sundström M. J. Biomol. NMR 2000;18:65.
18. Dalvit C, Fogliatto G, Stewart A, Veronesi M, Stockman B. J. Biomol. NMR 2001;21:349.
19. Ross A, Senn H. Drug Disc. Today 2001;6:583.
20. Diercks T, Coles M, Kessler H. Curr. Opin. Chem. Biol. 2001;5:285.
21. Weigelt J, van Dongen M, Uppenberg J, Schultz J, M. Wikström, J. Am. Chem. Soc. 2002;124:2446.
22. Weigelt J, Wikström M, Schultz J, van Dongen MJP. Comb. Chem. High Throughput Screen. 2002;5:623.
23. Jahnke W. Chem. Bio. Chem. 2002;3:167.
24. Dalvit C, Fasolini M, Flocco M, Knapp S, Pevarello P, Veronesi M. J. Med. Chem. 2002;45:2610.
25. Dalvit C, Flocco M, Knapp S, Mostardini M, Perego R, Stockman BJ, Veronesi M, Varasi M. J. Am. Chem. Soc. 2002;124:7702.
26. Jahnke W, Floersheim P, Ostermeier C, Zhang X, Hemmig R, Hurth K, Uzunov DP. Angew. Chem. Int. Ed. 2002;41:3420.
27. Siriwardena AH, Tian F, Noble S, Prestegard JH. Angew. Chem. Int. Ed. 2002;41:3454.
28. Dalvit C, Flocco M, Veronesi M, Stockman BJ, Comb. Chem. High Throughput Screen. 2002;5:605.
29. Dalvit C, Fagerness PE, Hadden DTA, Sarver RW, Stockman BJ. J. Am. Chem. Soc. 2003;125:7696.
30. Meyer B, Peters T. Angew. Chem. Int. Ed. 2003;42:864.
31. Zartler ER, Yan J, Mo H, Kline AD, Shapiro MJ, Curr. Top. Med. Chem. 2003;3:25.
32. Zartler ER, Hanson J, Jones BE, Kline AD, Martin G, Mo H, Shapiro MJ, Wang R, Wu H, Yan J. J. Am. Chem. Soc. 2003;125:10941.
33. Coles M, Heller M, Kessler H, Drug Discov. Today 2003;8:803.
34. Tengel T, Fex T, Emtenas H, Almqvist F, Sethson I, Kihlberg J. Org. Biomol. Chem. 2004;2:725.
35. Hajduk PJ, Sheppard G, Nettesheim DG, Olejniczak ET, Shuker SB, Meadows RP, Steinman DH, Carrera GM Jr, Marcotte PA, Severin J, Walter K, Smith H, Gubbins E, Simmer R, Holzman TF, Morgan DW, Davidsen SK, Summers JB, Fesik SW. J. Am. Chem. Soc. 1997;119:5818.
36. Hajduk PJ, Dinges J, Mikins GF, Merlock M, Middleton T, Kempf DJ, Egan DA, Walter KA, Robins TS, Shuker SB, Holzman TF, Fesik SW. J. Med. Chem. 1997;40:3144.
37. Hajduk PJ, Dinges J, Schkeryantz JM, Janowick D, Kaminski M, Tufano M, Augeri DJ, Petros A, Nienaber V, Zhong P, Hammond R, Coen M, Beutel B, Katz L, Fesik SW. J. Med. Chem. 1999;42:3852.
38. Hajduk PJ, Zhou MM, Fesik SW, Bioorg. Med. Chem. Lett. 1999; 9:2403.
39. Hajduk PJ, Boyd S, Nettesheim D, Nienaber V, Severin J, Smith R, Davidson D, Rockway T, Fesik SW. J. Med. Chem. 2000;43:3862.
40. Hajduk PJ, Gomtsyan A, Didomenico S, Cowart M, Bayburt EK, Solomon L, Severin J, Smith R, Walter K, Holzman TF, Stewart A, McGaraughty S, Jarvis MF, Kowaluk EA, Fesik SW, Med J. Chem. 2000;43:4781.
41. Liu G, Huth JR, Olejniczak ET, Mendoza R, DeVries P, Leitza S, Reilly EB, Okasinski GF, Fesik SW, von Geldern TW. J. Med. Chem. 2001;44:1202.
42. van Dongen M, Weigelt J, Uppenberg J, Schultz J, Wikström M. Drug Discov. Today 2002;7:471.
43. Szczepankiewicz BG, Liu G, Hajduk PJ, Abad-Zapatero C, Pei Z, Xin Z, Lubben TH, Trevillyan JM, Stashko MA, Ballaron SJ, Liang H, Huang F, Hutchins CW, Fesik SW, Jirousek MR, J. Am. Chem. Soc. 2003;125:4087.
44. Liu G, Xin Z, Pei Z, Hajduk PJ, Abad-Zapatero C, Hutchins CW, Zhao H, Lubben TH, Ballaron SJ, Haasch DL, Kaszubska W, Rondinone CM, Trevillyan JM, Jirousek MR. J. Med. Chem. 2003;46:4232.
45. Yu L, Oost TK, Schkeryantz JM, Yang J, Janowick D, Fesik SW, J. Am. Chem. Soc. 2003;125:4444.
46. Wyss DF, Arasappan A, Senior MM, Wang Y-S, Beyer BM, Njoroge FG, McCoy MA. J. Med. Chem. 2004;47:2486.
47. Dalvit C, Ardini E, Flocco M, Fogliatto GP, Mongelli N, Veronesi M, J. Am. Chem. Soc. 2003;125:14620.

48. McCoy MA, Wyss DF. J. Am. Chem. Soc. 2002;124:11758.
49. McCoy MA, Wyss DF, J. Biomol. NMR 2000;18:189.
50. Wang Y-S, Liu D, Wyss DF. Magn. Reson. Chem. 2004;42: 485.
51. Wyss DF. Cambridge Healthtech Institute's 4th Ann. Struct.-Based Drug Design. Boston (2004).
52. Lauer GM, Walker BD. Engl N., J. Med. 2001;345:41.
53. Reed KE, Rice CM. Curr. Top. Microbiol. Immunol. 2000;242:55.
54. Tan SL, Pause A, Shi Y, Nat. Rev. Drug Discov. 2002;1:867.
55. Levin MK, Patel SS, J. Biol. Chem. 277, 29377 2002.
56. Yao N, Reichert P, Taremi SS, Prosise WW, Weber PC. Structure 1999;7:1353.
57. Bartenschlager R. J. Viral Hepat. 1999;6:165.
58. McCoy MA, Senior MM, Gesell JJ, Ramanathan L, Wyss DF. J. Mol. Biol. 2001;305:1099.
59. Gesell JJ, Liu D, Madison VS, Hesson T, Y-S Wang, Weber PC, Wyss DF, Protein Eng. 2001;14:573.
60. Howe AY, Chase R, Taremi SS, Risano C, Beyer B, Malcolm B, Lau JY, Protein Sci. 1999;8:1332.
61. Lin C, Kim JL. J. Virol. 1999;73:8798.
62. Tai CL, Pan WC, Liaw SH, Yang UC, Hwang LH, Chen DS. J. Virol. 2001;75:8289.
63. Kim JL, Morgenstern KA, Griffith JP, Dwyer MD, Thomson JA, Murcko MA, Lin C, Caron PR, Structure. 1998;6:89.
64. Pause A, Sonenberg N, EMBO J. 1992;11:2643.
65. Oprea TI, Davis AM, Teague SJ, Leeson PD. J. Chem. Inf. Comput. Sci. 2001;41:1308.

NMR and Structural Genomics in the Pharmaceutical Sciences

Maša Čemažar and David J. Craik*
Institute for Molecular Bioscience, University of Queensland, Brisbane 4072, Queensland, Australia

Introduction

Structural genomics (or more correctly structural proteomics) involves the high-throughput determination of protein structures on a genome-wide scale. It is a new field of research that has the potential to add substantial value to the sequence information becoming available from genome projects. The purpose of structural genomics is to accelerate protein structure determination so as to obtain structural examples of all available protein folds. The derived information will be essential in providing new insights into relationships between protein sequence, structure, and function.

There have been a large number of structural genomics consortia established in recent years in many countries. In addition to their primary purpose of generating large numbers of protein structures many of these consortia are developing and improving techniques for high-throughput structure determination. Together with X-ray crystallography, NMR is one of the two main structure determination methods. The two techniques are complementary and have advantages over one another in the specific ways that they can be applied to the problem of structure determination. In the pharmaceutical sciences structural genomics is beginning to play an important role in drug design strategies. This chapter reviews current progress and trends in structural genomics, with an emphasis on NMR structure determination.

Strategies and Targets in Structural Genomics

The present stage of structural genomics has been referred to as the "low-hanging fruit" phase [1]. Presently the proteins that are being tackled mostly express well in *E. coli* in a soluble form, can be purified easily, and either crystallize readily or give high-quality NMR spectra. The structure determination of these proteins does not require any conceptual or technical advances, although such advances are being made and will accelerate progress. The reason for the choice of the "low-hanging fruit" is that researchers in the field are still seeking to improve the downstream technology, i.e. high-throughput data collection, processing, and structure determination, and to develop automated procedures for many of the steps before embarking on more extensive programs involving more challenging targets.

Table 1 lists some of the major structural genomics initiatives around the world that are using X-ray crystallography or NMR spectroscopy or, in a majority of cases, both techniques. There are several resources on the World Wide Web where links to the details of these initiatives can be found, including for example http://www.rcsb.org/pdb/strucgen.html#Worldwide. Many targets chosen by these initiatives are either from model organisms such as *E. coli*, *S. cerevisiae*, *C. elegans*, or are human proteins that have important functions in essential cell processes such as signal transduction or nucleic acid binding.

Sequence alignment strategies are frequently used to predict the folds of proteins from related families with high sequence homology from across different genomes. However, sometimes even though the homology between two sequences is rather low, they might have a high similarity in their folds and vice versa, so at the moment predictive methods have not replaced the need for experimental structure determination. Even in cases where the predictions of the fold work well, the atomic resolution of the predicted structures may not be sufficient for some applications. In particular, for pharmaceutical applications there is still often a need to experimentally determine the high-resolution structure of a particular target to be able to predict and study the interactions of this protein with ligands for drug design.

Advantages and Disadvantages of NMR for Structural Genomics

From a structural proteomics perspective, NMR studies are mostly limited because of the size of the proteins that can be successfully tackled and by the lengthy data collection and analysis steps. Even with recent developments in the field, complete structure determination of monomeric proteins has still not been pushed beyond the

* To whom correspondence hould be addressed.

Graham A. Webb (ed.), Modern Magnetic Resonance, 1411–1419.

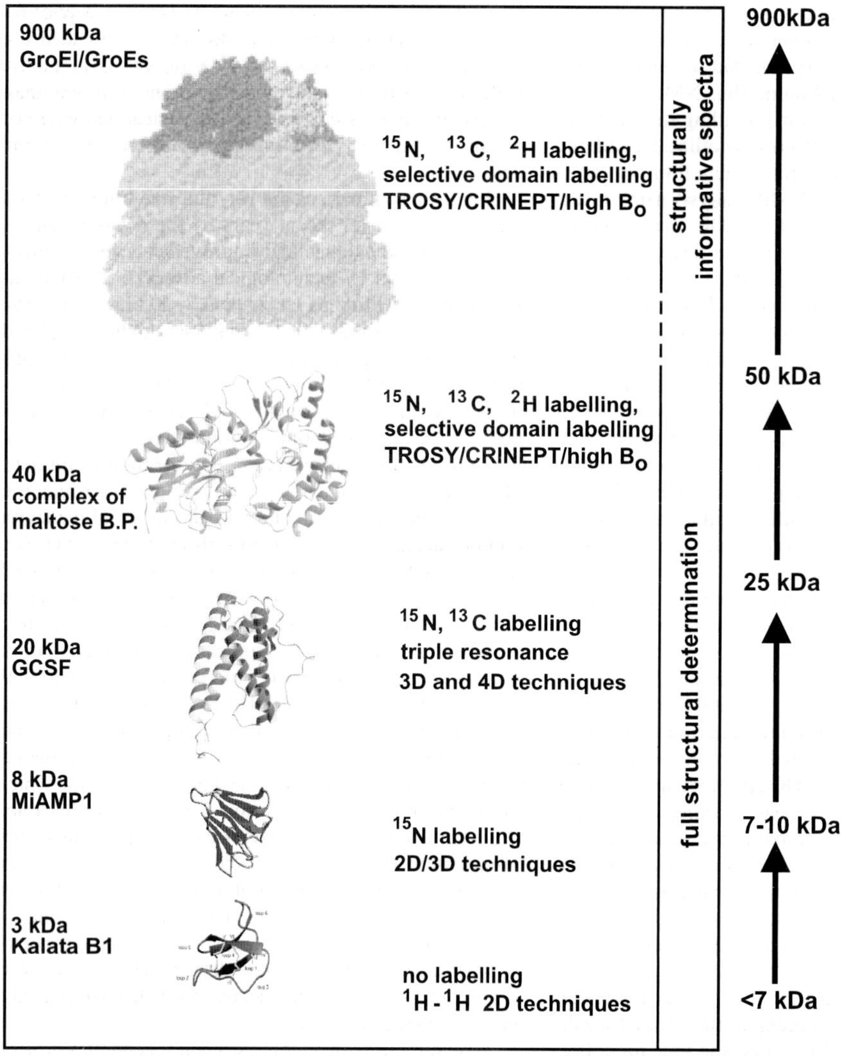

Fig. 1. NMR structural information and biomolecular size limits. A schematic representation of the relationship between various NMR structural determination techniques and biomolecular size limits. For each size range an example of a protein structure is given: 3 kDa macrocyclic peptide kalata B1, 8 kDa antimicrobial protein MiAMP1 from the nut kernel of *Macadamia integrifolia* [32], 20 kDa granulocyte colony stimulating factor [33], the 40 kDa complex of maltodextrin binding protein with β-cyclodextrin [34], and the 900 kDa complex of GroEl/GroEs [14]. On the right hand side of the figure the NMR techniques that can be applied to the specific mass range are summarized. So far full 3D structures have been determined up to the size limit of ~50 kDa, but this limit is set to expand over the coming years. (See also Plate 114 on page LXXII in the Color Plate Section.)

spectra that were judged too poor to allow NMR structure determination. It is useful to point out just what a central role HSQC spectra play in structural genomics programs. For proteins that are soluble and purifiable, HSQC spectra can be categorized as good, promising, or poor (or "none" for those where protein had been lost during the concentration procedure) and appropriate decisions made as to whether to proceed with structure determination [13]. A graphic representation of the distribution of spectral types is shown in Figure 2. The exact distribution varies for different organisms but the result is illustrative of general findings. It shows that almost half of the small (<20 kDa)